7th International Scientific Conference on Reliability and Durability of Railway Transport Engineering Structures and Buildings (Transbud-2018)

MATEC Web of Conferences Volume 230 (2018)

Kharkiv, Ukraine
14 - 16 November 2018

ISBN: 978-1-5108-7484-8

Printed from e-media with permission by:

Curran Associates, Inc.
57 Morehouse Lane
Red Hook, NY 12571

Some format issues inherent in the e-media version may also appear in this print version.

Printed by Curran Associates, Inc. (2019)

For additional information, please contact EDP Sciences – Web of Conferences at the address below.

EDP Sciences – Web of Conferences
17, Avenue du Hoggar
Parc d'Activité de Courtabœuf
BP 112
F-91944 Les Ulis Cedex A
France

Phone: +33 (0) 1 69 18 75 75
Fax: +33 (0) 1 69 28 84 91

contact-edps@webofconferences.org

Additional copies of this publication are available from:

Curran Associates, Inc.
57 Morehouse Lane
Red Hook, NY 12571 USA
Phone: 845-758-0400
Fax: 845-758-2633
Email: curran@proceedings.com
Web: www.proceedings.com

TABLE OF CONTENTS

Statement of Peer review

In submitting conference proceedings to MATEC Web of Conferences, I certify to the Publisher that I adhere to the **Policy on Publishing Integrity** of the journal in order to safeguard good scientific practice in publishing.

1. All articles have been subjected to peer review administered by the proceedings editors.
2. Reviews have been conducted by expert referees, who have been requested to provide unbiased and constructive comments aimed, whenever possible, at improving the work.
3. Proceedings editors have taken all reasonable steps to ensure the quality of the materials they publish and their decision to accept or reject a paper for publication has been based only on the merits of the work and the relevance to the journal.

Title, date and place of the conference

7th International Scientific Conference "Reliability and Durability of Railway Transport Engineering Structures and Buildings", Transbud-2018, November 14-16, 2018, Kharkiv

Proceedings editor(s):

Prof. Glib Vatulia, Prof. Andrii Plugin, Prof. Oleksandr Darenskiy

Date and editor's signature

October 24, 2018

Analysis of Vertical Irregularities and Dynamic Forces on the Switch Frogs of the Underground Railway

Volodymyr Boiko*, *Vitalii* Molchanov, *Volodymyr* Tverdomed, and *Olena* Oliinyk
State University of Infrastructure and Technologies, 9 Kyrylivska St., 04071 Kyiv, Ukraine

Abstract. This article presents the analysis of vertical irregularities and dynamic forces on the switch frogs in operating conditions on the main tracks of Kiev underground railway. For main railways the deterioration problem of tongues, framed rails, switch frogs and as a consequence formation of the character in interaction on the railroad switches is quite widely studied by different scientists, but for conditions of railway underground such researches are practically absent. However, in order to use metal elements of switches rationally and efficiently, the forecast of their service operation and replacement planning, it is necessary to estimate the level of dynamic forces within the most deteriorated elements of switches at different operational stages. This problem is particularly relevant in conditions of financial restriction for track facilities. The studies allowed determining the form, parameters and basic patterns of the irregularities formation in the rolling zone on switch frogs P65 and P50 mark of crossing 1/9 on wooden beams under different operating conditions, which is the basis for determining vertical dynamic forces of interaction within the switch frog subject to setting maximum permissible speeds of movement.

Introduction

Vertical irregularities on the rolling surface of switch frogs are one of the main disturbing factors in the formation of the oscillating process of the interaction between the wheels of the rolling stock and tracks within the most vulnerable zone of switches. This zone for ordinary switch frogs is located from the frog throat to the intersection of the triblet 20-40 mm, where the wheel completely goes to the rolling on the triblet surface. In this zone, due to the rapid deterioration of narrow contact areas in the check rail and the triblet, irregularities are quickly formed during the operation causing the formation of additional inertial forces of wheel and rail interaction. In this case, the emerging forces of interaction have a value, exceeding similar forces within the normal track significantly.

During the diagnostic of switch frogs, the control of deterioration only in certain sections does not provide the objective assessment of power loading. According to this, there is necessary to know the trajectory of wheel rolling in vertical. This allows simulating

* Corresponding author: zkks@ukr.net

the nature of the interaction, determining the level of dynamic forces and setting the permissible speeds of railway vehicles.

Therefore, the analysis of vertical irregularities on switch frogs is a necessary condition for studying the issue of power dynamics on switches.

Many scholars have conducted this scientific problem having carried out a number of experimental and theoretical studies and they have proposed different approaches to the definition of vertical irregularities in the switching section.

Researches of prof. Yakovlev V. F. [1] define main characteristics of the irregularity shape in describing of the wheel trajectory: the form and parameters of irregularity (length, depth and slopes of irregularities).

Prof. Danilenko E.I. [2] has carried out large experimental researches about trajectories of wheel rolling on the switch frogs of different designs and stamps that were set for different conditions in sections of main and industrial transport. A number of average irregularities has been established for various structures and types of switch frogs on timbers (1/5, 1/6, 1/7, 1/9 and 1/11) due to the different average deterioration and the deterioration of maximum.

There are well-known studies of switches on Ukrainian railways [3, 4, 5], where the formation of vertical irregularities on switch frogs and the estimation of power interaction of the "vehicle-track" system are considered.

Scientific works by Aniołek K., Herian J. [6] only consider a single cross section of the switch frog in the simulation of power interaction.

Zeng Z. and others [7] consider the wheel-rail contact parameters for undeteriorated switch frogs along the longitudinal direction of movement and deteriorated profiles of wheels.

J. Drozdziel and B. Sowinski [8] consider the irregularity formed on the base of the diameter geometry measured in the fixed sections of switch frogs.

In the scientific work of Jingmang Xu [9] the conditions of interaction within the arrow part of the switch with different deterioration characteristics have been modeled allowing to design the construction rationally.

Above-mentioned studies were mainly conducted for switches being operated on trunk transport, at the same time the conditions of switches operation in the underground railway have significant differences, and therefore require a separate research.

Measurement of vertical irregularities on the switch frogs

Investigations of vertical irregularities have been carried out in real conditions of the underground railway operation. In which case switches operated by the ground and underground zones of main underground railway tracks have been used.

Measurements have been carried out at 37 switches, including:

- 22 switches P50 mark of crossing 1/9;
- 15 switches P65 mark of crossing 1/9.

The selection of switches had been carried out according to the difference of characteristics in the tonnage missed during operation, and different operating conditions (the direction movement of trains – trailing or facing direction, different speeds, etc.).

The set speed of the movement V on switches is V_{max} = 80 km/h in the direction, V_{max} = 40 km/h in the lateral direction. Maximum axial loadings were up to P_o = 14 tons. The loading capacity G million tons per year of the underground sections was 21–25million tons per year for switch frogs P50 and 17 million tons per year for P65. The capacity tonnage T from the beginning to the time of investigations was from 13 to 133 million tons on the switch frogs P50, and from 25 to 106 million tons on P65. At which point switches with a preferred trailing movement direction of P50 and P65 there were investigated 13 and 10 correspondingly, and with facing movement direction 9 and 5.

Measurements of irregularities have only been performed in the direct direction of movement along the switching (main flow).

A mechanical device - a trajectory tracer – was used for measurements of rolling trajectories. The trajectory tracer consists of a semi-automatic device with a recording device that continuously records the position of the gravity center of the measuring wheel over the predominant direction of movement on the rolling surface along the switch frog.

To fix the characteristic points along the length of the switch frog, layout was made according to the scheme presented in Fig. 1

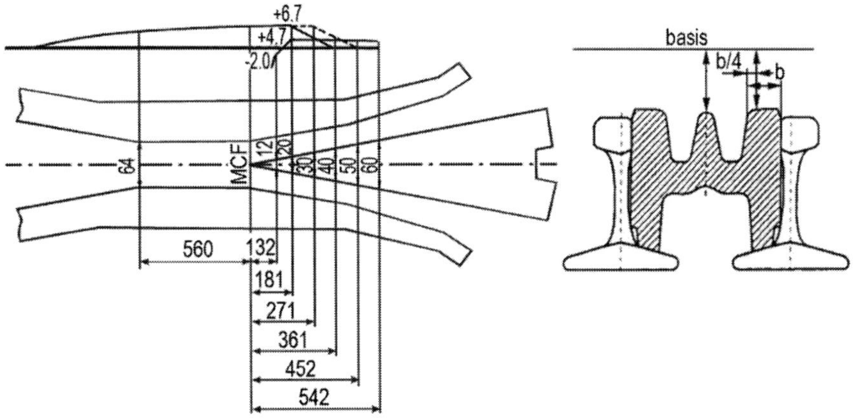

Fig. 1. The scheme of typical cross-sectional marking for measuring vertical irregularities of switch frogs, mark of crossings 1/9

The vertical irregularity on the switch frog represents the trajectory of the center movement of the measuring wheel, which has a rolling surface according to the shape of the wheel rim. For measurements there was used a roller sharpened under the profile of a new wheel (it is possible to use the wheel profile with any condition of deterioration).

Results of operational studies about vertical irregularities on the rolling surface of the switch frogs have been represented in the form of natural trajectories of rolling from the intersection of the cast part of the check rail to the core section of 60 mm and defining of tonnage which was skipped according to the design in the measurement period.

For the analysis of trends in the development of irregularities at different stages of operation, all received vertical irregularities have been grouped depending on the type of switch frogs, the capacity of tonnage and the preferred direction of the train movement. For each group there have been determined the shape and parameters of irregularities depending on the tonnage at different stages of operation.

Mathematical processing of the statistical material was performed after systematizing of irregularities according to operational characteristics, and groups of irregularities have been obtained according to different stages of frog deterioration:

1) the initial period of operation (approximately after 1–1.5 years of operation);

2) the medium deterioration condition (approximately after 60–75 million tons of capacity tonnage);

3) the deterioration condition closing to the maximum (approximately after 100-120 million tons of capacity tonnage).

The average trajectory of rolling by the center of the wheel along the switch frog is shown in Fig. 2, 3.

Obtained vertical irregularities for different conditions and stages of operation became the basis for calculating forces of interaction within switch frogs.

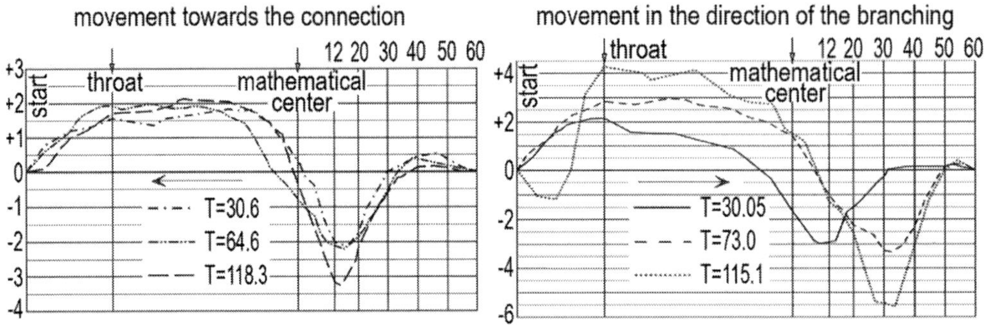

Fig. 2. Average vertical irregularities for the switch frogs P50, mark of crossings 1/9 at different stages of operation.

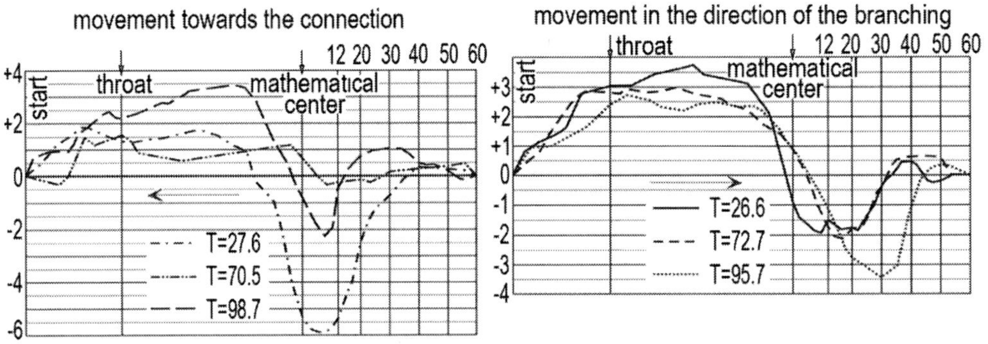

Fig. 3. Average vertical irregularities for the switch frogs P65, mark of crossings 1/9 at different stages of operation.

Estimation of vertical force dynamics on the switch frogs

Values of dynamic vertical forces acting on the switch frog from wheels of the rolling stock in the zone of irregularities can be determined by the formula:

$$P_{dyn} = P_{st} + \Delta P_k \tag{1}$$

The static force P_{st} is a known value and is taken from references depending on the type of rolling stock. The dynamic addition of vertical force from the influence of irregularity on the rolling surface of the frog sections ΔP_k is as a function from the parameters of irregularities – the depth and slopes of irregularity, and operating conditions – capacity tonnage, speed and the train direction:

$$\Delta P_k = f\,(i_z, \Sigma i, h, V) \tag{2}$$

In the formula (2): i_z – the opposite slope, ‰; Σi – the total slope on the fracture of irregularities, ‰; h – depth of the irregularity, mm; V – speed, km/h.

Parameters of irregularities i_z, Σi, h and conditions of operation of switch frogs: speed of movement V, capacity tonnage T, prevailing train direction - trailing or facing are the necessary information for calculating dynamic vertical forces.

As a calculation model, the flat longitudinal diagram of the interaction between the track and the 4-axle wagon, proposed in the article [2, 10] is presented in Fig. 4. As a disturbing factor in this scheme, irregularities on the surface of frog rolling have been determined.

The accepted calculation scheme of interaction describes the oscillatory process of the mechanical system of the vehicle-track sufficiently fully and with correctly choosing of design parameters of the track and the vehicle allows to obtain the result with the necessary accuracy, which satisfies the practical goals of the problem.

The peculiarity of the accepted calculation scheme is consideration of the mutual influence on the deflection under each of the considered wheels, the adjoining wheel in the trolley through the general support for wheels – the switch frog.

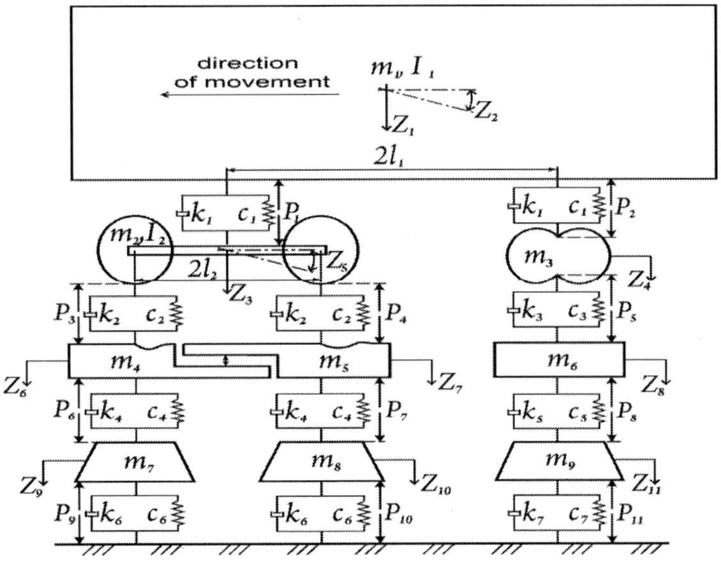

Fig. 4. The scheme of interaction of the switch frog and vehicle in the long vertical surface.

Differential equations of the oscillation process according to the described mechanical system «vehicle-track» on the surface of the rail thread (Fig. 4) due to d'Alembert's principle can be represented as a system of 11 equations (3) [2, 4]. The forces of vertical interaction P_i, included in this system of differential equations, are determined from dependencies (4).

$$\left.\begin{aligned}
&m_1Z_1''+P_1+P_2=0\\
&I_1Z_2''+P_2l_1+P_1l_1=0\\
&(m_{21}+m_{22})Z_3''-P_1+P_3+P_4=0\\
&m_3Z_4''-P_2+P_5=0\\
&I_2Z_5''-P_3l_2+P_4l_2=0\\
&m_4Z_6''-P_3+P_6+\delta_{6-7}P_7=0\\
&m_5Z_7''-P_4+P_7+\delta_{7-6}P_6=0\\
&m_6Z_8''-P_5+P_8=0\\
&m_7Z_9''-P_6-\delta_{6-7}P_7+P_9=0\\
&m_8Z_{10}''-P_7-\delta_{7-6}P_6+P_{10}=0\\
&m_9Z_{11}''-P_8+P_{11}=0
\end{aligned}\right\}(3)$$

$$\left.\begin{aligned}
&P_1=c_1(Z_1-Z_3-l_1Z_2)+k_1(Z_1'-Z_3'-l_1Z_2')\\
&P_2=c_1(Z_1-Z_4-l_1Z_2)+k_1(Z_1'-Z_4'-l_2Z_2')\\
&P_3=c_2(Z_3-Z_6-l_2Z_5-\eta_1(V_t))+k_2(Z_3'-Z_4'-l_2Z_5-\eta_1(V_t))\\
&P_4=c_2(Z_3-Z_7+l_2Z_5-\eta_2(V_t))+k_2(Z_3'-Z_7'-l_2Z_5'-\eta_2(V_t))\\
&P_5=c_3(Z_4-Z_8-\eta_3(V_t))+k_3(Z_3'-Z_8'-\eta_3(V_t))\\
&P_6=c_4(Z_6-Z_9)+k_4(Z_6'-Z_9')\\
&P_7=c_4(Z_7-Z_{10})+k_4(Z_7'-Z_{10}')\\
&P_8=c_5(Z_8-Z_{11})+k_5(Z_8'-Z_{11}')\\
&P_9=c_6Z_9+k_6Z_9'\\
&P_{10}=c_6Z_{10}+k_6Z_{10}'\\
&P_{11}=c_6Z_{11}+k_7Z_{11}'
\end{aligned}\right\}(4)$$

In the scheme (Fig. 4) and equations (3 and 4) there are following notations: m – mass of system elements; Z_i – vertical movements of masses, their first and second derivatives; c_i and k_i – corresponding deflection rate of elastic bonds and coefficients of viscous friction of dissipative viscous elements between the mechanical system; $\delta_{6-7} = \delta_{7-6}$ – ordinates of the influence line on the displacement under the force P_6 on the strength of P_7 and vice versa; η

(Vt) – functions of irregularities under the corresponding wheels of the vehicle; I – moments of inertia and trolley; $2l_1$ – distance between points of the body restraint on the trolley; $2l_2$ – the distance between axles of wheels in the trolley.

The practical implementation of this mathematical model can be performed in any software environment that allows solving mathematical and engineering problems.

Before the forces calculating, the basic elastic-dynamic parameters of the track and the rolling stock for the specific operating conditions have been determined, the speed of motion, initial conditions of the oscillation process and characteristics of irregularities were determined.

The solution of the equations systems (3), (4) is as a function of the distribution of all these interaction forces on the length of irregularities on the switch frog. The practical interest is in the interaction forces at the level of contact points "wheel-frog" P, which can be taken as an assessment criterion for the safe operation of the structure. Analysis of the calculation results showed that the largest values of dynamic forces arise under the first wheel of the calculation trolley.

One of the results obtained in calculating dynamic deflections and vertical dynamic forces P between the 1st wheel and the frog in rolling of the vertical irregularity with speed V = 40, 60, 80 km/h, for switch frogs P50, mark of crossings 1/9 with average deterioration (deterioration ≈ 3.0 mm) and trailing direction of movement is presented in Fig. 5.

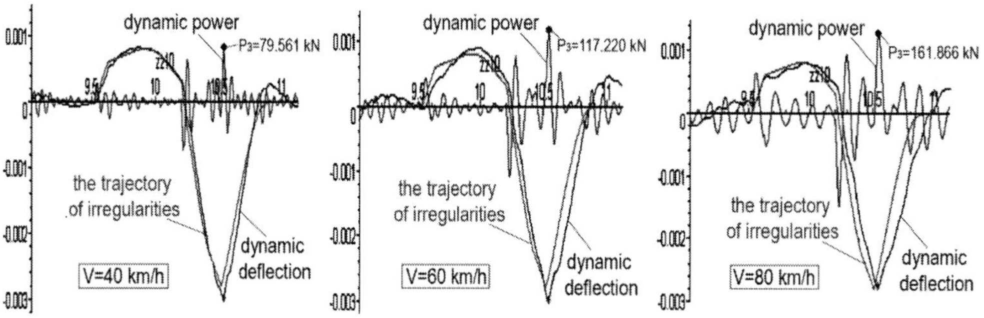

Fig. 5. Distribution of vertical dynamic forces and dynamic deflections of the rail thread in the contact of the wheel and rails on the switch frogs.

For practical implementation of results and the use of force dynamics on the switches frogs in predicting [2, 10] there had been proposed the algorithm for establishing functions of the connection for forces of interaction with the parameters of irregularities on the witch frogs at different stages of operation (2).

1. Finding equations of correlation connection $P = f(i_z)$, $P = f(\Sigma i)$ і $P = f(h_{max})$.

2. On the basis of the relation equations $P = f(i_z)$, $P = f(\Sigma i)$ і $P = f(h_{max})$ constructing the functional dependences $P = f(V)$ (differentiated for i_z and Σi) and formulating the equations of relation $P = f(V)$.

3. The cooperative solution of relation equations $P = f(i_z)$, $P = f(\Sigma i)$ і $P = f(V)$ for the case of the analysis of the counter i_z and the total slopes of the Σi irregularities. Combination the bifunctional equations $P = f(i_z, V)$, $P = f(\Sigma i, V)$ differentiated by i_z, Σi and V.

4. The graph-analytic analysis of equations $P = f(i_z, V)$, $P = f(\Sigma i, V)$. Definition of functional dependences of coefficients with arguments and free members of equations from arguments.

5. Solving equations of functional dependences of the universal type (2) separately for switch frogs with different operating conditions.

After the algorithm realization, there have been obtained ultimate universal equations of the form (2) establishing the functional connection of the dynamic interaction forces P with the parameters of the vertical irregularity (i_z, Σi, h) and the speed of the vehicle V for the switch frogs.

In the final view, the practical formula for determining the dynamic forces of switch frogs P50, mark of crossings 1/9 due to the medium deterioration condition is:

$$P=k_1\Sigma i+k_2\Sigma iV+k_3\Sigma iV^2+k_4V^2+k_5V+k_6i_z+k_7i_zV+k_8i_zV^2+k_9h+k_{10}hV+k_{11}hV^2+k_{12} \quad (3)$$

The comparative analysis of results in calculating the forces obtained by mathematical modeling and equations (3) shows a rather close correlation. The maximum deviations in the calculation results are 5–15%.

Conclusions

The comparative analysis of the average vertical trajectories of rolling on the switch frogs P65 and P50, mark of crossings 1/9 on wooden beams, laid down in the main tracks of the underground railway, allowed to establish main differences, trends, regularities of the development in the track operation and set up the following conclusions.

The shape and parameters of irregularities for the switch frogs P50 and P65 are not significantly different.

The basic form of the irregularity is formed during the initial period of exploitation after the 30-35 million tons capacity, the form is being changed insignificantly in the future but the depth, slope and length of the irregularity are changed.

The qualitative influence on the form and location of the deepest place on the switch frog is set by the prevailing direction of the train movement. Therefore, the switch frogs with trailing or facing movement directions should be considered separately.

According to the known form and parameters of vertical irregularities, additional dynamic forces can be determined at different speeds.

The suggested dependences (3) are universal and allow determining the dynamic forces P due to the complex of parameters of vertical irregularities on the switch frogs (2) and the established train speeds V for specific structures under given operating conditions. In practice, the use of the obtained equations does not require complex calculations of elastic-dynamic characteristics and the system solution of differential equations.

By the values of permissible vertical forces, it is possible to set permissible speeds of trains on the switch frogs and switches in general $[V]=f\,[P_{max}]$. The recommended values of permissible dynamic forces $[P_{max}]$ for the switch frogs under durability conditions of structures have been established [2, 10, 11] and the amount is $[P_{max}] \leq 450-490$ кН. Accordingly, the permissible level of the dynamic addition of vertical forces (1) from the irregularities influence on the rolling surface must be limited to the switch frogs by the meaning $[\Delta P_{max}]$ =370÷410 (кН).

Maximum forces did not exceed the permissible values for all above-mentioned switch frogs.

References

1. S.V. Amelin, V.I. Abrosimov, N. N. Elsakov, E. K. Smykov, L. N. Frolov, V. F. Yakovlev, *Improvement of the conduct of switch economy* (Transport, 1983)
2. E. I. Danilenko, *Theoretical bases and practical methods of calculation of durability and deterioration of intersections and connections of railways of industrial transport* (Kyiv, 1992)
3. E.I. Danilenko, V.D. Boyko, KUETT, **10**, 52-60 (2006)

4. V. Kovalchuk, M. Sysyn, Yu. Hnativ, O. Bal, B. Parneta, A. Pentsak, ETS, **1 (92)**, 33–42 (2018)
5. V. Boyko V. Molchanov T. Artiukhovych, UkrSURT, **169**, 39-47 (2017)
6. K. Aniołek, J. Herian, JTE, **139**, 533–539, (2013)
7. Z. Zeng, Z. Yu, X. Chen, JVS, **28**, 40–45 (2009)
8. J. Drozdziel & B. Sowinski, CR, **VII**, 673–682 (2000)
9. Jingmang Xu, Ping Wang, Li Wang, SAGE J, **7** (2016)
10. E.I. Danilenko, S.D. Taranenko, A.P. Kutah, *Switches of Ukrainian railways* (Transport of Ukraine, 2001)
11. E.I. Danilenko, V.V. Rybkin, *Rules for calculating the railway track for durability and stability* (Transport of Ukraine, 2005)

Method for the simulation and reconstruction of 3D particle size distributions in metal structures

Tetiana Danilenko*

State University of Infrastructure and Technologies, Ivan Ohienko str, 19, 03049 Kyiv, Ukraine

Abstract. The stereological method for studying grain structures is presented, it determines the interrelation between the size distributions of three-dimensional grains and their planar and linear sections. Method is applicable for both reconstruction and simulation of the 3D structures. The polyhedral shape of the grains is taken into account by means of the shape factors established for them. An original method to determine the distribution of plane sections from the certain grain-polyhedron to calculate the shape factors is presented. This method can be used to describe the shape of different structural components in many materials: steels, cast irons, concretes, rocks, ceramics, open-porosity materials, biofilms, foams, etc. The stereological method is also applicable to structures with spherical particles, for which the shape factors are equal to unity. The method makes it possible to create the standards of plane sections of grain structures and other structures with a previously known size distribution of 3D grains or particles.

Introduction

Material properties, including mechanical ones, depend on quantitative characteristics of their 3D structural components such as grains, voids, inclusions having a range of sizes and shapes. Examples of such materials are metals, concretes, rocks, ceramics, open-porosity materials, biofilms, foams, etc. Size distribution of their structural components is an important characteristic of materials [1-3] and it can be determined by reconstruction from a series of parallel sections planes [4-6] or by stereological methods that apply mathematical relations between 2D and 3D geometry [7, 8]. Stereological methods have an advantage over serial sections methods, because they are not so laborious and do not require such a significant material destruction as the parallel sections techniques.

The best known stereological methods for determining the distribution of spherical particles are described in [9, 10]. However, many structural particles have a non-spherical shape, for example in metals, ceramics, concretes and others. The particles can have the shape that differs significantly from sphere, e.g. cubic γ ' phase particles in nickel-based superalloys [11] or cement particles in concrete [12]. The study of the distributions of non-spherical particles, including grain-polyhedrons, is trying to apply stereological methods developed for spheres, but the difference between their shapes initially makes an obvious

* Corresponding author: danilenko.stereo.7@gmail.com

error in the result. Thus, the key problem of stereology in determining the 3D particle size distribution is the recognition of the shape of these particles [13].

The purpose of this work is to present the method of stereological reconstruction and simulation three diameter size distribution (3DSD) that is applicable for the grain polycrystalline metal structures and structures with spherical particles.

Materials and method

Initially, the method was developed for a spherical particle system (SPS) with a diameter distribution $P(D_k)$ in a scale of n discrete intervals. The main points of this method for the SPS are presented in the works [14, 15]. When the SPS is dissected by a plane, a system of plane sections with a diameter distribution $P_k(d)$ is obtained. Drawing on the resulting plane section of the lines gives a system of chords with the distribution of their lengths $P_k(l)$.

For particles of non-spherical shape it is proposed to use the shape factors of these particles [16]. If the shape of the particles is not uniform, then the shape factors are determined for a particle of the average shape, as will be done below for grain-polyhedrons. Shape factors α_i and β_i are intended to account for each size interval the differences in the distributions of $P_k(d)$ from the sphere and particle, and the differences in the distributions $P_k(l)$ from the circle and plane section of the particle, in accordance with that the shape factors need to be found from the equations:

$$\alpha_i = P_k(d)_{particle} / P_k(d)_{sphere} \tag{1}$$

$$\beta_i = P_k(l)_{particle} / P_k(l)_{circle} \tag{2}$$

The resulting equations establishing the relation between the relative frequencies $P_k'(d)$ and $P'(D_k)$; $P_k'(l)$ and $P'_k(d)$ are:

$$P_k'(d) = \alpha_a P'(D_k)\sqrt{D_k^2 - d_{k-1}^2} + \alpha_b P'(D_{k+1})\left(\sqrt{D_{k+1}^2 - d_{k-1}^2} - \sqrt{D_{k+1}^2 - d_k^2}\right) + ... + \alpha_n P'(D_n)\left(\sqrt{D_n^2 - d_{k-1}^2} - \sqrt{D_n^2 - d_k^2}\right) \tag{3}$$

$$P_k'(l) = \beta_a P'_k(d)\sqrt{d_k^2 - l_{k-1}^2} + \beta_b P'_{k+1}(d)\left(\sqrt{d_{k+1}^2 - l_{k-1}^2} - \sqrt{d_{k+1}^2 - l_k^2}\right) + ... + \beta_n P'_n(d)\left(\sqrt{d_n^2 - l_{k-1}^2} - \sqrt{d_n^2 - l_k^2}\right) \tag{4}$$

For the SPS shape factors are equal to unity. The obtained values of $P_k'(d)$ and $P_k'(l)$ should be normalized in accordance with equations:

$$P_k(d) = P_k'(d) \Big/ \sum_{k=1}^{n} P_k'(d) \tag{5}$$

$$P_k(l) = P_k'(l) \Big/ \sum_{k=1}^{n} P_k'(l) \tag{6}$$

The presented equations (3) and (4) allow to simulate different distributions $P(D_k)$ and calculate the corresponding distributions $P_k(d)$ and $P_k(l)$ and thus study the relation of the structure in volume and at its plane section. This is one of the important application of the method. The proposed simulation method can be widely used for the study of structures with spherical particles, of various surfaces that are formed by systems of hard spheres or which can be modelled by SPS. Examples of such systems are liquid-gas interfaces, the interfaces between fluid and porous substrate, surface cracks in concrete, where the hard spheres denote the aggregate particles, porosity profiles, where the pores are considered as spheres. Thus, the created method makes it possible to replace by simple calculation complex procedures of constructing models of spherical particle structures and their dissection by planes, as discussed in [3, 6, 17, 18].

For the reconstruction of 3DSD by this method two types of initial information can be used: $P_k(d)$ or $P_k(l)$, determined experimentally. In the first case we need to compile a system of equations (3), and solve it relatively unknowns $P'(D_k)$, then the results are normalized in accordance with the equation:

$$P(D_k) = P'(D_k) \Big/ \sum_{k=1}^{n} P'(D_k) \qquad (7)$$

If the initial information is $P_k(l)$, then first we need to compile a system of equations (4) and solve it relatively unknowns $P_k'(d)$, which are then normalized. Then compile the system of equations (3) using the found $P_k(d)$ as knowns and solve it relatively unknowns $P'(D_k)$, which then are also normalized.

Results and discussion

The presented method was applied to analysis of the grain recrystallized austenite structure. The distributions $P_k(d)_{grain}$ and $P_k(l)_{grain}$ were determined as the average for a statistically sufficient group of grains, i.e. for the 3D grain of average shape. The definition of $P_k(l)_{grain}$, necessary to determine the shape factors β_i, was not difficult, as will be discussed later. And to establish the $P_k(d)_{grain}$ from the 3D grain of average shape, necessary for calculating the shape factors α_i, the original method is proposed. According to the well-known principle of Cavalieri-Aker, the proportion of the structural component in the volume of the material is equal to its proportion on a plane section. In accordance to this principle it was assumed that the distribution of plane section diameters $P_k(d)$ from a certain 3D grain is close to the distribution of chords $P_k(l)$ at its maximum plane section, if the plane sections of 3D grain from each dimensional group are treated as separate structural components. The interrelation between the size of plane sections of 3D grain and the length of the chords is shown in Fig. 1. The plane sections Q, M and P, intersecting with the maximum for this 3D grain plane section G, produce on it chords, respectively, *EF*, *CD* and *AB*. The larger the plane section area, the longer the chord it gives on the maximum section G. That is, it should be expected that if the proportion of plane sections from the 3D grain in the *k*-th size interval is $P_k(d)$, then at the maximum plane section of this grain the proportion of chords in the *k*-th size chord interval $P_k(l)$ will be approximately equal to $P_k(d)$.

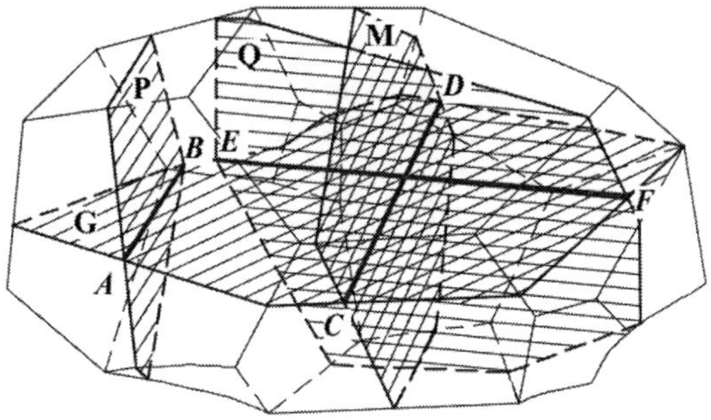

Fig. 1. Scheme of dissection of the metal grain model by planes (shaded) and obtaining chords EF, CD, AB as a result of intersections of planes Q, M, P with the maximum grain section G

The assumed assumption is verified for five polyhedrons, which can be considered as models of grains: pentagonal dodecahedron, tetrakaidekahedron, cubic octahedron, rhombic dodecahedron and truncated icosahedron that were studied in the works of other authors. As $P_k(d)$ used the data obtained by Adrian [19] for pentagonal dodecahedron, tetrakaidekahedron and cubic octahedron; by Hull and Houk [20] for rhombic dodecahedron and by Matsuura and Itoh [21] for truncated icosahedron. In [19] it is presented the distributions of the plane section perimeters for polyhedrons, which were adopted in this paper for the diameter distributions of the equal circles, based on the relation of the circle perimeter and diameter. In [20] and [21] done the distribution $P_k(d)$. To present the results of these works in the required size scale, $P_k(d)$ were determined as the area under the graphs for each size interval, as shown for (0.6 – 0.7) relative size interval for tetrakaidekahedron [19] in Fig. 2. This method was used to determine the distribution $P_k(d)$ for these five polyhedrons in scales of 10, 15 and 20 intervals.

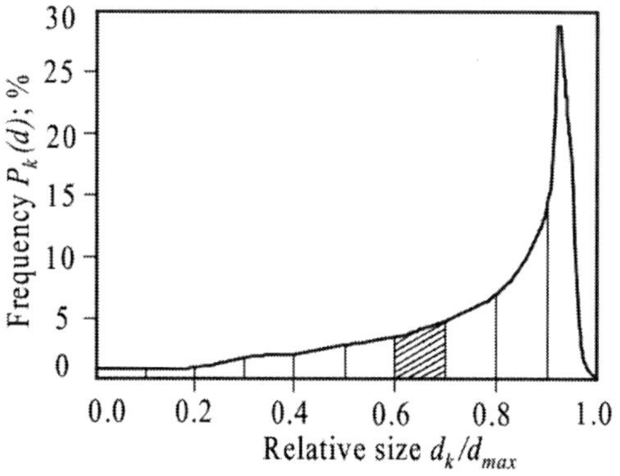

Fig. 2. The scheme for determining the area under the graph (taken from Adrian (1990) for the relative size interval (0.6 – 0.7) for tetrakaidekahedron

The distributions $P_k(l)$ were determined experimentally on the maximum plane sections of these polyhedrons using the computer program «Compass-3D», in which these plane sections were created, then the secant lines were applied in 32 directions at an equal distance between the lines in the same direction, which ensured the statistical uniformity of the secant lines. The number of chords for one plane section was 2500 – 3000 pieces, which, as shown by the additional studies, is sufficient. The results presented in Fig. 3 demonstrate the validity of the assumption that the distribution $P_k(d)$ of polyhedron is close to $P_k(l)$ obtained at its maximum plane section. These distributions are especially close for a size scale of 10 intervals (Fig. 3 *f, g, h, k, m*). For scales of 15 and 20 intervals, there are differences in the values of $P_k(d)$ and $P_k(l)$ in the two or three last size intervals. However, in general, these distributions are close enough, that gave grounds to applying this solution also for obtain $P_k(d)_{grain}$ of austenite grains, for which it is to be expected that the indicated differences in the distributions in scales of 15 and 20 intervals will be partially levelled due to the greater diversity of their shapes in comparison with the investigated polyhedrons.

To determine the shape factors of austenitic grains as outlined above, the shape of about 3000 plane sections of 3D grains was visually analysed and 20 maximum sections of the most representative shape were selected. Since these plane sections were of maximum sizes observed in the structure, they were considered to be the maximum plane sections of the grains from which they were obtained. On each of these plane sections, in accordance with the procedure described above for polyhedrons, $P_k(l)$ was determined using the computer program «Compass-3D» and took it as $P_k(d)$ from these 3D grains, and then determined the average value of $P_k(d)$ for each size interval for 20 grains. The obtained $P_k(d)$ adopted for $P_k(d)_{grain}$ from the 3D grain of the average shape. The number of grains chosen for analysis was considered sufficient, because $P_k(d)_{grain}$, established for two groups of 10 analysed grains, had insignificant differences between them.

To establish the $P_k(l)_{grain}$ the same 20 maximum grain sections were used, on which $P_k(d)_{grain}$ was found, and an additional 20 plane sections of the non-maximal sizes, since they should also represent the average distribution $P_k(l)_{grain}$.

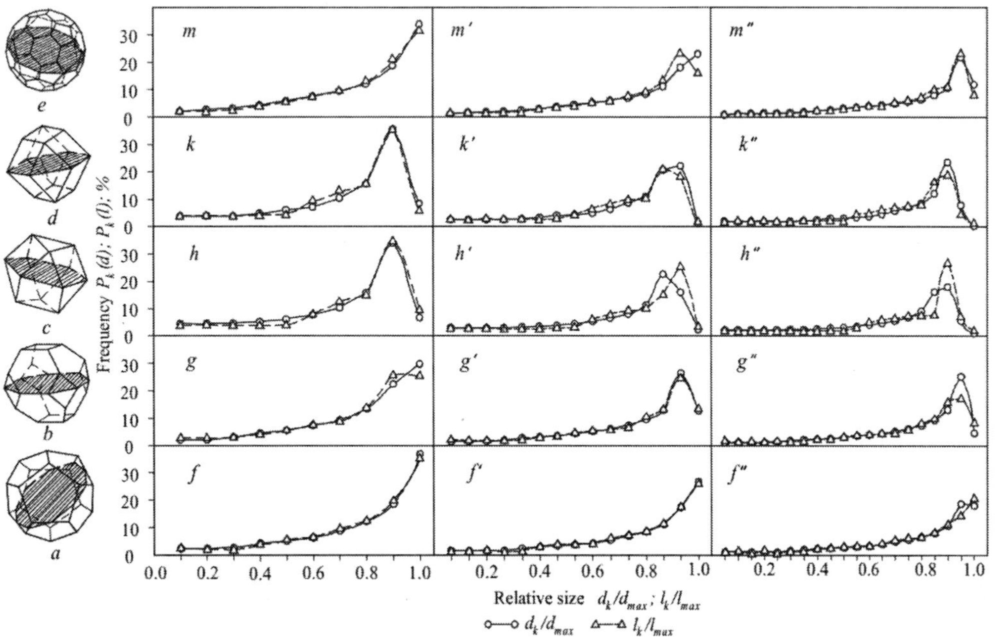

Fig. 3. Models of polyhedrons: pentagonal dodecahedron (a), tetrakaidekahedron (b), cubic octahedron (c), rhombic dodecahedron (d), truncated icosahedron (e) and their maximum plane sections (shaded) and distributions $P_k(d)$ from polyhedrons respectively (*f*, *f'*, *f"*, *g*, *g'*, *g"*, *k*, *k'*, *k"* taken from Adrian (1990); *h*, *h'*, *h"* taken from Hull and Houk (1953); m, m', m" taken from Matsuura and Itoh (1991) and distributions $P_k(l)$ obtained experimentally from the maximum plane section of polyhedrons

$P_k(d)_{sphere}$ and $P_k(l)_{circle}$ were calculated for a sphere with a relative diameter of 1 conventional unit by the equation:

$$P_k(d)_{sphere} = P_k(l)_{circle} = \sqrt{1-d_{k-1}^2} - \sqrt{1-d_k^2} \tag{8}$$

Further, according to equations (1) and (2), the shape factors α_i and β_i were calculated for uniform scales of 10 and 20 intervals, taking into account that from the plane section of the *k*-th size interval chords are obtained only in the size intervals from k-th to 1-th. A computer program has been developed for all calculations. The graphs of the distributions $P_k(d)_{grain}$ and $P_k(l)_{grain}$ obtained and $P_k(d)_{sphere}$ ($P_k(l)_{circle}$) calculated according with equations (8) are shown in Fig. 4, where the proximity of the distributions of $P_k(d)_{grain}$ and $P_k(l)_{grain}$ and the distinct difference between these distributions and $P_k(d)_{sphere}$ and $P_k(l)_{circle}$ are seen. The obtained result indicates that the stereological reconstruction of the 3D grain structure by the methods created for spherical particles can lead to erroneous results and conclusions. The results obtained agree with the data of [20] where was showed, that the maximum of distribution $P(d)$ for the regular polyhedrons is between 0.85 and 0.95 d/d_{max}, and concluded that it should be expected that the maximum frequency $P(d)$ for the average grain will not be as high and narrow as for the regular polyhedrons and an increase in the values of $P(d)$ will be observed at lower value of d/d_{max}. These assumptions are in good agreement with the results presented in Fig. 4. The resulting distribution $P_k(d)_{grain}$ has a maximum in the range 0.8 - 0.9 relative size d/d_{max}, i.e. is shifted to smaller values, and is wider, not so high, and has large values in the range of small d/d_{max}, for example, $P_k(d)$ is 10.3 for grains and (5.2 - 6.2) for polyhedrons at d/d_{max} equal to 0.6.

In the work [22] 3D simulation of grain structures and SPS in which grains and particles had different size distributions was done. The simulation of $P(D_k)$ showed significant differences of the corresponding distributions $P_k(d)$ and $P_k(l)$ for grain-polyhedrons and spheres, which indicates that the assumption that the grain-polyhedrons can be considered as spheres in stereological studies is incorrect. In addition, the distributions of sections have differences from the distributions of the 3D particles, the same applies to the distribution parameters (mean diameter, coefficient of variation, etc.).

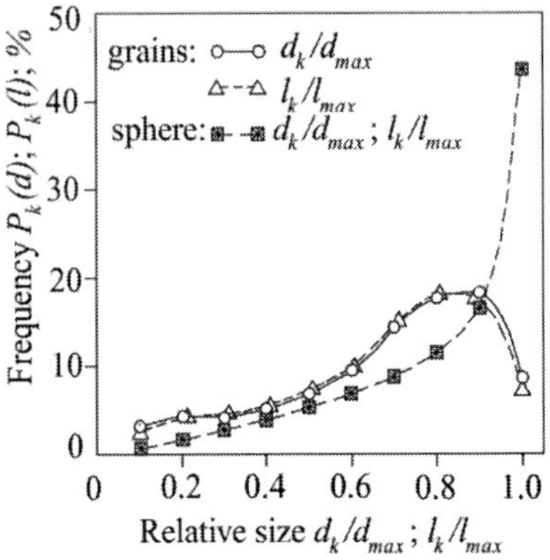

Fig. 4. The distributions $P_k(d)_{grain}$ and $P_k(l)_{grain}$ from the grain of average shape and $P_k(d)_{sphere}$ and $P_k(l)_{circle}$ from the sphere

Creation of the version of the structure standard

The important area of application of the proposed method is the possibility to create the plane section standards of the 3D structures with a known 3-D particles distribution [16]. In this paper a standard version of the plane section of the 3D grain structures was created in the form of the distribution $P_k(d)$ corresponding to the quantitative characteristics of the grain size index *G6* of ISO 643:2012 [23], but with a previously known distribution of 3D grains $P(D_k)$. To do this, first the existing standard *G6* [23] was analysed in the «Compass 3D» program, for this its scheme was created, the areas of plane grains were measured and the diameter distribution $P_k(d)$ of the equal circles was determined.

Then the distribution $P(D_k)$ was chosen, which gave a plane section distribution $P_k(d)$ with the quantitative characteristics corresponded to the distribution $P_k(d)_{ISO\ 643}$ of existing standard *G6*. The distributions $P(D_k)$ of the created version of the standard *G6* and the corresponding $P_k(d)$ and $P_k(l)$, as well as $P_k(d)_{ISO\ 643}$ of the existing standard *G6* are presented in Fig. 5, and the parameters of all distributions are in Table.

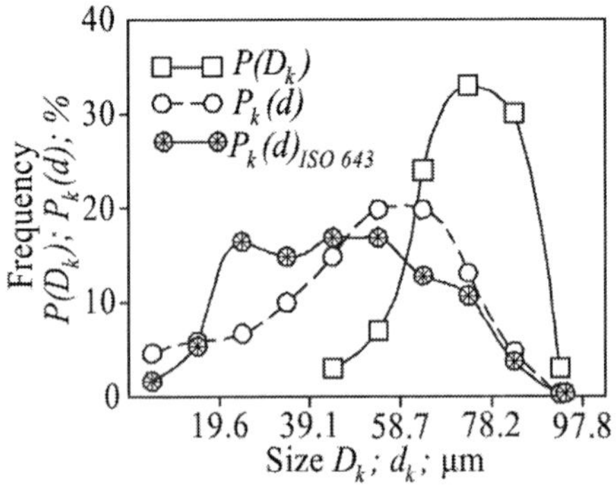

Fig. 5. Distribution of the plane section diameters $P_k(d)_{ISO\ 643}$ of the standard *G6* (ISO 643:2012) and the distribution of 3D grains $P(D_k)$ taken to create a variant of the plane section $P_k(d)$ close to the standard *G6*

These results showed that the plane section of the 3D grain structure which corresponds to the grain size index *G6* with the mean diameter of the plane grains d = 44.2 μm gives a 3D grain structure with a mean diameter D = 72.3 μm, i.e. much larger than observed on the plane section. The coefficient of variation, showing the heterogeneity of the structure, is much smaller for the 3D structure (0.16) than for plane section (0.42).

Table. The standard version parameters of the plane section of the grain structure with the previously known distribution $P(D_k)$, giving the distribution of plane sections $P_k(d)$, corresponding to the quantitative characteristics of the standard G6 of ISO 643:2012

Analysis object	Structure parameters	Values of parameters
Distribution of grain diameters, $P(D_k)$	Mean diameter of grains D, μm	72.3
	Mean square deviation from mean diameter, μm	10.7
	Coefficient of variation of the grain diameters distribution	0.14
	Modal diameter of grains, μm	73.3
	Maximum diameter of grains, μm	97.8
	Number of grains per 1 mm^3, pc.	5060
Distribution of plane sections diameters, $P_k(d)$	Mean diameter of plane grains d, μm	44.2
	Mean square deviation from mean diameter, μm	20.3
	Coefficient of variation of the diameter distribution	0.46
	Modal diameter, μm	58.7
	Mean area of plane grain a, μm^2	1953
	Maximum diameter of plane grains, μm	97.8
	Number of plain grains m per 1 μm^2, pc.	512
Chords length distribution, $P_k(l)$	Mean chord l, μm	39.0
	Mean square deviation from mean chord, μm	18.6
	Coefficient of variation of the distribution of chords	0.48
	Modal size chord, μm	44.0
	Number of chords per 1 μm, pc.	26
Structure parameters of the standard *G6* of ISO 643:2012	Mean diameter of plane grains d, μm	44.2
	Nominal number of plane grains m per 1 μm^2, pc.	512
	Mean area of plane grain a, μm^2	1950
	Mean chord l, μm	40.0
	Maximum diameter of the plane grains, μm	97.8
	Number of chords per 1 μm, pc.	25

After determining the distribution of $P_k(d)$, you can create the standard in the form of a grid, as is done in ISO 643:2012. This study showed that it is expedient to create standards of plane sections of 3D structures using the presented method.

Conclusions

The presented stereological method is applicable both for polyhedral structures and for structures with spherical particles. To take into account the shape of grain-polyhedron, the resulting equations of the method introduce the shape factors. An original method for determination of the shape factors is proposed, which can be applied to other polyhedral structures. The method can be used for creation standards of plane sections of 3D structure with a previously known size distribution $P(D_k)$ of 3D grains or other particles.

References

1. S. Berbenni, V. Favier, M. Berveiller. Int J Plast, **23**, 114 – 142 (2007)

2. T. Wejrzanowski, S. H. Ibrahimb, J. Skibinski, K. Cwieka, K. I. Kurzydlowski. Image Anal Stereol, **36**, 105-110 (2017)
3. B. Biswal, P-E Øren, R. J. Held, S. Bakke, R. Hilfer. Image Anal Stereol **28**, 23-34 (2009)
4. G. Liu, H. Yu. Image Anal Stereol **19**, 81-84
5. L. Bizet, J. Breard, G. Bouquet, J.-P. Jernot, M. Gomina. Image Anal Stereol **23**, 167-176 (2004)
6. A. Tewari, A. M. Gokhale. Mat Char **46**, 329-335 (2001)
7. H. E. Exner. Image Anal Stereol **23**, 73-82 (2004)
8. L. M. Cruz-Orive (2017). Image Anal Stereol **36**, 153-177 (2017)
9. S. A. Saltykov. *Stereometric metallography* (Moscow, Metallurgy, 1970)
10. E. E. Underwood. *Quantitative stereology* (Reading MA:Addison-Wesley, 1970).
11. C. Germain, R. Blanc, M. Donias, O. Lavialle, J.-P. Da Costa, P. Baylou. Image Anal Stereol **24**, 127-134 (2005)
12. H. He, Z. Guo, P. Stroeven, M. Stroeven, L. J. Sluys. Image Anal Stereol **29**,79-84 (2010)
13. H. E. Exner. Acta Stereol **6/III**, 1023-1028 (1987)
14. T. P. Danilenko. *Proc 1-st Int Conf on Devel Test Appl Mat.* Oct 2-5; Opatija, Croatia. 89-95 (1996)
15. T. P. Danilenko. Metallurgical and Mining Industry **4**, 82-88 (2016)
16. T. P. Danilenko, E. I. Danilenko. Patent for invention No 113112 Ukraine. Prom vlasnist **23**, **3**, 64-65 (2016)
17. D. Stoyan. Image Anal Stereol **33**, 225-229 (2014)
18. J. C. Tucker, L. H. Chan, G. S. Rohrer, M. A. Groeberc, A. D. Rollett. Scripta Mater **66**, 554–557 (2012)
19. H. Adrian. *Proc 3-d Conf on Stereol in Mat Science.* Oct; Krakow, Poland. 131-142 (1990)
20. F. C. Hull FC, W. G. Houk WG. J of Met. **April**, 565-572 (1953)
21. K. Matsuura, Y. Itoh. Mater Trans, JIM **32/11**, 1042-1047 (1991)
22. T. P. Danilenko. Metaloznavstvo ta obrobka metaliv **3**, 35-42 (2010)
23. BS EN ISO 643:2012 *Steels. Micrographic determination of the apparent grain size*

Results of theoretical and experimental studies on determining the coefficient of subgrade reaction of sleepers for the conditions of main railways with axial loads of 30-35 tons per axle

Oleksandr Darenskiy[1,*], *Eduard* Bielikov[1], *Olexii* Dudin[1], *Alina* Zvierieva[1], and *Anatolii* Oleshchenko[2]

[1]Ukrainian State University of Railway Transport, 61050, Feierbakh Square 7, Kharkiv, Ukraine
[2]National Academy of the National Guard of Ukraine, Defenders of Ukraine Square 3, 61001, Kharkiv, Ukraine

Abstract. The article considers obtaining numerical values of the coefficient of subgrade reaction of wooden and reinforced concrete sleepers with axial loads up to 30-35 tons per axle. It has been concluded that using the rolling stock with axial loads of up to 35 tons per axle is necessary in order to ensure sustainable development of the railway complex. The performance of the railway track thus should be investigated in order to predict its operation in such conditions. Generally, such studies are performed using numerical methods. One of the parameters that are required for such calculations is the parameter which is commonly called the coefficient of subgrade reaction. Empirical dependencies of the coefficient of subgrade reaction of wooden and reinforced concrete sleepers on the axial load and on the operating conditions of the track have been obtained. The obtained results can be used in studies of the interaction dynamics of the track of main railways with rolling stock with axial loads of 30-35 tons per axle, which will give an opportunity to provide well-grounded recommendations on the rules for the arrangement and maintenance of the track in such conditions.

Introduction

An increase in grain exports is the routine issue of the Ukrainian economy. Thus, despite logistical difficulties, 19.7 million tons of grain was exported by December 13, 2016, which is almost one million tons more than in 2015. However, maintaining such transportation volume becomes more difficult every year. About one thousand cars have been decommissioned (in 2011, the railways managed to dispatch 2,500 cars a day vs. about 2,100 cars per day now. The main factor of reduction is the physical wear of the fleet of grain hoppers, which in turn reduces the speed of railway operations and prevents from loading cars at full capacity). The number of available road vehicles is decreasing, as the

* Corresponding author: ppx_xiit@kart.edu.ua

fleet of vehicles is not renewed. Launching an automated car distribution system, which will also include grain hoppers, as well as an increase in axial load from 23.5 tons to 30-35 tons will help solve the problem.

The existing methods of calculating the effect of the rolling stock with such loads on the track cannot give an unambiguous answer to the question how the relationship of the track and rolling stock will change with such cars, how the dynamic interaction will change and how it will affect the performance of the track. Only dynamic calculation of the interaction forces between the track and the rolling stock can give the answer, for example, using the mathematical model of the 3D and dynamic system "vehicle-track" developed by A. M. Darenskiy [1].

One of the characteristics required for such calculations is the coefficient of subgrade reaction, which as shown by [1], depends on the load on the support, and, finally, on the value of the axial load. Thus, studying this parameter becomes a relevant task.

In the leading domestic and foreign studies, which will be outlined below, the coefficient of subgrade reaction is determined by experimental methods. Since in the global practice, the axial loads of 30-35 tons per axle are not used on main railways, the studies carried out by the authors for the conditions of non-public railways in conditions that are identical to those of the main lines were accepted as the basis.

The analysis of studies and publications

The effect of elastic properties of sleepers and their base on the formation of vertical stiffness of the track is very significant, which is noted by many researchers. For example, the calculations performed in [2] showed that the proportion of this parameter in the formation of the elastic modulus of the track with wooden sleepers in the summer is 77% to 90%. The experiments conducted at the German railways [3] determined that the specific weight of sleepers and their bases in the total vertical deformation of the track is 80% for wooden sleepers, and 95% for reinforced concrete sleepers.

It is noted that the stiffness of the track (taking into account the bending stiffness of the rails) and the stiffness of the rail supports depend essentially on the type of sleepers (wooden or reinforced concrete ones) and on the type of liners used in intermediate fastenings [4, 5, 6, 7]. In winter, when the ballast and roadbed are frozen, the stiffness of the track increases 1.5 times [7], and according to other data, even 2-3 times [4].

The elastic component of the displacement resistance forces of sleepers in the vertical plane is now assumed to be defined as:

$$R_y = C_{уш} \cdot Y_p = C_{y\sigma} \cdot \alpha \cdot \Omega_ш \cdot Y_p \text{, (kN)} \qquad (1)$$

where $C_{уш}$ – element of the vertical stiffness of the rail support, depending on elastic properties of the sleeper at bends under the action of vertical forces and elastic properties of the ballast layer and roadbed (kN/m); Y_p – vertical displacements of a sleeper in the subrail section under the action of force R_y (m); $C_{y\sigma}$ – coefficient of subgrade reaction (kN/m^3); α – bending coefficient of a sleeper; $\Omega_ш$ – area of a half-ties (m^2).

Many experiments [8-10, etc.] were devoted to the experimental determination of the values of the coefficient of subgrade reaction. Basically, values of $C_{уб}$ parameter are given for static loading of the track, since the methods for determining the dynamic stiffness are still neither well investigated nor defined. For example, methods for determining the dynamic parameters proposed in [9] are known. It is noted that in case of dynamic loads, the value of the coefficient of subgrade reaction can be 70%, and in some cases - 60%, of this value at static loading. Theoretical substantiation of this phenomenon as a result of ballast vibration is given in [5].

In the process of operation, as the ballast is compacted and contaminated, the coefficient of subgrade reaction of the sleepers increases by 40-80% [11].

For the conditions of main railways with loads up to 25 tons per axle, the value of the coefficient of subgrade reaction of wooden sleepers is recommended to be assumed in the range from 65 to 85 mPa/m for the summer and from 120 to 180 mPa/m for the winter conditions $6.5\text{-}8.5 \cdot 10^{-4}$ and $12\text{-}18 \cdot 10^{-4}$ kN/m^3. In case of reinforced concrete sleepers, the coefficient of subgrade reaction is [4, 9, 11] 180-240 mPa/m for the summer and 280-340 mPa/m for the winter conditions. Since there is no experience of operation of the rolling stock with high axial loads on main lines, the experience of operation of access ways should be used. For industrial railways, this issue requires special research.

The basic part

The coefficients of subgrade reaction of sleepers for the conditions of the railways of the Ukrainian industrial enterprises have been determined on the basis of the data of experimental studies conducted in 2010-2012.

The works were carried out at 21 sections of the track of metallurgical and mining enterprises. The studied sections were selected on the internal connecting and access roads of enterprises with the upper track structure on the reinforced (fastening KB) and wooden (fastening DO) sleepers.

At each investigated section, 3 to 5 sleepers were tested. Vertical loads on the sleepers from the rolling stock were recorded using dynamometric substrates, vertical displacement of sleepers were recorded using a digital video displacement metering system [12]. Vertical forces and displacements of sleepers were recorded continuously in digital form in the computer memory during the passing of the train.

The recorded information was copied and sent to MS Excel spreadsheet. Charts of elastic characteristics of sleepers for the mean values of displacements and loads for each section were constructed on the basis of these data. Elastic characteristics for both reinforced concrete and wooden sleepers are fading (rigid), and the vertical stiffness of sleepers increases with increasing loads (Fig. 1, Fig 2).

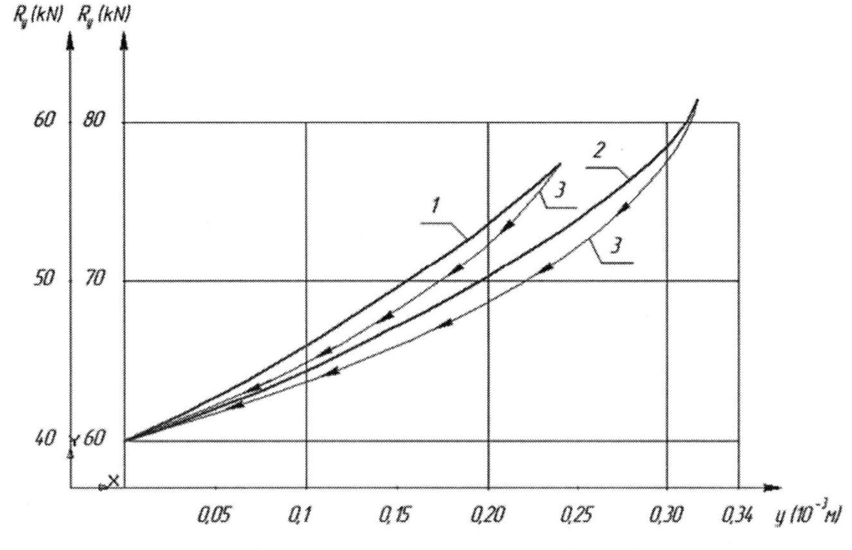

Fig. 1. Elastic characteristics of reinforced sleepers under vertical loads: 1 - section 14, hot-metal car 80t, *Poc* = 353 kN/axle, service life of the track – 5 years, tonnage – 60 million tons; 2 - section 19, dump car 6-BC-60, *Poc* = 215 kN/axle, service life of the track – 5 years, tonnage – 90 million tons; 3 - elastic characteristics at unloading

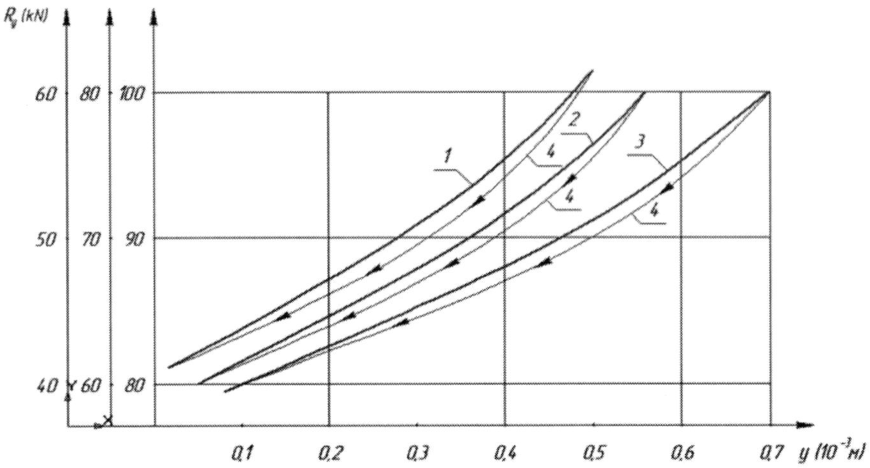

Fig. 2. Elastic characteristics of wooden sleepers under vertical loads: 1 – section 7, casting form bogies I-120-5500, *Poc* = 35.25 t/axle, service life of the track – 6 years, tonnage – 48 million tons; 2 – section 10, platform of hot-metal car 70 t, *Poc* = 245 kH/axle, service life of the track 12 years, tonnage –49 million tons; 3 – section 2, hot-metal car 50t, *Poc* = 201 kN, service life of the track – 7 years, tonnage – 50 million tons; 4 – elastic characteristics at unloading.

The stiffness of sleepers at vertical loads of a single rail line has been determined by linearizing the elastic characteristics using the chord method for the ranges 40-60, 60-80 and 80-100 kN, which corresponds to the actual loading on the sleepers when trains with axial loads up to 265 kN, 265 to 294 and more than 294 kN per axle move.

The obtained information suggests that the vertical stiffness of reinforced sleepers increases with the increase of operating load ranges with the comparable values of the tonnage passing the sections. This increase is 31-37% in the ranges 60-80 kN vs. the range 40-60 kN. For wooden sleepers this increase is from 22-27% to 52-59% (Fig. 1, Fig. 2) for different types of cars.

An increase in vertical stiffness was recorded for sections where identical rolling stock moves and which differ in terms of service life of the track. For example, Fig. 3 contains the values of the stiffness of reinforced concrete and wooden sleepers (single-rail line) obtained for the sections of loaded dump cars 6-BC-60 with permitted axial loads up to 216 kN/axle, with the service life of the track of up to 18 years, and the passing tonnage from 0 to 180 million tons.

It can be assumed, and this assumption is confirmed by the experiments, that it is the tonnage passing the sections as a generalizing factor of force impact on the track that has a determining influence on the change in the stiffness of sleepers under vertical loads and on the coefficient of subgrade reaction of sleepers. As the tonnage increases, density and contamination of the ballast also increase, which causes an increase in the stiffness of sleepers.

Coefficient of subgrade reaction of reinforced concrete and wooden sleepers was determined using the investigated data as

$$C_{y\sigma} = \frac{R_y}{\alpha \cdot \Omega_{ш} y_p} \qquad (2)$$

where R_y – vertical load effecting the sleeper (kN); $\Omega_{ш}$ – area of half-ties (m^2); y_p – vertical displacements of the sleeper in the subrail section (m); α – bending coefficient of the sleeper.

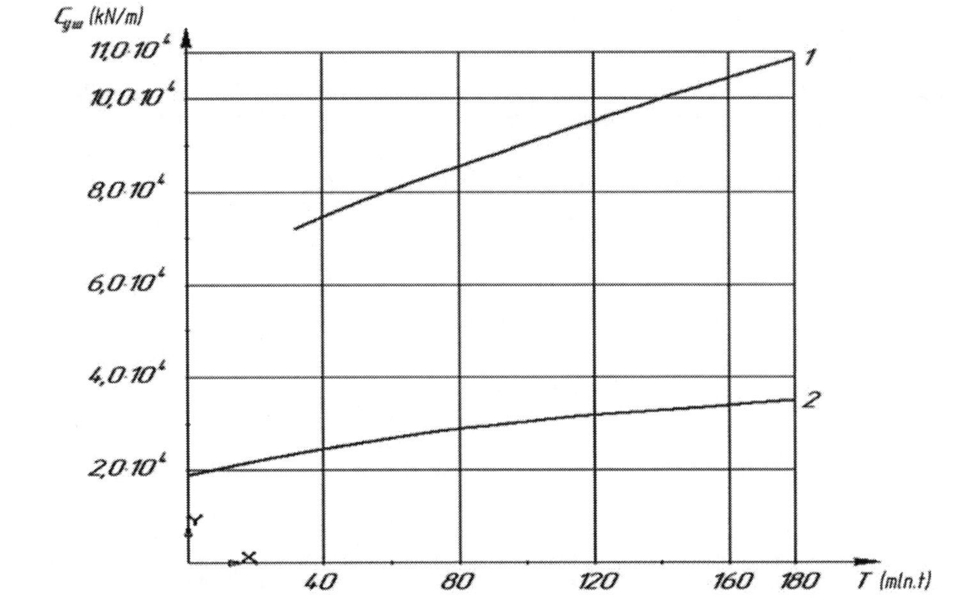

Fig. 3. Dependence of vertical stiffness of sleepers on the passed tonnage when dump cars 6-BC-60 move, 1 – reinforced concrete sleepers; 2 – wooden sleepers.

In equation (2), when the experimentally obtained data on changes in the values of R_y and y_p were used as arguments, two parameters remain unknown – the coefficient of subgrade reaction of sleepers $C_{y\sigma}$ (kN/m^3) and the bending coefficient of sleepers α. The bending coefficient of sleepers which is the ratio of the average subsidence of sleepers to the subsidence in the subrail section, in turn, depends on the loads affecting the sleepers, the geometric characteristics of the sleepers, elastic properties of its material, and its coefficient of subgrade reaction.

The values of coefficient α obtained in various studies vary greatly, they were obtained for the conditions of main tracks and are not determined in terms of their applicability – for example, according to the value of the coefficient of subgrade reaction.

Determination of parameters C_σ and α can be considered as identification of the object of the given structure. Here the mathematical model of the object is given by equation (2) and the method of calculating a sleeper as a variable section beam lying on a solid elastic basis which is characterized by the coefficient of subgrade reaction of the sleeper. The parameters C_σ and α of this model are selected so that at a given input (the value of the vertical forces R_y), the output of the model only slightly differs from the experimentally defined output.

Searching for unknown parameters in the form of semi-empirical function $\alpha = f(C_\sigma)$ was carried out by one of the hill-climbing methods for the objective function – the coordinate descent method [12]. A value equal to the sum of the squares of relative deviations of the experimental and calculated values of the bends of sleepers in the subrail section was assumed as an objective function:

$$\varepsilon = \sum_{i=1}^{i=n} \frac{(y_{pi} - y_{pэi})^2}{y_{pi}}; \quad y_{pi} \neq 0 \tag{3}$$

where y_{pi} – value of bending of the sleeper, determined using the calculation method at R_{yj}, (m); $y_{pэi}$ – same, determined experimentally (m); n – number of experimental points.

The conducted experiments, statistical processing of the obtained results and the calculations performed according to the stated method using the Mathcad software system allowed to determine the values of the coefficients of subgrade reaction of reinforced concrete and wooden sleepers and their coefficients of bending (Table 1). The values of the mean square deviations of the coefficients of subgrade reaction of reinforced concrete sleepers are within the range of 7%, of wooden sleepers – 13%.

For track sections with axial loads of 201-216 kN, the values of the coefficient of subgrade reaction and the bending coefficients of sleepers are consistent with the data given in [4, 5, 8].

Table 1. Coefficients of subgrade reaction and bending coefficients of sleepers in the studied section

Section No.	Sleeper type	Axial loading (kN)	Passed tonnage (mln. tons)	Coefficient of sbgrade reaction (kN/m^3)	Mean square deviations (kN/m^3)	Bending coefficient of sleeper
1	Wood. I A	201	14	$7.46 \cdot 10^4$	$0.64 \cdot 10^4$	0.815
2	Wood. I A	201	49	$8.97 \cdot 10^4$	$0.76 \cdot 10^4$	0.807
3	Wood. I A	353	60	$14.06 \cdot 10^4$	$1.15 \cdot 10^4$	0.787
4	Wood. I A	353	108	$15.87 \cdot 10^4$	$1.42 \cdot 10^4$	0.771
5	Wood. I B	211	25	$7.82 \cdot 10^4$	$0.81 \cdot 10^4$	0.812
6	Wood. I B	211	50	$8.91 \cdot 10^4$	$0.94 \cdot 10^4$	0.802
7	Wood. I A	346	48	$13.66 \cdot 10^4$	$1.45 \cdot 10^4$	0.785
8	Wood. I A	346	64	$13.92 \cdot 10^4$	$1.41 \cdot 10^4$	0.781
9	Wood. I B	245	2	$6.54 \cdot 10^4$	$0.61 \cdot 10^4$	0.824
10	Wood. I A	245	48	$8.67 \cdot 10^4$	$0.72 \cdot 10^4$	0.801
11	Wood. I A	216	9	$6.57 \cdot 10^4$	$0.67 \cdot 10^4$	0.820
12	Wood. I A	216	126	$10.72 \cdot 10^4$	$1.21 \cdot 10^4$	0.795
13	RC Sh-1	353	60	$35.42 \cdot 10^4$	$1.21 \cdot 10^4$	0.877
14	RC Sh-1	353	108	$37.64 \cdot 10^4$	$1.25 \cdot 10^4$	0.868
15	RC Sh-1	211	45	$21.25 \cdot 10^4$	$0.96 \cdot 10^4$	0.923
16	RC Sh-1	211	50	$21.53 \cdot 10^4$	$0.97 \cdot 10^4$	0.927
17	RC Sh-1	211	50	$20.14 \cdot 10^4$	$1.12 \cdot 10^4$	0.934
18	RC Sh-1	216	90	$24.35 \cdot 10^4$	$1.32 \cdot 10^4$	0.907
19	RC Sh-1	216	126	$26.25 \cdot 10^4$	$1.23 \cdot 10^4$	0.903
20	RC Sh-1	216	180	$29.43 \cdot 10^4$	$1.75 \cdot 10^4$	0.897
21	RC Sh-1	216	180	$29.48 \cdot 10^4$	$1.63 \cdot 10^4$	0.893

For wooden sleepers the parameters of this dependence are as follows:

$$C_\sigma = K_{oc}\,(6.5 \cdot 10^4 + 0.23 \cdot 10^4 \cdot T^{0.605})\ (\text{kN/m}^3), \quad (5)$$

where K_{oc} – 1, if axial loads <265 kN; K_{oc} =1.24, if loads is 265-294 kN; K_{oc} = 1.55, if loads >294 kN.

Semi-empirical dependence of the bending coefficient of sleepers on its coefficient of subgrade reaction has the form:

- for reinforced concrete sleepers: $\alpha = 1.2727 \cdot (C_\sigma \cdot 10^{-4})^{-0.1047}$ (dependence is applicable for the range of the coefficient of subgrade reaction C_σ =16.0-43.0·10^{-4} kN/m^3)

- for wooden sleepers: $\alpha = 0.9222 \cdot (C_\sigma \cdot 10^{-4})^{-0.0627}$ (applicable for the range C_σ =6.0-22.0·10^{-4} kN/m^3)

The works on determining the resistance of reinforced concrete and wooden sleepers to vertical movement, performed in winter conditions in 2012 at sections No. 11, 12, 20 and 21 showed that a 1.6 times increase in the coefficient of subgrade reaction for wooden sleepers and 1.8 times increase for reinforced concrete sleepers in winter for the Ukrainian industrial railways is substantiated.

Conclusions

1. On the basis of conducted experiments, the values of coefficients of subgrade reaction of reinforced concrete and wooden sleepers with axial loads of up to 35 tons per axle have been obtained. The dependence of the change of these parameters in the course of the operation of the track has been established taking into account the level of axial loads.

2. The obtained information will be used in further studies of interaction dynamics of the track of main railways with the rolling stock with axial loads of 30-35 tons per axle which will give an opportunity to provide substantiated recommendations concerning the norms of the arrangement and maintenance of the track in such conditions.

References

1. O.M. Darenskiy *Theoretical and experimental studies of the performance of railway tracks of industrial transport – monograph,* (Kharkiv: UkrSART, 2011)
2. V.P. Novichkov *Rational Methods of Strengthening the Railway Track,* (Zh. D. Transp., 1956)
3. F. Birmann, Eisenbahntechnische Rundschau, **7**, 229-246 (1977)
4. N.V. Ivolga, V.I. Klimov, Proceedings of DIIT, **167,** 73-81 (1975)
5. V.G. Albrekht, A.F. Zolotarsky, *Modern designs of the upper railway structure* (Transport, 1975)
6. A.F. Zolotarsky, B.A. Evdokimov, L.G. Isaev, L.G. Krysanov *Reinforced concrete sleepers for railways* (Transport, 1980)
7. E.I. Danilenko, V.V. Rybkin, *Rules for durability and stability calculation of the railway track,* (Transport of Ukraine, 2006)
8. O.P. Ershkov, Proceedings of the Central Research Institute of the MPS, **264**, 39-48 (1964)
9. V.F. Yakovlev, I.I. Semenov, V.I. Abrosimov, Proceedings of the LIIZhT, **326,** 66-85 (1971)
10. M.A. Frishman, L.Ya. Vorobeichik, R.S. Lipovskoy, Bulletin of the Central Research Institute of the Ministry of Transport, **8,** 31-35 (1970)
11. E. S. Varyzgin, *Contents of the ballast prism of the railway track,* (Transport, 1978)
12. O.M. Darenskiy, N.V. Bugaets, V.G. Vitolberg, Collection of sciences works of UkrDazt, **109,** 222-231 (2009)

A longer service life of intermediate elastic fastenings KPP-5 due to application of pads PRP-3.2

Oleksandr Darenskiy[1,*], *Nataliya* Bygaets[1], *Olga Horiyinova*[1], *Sergey* Voznenko[1], and *Stanislav* Kulik[2]

[1] Track and Track Facility Department, Ukrainian State University of Railway Transport, Feuerbach Sq. 7, 61050, Kharkiv, Ukraine
[2] The Kupiansk–Uzlovaya track maintenance department, Zheleznodorozhnaya Sq. 10, 63709 Kupiansk, Ukraine

Abstract. Since 2003 Ukrainian railways have used KPP-5 intermediate elastic fastenings. These fastenings unlike their predecessors (for example, terminal-bolted fastenings (KB) widely used on railways), first of all, have fewer elements. The KPP-5 fastening does not include screw joints, which considerably increases the pressing force between rail and tie, and resistance to track displacement forces. This may lead to reduction of the total labour input for routine track maintenance due to cutback of track works needed for mounting terminal and insert bolts. The article deals with forecasting the operational life of PRP 3.2 (9mm depth) and PRP 3.2.1 pads (10mm depth). In engineering the operational life is a time interval (for the railway track – tonnage passed) during which the element fulfils its functions. One of the main reasons why the assembly unit KPP-5 fails is low pressing forces between rail and tie. It occurs due to many processes; among the basic ones is the changed characteristic of underrail pads. And the application of new types of pads – PRP 3.2 (9mm) and PRP 3.2.1. (10mm) may solve this problem.

Introduction

The KPP-5 intermediate elastic fastenings have been applied on Ukrainian railways since 2003. The normative documents [1-6] stipulate laying down continuous welded rails on sections of various tonnages along straight lines and curves of the radii 350m and more. The length of continuous welded rails can be equal to the length of a block section or a railway haul (over length rails).

For continuous welded rails one of the main problems for intermediate fastenings is to create sufficient per-unit length resistance, i.e to provide the pressing force between rail and tie more than 23 kN. In operation resilient terminals and pads of the track suffer from residual deformations, liners wear out, the physical and mechanical characteristics of pads change, and all these factors lead to lower pressing forces. Application of pads PRP 3.2 and

* Corresponding author: ppx_xiit@kart.edu.ua

PRP 3.2.1 can prolong the operational life of the assembly unit, however, it is necessary to know dependencies of changes in the indicated parameters and, thus, to determine time when pads should be replaces and the operational life of the assembly as a whole [7-11].

The basic part

The mounting of the assembly unit KPP-5 provokes initial vertical deformation of the spring terminal. And each terminal presses a rail base to a tie with the force Q_{term}^{init} which can be defined as

$$Q_{term}^{init} = r_{term} \cdot \Delta Z_{init} \tag{1}$$

where r_{term} is the rigidity of the spring terminal at vertical deformations.

Under the terminal pressure $2 \cdot Q_{term}^{init}$ the underrail pad shrinks and creates the elastic (spring) resistance Q_{pad}. The shrinkage of the pad causes the inverse resilient deformation of the terminals leading to a decrease in the terminal pressing forces. Equilibrium of forces in the assembly unit occurs when the mounting forces of terminal pressure equal the elastic resistance of the pad

$$Q_{mount} = u_{pad}^{stat} \cdot \Delta Z_{pad}^{mount} \tag{2}$$

where u_{pad}^{stat} is the rigidity of a pad at static pressing;

ΔZ_{pad}^{mount} is the mounting pressing of the pad in an assembly.

As the structural peculiarity of the fastening KPP-5 is the equilibrium of the deformations of the pad and the inverse deformation of the terminal, the mounting pressing of the terminal can be presented as

$$Q_{mount} = Q_{init}^{mount} - \Delta Q = Ж_{pad} (\Delta Z_{init} - \Delta Z_{pad}^{mount}) \tag{3}$$

Or, considering expression (2), the mounting pressing on the pad can be defined as

$$Q_{mount} = \frac{2 \cdot Ж_{pad} \cdot Z_{init}}{u_{pad}^{stat} + 2 \cdot r_{pad}} \tag{4}$$

Thus, the mounting pressing on the terminal when laying down into the track equals to

$$Q_{mount} = r_{pad} (\Delta Z_{init} - \Delta Z_{pad}^{mount}) \tag{5}$$

On the basis of the analysis of the dimensions of structural elements of the KPP-5 fastenings (Fig. 1-3), the initial deformation of the elastic terminals depends on the thickness of the rail pad and is in a range of 11.24 – 13.24 mm.

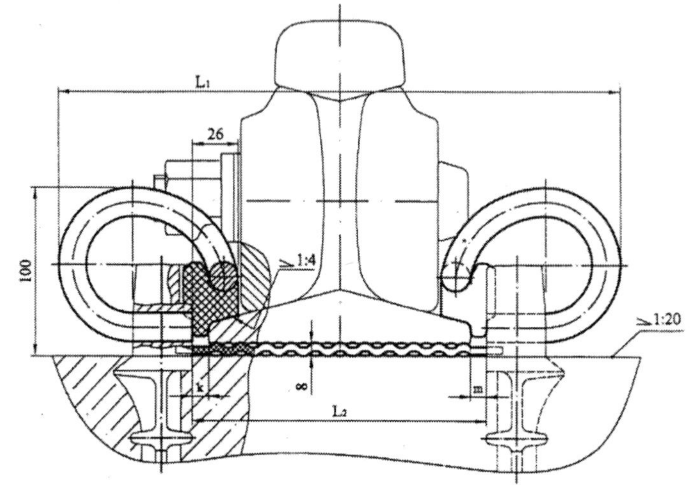

Fig. 1. Basic dimensions of the KPP-5 assembly unit

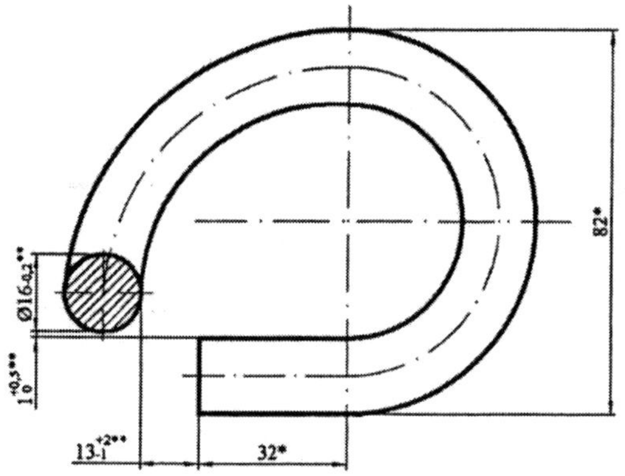

Fig. 2. Basic dimensions of the KP 5.2 terminal

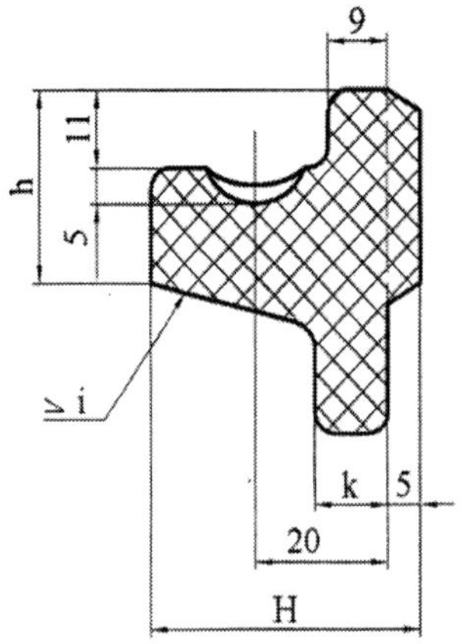

Fig. 2. Basic dimensions of the insulating liner VIP-65.11

The rigidity of the terminals, determined in the laboratory environment, is 1.4 kN/mm.

In track operation the mounting pressing of the terminals to the ties will be reduced due to residual deformations and deterioration of the underrail pad, final deformations of the terminals and deterioration of holes in the anchors and deterioration of the concrete of the underrail base, which increase the distance between the anchor's axle and the underrail base of the tie

$$Q_{mount} = r_{pad} (\Delta Z_{init} - \Delta Z_{pad}^{mount} - \Delta h_{pad} - \delta_h - \Delta h_{anch}) \tag{6}$$

where Δh_{pad} is the final deformation of the pad; δ_h is the final deformations of the terminal; Δh_{anch} is an increase of the distance between the underrail base and the anchor's top.

It should be mentioned that apart from the residual deformations and deterioration, in operation the assembly unit suffers from changes in rigidity (both dynamic and static) in the underrail pads.

The empirical dependency of increased static rigidity of the pad is

$$u_{pad}^{stat} = u_{pad} + 0.27 \cdot 10 \cdot t \tag{7}$$

Since the changes in some parameters depend not only on the service cycle, but also on the conditions of their power work, which can be generalized by the factor of the tonnage transported along the sections, the calculations were conducted for four load capacity values: 15, 30 45, 60 m tkm gross/km per year.

Fig. 4, 5 present generalized diagrams of changes in the pressing force between rail and tie in the assembly unit. The analysis of the diagrams showed that the pressing force was decreasing to a critical value of 20 kN. The reference periods for the pressing force in the assembly unit to be restored are given in Table 1.

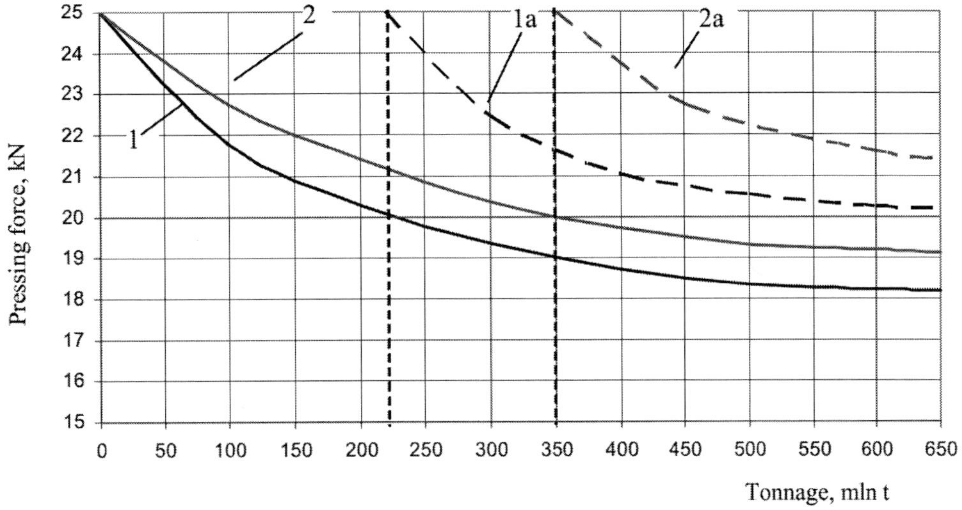

Fig. 4. Diagram of change in the pressing force between rail and tie in the assembly unit KPP-5: 1 – traffic density 15 mln tkm gross/km per year; 2 – traffic density 30 mln tkm gross/km per year; 1a, 2a – change in the pressing force after installation of PRP 3.2 pads

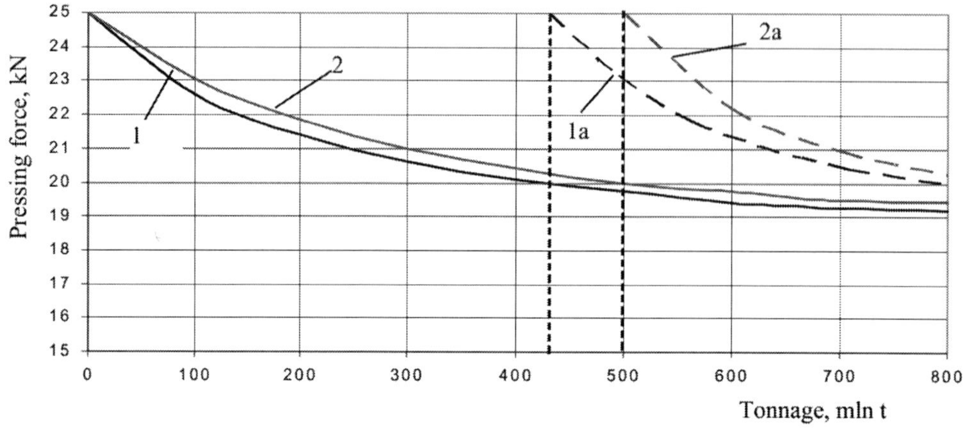

Fig. 5. Diagram of change in the pressing force between rail and tie in the assembly unit KPP-5: 1 – freight density 45 mln tkm gross/km per year; 2 – carrying capacity 60 mln tkm gross/km per year; 1a, 2a – change in the pressing force after installation of PRP 3.2.1 pads

Table 1. Reference periods for the pressing force to be restored in the assembly unit KPP-5 in dependence on the freight density

№	Freight density mln tkm gross/km per year.	Reference periods for the pressing force in the assembly unit to be restored	
		mln t	years
1	15	220	14.7
2	30	350	12.7
3	45	430	11.1
4	60	500	8.3

Considering the fact that pads of 10mm depth are more expensive, the final decision regarding the pad type should take into account the technical and economic assessment.

Conclusions

1. The formation of pressuring forces between rail and tie in the assembly unit KPP-5 is determined.

2. The dependencies of changes in residual deformations of terminals and pads, pad deterioration, changes in rigidity of PRP 2.1 pads in operation were obtained.

3. The periods when PRP 2.1 pads should be replaced for PRP 3.2 and PRP 3.2.1 pads under various operational conditions are determined. It will provide reliable work of KPP-5 fastenings during the whole overhaul period.

References

1. E.I. Danilenko, V.O. Yakovlyev, A.M. Orlovskyj, M.I. Karpov *Instrukciya z ulashtuvannya ta utrymannya koliyi zaliznyc Ukrayiny* (Transport Ukrayiny, 2006).
2. E.I. Danylenko, V.V. Rybkin. *Pravyla rozraxunkiv zaliznychnoyi koliyi na micznist i stijkist* (Transport Ukrayiny, 2006) .
3. V.G. Al'breht, E.M. Bromberg, N.B. Zverev, *Besstykovoj put'* (Moskow, Transport, 1982).

4. E.M. Bromberg, M.F. Verigo, V.N. Danilov, Vzaimodejstvie puti I podviznogo sostava (Moskow, Transport, 1956).
5. G.M. Shahunjants, *Zheleznodorozhnyj put', Monografija* (Moskow, 1987).
6. M.F. Verigo, A.J. Kogan, *Vzaimodejstvie puti i podvizhnogo sostava* (Moskow, Transport, 1986).
7. M.F. Verigo, Vestnik VNIIZhTa, **21 (5),** 25-37 (2001).
8. V.I. Klimov, V.V. Rybkin, Trudy DIITa, **32(244),** 72-84 (1987)
9. V.V. Govoruha. *Mekhanika deformirovaniya i razrusheniya uprugih ehlementov promezhutochnyh relsovyh skreplenij* (Dnepropetrovsk, 2005).
10. A. Darenskiy, D. Potapov, Y. Tuley, N. Bugaets, A. Malishevskaya, MATEC Web of Conferences. **116**, 03001 (2017).
11. S. Panchenko, T. Butko, A. Prokhorchenko, L. Parkhomenko, O. Zhurba, International Journal of Engineering & Technology. **7 (4.3)**, 217-221, (2018)

Specific feature of the restoration of operability of KPP-5 fasteners by using PRP-3.2 and PRP-3.2.1 rail pads

Denis Fast[1,*], *Natalia* Bugaets[1], *Volodymyr* Vitolberg[1], *Alexandr* Lichodey[2], and *Volodymyr* Stefanov[1]

[1]Ukrainian State University of Railway Transport, Feierbakh Square 7, 61050, Kharkiv, Ukraine
[2]National Academy of the National Guard of Ukraine, Defenders of Ukraine Square 3, 61001, Kharkiv, Ukraine

Abstract. Areas of rational use of the track construction with reinforced concrete sleepers have been defined, however, it requires improving and the problem of intermediate rail fastenings is particularly serious here. The most common intermediate rail fasteners used in Ukrzaliznytsya are lining terminal-bolted fasteners of KB type and direct fixation boltless fastening type KPP-5. Calculations have been made for the restoration of operability of KPP-5 fasteners using PRP-3.2 repair pads with a thickness of 9 mm and PRP-3.2.1 with a thickness of 10 mm. It has been established that in the sections with the traffic load of more than 15 million ton·km/km in year, restoration works should be performed after 13 years of operation. Using 10 mm thick PRP-3.2.1 rail pads will provide the necessary pressing force of the rail base against the sleeper until the expiration of the standard service life – no more than 30 years. When the traffic load is less than or equal to 15 million ton·km/km in year, recovery can be performed after 22 years of operation. The use of both types of pads will ensure the reliable functioning of KPP-5 fasteners until the expiration of its service life.

Introduction

To date, the railway track construction with reinforced concrete sleepers is quite common for the Ukrainian railways. Areas of rational use of the track construction with reinforced concrete sleepers have been defined, however, it requires improving and the problem of intermediate rail fastenings is particularly serious here. The maintenance and repair works on intermediate rail fasteners take a significant proportion of labour input in the track operation. This is caused by many reasons [1-5], the main of which include the absence of clear scientifically rationalized mechanical, technological and operational parameters of the intermediate rail fasteners, which significantly affect the reliability of the railway track.

The most common intermediate rail fasteners used in Ukrzaliznytsya, which have undergone complete long-term operational observations, are lining terminal-bolted fasteners of KB type and direct fixation boltless fastening type KPP-5.

* Corresponding author: ppx_xiit@kart.edu.ua

The main problems in the performance of these fasteners during operation can be determined based on the analysis of previous studies, the results of which are described in [6-10] devoted to the study of performance of KPP-5 fasteners and under conditions of non-public tracks:

1. The appearance of residual deformations in KPP-5 fastening elements due to wear of the rail pads, wear of the orifices in the anchoring beams, wear of the rail pads and the final deformation of the elastic terminals. This leads to a reduction in downforce of the rail base to the rail pad and, as a consequence, to violation of the normal temperature performance of the continuous welded track;

2. Changes in the stiffness of PRP-2.1 rail pads lead to the appearance of a spatial uneven stiffness of KPP-5 fasteners, which can lead to a significant increase in the spatial interaction forces between the track and the rolling stock, which in turn will increase the track defects.

PRP-3.2 and PRP-3.2.1 (A90 and A95) repair rail pads can compensate for these processes, but additional studies should be conducted in order to determine the rational areas of their application and their arrangement along the repaired section to ensure equal elasticity of the sub-rail base.

The basic part

To study the interaction dynamics of the rolling stock and the track in the vertical plane, it was convenient to use not absolute values of the interaction forces, but more convenient and informative dynamic coefficients. The dynamic coefficient is defined as the ratio of the value of dynamic forces P_{dyn} to its statistical value $P_{st.}$

$$K_{dyn} = P_{dyn} \,/\, P_{st.}. \tag{1}$$

It should be noted that when the rails are used as beams on a solid elastic foundation in dynamic calculations of the vehicle and the track when the vehicle moves along a straight section at a constant speed, the dynamic coefficient is 1. Meanwhile, under real conditions, vertical forces keep changing and this coefficient differs significantly from 1.

Fig. 1 shows the diagrams of the calculated changes in the vertical forces acting from the wheel to the rail when a 12-100 gondola car is moving along the track at a speed of 80 km/h. The distance travelled by the car is used as the abscissa, while the action force P is the ordinate axis. In this example, the spatial stiffness of the rail supports was assumed to be constant.

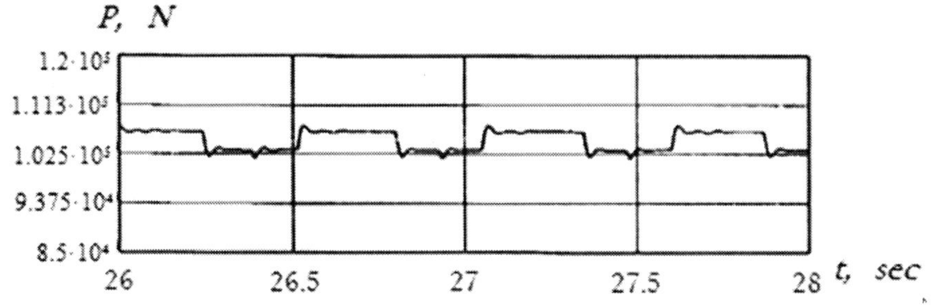

Fig. 1. Diagrams of dynamic vertical forces during the movement of a 12-100 gondola car

Introduction to the calculation of uneven elasticity of rail supports using the Monte Carlo method makes a very significant effect on the coefficients of dynamics. Diagrams of the dependence of changes in the dynamic coefficients on the uneven elasticity of the under-rail base with KPP-5 fasteners for PRP-2.1 rail pads for different service lives are shown in Fig. 2.

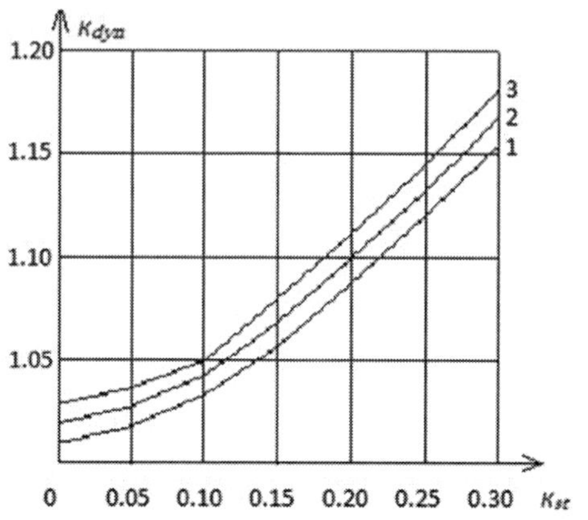

Fig. 2. Diagrams of dependence of the dynamics coefficients on the coefficients of the uneven elasticity of the under-rail base with KPP-5 fasteners and PRP-2.1 rail pads: 1 – service life 0 years; 2 – service life 13 years; 3 – service life 22 years

Further studies of the stress-strain state of rail track elements with SB-3-0 sleepers and KPP-5 fasteners [11] showed that the coefficient of uneven elasticity of the sub-rail base not exceeding 0.05 is permissible.

Table 1 shows the calculated values of the vertical stiffness of KPP-5 fasteners for various rail pads and their service life. The data were obtained from previous studies.

Table 1. Vertical stiffness of KPP-5 fasteners depending on service life and types of rail pads

Years	Types of rail pads					
	PRP-2.1	PRP-2.11	PRP-3.2 (9 mm A90)	PRP-3.2 (9 mm A95)	PRP-3.2.1 (10 mm A90)	PRP-3.2.1 (10 mm A95)
0	91.5	127.6	135.9	166.6	129.6	160.1
5	105	141.1	149.4	180.1	143.1	173.6
8.3	113.91	150.01	158.31	189.01	152.01	182.51
10	118.5	154.6	162.9	193.6	156.6	187.1
11.1	121.47	157.57	165.87	196.57	159.57	190.07
12.7	125.79	161.89	170.19	200.89	163.89	194.39
14.7	131.19	167.29	175.59	206.29	169.29	199.79
15	132	168.1	176.4	207.1	170.1	200.6
20	145.5	181.6	189.9	220.6	183.6	214.1
25	159	195.1	203.4	234.1	197.1	227.6
30	172.5	208.6	216.9	247.6	210.6	241.1

The results of previous studies enable to make a conclusion that the restoration of operability of KPP-5 fasteners is possible using PRP-3.2 rail pads with a thickness of 9 mm and PRP-3.2.1 with a thickness of 10 mm. Table 2 contains the calculated values of the stiffness gradients of fasteners and the coefficients of the uneven elasticity of the under-rail base are given after performing restoration.

Options of restoration section at which PRP-2.1 and PRP-2.11 rail pads were operated were considered. PRP-3.2 with a thickness of 9 mm and PRP-3.2.1 with a thickness of 10 mm was considered as rail pads being installed for recovery. Taking into account the earlier conclusion that the maximum permissible coefficient of the uneven elasticity of the under-rail base should be 0.05, the options with this and lower values do not require the formation of any transitional sections.

Also, in our opinion, the options of the restoration of operability of a fastener, in which the coefficient of uneven elasticity of the sub-rail base exceeds 0.25, are inadmissible. For this reason, they were excluded from further calculations.

Table 2. Stiffness gradients of KPP-5 intermediate rail fasteners and coefficients of uneven elasticity of the under-rail base when rail pads are replaced

Traffic load, million ton·km/km in year	Term replacement of pads, years	Types of pads until replacement	Stiffness of fastener at the moment of replacement, kN/mm	Stiffness gradient when repair rail pads are replaced $\frac{\Delta C_y kN/mm}{K_{un,el}}$			
				9 mm A90	9 mm A95	10 mm A90	10 mm A95
≤15	22	PRP-2.1	126.6	9.3 / 0.07	40 / 0.32	3.0 / 0.02	33.5 / 0.26
	22	PRP-2.11	162.7	-26.8 / 0.16	3.9 / 0.02	-33.1 / 0.2	-2.6 / -0.02
>15	13	PRP-2.1	150.36	-14.5 / 0.1	16.2 / 0.11	-20.8 / 0.14	9.7 / 0.06
	13	PRP-2.11	186.46	-50.6 / 0.27	-19.9 / 0.11	-56.9 / 0.31	-26.4 / 0.14

Other options, depending on the values of the coefficient of uneven elasticity, were divided into four groups, depending on the values of this coefficient, see Table 3.

Table 3. Options for restoration of operability of KPP-5 intermediate fastener

Option	Types of rail pads until restoration	Stiffness of fastener until restoration, kN/mm	Traffic load, million ton·km/km in year	Types of repair rail pads at replacement	Stiffness of fastener after restoration, kN/mm	Stiffness gradient and uneven elasticity, $\frac{\Delta C_y kN/mm}{K_{un,el}}$	Term replacement of rail pads, years
1	PRP-2.1	126.6	≤15	PRP-3.2 9 mm A90	135.9	9.3 / 0.07	22
2	PRP-2.11	186.46	>15	PRP-3.2 9 mm A95	166.6	-19.9 / 0.11	13
3	PRP-2.1	150.36	>15	PRP-3.2.1 10 mm A90	129.6	-20.8 / 0.14	13
	PRP-2.11	186.45		PRP-3.2.1 10 mm A95	160.1	-26.4 / 0.14	13
4	PRP-2.11	162.7	≤15	PRP-3.2.1 10 mm A90	129.6	-33.1 / 0.2	22

Four such options were accepted:

Option 1 – coefficient of uneven elasticity within >0.05÷0.09;

Option 2 – coefficient of uneven elasticity within ≥0.10÷0.13;

Option 3 – coefficient of uneven elasticity within ≥0.14÷0.16;

Option 4 – coefficient of uneven elasticity more than 0.16.

For these options, the pad arrangement schemes and the length of the transition sections to the recovery section were calculated using the software package of the dynamic "vehicle-track" system. The main criterion that was used as a basis for calculations is not exceeding the changes in stiffness of the under-rail base, at which the coefficient of uneven elasticity is 0.03. At such values, additional dynamic forces in the transition sections will not exceed 4 % of the static values of the wheel loads.

Conclusions

The results of the studies suggest the following conclusions:

1. Calculations have been made for the restoration of operability of KPP-5 fasteners using PRP-3.2 repair rail pads with a thickness of 9 mm and PRP-3.2.1 with a thickness of 10 mm. It has been established that in sections with the traffic load of more than 15 million ton·km/km in year, restoration works should be performed after 13 years of operation. Using 10 mm thick PRP-3.2.1 rail pads will provide the necessary pressing force of the rail base against the sleeper until the expiration of the standard service life – no more than 30 years. Use PRP-3.2 rail pads with a thickness of 9 mm will extent the operability for 10 years, then operability restoration works will be required again. Therefore, this option should be considered inappropriate. When the traffic load is less than or equal to 15 million ton·km/km in year, recovery can be performed after 22 years of operation. The use of both types of pads will ensure the reliable functioning of the KPP-5 fasteners until the expiration of its service life.
2. The results of the study of the influence of the uneven elasticity of the under-rail base on the dynamic interaction forces of the track and the rolling stock are given. The studies were conducted using the software package of the spatial dynamic "vehicle-track" system which was developed at the Department "Track and Track Facilities" of UkrSURT. It has been established that the uneven elasticity of the under-rail base of 0.05 should be considered permissible. In case of such values, the values of the additional dynamic forces will not exceed 1.035.
3. Possible options for performing restoration work using PRP-3.2 rail pads with a thickness of 9 mm thick and PRP-3.2.1 with a thickness of 10 mm (A90, A95) at the sections where PRP-2.1 rail pads were inserted in the track have been considered. The stiffness gradients and the coefficients of the uneven elasticity of the sub-rail base have been calculated. In addition, the replacement of PRP-2.1 pads with either PRM-3.2 pads with a thickness of 9 mm thick (A95) or PRP-3.2.1 pads with a thickness of 10 mm (A95) if a traffic load is up to 15 million ton·km/km in year, as well as replacement of PRP-2.11 pads with PRP-3.2 pads with a thickness of 9 mm (A90) or PRP-3.2.1 pads with a thickness of 10 mm (A90) if a traffic load exceeds 15 million ton·km/km in year is impermissible.
4. Four schemes of the arrangement of transition sections were adopted depending on the coefficient of uneven elasticity of the under-rail base for other options for the restoration of operability of KPP-5 fasteners. The procedure of laying rail pads and the length of transitional sections have been calculated.

References

1. E.I. Danilenko, *Railway Track. Arrangement, designing and calculations, interaction with the rolling stock* (Kyiv, 2010)
2. V.V. Govorukha, Geotechnical mechanics, **31**, 78-94 (2002)
3. A.N. Darenskiy, Zb. nauk. prats' UkrDAZT, **113**, 171-177 (2010)
4. I. A. Ivanov, I.V. Amelichev, Sb. nauch. trudov VNIIGT. «Application of reinforced concrete in the upper track structure», **118**, 56-59 (1984)
5. V.M. Tverdomed, Zb. nauk. prats' KUETT, **1-2**, 36-41 (2003)
6. V.V. Govorukha *Mechanics of deformation and destruction of elastic elements of intermediate rail fasteners* (Dnipropetrovsk, 2005)
7. F. Rezaie, M.R. Shiri, S.M. Farnam, Engineering Failure Analysis, **26**, 21-30 (2012)

8. V.V. Ribkin, I.O. Bondarenko, R.V. Markul, Praci Mignar. nauk.-tech. conf. «Problems of the interaction of the track and rolling stock» DNUZT, 39 (2013)
9. V.G. Vitolberg, *Forecasting of the service life of reinforced concrete sleepers SB 3-0 in conditions of non-public railways* (Kyiv, 2013)
10. A.M. Darenskiy, V.G. Vitolberg, D.A. Fast, A.V. Klymenko, Y.S. Leibuk, MATEC Web of Conferences, **116**, 03002 (2017)
11. N.I. Karpushchenko, N.I. Antonov, *Improvement of rail fasteners* (Novosibirsk, 2003)

Technical solutions for working elements of machines for efficient compaction of soil

Kazimir Glavatsky, *Volodymyr* Cherkudinov*, and *Olexandr* Posmitiuha

Dnipropetrovsk National University of Railway Transport named after Academician V. Lazaryan, Lazaryan St., 2, Dnipro, Ukraine

Abstract. The modernization of machines for compaction of soil on a modular basis has the following advantages: some of their structural elements (working equipment and working elements) can be performed as unified modular units, from which, depending on the technological requirements, it is possible to assemble the required configuration of the sealing machine; the initial state of the machine does not deteriorate, to which it is easy to return; the nomenclature of structural elements and machines in general decreases, as the machine can be completed with variable modules, which leads to an improvement in their quality. When consolidating the soil massive use static, dynamic and combined action on the soil in the form of rollers, vibro-and ramboards. Machines of the boot-type are classified into light, medium and heavy, which determines the order of its application. It is obvious that the reduction of technology for the process of soil compaction is possible due to the creation of machines that can change the intensity of the impact on the soil in a wide range.

1 Analysis of literary data and raising of problem

In modern conditions of ground works during the construction of soil embankments, in particular railway tracks, at the final stage, the issue of effective compaction of soil is relevant, since the stability of the soil structure depends to a large extent on the stability of the parameters of soil moisture and settlement load.

Known soil compacting machines (SCM) rollers, vibration plates and ramps traditionally comprise freshly ground soils in successive passages with the basic condition - not exceeding the surface pressure of the working element (WE) on the soil of the boundary of soil plasticity. To do this, there are certain technological sequences of the use of the Ministry of Emergency as a whole, and their WE [1 - 8].

In particular, the order of the SCM of the boot type involves, at the first stage, the consolidation of the soil with smooth, small and medium type, with subsequent use of their vibration systems (VS), followed by cam rollers with subsequent use of the aircraft. Further intensification of soil consolidation is possible with the use of trenches and vibration tracks of shock type.

The operation procedure of the SCM is characterized by either a gradual decrease in their contact area with the soil due to the change in the size and shape of their working

* Corresponding author: volodymyrcherkudinov@gmail.com

surface, for example, the gradual decrease in the diameter of the cylindrical metal roller, the installation on its surface of variable bandages with cam sizes and profile, gradual increase in pressure in the pneumatic boats, or a gradual increase in the load on the WE.

The general disadvantage of the work of all known SCM is that they do not limit the yield of soil from under their working surface (do not block the soil) in case of excess surface pressure of the soil to the soil of the boundary of soil plasticity.

Goal - the purpose of the work is to develop the principal variants of technical solutions of the SCM WE for accelerated blocked consolidation of soil.

2 Main part

The scientific hypothesis put forward by the authors is to exceed the surface of the soil's plasticity by blocking its movement from the working surface of the WE and using this effect to accelerate the process of soil consolidation by reducing the number of WE SCM passages and the use of the blocked and maximally compacted soil in the WE as the transmission link of pressure on the lower and, perhaps, insufficiently sealed layers of soil.

The structure of the blocking action on the soil proposed by the authors suggests that they create the conditions for blocking the soil from transverse dislocation from under the WE and during its movement.

This result can be achieved by executing the working surface of the WE roller of the roller or the working surface of a plate or a special form molding, for example, profile. The proposed technological series of profiles can be, for example, rectangular, triangular, trapezoidal, wavy, combined (Fig. 1).

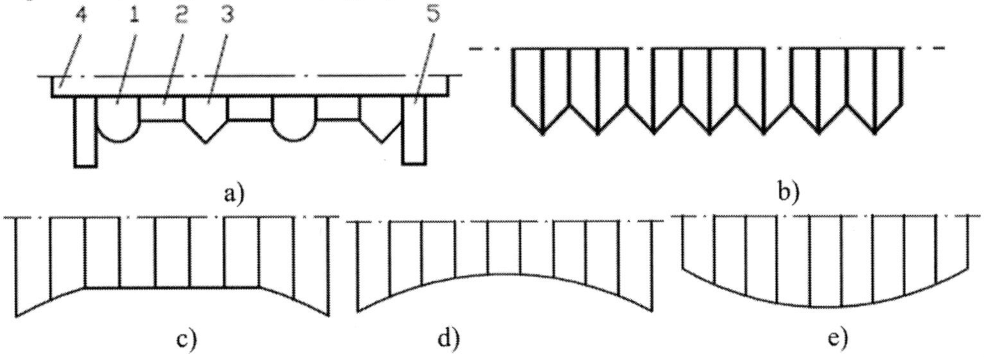

Fig. 1. Scheme of the working surface of the WE of special shape (blocking effect on the soil): 1, 2, 3 - variable discs; 4 - axle; 5 - side locking discs; a) example of a working elements with different types of disks; b) triangular profile roller; c) a trapezoidal combination roller; d), e) a roll with a convex and convex curved profile.

The authors proposed and protected by the patents a number of constructive decisions of the SCM, which provided for a change in the number of contacting with the soil WE, as well as developed technological sequences of the use of new SCM and machine tools in general with an emphasis on blocked seals [9, 10].

The proposed variants of the implementation of the variables WE of vibro-tramways and platforms provide a profile continuous or sectional execution of their working surface for the implementation of blocked consolidation of soil, both without moving the machine, and in the vertical and horizontal directions during the translational movement for the combined implementation of both blocked consolidation of soil and pressure changes at the point of contact of the working surface of the WE with the soil.

Proposed and investigated rollers with profiled working surface blocking action on the ground (Fig. 2).

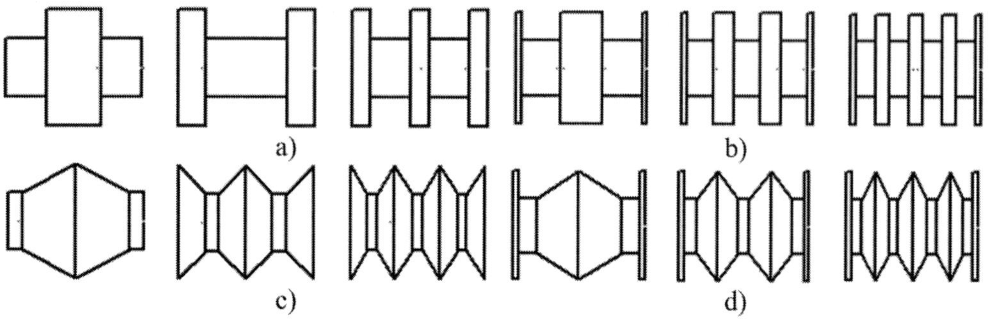

Fig. 2. Scheme of rolling rollers with a profiled working surface of blocking action on the soil with: a) is a cylindrical stepped form without side thin cylindrical disks; b) - cylindrical step form with lateral thin cylindrical discs; c) - a conic-cylindrical working surface without lateral thin cylindrical disks; d) - a conically cylindrical working surface with lateral thin cylindrical discs.

Thin lateral cylindrical disks are designed to limit the lateral slipping of soil from the working surface of WE rollers.

Also offered are variants of schemes of rollers of rollers with variable working surfaces, which are made to block the excavation of soil in the course of the rolling roller (Fig. 3).

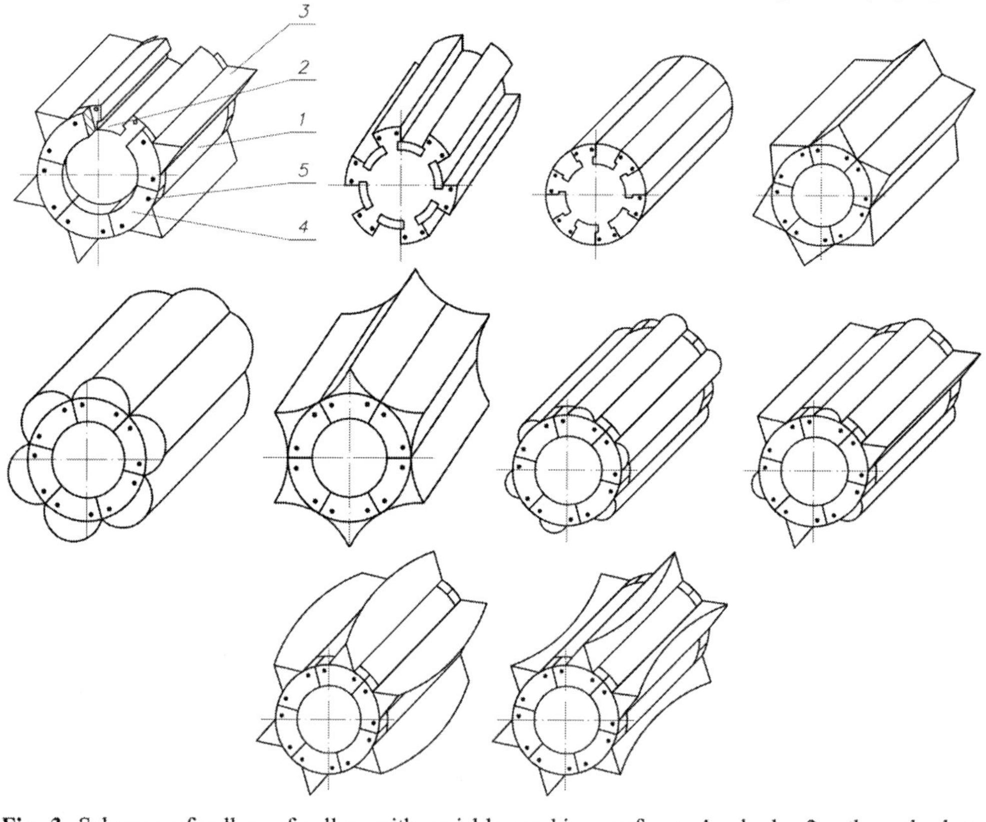

Fig. 3. Schemes of rollers of rollers with variable working surfaces: 1 - body; 2 - through slots; 3 - variable inserts; 4 - face fixing strips; 5 - threaded elements.

Variants of the profile of the working surface, blocking action on the soil (Fig. 4) are proposed to study WE ramps.

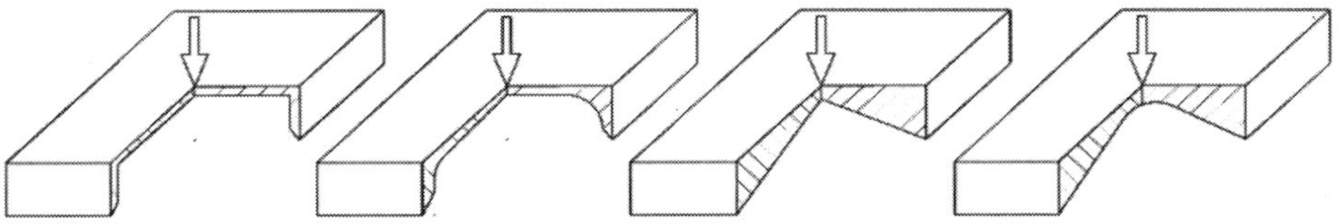

Fig. 4. Schemes of variants of working bodies of tractors of blocking action on the ground.

Variants of the profile of the working surface, blocking action on the soil are proposed to study the WE vibroplate (Fig. 5-8).

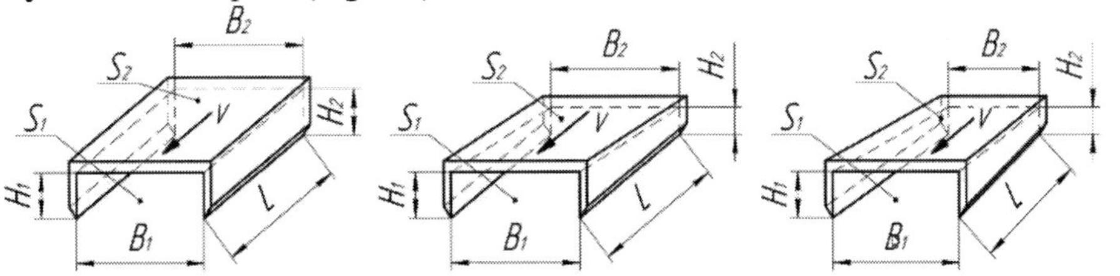

Fig. 5. Scheme of the П-shaped vibroplate profile.

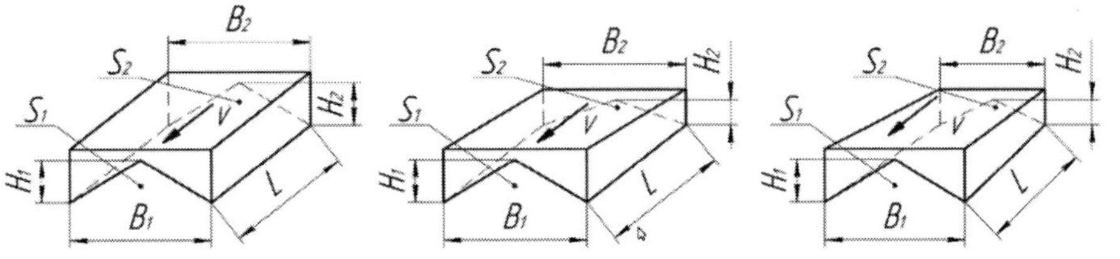

Fig. 6. Scheme of the triangular profile of the vibroplate.

In Fig. 5-8 possible value $H_1 = H_2$; $H_1 > H_2$; $B_1 = B_2$; $B_1 > B_2$; $S_1 = S_2$; $S_1 > S_2$ on the base length of the vibroplate L, and the direction of its movement is indicated by an arrow with the mark V.

In addition, to intensify soil consolidation, active use of vibration is foreseen in the form of integrated or modular vibration systems (VS) with the ability to regulate the disturbing force vector, type of oscillations, their amplitudes and frequencies.

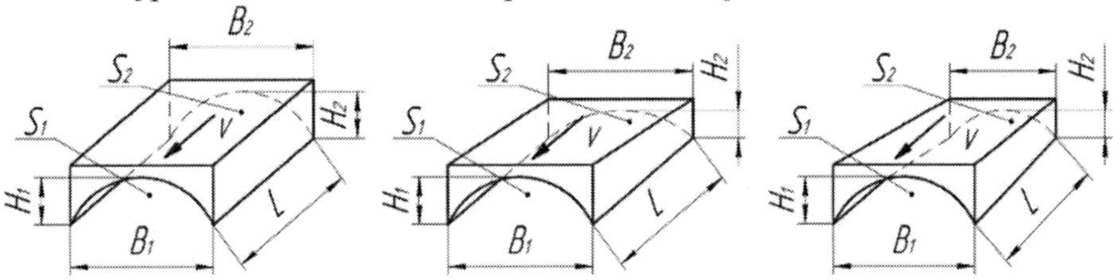

Fig. 7. Scheme of arched concave profile vibroplate.

The specified variants of the profiles of the working surfaces of the WE will restrict the transverse separation of the soil from the working surface of the WE and will allow:

- intensify the process of consolidation of soil by applying additional pressure on the soil, not limited to the limit of its plasticity;
- effectively perform seals of unconnected soils without disassembling them in the direction in relation to the direction of movement of the WE;
- to effectively perform seals of the surface layer of coherent soils, especially using the profile of the working surface WE in the variant (e) with the active application of the vibration system.

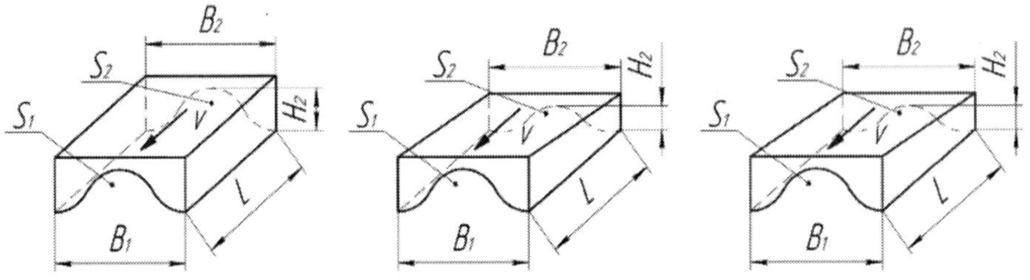

Fig. 8. Scheme of wavy Π-shaped vibroplates.

Execution of the working surface of the WE relief will increase its contact area with the soil, and, consequently, reduce the specific pressure on the soil. This will allow smoother removal of the gaseous and liquid phase of the soil from the compacted array.

The relief of the outer surface of the compacted layer of soil will create conditions for additional blockage of the next lumped soil layer with its consolidation.

Conclusions

The proposed development of the principal variants of technical solutions for the accelerated blocked consolidation of soil will ensure the achievement of the expected result by accelerating the process of soil consolidation by increasing the working stress of the soil below the surface of the WE due to its blocking.

The process of sealing the soil under the blocking surface of the SCM WE will be carried out not only in the vertical, but also in the horizontal and sloping direction, which will increase the efficiency and quality and sealing.

References

1. L. Khmara, S. Kravets, M. Skoblyuk. *Earthmoving Machines*: (KhNADU, Kharkiv, 2014)
2. L. Khmara, S. Kravets, L. Nichke, L. Nazarov, M. Skobul'k, V. Nikitin. *Earthmoving Machines* (Dnipropetrovsk – Kharkiv, 2010).
3. V. Blokhin, M. Malich. *Basic parameters of technological machines. Machines for earthworks. In 2 hours: Ch. 2.* (Higher School, Kyiv, 2009)
4. V. Blokhin, M. Malich. *Wheel tracked vehicles. In 2 hours: Ch. 1.* (IMA-press, Dnepropetrovsk, 2008)
5. V. Blokhin, M. Malich. *Wheel tracked vehicles. In 2 hours: Ch. 2.* (IMA-press, Dnepropetrovsk, 2008)
6. V. Blokhin, M. Malich. *Basic parameters of technological machines. Machines for earthworks. In 2 hours: Ch. 1.* (Higher School, Kyiv, 2006)
7. V. Blokhin, M. Malich. *Machines for earthworks. Conditions of half-way of their competitiveness* (IMA-press, Dnepropetrovsk, 2005)
8. E. Lokshyn, A. Rubaylov E. Lokshin, A. Rubaylov *Building and road cars. An overview of modern domestic self-propelled machinery* (Moscow, RIA Rosbiznes, 2004)
9. K. Glavatsky, M. Nebesnyi, O. Posmityukha, V. Proskurny. *Pattent 98823* (Ukraine (2012)
10. K. Glavatsky, M. Nebesnyi, O. Posmityukha, V. Proskurny, V. Cherkudinov. *Pattent 101518* (Ukraine, 2012)

Problems of providing international railway transport

Mykola Kurhan[1], and *Dmytro* Kurhan[2,*]

[1]Department of Roads Design and Construction, Dnipropetrovsk National University of Railway Transport named after Academician V. Lazaryan, Lazaryan St., 2, Dnipro, Ukraine
[2]Department of Track and Track Facilities, Dnipropetrovsk National University of Railway Transport named after Academician V. Lazaryan, Lazaryan St., 2, Dnipro, Ukraine

Abstract. Due to its geographic location and developed transport infrastructure, Ukraine has a significant potential in the development of cargo transportation, primarily in international traffic, in particular as a transistor country in the logistics chain of trade between Asia and Europe. The possibilities of railway transport for the organization of transportation between the countries of the European Union and Ukraine are not used to the full extent, since there is a number of technical reasons of the transport systems incompatibility between Ukraine and European countries. In the course of the research, methods of analysis and synthesis were used to study prospects of the European and domestic system of international railway transportations, the experience of creating, operating and optimizing these systems. To compare the above mentioned options, the authors developed a model for predicting and assessing the efficiency of railway transport from the border of one state to the border of the other one.

Introduction

Due to its geographic location and developed transport infrastructure, Ukraine has a significant potential in the development of cargo transportation, primarily in international traffic, in particular as a transistor country in the logistics chain of trade between Asia and Europe.

Ukraine's economy is dependent on the efficient shipment of bulk commodities such as coal, iron ore, steel, and grains – commodities that travel most economically over long distances by rail. The successful reform and restructuring of Ukrzaliznytsia will be a crucial part of Ukrainian economic reforms going forward [1].

According to the British Institute for Transport Studies, the transit rate of Ukraine is 3.75 (at a maximum of 5); it is the best indicator among European countries (for comparison, in Poland, which is second best, this figure is 2.92). Five international transport corridors pass through Ukraine; the expanded length of these routes is about 6.500 ths km, of which 3.500 are railways. In 2016, Ukraine joined the Coordinating Council for

* Corresponding author: kurhan.d@gmail.com

the Trans-Caspian International Transport Route (TITR) to China through Georgia, Azerbaijan and Kazakhstan.

The possibilities of railway transport for the organization of transportation between the countries of the European Union and Ukraine are not used to the full extent, since there is a number of technical reasons of the transport systems incompatibility between Ukraine and European countries, namely: different wheel gauge, characteristics of rolling stock, type of SSB, voltage in the contact network, dimensions, etc.

Future prospects for the integration of Ukraine's railways into the European transport network will depend on how successfully the tasks on the actual development of international transport corridors (ITC) will be solved [2, 3], on the availability of rolling stock ready to provide transport with governed speeds [4, 5], on political-economic and technical-technological problems of passenger turnover [6] and truck turn-around between Ukraine and Europe.

The basic part

Globalization of the economy and the development of modern supply chains contribute to the creating a transport product that would combine services of various transport modes in the most efficient and convenient way for consignors as well as it would be formed, first of all, on the basis of cargo interests, rather than individual participants in the process of transportation. Multimodal transportations have become such product, which are domestic or international transportation of products using several transport modes.

The object for the transfer in multimodal transportations, in principle, may be any cargo - bulk, unitized cargo. However, the most widespread are multimodal transportations using the so-called ITU (Intermodal Transport .Units, ITU) - containers, controllers, swap bodies. The cargo in ITU is all along the whole route, and all transport and cargo operations are not with heterogeneous packages, but with standard ITU (so-called reloading-free transport technology), which significantly speeds up and cheapens the technological processes, increases safety of cargoes and provides a wide range of other advantages.

The choice of the optimal transportation solution includes the rationale for the most profitable routes, the choice of rolling stock in accordance with the type of cargoes, the mode of cargo transportation (in which various modes of transport are used in stages), border stations technology, etc. [7].

Due to the intensive development of the integration processes of the Ukraine's railway transport into the European transport system, the following issues are of great importance: the functioning of the international transport corridors (ITC), the introduction of specialized rolling stock, which would allow operation both on the gauges of 1435 mm, and on 1520 mm ones, that is, application of European ASCS-technologies, based on 1520/1435 mm automatic gauge change systems [8]. This is development of "East-West" cars for direct cargo transportation by railways of various standards.

The study of international transportation technology can be summarized as following blocks: international transport corridors, rolling stock and operation of border stations (see Fig. 1).

In order to keep the transit volumes of cargo transportations in the direction of the sea trading ports, Ukraine has proposed to include additionally into ITC network directions leading to the Black Sea ports, Odessa, Nikolaev, and Kherson (Fig. 2).

Among the directions linking Lviv with the countries of Europe and the EU countries, two main ones can be delineated: one of them passes through the Mostyska II station and it is the shortest connection with Poland, Slovakia, the Czech Republic, Germany and other countries of Central Europe, as well as Scandinavia; The second direction passes through Chop and connects Ukraine with Slovakia, Hungary, Bulgaria, Romania and other countries of Central and Southern Europe. The second direction according to design parameters and the longitudinal profile does not allow to realize high speeds. Particular attention should be paid to the section Lavochne-Mukachevo, where gradients of the longitudinal profile larger than 15 ‰ are about of 28%, curves with a radius of up to 500 m - 29.3%, and the speeds do not exceed 60-70 km/h.

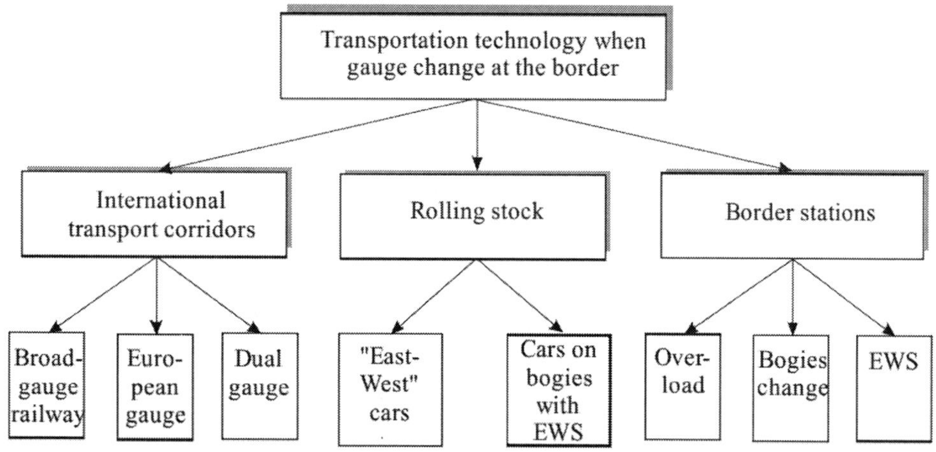

Fig. 1. The outline scheme of transportation technology at crossing the borders

Fig. 2. Passing of ITC to the Baltic Sea - the Black Sea through the territory of Ukraine

A significant obstacle in the ITC modernization No. 5 was the Beskidskiy single-track Tunnel built in 1886. The Beskidskiy Tunnel is of strategic importance, since it transits more than 40% of transit cargo to Western and Central Europe. The new Beskidskiy double-track railway tunnel was constructed at the expense of European Bank of Reconstruction and Development (EBRD) loan and own funds of railways in order to increase the transit potential of Ukraine, bringing to the relevant technical requirements of the International Transport Corridor and ensuring the safety of trains on the strategic direction of Kyiv-Lviv-Chop. The capacity of the new tunnel is three times larger than the existing one.

The carried out analysis of the technical state and parameters of the Lviv-Chop route showed that main reasons that restrain the increase in speed (except the tunnel) are the plan of the line with insufficient lengths of transition curves and curves of small radii, state of the permanent-way (excessive wear of rails and railroad switches), state of the artificial structures and road bed.

Plan and profile of the line (Table 1), speed limit at stations and a number of other reasons cause the need to reduce the speed with respect to its maximum level.

Issues relating to the poor state of infrastructure are proposed to be solved in the process of reconstruction of the railway. It is more complicated to change parameters of the plan and profile of the line.

Table1. Profile and plan indicators on international transport corridors within Ukraine (in %)

International Transport Corridor	Gradients on the longitudinal profile i>8 ‰	Radius curves in plan R<700 m
ITC No. 3	7.2	7.8
ITC No. 5	23.0	19.5
ITC Gdansk-Odesa	10.0	2.6

It is obvious that the integration of the transport system of Ukrainian railways is possible only under keeping of interstate requirements regarding maximum speed, time of cargo and passengers delivery, travelling comfort, etc. These challenges can not be solved today without bringing the ITC infrastructure to a proper state. For example, the monograph of H. N. Kirpa [9] presents an analysis of the technical state and infrastructure of international transport corridors within Ukraine. The features of each corridor are marked, the priority areas are identified, their restructuring will accelerate the cargo and passengers delivery, interest European and other countries in transit traffic through Ukraine.

In paper [10] features of the simulation model of the transport system are presented on the example of Poland. The model reflects those properties of the system that are important from the standpoint of research, namely: features of the transport network, the model of demand for passenger and cargo transport, distribution mechanism of cargo traffic and passenger flows into the network. The model allows performing computational experiments and drawing conclusions with respect to the distribution of vehicles in the national transport system, taking into account emissions from transport activities.

The issue of the compatibility of transport systems remains complicated one. In paper [11], the main aspects of the system approach to the issue of railway transport compatibility are presented. It has been shown that despite the difference in control systems, electricity supply and operation of railways, various standards of track infrastructure, there is a need to solve the problem of integrating the railway transport systems. The article covers the main directions of the railway system operation, which depend on the introduction of interoperability.

The analysis of the ways for organizing the cargo transportation in international traffic with the European Union countries showed that the following options are subject to comparison: cargo transshipment, including in containers from rolling stock, gauge of 1520 mm on a rolling stock, gauge of 1435 mm; replacement of bogies at the gauge-changing areas when crossing rail joints of a different standard; application of a special rolling stock equipped with bogies with extensible wheelsets; continuation (use of the existing) broad gauge of 1520 mm from the borders of Ukraine to the territory of Europe; continuation of European gauge of 1435 mm from the borders of Europe to the territory of Ukraine; Use of the dual gauge 1435/1520 mm.

To compare the above mentioned options, the authors developed a model for predicting and assessing the efficiency of railway transport from the border of one state to the border of the other one, taking into account all expenditures by the NPV indicator (Net Present Value of Discounted Cash Flow), which is the difference between total income (D) and all types of expenses taking into account the time factor (K_i – investments, costs in K_l – locomotive fleet, K_c - car fleet, C - current operating costs, and C_s - costs, depending on the type of technological operations and the time of freight cars presence at the station splicing of different gauges:

$$NPV(t) = \sum_{t=0}^{T_p} \left(D_t - K_{i\,t} - K_t - K_{l\,t} - K_{c\,t} - C_t - C_{s\,t} \right) \eta_t \qquad (1)$$

Uncertainty and risks in assessing the effectiveness of options are taken into account through the modified discount rate, which is included in the calculation of discount coefficient of the different-time expenditures (η_t).

To determine the transfer fee, the International Railway Transit Tariff (ITT) was used. It applies for the cargo transportation, as well as for transportation through border and port stations [12].

Conclusions

During the simulation (1), such source data as the initial volumes and the rate of their growth over time, length of haul, the type of cargo and cars for transportation (containers on platforms, tanks, universal cars) varied, the speed of delivery, which depended on the state of the railway infrastructure on the international transport corridors etc.

Let us consider, how the cargo traffic volume (G_t) and the rate of its growth (ΔG) affects the transport process at a various length of cargoes transportations. For example it is accepted: G_t=5,0 million tons/year. The length of transportation varied from 400 up to 2000 km.

At the same time, the largest total revenue is observed at length of haul of 1000 km (Fig. 3). If we set at 100% of the total revenue for the length of haul 400 m, then at a length of 600 km income increases by 26.1%, at a length of 800 and 1000 km, respectively, by 56.1 and 85.6%. At lengthes greater than 1000 km the profit decreases and at a length of 1700 km becomes negative. So, short routes of 600-1400 km are desirable.

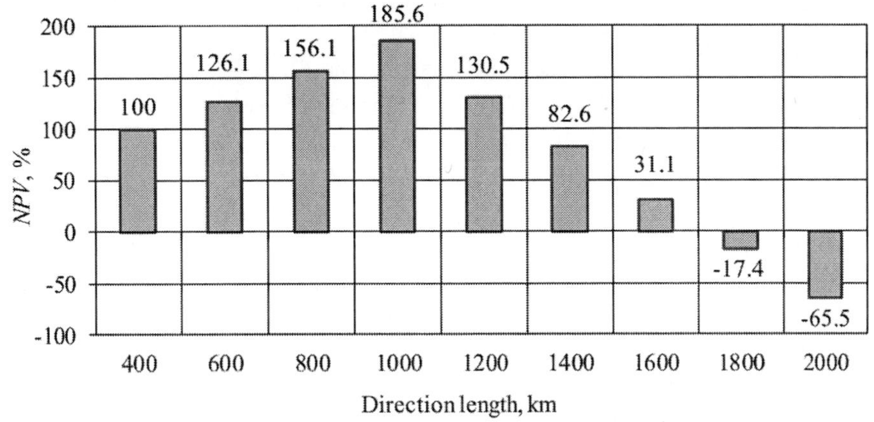

Fig. 3. Change in net present value (%), depending on on length of haul

To study the impact of the sectoral speed value on the final result, calculations are presented for a length of 1000 km at values of the sectoral speed from 30 to 60 km/h (for further extention) km/h. According to the obtained results, a diagram of the change in net present value $\sum_{t=1}^{15} NPV_t$ was constructed (Fig. 4) in comparison with sectoral speed (V_s) of 30 km/h.

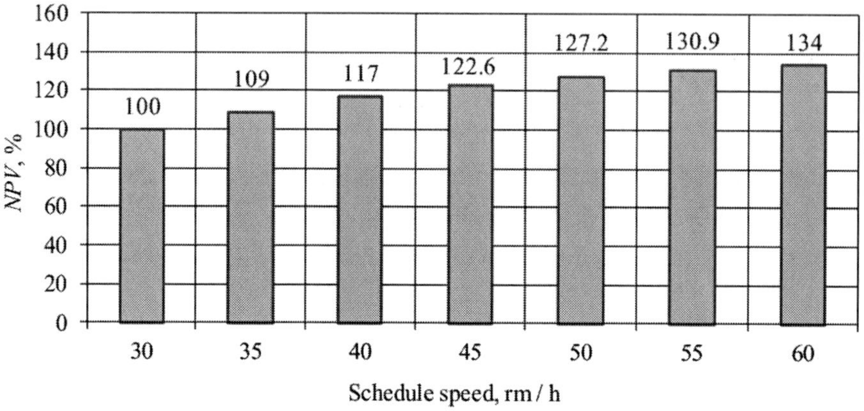

Fig. 4. Change in net present value, depending on the level of sectoral speed

The analysis of the diagram makes it possible to conclude that there is a need for modernization of international transport corridors in order to increase the speed affecting the circulation of rolling stock, the need for them to carry out transportations and, as a result, to obtain revenues from transportations. From analysis, it may be the recommended sequence of events. Thus, an increase V_s from 30 to 40 km/h leads to an NPV increase by 17.0%, with an increase V_s from 40 to 50 km/h - by 10.2%, and from 50 to 60 km/h - by 6.8%.

References

1. R. Pittman, Science and Transport Progress. Bulletin of Dnipropetrovsk National University of Railway Transport, **1(67)**, 34-50 (2017).
2. M. B. Kurhan, D. M. Kurhan, *Theoretical basis for the introduction of high-speed trains in Ukraine, DNUZT, Dnipro, Ukraine,* 283 (2016).
3. F. Horvat, S. Fischer, Közlekedésépítési Szemle, **59 (5)**, 33-37 (2009).
4. Ju. V. Demyn, Ju .V. Vereshhak, Visnik of the Volodymyr Dahl East Ukrainian National University, **5(147)**, 167-172 (2010).
5. S. Fischer, Acta Technica Jaurinensis, **8(3)**, 240-256 (2015).
6. M. B. Kurhan, D. M. Kurhan, Science and Transport Progress. Bulletin of Dnipropetrovsk National University of Railway Transport, **6(72)**, 40-48 (2017).
7. M. B. Kurhan, E. V. Voznaya, Ukraine Railway, **12(18)**, 24-33 (2014).
8. M. Szkoda, Eksploatacja i Niezawodnosc. **16 (3)**, 422–432. (2014).
9. G. M. Kirpa, *Integration of Ukraine's railway transport into the European transport system*, (DNUZT, 2004).
10. M. Jacyna, M. Wasiak, K. Lewczuk, M. Kłodawski, Archives of Transport, **(31)**, 23-35 (2014).
11. J. Szkopiński, Archives of Transport, **(29)**, 65-75 (2014).
12. *International railway transit tariff (Tariff 8100)* (Official Edition of OSJD, 2018).

Construction of the simulation model "car-railway track" for modeling the dynamic loading process of the boot node

Igor Martinov*, *Alyona* Trufanova, and *Vadim* Shovkun

Ukrainian State University of Railway Transport, Car Department, 61050, Kharkiv, Ukraine

Abstract. Using of software systems in the design of new samples is increasingly important, but there is a problem in adapting computer models to obtain adequate results of calculations. The modern development of the IT systems gives the opportunity to significantly reduce the timing of the introduction of new equipment into production by reducing the number of field experimental test samples of the design technology. Every year using of complexes software in designing new samples is becoming increasingly important, but there is a problem in adapting computer models to obtain adequate results of calculations. UM Universal Mechanism complex was used to simulate the dynamic loading process of the boot node. Modelling of dynamic loadings on the running gears of freight cars showed a sufficient convergence with the results of running tests, about 87%. Therefore, the proposed model can be used for assessing of disturbing loads when the calculating of the reliability of the bucket units of freight cars was done.

The safety of the movement of rolling stock depends on the reliability of aggregates, assemblies and parts of cars [1].

The reliability of the car is conditioned by its reliability, durability and the longevity of its parts. Therefore there is a need to increase the reliability of the rolling stock by allocating of sufficient resources for the current content and implementation of new principles of repair and elimination of the consequences in emergency situations, additional investment in new technologies, equipment modernization and development of methods aimed at improving operational the activity of cars.

The modern development of the IT systems gives the opportunity to significantly reduce the timing of the introduction of new equipment into production by reducing the amount of field experimental research samples of the design technology.

Every year, using of complex software in designing new samples is becoming increasingly important, but there is a problem in adapting computer models to obtain adequate results of calculations. UM Universal Mechanism complex was used to simulate the dynamic loading process of the boot node. The simulation model "railroad car" was developed by UM. The mathematical model includes geometric, inertial, stiffness and friction characteristics.

* Corresponding author: vadimshovkun62@gmail.com

The features of the model allow:

- to study indicators of dynamic characteristics (coefficients of dynamics of spring sets, vertical and transverse accelerations of separate points of a body, etc.);
- to investigate the safety indicators (coefficients of stability of wheels against the return of rails, frame and lateral forces, axial load, etc.);
- vary by main geometric, rigid, frictional and inertial parameters of the car and spring suspension;
- to investigate the movement of a car at different speeds with various wear of running parts, wheel profiles and rails, rigidity and inertia characteristics of the track with irregularities;
- to model the movement of cars with various structural modifications of trolleys.

A freight car consists of 19 solid bodies (4 wheel pairs with booms, 4 side frames, 8 friction wedges, 2 spindle beams, and 1 body) and their corresponding 114 degrees of freedom. **Fig.** 1 shows the design scheme of such a freight car.

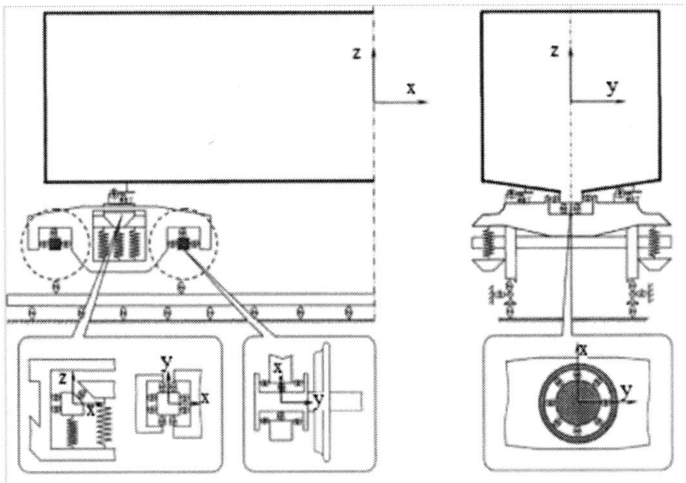

Fig. 1. Estimated power scheme of the car and railway track

In the mathematical model of UM software complex for each solid is given the mass *mi* and moments of inertia *Ji* to its own body coordinate system *(CKi) {Ci,xi,yi,zi}* Its beginning coincides with the centre of mass **Fig. 2**. Two general systems of coordinates are introduced to the scheme: the fixed base one *(CK00) {C0,X0,Y0,Z0}* and the base flexible one (track) *(CKO) {0,ξ,η,ζ}*.

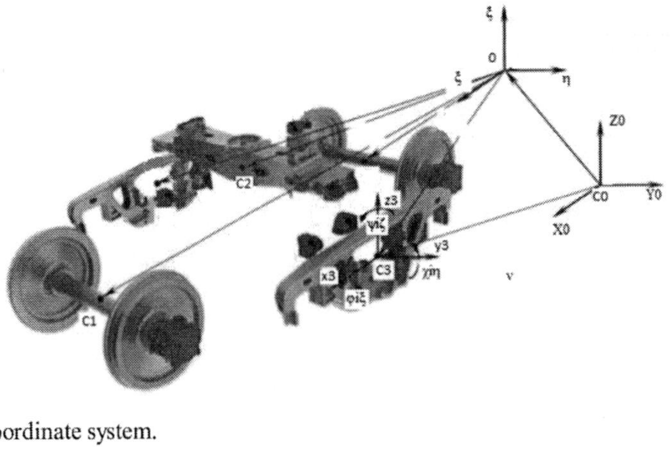

Fig. 2. Body coordinate system.

The movement of the car is presented as a complex movement. The moving base (track) coordinate system moves progressively along the axis of the railroad with the speed of the crew. The start of the JMA is associated with the centre of gravity of the bodies. The fixed system *CKO* determines the position of the rail path in space. The movement of the car is presented as a complex movement.

The moving base (track) coordinate system moves progressively along the axis of the railroad with the speed of the crew. The start of the JMA is associated with the centre of gravity of the bodies. The fixed system *CKO0* determines the position of the rail path in space.

The position of the *i-th* body in the *CKO* is determined by the position of the vector **ri** (0) (the lower index marks the body number for which the value is calculated, and the upper index shows in which coordinate system the calculation is made) connects the start of the *CKO* and *CKi,* and the angles between axes of *CKO* and *SKI.* These values can be represented as a vector tensor of the vector on the *CKO* (**Ri** (*0*) axis) and the rotation matrix (cosine paths) *(Ai (0)).*

In the UM complex, the kinematic characterization of the body system is described by a programmatic mathematical procedure and is implemented as a hinge linking this body with the base coordinate system [2,3].

The spatial orientation of each body included in the model of the freight car is given by its own system of coordinates parallel to the axes *CKO.* In the process of hinge description a vector **ri** (0) for each body in the initial position (for $t = 0$) is given. Thus, the position of the car bodies is set in the system of the railroad crew. If necessary, for any body, the number of degrees of freedom may be reduced by prohibiting the change of translational or rotational coordinates. At this stage, with the help of the problem of inertial and geometric quantities, parametrization of the mathematical model through various identifiers was carried out.

Based on the theorem about the motion of the centre of masses and the dynamic Euler equation, when the *CKs* coincide with the principal axes of inertia of the bodies the differential equations are obtained by UM software complex which describe the motion of the i-th body of the mathematical model, which have the form:

$$\begin{cases} mi\cdot aix(0)=\sum Fix(0)+\sum Rix(0);\\ mi\cdot aiy(0)=\sum Fiy(0)+\sum Riy(0);\\ mi\cdot aiz(0)=\sum Fiz(0)+\sum Riz(0);\\ Jix(i)\cdot d\omega ix(i)dt+\omega iy(i)\omega iz(i)\cdot\left(Jiz(i)-Jiy(i)\right)=\sum Mix(i)+\sum Lix(i);\\ Jiy(i)\cdot d\omega iy(i)dt+\omega iz(i)\omega ix(i)\cdot\left(Jix(i)-Jiz(i)\right)=\sum Miy(i)+\sum Liy(i);\\ Jiz(i)\cdot d\omega iz(i)dt+\omega ix(i)\omega iy(i)\cdot\left(Jiy(i)-Jix(i)\right)=\sum Miz(i)+\sum Liz(i)\end{cases}\tag{1}$$

mi- mass of the *i-th* body, *H*;

aix(0),aiy(0),aiz(0) of the acceleration vector of the *i-th* body on the coordinate axis *CK0*, m/s^2;

ΣFix(0),ΣFiy(0),ΣFiz(0) is the sum of the projections of all active forces applied to the *i-th* body, on the coordinate axes *CK0, H*;

ΣRix(0),ΣRiy(0),ΣRiz(0) the sum of the projections of all reactive forces applied to the i-th body, on the coordinate axes *CK0, H*;

Jix(i),Jiy(i),Jiz(i) moments of inertia of the *i-th* body relative to the axes *CKi, H/m2*;

ωix(i),ωiy(i),ωiz(i) projections of the angular velocity vector of the *i-th* body on the coordinate axes *SKi*, rad/s;

ΣMix(i),ΣMiy(i),ΣMiz(i) is the sum of the moments of all active forces applied to the *i-th* body, relative to the axes *CKi, H/m*;

*ΣLix(i),ΣLiy(i),ΣLiz(i)*is the sum of moments of all reactive forces applied to the *i-th* body, relative to the axes *CKi, H/м*.

The initial results of the simulation of dynamic modelling of loadings became the implementations which characterize the change in the coefficient of vertical and horizontal dynamics for a running wheel in a loaded and empty mode at a speed of 40-120 km/h, some implementations are shown in Fig. 3.

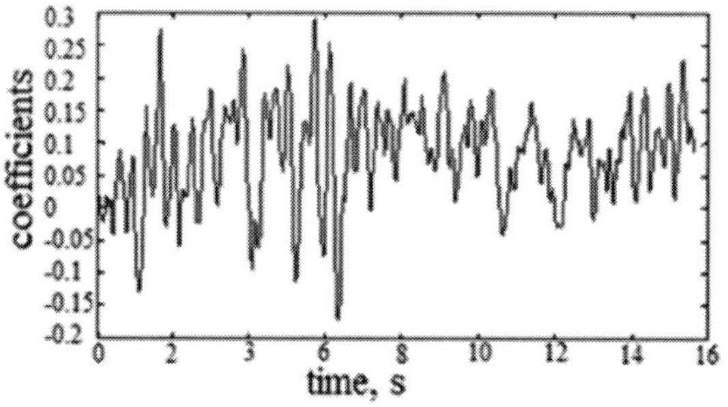

Fig. 3. Oscillogram of changes in the coefficient of vertical dynamics at a speed of 100 km/h.

Obviously, the process of time-varying coefficient of vertical dynamics is a random process with a pronounced sinusoidal component [4-10].

Similar conclusions can be drawn about the coefficients of horizontal dynamics: the simulation showed that there is a growth in accordance with the increase in speed. In load mode, these ratios did not exceed the normative values. In empty mode there was a slight excess of dependence on the velocity of the coefficients of vertical dynamics shown in Fig. 4.

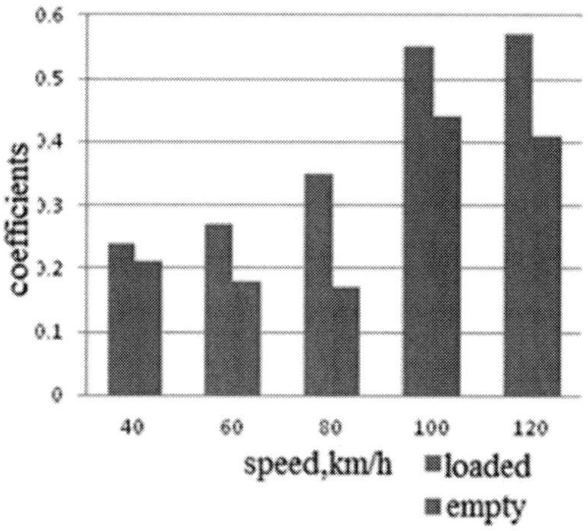

Fig.4. Dependence of the coefficient of vertical dynamics.

In the speed range from 11.1 to 33.3 m/s (40 ÷ 120 km/h), the ranges of the change in the trolley were from 0.24 to 0.54 for the empty mode, which is below the permissible value of 0.9.

Conclusions

1. Modelling of dynamic loadings on the running gears of freight cars showed a sufficient convergence with the results of running tests, about 87%.

2. Therefore, the proposed model can be used to estimate disturbing loads when calculating the reliability of the bucket units of freight cars.

3. The obtained data from the simulation can be used to calculate the reliability of the boot bearing nodes.

References:

1. I. Martynov, V. Ilchishin, East-European Journal of Advanced Technologies, **73**, (2015)
2. E. Bratchev, V. Kamayev, Abstracts of the report of the All-Union Conference. - Dnipropetrovsk From **97**, (1984)
3. A. Savoskin, A. Akishin, Computer modeling in railway transport: dynamics, durability, wear. A collection of abstracts, Bryansk **69**, (2016)
4. I. Martynov, A. Trufanova, V. Shovkun, Computer modeling in railway transport: dynamics, durability, wear. A collection of abstracts, Bryansk, **44**, (2016)
5. G. Mikhalchenko, D. Pogorelov, V. Simonov, Dnipropetrovsk National University of Railway Transport, **107**, (1996)
6. Simson, E. Bulletin of the NTU "KhPI", **37**, 142, (2010)
7. Averin, N. Vestnik VNIIZhT, **3**, 24, (2007)
8. Sanchugov V. Scientific journal "Izvestiya Samara Scientific Center of the Russian Academy of Sciences", **831**, (2007)
9. Gishvarov, A. Vestnik of the Ufa State aviation technical university, **1**, 43, (2007)
10. Haydamaka, A. Visnyk of Dnipropetrovsk National University of Railway Transport, **3**, 51, (2014)

Research of the dependence of pneumatic conveyor compressors power on project parameters of transport machines

Sergij Raksha[1], *Vladimir* Bohomaz[1,*], *Igor* Shcheka[1], *Volodymyr* Stefanov[2], and *Alex* Nesterenko[3]

[1]Department of Applied Mechanics and Materials Science, Dnipropetrovsk National University of Railway Transport named after Academician V. Lazaryan, Lazaryan St., 2, Dnipro, Ukraine
[2]Department of Construction, track and handling machines, Ukrainian State University of Railway Transport, Feierbakh Square 7, 61050, Kharkiv, Ukraine
[3]National Academy of the National Guard of Ukraine, Defenders of Ukraine sq. 3, 61001, Kharkiv, Ukraine

Abstract. The paper considers the algorithm for calculating the pneumatic transport systems of the main constructions: with high vacuum and with different pressures. The analysis of the dependence of all calculation elements on the design data of the transport system of the considered structures was carried out. In order to accelerate the calculation of such systems, the analytical dependence of the power of the compressor drive on the type of cargo, the coefficient of concentration of the transported mixture, the design productivity, and the geometric parameters of the route, the number and types of auxiliary devices (ells and shutters) are constructed. A graphical analysis of the dependence of compressor drive power of the pneumatic transport systems with high vacuum and with different pressures on the design data: the mixture concentration coefficient, the length of transportation, the design productivity for the transportation of cement with a given route configuration. The directions for further research have been determined.

1 Introduction

Pneumatic transport systems are very widely used in various industries, in construction, in transshipment work on rail and water transport. The load is transported through pipes and troughs due to the energy of the air flow in such systems. The main types of cargo that are transported by pneumatic conveyors are cement, mineral powders, crushed rock, ash and slag, chips and other types of goods.

The advantages of pneumatic conveyors in comparison with other transport vehicles include: tightness of the system; absence of the losses of transportable cargo; the convenience of conjugation of horizontal, vertical and inclined sections; the concentration of machine equipment in one place and the lack of need for complex maintenance throughout the route; the ability to move loads from several places to one and vice versa;

* Corresponding author: wbogomas@i.ua

the possibility of combining transport with several technological processes, such as cooling and drying.

One of the main elements of pneumatic systems is a compressor which is driven by an electric motor. Consequently, one of the important technical parameters of the pneumatic plant is the power of the drive of the compressor or blower machine.

The power of the compressor drive of a pneumatic transport installation depends on many factors. The main parameters influencing its value are: the cargo and its density; productivity of the plant for the cargo; horizontal and vertical dimensions of the transportation route; the number and types of auxiliary devices (ells and shutters); the given coefficient of the mixtures concentration.

2 Analysis of recent research and publications

The main publications describing the structure, design features, operational and design parameters of the modern machines of continuous transport are [1-8]. The analysis of publications shows that in order to determine the parameters of the drive of the compressor of the pneumatic transport unit, in particular its capacity, it is necessary to carry out a detailed calculation, namely: speed of air mixture (critical and actual), total pressure losses in the pneumatic system, air flow at the outlet of the compressor, working of the compressor and power drive compressor. The procedure for conducting such calculations is described in detail in the papers [1-4, 8]. Consequently, in order to accelerate the process of designing pneumatic transport units, it is necessary to construct a formula for determining the power of the compressor drive, which considers all output design data of the machine. There are papers in which the reduced algorithms for calculating the magnitude of the power of the drive of continuous transport vehicles are presented. In the paper [9] the authors constructed an algorithm for accelerated calculation of the belt elevator power with given design characteristics. The paper [10] is devoted to the construction of the dependence of the belt conveyors capacity on its design values.

3 Determine the purpose and objectives of the research

The purpose of the work is a graphical analysis of the dependence of the power of the compressor drive pneumatic transport units on the design productivity, type of cargo, the ratio of the mixture concentration and the length of the cargo transportation. For analysis, the task of the research is to construct an analytical dependence of the compressor drive power on the above design characteristics.

4 The main part of the research

The initial data for designing pneumatic transport machines are: the type of cargo and its density ρ_c; the air density ρ_a; the productivity of machine T_c; the length of the transportation route L; the lifting height H; the coefficient of the mixtures concentration μ; the number of the ells and shutters of the track.

The schemes of pneumatic transport machines are imagined in fig. 1.

The pneumatic transport machines, depending on the value of the mixture concentration ratio, are aspiration ($\mu \le 1$), with high vacuum ($\mu = 1...5$), pumping ($\mu = 8...25$).

For a further analysis of the dependence of all elements of the calculation from the initial design data, we present a general algorithm for calculating the pneumatic transport unit [2].

The weight productivity of the pneumatic machine by air:

$$T_a = \frac{T_c}{3.6\mu}. \tag{1}$$

The volume productivity of the pneumatic machine by air:

$$V_a = \frac{T_c}{3.6\mu\rho_a}. \tag{2}$$

The productivity of the pneumatic machine by air and the cargo:

$$V = V_c + V_a. \tag{3}$$

The diameter of the pipeline is taken d_p.

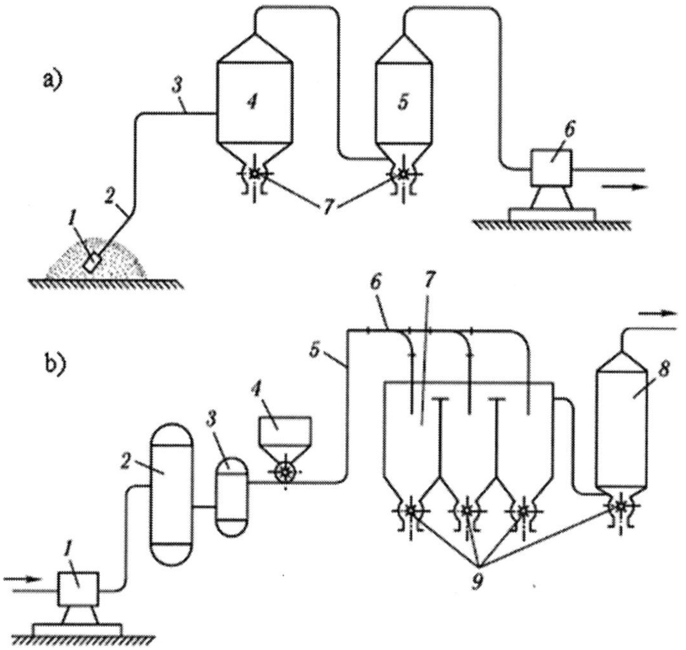

Fig. 1. The schemes of pneumatic transport machines: a) - suction machine (1 - nozzle; 2 - flexible pipeline; 3 - rigid pipeline; 4 - separator; 5 - filters; 6 - vacuum pump; 7 - locking ells); b) - pressure machine (1 compressor; 2 air sampler ; 3 - water separator; 4 - a loading device; 5 - a pipeline; 6 - two-way switch; 7 - separator; 8 - filters; 9 - locking ells)

The critical speed of the air mixture:

$$v_{cr} = n_2\sqrt{\mu a g d_p}, \tag{4}$$

where n_2 – the experimental constant; $n_2 = 0.1$ – for dusty, easy-to-handle cargoes; $n_2 \le 0.25$ – for dusty cargoes that are prone to felling; $n_2 \le 0.3$ – for granular cargoes; $n_2 \le 0.35$ – for lump cargo; in our calculations we accept $n_2 = 0.1$.

The coefficient a is found on the formula:

$$a = \frac{\rho_c - \rho_a}{\rho_a}. \tag{5}$$

The speed of the air mixture in the pipeline:

$$v_m = \frac{4V}{3600 \pi d_p^2}. \tag{6}$$

For the normal working of the machine, the speed of the air mixture in the pipeline must exceed the critical speed:

$$v_m > v_{cr}.$$

If $v_m < v_{cr}$ is valid, it is necessary to take another value of the pipelines diameter and do a recalculation.

The equivalent length of the pipeline:

$$L_f = L + H + L_{eq.el} + L_{eq.sh}, \tag{7}$$

where $L_{eq.el}$ – length equivalent to the number of ells (Table 1); $L_{eq.sh}$ – length equivalent to double-sided shutters (two-way switch is equivalent to a pipeline of 8 m in length).

The transverse section of the pipeline:

$$A = \frac{\pi d_p^2}{4}. \tag{8}$$

Table 1. The length of the pipelines, equivalent to the ells.

Bulk cargo	Value $L_{eq.el}$, m, in relation to R_{el}/d_p			
	4	6	10	20
Dusty	4…8	5…10	6…10	8…10
Grain homogeneous	-	8…10	12…16	16…20
Small-necked heterogeneous	-	-	28…35	33…45
The gluttonous is not homogeneous	-	-	60…80	70…90

Note. R_{el} – radius of curvature, d_p – inner diameter of the pipe.

The pressure losses in the pipeline in the case of clean air (isothermal air flow at pressure differences not exceeding $9.81 \cdot 10^4$ Pa):

$$p_0 = \sqrt{\frac{T_a^2 gRT\lambda L_f}{A^2 d_p} + p_e^2} - 9.81 \cdot 10^4, \tag{9}$$

where $R = 29.3$ – universal gas constant; $T = 283^\circ$ C – absolute temperature of the environment; $\lambda = 0.016...0.02$; $0.015...0.018$; $0.014...0.016$ – coefficient of resistance respectively for new and crowned pipes with diameters $d_p = 150$; 175; 200 mm; $p_e = 10.3 \cdot 10^4$ Pa – pressure at the end of the pipeline.

The losses of pressure in a pipeline of equivalent length L_f, taking into account rectilinear areas, rounding and two-way shutters:

$$p_n = p_0\left(1 + n_3\mu\frac{agd_p}{v_m^2}\right), \quad (10)$$

where $n_3 = 0.1\ldots0.075$ – the experimental constant.

The dynamic pressure losses:

$$p_d = \frac{v_m^2}{2}\rho_a(1+\beta\mu), \quad (11)$$

where β= 0.35…0.85 – an indicator of the relative velocity of the materials particles, β= 0.6…0.85 – for dusty cargoes.

The losses of pressure in short vertical pipelines:

$$p_v = (1+\mu)\rho_a Hg. \quad (12)$$

The losses of pressure in the feeders without a dosage device are equivalent to 90° rounding lengths p_Z=(1.96…2.94)·10^4 Pa. Consequently, we accept p_Z=2.45·10^4 Pa.

The total pressure losses in the pneumatic system:

$$p_t = p_n + p_d + p_v + p_z. \quad (13)$$

The air flow at the outlet of the blower machine:

$$V_t = (1.1\ldots1.15)V_a. \quad (14)$$

The end pressure in the pipeline:

$$p_e = p_t + p, \quad (15)$$

where p=9.81·10^4 Pa – the atmospheric pressure.

The work of the blower machine assigned to 1 m^3 of suction air by the isothermal pressure:

$$A_m = 225900 \cdot \lg\frac{p_e}{p}. \quad (16)$$

The power of the electric motor of the blower machine:

$$P = \frac{k \cdot A_m \cdot V_t}{10^3 \cdot \eta}, \quad (17)$$

where k=1.1 – the reserve coefficient; η=0.75 – coefficient of machine efficiency.

In order to construct the analytical dependence of the drive power of the blower unit on the design data of the transport unit, we will make some algebraic transformations.

Substituting (2) into (3), we have:

$$V = 1000T_c\left(\frac{1}{\rho_c} + \frac{1}{\mu\rho_a}\right). \quad (18)$$

Substituting (18) into (6), we have:

$$v_m = \frac{T_c\left(\rho_c + \mu\rho_a\right)}{0.9\pi d_p^2 \rho_c \mu \rho_a}. \tag{19}$$

For further studies, we denote $L_{eq.el} = \gamma$ and $L_{eq.sh} = \varepsilon$, where γ and ε – the corresponding values in table 1.

Substituting (1), (7), (8) into (9), we have:

$$p_0 = \sqrt{\frac{T_c^2 gRT\lambda\left(L + H + \gamma + \varepsilon\right)}{0.81\mu^2\pi^2 d_p^5} + p_e^2} - 9.81 \cdot 10^4. \tag{20}$$

Substituting (19) and (20) into (10), we have:

$$p_n = \left(\sqrt{\frac{T_c^2 gRT\lambda\left(L + H + \gamma + \varepsilon\right)}{0.81\mu^2\pi^2 d_p^5} + p_e^2} - 9.81 \cdot 10^4\right)\left(1 + n_3\mu^3 \frac{0.81agd_p^5\pi^2\rho_a^2\rho_c^2}{T_c^2\left(\rho_c + \mu\rho_a\right)^2}\right). \tag{21}$$

Substituting (19) into (11), we have:

$$p_d = \frac{T_c^2\left(\rho_c + \mu\rho_a\right)^2}{1.8\pi^2 d_p^4 \rho_c^2 \mu^2 \rho_a^2}\rho_a\left(1 + \beta\mu\right). \tag{22}$$

Substituting (12), (21), (22) into (13), we have:

$$\begin{aligned} p_t = &\left(\sqrt{\frac{T_c^2 gRT\lambda\left(L + H + \gamma + \varepsilon\right)}{0.81\mu^2\pi^2 d_p^5} + p_e^2} - 9.81 \cdot 10^4\right)\left(1 + n_3\mu^3 \frac{0.81agd_p^5\pi^2\rho_a^2\rho_c^2}{T_c^2\left(\rho_c + \mu\rho_a\right)^2}\right) + \\ &+ \frac{T_c^2\left(\rho_c + \mu\rho_a\right)^2}{1.8\pi^2 d_p^4 \rho_c^2 \mu^2 \rho_a}\left(1 + \beta\mu\right) + \left(1 + \mu\right)\rho_a Hg + p_z. \end{aligned} \tag{23}$$

Substituting (2) into (14), we have:

$$V_t = \frac{0.32T_c}{\mu\rho_a}. \tag{24}$$

We convert (17) and substitute (24), (16), (23) into (17), we have:

$$P = \frac{106.2T_c}{\mu\rho_a}\lg\left\{\left[\begin{aligned} &\left(\sqrt{\frac{T_c^2 gRT\lambda\left(L + H + \gamma + \varepsilon\right)}{0.81\mu^2\pi^2 d_p^5} + p_e^2} - 9.81 \cdot 10^4\right)\left(1 + n_3\mu^3 \frac{0.81agd_p^5\pi^2\rho_a^2\rho_c^2}{T_c^2\left(\rho_c + \mu\rho_a\right)^2}\right) + \\ &+ \frac{T_c^2\left(\rho_c + \mu\rho_a\right)^2}{1.8\pi^2 d_p^4 \rho_c^2 \mu^2 \rho_a}\left(1 + \beta\mu\right) + \left(1 + \mu\right)\rho_a Hg + p_z + 9.81 \cdot 10^4 \end{aligned}\right] / p\right\}. \tag{25}$$

For the graphical analysis of the dependence of the compressor drive power of the pneumatic transport unit, we accept: cargo – cement $\rho_c = 2600$ kg/m³; horizontal and vertical dimensions of the transportation route – $L = 200$ m and $H = 18$ m; air density $\rho_a = 1.2$ kg/m^3; a track with three-ell by the angle 90° and two double-sided shutters, that is, according to table 1 $\gamma = 30$ m and $\varepsilon = 16$ m.

By the given initial data, the dependence of the compressor drive power on the load productivity and the coefficient of the mixture concentration for systems with high vacuum and systems of various pressures are shown in Fig. 2.

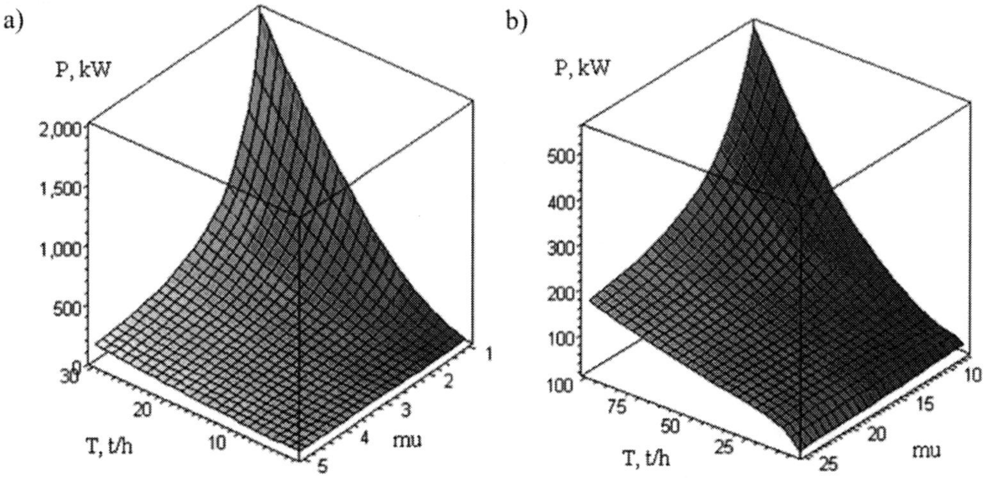

Fig. 2. Graphic dependence of the power of the electric motor of the blower machine on the productivity and coefficient of the mixture concentration: a) - machines with high vacuum by $T_c = 2...30$ t/h and $\mu = 1...5$; b) - machines with different pressure by $T_c = 2...100$ t/h and $\mu = 8...25$

In order to construct graphical dependence of the compressor drives power on the length of transportation we accept: $T_c = 30$ t/h. The graphic dependences constructed for the systems with high vacuum $\mu = 5$ and systems with different pressures $\mu = 25$ are shown in Fig. 3.

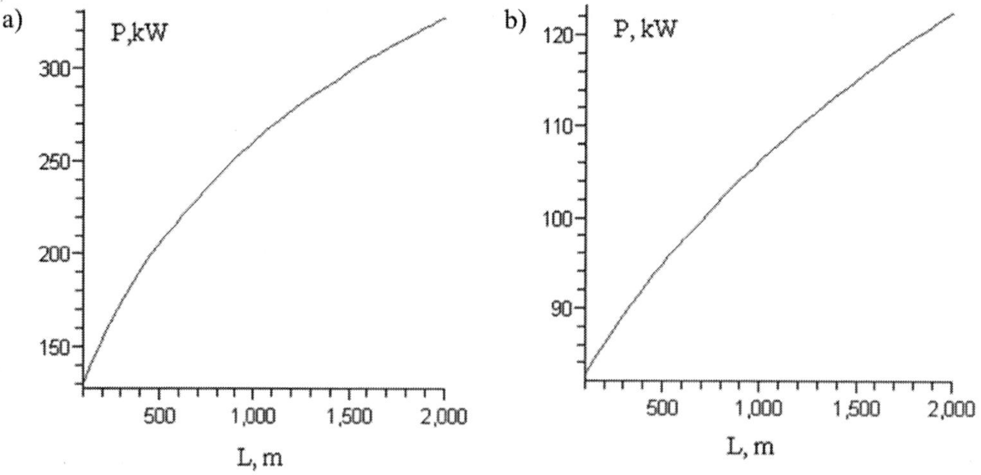

Fig. 3. Graphic dependence of the power of the electric motor of the blower machine on the transportation length: a) – machines with high vacuum by $T_c = 2...30$ t/h and $\mu = 5$; b) – machines with different pressure by $T_c = 2...30$ t/h and $\mu = 25$

5 Conclusions from research and prospects, further development in this direction

The analysis of the dependence of all elements of the calculation, given in [2], on the initial parameters for designing has been carried out for pneumatic conveyors. The analytical dependence of the compressor drives power on the loads density, the coefficient of the mixtures concentration, the productivity, the geometric dimensions of the route, the number and types of auxiliary devices have been constructed. Involving the constructed analytical dependencies, a graphical analysis of the change in the value of the compressor drive power during the variation of the design parameters values of the pneumatic machines was carried out. The analysis showed that the compressor drives power with the growth of the coefficient of the mixture concentration decreases, with increasing productivity - increases, with increasing length of transportation - increases. With the same output data, the compressor power of the systems with high vacuum is greater than the systems of different pressures. Further research is needed in the direction of developing an algorithm for optimizing energy consumption of pneumatic transport systems taking into considering the constructed dependencies.

References

1. N.E. Romakyn, *Mashyny nepreryvnoho transporta* (Akademiya, Moscow, 2008)
2. V.S. Bondaryev, O.I. Dubinets, M.P. Kolisnyk ta insh. *Pidyomno-transportni mashyny: rozrakhunky pidnimalnykh ta transportnykh mashyn* (Vyshcha shkola, Kyiv, 2009)
3. B.L. Vyshnya, B.S. Drozdov, V.T. Stefanenko. *Pnevmotransport. Raschoty, skhemy, oborudovanye* (Socrat, Yekaterinburg, 2010)
4. F.K. Ivanchenko *Pod'yemno-transportnyye mashiny* (Vyshcha shkola, Kyiv, 1993)
5. A.I. Voloshin, B.V. Ponomarev. *Mekhanika pnevmotransportirovaniya sypuchikh materialov* (Naukova dumka, Kyiv, 2001)
6. S.L. Litvinenko, G.I. *Nesterenko ta insh. Mekhanizatsiya i avtomatizatsiya pogruzochno - razgruzochnykh rabot* (Kondor, Kyiv, 2016)
7. N.N. Koval'nogov, A.S. Rtishcheva. *Raschet techeniya i soprotivleniya treniya potoka v soplakh Lavalya* (ULGTU, Ulyanovsk, 2007)
8. A.I. Baryshev, V.A. Budishevskiy, A.A. Sulima ta insh. *Raschet i proyektirovaniye transportnykh sredstv nepreryvnogo deystviya* (Nord-Press, Donetsk, 2005)
9. V.M. Bohomas, M.V. Borenko, S.V. Patsanovskyy, O.O. Tkachov, Nauka i prohres transportu **6**, 136-157 (2016)
10. V.M. Bohomas, Nauka i prohres transportu **1**, 131-146 (2016)

Frequency analysis of vehicle drive with cable traction

Serhii Raksha[1], *Oleksii* Kuropiatnyk[1,*], *Pavlo* Anofriev[1], *Dmytro* Onopreychuk[2], and *Ihor* Kovalov[3]

[1]Department of Applied Mechanics and Materials Science, Dnipropetrovsk National University of Railway Transport named after Academician V. Lazaryan, Dnipro, Ukraine
[2]Department of Construction, track and handling machines, Ukrainian State University of Railway Transport, Feierbakh Square 7, 61050, Kharkiv, Ukraine
[3]National Academy of the National Guard of Ukraine, Defenders of Ukraine Square 3, 61001, Kharkiv, Ukraine

Abstract. The paper outlines the basic principles of constructing frequency diagrams and specifies the features of their application for the analysis of the dynamic state of the vehicle drive with cable traction on the example of a ropeway. To draw up the frequency diagrams, the ropeway traction circuit was presented as a dynamic system containing lumped masses and elements with distributed parameters. In this case, the moments of inertia of the rotating masses of the drive were brought to the traction pulley as a lumped mass. The traction cable segments are represented as elements with distributed parameters. The mathematical model describing the dynamic state of the drive consists of equations in the form of deformations of the traction cable segments using the positions of wave mechanics. Studies have shown that in most cases there is a fairly wide inter-resonance band, limited by the first and second eigenfrequencies. This phenomenon may be the basis for justifying the rational values of the rotational speed of the pulley as one of the kinematic parameters of the ropeway drive.

1 Introduction

A lot of means of industrial transport have a flexible traction unit, among them belt and chain conveyor; vertical and inclined mine lifts. These machines also include tools with cable traction: hanging conveyors, monorail cable cars and ropeways. They differ in industries and technology, but the common feature is the presence of a drum drive (for belt conveyors), with one sprocket or a few sprockets (for chain conveyors and some hanging conveyors), or with a traction pulley (for mine lifts and ropeways). Instability of its operation (in particular, during transient modes) can lead to emergency situations. Therefore, the studies of the dynamic state of the vehicle drive with cable traction in order to prevent processes that disrupt its normal operation are relevant.

One of the measures to prevent the instability of the drive and to avoid emergencies

* Corresponding author: kuropyatnick@gmail.com

when using the means of industrial transport is to limit the speed of the traction unit and the associated working elements (scrapers for the chain conveyor, overhead conveyor car, cable car, etc.). However, this does not allow controlling the dynamic processes during the start-up period, in particular, to prevent and limit resonant phenomena.

To analyze the eigenfrequency spectrum of a vehicle drive with a cable traction under which there are resonance phenomena, it is convenient to use frequency diagrams that reflect the eigenfrequency as a function of moving the working element. By imposing on such diagrams a tachogram of the drive, it is possible to conclude about the safety of the vehicle as a dynamic system, determining the number of hazardous resonant zones that the drive pass during acceleration and deceleration, as well as the frequency and (for certain additional calculations) amplitudes of oscillations in the resonant zones. In addition, the inverse problem solution allows the development of rational tachograms of the drive to prevent the occurrence of resonant phenomena.

Consider the features of the frequency analysis of the vehicle drive with cable traction on an example of a ropeway.

2 Analysis of publications

Some results of research on rational design and effective use of ropeway elements are given in [1-3]. However, most of them are due to calculations based on static formulas.

The work [4] is devoted to study on the dynamic load of ropeway elements. That paper considers the processes in traction ropes, but only transverse oscillations are taken into account; the effect of these processes on the dynamics of the drive is not taken into account.

The importance of applying an integrated approach to evaluate the load of ropeway elements is confirmed by foreign experience [5-10]. However, the recommendations based on the analysis of the eigenfrequency spectrum of the drive have not been detected.

The formation of basic approaches to the construction and analysis of the eigenfrequency spectrum of the drive by frequency diagrams was begun in the works [11, 12]. However, the materials contained therein don't apply to ring-type ropeways.

3 Purpose of work

The purpose of this work is to develop the basic principles of constructing frequency diagrams and to formulate approaches to the application of such diagrams during the analysis of the dynamic state of vehicle drives with cable traction on the example of ropeway.

4 Theoretical studies

To compile frequency diagrams of the drive, we consider the traction circuit of cableway as a dynamic system, which includes a drive, a traction cable with a tensioner and cars (Fig. 1). In this case, we replace the components of the drive by three lumped masses – the rotor of the electric motor, the gear unit and the pulley. These masses are interconnected by elastic weightless connections. The cars and the tensioner are interconnected by segments of a traction cable as elements with distributed parameters.

In this problem statement, a mathematical model describing the dynamic state of a system consists of differential equations in ordinary derivatives and partial differential equations. Solution the system of this equation in order to obtain the frequency function is complicated, so within the scope of this work we perform the replacement the schema shown in Fig. 1 by simplified schema (Fig. 2). The peculiarity of this approach is to bring the masses of the drive to the pulley, which allows us to restrict the use of differential equations in partial derivatives during the compilation of a mathematical model.

In order to formulate the description of the mathematical model, we introduce the following concepts:

– number of cars in one direction of travel (n) – the number of cars that are simultaneously located on the area between the pulley and the tensioner as lumped masses (constructively between the drive and tension stations) and move in one direction; thus the total number of cars is $2n$;
– lower car – a car moving from the lower station to the upper;
– upper car – a car moving from the upper station to the lower one;
– lower direction of movement – the section of the route, along which the lower cars move;
– upper direction of movement – the section of the route, along which the upper cars move.

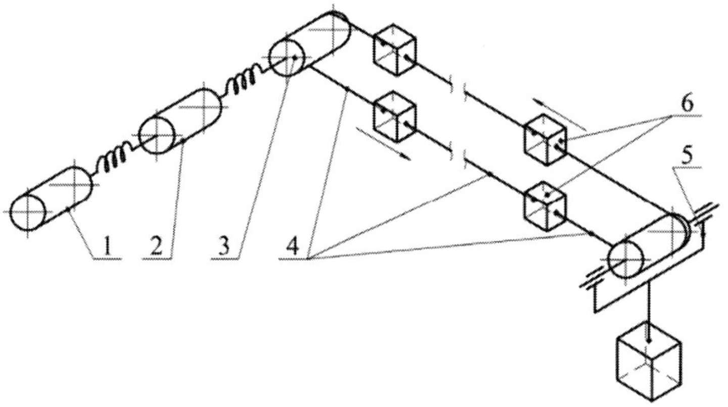

Fig. 1. Ropeway traction circuit: 1 – the rotor of the electric motor; 2 – mass of the gear unit; 3 – pulley; 4 – the traction cable segments; 5 – tensioner; 6 – cars

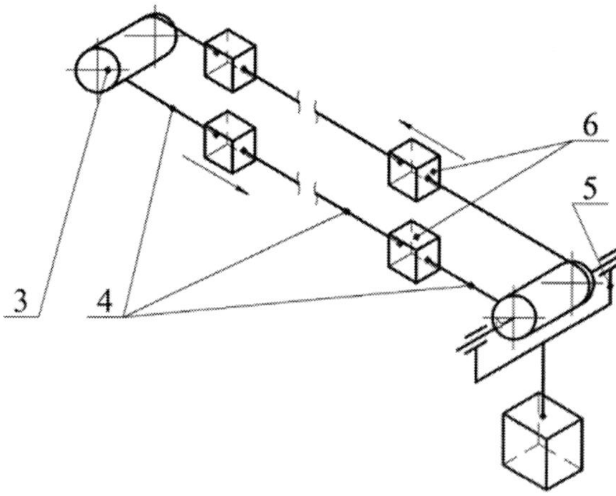

Fig. 2. Simplified model of the ropeway traction circuit (designation according to Fig. 1)

In absolute terms these statements are valid for ropeways, whose profile changes monotonically; that is the difference in elevation between two adjacent reference points in the upper direction of motion is always positive, and in the lower direction of motion it is negative. In the case of changes in the profile curvature, these concepts can be used for each span individually, and then combine into a common simulation system.

For the convenience of compiling a mathematical model we use a substitution scheme in which all masses make translational movement [13] (Fig. 3, a). In this case, the feature of such a scheme for the traction circuit of ropeway is its closeness, which determines a certain specificity of the formation of equations. It is reflected in the application of additional forces that characterize the mutual influence of finite masses (Fig. 3, b).

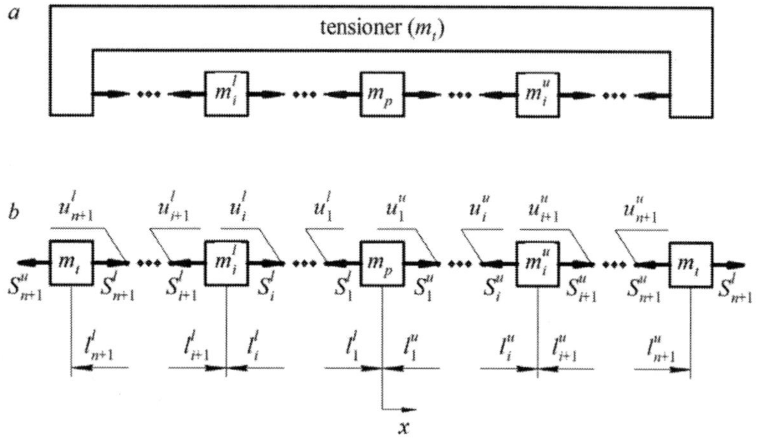

Fig. 3. Substitution scheme: a – general view; b – with additional force applied

In the substitution scheme we use the following notation:

m_p – mass of the pulley (equivalent weight of all elements of the drive);

m_t – mass (equivalent to the working force) of the tensioner;

m_i^l – mass of the i-th lower car ($i = \overline{1, n}$);

m_i^u – mass of the i-th upper car ($i = \overline{1, n}$);

u_1^l – elastic deformation of the traction cable segment between the pulley and the first car in the lower direction of movement;

u_i^l – elastic deformation of the traction cable segment between (n-1)-th and i-th cars in the lower direction of movement ($i = \overline{2, n}$);

u_{n+1}^l – elastic deformation of the traction cable segment between the i-th car in the lower direction of movement and the tensioner;

u_1^u – elastic deformation of the traction cable segment between the pulley and the first car in the upper direction of movement;

u_i^u – elastic deformation of the traction cable segment between (n-1)-th and i-th cars in the upper direction of movement ($i = \overline{2, n}$);

u_{n+1}^u – elastic deformation of the traction cable segment between the i-th car in the upper direction of movement and the tensioner;

S_1^l , S_i^l , S_{n+1}^l , S_1^u , S_i^u , S_{n+1}^u – elastic forces that correspond to deformations u_1^l , u_i^l , u_{n+1}^l , u_1^u , u_i^u , u_{n+1}^u ;

$$S_i^l = EF\frac{\partial u_i^l}{\partial x}; \qquad S_i^u = EF\frac{\partial u_i^u}{\partial x}, \qquad i = \overline{1, n+1}, \tag{1}$$

E, F – the elastic modulus and the area of the traction cable cross-section.

According to the indicated symbols, the mathematical model describing the dynamic state of the drive consists of equations in the form of deformations of the traction cable segments using the positions of wave mechanics [13]:

$$\begin{cases} \frac{\partial^2}{\partial t^2} U^l = a^2 \frac{\partial^2}{\partial x^2} U^l; \\ \frac{\partial^2}{\partial t^2} U^u = a^2 \frac{\partial^2}{\partial x^2} U^u, \end{cases} \tag{2}$$

$U^l = \left[u_i^l \right]$ – vector of deformations of the traction cable segments on the lower direction of the car movement ($i = \overline{1, n+1}$); $U^u = \left[u_i^u \right]$ – vector of deformations of the traction cable segments on the upper direction of the car movement ($i = \overline{1, n+1}$); a – the propagation velocity of the elastic wave in the longitudinal direction of oscillation.

In order to simplify the recording, we introduce the following notation:

$x_i^l = -\sum_{j=1}^{i} l_j^l$ – coordinate of the i-th car in the lower direction of movement;

$x_i^u = \sum_{j=1}^{i} l_j^u$ – coordinate of the i-th car in the upper direction of movement;

$\alpha_i^l = ql / m_i^l$ – mass coefficient of the i-th car in the lower direction of movement;

$\alpha_i^u = ql / m_i^u$ – mass coefficient of the i-th car in the upper direction of movement;

$\alpha_p = ql / m_p$ – mass coefficient of the pulley;

$\alpha_t = ql / m_t$ – mass coefficient of the traction cable tensioner;

$\lambda = kl$ – the actual number of the frequency function (k is a certain constant used in the transformation of the system of equations [13]);

A, B (with any indexes) – amplitude coefficients.

Also consider the following parameters of the traction cable: E – elastic modulus; F – cross-section area; q – linear mass; l – the length of the traction cable segments between the pulley and the tensioner as lumped masses in one direction of movement.

For equations in the form of the first expression of the system (2), which describe the movement of the i-th car in the lower direction, the boundary conditions have the form ($i = \overline{1, n}$):

$$m_i^l \left. \frac{\partial^2 u_i^l}{\partial t^2} \right|_{x=x_i^l} = EF \left. \left(\frac{\partial u_i^l}{\partial x} - \frac{\partial u_{i+1}^l}{\partial x} \right) \right|_{x=x_i^l}; \qquad u_i^l \left(x_i^l, t \right) = u_{i+1}^l \left(x_i^l, t \right). \tag{3}$$

For equations in the form of the second expression of the system (2), which describe the movement of the i-th carriage in the upper direction, the boundary conditions have the form ($i = \overline{1, n}$):

$$m_i^u \left. \frac{\partial^2 u_{i+1}^u}{\partial t^2} \right|_{x=x_i^u} = EF \left. \left(\frac{\partial u_{i+1}^u}{\partial x} - \frac{\partial u_i^u}{\partial x} \right) \right|_{x=x_i^u}; \qquad u_i^u \left(x_i^u, t \right) = u_{i+1}^u \left(x_i^u, t \right). \tag{4}$$

To combine the equations (3) and (4) with the system and to ensure the closeness of the traction circuit, we use the conjugation conditions at the traction cable segments near the pulley and the tensioner. We make such expressions in the form of boundary conditions to differential equations which describe the movement of the pulley and the tensioner as lumped masses m_p and m_t respectively.

According to the above notation, the boundary conditions for the equation describing the movement of the pulley have the form:

$$m_p \frac{\partial^2 u_1^u}{\partial t^2}\bigg|_{x=0} = EF\left(\frac{\partial u_1^u}{\partial x} - \frac{\partial u_1^l}{\partial x}\right)\bigg|_{x=0}; \qquad u_1^l(0,t) = u_1^u(0,t), \tag{5}$$

$x = x_1^u = x_1^l = 0$ – pulley coordinate.

The boundary conditions for the equation describing the motion of the tensioner are as follows:

$$m_t \frac{\partial^2 u_{n+1}^l}{\partial t^2}\bigg|_{x_{n+1}^l} = EF\left(\frac{\partial u_{n+1}^l}{\partial x}\bigg|_{x_{n+1}^l} - \frac{\partial u_{n+1}^u}{\partial x}\bigg|_{x_{n+1}^u}\right); \quad u_{n+1}^u\left(x_{n+1}^u, t\right) = -u_{n+1}^l\left(x_{n+1}^l, t\right). \tag{6}$$

After separating the variables in equations (3) – (6), we combine the results obtained with the system (7).

$$\begin{cases}
\left[\sin\left(\lambda\frac{x_i^l}{l}\right) - \frac{\lambda}{\alpha_i^l}\cos\left(\lambda\frac{x_i^l}{l}\right)\right]A_i^l - \left[\frac{\lambda}{\alpha_i^l}\sin\left(\lambda\frac{x_i^l}{l}\right) + \cos\left(\lambda\frac{x_i^l}{l}\right)\right]B_i^l - \\
\qquad - \sin\left(\lambda\frac{x_i^l}{l}\right)A_{i+1}^l + \cos\left(\lambda\frac{x_i^l}{l}\right)B_{i+1}^l = 0; \\
\cos\left(\lambda\frac{x_i^l}{l}\right)A_i^l + \sin\left(\lambda\frac{x_i^l}{l}\right)B_i^l - \cos\left(\lambda\frac{x_i^l}{l}\right)A_{i+1}^l - \sin\left(\lambda\frac{x_i^l}{l}\right)B_{i+1}^l = 0; \\
\sin\left(\lambda\frac{x_i^u}{l}\right)A_i^u - \cos\left(\lambda\frac{x_i^u}{l}\right)B_i^u - \left[\sin\left(\lambda\frac{x_i^u}{l}\right) - \frac{\lambda}{\alpha_i^u}\cos\left(\lambda\frac{x_i^u}{l}\right)\right]A_{i+1}^u + \\
\qquad + \left[\frac{\lambda}{\alpha_i^u}\sin\left(\lambda\frac{x_i^u}{l}\right) + \cos\left(\lambda\frac{x_i^u}{l}\right)\right]B_{i+1}^u = 0; \\
\cos\left(\lambda\frac{x_i^u}{l}\right)A_i^u + \sin\left(\lambda\frac{x_i^u}{l}\right)B_i^u - \cos\left(\lambda\frac{x_i^u}{l}\right)A_{i+1}^u - \sin\left(\lambda\frac{x_i^u}{l}\right)B_{i+1}^u = 0; \\
\frac{\lambda}{\alpha_p}A_1^u + B_1^u - B_1^l = 0; \\
A_1^u - A_1^l = 0; \\
\sin\lambda\cdot A_{n+1}^u - \cos\lambda\cdot B_{n+1}^u + \left[\sin\lambda + \frac{\lambda}{\alpha_t}\cos\lambda\right]A_{n+1}^l - \left[\frac{\lambda}{\alpha_t}\sin\lambda - \cos\lambda\right]B_{n+1}^l = 0; \\
\cos\lambda\cdot A_{n+1}^u + \sin\lambda\cdot B_{n+1}^u + \cos\lambda\cdot A_{i+1}^l - \sin\lambda\cdot B_{n+1}^l = 0.
\end{cases} \tag{7}$$

The frequency function is established as a determinant of the matrix of the coefficients of the system (7). The zero arguments of this function (such that the function acquires a null value) are the eigenvalues λ, by which we determine the eigenfrequencies of the ropeway traction circuit as a dynamic system:

$$\omega = \frac{\lambda}{l}\sqrt{\frac{EF}{q}}\,. \tag{8}$$

Taking into account that the ropeway traction circuit is modeled as a system of lumped masses (pulley, cars and tensioner) connected by elastic elements with distributed parameters (traction rope segments), the frequency function has a set of zero arguments, and hence the drive as the dynamic system has a set of eigenfrequencies ω, which are determined by their eigenvalues λ.

We note that the frequency diagram shows the change of the eigenfrequencies of the drive during the movement of cars between stations. Examples of frequency diagram under the condition of the same loading of cars are shown in Fig. 4 (shows the first three eigenfrequencies). Compared to the second and third, the first frequency is relatively constant, since it is largely determined by the pulley mass. The coordinates of cars have a significant influence on the second and third eigenfrequencies, which explains the substantial nonlinearity of the corresponding curves. For a ring-type ropeway, there is a certain cyclic change in the third frequency, with the number of cycles corresponding to the number of cars in one direction of movement.

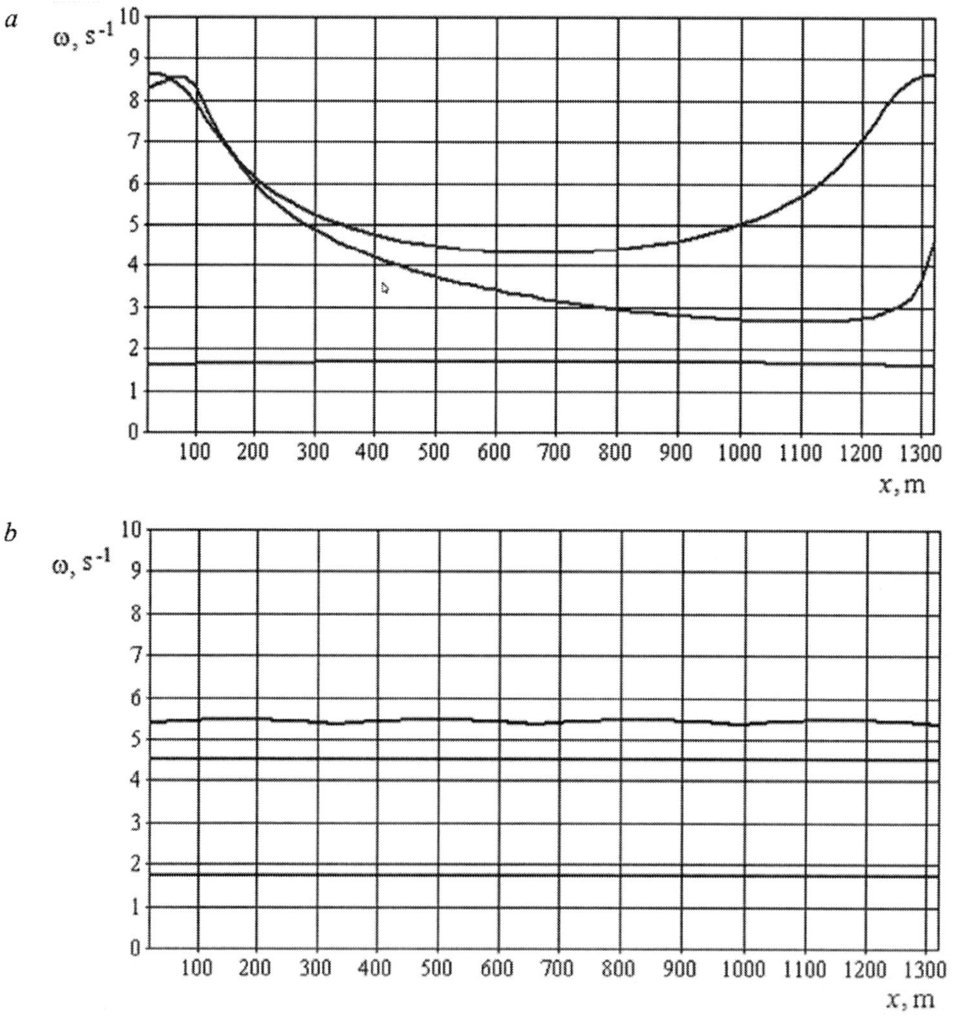

Fig. 4. Frequency diagrams of the ropeway drive: a – pendulum-type ropeway; b – ring-type ropeway

5 Conclusions

In most cases, there is a fairly wide interresonance band, limited by the first and second eigenfrequencies. This phenomenon may be the basis for justifying the rational values of the rotational speed of the pulley as one of the kinematic parameters of the ropeway drive. The results obtained are valid for all ropeways (both pendulum-type and ring-type) by the nature of the frequency diagram; however, the frequency values may vary, depending on the ropeway technical characteristics. For other vehicles with cable traction: for belt and chain conveyors, the degree of nonlinearity of frequency diagrams depends on the uniform distribution of cargo on the carrying surface; for hanging conveyors, the nature of the frequency diagrams is similar to that for a ring-type ropeways; frequency diagrams for mine lifts drives are similar to the pendulum-type ropeway diagrams but they are characterized by greater symmetry of the curves of the second and third eigenfrequencies for vertical lifting.

References

1. M. Berkman. *Suspended cable cars* (Mechanical Engineering, Moscow, 1984)
2. S. Raksha, O. Kuropiatnyk, A. Kurka. Science and Transport Progress. Bulletin of Dnipropetrovsk National University of Railway Transport **1 (49)**, 125–131 (2014)
3. Yu. Goriachev, O. Kuropiatnyk, M. Izmailov. Science and Transport Progress. Bulletin of Dnipropetrovsk National University of Railway Transport **3 (51)**, 109–116 (2014)
4. D. Pataraia. *Calculation and design of cable systems on the example of ropeways* (Mezniereba, Tbilisi, 1991)
5. J. Nejez. ISR **6**, 47 (2011)
6. G. Kopanakis. ISR **6**, 48–50 (2011)
7. G. Kopanakis. ISR **1**, 46–49 (2012)
8. G. Kopanakis. ISR **3**, 63–66 (2012)
9. J. Kowal. AME vol. LV **4**, 357–368 (2008)
10. M. Knawa. PAMM vol. 8 **1**, 10297–10298 (2008)
11. S. Raksha, Yu. Goriachev, O. Kuropiatnyk. Construction, materials science, mechanical engineering **66**, 249–256 (2012)
12. S. Raksha, Yu. Goriachev, O. Kuropiatnyk. Herald of the DSEA **1 (30)**, 110–116 (2013)
13. A. Stepanov. *Dynamics of Machines* (UBRAS, Ekaterinburg, 1999)

Determination of equivalent and optimal sizes of wedge tip from flange for the static perforation of soil

Olexander Posmituha[1,*], *Svyatoslav* Kravets[2], *Volodymyr* Suponyev[3], and *Kazimir* Glavatsky[1]

[1]Dnipropetrovsk National University of Railway Transport named after Academician V. Lazaryan, Dnipro, Ukraine
[2]National University of Water and Environmental Engineering, Rivne, Ukraine
[3]Kharkiv National Automobile and Highway University, Kharkiv, Ukraine

Abstract. The purpose of the study is to create a new design of the working tip for the perforation of the soil when laying two or more buildings trenchless way and determine its rational parameters. Method. It is known that for static perforation in the formation of cavities in the soil for communication is traditionally used cone-shaped working elements, which is the most common, cheap and reliable. However, the study revealed inefficiency. The design of the working elements of the soil perforation for trenchless laying of two or more buildings simultaneously in the form of a tip, which consists of a wedge-shaped part and a control part in the form of a rectangle with half-cylinders. The basic principles and mathematical dependences of determining the size of the working body are described. Scientific novelty. Regularities of changes in soil pressure on the lateral, conical and wedge surface, flat and cylindrical part of the tip, allowed to obtain an analytical dependence for the calculation of the resistance force depending on the physical and mechanical properties of the soil, the diameter and the number of cases established simultaneously. Also, when comparing the cone-cylindrical and flat tips, it can be argued that the breakdown force of the flat tip decreases from 1.2 to 1.7 times.

1 Analysis of literary data and problem statement

General conformities to law of processes of perforation of soil and formation horizontally-directed mining holes expounded in works [1, 3, 4, 6, 9, 10], researches of influence of structural parameters working elements on the process of perforation in works [2, 3, 6, 9]. Changes of descriptions of soil are in surrounding space of mining hole (well space) with static puncture of soil and the influence on the formation of a wellhead leader [2, 5, 6]. Researches on determination of rational and optimal sizes tip for to puncture the soil of working element of cone form for the decision of this task in works [4, 6].

For the first time such a form of cross section of the hole is considered that analytical dependence of resistance of moving of working organ is create from the form of working

* Corresponding author: AleksandrP@3g.ua

organ (Fig. 1.) and the number of linearly lengthy objects, what laid simultaneously. The got dependences give an opportunity to assert, that with this shape of the hole depending on the number of communications, which are laid simultaneously it is possible to decrease effort to 1.7 parts, and also to decrease the zone of structural changes in soil round the got cavity in comparing to the traditional cone tip. From work [6] it is known that the puncture of the soil by a cone tip it can be effective within the limits of diameter of mining hole, what is got from the condition of optimal resistance. It was so set, that the maximum reduction in the resistance to piercing takes place: for hard sandy loam at diameters $D=0.2...0.25$ m, for semi-solid loam with diameters $D=0.13...0.17$ m, for taut plastic clay at diameters $D=0.12...0.13$ m. Obvious is that through such cavities the number of pipe laying is limited, especially paying attention to that, that their diameters fold not less than 100 mm.

2 Methodology of calculation of resistance of wedge tip

We propose the construction of a groundbreaking working element (Fig. 1) for a simultaneous gasket two and more cases 3, what is folded pricking wedge-shaped part 1 with the halves of cone and the calibration part 2 in form a rectangular cavity with half the cylinder at the edges.

Will define the law of change of closeness of soil on a thickness d tip based on the equality of masses [6], what according to the chart of action of normal pressure of soil on a working elements (Fig. 1) looks like:

$$(n-1)d^2\rho_{nat} = (n-1)d(d-2x)\rho , \tag{1}$$

where d – the diameter of the linear-extending object (LEO); n – quantity of LEO, which are simultaneously laid; ρ_x – a law of change of closeness of soil is on the lateral verges of wedge on the height of cone surface; ρ_{nat} – a closeness of soil is in the natural state; x – deformation of soil by the frontal surface of tip.

Then law of change of normal pressure of soil on the frontal surface of tip in accordance with methodology [4] written down in a kind:

$$q_x = E_{soil}\frac{2x}{d}, \tag{2}$$

where $E_{soil} = \frac{1+\omega}{c_k}\frac{\rho_{hard}}{\rho_{nat}}$ – compression module of deformation of soil; c_k – coefficient of compression of the soil [8], MPa^{-1} ($c_k = 0.07...0.09(\omega_T - 10)$); ρ_{hard} – density of solid phase of soil (without pores); ω – humidity of soil, %; ω_T – limit of fluidity of soil.

The force of immersion of the tip consists of the effort of immersion of the wedge in height $h_{wedge} = (n-1)d$ that effort of immersion of two extreme halves of cones equivalent to effort of immersion of one cone diameter d.

$$P_{comeback.} = P_{wedge} + P_{cone.} . \tag{3}$$

Will define resistance of wedge:

$$dP_{wedge} = 2q_x\partial F\sin\beta(1+f\cdot ctg\beta). \tag{4}$$

where ∂F – elementary area of frontal surface of wedge ($\partial F = \partial h \cdot \frac{\partial x}{\sin\beta}$); 2β – corner of intensifying of wedge; f – a coefficient of friction of soil is on a wedge.

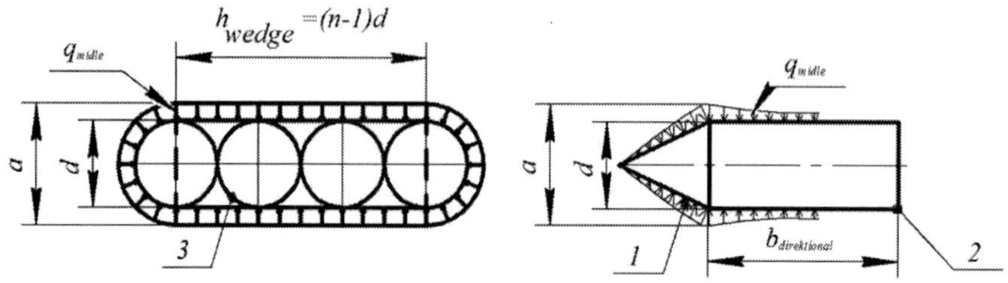

Fig. 1. Cross section of working elements: 1 – wedge piercing part with half cone sideways; 2 – calibrating part; 3 – cases that is laid.

Then

$$\partial P_{wedge} = E_{soil}\frac{4x}{d}\partial h\frac{\partial x}{\sin\beta}\sin\beta\left(1 + fctg\beta\right), \tag{5}$$

$$P_{wedge} = E_{soil}\frac{4h_{wedge}}{d}\left(1 + f\cdot ctg\beta\right)\int\limits_{0}^{d/2} x\partial x = \frac{d\cdot h_{wedge}}{2}E_{soil}\left(1 + f\cdot ctg\beta\right) \tag{6}$$

$$P_{wedge} = \frac{\left(n-1\right)d^{2}}{2}E_{soil}\left(1 + fctg\beta\right). \tag{7}$$

Obviously, that total resistance the puncture consists of the resistance of the conical and wedge parts of the tip (Fig. 1) for a gasket n-casing. According to previous work [4, 6] authors, the theoretically determined resistance to penetration of the conical tip into the soil equals:

$$P_{comeback.cone.} = \frac{\pi}{8}D^{2}E_{soil}\left(1 + f\,ctg\beta\right). \tag{8}$$

where D – the size of the hole in which fit n-casing, what laid simultaneously (Fig. 2).

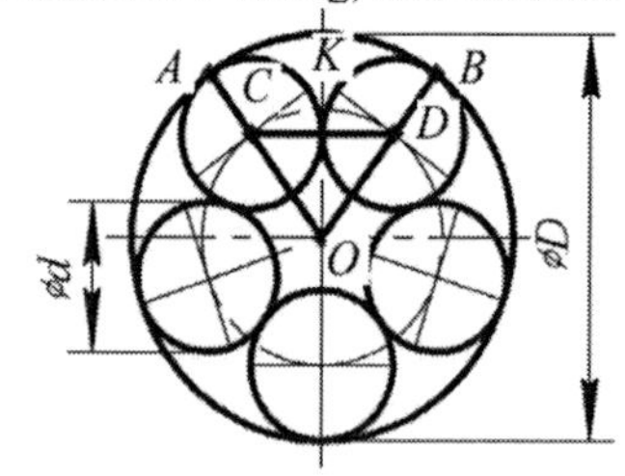

Fig. 2. Determination of opening diameter is at a group gasket n- casing.

Consider the triangles ΔAOB , ΔCOD та ΔKOD .

Angle $\angle KOD = \frac{\pi}{n}$, from where

$$\sin\frac{\pi}{n} = \frac{KD}{OD}, \quad OD = \frac{KD}{\sin\left(\pi/n\right)} = \frac{d}{2\sin\left(\pi/n\right)}, \tag{9}$$

$$OB = OD + DB = \frac{d}{2\sin\left(\pi/n\right)} + \frac{d}{2} = \left(1 + \frac{1}{\sin \pi/n}\right)d. \quad (10)$$

In the general case for any quantities casing:

$$D = \left(1 + \frac{1}{\sin \pi/n}\right)d. \quad (11)$$

The total frontal resistance is equal to:

$$P_{come back.wedge} = \frac{\pi + 4(n-1)}{8} d^2 E_{soil}\left(1 + fctg\beta\right). \quad (12)$$

Determine the ratio of frontal perforation resistance by cone and wedge tips with taking into account (11):

$$\frac{P_{come back.cone.}}{P_{come back.wedge}} = \frac{\pi D^2}{\left[\pi + 4(n-1)\right]d^2} = \frac{\pi\left(1 + \frac{1}{\sin \pi/n}\right)^2 d^2}{\left[\pi + 4(n-1)\right]d^2} = \frac{\pi\left(1 + \frac{1}{\sin \pi/n}\right)^2}{\pi + 4(n-1)}. \quad (13)$$

Will define dependence of size of circle, what is described round a few casing fixed alongside. Will consider a few variants of reaching of pipes an amount: 2, 3, and *n* piece.

Putting a value in a formula (8) at an identical size one casing will define the relation of resistance of pricking cone and wedge tips.

If $n=2$ then $\frac{P_{come back.cone.}}{P_{come back.wedge}} = 1.76$. If $n=3$ then $\frac{P_{come back.cone.}}{P_{come back.wedge}} = 1.309$.

If $n=4$ then $\frac{P_{come back.cone.}}{P_{come back.wedge}} = 1.209$. If $n=5$ then $\frac{P_{come back.cone.}}{P_{come back.wedge}} = 1.2$.

If $n=10$ then $\frac{P_{come back.cone.}}{P_{come back.wedge}} = 1.44$. If $n=15$ then $\frac{P_{come back.cone.}}{P_{come back.wedge}} = 1.79$.

3 Methodology of calculation of total resistance of moving of wedge tip and flat directing in soil

The lateral pressure of the soil acts on the entire perimeter of the body of the tip, which creates additional resistance from forces of friction of soil on lateral walls and edges of tip with rounding. Will define this resistance.

The average pressure of the soil on the side walls and the tips of the tip is equal [4, 6]

$$q_{midle} = 2c \cdot tg\left(\frac{\pi}{4} + \frac{\varphi_0}{2}\right) + \frac{a_q}{2}h. \quad (14)$$

where c – coefficient of soil adhesion; φ_0 – corner of internal friction of soil; a_q – coefficient of proportion (a_q = 0.026 MPa/m); h – depth of placing of tip (the depth of laying of communications).

The length of the perimeter of the body of the tip is equal to

$$L_{perimeter} = 2(n-1)d + \pi d \ . \tag{15}$$

Resistance is from forces of friction it is determined through the plaza of side of tip that the average pressure, and equals:

$$F_{friction} = \left[\pi d + 2(n-1)d\right] b_{direktional} f \left[2c \cdot tg\left(\frac{\pi}{4} + \frac{\varphi_0}{2}\right) + \frac{a_q}{2} h \right]. \tag{16}$$

where $b_{direktional}$ – width of the lateral surface of the tip (fig. 1); f – a coefficient of friction of soil is on material of tip.

Then total resistance of tip can differ to 2 times from the type of soil and from 1.2 to 1.76 times depending on the amount of casing, and finally has a look:

$$\begin{aligned} P_\Sigma = P_{comeback.wedge.} + F_{friction} = \frac{\pi + 4(n-1)}{8} d^2 E_{soil}\left(1 + fctg\beta\right) + \\ + \left[\pi d + 2(n-1)d\right] b_{direktional} f \left[2c \cdot tg\left(\frac{\pi}{4} + \frac{\varphi_0}{2}\right) + \frac{a_q \cdot h}{2} \right] \end{aligned} \tag{17}$$

4 Methodology of calculation of equivalent and optimal sizes of wedge tip is with performances

The construction of tip is offered in that wedge part is done with performances, what removes operating of resilient deformations and accordingly force of friction under the perimeter of calibrating part.

Will define the minimum thickness of wedge (Fig. 3) with performances a_{wedge}^{min}, what removes operating of resilient deformations under the perimeter of tip:

$$a_{wedge}^{min} = a + \Delta a \ . \tag{18}$$

what removes operating of resilient deformations under the perimeter of tip Hooke [1].

$$\sigma_1 = E_V \varepsilon \ . \tag{19}$$

where E_V – modulus of resiliency under volume compressed soil (module of by volume deformation); ε – relative deformation of the soil ($\varepsilon = \Delta a/a$).

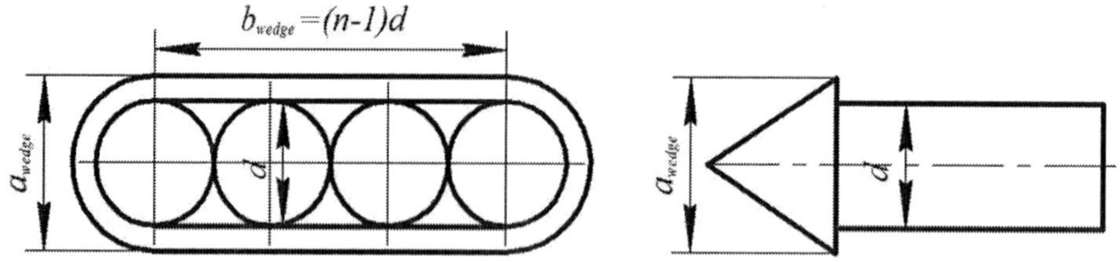

Fig. 3. Scheme of the working elements in the form of a wedge with protrusions.

Then in accordance with [1, p. 247].

$$\Delta a = \frac{\sigma_1}{E_V} a . \tag{20}$$

For solid sandy loam $\Delta a = \frac{0.108}{0.95} a = 0.114a$ ($a_{min} = 1.114a$).

For noncompressed clay loam $\Delta a = \frac{0.157}{0.64} a = 0.245a$ ($a_{min} = 1,24a$).

For plastic clay $\Delta a = \frac{0.167}{0.477} a = 0.35a$ ($a_{min} = 1.35a$).

Thus

$$a_{wedge}^{min} = a + \frac{\sigma_1}{E_V} a = \left(1 + \frac{\sigma_1}{E_V}\right) a . \tag{21}$$

A decline of resistance of wedge tip is with performances at a minimum thickness a_{wedge}^{min} it is resulted on (Fig. 4).

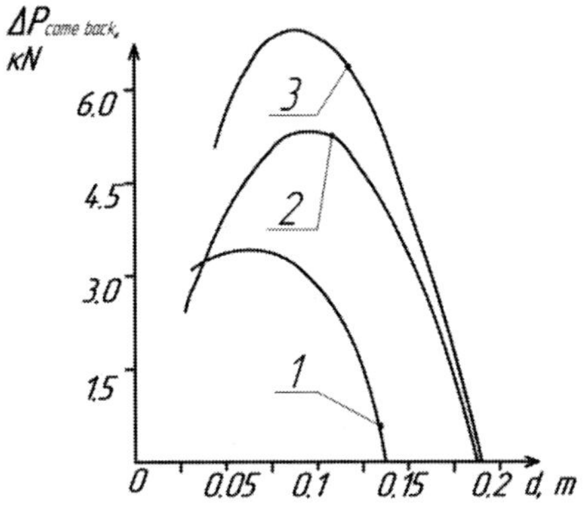

Fig. 4. Dependence of reduction of resistance to perforation wedge-shaped tip with performances from the diameter of the communications (n=2): 1 – plastic clay; 2 – noncompressed clay loam; 3 – solid sandy loam.

From (Fig. 4) it is evident that the maximum reduction in resistance pricking of wedge tip with performances depends on the type of soil, and diameter linear-extending object LEO, what are simultaneously laid.

For determination of maximal size casing will consider a difference between total resistance of wedge tip without performances (17) and the ground resistance of the tip with the projection, what takes into account resistance of frontal surface of wedge tip only, but eliminates the action of resilient deformation of soil, but accordingly force of friction for the surfaces of him calibrating part. For implementation of condition of absence of forces of friction on the surface of calibrating part and determination of force of resistance of soil tip with protrusion at calculation dependence instead of dependence of diameter LEO insert the minimum wedge size (21) and get the following expression:

$$\Delta P_{comeback.wedge} = P_{\Sigma} - P_{comeback.wedge} = \frac{\pi + 4(n-1)}{8} E_{soil}(1 + fctg\beta)d^2 + \\ +[\pi + 2(n-1)] fb_{direktional}\left[2c \cdot tg\left(\frac{\pi}{4} + \frac{\varphi_0}{2}\right) + \frac{a_q}{2}h\right]d - \\ -\frac{\pi + 4(n-1)}{8} E_{soil}(1 + fctg\beta)\left(1 + \frac{\sigma_1}{E_V}\right)^2 d^2 = \\ = \frac{\pi + 4(n-1)}{8} E_{soil}(1 + fctg\beta)\left[1 - \left(1 + \frac{\sigma_1}{E_V}\right)^2\right]d^2 + \\ +[\pi + 2(n-1)] fb_{direktional}\left[2c \cdot tg\left(\frac{\pi}{4} + \frac{\varphi_0}{2}\right) + \frac{a_q}{2}h\right]d \quad (22)$$

Differentiating the got expression (24) relatively d will get:

$$\frac{\partial \Delta P_{comeback.wedge}}{\partial d} = 2\frac{\pi + 4(n-1)}{8} E_{soil}(1 + fctg\beta)\left[1 - \left(1 + \frac{\sigma_1}{E_V}\right)^2\right]d + \\ +[\pi + 2(n-1)] fb_{direktional}\left[2c \cdot tg\left(\frac{\pi}{4} + \frac{\varphi_0}{2}\right) + \frac{a_q}{2}h\right] = 0 \quad .. (23)$$

Deciding the got equalization (23) relatively d will get calculation of optimal size of diameters of pipes for their gasket by a wedge working elements:

$$d_{onm} = \frac{4 fb_{direktional}(\pi + 2(n-1))\left(2c \cdot tg\left(\frac{\pi}{4} + \frac{\varphi_0}{2}\right) + \frac{a_q}{2}h\right)}{(\pi + 4(n-1))\left(\left(1 + \frac{\sigma_1}{E_V}\right)^2 - 1\right)E_{soil}(1 + fctg\beta)}. \quad (24)$$

Thus, for example for ($n=1$) most effectively to lay communication by a wedge tip with performances in plastic clay diameter $d_{optimal} = 0.096$ m; in noncompressed clay loam – $d_{optimal} = 0.127$ m.; in solid sandy loam – $d_{optimal} = 0.129$ m. this diameter depends on the type of soil, geometrical parameters of tip, depths of gasket and from an amount LEO, what are simultaneously laid.

Determine the diameter of the communications, at that a tip with performances is not effective, id est $\Delta P_{comeback.wedge} = 0$. From equalization (22) will get:

$$d_{max} \geq \frac{8 fb_{direktional}(\pi + 2(n-1))\left[2c \cdot tg\left(\frac{\pi}{4} + \frac{\varphi_0}{2}\right) + \frac{a_q}{2}h\right]}{(\pi + 4(n-1))\left[\left(1 + \frac{\sigma_1}{E_V}\right)^2 - 1\right]E_{soil}(1 + fctg\beta)} \quad (25)$$

$$d_{optimal} = \frac{d_{max}}{2}. \quad (26)$$

Calculations are defined (for example for $n=1$), that the maximum diameter of the effective use of the wedge tip with the protuberance is in solid sandy loam $d_{max} \geq 0.258$ m, in noncompressed clay loam $d_{max} \geq 0.254$ m and in plastic clay $d_{max} \geq 0.192$ m.

Conclusions

Installed, that for an effective without a trench gasket simultaneously two or more underground linear-extending object communications by the method of static puncture of soil, it follows to remove the action of him resilient deformations on the perimeter of calibrating part of tip wedge working element due to the performance of basis of wedge outside the calibrated part.

Got dependence for determination of minimum thickness of wedge allowed to set optimal diameters LEO what are laid by a wedge working element. Installed(for example for $n=1$), that most effectively to lay communications a wedge tip with performances in plastic clay diameter $d_{optimal} = 0.096$ m; in noncompressed clay loam – $d_{optimal} = 0.127$ m.; in solid sandy loam – $d_{optimal} = 0.129$ m.

It is also educed that this diameter depends on the type of soil geometrical parameters of tip, depths of gasket and from quantity LEO, what are simultaneously laid.

Calculations were made to determine the maximum diameter effective use of a wedge tip with a projection, which makes up (for example for $n=1$) in solid sandy loam $d_{max} \geq 0.258$ m, in noncompressed clay loam $d_{max} \geq 0.254$ m and in plastic clay $d_{max} \geq 0.192$ m.

References

1. V. Kovanko, O. Lukianchuk. *Scientific bases of creation of earth-moving-tier machines and underground movable devices* (NUWEE, Rivne, Ukraine, 2015)
2. K. Glavatsky, A. Gudenko. Herald of the DSEA, **3**, 101-105, (2015)
3. S. Balezny. Bulletin KNAHU, **76,** 138-141, (2017)
4. S. Kravets, O. Posmituha, V. Suponyev. Construction, materials science, mechanical engineering, PSACEA, **103,** 91-98, (2017)
5. S. Khachaturian, V. Suponyev, V. Oleksin. Bulletin KNAHU, **73**, 196-202, (2016)
6. S. Kravets, O. Posmituha, V. Suponyev, Science and transport progress. Bulletin DNURT named after Academician V. Lazaryan, **70**, 89-98, (2017)
7. Zhao Jun Ling Bian. *Trenchless technology underground pipes* (Machinery Industry Press., China, 2014)
8. P. Goldstein, *Mechanical properties of soils: stress-strain characteristics of soils.* (Stroyizdat, Moscow, 1979)
9. V. Ziemskov, A. Sudakov. Sciences of Earth. Tidings of the TSU. **6**, 35-38, (2005)
10. A. Pridmore, J. Geisbush. *Planning and Design Pipelines* (Phoenix, Arizona, 2017)

MATEC Web of Conferences **230**, 01012 (2018) https://doi.org/10.1051/matecconf/201823001012
Transbud-2018

Effect of joint and isolated irregularities of the track on the wear of rails in curves

Dmytro Potapov[1,*], *Sergij* Panchenko[1], *Yaroslav* Leibuk[1], *Yuseph* Tuley[2], and *Pavel* Plis[3]

[1]Ukrainian State University of Railway Transport, Feierbakh Square 7, 61050, Kharkiv, Ukraine
[2]Chief Engineer - First Deputy Director of regional branch "Pivdenna Zaliznytsia" PJSC "UKRZALIZNYTSIA", str. Evgena Kotlyara, 7, 61052, Kharkiv, Ukraine
[3]Kupianskaya Railway Machine Station 133, Str. Sovetskaya, 14, 63709, Kupiansk, Ukraine

Abstract. The purpose of the work is to identifying causes of intensive lateral wear of rails and other track failures in short radius curves using numerical methods. The study has been conducted for the most common type of rolling stock: four-axle freight cars on TsNII-Ch3 bogies, transportation of which makes 75-90% of freight load of the track sections. Thus, it is the effect of these cars which is the main cause of the above disorders. To solve the problem, effect of the following parameters of the track plan on dynamic processes of interaction of the track and rolling stock was investigated in the work: effect of the curves radius, the unbalanced accelerations and the width of the track; effect of local joint and smooth irregularities of the track in plan view. The results of numerical studies. A mathematical model of the spatial dynamic system "vehicle-track" has been used to perform the study, which is based on the design diagrams of the track in the form of long rail beams resting on multiple dissipative elastic supporting sleepers with nonlinear characteristics. This model is implemented in the MathCAD software system.

Introduction

The sections of the track in short radius curves (less than 400 m) differ significantly from other sections by multiple technical failures, including lateral wear of the rails, derangements in plan view and widening of the railway track, although the total length of curves with radii less than 400 m is about 2.5% of the total length of main tracks of the Ukrainian main railways.

The main applicable regulatory document [2] is based on the quasi-static calculation method for the effect of vertical forces. The effect of horizontal lateral forces, including in curves, on the value of normal stresses in the edges of the top and base of the rail is taken into account using correction factors depending on the type of rolling stock and the radii of the curves.

Methods for determining horizontal lateral forces in the curve were further developed in [3, 5, 6, 8], in which vertical and horizontal lateral forces of interaction of rolling stock and

* Corresponding author: ppx_xiit@kart.edu.ua

tracks were determined using methods of statistical dynamics and the theory of random functions. However, more recent studies [1] have shown that in certain operative conditions, for example, in short radius curves, the principle of superposition is not applicable.

It should be emphasized that, according to some authors [2, 4, 7, 9], the reason of using design diagrams in the form of a beam on elastic foundation was problems in computations when sufficiently powerful computing means were unavailable.

In [1] a mathematical model of the spatial dynamic system "vehicle-track" is presented, which is based on the design diagrams of the track in the form of long rail beams resting on multiple dissipative elastic supporting sleepers with nonlinear characteristics. In addition, the principle of superposition in the dynamic subsystem "vehicle" was not used. The mathematical model is implemented in the MathCAD software system.

The purpose of the work is identifying causes of intensive lateral wear of rails and other track failures in short radius curves using numerical methods. The study has been conducted for the most common type of rolling stock: four-axle freight cars on TsNII-Ch3 bogies, transportation of which makes 75-90% of freight load of the track sections. Thus, it is the effect of these cars which is the main cause of the above disorders.

To solve the problem, effect of the following parameters of the track plan on dynamic processes of interaction of the track and rolling stock has been investigated in the work:

- effect of curve radius, unbalanced accelerations and track width;
- effect of local joint and smooth irregularities of the track in plan view.

In the short radius curves, due to insufficient flexural stiffness of the joint pads, angle-shaped indentations are present practically everywhere in plan view. These "angles" vary from 0.2^0 to 2^0, depending on the type of joint pads (with four or six-bolt joint) and the bolt tension.

Numerical studies were carried out when a four-axle freight car was moving on a curve with a radius of 300 m in order to detect the effect of these indents. The adopted value of unbalanced acceleration in all cases was 0.4 m/s^2, speed of movement – 17.7 m/s with the rise of external rails of 100 mm and 20, 25 m/s with the rise of 150 mm. The angle at the joint was taken to be equal to 0.25^0 to 2.0^0, which allowed detecting the effect of this parameter on the dynamics processes. The results are shown in Table 1. An example of a diagram of changes in horizontal transverse lateral forces of the first wheel pair in the direction of travel is shown in Fig. 1.

Table 1. Extreme values of directive and lateral forces in the presence of joint irregularities.

Rail base	Speed, m/s	Rise of external trail, mm	Angle in the joint, degree.	Lateral forces, kN			Total wear coefficient kN rad
				Directive forces	External rail	Internal rail	
Wooden sleepers, fasteners DO	17.7	100	0.25	15.5	20.1	-8.6	2.340
	17.7	100	0.5	25.5	29.5	-14.6	2.5703
	17.7	100	0.75	50.2	41.3	-16.6	3.4705
	17.7	100	1.5	64.3	65.4	-20.3	4.1704
	17.7	100	1.75	84.6	87.7	-27.3	5.4062
	17.7	100	2.0	106.6	120.2	-39.8	5.4067
	20.25	150	2.0	52.3	61.3	-26.2	4.167
Reinforced concrete sleepers, Sh 1-1, SKD-B	17.7	100	0.25	19.6	27.6	-9.2	2.750
	17.7	100	0.5	29.2	38.4	-16.5	2.985
	17.7	100	1.5	79.6	185.4	-24.5	5.112
	17.7	100	2.0	120.5	137.4	-41.5	6.65

The data in the Table 1 show that the increase of the angles in joints in the track with wooden sleepers under other equal conditions causes the growth of lateral forces and the total wear coefficient. When the angles change from 0.25° to 2.0°, directive forces increase 6.87 times, lateral forces – 5.98 times, and the total wear coefficient – 2.37.

When reinforced sleepers with SKD65-B fasteners are used, the horizontal lateral forces increase up to 120 kN for directive forces and up to 137 kN for horizontal lateral forces. The total wear coefficient of rails increases by 38%. The interaction of the wheel flange with the top of the rail is definitely of impact nature (Fig. 1).

Thus, in the joints, even with six-bolt joints and high performance bolts, angles of 0.25° occur. In other conditions, these angles can reach 2.0°. At speeds of 60-70 km/h, the wear coefficient increases by almost 3 times. These values can be reduced by decreasing flexural stiffness of the joint. Reducing the horizontal transverse rigidity of rail supports is also useful.

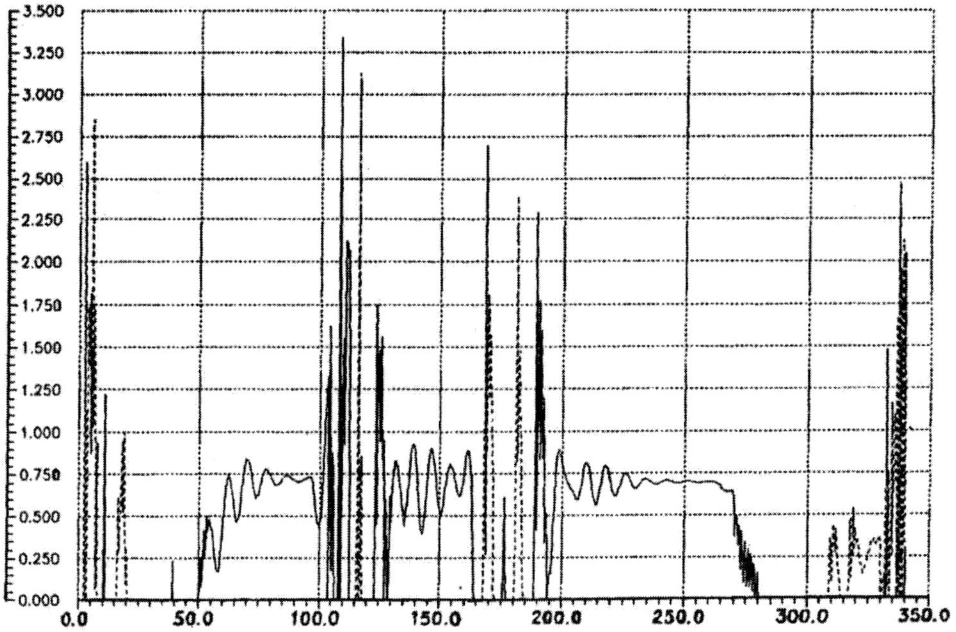

Fig. 1. Diagram of dependences of directive forces on the first wheel in the direction of travel in the areas of joint irregularities

The study of the effect of smooth isolating tracks in plan view was carried out using inequality models that were given in the form of a displaced cosine curve:

$$\eta(x) = \frac{a_{ner}}{2}\left(1 - \cos\frac{2 \cdot \pi \cdot x}{L}\right) \tag{1}$$

where a_{ner}– amplitude of irregularity, (m); L – its total length, (m); x – distance from the beginning of the irregularity to ordinate $\eta(x)$.

Numerical studies have been conducted on the effect of different values of amplitude (from 6 to 24 mm) for irregularity of this kind. The length of the inequalities was assumed constant - 20 m. The curve radius was adopted equal to 300 m, the rise of the external rail – 110 mm, the traveling speed of the car – 20.3 m/s.

Moreover, additional calculation has been performed for the option of the curve radius of 600 m to compare the results. In this case, the speed was adopted at 27.7 m/s. This option was considered because in the actual operating conditions in curves with radii $R \geq 600$ lateral wear is very slight or absent.

Track irregularity in plan view increase significantly the curvature of rails in plan view. This leads to an increase in wear coefficients 8.74 times and the emergence of impact forces if up to 118-122 kN for wooden sleepers and 123-135 kN for reinforced concrete with SKD65-B fasteners.

In addition to studies on determining the allowable values of the parameters of smooth isolated irregularities in plan view, studies have been conducted to identify the most favorable modes of movement of cars on smooth isolated irregularities. For this purpose, studies have been carried out on the effect of the magnitude of unbalanced accelerations on horizontal lateral forces and the total wear coefficient, with constant values of length of irregularity and its amplitudes, and variables of unbalanced accelerations.

In Table 2, the results of the study of parameters during the movement of the calculated car on the short (1-2 m) isolated irregularities in the curve of a radius of 300 m are given.

During conducting these studies, the curve radius was adopted equal to 300 m, unbalanced acceleration – 0.4 m/s^2, traveling speed – 17.5 m/s.

Table 2. Effect of short isolated irregularities on the extreme values of horizontal lateral forces and total wear coefficient

Length of irregularities, m	Amplitude of irregularity, mm	Extreme values of horizontal lateral forces, kN		Total wear coefficient kN · rad
		directive	lateral	
1.0	4	27.3	-9.38	9.616
2.0	4	12.23	-15.15	9.381
2.0	2	31.2	-12.74	2.482

Based on the results of presented calculations, the following conclusions can be drawn. The factor that mainly determines the dynamic processes during movement of a vehicle in the short radius curves, which have track irregularities in the plan, is the radius of the total curve.

$$R_c = \frac{R_{cr} \cdot R_{ner}}{R_{cr} + R_{ner}} \tag{2}$$

where R_{cr} – curve radius (m); R_{ner} – radius of the irregularity (m).

If irregularity in plane view, which is described by equation (1) is adopted, value R_{ner}, can be determined as follows:

$$R_{ner} = \frac{L^2}{2 \cdot a_{ner} \pi^2} \tag{3}$$

Conclusions

According to the results of the studies, if a total length of irregularity is 20 m and an amplitude is 23.3 mm (inequality of grade 4), its minimum radius will be 217 m, and if these irregularities are in a circular curve of 300 m, the total curve radius will be 126 m. When a freight car of type 18-1000 is moving at a speed of 70 km/h, the total value of the lateral wear coefficient increases from 0.252 kN/rad (track without irregularities, Table 1) to 17.219 kN · rad, i.e., 68.3 times, while when the wheel flange hits the rail, the value of the directive forces will exceed 380 kN for wooden and 425 kN for reinforced concrete sleepers. However, no speed limits are established for such inequalities. According to the data, dynamic effects on the track with the irregularities in plan view depend not only and not so much on the value of unbalanced acceleration, but on the impact force and the angle of attack of the wheel flange on the rail.

References

1. O. M. Darenskiy, *Theoretical and experimental studies of operation of railway tracks of industrial transpor,* (Kharkiv, 2011)
2. E. I. Danilenko, *The rules of strength and stability calculation of the railway track*, 168 (Transport Ukrainy, 2006)
3. M.F. Verigo, *Vzaimodeystviye puti i podvizhnogo sostava v krivykh malogo radiusa i bor'ba s bokovym iznosom rel'sov i grebney koles* (Moscow, 1997)
4. O.P. Ershov, *Dinamicheskaya otsenka otstupleniy v soderzhanii zheleznodorozhnogo puti i dal'neysheye yeye sovershenstvovaniye* (Moscow, Transport 1989)
5. *J.* Herian, *Achievements in Materials and Manufacturing Engineering*, **49**, 71-81 (2011)
6. H. Baluch, *Diagnostyka nawierzchni kolejowej* (Warszawa, 1978)
7. B. Lichtberger, *Track Compendium. Formation, Permanent Way, Maintenance, Economic* (Hamburg, Eurailpress, 2005)
8. J. Xu, P. Wang, L. Wang, Advances in Mechanical Engineering , **8**, 1-14 (2016)
9. R. Matching, Railway Track & Structures, **10**, 2-3 (2016)

Reused rails for underground systems

Dmitry Potapov[1,*], *Volodymyr* Vitolberg[1], *Danylo* Shumyk[1], *Viacheslav* Ovcharenko[2], and *Viktor* Bulgakov[3]

[1]Ukrainian State University of Railway Transport, Feierbakh Square 7, 61050, Kharkiv, Ukraine
[2]National Academy of the National Guard of Ukraine, Defenders of Ukraine sq. 3, 61001, Kharkiv, Ukraine
[3]Utility Company "Harkovvodokanal", st. Kontorskaya, 90, Kharkiv, Ukraine

Abstract. The structural components of the rail track, including rails, deteriorate in operation. However, some of them retain their operational characteristics and can be reused. This practice has found wide application on Ukrainian railways and shown a lot of self-evident advantages, among which the basic one is a rational use of material and technical means. Besides, development of a used rail system for underground railways may also optimize the track facility system for such specific conditions. The article presents results of the research into the change of the crumbling depth on the roll surface (the main rail defects of underground railways); it allowed the authors to conclude about the possibility to reuse rails for underground systems. Moreover, assessment of interacting forces in the wheel/rail system and the crumbling volume made it possible to determine rational application for used rails. The proposed system takes into account all the features of operating the railway track in underground systems, and also allows more efficient use of the residual rail resource without reducing the necessary level of traffic safety.

Introduction

A long experience in reusing materials of the permanent way on Ukrainian railways has demonstrated a lot of obvious advantages of these engineering solutions. One of the basic advantages of used materials is a more rational use of material and technical means, which has allowed optimizing the whole track facility system including the rail system. Therefore, implementation of the reused rail system for underground railways is one of the key prospective ways to provide operation of the enterprise under limited resources.

The research conducted [1 - 5] contributed to a considerable increase in the service life of rails due to a detailed study of defect development considering the crumbling dynamics on the rail roll surface (digital code of defect 11.1-2-3 according to [6]). Determination of main groups of defects and damages in rails during 1986-2012 contributed to safe traffic operation due to timely detection of the second group of defects (digital code of defect 21.1-2-3 according to [6]).

* Corresponding author: ppx_xiit@kart.edu.ua

Therefore, in order to solve the problem to reuse 12.5-m and 25-m rails for underground railways the authors chose one of the Ukrainian metro systems as the basic enterprise.

The basic areas of research were:

- To determine possibility to reuse rails for main lines considering their operational peculiarities and taking into account the crumbling depth development.
- To determine groups by serviceability and fields of application for used rails by assessing interacting forces of the wheel/rail system, and considering the volume of crumbling on the roll surface.

1 Determination of the crumbling depth on the roll surface

The technique is based on the fundamental concepts of the reliability theory, the probability theory and the mathematical statistics. According to [7, 8] the failure is an event leading to full or partial inability of the object (or structural element) to function properly. Thus, all rail failures can be divided into full and partial ones in accordance with their impact on the general state of the system (the whole permanent way structure).

The most harmful rail defects (digital code of defect 21, 30Г, 30В) requiring stoppage of the traffic can be considered, for example, as full failures.

The main defects (about 75%) requiring rail removal from the metro tracks are the defects by code 11; they can be regarded as partial failures. Partial failures worsen the general serviceability of the system (as requiring a lower traffic speed) and sometimes may cause full failures.

According to [4] the deepest the crumbling is, the lower is the speed, on the condition that the rail traffic safety should be guaranteed. If the crumbling depth on the roll surface exceeds 8mm, the rail should be removed.

The progressive rail failure by code 11 is graphically presented in Fig.1.

Thus, four states of rail damages (I, II, III, IV) can conditionally be distinguished; they are located in the middle of the corresponding crumbling depth intervals on the roll surface. Since the passenger transportation should be provided (an established speed along the section), the rails can be replaced in any of these states.

The intensity of depth development on the corresponding intervals Δt_1, Δt_2, Δt_3, Δt_4, first of all, depends on the rail track layout. The lower the radius of the corresponding section of the track is, the higher is intensity in defect development.

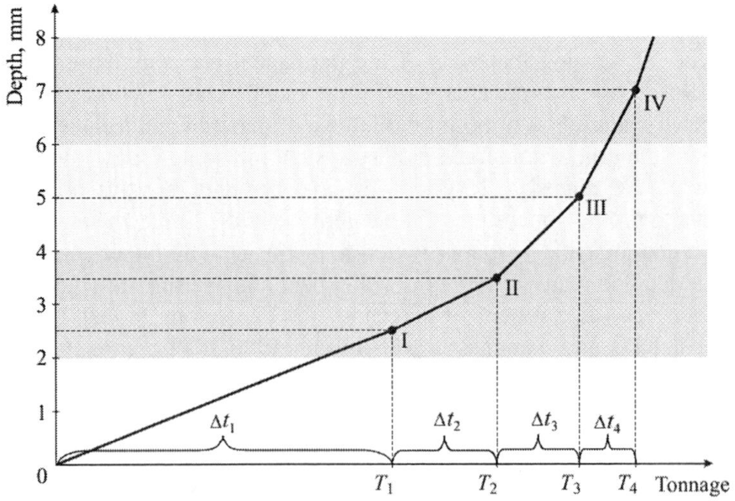

Fig.1. Stages in development of the crumbing depth in rails by code 11.

According to earlier research the maximum depth, at which the rails are withdrawn, lies in a range of 2.6 – 6.10 mm, i.e. for states I – III (Fig.1).

Therefore, section can serve as a reserve to increase the service life of a rail. It should be mentioned that the authors do not consider it reasonable to transcend the upper limit of the red zone as at a crumbling depth exceeding 8 mm it is sufficient (for underground railways) to considerably decrease the traffic speed up to 25 km/h; besides, there is a possibility for it to grow into a more harmful defect (code 11).

It should also be mentioned that the number of rails with a crumbling depth exceeding 6 mm was insignificant.

By conducting full-scale instrumental research according to the instrument ratings, all curves of one of the Ukrainian metro systems were divided into groups by radii.

And some sections were chosen for measuring the crumbling depths with an electronic digital calliper. The measurement pitch was set separately for each defect and was within a range of 15-30 mm. Then the results obtained were processed in order to compute the average depth.

It allowed establishing that for each category the value of a random crumbling depth followed the normal distribution law.

Thus, for curves with a radius of 500 m and less the average crumbling depth was in a range from 2.5mm to 4.25 mm, which is an absolutely high value among all the categories of radii. The lowest value was 1.8 mm (for straight sections and curves of a radius exceeding 1001 m). In curves of other radii the average depth changed within an approximately equal limits from 2.5 to 3.87 mm (for radii of 501-650 m), and from 2.65 to 4.02 mm (for radii of 651-800 m).

On the basis of the calculation the dependencies between the average crumbling depth and the tonnage passed were obtained (Fig. 2).

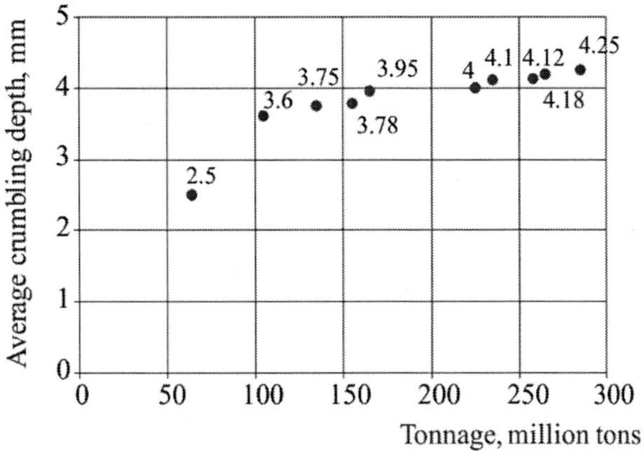

Fig. 2. Average values of crumbling depths by code 11 in curves of the radii 500m and less.

It should be mentioned that the previous assumption (Fig. 1) on the nature of defect development in depth (code 11) was not absolutely right. By generalizing the results obtained, an increase in the depth can be described as follows (Fig. 3):

Three stages of the development can be conditionally distinguished. The first stage is characterized by a rapid growth of the defect's depth. It is explained by mashing the surface layers of the metal, which leads to creation of longitudinal and linear-inclined cracks in the undersurface zones; these cracks while growing encounter various stress concentrators, and, most likely, turn into transversal cracks with consequent crumbling. The undersurface microcracks can be observed as dark spots on the roll surface (Fig. 4).

In Fig. 4 the crumbling depth is within 2.56 – 3.3 mm. And concentration of dark spots can be observed on both sides of the defect.

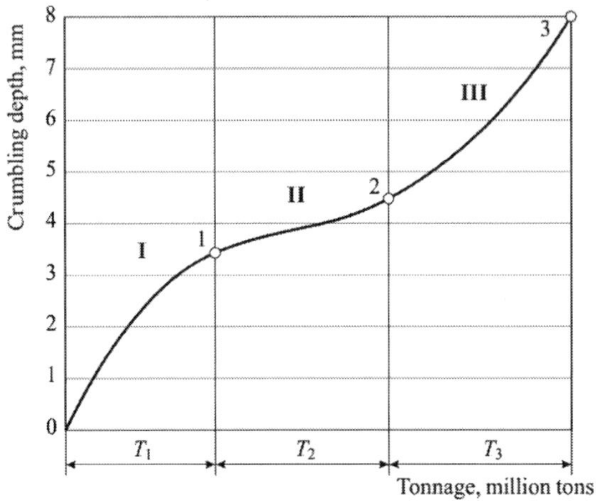

Fig. 3. The process of the crumbling depth development by code 11 based on the calculation.

Fig.4. Dark spots on the roll surface.

The second stage is characterized by a considerable decrease in crumbling depth development (for example, in curves of the radius 500 m and less the depth does not virtually change within a range of 3.95 – 4.12 mm). In addition to that, taking into account an increase in the number of microcracks in undersurface layers of the metal, the microcracks confluence with the existing ones and that leads to a greater crumbling on the roll surface (Fig. 5).

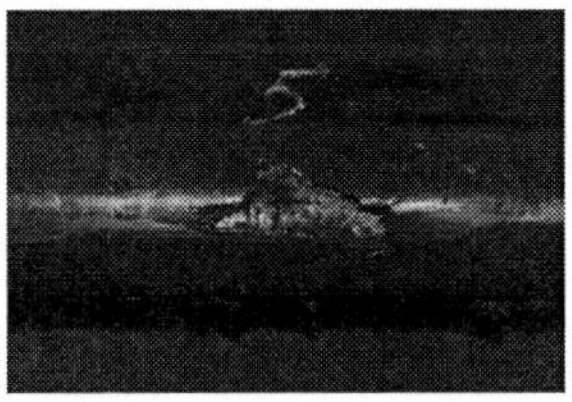

Fig.5. Growth of the crumbling area on the roll surface.

The confluence of microcracks in area 1 is clearly seen; it causes an expension of the defect area. The depth of the defect is within a range of 2.94 – 4.60 mm.

At the final (third) stage irreversible processes in the metal rail head (including structural changes) run to deeper layers, thus repeatedly increasing the speed of the crumbling depth development.

The average crumbling depths obtained (H) were consequently approximated in the MathCAD as follows

$$H(t) = A_1 \cdot t^3 + A_2 \cdot t^2 + A_3 \cdot t + B \quad (1)$$

where t is the tonnage passed, million tons. A_1, A_2, A_3, B are the empirical coefficients calculated with the least-squares method.

And on the base of the experimental data the theoretical curves to define the increase intensity of a crumbling depth was obtained. Fig.6 gives the diagram of curves of the radii 500 m and less as an example.

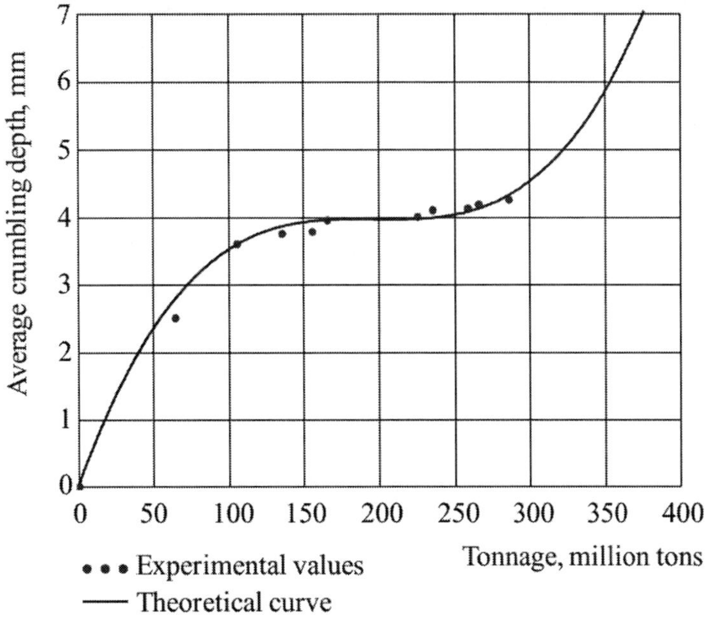

Fig. 6. Approximation of crumbling depths by code 11 in the radii 500 m and less.

2 Determination of crumbling volumes

While planning the rail observation procedure, the authors selected track sections with the tonnage (T_{ton}) close to the normative (T_{norm}), in order to more accurately assess the geometry of rail defects in the pre-critical state.

To this effect, the dimensionless coefficient K_{res} was calculated:

$$K_{res} = \frac{T_{ton}}{T_{norm}} \tag{2}$$

where T_{ton} –value of tonnage passed along the section, million ton; T_{norm} – normative tonnage, million tons.

According to the requirement $0.6 \leq K_{res} \leq 1$ the sections for observation were chosen.

During the observation the basic geometry (length, width, depth) of the crumbling on the roll surface was measured with the caliper and the crumbling areas exceeding 25 mm in length were taken into considerations.

The measurement pitch of a depth for each crumbling zone was separately established and depended on the length and width.

The following parameters were calculated for each testing section:

$\bar{L}$ is the average length of the crumbling areas, mm;

$\bar{A}$ is the average width of the crumbling areas, mm;

$\bar{h}$ is the average depth of the crumbling areas, mm.

Besides, the maximum crumbling depth h_{max} was established.

The volume of each crumbling (V) was calculated on the basis of its outline by one of the variants (in some cases several variants were applied simultaneously): layer segment, half of the oblate spheroid, paraboloid and spherical sphere [9].

The total volume of crumbling (V_{total}) for a specified testing section was calculated as:

$$V_{total} = \sum_{i=1}^{N} V_i \tag{3}$$

where V_i – volume of the corresponding crumbling, mm^3; N – total number of crumbling areas along the testing section.

The authors believe that as far as the number of testing sections for certain radius categories was not sufficient and, moreover, several parts of the curves had already been replaced to provide the safe operation on (in some cases the whole track line, which meant that the volume of crumbling could not be assessed). Therefore, the authors decided to introduce a single factor – a specific crumbling volume ($V_{specific}$). It was defined as

$$V_{specific} = \frac{V_{total}}{T_{ton} \cdot K_{res}}, \text{ (mm}^3\text{/million tons)} \tag{4}$$

Then the average value of it $\overline{V_{specific}}$ was calculated for each radius category.

In order to define the serviceability groups and spheres of application for used rails in underground systems the authors investigated into the action of a vehicle to the track on the base of [10, 11, 12].

By comparing the level of acting stresses in the rail head with the admissible ones, it is recommended to use for main metro lines those used rails which were removed from all radius categories along main lines, both during complete replacement and routine maintenance procedures, with the exception of curves of the radius 500 m and less.

In terms of the average depth and specific crumbling volume on the roll surface, it is advisable to divide used rails into three groups by serviceability:

I group – used rails for main lines on straight sections and curves of the radius 501 m and more.

II group – used rails for main lines on straight sections and curves of the radius 651 m and more.

III group – used rails which cannot be used for main lines.

Conclusions

1. The average crumbling depth on the rail roll surface with the tonnage close to normative is in a range of 3.24 – 3.97 mm.

2. The average specific crumbling volume on the rail roll surface with the tonnage close to normative is 98.92 – 141.15 mm^3/million tons. It increases when the radii decreases.

3. Three serviceability groups for used rails were established by the results of the research with the application of numerical methods for the action of the vehicle to the track with consideration of the average depth and specific crumbling volume on the roll surface.

References

1. E.I. Danilenko, R.M. Yosifovich, O.A. Oliynik, O.O. Soroka, Zb. nauk. prats' DETUT. Seriya «Transportni sistemi i tekhnologii», **(22)**, 10-20 (2013)

2. V.V. Kosarchuk, A. V. Agarkov, Zb. nauk. prats' DETUT. Seriya «Transportni sistemi i tekhnologii», **(20)**, 77-89, (2012)
3. J.W. Ringsberg, H. Bjarnehed, A. Johansson, B.L. Josefson, *Proceedings of the Institution of Mechanical Engineers, Part F: Journal of Rail and Rapid Transit*, **1(214)**, 7-19 (2000)
4. A.Yu. Abdurashitov, Put' i putevoe khozyaystvo, **11**, 16-20 (2002)
5. M.Kh. Akhmetzyanov, Vestnik VNIIZhT, **2**, 36-43 (2003)
6. *Classification and Catalogue of Defects and Damages in Rails on Ukrainian Railways* (Kyiv, 2013)
7. M.N. Stepnov, *Statisticheskie metody obrabotki rezul'tatov mekhanicheskikh ispytaniy* (Moskow, 1985)
8. K.A. Braunli, *Statisticheskaya teoriya i metodologiya v nauke i tekhnike* (Moskow, 1977)
9. G.A. Korn, T.M. Korn, *Mathematical handbook,* (Moskow, 1974)
10. O.M. Darenskiy, Yu.L. Tuley, D.O. Potapov, A.S. Malishevs'ka, «Science and Transport Progress». Bulletin of Dnipropetrovsk National University of Railway Transport, **6(66)**, 96-104 (2016)
11. O.M. Darenskiy, A.S. Malishevs'ka, Information and control systems at railway transport, **1(122)**, 43-51 (2017)
12. A. Darenskiy, D. Potapov, Yu. Tuley, N. Bugaets, A. Malishevskaya, MATEC Web of Conferences, **116**, 03001 (2017)

Ways of introduction of the high-speed movement of passenger trains in Ukraine

Anna Shevchenko[1,*], *Oleksander* Matviienko[2], *Vitalii* Lyuty[1], *Vladimir* Manuylenko[1], and *Mykhailo* Pavliuchenkov[1]

[1]Department of Track and Track Facilities, Ukrainian State University of Railway Transport, Feierbakh Square 7, 61050, Kharkiv, Ukraine
[2]Engineering systems "El-term". Vosstaniya Square 7/8, 61000 Kharkov. Ukraine

Abstract. The article describes the historical background of the introduction and development of high-speed traffic in the world. The main documents regulating the joint work of high-speed trains of the Commonwealth countries are considered. The main problems of high-speed traffic in Ukraine, and the railway as a whole are studied, analyzed, and described. Recommendations of the company SYSTRA, and the results of developments of Khargiprotrans JSC on the design, implementation and operation of high-speed highways in Ukraine are considered. In conclusion, the current state of the railways, the problems of the current track maintenance has been studied. For Ukraine, first of all, the fourth stage of construction which will connect Lviv with Kiev should be put into operation, based on the analysis of the passenger traffic of the State Committee of Statistics. In this direction, the roadbed and artificial structures are in satisfactory condition. The upper track structure of has sufficient strength. The effect of the profile on the speed of motion is compensated by the power of the locomotive.

Introduction

Mass transportation of goods and passengers is currently carried out by rail, road, air and water transport.

The formation of the transport market in Ukraine and the possibility of a free choice of means of transportation naturally causes competition between the above types of transport.

On foreign railways, competition for various modes of transport is assessed on the basis of the following factors [1]:

- delivery time of goods;
- frequency of departures;
- compliance with the traffic schedule;
- cost of cargo delivery;

* Corresponding author: Annshevc@gmail.com

- ability to transport different goods;
- cost of transportation.

A serious competitor to rail transport is road transport, which has the following advantages:

- high speed of transportation;
- delivery of goods "from door to door";
- mobility;
- the possibility of delivery in small batches;
- lower costs for road infrastructure.

From statistical data it is known that the largest number of goods and passengers in our country is transported by rail.

Following the commissioning of high-speed railway which operated at a speed of 200 km/h in Japan in 1964, in 1967 SNCI (French Railways) began a study of the creation of a railway network at a higher speed, namely, more than 250 km/h. The first results of this research appeared in 1970 and included the proposal to create a new line between Paris and Lyon. This proposal was based on the following three principles: a new line exclusively designed for passenger transportation, line compatibility with the existing network, operation based on high frequency of traffic, short journey duration. In 1986-1987, a plan was also developed in the Soviet Union to ensure the introduction of high-speed (up to 160 km/h) passenger trains on the main railways of the USSR by the leading railway universities [1-2].

For this purpose, on July 23, 1996, the European Community adopted Directive 96/48/EC on the Interoperability of the Trans-European High-Speed Railway Network. It should be noted that this directive combines the main technical characteristics of the infrastructure, rolling stock and operation of French high-speed technologies. This Directive was amended and supplemented in 2004 and then in 2007 by Directives 2004/50/EC and 2007/132/EC, which update its content in accordance with the 2001 Directive on the Interoperability of the Traditional Trans-European Railway System. This new impetus is part of the strategy for revitalizing rail transport in Europe and helping to cope with the harmful consequences associated with excessive use of road transport [3-6].

France is the first European country that introduced high-speed railways. The total length of the lines currently operating (2002) is 1.520 kilometres.

Thus, a unique high-speed line with a length of 1.070 km has been built from Calais to Marseilles with a train speed of 300 km/h to create a link between the far north (tunnel under the English Channel) and the far south of France (Mediterranean Sea). The duration of the trip by the railroad system is a bit over 3 hours.

Considering that the geography of transport corridors is constantly changing due to many factors, the development of intercontinental communication along the main railway routes is constantly adjusted and improved.

The basic part

In the directives of the possibility of interaction of the trans-European high-speed railway network [3, 7], the main technical characteristics of the infrastructure, rolling stock and operation of French high-speed technology are combined, which are based on the following principles:

1. The new high-speed line is intended exclusively for passenger traffic.
2. High-speed line compatibility with the existing network.

3. The operation is based on high frequency of movement and short duration of the trip.

The new line, intended for passenger transportations, together with the high capacity of the composition with respect to the mass speed allows rolling stock operation at steeper tilt angles [8, 9]. The advantages of this option are listed below:

- a significant reduction in the cost of infrastructure by creating steeper slopes, reducing the number and size of necessary structures (tunnels, overpasses, etc.) and ensuring the most direct line laying. Thus, the length of new lines that will be built is reduced, and accordingly, the duration of trips is reduced.
- maximum utilization of the capacity of the new line by uniform distribution of the speed of the run of high-speed trains on new lines.
- providing a new line exclusively for the passenger movement. Due to this, the load on traditional railway lines in favor of freight traffic is substantially reduced and optimal operation of both types of railways is ensured, since the trains will run at approximately the same speeds.

Provided it is compatible with the existing railway network, the high-speed trains will be able to continue along existing routes and penetrate the city centers, thus creating extensive access to the new network throughout its territory. This is achieved through the following factors:

- penetration into the downtown area of cities by using existing lines, without the need to build an additional railway within the city, which would entail significant costs;
- no need to change trains at each end of the new line;
- gradual introduction and expansion of high-speed rail services.

Such an approach will help to reduce the cost of building a new line, expand the potential coverage of the territory and increase traffic. The rolling stock of the high-speed railway is fully compatible with the existing network, and will be able to continue on traditional railways at speeds acceptable for local lines. Thus, the high-speed railroads will be able to serve cities located several hundred kilometres from the new line. In countries that have different track widths on traditional and high-speed rail, new lines can be brought to existing stations by modifying the existing track width on station tracks. However, the decision to significantly cover the territory can be the manufacture of rolling stock, which automatically changes the width of the track, as it is in Spain and Japan [7].

High-speed compositions are light weight convoys with two locomotives and a high rated power that can be coupled. As a result, these compositions are capable of performing the following functions:

- to achieve a commercial operation speed of 300-350 km/h and a high average income rate in some regions of 250 to 300 km/h (including the use of existing lines and stations). Such income speed significantly shortens the duration of trips, due to which absolutely competitive in comparison with air travel the duration of trips at distances of up to 1.500 km [9, 10].
- run at high speed due to rapid turnover (design with two locomotives, which makes it possible to easily change the direction at the stations), despite the limited number of trains.
- to reach a full load in periods of maximum passenger turnover on new lines by creating a longer train by linking the two trains.

These principles have proven their reliability on high-speed rail in France.

Based on the successful experience of operating a high-speed railway in France, other European countries, such as Germany and Italy, have developed their own railways. These national high-speed railways have various technical features and, in general, do not have the possibility of interaction. The European Community promotes the development of a trans-European network of high-speed railways. Therefore, it is necessary to guarantee the possibility of interaction between individual railways and their individual systems.

Proceeding from the tasks of creating a high-speed rail network, the total population and the economic situation in Ukraine, and also based on the basic principles of a high-speed railway, Ukraine proposes to create the following system of high-speed railways.

General information about Ukraine was analyzed in terms of geographical location, administrative regions, urban population and economic situation.

All the positive changes in the economy of Ukraine, the growth of the gross domestic product (as of August 2018 - the GDP of Ukraine is 2.9% according to the World Bank), the activation of economic activity, the welfare and living standards of the population contribute to the increase in mobility and the volume of long-distance transportation over long distances.

In terms of population size, economic activity, cultural development, and tourism development, the following most important regions of Ukraine can be identified: Kiev and the Kiev region; Kharkiv and Poltava; Dnepropetrovsk, Dneprodzerzhinsk, Zaporozhye and Krivoy Rog; Region of Donbass (Donetsk, Lugansk, Mariupol, Gorlovka and Makeyevka); Odessa and Crimea peninsula (Odessa, Nikolaev, Kherson, Simferopol and Sevastopol); Western region (Lviv, Ternopil, Khmelnitsky and Vinnitsa) [4, 6].

Based on the experience of the networks of high-speed railways in Europe and in the world, the following tasks of creating a high-speed railway can be distinguished:

- the need for modern society is the creation of communication between cities, especially between large cities with economic, cultural and tourist potential;
- the movement of people: people always try to shorten the duration of travel as much as possible. The high-speed railway, a symbol of modern society, offers a fast, safe, convenient and efficient way of traveling with a high frequency of traffic, accessible to everyone;
- development of the country: high-speed railways do not harm the environment and are a safe mode of transport. They have a significant positive impact on economic activity (trade, tourism, hotel business, various services, etc.), urban development, real estate, employment, image of the country, etc. Thus, the high-speed railway greatly contributes to the development of the country.

Based on the successful experience of networks of world high-speed railways, the following main principles of their operation can be singled out: the lines are intended exclusively for passenger transportation; compatibility of the high-speed rail system with the existing rail network, as well as functioning based on the principles of high-frequency traffic and a competitive short duration of travel [3, 6, 9].

Based on the tasks and basic principles of creating a network of high-speed railways (creating a communication between cities with a population of more than 300 thousand inhabitants, as well as constructing lines where possible on the most direct routes, the proposed high-speed lines are a long-term project for the future Table 1.

The length of the proposed high-speed network is 3.071 km.

The width of the track on the traditional railway of Ukraine differs from the standard gauge of the international network of transport corridors. The choice of the gauge width of 1520 mm (CIS) or 1435 mm (the international network of transport corridors standard) for use on the Ukrainian high-speed network is one of the most important issues of network implementation.

Thus, for the high-speed lines, three scenarios were developed: the use of the standard track gauge for the CIS countries (1520 mm), the use of the standard gauge of the international network of transport corridors (1435 mm), and the use of rolling stock adapted to both types of track (standard gauge of UIC for high-speed lines and standard gauge for traditional railways of the CIS countries, due to the variable width of wheel sets of rolling stock).

Table 1. Suggested routes for high-speed lines

Point of departure / arrival		Distance, km		
		Road	Railway	High-speed railway
Kiev	Poltava	337	351	310
Poltava	Kharkiv	150	140	130
Amount		487	491	440
Poltava	Dnepropetrovsk	196	214	130
Dnepropetrovsk	Zaporozhye	97	126	80
Amount		293	340	210
Dnepropetrovsk	Donetsk	253	269	220
Donetsk	Lugansk	148	152	140
Donetsk	Mariupol	121	110	120
Amount		522	531	480
Kiev	Belaya Tserkov	67	90	65
Belaya Tserkov	Nikolaev	409	512	340
Nikolaev	Odessa	120	242	120
Nikolaev	Kherson	66	62	60
Kherson	Simferopol	251	275	240
Amount		913	1181	825
Belaya Tserkov	Vinnitsa	150	200	130
Vinnitsa	Khmelnitsky	120	146	110
Khmelnitsky	Ternopil	114	119	105
Ternopil	Lviv	128	141	120
Amount		512	606	465
Dnepropetrovsk	Krivoy Rog	151	165	150
Krivoy Rog	Nikolaev	159	217	150
Amount		310	382	300
Zaporozhye	Melitopol	113	115	110
Melitopol	Simferopol	248	241	241
Amount		361	356	351

Each scenario has a different impact on rolling stock, networking, infrastructure, and financial issues. These results are presented in the following tables.

Scenario 1: Ukrainian high-speed network with a track width of 1520 mm. There is no compatibility with European high-speed lines.

The total amount of fixed costs for research, approval of rolling stock, design of tracks and construction works is estimated at 100 million Euros. Typically, in Western Europe, the design of a new high-speed composition takes 3 to 5 years.

Scenario 2: Ukrainian high-speed network with a gauge of 1435 mm. There is no compatibility with the traditional network of Ukrainian railways; there is no compatibility with high-speed networks of the CIS countries (in case they take a gauge of 1520 mm) Minimum costs for adapting existing stations in comparison with the total costs for this project.

Scenario 3: Rolling stock, adapted for two types of track width. Designing a new type of rolling stock with an automatically variable width of wheel pairs 1435/1520 mm. A more complex rolling stock design entails an increase in maintenance costs and a lengthy approval process for this solution. The total amount of fixed costs for the study and approval of the rolling stock system is estimated at 50 million Euros. Typically in Western Europe, the design of a new high-speed rolling stock takes 3 to 5 years.

Taking into account the presented information, the choice of track width for the Ukrainian high-speed network is more operational, financial and political, rather than a technical issue [5, 7, 10].

The introduction of the high-speed railway network proposed for Ukraine should be carried out in stages. A program for the development of high-speed lines should be developed, in which priority should be given to the introduction of these lines. The following aspects should be analyzed in order to determine whether it is expedient to continue work on any component: financial profit, socio-economic aspects, environmental impact and political criteria. This study showed that the entire network from an economic point of view is appropriate. Therefore, further feasibility study of the project is recommended. Such a study is aimed at detailed development.

From the point of view of economic profitability, according to calculations of the company SYSTRA, the following sequence of construction of the high-speed railway, presented in Table 2, is proposed.

Table 2. Stages of construction of railways.

Stages	Point of departure	Point of arrival	High-speed railway
I	Kiev	Poltava	310
	Poltava	Kharkiv	130
	Poltava	Dnepropetrovsk	130
	Amount	5701	
II	Dnepropetrovsk	Donetsk	220
	Dnepropetrovsk	Zaporozhye	80
	Amount	300	
III	Kiev	Belaya Tserkov	65
	Belaya Tserkov	Nikolaev	340
	Nikolaev	Odessa	120
	Amount	525	
IV	Belaya Tserkov	Vinnitsa	130
	Vinnitsa	Khmelnitsky	110
	Khmelnitsky	Ternopil	105
	Ternopil	Lviv	120
	Amount	465	
V	Nikolaev	Kherson	60
	Kherson	Simferopol	240

	Amount	300	
VI	Dnepropetrovsk	Krivoy Rog	150
	Krivoy Rog	Nikolaev	150
	Amount	300	
VII	Zaporozhye	Melitopol	110
	Melitopol	Simferopol	241
	Amount	351	
IIX	Donetsk	Lugansk	140
	Donetsk	Mariupol	120
	Amount	260	

When operating the entire network, starting from 2035 - 2040, the network will be able to use about 130 million passengers annually or 356 thousand passengers daily. This amount of traffic means an annual passenger turnover of 54 billion pass-km [7].

About 34 different routes and 200 trains daily throughout the network will be put into operation. Thus, the entire network will be covered to meet demand.

Taking into account the recommendations of the company SYSTRA of Khargiprotrans JSC, a feasibility study of the high-speed railway Kiev-Poltava-Kharkov was developed for the speed of passenger trains of 250 km/h. Khargiprotrans JSC has designed several options and the main indicators are given in Table 3.

Table 3. Key indicators of high-speed railroad options Kyiv - Poltava – Kharkov

Options	Units	Variant		
		I	II	III
Variant length	km	463	454	457
Length of the new line	km	440	404	407
Length of the existing railway	km	23	50	50
Profile volume of excavation works	mln.m^3	49,7	45,6	46,0
Area of the right-of-way	hectare	1760	1600	1650

In order to evaluate this or that alternative, you must choose the optimality criterion. In modern conditions, the integral effect is taken as the criterion of optimality E_{int}, which is determined by the formula

$$E_{\text{int}} = \sum (R_t - K_t)\eta_t \, , \tag{1}$$

where T_p – settlement period, R_t – result for the year, K_t – outgoings for the year;

η_t – the discount rate is determined by the formula

$$\eta_t = \frac{1}{(1+E)^t} \, , \tag{2}$$

where E – discount rate.

With the optimality criteria (1), a passive search method can be used to find the optimal variant, which is used with a small interval of uncertainty.

In the presence of a large number of possible options for finding the optimal solution, apply directional search. There are a lot of targeting methods, they have their drawbacks and advantages.

For today, 10 high-speed trains "Intercity" of Korean production "Hyundai" and two trains of the company "Skoda" run in Ukraine. The maximum speed of these trains is 160 km/h, [5].

Conclusions

The introduction of the high-speed railway network proposed for Ukraine should be carried out in stages. A program for the development of high-speed lines should be developed, in which priority should be given to the introduction of these lines. As a result of the conducted research, it was established that the speed limit on the hauls was caused by the following main reasons: a) unsatisfactory condition or presence of defects in the earthen linen and artificial structures; b) insufficient capacity of the upper structure of the track; c) unfavorable conditions of the plan and profile.

For Ukraine, first of all, it is necessary to put into operation the fourth stage of construction, based on the analysis of the passenger traffic of Derzhstat, which will connect Lviv with Kiev.

In the considered direction, the roadbed and artificial structures are in satisfactory condition. The upper structure of the track has sufficient power. The effect of the profile on the speed of motion is compensated by the power of the locomotive.

Thus, the main reason for the restriction of the speed of the passenger trains on the runway is the conditions of the line plan. The line plan for this direction is relatively favorable, but nevertheless there are the following drawbacks: a) the presence of curves of small radii (600-700 m); b) insufficient length of straight inserts between the curves; c) insufficient elevation of the outer rail in the curved sections of the track; d) a large amount of elevation slopes due to insufficient length of transitional curves; e) location of turnouts in curved sections of the track; e) availability of stations on main routes; g) on the main tracks of large stations, shunting routes with human flows in one level often

References

1. *Regulations on carrying out planned preventive repair and road works on the railways of Ukraine* (Ministry Transport of Ukraine, 2004)
2. M. B. Kurhan, D. M. Kurhan, *Theoretical basis for the introduction of high-speed trains in Ukraine*, (DNUZT, Dnipro, 2016).
3. Korzhenevich I.P. Railway transport of Ukraine, **5**, 79-82 (2015)
4. *Guidance on the main issues of the arrangement and maintenance of the Railroad.* (Transport of Ukraine, 2007)
5. A.A. Shevchenko. Conference materials 80th International Scientific and Technical Conference, **177**, 112-113 (2018)
6. Vytautas Dzerkelis, Žilvinas Bazaras, Vaidas Lukoševičius. Transport, **30(2)**, 194–201 (2015)
7. Marijonas Bogdevičius, Rasa Žygienė, Stasys Dailydka, Vilius Bartulis, Viktor Skrickij, Saugirdas Pukalskas. Transport, **30(2)**, 217–232 (2015)
8. Paola Carolina Bueno Cadena, José Manuel Vassallo Magro. Transport, **30(3)**, 298–306 (2015)

9. G.M. Kirpa, *Integration of Ukraine's railway transport into the European transport system*, (DNUZT, 2004)
10. M.B. Kurhan, D.M. Kurhan, Science and Transport Progress. Bulletin of Dnipropetrovsk National University of Railway Transport, **6(72)**, 40-48 (2017)

Formation of the stress-strain state of rail bars during welding with preliminary vertical bending

Volodymyr Shramenko[1*], *Nataliia* Bielikova[1], *Olexandr* Ovchinnikov[1], *Dmytro* Onopreichuk[1], and *Stanislav* Horielyshev[2]

[1]Ukrainian State University of Railway Transport, Feuerbach sq., 7, Kharkov, Ukraine, 61050
[2]National Academy of National Guard of Ukraine, Zakhysnykiv Ukrainy sq., 3, Kharkiv, Ukraine, 61001

Abstract. The article defines the parameters and working conditions for the final restoration of the integrity of the rail beams of a continuous gauge rail by electrocontact welding, as well as when welding rails with each other when they are elongated in the field by the method of preliminary bending. This involves the process of welding rails at temperatures above the temperature of their fixation. In this connection, an appropriate calculation scheme has been considered in relation to which the necessary parameters have been determined to ensure the conditions for straightening the residual curvature and, as a consequence, the formations of the corresponding stress state. To ensure the self-straightening conditions, the resulting formula for determining the residual bending arrow, the elimination of which under the influence of its own weight leads to the appearance of compressive stresses in the weld zone. In addition, a formula has been derived for determining the compressive stresses in the weld zone arising in the rail after the elimination of residual curvature. This means that when performing welding operations without forced straightening of the rails, you can restore not only their integrity, or weld them together. Thereby form a corresponding temperature-stress state.

1 Introduction

One of the ways to increase the efficiency and reliability of the continuous welded rail track is to increase the length of rail bars [1]. With the advent of an efficient welding technology for short rail bars – electrical resistance welding – manufacture of block-long and section-long rail bars became possible [2].

1.1 Analysis of recent researches and publications

Scientific studies [3-8] on the integrity restoration and manufacture of long rail bars have shown that bar bending can be mechanized, and within certain limits the conditions for self-correction can be created with the provision of the appearance of a compressive force in welding area and, as a result, improving the quality of the welded joint. Obligatory

* Corresponding author: dek_str@ukr.net

conditions ensuring the quality of the welded joint are matching of the welded ends of the bars excluding both any angle between them, and bending moments in the support sections at the ends of the curved part of the bar.

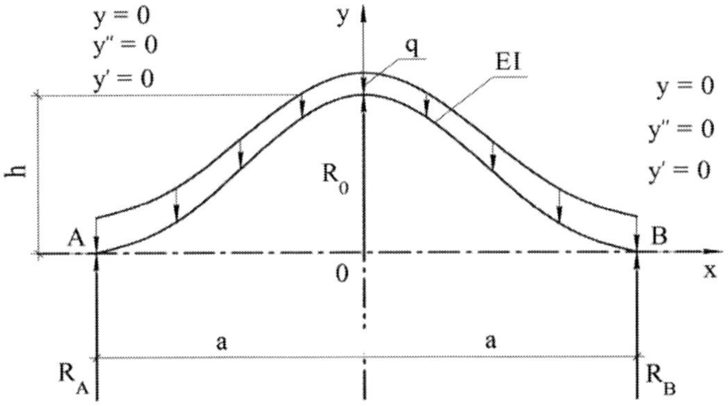

Fig.1. Computational model for determining the parameters of the stress-strain state of the bar during vertical bending

Based on these two conditions, the length of the curved part of the bar, or rather of the half-chord, is obtained

$$a = \sqrt[4]{\frac{72hEI}{q}} \quad \text{or} \quad a = 2.91\sqrt[4]{\frac{hEI}{q}}; \tag{1}$$

elastic line equation:

$$y_x = -h\left[3\left(\frac{x}{a}\right)^4 - 8\left(\frac{x}{a}\right)^4 - 3\left(\frac{x}{a}\right)^2 - 1\right]; \tag{2}$$

bending moment equation:

$$M_x = -\frac{12hEI}{a^2}\left[3\left(\frac{x}{a}\right)^2 - 4\frac{x}{a} + 1\right]; \tag{3}$$

transverse force equation:

$$Q_x = -\frac{48hEI}{a^3}\left(\frac{3x}{2a}\right); \tag{4}$$

reactions of the supports

$$R_A = R_B = \frac{24hEI}{a^3}, \quad R_o = \frac{48hEI}{a^3}, \tag{5}$$

where h – the lifting height (bending arrow) of the rail bar; a – the half-chord length; q – linear weight of the rails; EI – flexural stiffness of the rail bar in the vertical plane.

1.2 Determine the purpose and objectives of the study

One of the advantages of the method with the vertical pre-bending of a part of the bar is the use of the own weight of the rails to straighten the residual curvature. Self-straightening can take place only with a residual bending arm of a certain value determined from the equilibrium conditions of all forces arising as a result of their redistribution after the middle part of the curved part of the rail bar is released from the support. Accordingly, the limiting value of the residual bending arm should be determined starting from which, under the influence of its own weight, the bar can be straightened, as well as other parameters of the stress-strain state during welding.

2 The basic part of the research

After the rail bar from is released form the support in the middle part of the curved part of the bar (Fig. 2), a longitudinal force (distance) appears on its ends, the value of which is determined from the equilibrium equation.

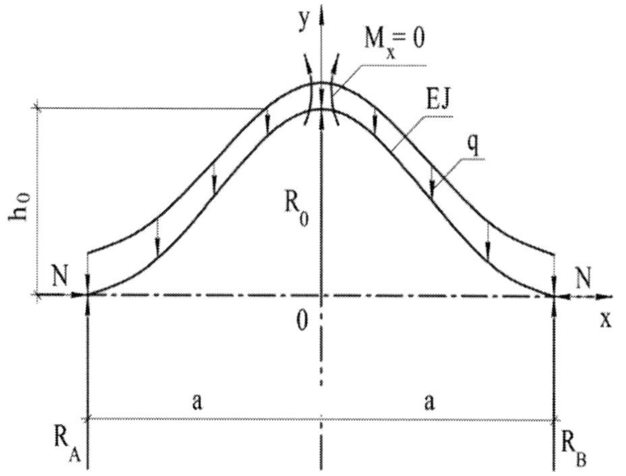

Fig. 2. Computational model for determining the maximum allowable bending arm at which self-straightening of residual curvature is provided

$$M_{x=0} + \frac{qa^2}{2} + Nh - qa^2 = 0, \quad (6)$$

or

$$\frac{12hEI}{2^2} - \frac{qa^2}{2} + Nh = 0, \quad (7)$$

then

$$N = \frac{qa^2}{2h} - \frac{12EI}{a^2} \quad (8)$$

After substituting the length of the half-chord a (1) into equation (8), an expression is obtained for determining the magnitude of the longitudinal force that arises in the bar when it is released from the support in the middle of the curved part.

$$N_h = 2\sqrt{\frac{2qEI}{h_o}}, \quad (9)$$

where h_o – the value of the residual arm bending of the rail bar after it is released from the support in the middle of the curved part.

At the same time, to eliminate the residual bending arm, the longitudinal force N_o should be overcome, which according to Hooke's law is defined as

$$N_o = \frac{\Delta l}{2a} EF, \tag{10}$$

where Δl – difference between the length of the curved part of the bar; l – length of the chord $2a$; E – elastic modulus of rail steel ($E = 2.1 \cdot 106$ kg/cm^2); F – cross-sectional area of the rail.

The length of the curved part of the bar is determined by the formula

$$l = 2\int_o^a \sqrt{1 + \left(\frac{dy}{dx}\right)^2} \cdot dx, \tag{11}$$

or in relation to equation (2)

$$l = 2\int_o^a \sqrt{\left[\frac{12hx(x-a)^2}{a^4}\right]^2 + 1} \cdot dx, \tag{12}$$

The length of the curved part of the bar is determined by numerical integration using Simpson's rule [9] with respect to rails of the R65 type, for which q = 0.65 kg/cm, for h values of 10 cm to 100 cm in 10 cm pitch.

Table 1. Results of the calculation

h, cm	10	20	30	40	50	60	70	80	90	100
Δl, cm	0.081	0.272	0.553	0.915	1.352	1.861	2.436	3.078	3.782	4.547

The results of calculating the difference Δl between the length of the curved part of the bar and the corresponding chord, depending on the lifting height (bending arm), h are shown in Table 1, and in the form of a diagram (Fig. 3).

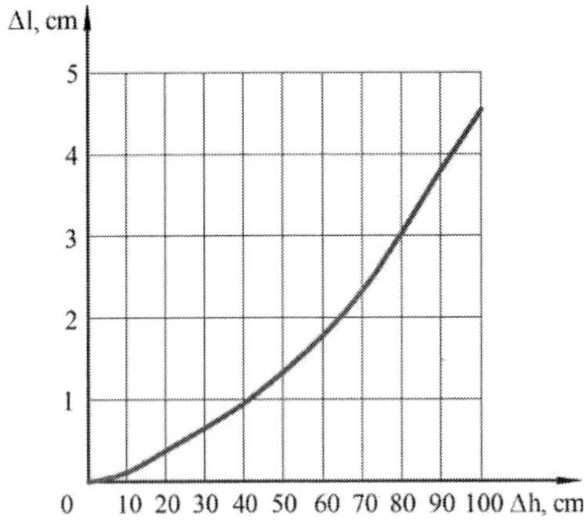

Fig. 3. Dependency diagram of the length of the curved part of the bar and the corresponding chord

The diagram has the form of the power function, which, after processing by the least square method [10], adopted a specific form for rails of type R65

$$\Delta l = 0.00144\text{h}^{1.75} \tag{13}$$

After further transformations in relation to all rail types, this formula took the form

$$\Delta l = 0.4706\sqrt[4]{\frac{qh^7}{EI}}. \tag{14}$$

After using equations (1) and (14) in equation (8), a formula was obtained for determining the force N_0, overcoming which leads to straightening the residual curvature

$$N_0 = 0.08075 \cdot F\sqrt{\frac{qh^3E}{I}}. \tag{15}$$

For self-straightening of the residual curvature, the condition should be met $N_h \geq N_0$.
That is, using equations (9) and (15), inequality was made

$$2\sqrt{\frac{2qEI}{h}} \geq 0.08075F\sqrt{\frac{qh^3E}{I}}, \tag{16}$$

from which

$$h_0 \leq 5.918\sqrt{\frac{1}{F}}, \tag{17}$$

where I and F – moment of inertia and the cross-sectional area of the corresponding rail type, respectively.

3 Conclusion

Thus, to ensure the conditions for self-straightening of the residual curvature, the residual arm should not exceed the value determined from formula (17).

Based on equation (15), a formula was obtained for determining the compressive stresses, arising in a rail after the elimination of residual curvature.

Table 2 shows the values of such a bar for different rail types.

Table 2. Maximum residual bending arm for different rail types

Rail type	Moment of inertia of the trail section I, cm	Cross-sectional areas of the rail F, cm^2	Residual arm h_o, cm
R50	2018	65.9	32
R65	3548	82.6	38
R75	4490	96.1	40

So, if the residual curvature is self-straightened, compressive stresses can be created, e.g. in rails P65, which are equal to 371 kg/cm^2 (≈37 MPa), which, expressed in terms of temperature difference, will be about 15 °C. This means that when welding operations are performed without forced straightening of the rails, not only the integrity of the rail bars, but also their temperature mode within the limits of up to 15 °C can be restored.

References

1. V. Shulga, V. Novakovich, V. Laptev. Track and track facilities. **9**, 28-31. (1981)
2. V. Novakovich *Besstykovoy zheleznodorozhnyy put' s rel'sovymi pletyami neogranichennoy dliny* (Lvov, Vysshaya shkola, 1984)
3. N. Vinogorov, A. Lugovoi, V. Sushkov, V. Shramenko *Track and track facilities*. **9**, 23-24. (1988)
4. V. Sushkov, V. Shramenko *Collection of scientific papers HIIT*. **5**, 33-37. (1988)
5. *Instruktivnyye materialy po svarochno-naplavochnym rabotam v putevom khozyaystve* (Moskow, Transport, 1991)
6. I. Voznenko *Bezstikova koliya with rakami neobmezhenoї dovzhini* (Khakiv, KharDAZT, 1998)
7. Z. Kreynis, N. Seleznyova *Besstykovyy put'. Kak remontirovat' besstykovyy put'* (Moskow, Marshrut, 2005)
8. *Tekhnichni vkazivky po ulashtuvannyu, remontu i utrymannyu bezstykovoyi koliyi na zaliznytsyakh Ukrayiny* (Kyiv, 2012)
9. G. Dwight *Tables of integrals and other mathematical formulas* (1997)
10. I. Bronshtein, K. Semendyaev *For engineers and students of universities* (1962)

Determination of the level of separate rail failure using the indicator of their reliability

Anatolii Shtompel[1], *Oleksii* Skoryk[1], *Vadym* Novikov[1], *Yuliia* Kravchenko[2], and *Yevhen* Korostelov[1*]

[1]Ukrainian State University of Railway Transport, Feierbakh Square 7, 61050, Kharkiv, Ukraine
[2]Kharkiv Petro Vasylenko National Technical University of Agriculture, Department of Economics and Marketing, Alchevskikh st., 44, 61002, Kharkov, Ukraine

Abstract. During the operation, the upper railway track structure constantly tends to age, which includes the increasing number of failures. This process is controlled by the track maintenance system, which provides for the current maintenance of the track in a certain railway section. As part of the current maintenance of the track, in particular, works are performed to eliminate defective elements of the upper railway track structure, which appeared during functioning of the track structure. The dependence of the operational cargo turnover on the amount of transported cargoes has been analyzed. It is directly related to the issues of organization of maintenance of the railway track structure taking into account the current track operation conditions. Regression analysis of the reported data allowed establishing a mathematical model of the dependence of the value of operational cargo turnover on the value of the transported cargo. The service life of rails (rail bars) in the track section has been established to be related to the indicator separate rail failure during the period of operation of the upper railway track structure by defects. The safety level of rails (rail bars) can be determined based on the value of this indicator.

1 Introduction

Appropriate mathematical models to predict the breakdown of elements of the upper railway track structure should be available (taking into account operational conditions of the railway section) to plan the rail repair during the current maintenance of the track. These models require the existence of an appropriate method for assessing the probability of failure of elements of the upper railway track structure, including rails.

The state of all elements of the upper railway track structure (within the life cycle of the upper railway track structure) should «ensure the safe and smooth movement of trains with the speeds established for this section» [1]. The technical condition of the track structure should be relevant to the level of the train traffic intensity, i.e., the railway track maintenance system should be susceptible to possible changes in the operational cargo

* Corresponding author: dek_str@ukr.net

turnover. At the same time, these studies were conducted for the conditions of both main railways [2-5] and subways [6-7].

The issue of the numerical value of the indicator m_p – separate rail failure (during the period of operation of the upper railway track structure) by defects has constantly been in the focus of the track studies [8-14].

2 The basic part of the research

One of the railway performance indicators is operational cargo turnover $Q_{gr.oper.}$, mln. t. km gross. This indicator characterizes the amount of the train flow, which determines the power load on the track structure.

Thus, establishment of dependence $Q_{gr.oper.} = f(t)$ for a certain period of time (t - the current year of this period) is of practical importance for the track facilities and is directly related to the issues of organization of maintenance of the railway track structure.

The value of $Q_{gr.oper.}$ is directly related to the volume of transported cargo $Q_{cargo.}$.

Using regression analysis of the corresponding reporting data, a mathematical model of dependence $Q_{gr.oper.} = f(Q_{cargo})$, was established which has the following form:

$$Q_{gr.oper.} = 0.987 Q_{cargo}, \tag{1}$$

where the units of measurement of Q_{cargo} and $Q_{gr.oper.}$ are mln.t and mln. t.km gross, respectively.

The relevance of use of this mathematical model in confirmed with coefficient of correlation $r = 0.998$ and coefficient of determination $r^2 = 0.995$.

In [3], the change of operational cargo turnover over the railway network in 2012-2016 was considered. It was found that the indicator $Q_{gr.oper.}$ varied linearly

$$Q_{gr.oper.} = 493.97 - 34.05\ t \tag{2}$$

Coefficient of correlation $r = 0.967$ and coefficient of determination $r^2 = 0.935$.

Using the operational cargo turnover, average (over the railway network) traffic capacity of the main track can be determined

$$C_{aver.} = Q_{gr.oper.} / L_{main} \tag{3}$$

where L_{main} – main track length (mln. t.km gross/km per year).

I.e. for the specified period, dependence $C_{aver.} = f(t)$ has the form as follows

$$C_{aver.} = 16.7 - 0.66\ t, \tag{4}$$

coefficient of correlation $r = 0.986$ and coefficient of determination $r^2 = 0.972$.

The service life of rails (rail bars) in a section is related to such an indicator as m_p – individual overage the operating life of the upper railway track structure by defects (rail failure during the period of operation of the upper railway track structure by defects (for the continuously welded rail track - the number of defective spots on a rail bar).

Some data on numerical value of this parameter (for new heat hardened rails R65) with the certain historical aspect are given in Table 1.

According to the analysis of data in Table 1, a unique estimate of the value of the indicator m_p is unavailable. The lower limit of the indicator varies from three to ten rails per 1 km of track. Therefore, additional studies are required to define m_p taking into account the current operational conditions of the track.

The value of the indicator m_p is directly related to the reliability of the rails (rail bars), and therefore has a high practical value.

Estimation of the reliability of rail bars of the continuously welded track during the operation involves the conditional division into segments each being 12.5 m long (such is the length of the rail insertion being welded into the bar during the final restoration of its integrity). In this case, rails of equalizing spans are not considered.

Table 1. Values of m_p

№	Period, source	Traffic capacity, mln. tkm gross / km a year	Value of the indicator m_p, pcs/km
1	1960-s [10]	25-50	5
2	1970-s [11]	25 37.5 65	6,53 5,92 5,21
3	1980-s [12]	50-80	9
4	2004 [13]	15 and more	3 and more
5	2008 [14]	30 50 80	8 6,5 5,6
6	2010 [8]	15 and more	10 and more
7	2014 [9]	15 and more	3 and more

In [13], the dependence of the total separate failure m_p of the heat hardened rails R65 (within a rail bar) on the total transported tonnage T, million tons gross (for the sections with an average weighted axial load of 155 kN and the train speed of 100 km/h) is given.

$$m_p = 0.56 \cdot 10^{-8} T^3. \tag{8}$$

Probability of rail failures during the operation can be determined using the formula:

$$F_p(t_i) = m_p(t_i)/N_p, \tag{9}$$

where N_p – number of conditional 12.5 m long rails per 1 km track.

In this case this indicator is

$$F_p(t_i) = 0.35 \cdot 10^{-10} T^3. \tag{10}$$

Failure intensity for rails in the track section correlates with the dependence

$$\lambda_p(t_i) = m_p(t_i) / [N_p - m_p(t_i)] \, \Delta T, \tag{11}$$

where ΔT –transported tonnage in the period of time being considered.

Also, function $\lambda_p(t_i) = f(T_i)$ is described by the following mathematical model

$$\lambda_p(t_i) = 0.35 \cdot 10^{-10} T^4. \tag{12}$$

This model is the result of the corresponding processing of paired values λ_p and T within interval T = 50 - 750 mln. t gross.

Verification of the dependence (12) using the Fisher test confirmed the adequacy of the established model.

In Table 2, the results of calculation of some indicators are summarized.

Table 2. Results of calculation

T, mln. t	200	300	400	500	600	700	800	900
m_p, pcs/km	0.045	0.151	0.358	0.700	1.210	1.921	2.867	4.586
$F_p(t_i)$	0.0003	0.0010	0.0022	0.0044	0.0076	0.0120	0.0179	0.0255
$1 - F_p(t_i)$	0.9997	0.9980	0.9978	0.9956	0.9924	0.9880	0.9821	0.9745

We can note as follows:

- after transporting 800 mln.t gross, as reference value T_{ref} for this upper railway track structure, appearance of three failed rails (m_p = 2.867 pcs/km) for 1 km of track with probability $F_p(t_i) = 0.0179$ should be expected, probability of failure-free operation of rail bars decreases from 1 to 0.9821, i.e. approximately by 2 %;
- the results of calculations correlate with previous study [13], where a similar upper railway track structure was considered (bars of P65 rails, steel and concrete sleepers with KB fastening) and the probability of failure-free operation of rail bars (after transporting 600 mln. t gross) was determined being 0.9765 (in this case this indicator is 0.9924, i.e. divergence is 1.6 %).

The latter confirms the relevance of use of the mathematical model (8) to determine the value of indicator m_p.

Thus, the value $m_p = 3$ pcs/km can be considered as the lower limit in the estimation of total separate rail failure in a certain track section.

Study [13] has established the admissible (provided rail operation is economically reasonable) level of the total separate rail failure m_p for sections with different traffic capacity C (Table 3):

Table 3. Values of the indicator m_p for sections with different traffic capacity C

Indicator №	Traffic capacity of the section T, mln. t-km gross/km per year	m_p, pcs./km
1	30	8
2	50	6.5
3	80	5.6

The above data were processed to determine the corresponding dependence $m_p = f(C)$, which is described by the equation:

$$m_p = 0.0009\,T^2 - 0.147T + 11.6 \tag{13}$$

Thus, the value of the indicator m_p can vary from 3 to 11 pcs/km depending on parameter *C*, i.e. the track category.

The experience of the operation of the railway track suggests that the condition of the rail facilities in a certain section of the road is characterized by two main indicators:

- volume of the transported tonnage T;
- total number of trail failure m_p.

In Table 4, recommendations concerning the indicator m_p at the track section are given:

Table 4. Values of indicator m_p

Track category	Transported tonnage (service life, years), % of the reference one (at least)	Total separate rail failure (number of defect spots) after track laying (in average at the section), pcs./km	
		welded rail track	jointed track
I-III	100	3 and more	5 and more
IV	100	4 and more	6 and more
V-VI	100	8 and more	10 and more
VII	100	10 and more	

Thus, provided rail operation is economically reasonable, the level of the total separate rail failure m_p for sections with different traffic capacity *C* was determined.

3 Conclusion

The performed regression analysis allowed determining the dependence of the operational cargo turnover on the volume of the transported cargo.

Also, determination of dependence $Q_{gr.oper.} = f(t)$ for the certain period of time is of practical importance for the track facilities and is directly related to the issues of organization of maintenance of the railway track construction.

Average (for the railway network) traffic capacity of the main track was determined using the cargo turnover. However, it should be noted that parameter $C_{aver.}$ changed not just in singe digit during the calculation period.

Dependence of rail failure intensity in a track section taking into consideration the current operation conditions were determined. For the considered operation conditions, after transporting 800 mln.t gross, appearance of three failed rails ($m_p = 2.867$ pcs/km) for 1 km of track with probability $F_p(t_i) = 0.0179$ should be expected, probability of failure-free operation of rail bars decreases by about 2 %;

The obtained results of calculations correlate with the finding of the previous studies for similar operation conditions of the railway track and the probability of the failure-free operation of rail bars (after transporting 600 mln. t gross) of 0.9765 was determined (in this case this indicator is 0.9924, i.e. divergence is 1.6 %).

References

1. *Rules of technical operation of the railways of Ukraine* (Kyiv, 1995)
2. A. Stompel. Collection of scientific works on the materials of the international scientific and practical conference «Modern areas of theoretical and applied research». **27(1)**, 27-32. (2011)
3. A. Stompel Scientific view to the future. **6(2)**, 89-91. (2017)
4. A. Stompel, O. Skoryk, O. Ovchinnikov. Collection of scientific works of Sworld. **36(1)**, 64-69. (2014)
5. M. Kurgan, D. Kurgan, N. Khmelevskaya. Ukrainian Railway. **9-10(51-52)**, 14-21. (2017)
6. S. Voronin, O. Skoryk, Ye. Korostelov. Eastern-European Journal of Enterprise Technologies. **4(80)**. 11-17. (2016)
7. S. Voronin, O. Skoryk, V. Stefanov, D. Onopreychuk, Ye. Korostelov. MATEC Web of Conferences Transbud. **116**, 03005. (2017)
8. E. Danilenko, M. Karpov, V. Yakovlev. *Regulation on the system of conducting a track-and-field economy on the railways of Ukraine.* (Kyiv, 2011)
9. *Regulations on carrying out scheduled maintenance and repair works on the railways of Ukraine.* (Kyiv, 2015)
10. P. Tsukanov, A. Zolotarsky. Railway Transport. **6**, 27-33, (1961)
11. L. Krysanov, M. Klokov. Railway transport. Series «Traveling economy». **4**, 18-26. (1986)
12. A. Bolotin, V. Shulga. Interuniversity Collected Works. **805**, 94-97. (1989)
13. A. Shtompel, B. Nosenko, T. Stomp. A scientific glance into the future. **2(1)**, 72-75. (2016)

Probabilistic evaluation of the railway track infrastructure components failure risk

Anatoliy Shtompel, *Liudmyla* Trykoz*, *Dmytro* Borodin, *Andrii* Ismagilov, and *Yaroslav* Chmuzh

Ukrainian State University of Railway Transport, Feierbakh Square 7, 61050, Kharkiv, Ukraine

Abstract. The permanent way components are of key importance for safe operation of a rail way. The country regulations, in particular in Ukraine, specify the operational life limits for the permanent way but they do not define any tool or method to predict deterioration of the permanent way condition over time. The study is aimed to develop a method for assessing failure risk of the permanent way components in operation. There was a method offered to evaluate risk of failure of the permanent way components of the welded tracks, which considers accumulated freight load on a rail section. Each element of the permanent way, such as rails, fasteners, sleepers, ballast layer, accumulates defects and deformations. The accumulation rate is different for the above components and depends on freight traffic. There was probability of failure-free operation calculated for each component for the first time and an integral fatigue index of the construction has been offered which considers freight traffic accumulated load. There was a mathematical failure forecast model developed which allows planning of track maintenance. The model allows to take into account operating conditions of a railway section. The results of simulation are presented in various diagrams.

1 Introduction

Risk is not avoidable in railway operation. It is defined as a measure combination of hazard likelihood and severity of consequences of an operational failure. Transport safety implies several aspects: railway infrastructure safety, transport system operation safety, goods safety, health and life safety of the passengers and employees. The term "safety" does not mean complete absence of any hazard in railways operation. It means an ability to ensure safe cargo and passenger traffic in case of an emergency of failure. The reliability theory defines emergency prevention as ability to ensure long-term fail-free operation of each component of the railway infrastructure. It is not enough to use those tools for acquisition and processing of data related to failure rates of the infrastructure components. There is a more important task in railway transport operation, which is risk management involving timely detection of negative trends in each infrastructure component, prediction of possible consequences and decision making on how to prevent or minimize negative impact.

* Corresponding author: lvtrikoz@ukr.net

Such a probabilistic simulation can be used in the railways engineering and construction [1, 2]. As it is shown in [1], there are numerous risk factors in railway construction, such as decision making, engineering, construction, technology, quality, investment, natural disasters, force majeure circumstances, etc. It has been is proven that risk management is a practical method in the railway construction and an important tool for design risk minimization, construction quality improvement and efficient project cost management. A quantitative risk analysis can be performed for various railway infrastructure components such as tunnels [3], bridges [4, 5], rolling stock [6]. Some authors introduce the concept of "critical element" to evaluate the failure risk [7-9]. The above papers present the results of risk assessment models development based on railway accident scenarios. The researchers also have tried to create an efficient method that allows to evaluate significance of railway infrastructure elements. The road junctions, bus and train stations, bridges and tunnels were defined as the potentially significant elements of the infrastructure. This approach is based on events that have already happened (accidents). In our opinion, it does not allow to predict the risks of daily "aging" of the whole railway and, accordingly, to plan preventive repair of the track to avoid accidents.

Some researchers propose to create an information system for railway risk management. They consider using of a large amount of daily data: results of dynamic and static monitoring of railway lines and bridges, overhaul and medium repairs data, maintenance data and records about bridges and tunnels failures [10]. Such systems should be able to acquire, store and manage huge amounts of data related to condition of railway, bridges and tunnels. The data management and processing should be combined with daily operations and monitoring to obtain information about the status of the permanent way. Based on big data collected over long period of time, the authors tried to define how condition of track changes with time and to develop a model to predict it. They want to change a concept of track maintenance. The new approach should be based on preventive and condition based maintenance which makes it absolutely different from the existing approach based on scheduled maintenance or repair-after-failure concept. If this new approach will be implemented than the managers in all levels will be provided with tools to monitor railway equipment in real time and increases efficiency of management.

The study [11] was aimed to improve railroad drift simulation based on comprehensive field research. The authors apply both engineering and statistical data processing approach. They have collected and processed comprehensive data about track condition related to approximately 100 km of a railway line. The data acquisition was performed during two years. They have studied the factors influencing track deterioration, including load traffic, inspection monitoring, track quality. This allowed them to find correlations between the influencing factors and road deterioration. The forecasting approach offered in [12] offers a method to evaluate track segment drift. The evaluation results could be used to make decisions on allocation of limited budget. Probabilistic safety assessment becomes quite complicated in complex systems with several failure modes. Such systems can be simulated by Marcov processes which suppose multiple states and transitions [13]. There is a probabilistic estimate approach to risk assessment for infrastructure networks with high risk of flood presented in [14]. The proposed method of risk analysis combines both probability of the infrastructure failure and expected efficiency of the network infrastructure or loss of capacity due to component failure.

All the above mentioned studies do not consider risk of failure of the components of the permanent way structure. However, the permanent way condition determines availability of any infrastructure component. The purpose of this study is to establish a method of evaluation of failure risk of the permanent way with welded railway tracks subjected to load traffic.

The track permanent way experiences load from rolling stock. The level of this load significantly affects operation of the track and causes the changes in its technical condition. The more load traffic passes the section the higher is tendency of deterioration of permanent way technical condition due to the accumulation of residual deformations in it and failure of the components. All this leads to decrease of operational reliability of the permanent way structures and the deterioration of the train traffic safety. Maintenance is struggling to slow down this process of "aging" of the track. The regular maintenance includes, in particular, works to eliminate failures of permanent way components which happened in operation. In order to be able to plan maintenance and repair of a railway, it is necessary to have appropriate mathematical forecasting models of permanent way components failure. Those should take into account operational conditions of the railway section. The models require an appropriate method of permanent way failure probability evaluation.

2 Development of the mathematical model of the probability of failure-free operation of the track permanent way

In order to remove negative effects of deterioration of operating conditions at a certain site, an appropriate system to manage railway availability is required. This system should be based on technical condition status of the way section derived from information about failure of the track permanent way (TPW) components. The applicable regulations define the operational limits (performance criteria) of the TPW construction at certain stages of its life. The assembled rails and sleepers are considered as basis (welded heat-treated rails, prestressed concrete sleepers with a sleeper density N_{sl} = 1840 pcs per km, an intermediate bonded-bolted fastening, crushed stone ballast). This type of rails operates in almost 75% of the main railways of Ukraine.

It is accepted that the process of accumulation of single defects of rails as a function of traffic load is the one defining change of railway track condition in a particular section. If replacement of a defect rail is assumed to be a failure, then the failure probability of $F_R(T)$ and faultless operation of rails $P_R(T)$ by the moment of traffic load accumulation T is

$$F_R(T) = r_R(T) / N_R; \quad (1)$$

$$P_R(T) = 1 - r(T) / N_R, \quad (2)$$

where $r_R(T)$ is the number of rail failures by the moment of traffic load accumulation T; N_R is the number of rails per 1 km of track. For N_R = 80 pcs / km and $r_R(T)$ = 10 pcs / km (for R65 rail type), $F_R(T)$ = 0.125, and $P_R(T)$ = 0.875.

Practical experience shows that one of hundred defective rails can not be detected by existing flaw detectors and it gets damaged by the train running over. Let assume the probability of breaking a defective rail F_R = 0.01, then the probability of failure after replacing 10 rails (per 1 km of track) is $F_R = 0.125 \cdot 0.01 = 0.00125$, and the probability of fail-safe operation is P_R = 0.99875. It is known that not every break of a defective rail under train causes its get off from track or accident. Therefore, the actual reliability of the rails (under the condition that the defective rails are replaced in scope of 10 pcs per km during regular maintenance of the TPW) can be estimated as P_R = 0.999, which allows to classify the rails as highly reliable components according to the valid classification.

A track is assumed to be defective if there are 100 defective concrete sleepers found per 1 km of track. Taking the sleepers density N_{sl} = 1840 sleepers per km, the failure probability of this component is $F_{sl}(T)$ = 100/1840 = 0.054, and probability of its failure-free operation is $P_{sl}(T)$ = 0.946. These values F_{sl} and P_{sl} are set for a case when the defective sleepers are not replaced in the section. This is a hypothetical case and it is not realistic. If $F_{sl} = F_R = 0.01$, then the permissible number of defective concrete sleepers can be determined per 1 km of track $r_{sl} = N_{sl} \cdot F_{sl}(T) = 1840 \cdot 0.01 \approx 20$ sleepers. In this case, probability of failure-free operation of the specified component is $P_{sl} = 0.99$.

The limit amount of defective rails fastenings allowed per 1 km (for the life cycle of the TPW) is set by regulations at 25% of the total number of sleepers (the total amount of defective rail pad plates and fixing bolts is considered). Then, probability of failure in the operation of the fastenings is equal to $F_{fas}(T) = 0.25$, and the probability of its failure-free operation is $P_{fas}(T) = 0.75$. As in the previous case (concrete sleepers), the obtained values of F_{fas} and P_{fas} are calculated for the unrealistic hypothetical situation, when the defective rail pad plates and fixing bolts are not replaced regularly in a site. Taking $F_{fas} = F_R = 0.01$, it is possible to set the permissible number of defective fastening components per 1 km of the track in operation: $r_{fas} = N_{fas} \cdot F_{fas}(T) = 1$ % (from the total number of rail pad plates and fixing bolts). The probability of failure-free operation of fastening is $P_{fas} = 0.99$. Indicators r_R, r_{sl} and r_{fas} should be considered as criteria for making decision about rehabilitation of the TPW in a certain section. They provide quantitative evaluation of technical condition of the rail and sleeper grid (RSG).

Considering the known values of the probability of failure-free operation of the i-th element (for a specific section of the track), the integral indicators for the TPW are defined for accumulated traffic load

$$P_{RSG} = P_R \cdot P_{sl} \cdot P_{fas}; \tag{3}$$

$$F_{RSG} = 1 - P_{RSG}. \tag{4}$$

One can calculate for the above considered cases as follows:
- in case no maintenance done in the track of a railway section
$P_{RSG} = 0.875 \cdot 0.946 \cdot 0.75 = 0.62$; $F_{RSG} = 1 - 0.62 = 0.38$;
- in case the defective TPW components are regularly replaced in the section
$P_{RSG} = 0.999 \cdot 0.99 \cdot 0.99 = 0.98$; $F_{RSG} = 1 - 0.98 = 0.02$.

The value P_{RSG} is used to rank track sections (in a certain area) by their reliability and traffic load accumulated. This ranking is a part of the system of railroad maintenance planning.

During track operation, polluted ballast is mixed with water and pressed out from under the sleepers when the train rolls over. This phenomenon is named "track pumping". There is the relationship between the number of track pumpings m_{pum} (sleepers per km of track) and the volume of the load traffic T (million tons gross) passed over the welded rails section (welded R65 type rails, prestressed concrete sleepers, gravel ballast) with averaged operating conditions [15]

$$m_{pum} = 1.6 \cdot 10^{-6} T^3. \tag{5}$$

A single track pumping (per 1 km of track) is not considered as a failure of the ballast layer. This defect is usually fixed during regular maintenance of the track structure at the site. The regulations specify the number of ballast pumpings per 1 km of track as the main criterion planning maintenance in the section. The maximum permissible number of sleepers with local pumpings is 3-15% of the number of sleepers per 1 km and the absolute value depends on sleepers density. Obviously, this number of pumping per 1 km of track can be considered as a failure of the ballast layer. The probability (risk) of occurrence of pumpings of crushed stone ballast during the accumulation load traffic is based on the formula

$$F_{pum}(t_i) = m_{pum}(t_i) / N_{sl}, \tag{6}$$

where $m_{pum}(t_i)$ is the number of pumpings at time t_i; N_{sl} is the sleeper density.

Taking into account the formula (5) for the continuous welded rails (with N_{sl} = 1840 sleepers per km), the mathematical model for estimation of the failure probability $F_{pum}(t_i) = f(T_i)$ will be as follows

$$F_{pum}(t_i) = 8{,}7 \cdot 10^{-10} T^3. \qquad (7)$$

The intensity of occurrence of ballast pumpings within the track segment is determined by the formula

$$\lambda_{pum}(t_i) = m_{pum}(t_i) / N(t_i) \Delta T, \qquad (8)$$

where $N(t_i)$ is the number of sleepers without ballast pumpings at time moment t_i; ΔT is the value the passed gross load at the considered time interval. In this case, the function $\lambda_{pum}(t_i) = f(T_i)$ is presented as follows:

$$\lambda_{pum}(t_i) = 4.3 \cdot 10^{-10}\, T^{2.14}. \qquad (9)$$

This model is received by processing the pairs of values λ_{pum} and T in load traffic accumulated interval T=50 ÷ 700 million tons gross. The verification of formula (9) by Fisher's criterion has confirmed that the offered model is correct. Thus, the number of pumpings of crushed stone ballast for 1 km of welded rail corresponding to 3% (from N_{sl}) and 15% (from N_{sl}) should be expected at 330 and 560 million tons of accumulated load traffic gross respectively. This corresponds to the probability of occurrence 0.03 and 0.15, respectively.

The probability of failure-free operation $P_{TPW}(t_i)$ is used to evaluate reliability of the TPW. It is calculated at a specific moment of time t_i (after the accumulation of a specific load traffic T_i, in million tons gross) according to the formula

$$P_{TPW}(t_i) = P_R(t_i) \cdot P_{sl}(t_i) \cdot P_{fas}(t_i) \cdot P_{pum}(t_i), \qquad (10)$$

where $P_R(t_i)$; $P_{sl}(t_i)$; $P_{fas}(t_i)$; $P_{pum}(t_i)$ are the probability of failure-free operation (at the time t_i), respectively, of rails, sleepers, intermediate fastenings and ballast layer.

The welded rails are split into segments of 12.5 m for the purpose of evaluation of their reliability (this is length of the rail insert that is welded into the rail string when its integrity is finally restored). Rails of equalizing segments are excluded. The total number of single defects m_R (pcs per km) of thermally reinforced rails of the R65 type (within the rail string) as a function of accumulated load traffic T, million tons gross (for areas with an average axle load of 155 kN and a train speed of 100 km/h) can be defined as follows

$$m_R = 0.56 \cdot 10^{-8} T^3. \qquad (11)$$

The probability of rail failures in operation is defined as follows

$$F_R(t_i) = m_R(t_i) / N_R, \qquad (12)$$

where N_R is the number of 12.5 m long rail sections per 1 km of track.

In this case, this value is calculated as

$$F_R(t_i) = 0.35 \cdot 10^{-10} T^3. \qquad (13)$$

Based on the materials [15], the probability of sleeper failure (for the considered construction of the TPW) can be described by the mathematical model

$$F_{sl}(t_i) = 7.93 \cdot 10^{-10} T^2. \qquad (14)$$

It is considered that a system consisting of connected components (the node of fastening) is operational in the case all its components are operational (not failed). The corresponding data processing allowed to define the following mathematical models for evaluation:

- the probability of fastening failures (for the construction of the TPW with N_{sl} = 1840 pc / km) as a function of accumulated traffic load

$$F_{fas}(t_i) = 28.3 \cdot 10^{-10} T^3, \tag{15}$$

- total fastening failure units due to defects during operation

$$m_{fas} = 10.4 \cdot 10^{-6} T^3. \tag{16}$$

The verification of formulas (15) and (16) in accordance with Fisher criteria confirmed the proposed models are correct.

The rate of fastening failure in a track section is defined by the formula

$$\lambda_{fas}(t_i) = m_{fas}(t_i) / N(t_i)\Delta T, \tag{17}$$

where $N(t_i)$ is the number of fastenings without failure at the time t_i.

In this case, the function $\lambda_{fas}(t_i) = f(T_i)$ is described by following mathematical model

$$\lambda_{fas}(t_i) = 3.5 \cdot 10^{-10}\, T^{2.4}. \tag{18}$$

This model is received by processing pairs of values λ_{fas} and T in the interval T = 100÷600 million tons gross. It is verified by Fisher criteria. A failure of fixing assembly does not mean a failure of the system of rail links to supports. According to the valid regulations, the transversal stability of the rail line is considered to be partially failed when 3 or more neighboring linking assemblies fail. The researches defined the value of probability of failure-free operation of a system of transverse links of the rails with supports through intermediate fastenings. The data processing allowed to define how failure probability of the system changes (the probability of the occurrence of the critical sets of failed assemblies) $F_{fas}^{bush}(t_i)$ as a function of traffic load accumulated

$$F_{fas}^{bush}(t_i) = 5.5 \cdot 10^{-20} T^7. \tag{19}$$

Based on the calculations made the following mathematical model has been received for the function $P_{TPW}(t_i) = f(F_R(t_i), F_{sl}(t_i), F_{fas}(t_i), F_{pum}(t_i))$

$$P_{TPW}(t_i) = -2.7 \cdot 10^{-6} T^2 + 0.5 \cdot 10^{-3} T + 0.99. \tag{20}$$

The formula has been verified in accordance with Fisher criteria. The correlation factor and determination factor are calculated respectively as r = 0.9996 and r^2 = 0.9992. The formula (20) can be applied within the range from 200 million tons gross to 700 million tons gross. Calculation results by using formula (20) is given in the table 1.

Table 1. The probability of failure-free operation of the TPW as function of accumulated traffic load

Accumulated traffic load, million tons gross	0	200	300	400	500	600	700
Probability of failure-free operation of the TPW	0.990	0.982	0.897	0.758	0.565	0.318	0.017

3 Application of the mathematical model of the to the section of the railway of Ukraine

The formula (20) is used for a hypothetical case, which assumes that the failed components of TPW are not fixed regularly during maintenance. This explains the sharp falling of the function $P_{TPW}(t_i)=f(T)$. In real conditions, tracks are maintained regularly. The simulation results processing allowed to establish the following mathematical model of the relationship

$$P_{TPW}(t_i)= 0.9975 - 2.3\cdot10^{-4}T. \quad (21)$$

The correlation factor is $r = 0.74$, the determination factor is $r^2 = 0.54$. The formula (21) can be applied in a range from 100 to 800 million tons gross.

To illustrate the application of the developed mathematical model for specific conditions, let's take a section of one of the railways of Ukraine. There are certain areas in the selected section where traffic load accumulated after TPW rehabilitation differs. We have used formula (21) to calculate probability for various values of the accumulated traffic load. The sections were ranked based on the calculated values. Let's assume that the probability of failure-free operation is $P_{TPW} \geq 0.95$, then all components of the TPW operate reliably in the selected site. If the probability of failure-free operation is $0.95 > P_{TPW} \geq 0.85$, then one can assume that failure of defective individual components occurred and this is an indication that rehabilitation should be planned and scheduled. When $P_{TPW} < 0.85$, the related areas should be included in rehabilitation plan as soon as possible. For a better visualization of above ranking there is a layout of a railway area provided below where each category of the failure-free operation is highlighted by a specific color (Fig. 1, *a*).

a) *b)*

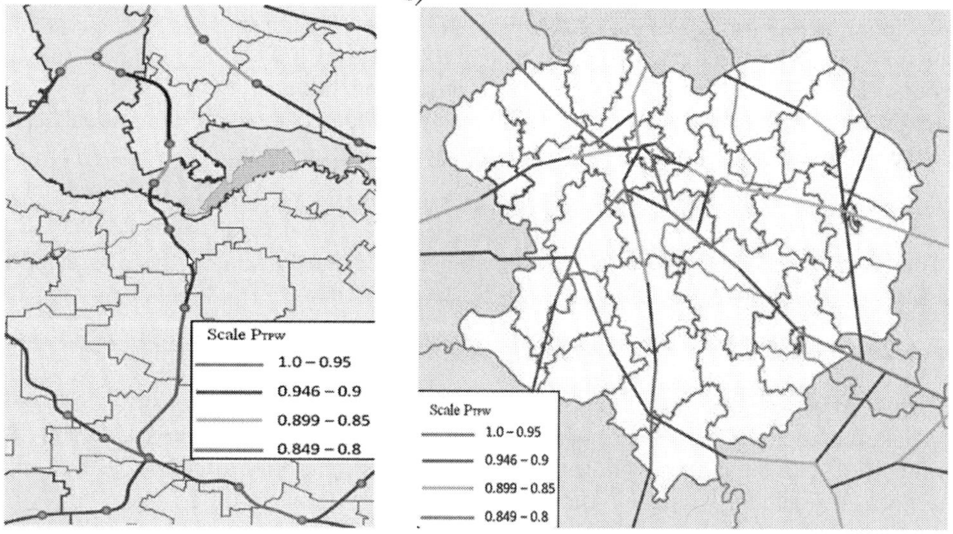

Fig.1. Classification of the railway track sections according to the probability of failure-free operation P_{TPW}

The management of this area is able to carry out long-term planning of the different types of repair and maintenance, taking into account the technical condition of the TPW. This allows to schedule maintenance, specify location, plan resources. A fairly simple software can calculate the probabilities daily, do classification and highlight the railway sections by color accordingly. Such daily monitoring will increase the operational reliability of this area and the entire railway in general (Fig. 1, *b*).

4 Conclusions

This paper is the first step towards creating a risk management system in the railway industry. The developed mathematical model allows to determine probability of failure-free operation of all the components of the track permanent way: rails, sleepers, fastenings and ballast layer, which is new for the railway industry. The model allows ranking of the track sections by a degree of their "aging" and provides basis for timely planning of various types of tracks maintenance or repair.

In order to increase reliability of the track sections, the mathematical model can be used as a basis for technical condition monitoring of railways. This model considers the maximum possible number of factors that affects reliability of the transportation. Monitoring allows sorting track sections according to their fail risk degree and identify the most dangerous places that require additional inspection or repair. The model can be for rational planning and allocation of funding and resources.

References

1. L. Wang, Y. Li, E. Wang, Systems Engineering Procedia, **1**, 174-180 (2011).
2. S. Lee, International Journal of Railway, **6 (2)**, 45-52 (2013).
3. J. H. Park, C. S. Shim, Journal of the Korean Society for Railway, **20 (3)**, 400-412 (2017).
4. J. M. Rocha, A. A. Henriques, R. Calçada, A. Rönnquist, Engineering Structures, **101**, 138-149 (2015).
5. J. M. Rocha, A. A. Henriques, R. Calçada, Structure and Infrastructure Engineering, **12 (1)**, 78-92 (2016).
6. D. Lee, H. S. Lee, Journal of The Korean Society For Urban Railway, **5 (3)**, 867-871 (2017).
7. B. Leitner, Procedia Engineering, **187**, 150-159 (2017).
8. B. Leitner, L.Môcová, M. Hromada, Procedia Engineering, **187**, 143-149 (2017).
9. Z.Dvořák, E.Sventeková, D.Řehák, Z.Čekerevac, Procedia Engineering, **187**, 548-555 (2017). doi: 10.1016/j.proeng.2017.04.413.
10. C.Jia, W.Xu, Y.Wang, Procedia Engineering, **15**, 1293-1297(2011).
11. J. Sadeghi, H. Askarinejad, Structure and Infrastructure Engineering, **6 (6)**, 677-688 (2010).
12. S-H. Kim, Journal of the Korean Society for Railway, **14 (6)**, 569-574 (2011).
13. D.Ionescu, N.Brinzei, J-F. Petin, IFAC-PapersOnLine, **48 (21)**, 314-319 (2015).
14. K.Liu, M.Wang, W-H. Zhu, DEStech Transactions on Computer Science and Engineering, Jun 21; (mcsse), 421-425 (2017).
15. A.Shtompel, SWorld Journal, **13**, 271-277 (2017).

Diagnostics of railway turnout systems using tools of centralization blocking system

Ivan Siroklyn*, *Serhii* Zmii, *Artem* Maslii, and *Serhii* Buriakovskyi

Ukrainian State University of Railway Transport, Feierbakh Square 7, 61050, Kharkiv, Ukraine

Abstract. Turnout systems are one of the few most important components of the railway infrastructure. Recently there has been a revival of research aimed at prognostics and health management of railway turnout systems. The works of the researchers are devoted to evolution timely diagnosis of damage of mechanical components or failures electrical components of switch motor, and contaminated by turnout system. In this article, the approaches of the most common specialized systems of diagnosing and predicting the technical condition of railway turnout systems elements such as POSS, SIDIS W, and APK-DK. The work presented in this article are followed by available approaches and studies of centralization blocking system tools' informativeness in the field of turnout systems failure diagnostic. The main point of view are methods of engine current analysis. The analysis of the motor current curves of the railway turnout obtained by the experimental method for different values of the ADC operating frequency is presented. Similar current curves are obtained by mathematical modeling of the normal operating mode of the elements of the railway turnout and under the condition of working with developing damage.

1 Introduction

An analysis of recent works in terms of diagnosis and prediction of the technical state of the railway turnout's elements indicates an increase of interest in this topic [1]. In [2] is shown, that 19% of all cases of train delays in Germany are related to switches' denials.

The possibility of contamination and other failures prediction for a railway switches was investigated in paper [3, 4]. For determination of the most effective degradation indicators five different type of sensors are used [4] (force, current, voltage, linear speed and distance sensor). However, the mathematical model of object was not investigated, and the experiment concerned only the contamination. The more extensive researches in the field of turnout systems diagnosis, based on analysis of power curve which is consumed by the engine during a switch passing is shown at [4-10]. For the most part, a time analysis of the curve and power levels are used in studies and the failure prediction has been neglected.

Turnout system failures can be divided into two categories: mechanical failures and electrical ones. The first category includes degrading of switches, stockrails, condition of slide chair and other failures of elements related to the track-and-field economy. Second

* Corresponding author: seroklin.iv@gmail.com

category includes failures of switch motor and its elements. The analysis of security divisions of railway transport of Ukraine shows that the share of the first category account for about 53% of turnoute systems damage.

According to switches of the track 1520 and the most common switch motors with engine MSP and MST types there are a number of works, which describe the diagnostic signs of damages [8]. Diagnostic signs of deterioration of the technical condition of the main elements separately for the mechanical part of the railway turnout are presented in Table 1.

Table 1. Diagnostic signs of railway turnout systems damage.

Failures	Diagnostic signs
Absence of distance in the heel spread	Gradual or sharp increase in the amplitude of the signal (current) at the end of the time dependence
Shift of stockrails and/without clamp of switch under the stockrail	Gradual or sharp increase in the amplitude of the signal (current) at the end of the time dependence
Contaminated slide chair	Gradual accretion of the current amplitude during each subsequent pass of switches
Contaminated turnout system of loose materials	Curt accretion of the current amplitude during the pass of the current switch compared with previous one
Slacks on connecting rods	Operating time increasing before switches start moving. Increasing of the operating time in general

2 Specialized systems for diagnosing the technical state of the turnout system elements

In Europe there are several systems which are capable to diagnose failures of turnout systems: Roadmaster 2000 (VAE), POSS (Strukton), and SIDIS W (Siemens). In POSS systems (Germany, Netherlands) a fairly simple approach is used in which, each time the switch is moved, the charts of the main power parameters changes are formed and compared with the reference for a particular switch [9]. The SIDIS W system uses the proportionality ratio between the switch shift effort and the active power consumed by the engine. The change in active power curve whilst moving the switch is used to diagnose bounce. It was the data of the SIDIS W system used in [2] to analyze the most efficient mathematical model of the prediction. The paper substantiates the expediency of choosing an interval of 3-5 days to warn staff of the high likelihood of occurrence of a particular type of failure (RUL). The basis for failure prediction is the principle of monitoring the power level, time characteristics of the signals and the climatic parameters has been taken into account. According to the author's assurances, the best methods for predicting the residual resource allowed to predict about 50% of documented failures.

On the territory of Ukraine, "Kaskad", "Prohnoz", "APK-DK", "ASDK" systems are operated. In one way or another, they provide information on the operating modes of drives, but information is stored only in the form of discrete signals. The market offers "APK-DK" system, which is specially designed for diagnostics of the elements of the switch according to the current curve. The ability to store an analog signal (for example, motor current curves) is only available in new microprocessor centralization blocking systems. Intelligent processing in the part of the system is performed for diagnostic tasks mostly often in the form of exceeding control for the maximum permissible value of current (or power), or identification of engine operation on friction. The system of electricity supply control, which, as a rule, is an integral part of the equipment of the centralization blocking system, continuously records the main parameters.

In addition to the analysis of currents levels, in work [8] the synthesis of two other approaches in the diagnostics of the electric drive: time principles and spectral ones are proposed, where it is proposed to analyze the spectrum within the first five harmonics.

3 Control of the current curve by means of centralization blocking systems

Most centralization blocking systems produced in Ukraine use controllers of switches, which are capable of removing the current values of an engine with a period which is close to 1 ms. he step (quantum) of fixation with the value of current is less than 0.01 A. For comparison, the specialized system of diagnostics of the railway turnout systems "APK-DK" uses the ADC with a period of 0.001s but averages the values of ten measurements to reduce the nuisance impact. Thus the informative period is 0.001s. However, in centralization blocking systems, the main limiting factor is not the parameters of the ADC of the object controller, but the data transmission network to the central processor of the system. Bandwidth of a network of existing systems rarely uses a period of 0.05 with more frequent use of 0.1 s or 0.4 s. For analyzing the informativity of measurements in laboratory conditions, with digital measurement means, the current curve with a period of 0.001s was removed and the average value of the current of each 10 measurements was calculated. (Curve a), Fig. 1). The average of ten measurements of the current curve with a period of 0.02 s, 0.05 s and 0.1 s were made (Curves b), c) and d), Fig. 1). By expression a), the value of the mean-square deviation of the current values of curves b), c) and d) from curve 1 is calculated, which is taken as the reference:

$$\sigma^2 = \sum_{n=1}^{j} (A'_n - A''_n)^2 , \tag{1}$$

where σ – the mean-square deviation of the results of measurements of the values of the engine current at the measuring unit n, j – number of measurements for constructing the current curve, A'_n – value of the current of the reference curve at the measuring unit n, A''_n – value of the current curve at the n-measurement.

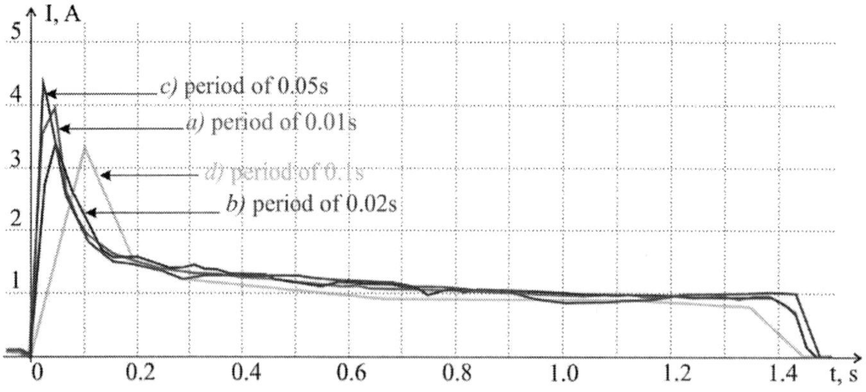

Fig. 1. Measurements of the current in laboratory conditions.

Using the methods of linear extrapolation and interpolation, the number of points of current measurements (*j*) on which the deviation was calculated is taken $j = 71$ with a uniform distribution along the curve. The results of calculating deviations of values from the reference are presented in Table 2. One can see that the deviation is not significant, given the fact that for the diagnosis the deviation of the values of the controlled value by 10% or 20% will be informative.

Table 2. Results of calculating deviations of values from the reference.

Period of measurements, c	0.01	0.02	0.05	0.1
Medium-square deviation of results, A	-	1.102	1.26	3.341

3 Theoretical studies

For theoretical studies, a mathematical model of the operation of switching was used [10]. Particular attention was paid to the description of the turnout kinematic line in the nature of the movement of the switch points, as the 3 masses mechanical system. Since the process of transformation of the energy of an electric field into a rotating moment on a machine shaft (and, consequently, in the motion of a working body) is interconnected, the junction was considered as a single electromechanical system. In Fig. 2 shows the diagram of the motor current in the absence of load.

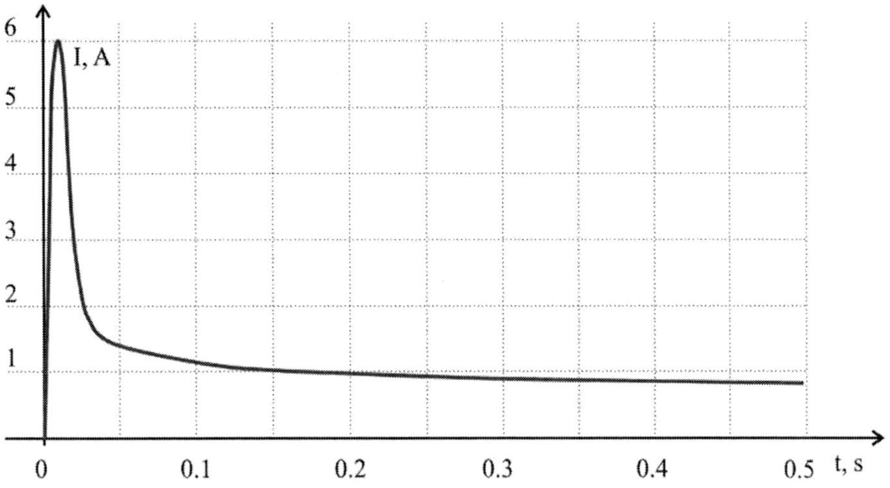

Fig. 2. Diagram of motor current without load

Load parameters - the weight of the switch points and the strength of their resistance to movement during the transfer, as well as the features of the turnout work (for example, operating mode at contaminated transfer surfaces or other faults) affect the electric engine, which, in turn, changes any processes affects the nature of the movement of rails. On the mathematical model [10] an experiment was conducted in the mode of contaminated transfer surfaces. From the current diagram (Fig. 3) it is clear that exactly this factor causes a variation of the force that affects the engine current. In this case, the frequency of such oscillations is within 70-100 Hz. That is, to imagine the behavior of the motor current, it is desirable to measure its dimensions with a frequency of not less than 1 kHz.

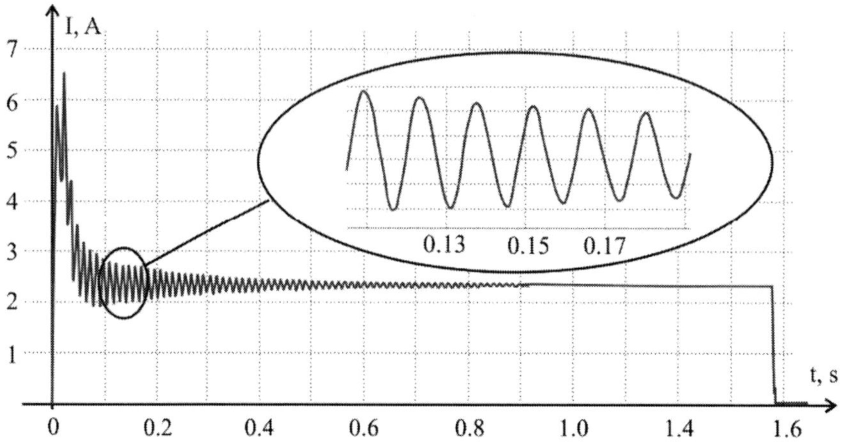

Fig. 3. Diagram of engine current during transfer with contaminated transfer surfaces

4. Conclusions

Available technical means of modern centralization blocking systems enable to realize the functions of diagnostics of the technical condition of switching elements using methods of analysis of the current curve of the electric drive motor. This relates to controlling the current level, character, and time of the switch transition.

In most cases, the bandwidth of data transmission systems restricts the use of some of the methods requiring an increase in the frequency of measurements. As a way out of the situation, one can propose to consider the option of embedding diagnostics in general, or certain types of diagnostics on the object controller itself, with the transfer of the signs of the switch technical state to the CPU of the codes system.

References

1. V. Atamuradov, K. Medjaher, P. Dersin, B. Lamoureux, N. Zerhouni, Int. J. of PHM, **060**, 1–31 (2017)
2. T. Bohm, Int. J. of PHM, **059**, 1–15 (2017)
3. U. Oßbergera, W. Kollment, S. Eck, Procedia Structural Integrity, **4**, 106–114 (2017)
4. F. Camci, O. Faruk Eker, S. Başkan, S. Başkan, Proceedings of the Institution of Mechanical Engineers, Part F: Journal of Rail and Rapid Transit, **230**, 24–42 (2016)
5. V. Atamuradov, F. Camci, S. Baskan, M. Sevkli, Ieee international symposium on diagnostics for electric machines, power electronics and drives, Piscataway (NJ), 1–5 (2009)
6. O. Eker, F. Camci, U. Kumar, U. Kumar (Ed.), emaintenance Lulea (Sweden): Univ, 248–251 (2010)
7. T. Asada, C. Roberts, T. Koseki, Transportation Research Part C: Emerging Technologies, **30(0),** 81–92 (2013)
8. S.Iu. Buriak, V.I. Havryliuk, O.A. Hololobova, Bulletin of the Dnipropetrovsk National University of Railway Transport, **4 (58)**,7–25 (2015)
9. K. Seidl, Signal + Draht, **12**, 39–40 (2003)
10. S.H. Buriakovskyi, A.S. Maslyi, Bulletin of the National Technical University "Kharkiv Polytechnic Institute", **28**, 392–393 (2010)

Evaluation of wear reduction measures for wheelsets and rails operated in high mountain conditions

Sergei Voznenko[1], *Anatoliy* Falendysh[1,*], *Andrey* Sumtsov[1], *Olha* Kletska[1], and *Miroslav* Blatnicky[2]

[1]Ukrainian State University of Railway Transport, 61050, Feuerbach square 7, Kharkiv, Ukraine
[2]University of Zilina, Univerzitná 8215/1 010 26 Žilina, Slovakia

Abstract. There are about 20 major factors influencing wear of railway system of a wheel, but the most critical include service and lubricant of rails. On the basis of locomotive depot Rodakovo the analysis of regularity of wear of a surface of wheels of the locomotive during the long period of time has been made and the method of correction of wheels has been offered and calculations which have shown that during operation of wheel couples there can be several corrections of a profile of a wheel before full grinding that leads to increase in life cycle of a wheel have been carried out. Also the approximate method for assessment of efficiency of introduction of methods of maintenance and repair of a wheel which includes the cost of life cycle of a wheel of the locomotive which will be developed further has been offered, estimating at the same time several methods, comparing their efficiency and choosing the most preferable options in the set parameters.

Introduction

Today, the rated service life of rails at the Ukrainian railways is 700 million tons gross, while at the American and European railways it is 2-3 times higher. The main factors that determine the service life of rails include the quality of rail steel, the quality of the rail production, the design of rails, the rail maintenance system, and the operation conditions of rails [1, 2].

According to [3, 4], there are about 20 main factors affecting the wear of the wheel-rail system:

- variation between wheel diameters;
- wheel profile;
- skewness of the axles of wheelsets;
- Track gauge;
- rail canting;
- rail profile;
- amplitude of irregularities;

* Corresponding author: fap_hiit@ukr.net

- slope and curved track sections;
- tread of wheels on rails;
- lubrication of rails and a type of grease;
- climate;
- construction and type of the rolling stock bogie;
- weight of the train.
- speed and length of the train;
- direction of the train;
- type of locomotives and cars;
- shape of the wheel flange;
- type of the brake system;
- train control by the driver;
- periodicity and repair time of tracks and wheelsets.

Article [5] presents operational observations and experimental studies, the results of mathematical simulation without regard to lubrication are summarized in Table 1. All the above factors affect the wear of rails and wheels, but the most critical ones include maintenance and lubrication of rails.

To prolong the service life of rails and reduce the damage with contact wear defects, polish the skid wears, pits, roughness and microdamaged surface layer of the metal should be ground away, both in the middle part of the top of the rail and in its lateral curls, forming a more convex outline of the upper surface of the top of rail, which shifts the wheel-to-rail contact area to the axle of the top, increases the strength of rails from 15-45 to 80-95%. Restoring of performance of rails in the track by forming a given convex shape of the lateral profile of the top is possible by using a grinding train with rotating grinding wheels located at different angles to the tread surface of the rails. Such grinding trains used abroad, at speeds up to 8 km/h, enable forming a convex profile of the top, removing the damaged metal layer up to 0.38 mm for one run [1, 2].

Table 1. Effect of different factors on the wear of the wheel-rail system

Factor	Range of measurements	Degree of effect, %	Grade
variation between wheel diameters	0-4 mm	+ (116-128)	1st
wheel profile	New of average wear	(47 98)	2nd
skewness of the axles of wheelsets	0-4 mm	+ (35)	3rd
track gauge	1510 1530 мм	-(21 28)	4th, 5th
Rail canting	1/20 1/12	+ (27)	4th, 5th
Rail profile	New of average wear	(8 - 160)	6th
Amplitude of irregularities	1-1.2	+ (3-4)	7th

According to [8], the wheel should be inserted into the contact surface of the rail to transfer the torque and ensure the adhesion of the wheel to the rail. In order to ensure the normal wear in the wheel-rail system with a slippage of up to 10%, the stiffness ratio should be at least 1.2: 1, that is, with hardness of the wheel is 20% higher than that of the rails. However, since the late 1950s when bulk hardened rails began to be used, the ratio in the wheel-rail pair was set at 1:1.4, which today leads to poorer adhesion, increases slippage, and causes intensive wear of wheelsets.

In [6, 7, 10, 11], based on the analysis of the global experience and experimental works, the conclusion has been made that an increase in the hardness of the wheels by 1 HB within the operating range of hardness increases their wear resistance by 1-2%. In addition, an increase in the hardness of the wheels from 250 to 600 HB practically does not affect the wear and contact design life of the rails, and the contact fatigue design life of the wheels increases in proportion to the square of the increase in their hardness. Thus, an increase in the hardness of rails and the tread surface of wheelsets leads to an increase in their service life.

In [9], it was shown that the surface layer in the wheel-rail tribosystem in the actual operating conditions, especially the flanges and lateral surfaces of the rail, is strengthened. Moreover, the depth of the frictional layer reaches 0.01-0.03 mm, and the increase in hardness at the surface is 220 to 1200 HV. According to the authors of [9], the strengthening mechanism of the friction layer is more complicated than simply cold work hardening or phase conversion hardening, although high contact loads up to 1000-1500 MPa and high temperatures up to 1000°C exist in small volumes of the actual wheel-rail contact. However, they exist for thousandths of a second. Therefore, it would be more correct to refer not to heating or cold work hardening of the surface layer, but to its high energy saturation. With such energy density, the metal in the shear layer flows amorphously, like liquid glass. After leaving the friction area, the metal in the slip band cools instantly, while retaining the amorphous structure of metal glass with a hardness of 800 HV or more.

Currently, there are several techniques to increase the hardness of wheelsets, which are classified by the heating and cooling method:

- bulk hardening in furnaces;
- HFC hardening;
- laser, electron beam hardening;
- plasma arc (jet= hardening;
- electrocontact hardening;
- oxygen-acetylene hardening;
- laser cladding;
- plasma coating.

On the basis of Rodakovo locomotive depot of Ukrzaliznytsia, the analysis of regularities of wear of the tread surfaces of locomotive wheels has been carried out. The wear parameters were monitored on 10 diesel locomotives 2TE116 (120 wheelsets) for 30 months. During the period of the study, the locomotive mileage averaged 220-300 thousand km. Also, 360 tyre grindings have been performed. The statistical analysis of operative wear parameters of tyres showed that by the time of grinding, neither clearance nor flange thickness parameters reach their maximum permissible values. As a rule, with a 2-3-fold reserve of the tyre body for clearance and flange thickness, the tyres are completely ground due to the value beyond the permissible limits of the slope parameter. The average clearance value before grinding was 2.0-2.5 mm, and the average flange thickness was 27.8 mm. And this is with allowable values of rolled products – up to 7 mm, flange thickness – up to 25 mm. In 77% cases, tyres were ground due to reaching the limiting value of the flange slope (6.5 mm) [12].

Figure 1 shows a typical example of a diagram of grinding tires with GOST 11018-87 profile and the following values of the geometric parameters of wear: flange thickness 27.4 mm; clearance 0.9 mm; flange slope 6.4 mm.

A wheelset with such geometric parameters of wear should be rejected, and the tyres are ground to the initial profiles. In this example, a layer of metal weighing more than 35 kg is removed by grinding, and the tyre becomes 11 mm thinner. In this case, the ground profile acquires the geometric parameters of the new profile, namely: flange thickness 31.9 mm; clearance 0.0 mm; flange slope 10.3 mm.

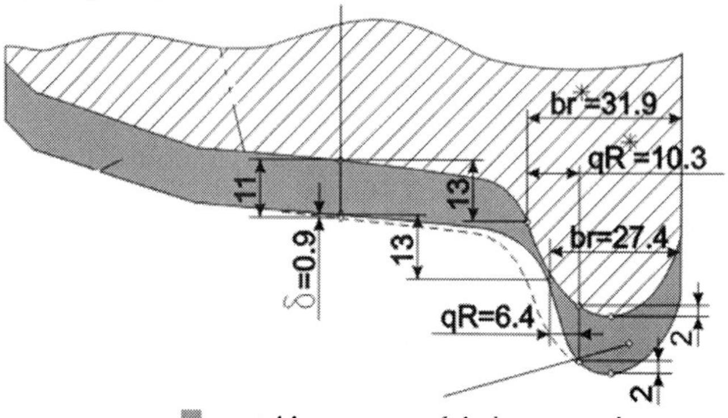

Fig. 1. Diagram of profile restoration of tyre GOST 11018-87 using the existing technique

Since in this example the tyre before grinding had a significant clearance and flange thickness margin, it is expedient to perform not complete grinding, but only partial grinding, which would correct only by the wear indicator, the value of which are beyond the permissible limits, namely, the flange slope. The operation of the partial grinding of the tyre is called the correction of the flange profile. Figure 2 shows the profile correction diagram.

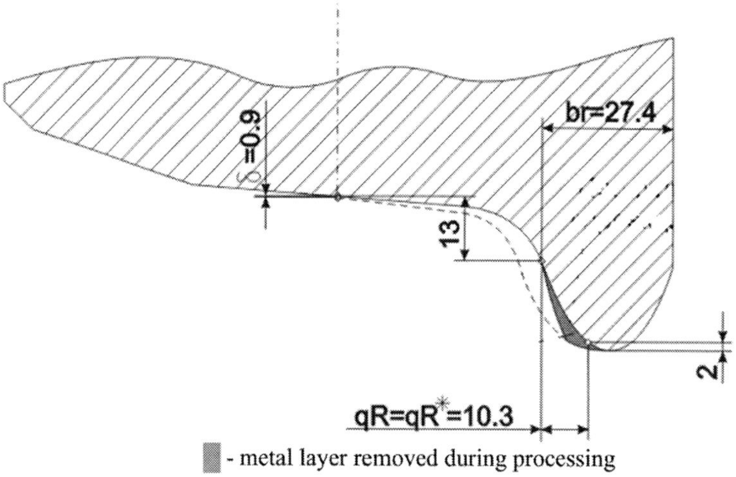

Fig. 2. Diagram of correction of the tyre flange profile

After correction, this profile retains the clearance of 0.9 mm and the flange thickness (27.4 mm) the same as before grinding, which is allowed by the "Instruction on formation, repair and maintenance of wheelsets of traction rolling stock of the Ukrainian railways of 1520 mm gauge" (VND 32.0 .07.001.2001) for further operation of the wheelset. In this case, the flange slope increases to the normative value of the slope of the initial profile - 10.3 mm.

When the tyre grinding technique described in this example is used and replaced with the flange profile correction, the metal layer removed when the profile of the wheels of one wheelset is restored, decreases from 70 kg (with the conventional methodology) to 1.8 kg (with the new methodology), and the tyre thickness remains unchanged.

The conducted calculations show that during the operation of wheelsets, one to three flange profile corrections can be performed between complete grinding of tyres.

The above methodology of restoring the profile of the bandages, which involves replacing of complete grinding of the tyre with a flange profile correction allows us to hope for extension of the service life of tyres probably in 2.0-2.5 times.

The brief analysis suggests that increasing the wear resistance of the wheel-rail friction pair is a complex multifactor process that requires an integrated approach. An important direction of further studies is the rationale for searching for optimal combinations of various techniques by using an integrated approach to estimation of the wheel-rail problem.

To implement this approach, a simple principle of qualimetry, based on a direct comparison of indicators is applicable. Besides, comparable indicators should be used as indicators of evaluation. The most expedient method of estimation to be used is the weighting factors method, which is generally determined by the following formula

$$K = \frac{\sum_{i=1}^{i=5} k_n \varphi(i)}{\sum_{i=1}^{i=5} \varphi(i)} \tag{1}$$

where K – indicator; k – weighting factor; φ_p – value of the parameter after the implementation; φ – value of the parameter before the implementation.

The following parameters were chosen as the basic comparison parameters using the expert method:

- service life of the rail;
- service life of the tyre;
- life cycle cost of the rail and the wheelset.

The above indicators fully reflect the effectiveness of the use of individual methodologies to increasing the service life of the wheel-rail pair. Moreover, the integrated methodology takes into account the influence both on the tyre and on the top of rails.

Ranking of weighting factors using the expert method produced following results: $k_1 = 0.3$; $k_2 = 0.3$ and $k_3 = 0.4$. The first two indicators are equivalent and reflect the service life of the rail and the wheelset, and the last, more significant, the costs for its provision. Thus, both technical and economic components are taken into account. It should also be noted that the service life is a rational indicator, while the cost is the irrational one. In view of this, the formula for calculating the complex indicator takes on the following form.

$$K = 0.3\ \varphi_{p\,1} / \varphi_1 + 0.3\ \varphi_{p\,2} / \varphi_2 + 0.4\ \varphi_3 / \varphi_{p\,3} \tag{2}$$

The implementation of the proposed methodology for evaluating the measures to extend the lifespan of the wheel-rail pair has been conducted using the evaluation of two measures: plasma hardening of the bandage surface and the use of corrective grinding.

The complex estimation of parameter K for the effect of the use of plasma hardening of wheelsets according to the data on operation of locomotives in mountain conditions is as follows. When bulk hardened rails are used, the application of reinforced tyres does not affect their service life, since the hardness of both surfaces is comparable. In this case, the service life of the wheelset before is 106.3 thousand km before hardening, and 137.2 thousand km afterwards. The life cycle cost was 42 UAH,860.4 before the implementation and UAH 49,725.1 afterwards. As a result of the calculation, we obtain

$$K = 0.3 + 0.3 \cdot 137.2 / 106.3 + 0.4 \cdot 42\,860.4 / 49\,725.1 = 1.03.$$

The methodology for determining the effect of use of corrective grinding involves determination of the service life of the bandage and the life cycle cost. The wear of the rails does not change, and may be neglects in the evaluation. Thus, the first element of the indicator will be 0.3. The service life of the wheelset is 106.3 thousand km before the implementation, and 149.1 thousand km afterwards.

According to the data provided by the locomotive depot, the life cycle cost will be UAH 42,860.4 before the implementation of the methodology and UAH 38,657.9 after the implementation. Thus, the complex indicator will be

$$K = 0.3 + 0.3 \cdot 149.1 / 106.3 + 0.4 \cdot 42\,860.4 / 38\,657.9 = 1.16.$$

Conclusions

Comparison of the results of the first and second method of increasing the service life of the wheel-rail pair demonstrates that the corrective grinding is more advantageous vs. plasma hardening.

The proposed method for estimation of the measures to increase the service life of the wheel-rail pair will be further developed by estimating the simultaneous implementation of several measures with subsequent evaluation and comparison of their effectiveness and choosing the most preferable options.

References

1. Lapidus B.M. Zheleznodorozhnyy transport, **6**, 46 – 48 (2013)
2. Karnushchenko N.N. Velichko D.V, Bobovnikova P.A., Transport Rossiyskoy Federatsii, **3**, 10-14 (2010)
3. Sazonov V N., Zagitov E.D., Zheleznyye dorogi mira, **4** (2008)
4. Senkin P. L., Put i putevoye khozyaystvo, **5**, 13-14 (2007)
5. Gozbenko V. Ye., Tuvshintur B., Sovremennye tekhnologii. Sistemniy analiz. Modelirovanie, **3**, 191 – 197 (2015)
6. Lokomotiv-inform, **11**, 20-22 (2008)
7. Moroz B.A., Maryutin K.A., Balanovskiy A.Ye., Lokomotiv, **19**, 19-22 (1998)
8. Balanovskiy A.Ye., Glazkov V.S., Moroz B.A. et al., Sb. tr. SGUPS "Novye tekhnologii na VSZHD", 57-65 (1999)
9. Balanovskiy A.Ye., Khayasi S.M. *Problema iznosa pary treniya koleso-rels (kratkiy analiz i predlozheniya)* (Irkutsk: Plazmoprotek, 1997)
10. Potakhov D.A., Izvestiya Peterburgskogo universiteta putey soobshcheniya, **1**, 139 – 147 (2013)
11. Buynosov A.P. *Metody povysheniya resursa kolesnykh par tyagovogo podvizhnogo sostava: Monografiya* (M.: GOU "Uchebno-metodicheskiy tsentr po obrazovaniyu na zheleznodorozhnom transporte", 2010)

The impact of the reinforcement percentage on the stress-strain state of the bending steel fiber reinforced concrete elements

Oleksandr Andriichuk[1,*], *Volodymyr* Babich[2], *Ivan* Yasyuk[1], and *Serhii* Uzhehov[1]

[1] Lutsk National Technical University, 43018 Lutsk, Lvivska St. 75, Ukraine
[2] National University of Water Management and Nature Resources Use, 33000 Rivne, Soborna St. 1, Ukraine

Abstract. The use of concrete with traditional reinforcement pose the problem of finding ways to increase the crack resistance, impact strength, frost resistance and other characteristics, which in turn depend on the strength properties of the material. One of the solution is the use of dispersion reinforced concrete - steel fiber concrete (with short steel fibers of 30–50mm). The combination of rigid fibers with significant strength reserves and the concrete could increase the crack resistance of the matrix and other strength characteristics. The paper presents the results of theoretical studies of the reinforcement percentage effect on the stress-strain state of the rectangular cross-section elements of steel fiber reinforced concrete (SFRC) in bending from at repeated loads with levels of application $\eta = 0.5$ and $\eta = 0.7$. It is found that the limit deformations of the SFRC cross-section in bending is recommended to calculate using a polynomial function, which gives the greatest convergence between the true values and approximations in comparison with the logarithmic and exponential functions.

1 Introduction

During the authors' previous studies, it was found that steel fiber reinforced concrete (SFRC) as an effective material should be used for the production of gutter drainage systems [1-2]. The use of SFRC gutters instead of standard reinforced concrete gutters allows to completely avoid the use of rebar; reduce the consumption of materials of construction, and as a consequence, their weight; reduce labor costs for production of structures and significantly increase the durability of constructions. These studies were conducted in the building laboratory of Lutsk NTU. It should be noted that the effectiveness of the use of dispersion-reinforced concrete is confirmed by the results of studies and other authors [3-5].

In recent years, the study of dispersion reinforced concrete and structures based on them, working under the influence of repeated static and dynamic loads, has gained significant relevance. Leaders in these studies are the Russian Federation [6, 7] and Israel

* Corresponding author: aleklutsk@gmail.com

[8] because high-strength fiber reinforced concrete hold a great promise in the development of modern protective structures for defense purposes. It should be noted that at present, in other countries, researches are carried out to increase the strength properties of structural elements by using steel fibers for their dispersed reinforcement [9-12].

To obtain steel fiber reinforced concrete as a dispersed reinforcement of experimental gutters, it is used a wave fiber with a diameter of 0.8 mm and length of 50 mm (Fig. 1).

The geometric dimensions of the gutters were as follows: wall thickness $t = 40$ mm; an inner diameter $d_{in} = 300$ mm; outer diameter $d_{out} = 380$ mm; length of the elements $l = 300$ mm. General view of SFRC gutters is shown in Fig. 1.

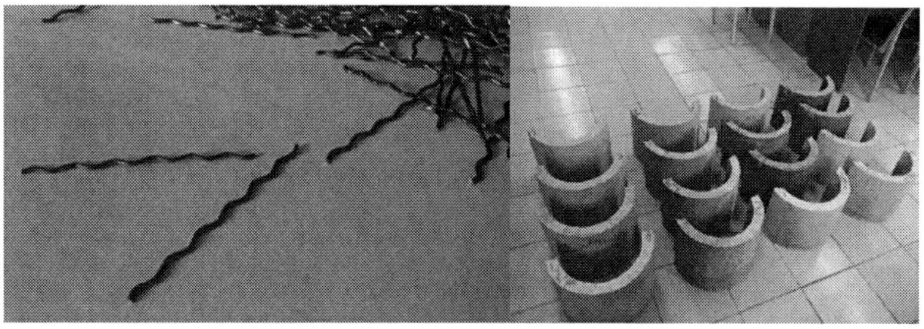

Fig. 1. General view of steel anchor fibers for the production of steel fiber reinforced concrete.

During the experimental studies [2], it was determined the actual stress-strain state of SFRC gutters with reinforcement coefficients $\mu = 1$, $\mu = 2$ and $\mu = 3$ from the effect of repeated loads on them with application levels $\eta = 0.5$ and $\eta = 0.7$ (within ten cycles). At the 11th cycle the load increased to destructive for determining the bearing capacity of the elements. The main factors that are determined the stress-strain state of structures should be: linear dimensions of the structure and its cross sections that deform; type of applied load; modulus of elasticity of material.

The moments of resistance (W) of the rectangular cross-section for three different percent of reinforcement with a variation of 1% ($\mu = 1$, , $\mu = 2$ and $\mu = 3$) of the SFRC gutters are characterized by the dependence:

$$W = \frac{b \times h^2}{6} \tag{1}$$

b - width of section; h - height (thickness) of section;

The resulting stresses in concrete and SFRC are characterized by the expression:

$$\sigma = M / W \tag{2}$$

Due to the fact that the geometry of the gutters cross-sections ($\frac{b \times h^2}{6}$) at different percentages of reinforcement ($\mu = 1$, $\mu = 2$ and $\mu = 3$) is constant, the stress-strain state of the cross section can be estimated on the basis of the limiting moment that the research elements are takes at different percentages of reinforcement and, respectively, the emerging limit stresses in concrete and SFRC (σ).

On the basis of experimentally obtained data of limit stresses in the concrete section of the gutter (σ_c) and SFRC cross-sections (σ_{sf}), depending on the percentage of reinforcement (μ), it is advisable to describe the functional relationship between them $f(\mu)$ given by the expression:

$$\sigma_{sf} = f(\mu) \cdot \sigma_c \quad (3)$$

It should be noted that at μ=0 (concrete element) the function $f(\mu) = 1$.

Then the dependence of the limit stresses in the SFRC section (σ_{sf}) on the experimentally determined limit stresses in the base concrete section (σ_c) can be found using the determined function, which depends on the percentage of reinforcement $f(\mu)$.

Experimental results of testing samples according to the experimental plan [2] are presented in Table 1.

Table 1. Bearing capacity of steel fiber reinforced concrete cross-sections of gutters.

The level of applied repeated loads	Limit stress	The percentage of reinforcement, %			
		$\mu = 0$	$\mu = 1$	$\mu = 2$	$\mu = 3$
$\eta = 0.5$	*σ, MPa*	15.88	19.40	27.72	30.64
$\eta = 0.7$			17.03	27.72	29.69

To determine the intermediate values of the stress limits for other percentages of reinforcement, we apply the least squares method, which is one of the methods of regression analysis, that is used for statistical evaluation of the regression model parameters from empirical data.

According to this method, the parameters of the model must correspond to a regression equation which provides the smallest value of the sum of the squares of deviations of empirical data from those calculated by the regression equation. Thus, of two different approximations of the same empirical function, which is given in the form of a table, the best is the one for which the sum of the squares of the deviation has the smallest value. A generalization of the conditions for the application of the least squares method is formulated in the Gauss–Markov theorem [13].

Dependences of limit stress on the reinforcement ratio ($\sigma - \mu$) for cross-section of SFRC gutters with the intermediate values are presented in Fig. 2, with $\eta = 0.5$ and $\eta = 0.7$.

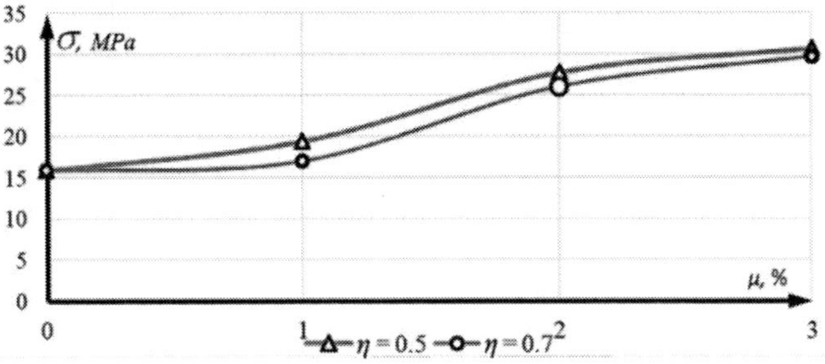

Fig. 2. Dependences of limit stress on the reinforcement ratio of cross-section of SFRC gutters.

Let approximate the obtained curves at the load levels of SFRC gutters $\eta = 0.5$ and $\eta = 0.7$ using three functions (logarithmic $y = \log_a x$, exponential $y = e^x$ and polynomial $y = c_0 + c_1 x + ... + c_n x^n$) and build the corresponding trend lines. At the level of repeated loads $\eta = 0.5$ the trend lines for the approximated curve are shown in Fig. 3 (a, b, c), and at the load level $\eta = 0.7$ are shown in Fig. 4 (a, b, c).

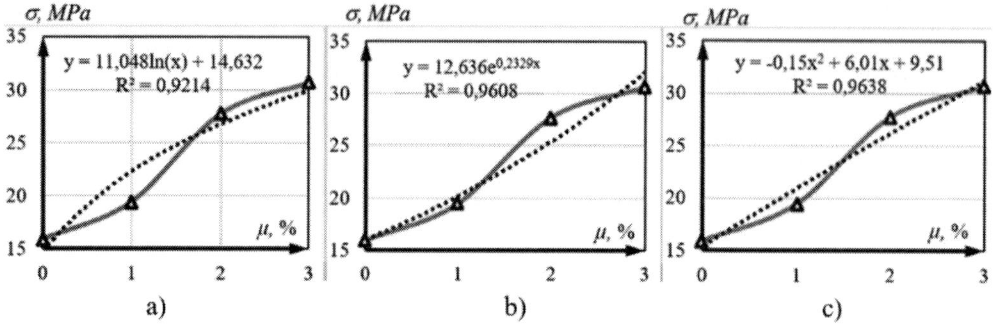

Fig. 3. Obtained by approximating the trend line at $\eta = 0.5$ using (a) the logarithmic function; (b) the exponential function; (c) the polynomial function.

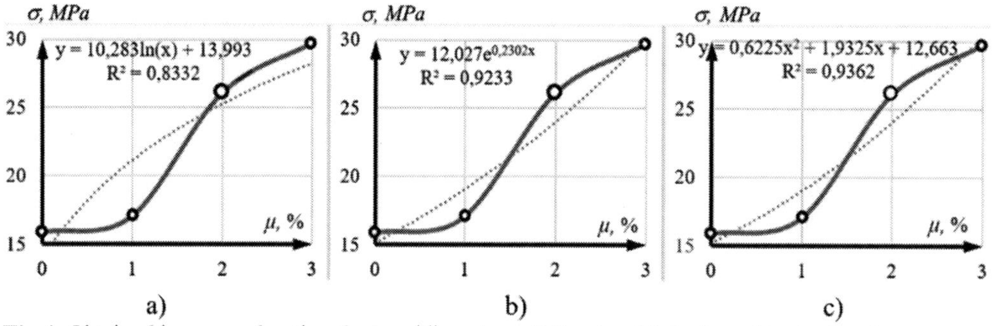

Fig.4. Obtained by approximating the trend line at $\eta = 0.7$ using (a) the logarithmic function; (b) the exponential function; (c) the polynomial function.

Let also determine the deviation – correlation coefficients (R^2) for each of the functions obtained by approximation at $\eta = 0.5$ and $\eta = 0.7$.

At the approximation of data using the logarithmic functions and given that $f(\mu) = \sigma_{sf} / \sigma_c$ (according to (3)), limit strain of SFRC sections are:

at $\eta = 0.5$ $$\sigma_{sf} = (11.048\ln(\mu) + 14.632) \cdot \sigma_c \quad (4)$$

at $\eta = 0.7$ $$\sigma_{sf} = (11.048\ln(\mu) + 14.632) \cdot \sigma_c \quad (5)$$

At the approximation of data using the exponential function and given (3), limit strain of SFRC sections are:

at $\eta = 0.5$ $$\sigma_{sf} = 12.636e^{0.2329\mu} \cdot \sigma_c \quad (6)$$

at $\eta = 0.7$ $$\sigma_{sf} = 12.027e^{0.2302\mu} \cdot \sigma_c \quad (7)$$

At the approximation of data using the polynomial function and given (3), limit strain of SFRC sections are:

at $\eta = 0.5$ $$\sigma_{sf} = (-0.15\mu^2 + 6.01\mu + 9.51) \cdot \sigma_c \quad (8)$$

at $\eta = 0.7$ $$\sigma_{sf} = (0.6225\mu^2 + 1.9325\mu + 12.663) \cdot \sigma_c \quad (9)$$

The correlation coefficients at the level $\eta = 0.5$ are determined as follows: $R^2_{\log} = 0.9214$; $R^2_{\exp} = 0.9608$; $R^2_{pol} = 0.9638$. The correlation coefficients at the level $\eta = 0.7$ are determined as follows: $R^2_{\log} = 0.8332$; $R^2_{\exp} = 0.9233$; $R^2_{pol} = 0.9362$.

Conclusions

The studies that are presented in the article provide the possibility of determining the limit stresses in the cross section of the steel fiber reinforced concrete element from its reinforcement percentage. This requires only the presence of the strength characteristics of the concrete section.

After analyzing the obtained correlation coefficients, it can be argued that the greatest convergence between the true values and approximations is in the application of the polynomial function in accordance with the equations 8 – 9 (at $\eta = 0.5$ and at $\eta = 0.7$).

It is found that the limit deformation of the cross-section of SFRC in bending is recommended to calculate using a polynomial function, which gives the greatest convergence (95%) among the true values and approximations in comparison with the logarithmic and exponential functions.

References

1. E.M. Babych, O.V. Andriichuk. *Strength of elements with annular cross sections made of steel-fiber-reinforced concrete under one-time loads.* Materials Science, New York, **52**, 509–513 (2017)
2. O. Andriichuk, V. Babich, I. Yasyuk, S. Uzhehov. MATEC Web of Conferences, **116**, 02001, 1-9 (2017)
3. L. Dvorkin, O. Dvorkin, Y. Ribakov. *A method for optimal design of steel fiber reinforced concrete composition.* Materials and Design (2011)
4. R. Kinash, V. Bilozir. *Deformational calculation method of bearing capability of fiber-concrete steel bending elements.* Czasopismo Techniczne (2015)
5. R. Shmyh, V. Bilozir, A. Vysochenko, V. Bilozir. *Carrying capacity of bending concrete elements reinforced by fibro and stripes taken from used PET bottles.* International Scientific and Practical Conference World science (2018)
6. D.A. Lamzin. *High-speed deformation and destruction of fine concrete.* Thesis. (2014)
7. L.A. Avetisyan. *Strength and deformability of compressed reinforced concrete elements under dynamic loading under fire conditions.* Thesis. (2015)
8. S.U. Baloch. *Steel fiber reinforced concrete.* Science Technology and Development **24**, 21-32 (2015)
9. S.V. Klyuyev, A.V. Klyuyev, R.V. Lesovik and A.V. Netrebenko. World Applied Sciences Journal **24**: 1280-1285 (2013)
10. A. Winkler, C. Edvardsen, T. Kasper. *Examples of bridge, tunnel lining and foundation design with steel-fibre-reinforced concrete.* Fibre-reinforced concrete from design to structural applications. **310**, 451-460 (2017)
11. J. Ming, J. Shi, W. Sun. *Effect of mill scale on the long-term corrosion resistance of a low-alloy reinforcing steel in concrete subjected to chloride solution.* Construction and building materials. **163**, 508-517 (2018)
12. J. Yuan and B.A. Graybeal. *Bond behavior of reinforcing steel in ultra-high performance concrete.* Federal highway administration (2014)
13. H. James, H. Stock, W. Mark. *Regression with a single regressor: hypothesis tests and confidence intervals.* Introduction to Econometrics, **3**, Addison-Wesley, 163-164. – 785 (2011)

Computerized method for estimation of ultimate load of piles

Basheer Younis[1]*, *Khudhair* Thamer[2], and *Firas* Kazimahomedov[1]

[1]Kharkiv National University of Civil Engineering and Architecture, Department of Building Materials and Products, 61002 Sumska st, 40, Kharkiv, Ukraine
[2]Al-Maaref University College, 330 Rmadi, Anbar, Iraq

Abstract. A new method was established to estimate the ultimate load, which is utilized in the assessment of pile test. In the present study twenty five reports of failure test piles (28.5 x 28.5 cm in dimensions, 12 m in depth and the working load was 40 ton) of Um-Alqura Mosque project were studied in according to their loading and settlement. The (mean ± SD) of the pile settlement for each load of all piles studied were determined, and the results obtained were utilized to generate the equation that correlates between the load and settlement using (4th degree polynomial fit graph). The equation obtained was applied to estimate the ultimate load for the maximum settlement (10% of the pile diameter). The estimated ultimate load (130 ton) in the present study was compared with the ultimate load (132 ton), which was determined previously using (Fellenius and Decourt presentation, 2001). The deviation between the ultimate load of the previous and the present method was (1.5%), but the present method depends on two working load, so it saves time and efforts and suitable to implement in the assessment of the test pile.

1 Introduction

A pile is structural member of timber, concrete or steel used to transmit structure load through water soft soil to a good bearing stratum by means end bearing resistance and skin friction. The piles can be classified according to pile material (timber, steel and composite piles), their effect on an adjacent soil (non displacement, displacement piles), and according to method of execution (driven and bored piles) [1-6]. The purpose of the static load test is to find the allowable load, which is established by dividing the ultimate load with a factor of safety. The factor of safety normally applied in the industry ranges from a low of 1.8 through a high of 2.5. The factor of safety is not a singular value applicable at all time. Its value depends on the desired freedom from unacceptable consequence of failure, as well as on the level of knowledge and control of the aspects influencing the variation of capacity at the pile. Most codes specify a single factor regardless of condition usually, frequently larger. There are many criterions in assessment the test pile and calculate the ultimate load. All the previous criterions extrapolated from later part of the load – settlement curve [7-10]. In the present study new equation was generated from the load – settlement curve to

* Corresponding author: docbasheer01@gmail.com

estimate the ultimate load of pile which was related to the maximum settlement of piles utilizing (ASTMD 1143-1987-5.1) [11], and computer algorithms that concerning the graphic purpose.

2 Materials and methods

Twenty-five precast concrete piles test reports of (Um-Alqura Mosque project / Iraq) were studied (2% of the total number of the executed piles). The piles were (28.5x28.5 cm and 12 m) in dimensions and depth respectively. The piles were tested by (ASTM D1143-1987-5.1), and the working load of each pile was 40 ton.

The following tools were used in the pile test:

1. Kentledge (collection from the steel members).
2. Hydraulic jack.
3. Cell load.
4. Concrete block (collection from uniform concrete sections
5. Hydraulic pump and dial gage.
6. Gages for settlement reading.
7. References beam for settlement gages.
8. The survey equipments.
9. Bearing plate.
10. Crane and shovel.

In this study, new equation was established to estimate the ultimate load according to the following steps:

1. Determination of the mean settlement of all piles studied, listed in Table 1.
2. Draw the relation between load and mean settlement, (Fig. 1).
3. Determination of the equation that correlates between load and settlement utilizing curve expert program (4th degree polynomial fit graph).
4. The equation established is used to estimate the ultimate load that causes settlement of (10% of the diameter of the piles studied).

3 Results

The results obtained depend on practical results that were excluded from twenty-five pile test reports of the assessment of the piles studied. Table 1 showed the load up to double working load with the mean settlement for the each load of all piles studied.

The results that listed in Table 1 were utilized to generate Fig. 1, and the following equation was obtained:

$$Y = 0.026 + 0.0042x - 0.002x^2 + 4.765e - 005x^3 - 3.789e - 007x^4, \quad (1)$$

The equation that mentioned above, was to estimate the ultimate load (that related to the maximum settlement not exceed 10% of the pile diameter).

The result obtained of the estimated ultimate load is compared with the calculated ultimate load of eight piles of the same project, using another method of assessment of pile test (depends on three working load) [12], the results are listed in Table 2, and shows that the deviation between the ultimate load of the present and the previous method does not exceed (1.67%).

Table 1. The load (ton) and settlement of the piles studied.

Load (ton)	Settlement (mm) mean	±SD *
0	0	0
10	0.2344	0.16981
20	0.5428	0.32027
30	0.8796	0.51923
40	1.2676	0.78756
50	1.74	1.07943
60	2.3864	1.57431
70	3.1292	2.17814
80	4.89	5.19013

*SD = Standard deviation.

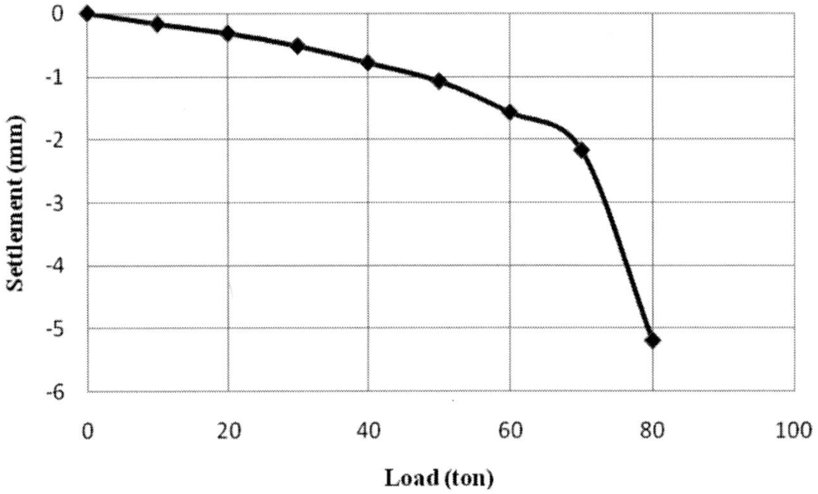

Fig. 1. Load – settlement curve of the piles studied.

Table 2. Comparison between the estimated ultimate load utilizing the present method and the method of (Fellenius and Decourt presentation, 2001).

Methods	Working load (ton)	Ultimate load (ton)	Deviation (percent)
Present method	2 working load (80 ton)	130.2	1.67%
Fellenius &Decourt method, 2001	3 working load (120 ton)	132.0	

4 Discussion

Many methods were applied in assessment of pile test. Such methods were arranged and used according to the types of piles, soils and the technique used. The advancement in this field added some new practical applications to the assessment of the pile test that depends on different facilities and equipments. The assessment methods of the pile test depend on work loading, value of settlement, and special technical management for pile loading, in according to the type of piles (precast, bored, timber, etc), types of soils, diameter of pile, depth of excavating, working load, geological properties, topographic properties and the nature of the material used in manufacturing of the piles. Also, all these methods of assessment of pile test depend on the final settlement of the executed piles that is not exceeded 10% of the pile diameter, (Davisson, 1972, Hanssen, 1970, Decourt, 1999, Fellenius, 2001). In the present study to overcome all the parameters that effect on the assessment of the pile test, new equation was generated and applied in the assessment of piles that (previously tested and assessed as succeeded piles, by method of ASTMD1143-1987-5.1). The equation established was utilized to estimate the ultimate load of the piles studied and depends on the real load-settlement curve (two working load) that related to the maximum pile settlement, bearing capacity, skin friction, water content of soil, volume of the soil, that effect on the pile settlement. Also the present method was applied for two working load to save the time and efforts in the assessment of the executed piles. The comparison between the present method and the method of (Fellenius and Decourt presentation, 2001), showed that the deviation of the estimated ultimate load of the two method was not exceed (1.67%). Also the applications of the present method in different projects and area will be used as reference for the future projects for soils of the same nature.

5 Conclusions

New method was established to estimate the ultimate load for assessment of pile test, utilizing computer programs for fitting load - settlement curve. The present method depends on two working load, concerning maximum settlement. The application of the present method will give new reference of ultimate load and settlement of different types of soils. So the present method save time and efforts and suitable to implement in the assessment of pile test.

References

1 J.E. Boweles, *Foundation analysis and Design,* (Second Edition, 1977)
2 BSI. CP2004, *Code of practice-foundations,* (1971)
3 B.H. Fellenius, *Basics of Foundation Design,* (Canada, 2009)
4 R. Rajapakse, *Pile Design and Construction Rules of Thumb,* (Elsevier, (2008)
5 ICE. *Behaviour of piles,* (London, 1971)
6 K. Karlsrud, *PhD Thesis,* (Trondheim, Norwegian University of Science and Technology, 2012)
7 M.T. Davisson, *Proceedings Lecture Series,* (Illinois Section American, **52** (1972)
8 J.B. Hansen. *A Revised and Extended Formula for Bearing Capacity*, Denmark, **28** (1970)
9 B. Fellenius, Geotechnical Engineering Journal of the SEAGS & AGSSEA, **42** (2), 70-77 (2011)
10 B.H. Fellenius, *Proceedings of the Deep Foundation Institute. Bearing capacity-a delusion,* (Dearborn, Michigan, 1999)
11 ASTM D1143-87, 5.1. *Testing Piles Under Static Axial Compressive load,* (1987)
12 L Decourt, Felleniuse, *Deep Foundation Institute. Extrapolation method, Fulcrum,* (New Jersy, 2001)

Bearing capacity of RC beams reinforced with high strength rebars and steel plate

Taras Bobalo[1], *Yaroslav* Blikharskyy[2,*], *Rostyslav* Vashkevich[1], and *Myhailo* Volynets[1]

[1]Lviv Polytechnic National University, Department of Building Constructions and Bridges, 79013, 6, Karpinskogo str., Lviv, Ukraine
[2]Lviv Polytechnic National University, Department of Highways and Bridges, 79013, 6, Karpinskogo str., Lviv, Ukraine

Abstract. Nowadays, reducing the material content of not only buildings and structures in general, but also individual constructions is a topical task that can be realized through the use of high-strength concrete and reinforcement, as well as with the use of external reinforcement. The concentrated location of sheet reinforcement on the external the most tense facets of steel and concrete structures increases the operating height of the cross-section, makes it possible to more effectively use the strength properties of steel in comparison with conventional reinforced concrete, and with the same bearing capacity to economize on expenses. Composite and monolithic reinforced concrete structures with external reinforcement are used in various construction sectors around the world. This contributed to the expansion of the use of reinforced concrete for special buildings of power-engineering and hydrotechnical construction. The technical and economic efficiency, as well as the possibility of using external rebar as formwork for monolithic concrete construction, have been proved. Therefore, there is a need for the study of structures with combined reinforcement, in which high rigidity of steel and concrete structures is combined with an effective use of high-strength reinforcing bars (rebar) without prior tension.

Introduction

The use of composite steel-concrete beams and other combined systems in construction provides for more efficient use of the characteristics of the strength of materials and increase the bearing capacity of structures [1-4]. The concentrated location of band reinforcement on the external facets of the structural intersections allows to reduce their mass, decrease the size of the cross-section compared to the reinforced concrete or to reduce the use of steel at the same height of the cross-section. However, there is a danger of corrosive effects [5], therefore, it is necessary to provide for a qualitative protection against the influence of the environment [6, 7].

The advantage of high-strength steels is their durability and crack resistance. This allows to reduce the amount of metal used in construction, without losing the reliability of

*Corresponding author: yaroslav.z.blikharskyy@lpnu.ua

the structure [8, 9]. When increasing the bearing capacity of reinforced concrete samples using high-strength rebar, it is also necessary to provide for the strength of sloping cross-sections [10]. However, if it was not envisaged in the design, then there are ways to enhance them [11].

Moreover, in addition to strength, another important characteristic of such steel is its deformability. In order to exhaust excess deformability, high-strength reinforcement is usually used in pre-stressed reinforced concrete structures [12], but the pre-stressing of the bearing bars leads to additional costs and is difficult to do under normal conditions. Therefore, the use of combined reinforcement allows to find the optimal ratio of different types of steels, which will reduce the cost of metal, while ensuring compliance with requirements for the strength and deformability of such structures.

The study of the strength and deformability of composite steel-concrete beams using sheet external band reinforcement as well as high-strength rebar, is relevant in terms of expediency and practical application of such structures in construction.

1 Research methodology

1.1 Test specimen description and material properties

For the experiment, eight composite steel-concrete beams were made with a cross-section of $0.12{\times}0.24$ m, a calculated span of 2.4 m, in which the adhesion of the external band reinforcement was provided with U-shaped anchors (Fig. 1, 2) welded to it.

a) b)

Fig.1. Skeletons of steel-concrete experimental beams with adhesion of band reinforcement with concrete: a) view of the flexion area; b) view of the bearing area.

To reduce the absolute error, the study was conducted using twin beams, that is, two beams with the same reinforcement were made from the same batch of concrete. This approach made it possible to obtain qualitative output material for the experiment.

The concrete class was decided after testing concrete cubes of the size $150{\times}150$ mm, and prisms of $150{\times}150{\times}600$ mm. Thus, it was determined that the characteristic strength of the concrete corresponded to the class C30/40-C35/45. The calculated strength was accepted taking into account the coefficient of reliability γ_c, which for heavy concrete is 1.3. Concrete was made using Portland cement with the activity of M500 and superplasticizer HK-1 (BM) to facilitate the laying of the mixture, and to reduce the water-cement ratio.

Notation for the beams was selected as follows: B - beam; I to IV - designation of the type of reinforcing skeleton; 1, 2 - serial number of the beam.

Compaction of concrete mix was vibrational. For all beams, the cross-reinforcement was adopted from the bars Ø5 mm of A240C class installed with a spacing of 70 mm in the zone of transverse forces acting on the support section (Fig. 1b), and with a spacing of 100 mm in the span (Fig. 1a).

The operating longitudinal reinforcement of the stretched zone was formed with smooth steel band rebar C275 of thickness t=8 mm, given that the adhesion with concrete was provided by U-shaped anchors, and with deformed reinforcement Ø 10 mm of A1000 class. The made beams varied by different types of reinforcing skeleton and different ratios of band and bar reinforcement. Physical and mechanical characteristics of materials of the experimental beams are given in Table 1. Steel stretch was tested by the standard method, however the method using digital image correlation becomes more popular nowadays [13].

Table 1. Physical and mechanical characteristics of materials of the experimental beams.

Notation of the beams		B - I - 1, B - I - 2	B - II - 1, B - II - 2	B - III - 1, B - III - 2	B - IV - 1, B - IV - 2
Heavy concrete	$f_{ck,cube}$ /$f_{ck,prism}$, MPa	48.7 / 29.6	46.3 / 28.1	46.3 / 28.1	49.2 / 29.9
	f_{cd}, MPa	22.8	21.6	21.6	23.0
	E_{cm}×103, MPa	38.00	37.76	38.04	38.50
Reinforcement of the stretched zone–longitudinal band	$B_s \times t_s$, mm	82×8	46×8	-	114×8
	f_{yk}, MPa	287	287	-	287
	f_{yd}, MPa	273	273	-	273
	E_p×10^5, MPa	2.05	2.05	-	2.05
	Brand	C275	C275	-	C275
Reinforcement of the stretched zone–longitudinal bar	Ø, mm	1ø10	2ø10	3ø10	-
	f_{yk}, MPa	1080	1080	1080	-
	f_{yd}, MPa	900	900	900	-
	E_p×10^5, MPa	1.85	1.85	1.85	-
	Class	A1000	A1000	A1000	-
Reinforcement of the compressed zone–longitudinal bar	Ø', mm	8	8	8	8
	f_{yk}', MPa	594.5	594.5	594.5	594.5
	f_{yd}', MPa	495	495	495	495
	E_p'×10^5, MPa	2.05	2.05	2.05	2.05
	Class	A400C	A400C	A400C	A400C
Bar cross-reinforcement	Ø, mm	5	5	5	5
	f_{ywd}, MPa	296	296	296	296
	E_p×10^5, MPa	2.05	2.05	2.05	2.05
	Class	A240C	A240C	A240C	A240C

For the reinforcement of beams, types of skeletons with different percentage reinforcement of the operating rebar of the lower stretched zone were used. The main task for the layout of skeleton type was to provide a close total tensile force that can be taken simultaneously by all the reinforcement of the stretched zone. Thus, in the skeleton type I, the total calculated tensile force taken by the reinforcement of the stretched zone was 2497.4 MPa, 68.8% of which was taken by the steel rebar C275 and 31.2% by the A1000 rebar. In the skeleton type II, the total calculated tensile force taken by the reinforcement of the stretched zone was 2417.6 MPa , 38.1% of which was taken by the steel rebar C275 and 61.9% by the A1000 rebar. In the skeleton type III, only the reinforcement of A1000 class was used, and the calculated tensile force was 2119.5 MPa . In the skeleton type IV, only the reinforcement of C275 class was used, and the calculated tensile force was 2489.8 MPa . Characteristics of the experimental beams is given in Table 2.

Table 2. Characteristics of the experimental beams.

Notation of the beams		B - I - 1, B - I - 2	B - II - 1, B - II - 2	B - III - 1, B - III - 2	B - IV - 1, B - IV - 2
Beam cross-section parameters	width b, mm	120	120	120	120
	height h, mm	235	236	238	235
	area A, cm^2	282	283	286	282
Reinforcement of the stretched zone–longitudinal band of C275 class					
Area As, cm^2		6.56	3.68	-	9.12
Reinforcement of the stretched zone–longitudinal bar of A1000 class					
Quantity and diameter of bars, mm		1ø10 A1000	2ø10 A1000	3ø10 A1000	-
Area As, cm^2		0.785	1.570	2.355	-
$f_{yk}{\cdot}A_s$, A1000 $f_{yk}{\cdot}A_s$, of a sheet		31.2% 68.8%	61.9% 38.1%	100% 0%	0% 100%
Total percentage of reinforcement by operating rebar		2.77%	1.97%	0.89%	3.45%

1.2 Experimental program

A peculiarity of the experiment was that with different percent reinforcement by the operating rebar in the lower zone from 0.89% to 3.45%, all experimental samples were designed equal, that is, the destructive load for them, according to the calculation of strength was the same. This was ensured by different ratio of areas between the band reinforcement made of steel plate C275 class and the high-strength reinforcement of A1000 class. The beams of types B-III and B-IV were adopted as reference, without combined reinforcement, they have one class reinforcement for the operating rebar of the lower zone. The B-III beam in the lower stretched zone was reinforced only with A1000 class reinforcement, and the B-IV beam was reinforced only with steel plate C275 class.

The loading of the experimental beams was carried out by stages of 10% of the destructive calculated according to [14], carried out with the help of a hydraulic jack with a power of 1000 kN . Through the distributive traverse, force was applied to the top surface of the beam in the form of two concentrated forces applied symmetrically to the middle of the beam at a distance from the supports 1/3 of the span. The loading time at each stage was 30 minutes, after which the device readings were taken.

2 Results of the experimental study

2.1 Bearing capacity of experimental beams

The onset of the flow started earlier for the band reinforcement than for high-strength reinforcement. This was accompanied by an increase in the growth of deformations and deflections in the subsequent loading of the beam.

When the flow of the band reinforcement was achieved, the tension in it did not decrease, but it continued to take the load during the flow, with the constant tensile force

$f_{yk} \times A_{s(band)}$. Strengthening of the band reinforcement after passage of the site of flow was not achieved, because of the considerable length of this area in the diagram of band steel stretch.

The bending moment corresponding to the flow tension in the band reinforcement was conventionally indicated M_{dr1}, and the moment corresponding to the conditional flow of the high-strength rebar was called the "marginal moment" M_{dr2}.

Since the calculation of the strength of the bending structures is carried out in terms of marginal state, then at the basis of the further analysis was laid the "marginal moment" *Mdr2*, which corresponds to the onset of the flow in high-strength rebar.

The reliability coefficient γ_s for A1000 class reinforcement according to [14] was $\gamma_s - 1.2$, for band reinforcement C275 $\gamma_s - 1.05$.

The results of the experiment are presented in Table 3.

Table 3. Results of the experimental study.

Notation of the beams	*Exp. M_{dr1}, kNm	*Th. M_{dr1}, kNm	Exp. deflection, mm	Th. deflection, mm	Bearing capacity, kNm			Deviation,%	
					Exp. M_{dr2}, kNm	Th. M_{max}^{DBN}, kNm	**Th. M^{DBN}, kNm	$(M_{dr2}-M^{DBN})/M_{dr2}$, %	** $(M_{dr2}-M^{DBN})/M_{dr2}$, %
B - I - 1	42.70	39.9	6.87	6.71	51.52	51.38	46.39	0.3	10.0
B - I - 2	43.60		7.98		52.64			2.4	11.9
B - II - 1	31.10	27.2	8.01	10.6	55.20	52.10	46.95	5.6	14.9
B - II - 2	31.10		7.36		52.40			0.6	10.4
B - III - 1	-	-	-	-	55.36	52.85	47.82	4.5	13.6
B - III - 2	-		-		57.04			7.3	16.2
B - IV - 1	51.49	50.25	7.91	5.39	51.49	50.25	49.79	2.4	3.3
B - IV - 2	50.57		6.72		50.57			0.6	0.6

Exp. – experimental, th. – theoretical according to DBN B.2.6-98:2009 [14];
*moment at yielding of steel rebar;
**bearing capacity with the use ofreliability coefficients for materials γ_s, γ_c.

For the beam B-I-1, B-I-2, the percentage of beam reinforcement was 2.77%. The flow of the band reinforcement was achieved at load $M_{dr1} = 42.7$ kNm and 43.6 kNm. The flow of the high-strength rebar was achieved at the bending moment $M_{dr2} = 51.52$ kNm for the beam B-I-1, and 52.64 kNm for B-I-2.The theoretical calculation assumed the magnitude of the destructive torque $M_{max}^{DBN} = 51.38$ kNm, $M^{DBN} = 46.39$ kNm. Thus, the margin of safety when using the calculated characteristics of materials for this series is 10.0–11.9%.

For the beam B-II-1, B-II-2, the percentage of beam reinforcement was 1.97%.The flow of the band reinforcement was achieved at load $M_{dr1} = 31.1$ kNm. The flow of the high-strength rebar was achieved at the bending moment $M_{dr2} = 55.2$ kNm, for the beam B-II-1, and 52.40 kNm for B-II-2.The theoretical calculation assumed the magnitude of the destructive torque $M_{max}^{DBN} = 52.10$kNm, $M^{DBN} = 46.95$ kNm.Thus, the margin of safety when using the calculated characteristics of materials for this series is 10.4–14.9%.

For the beam B-III-1, B-III-2, the percentage of beam reinforcement was 0.89%. The flow of the high-strength rebar was achieved at the bending moment $M_{dr2} = 55.36$ kNm, for the beam B-III-1, and 57.04 kNm for B-III-2. The theoretical calculation assumed the magnitude of the destructive torque $M_{max}^{DBN} = 52.85$ kNm, $M^{DBN} = 47.82$ kNm. Thus, the margin of safety when using the calculated characteristics of materials for this series is 13.6–16.2%.

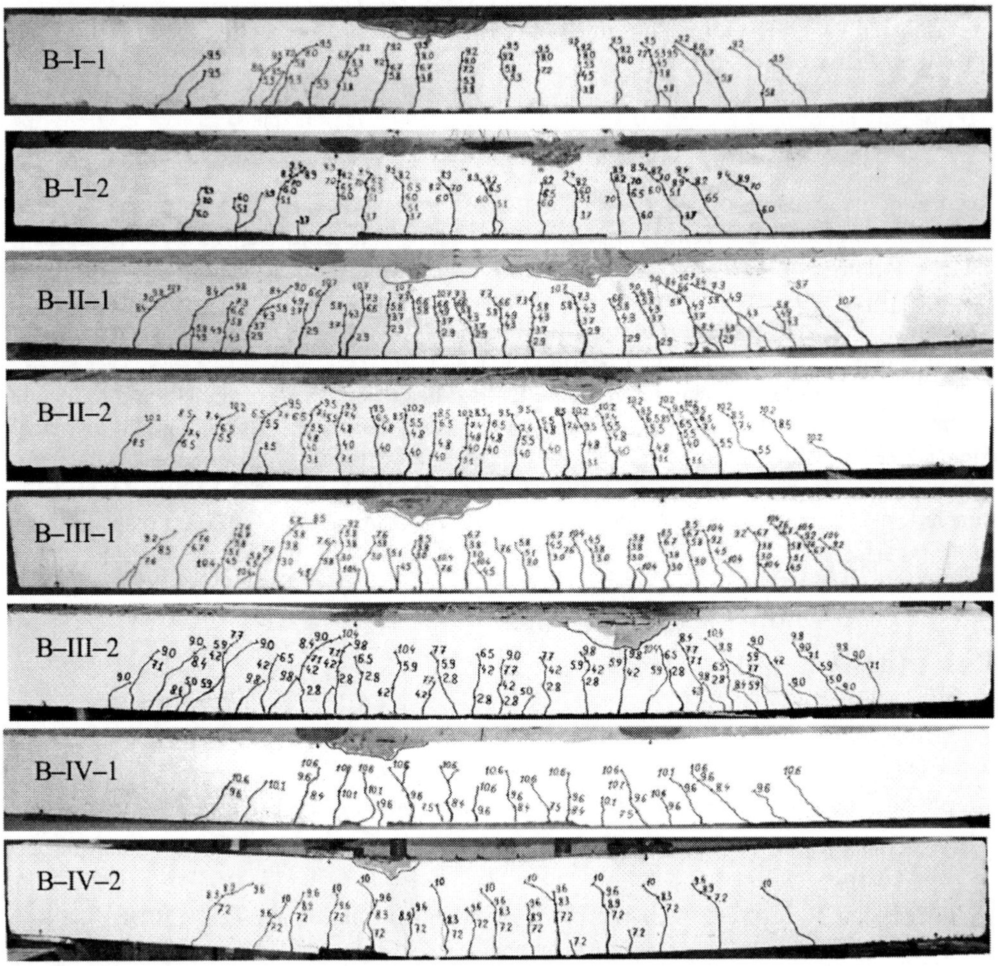

Fig. 2. Destruction pattern of the experimental beams.

For the beam B-IV-1, B-IV-2, the percentage of beam reinforcement was 3.45%. The flow of the band reinforcement was achieved at load M_{drl} = 51.49 kNm for the beam B-IV-1, and 50.57 kNm for B-IV-2.The theoretical calculation, according to [12], assumed the magnitude of the destructive torque $M_{max}^{DBN} = 50.25$ kNm, $M^{DBN} = 49.79$ kNm. Thus, the margin of safety when using the calculated characteristics of materials for this series is 0.6–3.3%.

From the results of the research, it follows that the current regulatory instrument DBN B.2.6-98: 2009 [14], when using the characteristic values of the strength of materials, allows to estimate with sufficient accuracy the strength of both composite steel-concrete reinforcement and reinforced concrete. Deviation from the experimental data does not exceed 7.3%.

Composite steel-concrete beams which have a higher percentage of high-strength rebar, compared with the calculation using the reliability coefficients for materials γ_s, γ_c, have a greater margin of safety. Thus, in composite steel-concrete beams without the high-

strength rebar, the margin of safety was up to 3.3 %.In the beams where the percentage of reinforcement with high-strength rebar was 31.2%, the margin of safety was within the range of 10.0–11.9%. With a percentage of high-strength rebar on the level 61.9%, the safety margin was 10.4–14.9%. In reinforced concrete beams reinforced only by high-strength rebar of class A1000, the safety margin was 13.6–16.2%.
In reinforced concrete beams reinforced only by high-strength rebar's of class A1000, the safety margin was 13.6–16.2%.

Conclusions

According to the results of experimental and theoretical studies, it has been established that the beams with combined reinforcement by high-strength and ordinary steel, after the flow of less stable reinforcement, continue to take the load until a conditional margin of safety for the high-strength rebar is reached. With a decrease in the percentage of high-strength rebar, the deformability of the beam also decreases, but the percentage of reinforcement increases at constant bearing capacity.

The performed experimental and theoretical studies of the work of the stress-strain state of structures allowed to determine that, when using combined reinforcement by steel concrete beams with a span of 2.4 m , the optimal percentage of reinforcement with high-strength rebar of class A1000 is achieved, provided that 25–55% of tensile force is transferred on it. This reduces the percentage of reinforcement by 15–30% correspondingly, while ensuring compliance with requirements for deformability, crack resistance and durability.

References

1. Lu, Y & Hu, L & Liang, H & Li, W. Journal of Building Engineering (2015)
2. Thamrin, Rendy. *Matec Web of Conferences.* **103** (2016)
3. Y. Blikharskyy, R. Khmil, Z. Blikharskyy. *Matec Web of Conferences.* **174** (2018)
4. P. Krainskyi, Y. Blikharskyy, R. Khmil, Z. Blikharskyy. *Matec Web of Conferences.* **183** (2018)
5. J. Selejdak, R. Khmil, Z. Blikharskyy. *Matec Web of Conferences.* **183** (2018)
6. Gene Corley, W. MP, **34** (1995)
7. Abbas Abdulsada, Shaymaa & Al-Mosawi, Ali & Amer, Ali. Biopr. Eng. **1** (2017)
8. Yoon, Seok-Kwang & Lee, Su-Chan & Lee, Do-Hyeong & Lee, Jung-Yoon.. J Korea Concr. Inst., **26** (2) (2014)
9. Dios Garay, Juan & Lubell, Adam. *ASCE* (2008)
10. P. Vegera, R. Vashkevych, Z. Blikharskyy. *Matec Web of Conferences.* **174** (2018)
11. Z. Blikharskyy, R. Khmil, P. Vegera. *Matec Web of Conference.* **116** (2017)
12. Li, J. CEt, **51** (2016)
13. Q. Zhang, Y. V. Mol'kov, Y. M. Sobko, Y. Z. Blikhars'kyi, R. E. Khmil. *Materials Science*, **50** (6) (2015)
14. National Standard of Ukraine, DBN B.2.6 - 98: 2009, 84 (2011)

Development and implementation of algorithms of building structure engineering calculations for shear fraction under pressing-through.

Tetiana Dmytrenko[1,*], *Oleksandr* Semko[1], *Andrii* Dmytrenko[1], *Tetiana* Derkach[1], and *Olena* Voskobiinyk[1]

[1] Poltava National Technical Yuri Kondratyuk University, Pershotravnevyj Ave, 24, Poltava, 36011, Ukraine

Abstract. Development of calculating algorithms for designed and patented constructive solutions for beamless monolithic slab and concrete reinforced column connection in the form of software program is presented in this article. Also a mathematical model for the constructive solutions was created using finite elements method. A theoretical part of the research includes engineering technique for the connection nodes between monolithic concrete slab and reinforced concrete column in stress strain state. A technical goal is simplifying and reliability increasing of loading capacity of the junction and assembly work facilitation along with reducing building frame costs. To perform testing of connection nodes between monolithic reinforced concrete beamless non-bearing slab and reinforced concrete column for shear existing calculating algorithm was improved using the designed formula. A connection node for a shear along the column body calculation according to the technique was also implemented as a software program. A technique for calculating a connection node between monolithic beamless non-capital concrete reinforced slab and concrete reinforced columns was developed. Visual Basic for Applications was used to automate calculating of the connection nodes for shear. The research results have been implemented into practical design and calculation during extension of manufacturing building of Skvira confectionary shop in Skvira settlement.

1 Introduction

Nowadays there are a great number of junction's variations between columns and slabs. The most popular are usage of confinement or rigid reinforcement in the slab and junctions with capitals. However these approaches have some weaknesses, as high material capacity, labor intensity, complexity of calculation and design. [1, 2].

Computer technology using during the engineering calculations performing gives an opportunity to increase the structure's reliability and reduce a designing period. It is necessary to develop a calculation algorithm and it's automation to perform beam-less monolithic slabs calculations.

Usage of steel rectangular pipes, filled by concrete, allows combining of concrete (reinforced concrete) and steel constructions advantages with mutual deficiencies compensation [1], [3].

* Corresponding author: dmitr_tat@ukr.net

2 Analysis of the recent research

Development and application issues of building constructions of monolithic slabs nodal connection calculations using up-to-date computer technologies were investigated by Horodetskyi D.O., Horodetskyi O.S. [2], Perelmuter A.V. [4], Shymkovych D.H. [5] and others [3], [6, 7].

It is necessary to develop a calculation algorithm for monolithic slab under forcing through and automation of the process to provide engineering calculations accuracy.

3 Problem

Punching shear may occur at a column and ceiling junction under load at failure.

Creating a real calculation model, that allows to describe construction resistance against shear effect along with inflection is one of the most complex tasks in reinforced concrete theory. That is why during monolithic beamless slabs design bursting calculation was done using methods for splitted calculating of oblique sections on shear forces and bending moment's performance [8].

There for an assumption about elastic behavior of concrete at forcing through zone under loads close to the boundary shows real work of a junction.

4 Main part

While solving the task a new construction solution of composite reinforced concrete columns nodes connection with a monolithic reinforced concrete beamless slab was designed and patented:

– design of knot in which monolithic reinforced concrete overlapping joins to composite reinforced concrete columns with use of anchor bending's which are welded to composite reinforced concrete columns (Figure 1) [9].

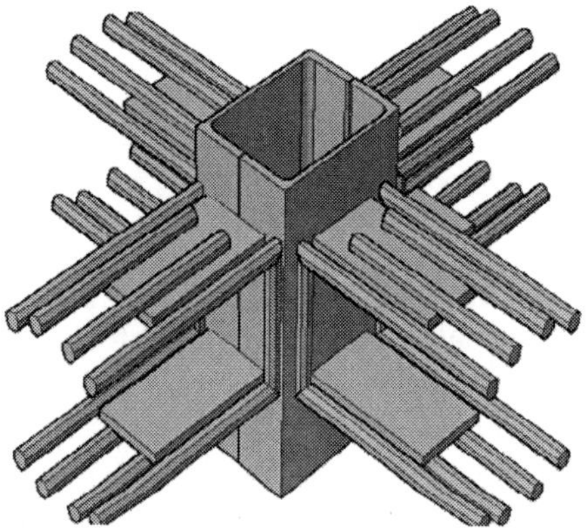

Fig. 1. Metal constructions of a monolithic girder beamless node connection with columns made of beam channels.

– node construction, where monolithic reinforced concrete floor slap is connected to composite column using anchoring bendings, which are welded to composite columns. It provides local stability of a wall of a channel (figure 2) [10].

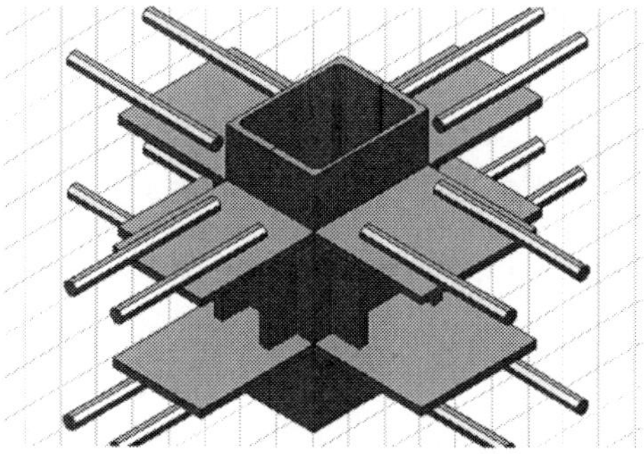

Fig. 2. Metal designs of knot of connection the composite reinforced concrete of columns from channels from a monolithic reinforced concrete girder less a plate

– design of knot in which column bore through openings for drawing of fittings. Diameter of openings is at 3 mm more than the size of fittings. Fittings settle down in two directions, and one above other (Figure 3) [11].

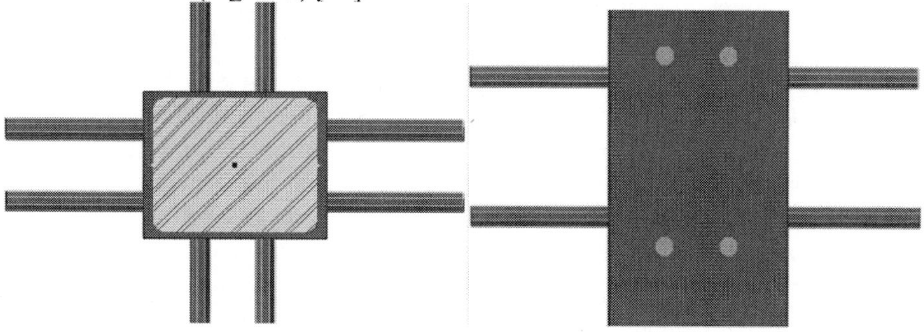

Fig. 3. The scheme of knot of connection monolithic girder less overlapping's with composite reinforced concrete columns from channels

Development of an algorithm of calculation of knots of connection monolithic reinforced concrete girder less overlapping's with composite reinforced concrete a column was carried out step by step:

Stage 1 - Theoretical research

– Analysis of existing design solutions of connection nodes between beamless monolithic slab and reinforced concrete column

– Defining features of connection nodes between beamless monolithic slab and reinforced concrete column

– Research for new design solutions of connection nodes between beamless monolithic slab and reinforced concrete column

– Calculation of reinforcing rods in connection nodes between beamless monolithic slab and reinforced concrete column

– Calculation and design slabs under bursting at the location of the columns within the slab

– Analysis of existing calculation procedures solutions of connection nodes between beamless monolithic slab and reinforced concrete column

– Analysis calculation procedures for the shear under bursting

– Calculation of connection nodes between beamless monolithic slab and reinforced concrete column using finite element method

– Calculation model design by finite element method

– Stress-strain state of the connection nodes numerical research by the finite elements method
– Gusset calculation for composite girder connection node between beamless monolithic slab and reinforced concrete column under bursting
Stage 2 - Experimental research
– The nodal connections durability research for the tension with filling the columns with concrete
– Research of the connection node durability for the tension without filling the columns with concrete
– Research of the reinforcement influence on connection durability without using gussets
– Investigation of bearing capacity of the node connection without using gussets
– Investigation of bearing capacity of the node connection using gussets
– Design of the calculation procedure of the node connections monolithic reinforced concrete beamless overlap with the composite reinforced concrete column
Stage 3 - Calculation algorithm development
– Algorithm of the monolithic reinforced overlap for the share under bursting calculation according to Euro code.
– Algorithm of connection nodes test between monolithic beamless non-capital overlap and reinforced concrete column for the share (according to the introduced procedure)
– To design calculation algorithms of monolithic slab experimental research of suggested new connection nodes constructions (Figure 4) was performed to detect the nature of their destruction and load capacity. A mathematical model of stress-strain state of the connections was also created along with calculating by finite elements method.

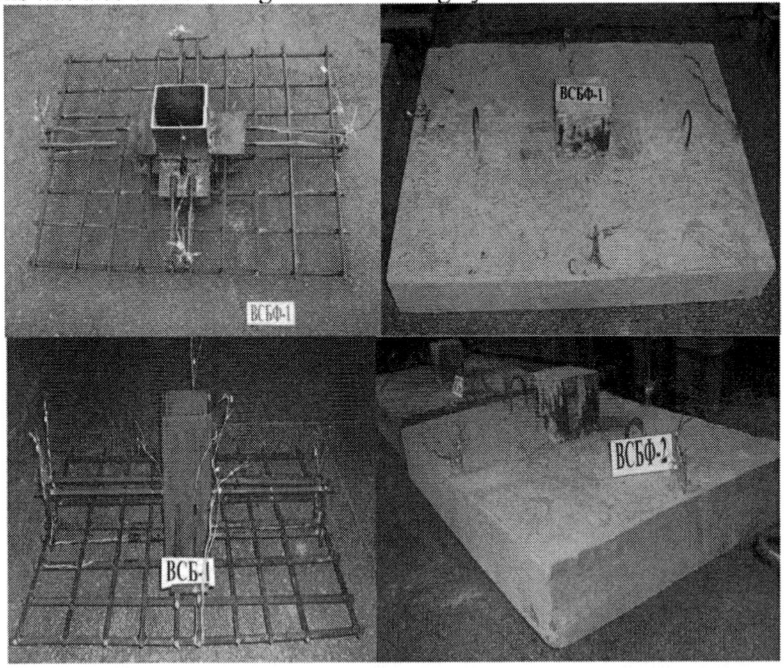

Fig. 1. Introduced new design decisions for connection nodes between monolithic concrete slab and reinforced concrete columns

To compare experimental models with mathematical ones of the nodal connections between a beamless monolithic slab and a composite column computer programs were used. They allowed to decrease a number of full-scale experiments (Femap version 8.1 (2001) as a pre- and postprocessor for MSC.Nastran. Educational demo version SDRC-

Femap 8.1a S/N 000-00-00-demo-406F-00000000) was also used [12].

To perform constructions stress-state analysis by finite element method a calculation schema of nodes junction between beamless monolithic reinforced concrete slab and reinforced concrete column was introduced (figure 5).

As a boundary condition for the connection node, an embedding is done at the base with the command Model → Constraint On Surface. Applied load is equal to 100 кН.

Necessary to mention, that the model isn't enough accurate because different type elements are connected in the node [13].

4-nodes and tetrahedron type finite elements were used to calculate reinforced concrete constructions (figure 6).

Beamless monolithic reinforced concrete slab and reinforced concrete column connection node stress-state calculation was used to compare numerical and experimental research results. The construction destruction is supposed to happen at pyramid bursting area [5].

During experiment was observed a destruction along the compressed area caused by crushing concrete over the cracks that formed the body in the form of a truncated pyramid. The element stress-state under crushing concrete over the cracks features are reinforcement and concrete deformation concentrations zones (figure 7). Concrete deformation zone is could be found next to the top of the crack. Concentration of longitudinal reinforcement deformation zone is at intersection points with the crack, that makes bursting pyramid.

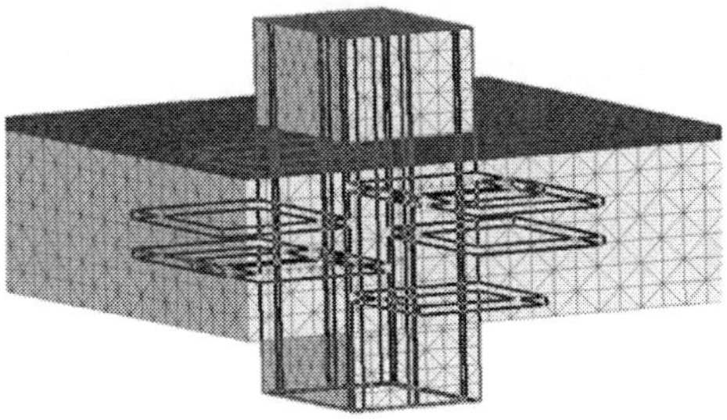

Fig. 5. A calculation schema of nodes junction between beamless monolithic reinforced concrete slab and reinforced concrete column was introduced

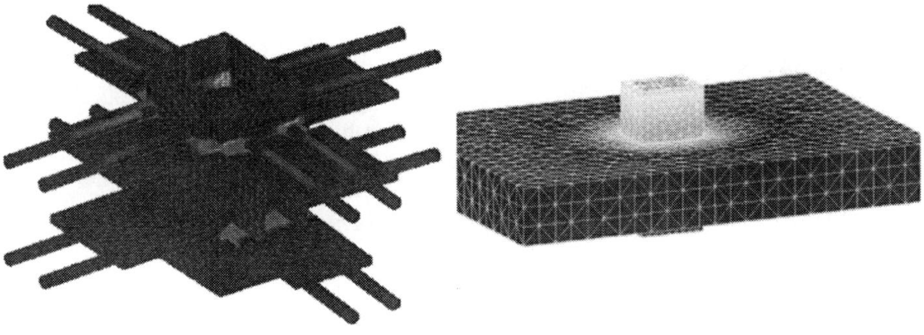

Fig. 6. A graphic representation of the deflected mode calculations results using final elements procedure

During the experiment c/h_0 ratio influence at load bearing capacity of the sample was analyzed. It was performed in planes ZX and ZY. The sample performance in plane ZX is better than in ZY plane. Accordingly c/h_0 ratio in ZX plane is less. In other words load bearing capacity is decreased for 22 % under c/h_0 ratio increasing (table 1).

Based on the performed theoretical and field experiments calculation methodology for connection nodes between beamless monolithic concrete slab and reinforced concrete column was created.

Table 1. Experiment input according to VSB series samples

Plane	Gage	c/h_0	The force at which fracture appeared Q, kN	Destructive power Q_u^e, kN	Burst occurring level Q/Q_u^e
ZY	31-46	4.19	140	300	0.47
ZX	44-45	2.29	180	296	0.61

Where h_o – plate's working length; c – is the projection length on the longitudinal axis of the most dangerous oblique section element.

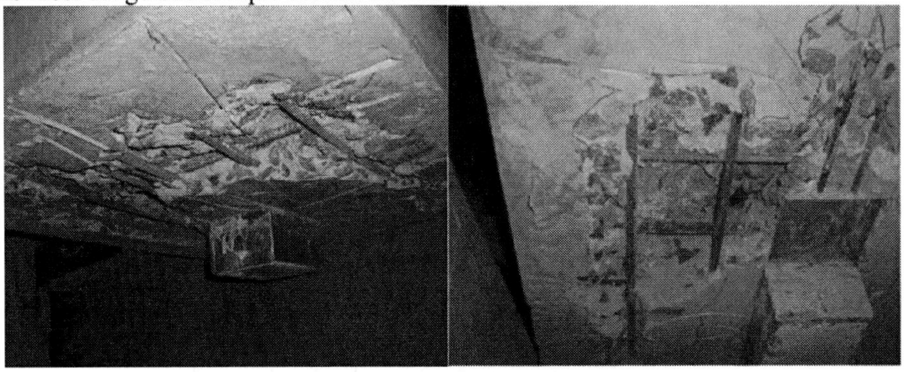

Fig. 7. The samples after testing

Based on the performed theoretical and field experiments calculation methodology for connection nodes between beamless monolithic concrete slab and reinforced concrete column was created.

An introduced procedure of a beamless monolithic slab and a composite column connection node includes two stages:

To check a work of the connection node of a beamless monolithic slab and a composite column during forcing through was developed an improvement of the calculation algorithm according to operative rules using the formula;

A nodal connection between beamless monolithic slab and a composite column calculation for a shear along the column body according to the introduced procedure.

To check a work of a nodal connection for a shear during forcing through it is necessary to determine a length of a design perimeter, using the net formula:

$$u_i = 2(a+b) + 4\pi h_o \quad (1)$$

Where a, b – reinforced concrete column geometric sizes; h_o –working length of the plate.

If applied loading is uneven, a bending moment of plates occur. It is counted in formulas to determine eccentricity of axes Y and Z.

$$\left.\begin{aligned} e_y &= \frac{M_{edy}}{V_{ed}} \\ e_z &= \frac{M_{edy}}{V_{ed}} \end{aligned}\right\} \quad (2)$$

Where e_y and e_z– eccentricity of axes Y and Z;
M_{edy} – bending moment, acting in the plate on axis Y;
M_{edz} – bending moment, acting in the plate on axis Z;
V_{ed} – total lateral force acting on the estimated length of the perimeter.

To determine geometric parameters of the forcing through pyramid perimeter the next formulas can be used:

$$\left.\begin{aligned} b_X &= a + 4 \cdot h_O \\ b_Z &= b + 4 \cdot h_O \end{aligned}\right\} \tag{3}$$

Where *a, b* – geometric sizes of reinforced concrete column;
h_o –working length of the plate.

Loading eccentricity for columns with a rectangular cut is determined using the next formula:

$$\beta = 1 + 1.8\sqrt{\left(\frac{e_y}{b_z}\right)^2 + \left(\frac{e_z}{b_y}\right)^2} \tag{4}$$

If the loading eccentricity is less than 1.15 it is necessary to assume 1.15 and determine total shear force acting on the critical perimeter length:

$$\vartheta_X = \beta \frac{V_{Ed}^2}{u_i d} \tag{5}$$

To determine a ratio of reinforcement assume the next coefficient:

$$k = 1 + \sqrt{\frac{200}{d}} \tag{6}$$

The coefficient shouldn't be more than 2.

Ratio of reinforcement is equal to:

$$\rho_1 = \sqrt{\rho_{1x} + \rho_{1y}} \le 0.02 \tag{7}$$

To determine hardness of a plate without shear reinforcement against forcing through By reinforcement (using the formula introduced by author) [13]:

$$V_{Rd,s} \le 1.125 \cdot f_{ctd} \cdot u \cdot h \cdot \sqrt{1.1 + 0.7\rho} \tag{8}$$

By concrete:

$$V_{Rd,c} = 0.4 \cdot f_{ctd} \tag{9}$$

The larger value is accepted as calculated.

The calculation algorithm for ceiling slabs under bursting by a column was performed in high level programming language Object Pascal in Delphi environment.

According to the algorithm a computer program was developed using programming language Visual Basic for Applications (VBA). The language is used for applications intended to manipulation of databases and users' interface setup development. VBA is a structured high-level programming language. General principles of object-oriented programming are implemented in VBA language.

Software development stages:

1. Issue statement

– The purpose of the software is to check the connection between monolithic beamless non-capital overlap and reinforced columns for the shear using the proposed formula.

2. Issue analysis and design

– At this step inputs and results were determined, revealed constraints on their values, made a formal description of the problem and the mathematical model suitable for solving the problem using computer.

3. Development and the algorithm selection

– Based on the mathematical description of the problem the solving algorithm was created.

4. Design if the general program structure

– The solving model was created with further specification and division into subprograms, defined the software architecture and the way of data storage.

5. Coding

– According to the algorithm software was developed using Visual Basic for Applications (VBA). This programming language is used for applications development, used for interacting with data bases and adjusting the user's interface. VBA is a structured high level programming language. In the VBA language the general principles of object-oriented programming are implemented.

6. Software adjustment and testing

– Performed a verification check of the calculation of connection nodes between monolithic beamless non-capital overlap and reinforced columns for the shear.

7. Results analysis

– Actual results meet the experiment data results.

A programming module used for junction of beamless monolithic slab and a composite column during forcing through testing is shown below (using introduced procedure:

As stated above, it is necessary to perform a calculation of the junctions along the column body according to the procedure.

To determine optimal field performances of rebar the europium method of bracing calculations was used. The method is known as a Concrete Capacity Method.

It includes a concept of hardness estimation using partial safety coefficients.

The main differences of the method are:

Differentiation between collapses mode: an extraction form concrete or steel collapse.

Differentiation between safety coefficients, based on different types of collapse;

According to the method an algorithm of junction between a beamless monolithic slab and a composite column during forcing through calculation along the column body was introduced (figure 8).

According to the algorithm a software program was developed to automate the calculation process (figure 9).

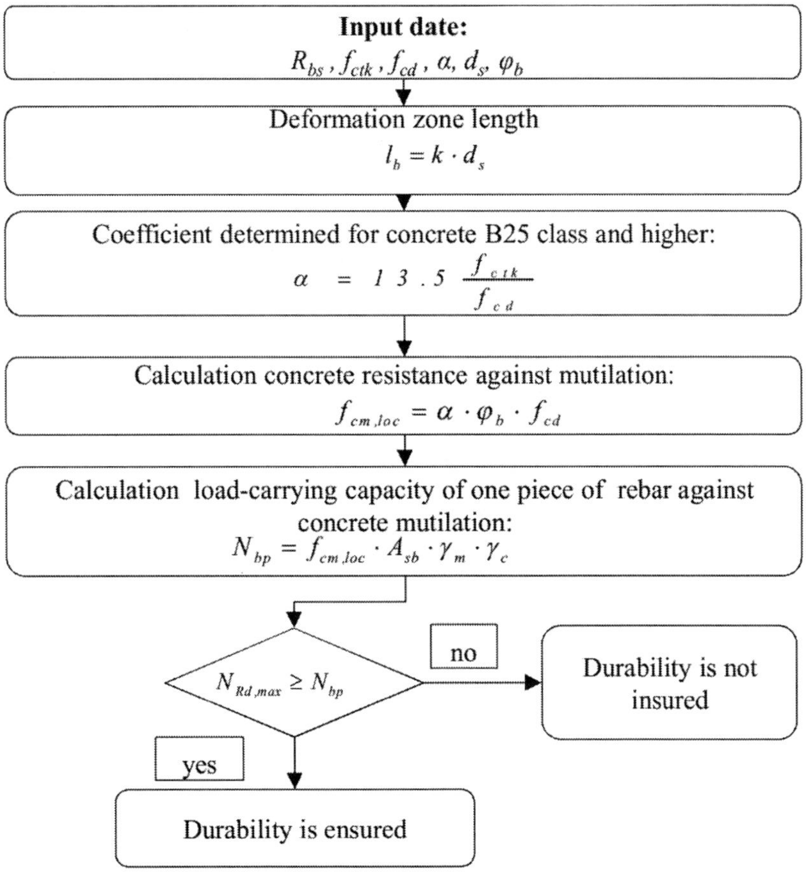

Fig. 8. The algorithm of a junction between a beamless monolithic slab and a composite column during forcing trough calculation (according to the introduced procedure)

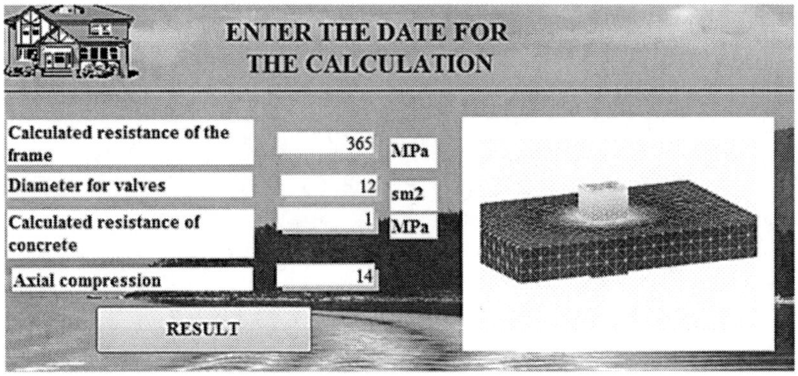

Fig. 2. Software interface of a junction between a beamless monolithic slab and a composite column during forcing through calculation (according to the introduced procedure) calculation

5 Conclusions

1. New constructive solutions of the junction of the monolithic reinforced concrete beams with non-beam overlap with steel reinforced concrete columns offer the possibility of redistribution of the bending moment in comparison with the acting analogues.

2. The model of the stress-strain state is developed and made calculations of the joints between monolithic reinforced concrete non-beam overlap with steel reinforced concrete column by the finite element method is implemented in the Femap software environment. It takes into account the concrete and reinforcement work characteristics, which allows to increase the bearing capacity by 10%.
3. The proposed engineering method for calculating the junction points of monolithic bezel-faced ceilings with steel-reinforced concrete columns, taking into account the features of the stress-strain state, gives more accurate calculation results in comparison with the current norms. The difference between the calculation results for the current and proposed method was 5%.
4. The methodology of calculating the proposed connections includes two stages of calculation: calculation for shear fraction under pressing-through and shearing action along the scape provides the structure load-bearing capacity.
5. The calculation algorithm implemented in Visual Basic for Applications is developed to ensure the accuracy of the engineering calculations based on the method developed by the authors.

References

1. M.A. Eder, R.L. Vollum, A.Y. Elghazouli, Eng. Struct., **36**, 239 – 257 (2012).
2. A.S. Gorodetskiy, I.D. Evzerov, *Computer models of structures*, K.:Fakt, 344 (2006).
3. R.A. Kanishchev, Mag. of Civ. Eng., **4(64)**, 59–68 (2016).
4. A.V. Perelmuter, V.I. Slivker, *Calculation models of structures and the possibility of their analysis*, Stal, 710 (2011).
5. D.G. Shymkovich, *Calculation of structures in MSC Nastran for Windows*, M: DMK Press. 447 (2003).
6. A.A. Pisano, Fuschi, P., D.D. Domenico, Comp. & Struct., **160**, 42–55 (2015).
7. N.I. Vatin, A.D. Ivanov, *Pairing of columns and slabs without edges and without capitals monolithic in a reinforced concrete frame building*, Sankt-Peterburg, 82 (2006).
8. D. Zenunovica, R. Folic, Eng. Struct., **40**, 466–478 (2012).
9. V.O. Semko, T.A. Dmitrenko, *The unassembled knot of the connection of a monolithic bulkhead floor with columns of channels*, Pat. 41483 Ukrayina, MPK E 04 B 5/32, Bul. **6**, 4 (2009).
10. V.O. Semko, T.A. Dmitrenko, *Connection unit of monolithic unbaked bulkhead overlap with columns of channels*, Pat. 48566 Ukrayina, MPK E 04 B 5/32, Bul. **10**, 4 (2009).
11. V.O. Semko, T.A. Dmitrenko, A.O. Dmitrenko, *Connection unit of monolithic overlap with steel-reinforced concrete columns*, Pat. 59155 Ukrayina, MPK E 04 B 5/32 (2006.1), Bul., **9**, 4 (2011).
12. S.Yu. Fialko Mag. of Civ. Eng, **5 (40)**, 116–124 (2013).
13. T. A. Dmitrenko, A. O. Dmitrenko, T. M. Derkach, *Computer simulation of robots joints for design on the punching shear*, Resource-efficient technologies in the construction complex of the region, 232–237 (2014).

Bearing capacity calculation of reinforced concrete corbels under the shear action

Oksana Dovzhenko[1,*], *Volodymyr* Pohribnyi[1], *Volodymyr* Pents[1], and *Dmytro* Mariukha[1]

[1] Poltava National Technical Yuri Kondratyuk University, Pershotravneviy sq., 24, Poltava, 36011, Ukraine

Abstract. The necessity of creating a general methodology for concrete and reinforced concrete elements strength calculation under the shear is established. Three failure cases of reinforced concrete corbels under the shear are considered. The solutions of problems of corbels strength with failure along the whole section, close to the normal, in the compressed zone under an inclined crack and within the compressed inclined strip are given. A variational method in the plasticity theory, the virtual velocities principle and the characteristic lines method are used for concrete and reinforced concrete elements calculations. In the ultimate state, concrete is considered as a rigid-plastic body. The shear is realized in case when the plastic deformation is localized in the compressed zone. The calculating ultimate load results for different failure cases are given. Such a design scheme is implemented, in which the console strength is minimal. This corresponds to the minimum of power of plastic deformation in concrete compressed zone. Reinforced concrete corbels calculation engineering methods are offered. The elements obtaining effective constructive decisions direction based on the plasticity theory use is determined.

1 Introduction

Concrete and reinforced concrete elements that work on the perception of joint action of transverse forces and bending moments are widely used in practice. These include the beam and slab elements supporting areas, column corbels, T-beams and I-beams flanges, base plates cantilevered parts, pile grillage, double-girder columns lintels and other structures. The reinforced concrete supporting corbels are used in the railway transport structures for the superstructure support.

An important task is to increase the efficiency of their work and reduce the material consumption of constructive solutions used in the construction, reconstruction and structures restoration. One of the components in this direction is the methods for calculating bearing capacity improvement.

Research articles [1-4] are devoted to the strength problem of reinforced concrete corbels. The ultimate stress-strain state is investigated, experimental data are presented and their calculation methods analysis is performed in these articles. Existing calculation methods have a significant number of disadvantages, namely: the elasticity theory methods

* Corresponding author: o.o.dovzhenko@gmail.com

do not correspond to the elements actual work in the ultimate state; the empirical approach is cumbersome, the obtained results are influenced by the experimental conditions, it is not always possible to take into account the full number of determinants; frame-core model does not consider all possible cases of failure at the various geometric parameters of elements, loading conditions, reinforcement, etc.

2 Unsolved part of the problem

The creation of general methodology for calculating bearing capacity of reinforced concrete corbels is an urgent task. This methodology should cover various failure forms, determine their implementation boundaries, take into account the operation specifics in the ultimate state, allow to obtain constructive solutions with optimal longitudinal and transverse elements reinforcement equal to the normal and inclined sections, and allow to reduce material capacity of structures and to exclude cases of strength revaluation.

The specificity of work of these elements lies in the fact that fields, strips, intense strains and stresses concentration lines are formed in the power flows action direction [5, 6], the which presence is confirmed by experimental research [7-9]. Their placement is determined by the external forces and supporting reactions of the place of application. The plastic deformation concentration and the biaxial compression zones presence permits to use the variational method of concrete plasticity theory [10, 11] and the method of characteristics [12] as the general methodology basis for calculating the concrete and reinforced concrete elements bearing capacity.

The work purpose is to improve the methods for calculating the reinforced concrete corbels bearing capacity under the shear on the unified approach basis with the plasticity theory use.

3 Calculating methods for corbels based on the plasticity theory

In this paper, three failure cases of reinforced concrete corbels under shear are considered: for an integral section close to the normal, in a compressed zone under an inclined crack and for a sloping compressed strip.

In case of $a/h \leq 0.15$ and shear spans $c/h \leq 0.3$, simultaneous failure of the reinforced concrete corbel in the compression and tension areas is observed – failure along the whole cross-section (Fig. 1 a). In this case, the concrete plastic deformation is localized in a compressed zone in a thin layer on the displacement surface, and the stress in the tensile zone reaches the tensile strength values.

With the increase of parameter a/h and the impact of bending moment, the stress level of relative limit values in the tensile zone of corbel is ahead the stress level in the compressed zone. First a crack appears near the edge of the loading platform, which is close to the corbel fixed and. It grows and at a transverse reinforcement slight intensity and yield point stresses in the reinforcement, a compressed concrete zone shear is realized (Fig. 1 b).

In the case of a strong transverse reinforcement sufficient to prevent the first crack development, and with the c/h that are short corbels characteristic, a second crack emerges that extends from the outer facet of the loading platform to the corbel lower corner. There is a compressed inclined strip and a shear within the strip is implemented (Fig. 1 c).

To determine the reinforced concrete corbels bearing capacity, the kinematic schemes and the stresses distribution on the failure surface along the whole section (Fig. 2 a) are considered. The strength problem of concrete wedges, which simulate the compressed zone under an inclined crack (Fig. 2 b), and the strength problem of compressed inclined strip (Fig. 2 c) are solved.

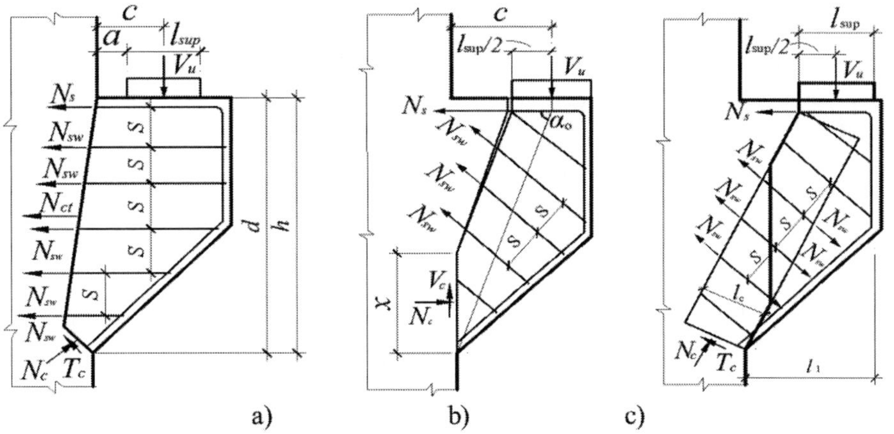

Fig. 1. Design scheme of reinforced concrete corbel: a) with failure for whole section; b) with compressed zone under inclined crack shear; c) with compressed strip shear.

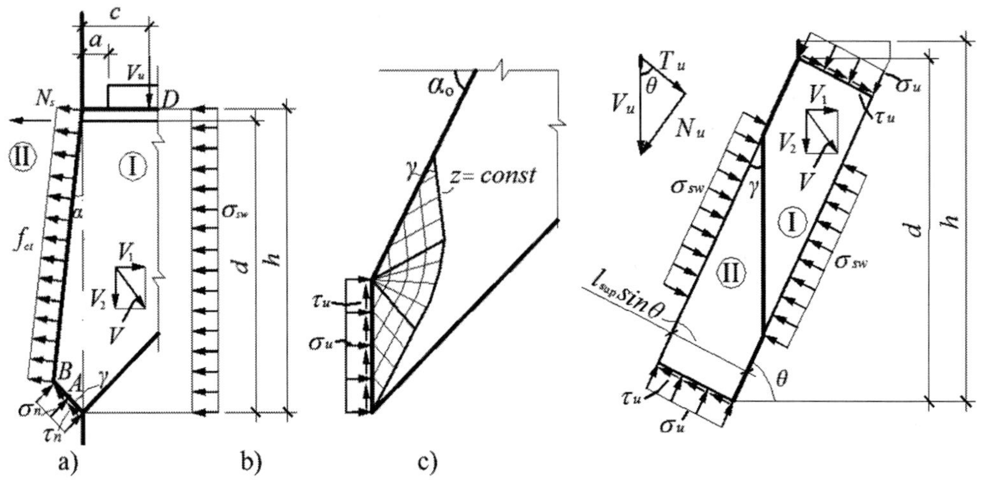

Fig. 2. For calculating the reinforced concrete corbels strength: a) with whole section failure; b) with compressed concrete wedge under an inclined crack shear; c) with shear within the compressed inclined strip.

A variational method is used in the plasticity theory and functional of principle of virtual velocities [5], which, in the inertial and mass forces absence, has the form

$$J = \int_{S_l} (W_c + W_s)\, dS - \int_{S_f} f_i^* V_i \, dS, \tag{1}$$

where W_c and W_s – concrete and reinforcement deformation power; S_l – failure surface; S_f – surface on which the forces (stresses) f_i^* are given at velocities V_i. The reinforcement effect is considered as a pre-stressing.

The functional is investigated for the stationary state using the equation $\delta J = 0$. Parameter $k = V_1 / V_2$ is varied in the orthogonal directions and the inclination angles of areas of failure surface α and γ : $\partial k / \partial V_u$, $\partial tg\alpha / \partial V_u$ and $\partial tg\gamma / \partial V_u$ [13, 14]. Balandin–Geniev condition for the concrete strength [12, 15] is accepted. The failure occurs by moving one rigid disk I to another II (Fig. 2 a, b). The ultimate load magnitude corresponds to the deformation power minimum, taking into account the reinforcement work.

By the failure along the whole cross-section, the relative transverse force that the reinforcement concrete corbel perceives is determined by the formula

$$\frac{V_u}{f_c bh}=\frac{\left[\sqrt{(1-\chi+\chi^2)/3}\sqrt{4(k-tg\gamma)^2+(1+ktg\gamma)^2}-(1-\chi)(k-tg\gamma)\right]tg\alpha+\chi tg\gamma(k+tg\alpha)}{tg\alpha+tg\gamma}+ \quad (2)$$
$$+k(\sigma_s+\sigma_{sw})/f_c,$$

where $\chi = f_{ct}/f_c$; $\sigma_s = \rho_f f_y$, here $\rho_f = \frac{A_s}{bh}$; $\sigma_{sw} = \rho_w f_y$, $\rho_w = \frac{A_{sw}}{bs}$,

when applying equilibrium equations

$$\sum M_A = 0; \sum M_B = 0; \sum M_D = 0. \quad (3)$$

Normal and tangential stresses that act on the failure surface compressed section and are used in equilibrium equations, are equal to

$$\frac{\tau_n}{f_c} = \pm\frac{\sqrt{(1-\chi+\chi^2)/3}\left(1+ktg\gamma\right)}{2\sqrt{\left(k-tg\gamma\right)^2+0.25\left(1+ktg\gamma\right)^2}}, \quad (4)$$

$$\frac{\sigma_n}{f_c} = 1-\chi-\frac{2\sqrt{(1-\chi+\chi^2)/3}\left(k-tg\gamma\right)}{\sqrt{\left(k-tg\gamma\right)^2+0.25\left(1+ktg\gamma\right)^2}}. \quad (5)$$

The obtained results in solving the problems of corbel strength in the failure case in the whole section are shown in Table 1.

Table 1. Ultimate value of relative transverse force that is perceived by the corbel under the shear across the whole cross-section.

$\frac{f_{ct}}{f_c}$	$\rho_s\frac{f_y}{f_c}$	$\frac{V_u}{f_c bh}$ by c/h								
		0.15			0.2			0.25		
		$\sigma_{sw}=\rho_w f_y/f_c$								
		0.025	0.05	0.075	0.025	0.05	0.075	0.025	0.05	0.075
0.05	0.075	0.353	0.375	0.395	0.314	0.335	0.349	0.281	0.301	0.319
	0.1	0.4	0.42	0.438	0.359	0.378	0.395	0.324	0.342	0.358
	0.125	0.443	0.46	0.476	0.401	0.417	0.433	0.364	0.38	0.394
0.1	0.075	0.394	0.412	0.429	0.353	0.371	0.387	0.318	0.335	0.351
	0.1	0.435	0.452	0.467	0.393	0.409	0.423	0.357	0.372	0.386
	0.125	0.474	0.488	0.502	0.431	0.445	0.457	0.393	0.406	0.418
0.15	0.075	0.427	0.442	0.456	0.385	0.4	0.413	0.349	0.363	0.376
	0.1	0.464	0.478	0.49	0.422	0.435	0.447	0.384	0.367	0.408
	0.125	0.499	0.511	0.522	0.455	0.467	0.478	0.417	0.428	0.438

Unknown variables of strength problem when shearing a compressed zone under an inclined crack: V_u, N_c, V_c, N_s, x.

Equation 1) $\sum X = 0$; 2) $\sum Y = 0$; 3) $\sum M = 0$; 4) $x = 0,5 l_{\sup} d / c$;

$$5)\ V_c = \varphi_1 f_c bx + \varphi_2 N_c, \tag{6}$$

where φ_1 and φ_2 are parameters that depend on the concrete class and the ratio d / c.

The truncated wedge strength can be estimated by the method of characteristic lines [13]. Using the known uniaxial compression stress state in the plasticity triangular area near the facet, formed by an inclined crack, the characteristics properties $z = \text{const}$, force dependence (6) between V_c and N_c on the truncated facet is obtained.

In the case of application of virtual velocities principle [10, 16], the dependence (6) has the following form

$$N_c = \frac{f_c bx\left[\sqrt{(1-\chi+\chi^2)/3}\sqrt{4(k-tg\gamma)^2+(1+ktg\gamma)^2}-(1-\chi)(k-tg\gamma)\right]\left[1+tg\alpha_0/(tg\gamma-tg\alpha_0)\right]}{tg\gamma} - kV_c. \tag{7}$$

The forces ultimate values that are perceived by the compressed zone concrete wedges, are given in Table 2.

Table 2. Ultimate value of relative transverse and longitudinal forces that are perceived by concrete wedges at $\alpha_0 = 45°$ and $f_{ct} / f_c = 0.1$.

Relative efforts	Loading direction angle $\beta = arctg\ \tau_n / \sigma_n$, °							
	10	15	20	25	30	35	40	45
$\frac{N_c}{f_c bx}$	1.8	1.5	1.27	1.05	0.887	0.737	0.615	0.516
$\frac{V_c}{f_c bx}$	0.317	0.403	0.461	0.495	0.512	0.515	0.516	0.516

Unknown variables of strength problem under the shear in the inclined strip: V_u, N_c, T_c.
Equation:

$$1)\ \tau_u = \frac{T_c - N_s \sin\theta}{bl_c} = \frac{T_c}{bl_c} - \frac{A_s E_s}{bl_{\sup} E_c} f_c, \tag{8}$$

where $T_c = \tau_n bl_c = \left\{\frac{f_c\left[\sqrt{(1-\chi+\chi^2)/3}\sqrt{4(k-tg\gamma)^2+(1+ktg\gamma)^2}-(1-\chi)(k-tg\gamma)\right]+k\sigma_{sw}}{(tg\theta+k)tg\gamma}\right\} bl_c$;

$$2)\ N_c = \sigma_u bl_c = \left\{\frac{f_c\left[\sqrt{(1-\chi+\chi^2)/3}\sqrt{4(k-tg\gamma)^2+(1+ktg\gamma)^2}-(1-\chi)(k-tg\gamma)\right]+k\sigma_{sw}}{tg\gamma} - k\tau_n\right\} bl_c; \tag{9}$$

$$3)\ V_u = N_c l_{\sup} / l_c. \tag{10}$$

The ultimate values of forces that are perceived by the inclined compressed element are given in Table 3.

Of the above kinematic schemes of failure, the one in which the bearing of corbel is minimal is implemented.

Table 3. Relative effort ultimate value perceived by reinforced concrete corbel under the shear by inclined strip.

$\frac{f_{ct}}{f_c}$	$\frac{\sigma_{sw}}{f_c}=\frac{\rho_w dE_{sw}}{l_c E_c}$	$\frac{V_u}{f_c b l_{\sup}}$ by								
		Loading direction angle $\beta = arctg\ \tau_n / \sigma_n$, °								
		5	10	15	20	25	30	35	40	45
0.1	0.05	0.969	0.866	0.771	0.684	0.605	0.532	0.464	0.403	0.346
	0.1	1.04	0.933	0.834	0.742	0.657	0.578	0.505	0.437	0.375
	0.15	1.11	0.995	0.891	0.794	0.704	0.62	0.541	0.468	0.401

The engineering calculation of bearing capacity of reinforced concrete corbels is proposed to be carried out using formulas (11) and (12), based on the results of solving the strength problems of the truncated wedge and the compressed inclined strip.

Output data for solving the problems of bearing capacity of reinforced concrete corbels: loading area length $l_{\sup}$; distance from the column facet to the force application place c; width, height, working height and corbel span b, h, d, l_1; concrete class C (f_{cd}, E_c); longitudinal and transverse reinforcement (E_s, f_{ywd}, A_s, A_{sw}, s).

Bearing capacity of the corbel under the shear of the compressed zone under an inclined crack is equal to

$$V_{Rd} = \frac{f_{cd} bdn(0.5\varphi_1 l_{\sup} / c + \varphi_2 \alpha_{sw} + \alpha_s)}{n - \varphi_2 c / d}, \tag{11}$$

where $n = 1 - 0.25 l_{\sup} / c$; $\varphi_1 = \overline{\varphi}_{c1} / (1 + \overline{\varphi}_{c1} tg\alpha)$, here $\overline{\varphi}_{c1} = 0.75$ at the concrete class C20/25 and less and $\overline{\varphi}_{c1} = 0.75 f_{c,C20/25} / f_c$ at the concrete class more C20/25, $tg\alpha_o = d / c$; $\varphi_2 = tg(\alpha_o - \gamma)$, where $\gamma = arctg\,\overline{\varphi}_{c1}$; $\alpha_{sw} = \frac{f_{ywd} A_{sw}}{f_{cd} bs}$; $\alpha_s = 0.9\sqrt{\frac{A_s}{bd} - 3.5\times10^{-3}} \geq 0$.

Bearing capacity under the shear within a compressed strip

$$V_{Rd} = \alpha_{c,s} f_{cd} b l_{sup} (1 + \alpha_{sw}), \tag{12}$$

where $\alpha_{c,s} = 1 - \frac{l_1}{d} + t\frac{A_s}{bd}\sin\theta$, here $t = 1.5\frac{E_s d}{E_c l_c}$, $l_c = l_{sup}\sin\theta$, $\sin\theta = \sqrt{\frac{d^2}{d^2 + l_1^2}}$; $\alpha_{sw} = t\frac{A_{sw}}{bd}$.

4 Conclusions

There is a need to create a sufficiently general methodology for calculating the bearing capacity of concrete and reinforced concrete elements under the shear that would cover various cases of failure and allow obtaining effective structural solutions.

In reinforced concrete corbels, the shear is realized at failure along the whole section, close to the normal, in a compressed zone under an inclined crack, within the inclined strip.

It is proposed to use the mathematical apparatus of plasticity theory as the basis of the determining methods of bearing capacity of reinforced concrete corbels. In this case, concrete is considered as a rigid-plastic body, a variational method using the virtual velocities principle and the characteristic lines method is used. Plastic deformations are considered to be localized in thin layers on the failure surface. The ultimate load corresponds to the plastic deformation power minimum.

References

1. Layla A. Gh. Yassin, Eyad K. Sayhood, Qais Abdul Majeed Hasan, JMES, **2**, 180-205, (2015)
2. D. L. Araújo, A. P. Silva Neto, F. A. Lobo, M. K. El Debs, IBRACON Structural and Materials Journal, **9(3)**, 435-470, (2016)
3. R. M. F. Canha, D. A. Kuchma, M. K. El Debs, R. A. Souza, Engineering Structures, **74**, 130-144, (2014)
4. M. Lachowicz, K. Nagrodzka-Godycka, Engineering Structures, **108**, 1-11, (2016)
5. L. O. Doroshkevych, B. Iu. Maksymovych, Mizhvidomchyi nauk.-tekhn. zb. nauk. prats (budivnytstvo), Kyiv, **59(1)**, 57-63, (2003)
6. V. Pohribnyi, O. Dovzhenko, L. Karabash, I. Usenko, Web of Conferences, **116**, 02026, (2017)
7. L. B. Kriz, C. H. Raths, Journal of the Prestressed Concrete Institute, **10(1)**, 16-61, (1965)
8. O. Dovzhenko, V. Pohribnyi, I. Yurko, I. Shostak, Web of Conferences, **116**, 02011, (2017)
9. O. Dovzhenko, V. Pohribnyi, L. Karabash, Journal of Engineering & Technology (UAE), **7 (3.2)**, 354-359, (2018)
10. V. L. Kolmogorov, *Mehanika obrabotki metallov davleniem*, Moskow, 688, (1986)
11. V. P. Mitrofanov, Stroitelnaya mehanika i raschet sooruzheniy, Moskow, **6**, 23-28, (1990)
12. G. A. Geniev, V. N. Kissyuk, G. A. Tyupin, *Teoriya plastichnosti betona i zhelezobetona,* Moskow, 316, (1974)
13. O. A. Dovzhenko, V. V. Pogrebnoy, Ju. V. Chursa, Promyshlennoe i grazhdanskoe stroitel'stvo, Moscou, **2**, 70-74, (2017)
14. O. A. Dovzhenko, V. V. Pohribnyi, L. V. Karabash, Science & Technique, **17(2)**, 146-156, (2018)
15. O. Dovzhenko, V. Pogrebnyi, I. Yurko, News NAS RK. Series of geology and technical science, **2 (428)**, 212-219, (2018)
16. V. Pohribnyi, O. Dovzhenko, O. Maliovana, International Journal of Engineering and Technology(UAE), **7 (3.2)**, 19-26, (2018)

Increasing the reliability of engineering buildings and strengths of use of new small-based equipment for concrete works

Inga Emeljanova[1], *Mykola* Derevyanko [1], *Sergiy* Guzenko[1], *Denys* Chayka[1,*] and *Dmytro* Subota[1]

[1]Kharkiv National University of Civil Engineering and Architecture, Sumska st. 40, 61002 Kharkov, Ukraine

Abstract. The article considers the possibility to increase the reliability of engineering buildings and facilities due to the use of new small-based equipment both in the operating conditions of technological complexes and in the conditions of technological sets. An example of the use of a technological complex with a universal hose concrete pump for strengthening the supporting columns of a medical centre being built by the wet concrete spraying method is given. The structural scheme of energy consumption by technological sets of small-based equipment is given. The variants of using the technological set of small-sized equipment for concrete work in various ways. One of the universal technological kits of effective small-based equipment is shown. As part of the proposed kit, the base machine can successfully be used universal non-piston hose concrete pump. Dependencies for determining the productivity of a hose concrete pump and the cost of its power for transporting a concrete mixture are proposed.

1 Introduction

Up-to-date construction assumes various types of equipment which are used as separate machines [1 – 5] when performing concrete works. A rational solution in the performance of concrete work in the construction site [6] is the use of certain types of equipment in both technological complexes and technological kits.

Technological complexes consist of separate types of machines (equipment), which consistently perform operations of the technological cycle in the manufacture of reinforced concrete structures or the erection of buildings or structures from monolithic reinforced concrete directly in the construction site.

Considering the concreting processes under such conditions [7, 8], a new universal small-based equipment, which is patented in Ukraine [9], attracts attention and passed approbation at many construction sites when performing works by the wet concrete spraying method.

New small-based equipment that works in the conditions of technological complexes or kits includes the following:

- concrete mixers operating in a cascade mode (trivalent and gravity-forced action);
- mortar-concrete pumps (with ball and poppet valves);
- concrete pumps (universal non-piston hose);
- robotic nozzles for concrete spraying method with a ring nozzle.

* Corresponding author: d.chayka93@gmail.com

One of the factors determining the efficiency of small-based equipment is its energy consumption. The completeness of a particular technological complex or kits in the conditions of a construction site, first of all, should ensure its operation with minimal energy consumption in the presence of a basic machine.

One of the last facilities, where the efficiency of the technological complex with the new equipment mentioned above was confirmed, is the construction of the medical center "OnClinic" (Kharkov).

Shotcrete works were carried out according to the technological scheme shown in Fig. 1.

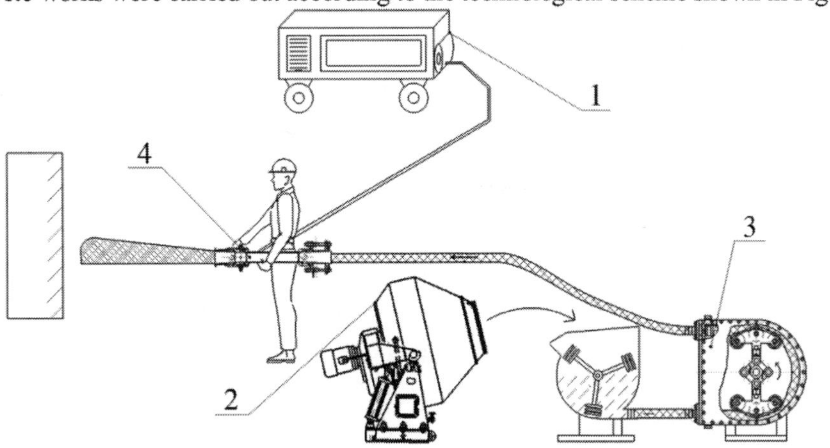

Fig. 1. Technological scheme of performance of concrete gun works by wet method when using a complex of small-based equipment with universal non-piston hose concrete pump. 1 – compressor mobile installation; 2 – gravitational concrete mixer; 3 – non-piston hose concrete pump; 4 – gunned nozzle with a ring nozzle.

The basic machine in this complex is a non-piston hose concrete pump which is shown below on the construction site of the medical centre "OnClinic".

Fig. 2. Universal non-piston hose concrete pump with hydraulic drive on the territory of the "OnClinic" medical centre under construction.

According to the method of wet concrete spraying, the supporting columns of the building were reinforced. At the same time, the following consumption of constituents per 1 m^3 of concrete was used: sand – 640 kg, cement – 340 kg, crushed stone of fraction up to 5 mm – 1240 kg, water – 170 l. Polypropylene fiber - 1.5% (fiber length l = 12 mm); plasticizer Sika plus Mix – 0.2%.

Figure 3 shows the process of reinforcing columns using a nozzle with a ring nozzle. It is seen that the nozzle also produces a concentrated jet of concrete mixture practically without rebound.

Fig. 3. Strengthening of the column during concreting by the wet concrete spraying method.

More rational is the use of small-based equipment in the conditions of the technological kit.

The structural scheme of energy consumption by technological kits of small-based equipment is shown in Fig. 4 [10].

The technological kit has a common frame on which certain types of equipment are mounted, which are independent modules, which, when performing concrete works, allow all operations of the cycle to be combined in time.

It can be considered that the technological kits are assembled from separate modules, which are machines (equipment) presented in this scheme. Each of the modules is characterized by separate capabilities that are used in specific construction conditions [11].

The use of such modular kits for the manufacture of reinforced concrete structures and products in the construction site using the method of off-shuttering concreting is of particular use.

Thus, new small-based equipment in various combinations in the form of modules can effectively work as part of universal technological kits and be used for various types of works. For example, for:

- performing spray-concrete works using the wet concrete spraying method: "metering unit → concrete mixer operating in cascade mode → two-piston or hose-type concrete pump → working nozzle with annular nozzle → concreting surface".
- off-shuttering concreting by wet concrete spraying using fibrous concrete mixtures for the manufacture of reinforced concrete structures of complex geometric shapes: "Fiber

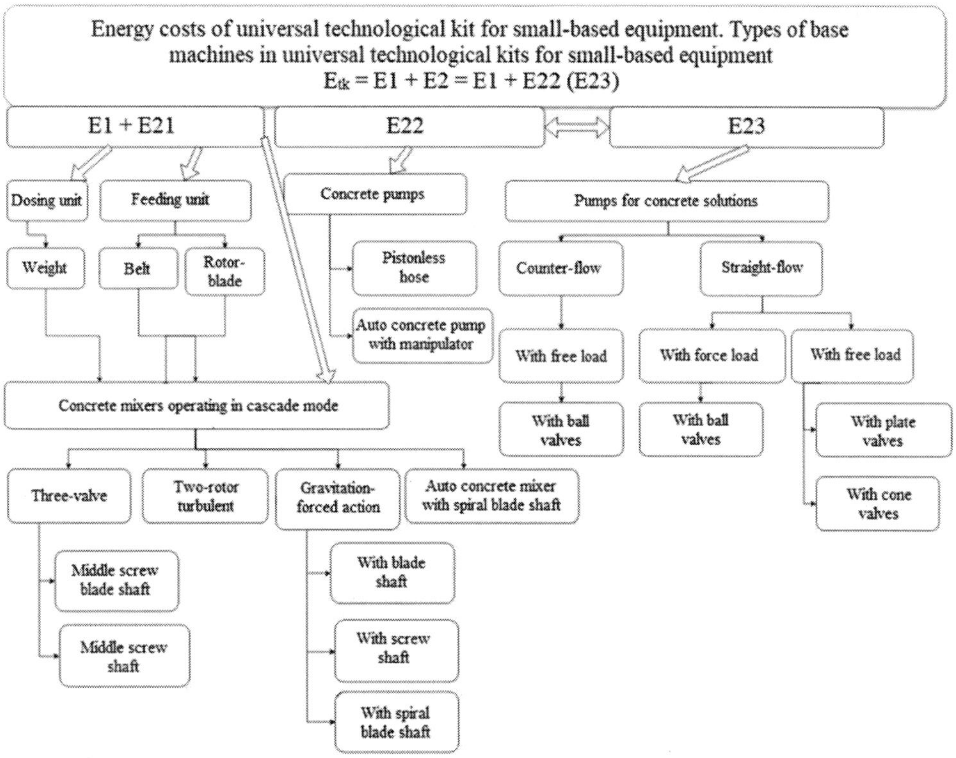

Fig. 4. Structural scheme for determining the power costs of a universal technological kit of small-based equipment for the conditions of the construction site.

automatic cutter → dosing unit for concrete mix components → three-shaft concrete mixer → non-piston hose concrete pump → concrete spraying nozzle with ring nozzle → concrete frame with embedded structures details".

Fig. 5 shows one of the universal technological kits of effective small-based equipment.

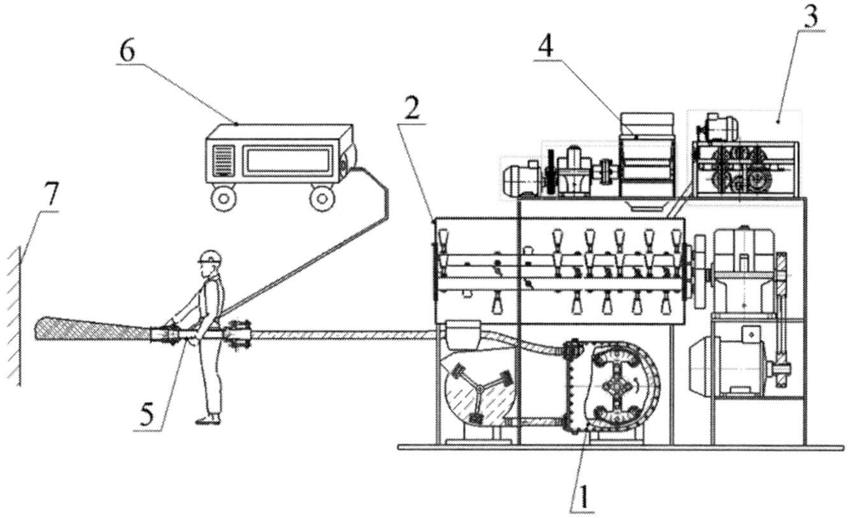

Fig. 5. Universal technological kit for small-based equipment with a pistonless hose concrete pump. 1 – universal pistonless hose concrete pump; 2 – three-shaft concrete mixer; 3 – machine-cutter of fiber; 4 – belt feeder; 5 – concrete spraying nozzle with a ring nozzle; 6 – mobile compressor unit; 7 – concrete spraying surface.

Basic concrete pump with hydraulic drive, in comparison with two-piston concrete pumps, has a number of advantages: minimum weight with a wide range of performance, reliability of construction and improved operating conditions due to the improvement of its design in comparison with existing hose concrete pumps, economic energy consumption.

2 Theoretical part

Performance of a hose concrete pump which, as a basic machine, determines the productivity of the entire technological kit is defined as [12]:

$$\Pi_{tech} = 3600 \cdot F_{hose} \cdot v_{av} \cdot k_1 \cdot k_2 \cdot k_3, \tag{1}$$

where, F_{hose} – cross-sectional area of the hose in the concrete pump housing, m^2; v_{av} – average speed of the concrete mixture through a flexible hose, m/s; k_1 – coefficient that takes into account the gradual build up of the force created by the pinch rollers of the rotor when compressing the outside of the hose in the pump casing; k_2 – coefficient that takes into account the reliability of the hose in the pump casing, taking into account the ultimate stress state; k_3 – coefficient that takes into account the conditions for the mixture to be delivered by a concrete pump via a flexible hose, taking into account the presence of its reverse currents.

Coefficient
$$k_1 = \frac{G_1 + G_2 + G_3}{G_1 + G_3}, \tag{2}$$

where, G_1, G_2, G_3 - respectively, the weight of the concrete mixture resisting the influence of the lateral and central rollers of the rotor on the outer surface of the deformable hose, through which the mixture is fed into the transport main.

Coefficient
$$k_2 = \frac{\sigma_{eq}}{[\sigma]_g}, \tag{3}$$

where, σ_{eq} – total all-round voltage that the hose in the concrete pump casing undergoes as the next portion of the concrete mix moves under it under the action of the pinch rollers of the rotor, MPa; $[\sigma]_g$ – allowable stress on hose breakage, MPa.

Coefficient
$$k_3 = \frac{m_{outl}}{m_{inl}}, \tag{4}$$

where, m_{outl} – weight of concrete mixture at the outlet of the working part of the concrete pump hose; m_{inl} – weight of the concrete mixture that enters the concrete pump from the hopper.

The following dependence can be used to determine the cost of power universal hose concrete pump:

$$P_{dr} = k_{p.r} \cdot \Pi_{tech} \cdot \rho_{0c.p} \cdot v_{av}^2 / 1000 \cdot \eta_{gen}, \tag{5}$$

where, $k_{p.r}$ – coefficient that takes into account the power reserve for overcoming resistance to movement of the mixture through the pipeline; $\rho_{0c.p}$ – average density of transported concrete mix; η_{gen} – overall efficiency of the concrete pump.

If there is a gradual loading of the hose in the working part of the pump casing with the rotating rotor rolls [13, 14], the service life of the new machine exceeds the life of the hours of modern hose-type concrete pumps by 25%.

Thus, technological sets of small-sized equipment with a new universal hose concrete pump allow us to carry out concrete work in the construction of reliable engineering buildings and structures in stable construction conditions.

3 Conclusions

1. The list of new equipment protected by patents of Ukraine is given which can be used as separate modules of universal technological complexes and sets of small-sized equipment for construction site conditions.
2. The use of a non-piston hose concrete pump of a new constructive solution is shown with the reinforcement of support columns by the wet concrete spraying method, the medical centre under construction.
3. The structural scheme of energy consumption by technological sets of small-sized equipment is given.
4. One of the universal technological kits of effective small-sized equipment is shown.

References

1. S. Moran, Process plant layout (Chapter 25), 369–376 (2017)
2. M. C. Valigi, I. Gasperini, Simulation Modelling Practice and Theory, **15 (10)**, 1211-1223, (2007)
3. J. Henikl, W. Kemmetmiller, A. Kugi, Automation in Construction, **61**, 112–123, (2016)
4. J. Klespitz, L. Kovács. *SAMI 2014, IEEE 12th international symposium on applied machine intelligence and informatics*, 191–194, (2014).
5. A. Daszczenko J. Glinski, E. Krasowski, et al., Hydraulika: Maszyny hydraulicznu. 385, (2010).
6. S. A. Hamid, A. Nouh, N. Y. Zabel, HBRC Journal, (2017).
7. Ö. Sjöström, Storage in Excavated Rock Caverns: Rockstore **77**, 585-591, (1978)
8. J. Wang, D. Niu, Y. Zhang, Construction and Building Materials, **95**. 312-328, (2015)
9. I.A. Emeljanova, A.A. Zadorozhny, S.A. Guzenko, N.A. Melentsov. Two-piston mortar-concrete pumps for construction site conditions, (Kharkov: Timchenko AN), 196, (2011)
10. I.A. Emeljanova, V.V. Blazhko, A.I. Anischenko, Scientific Bulletin of Civil Engineering **91 (1)**, 238-243, (2018)
11. I.A.Emeljanova, D.O. Chayka, D.S. Kabanets Utilities of cities, **137**, 98-103, (2017)
12. I. Emeljanova, V. Virchenko, D. Chayka, IJET, **7(3,2)**, 97-101 (2018)
13. D. Chayka, I. Emeljanova, A. Grigoryev, P. Andrenko, IJERA, **38**, 46-59, (2018)
14. I. Emeljanova, *V.* Blazhko, V. Shatobhin, D. Chayka, D. Kabanets, Scientific Bulletin of Civil Engineering, **90 (4)**, 136–145, (2017)

Calculation of composite elements of slabs in buildings, structures and fragments of spans of bridges

Stanislav Fomin[1,*], *Yuriy* Izbash[1], *Serhii* Butenko[1], *Maryna* Iakymenko[1], and *Karina* Spirande[1]

[1]Kharkiv National University of Construction and Architecture, 61002, Kharkiv, vul. Sumska, build. 40, Ukraine

Abstract. The calculation consists of two stages. The first one begins with the definition of their class, bearing capacity at temperature of 20 °C, according to EN 1992-1-1. At the second stage, the calculation at high temperatures shall be carried out in accordance with Eurocode 4 part 1-2. Comparison of the "stress-strain" diagram of concrete of class 30 under compression and temperature of 20 °C in two formulas showed their difference. That is, the designers do not have the opportunity to continue the calculation of diagrams at different heating temperatures. There was a need to improve the mathematical model of the "stress-strain" ratio of concrete high temperatures, clarification of the criteria of the bearing capacity of concrete in calculation of the fire resistance of composite structures in EN 1994-1-2:2005. In this paper, the method of determination of $\varepsilon_{cu1,\theta}$ developed has allowed, based on the energy approach, to formulate the corrected dependence of the limit deformation on temperature, dependence of the maximum deformation on temperature, and the value of the parameters of the "stress-strain" diagram. According to these data, using the formulas of the first stage, the "stress-strain" diagrams of the concrete of class 30 are calculated at the compression and heating according to EN 1992-1-2:2004.

1 General provision

The calculation of the fire resistance of composite structures in the static part should take into account the following steps: the calculation of mechanical characteristics under fire conditions, the effects for thermal analysis and impacts for mechanical analysis.

Since the steel-and-reinforced concrete structures consist of reinforced concrete slabs and rolling steel beams, it is necessary to know the complete "stress-strain" diagrams of concrete, reinforcement and rolling beams to calculate their fire resistance.

Steel girders, which work together with reinforced concrete slabs, are widely used in bridge building.

* Corresponding author: sfomin@ukr.net

2 Improvement of mathematical model of deformation diagram of compressed concrete of steel-and-reinforced concrete structures under heating

The paper [1] describes defects in the strength and deformation properties of concrete at higher temperatures that are given in EN 1994-1-2:2005 Eurocode 4, which during harmonization were included in DSTU-N-P B B.2.6-159:2010. The method of determination of the specified diagrams "$\sigma_{c,\theta} - \varepsilon_{c,\theta}$" for the design of steel reinforced concrete structures under the fire conditions was developed. The values of the parameters of the diagram for concrete based on silicate filler at higher temperatures were specified, and the mathematical model of the "stress-strain" ratio of the concrete at compression and higher temperatures was refined.

Despite such a significant refinement of the mathematical model of the "stress-strain" diagram of concrete under compression and higher temperatures, designers do not have sufficient information to calculate the fire resistance of the structure.

The calculation of elements of reinforced concrete and composite structures consists of two stages. The first stage begins with determination of their bearing capacity at a normal temperature of 20 °C, that is, when using Eurocode 2 EN 1992-1-1:2005 [2] or DBN V.2.6-98:2009 [3]. These standards propose to use two equations to describe the relationship between σ_c and ε_c for short-term axial load. Equation (1) that is used in Eurocode 2 EN 1992-1-1:2005 [2]:

$$\frac{\sigma_c}{f_{cm}} = \frac{k\eta - \eta^2}{1+(k-2)\eta} \tag{1}$$

where $\eta = \varepsilon_c / \varepsilon_{c1}$, ε_{c1} - deformations at maximum loads according to table 1,
$k = 1.05E_{cm} \mid e_{c1,cd} \mid / f_{cm}$ (f_{cm} - according to table [2]);
and equation (2) in the form of a quintal polynomial, which is based on the results of numerous experimental researches of the State Research Institute of Building Structures, the statistical processing of which allowed to propose a more complete normative basis:

$$\sigma_c = f_{(ck),(cd)} \sum_{k=1}^{5} \alpha_k \eta^k \tag{2}$$

For both formulas, use limits were defined:

$$0 < |\varepsilon_c| < |\varepsilon_{cu1}|, \tag{3}$$

where ε_{cu1} - nominal boundary deformations of concrete.

The paper [1] also shows that graphs practically coincide on the ascending branch according to DBN and EN92-1-1. Both normative documents contain the concrete classes, values of the relative deformation of the compression of concrete ε_{c1} at maximum stresses f_c, nominal boundary deformations of concrete ε_{cu1} (the value of relative boundary deformation of concrete compression) and the mean value of the initial elastic modulus of concrete E_{cm} (GPa). The data given in tables 1 and 2 is sufficient for the calculation of reinforced concrete structures at a temperature of 20 °C.

So, using the base Eurocode 2 EN 1992-1-1:2005 or the national building codes of Ukraine DBN V.2.6-98:2009, which take into account the main provisions (principles) of

Eurocode 2 EN 1992-1-1:2005, the calculations of reinforced concrete structures at temperature 20 °C are conducted.

Table 1. Classes of concrete strength according to Eurocode 2 EN 1992-1-1:2005.

$f_{ck,cube}$ (MPa)	15	20	25	30	37	45	50	55	60	67	75	85	95	105
ε_{c1} (‰)	1.8	1.9	2.0	2.1	2.2	2.25	2.3	2.4	2.45	2.5	2.6	2.7	2.8	2.8
ε_{cu1} (‰)	3.5	3.5	3.5	3.5	3.5	3.5	3.5	3.5	3.5	3.2	3.0	2.8	2.8	2.8
E_{cm} (GPa)	27	29	30	31	33	34	35	36	37	38	39	41	42	44

Table 2. The concrete strength classes according to DBN B.2.6-98:2009.

	C 8/10	C 12/15	C 16/20	C 20/25	C 25/30	C 30/35	C 32/40	C 35/45	C 40/50	C 45/55	C50/60
$f_{ck,cube}$ (MPa)	10	15	20	25	30	35	40	45	50	55	60
$\varepsilon_{c1,cd}$ (‰)	1.56	1.58	1.62	1.65	1.69	1.72	1.76	1.80	1.84	1.87	1.91
$\varepsilon_{cu1,cd}$ (‰)	3.75	3.70	3.59	3.44	3.28	3.10	2.93	2.72	2.57	2.43	2.29
E_{cm} (GPa)	18	23	27	30	32.5	34.5	36	37.5	39	39.5	40

Simultaneously, the concrete strength classes, deformation of concrete under compression ε_{c1} at maximum stresses f_c, nominal limit deformations of concrete ε_{cu1}, mean value of the initial elastic modulus of concrete E_{cm} and complete diagram of stress-strain of concrete were determined.

For example, for concrete strength class 30 according to Eurocode 2 EN 1992-1-1, the concrete deformations at maximum stresses $f_c = 30\text{MPa}$ are equal to $\varepsilon_{c1} = 2.1‰$, nominal limit deformations of concrete - $\varepsilon_{cu1} = 3.5‰$, mean value of the initial modulus of elasticity of concrete $E_{cm} = 31\text{GPa}$, and complete "stress-strain" diagram of concrete shall be determined by the formula (1).

For concrete strength class C 30/35 according to DBN V.2.6-98, the concrete deformation $f_c = 35\text{MPa}$ are equal to $\varepsilon_{c1} = 1.72‰$, nominal limit deformations of concrete - $\varepsilon_{cu1} = 3.1‰$, mean value of the initial modulus of elasticity of concrete $E_{cm} = 34.5\text{GPa}$, and complete "stress-strain" diagram of concrete shall be determined by the formula (2).

If the calculation showed the necessity of the strength class of concrete 30 according to Eurocode 2 EN 1992-1-1, the deformation of the concrete at maximum stresses $f_c = 30\text{MPa}$ are equal to $\varepsilon_{c1} = 0.0021{=}2{,}1‰$, the limit deformation of the concrete $\varepsilon_{cu1} = 0.0035 = 3.5‰$, concrete elastic modulus $E_{cm} = 31\text{GPa} = 31.000\text{MPa}$, complete "strain-deformation" diagram of concrete at a temperature of 20 °C shall be determined by the formula (1).

2.1 Analysis of recent research

In the course of the study [4], a method for determination of the criterion of concrete bearing capacity $\varepsilon_{c1,\theta}$ was developed for calculation of the fire resistance of reinforced concrete structures, on the basis of which data of characteristics of strength and deformation properties of concrete at higher temperatures given in Eurocodes EN 1992-1-2:2004 [9] and ENV 1992-1-2:2005 Eurocode 2 were refined. The comparison of the diagram for DBN B.2.6-98 [3] at the normal temperature (2) with a similar standard diagram [5] on the ascending branch for $\varepsilon \le \varepsilon \le \varepsilon_{c1,\theta}$ the diagram under the formula (4) shows a significant overstatement of the value $\varepsilon_{cu1,\theta=20^{\circ}C} = 0.0200 = 20‰$ (on the [3] $\varepsilon_{cu1,ck} = 2.4 \ldots 4.5‰$ depending on the concrete class).

$$\sigma_{c,\theta} = f_{c,\theta}\left[3\left(\frac{\varepsilon_{c,\theta}}{\varepsilon_{cu,\theta}}\right)\Bigg/\left\{2+\left(\frac{\varepsilon_{c,\theta}}{\varepsilon_{cu,\theta}}\right)^{3}\right\}\right] \quad (4)$$

Is is shown that these diagrams are obtained on special presses equipped with a heating device and a servo-control of pressure in the press cylinder, which allows passing through the descending branch practically zero stress. In this case, the branch descending to zero is no longer the area of deformation of the integral concrete sample, but the area of deformation of individual parts of the fragmented concrete. So, the values of deformations in columns 4 and 7 of Table 3.1 [5] are not limit deformations of concrete $\varepsilon_{cu1,\theta}$. The deformations $\varepsilon_{c1,\theta}$ in this table are taken to be the same for all classes of concrete, including for silicate (column 4) and carbonate (column 7) fillers. So, depending on the class of concrete in formula (4), it is possible to vary only the value of compressive strength f_c, therefore these diagrams, even on the ascending branch, incorrectly represent the actual properties of the physical nonlinearity.

Fig. 1 shows a "stress-strain" diagram of concrete of class 30 under the formula (1) in comparison with the formula (4).

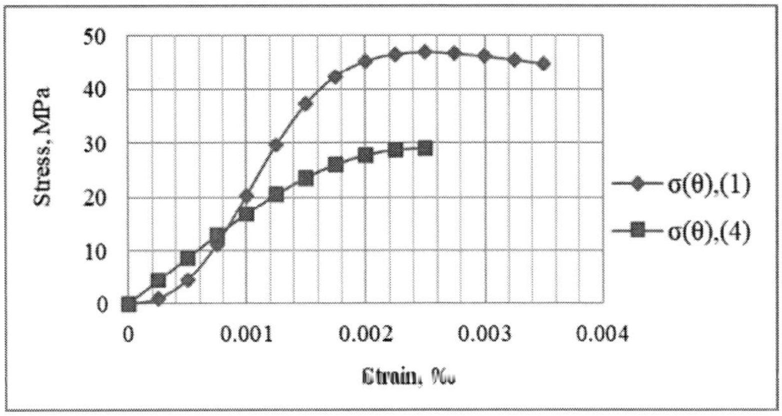

Fig. 1. The "stress-strain" diagram of concrete of class 30 under compression and temperature of 20 °C by formula (1) and (4).

This difference between the diagrams occurs because of the class of concrete is not taken into account. That is, there is no possibility for designers to continue the calculation of diagrams at different heating temperatures because they are offered a new algorithm with the absence of the concrete classes, criterion of the bearing capacity of concrete $\varepsilon_{cu1,0}$ of the elasticity of concrete E_{cm}, new characteristics of concrete based on a silicate and carbonate filler, and so on.

The defects in the standard DSTU-N-P B V.2.6-159:2010 (EN 1994-1-2:2005 Eurocode 4. MOD) [7], [8] harmonized with EN 1994-1-2:2005 [6] are also found.

So, it is necessary to improve the mathematical model of the "stress-strain" ratio of concrete under compression and elevated temperatures, refinement of the criteria of the bearing capacity of concrete when calculating the fire resistance of composite structures in EN 1994-1-2:2005.

2.2 Results of theoretical studies

Section 3 of the standard [5] states that the strength and mechanical properties of concrete at higher temperatures can be determined from the stress-strain dependencies given in EN 1992-1-2 [7].

In fact, the deformations $\varepsilon_{c1,\theta}$ given in Table 3.1 are taken to be the same for all classes of concrete, including for silicate (column 4) and carbonate (column 7) fillers. So, depending on the class of concrete in formula (4), it is possible to vary only the value of compressive strength f_c, therefore these diagrams, even on the ascending branch, incorrectly represent the actual properties of the physical nonlinearity.

Eurocode EN 1992-1-2:2004 is definitely the leading normative document in the theory of concrete, reinforced concrete and composite structures, their fire resistance, and is based on high-level experiments, with new test devices, which allowed the implementation of progressive nonlinear computational models.

At the same time, the main results of the conducted studies are not consistent with the basic Eurocode 2 EN 1992-1-1:2005 [2]: Design of reinforced concrete structures – Part 1-1: General norms and rules for structures, that is shown, for example, by comparing deformation diagrams at a temperature of 20 °C and other parameters.

The correction in the indicated tables and the method for determination $\varepsilon_{cu1,\theta}$ on the basis of the energy approach [10] have made it possible to formulate the corrected dependence of the limit deformation on temperature (Fig. 2), the dependence of the maximum deformation on temperature (Fig. 3) and Table 3 - Value of the parameters of the "stress-strain" diagram.

According to these data, using the formula (1), the "stress-strain" diagrams of concrete of class 30 were calculated under compression and heating according to EN 1992-1-2:2004 (Fig. 3).

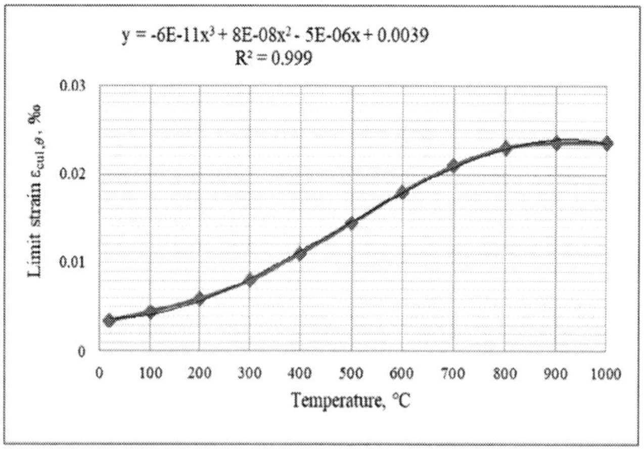

Fig. 2. The dependence of the limit deformation on temperature was adjusted.

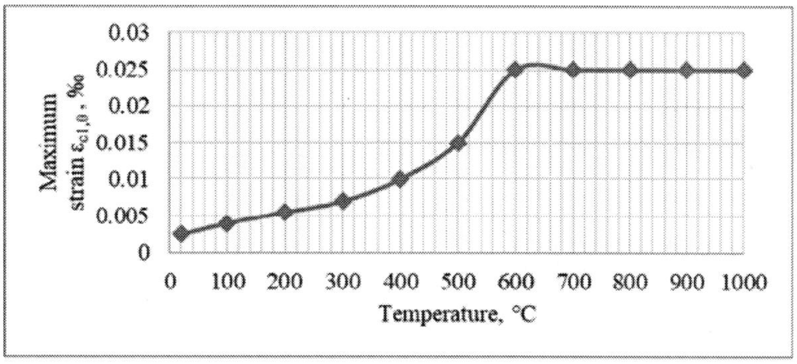

Fig. 3. The dependence of the maximum deformation on temperature.

Table 3. - The value of parameters of the "stress-strain" diagram for silicate filler based concrete at higher temperatures.

Concrete temperature θ, °C	Silicate filler					
	$f_{c,\theta}/f_{ck}$	$\varepsilon_{c1,\theta}$	$\varepsilon_{cu1,\theta}$	$f_{c,\theta}$, MPa	$E_{m\theta}$, MPa	$E_m/E_{c,\Theta}$
20	1	0.0025	0.0035	30	31000	1
100	1	0.004	0.0045		21328	0.688
200	0.95	0.0055	0.006		15283	0.493
300	0.85	0.007	0.008		11067	0.357
400	0.75	0.01	0.011		7812	0.252
500	0.6	0.015	0.0145		4805	0.155
600	0.45	0.025	0.018		2790	0.09
700	0.3	0.025	0.021		1705	0.055
800	0.15	0.025	0.023		868	0.028
900	0.08	0.025	0.0235		536.3	0.0173
1000	0.04	0.025	0.0235		127.1	0.0041

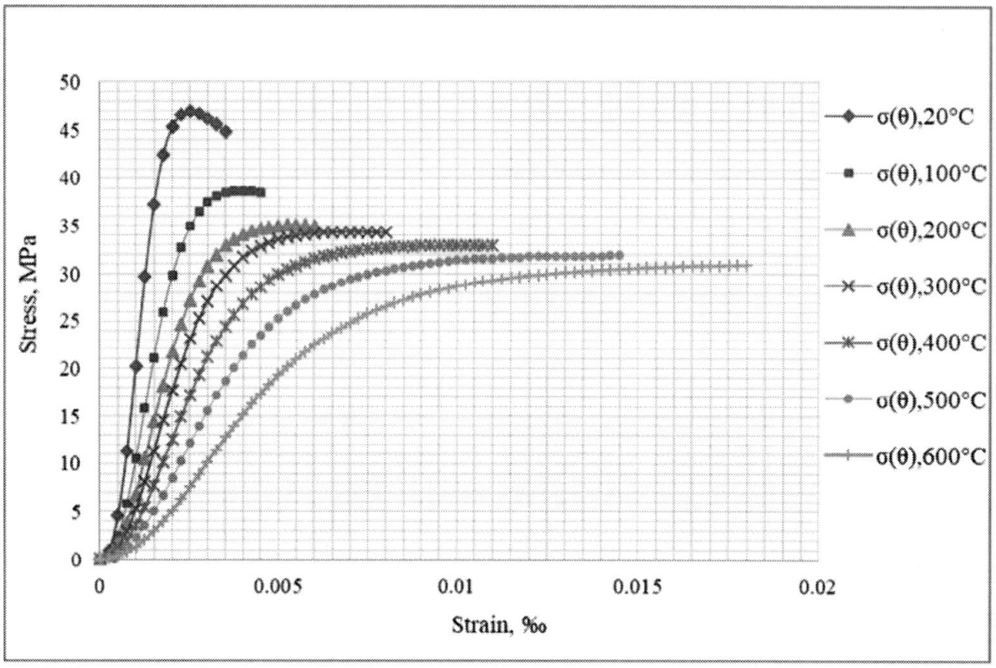

Fig. 4. Stress-strain diagrams of concrete of class 30 under compression and heating according to Eurocode EN 1992-1-2:2004.

Conclusions

1. The paper [1] revealed defects in strength and deformation properties of concrete at higher temperatures given in EN 1994-1-2:2005 Eurocode 4. Despite the refinement of the mathematical model of the "stress-strain" diagram of concrete under compression and higher temperatures, the designers lack information to fully calculate the fire resistance of the structure.

2. It is shown that the calculation of elements of reinforced concrete and composite structures consists of two stages. The first stage begins with determination of their bearing capacity at a normal temperature of 20 °C, that is, when using Eurocode 2 EN 1992-1-1:2005 [2] or DBN V.2.6-98:2009 [3]. These standards propose to use two equations to describe the relationship between σ_c and ε_c for short-term axial load. Equation (1), which is used in Eurocode 2 EN 1992-1-1:2005 [2], and equation (2) in the form of quintal polynomial, which is based on the results of numerous experimental researches of the National State Institute of Concrete Structures (NSICS), the statistical processing of which allowed to offer a more complete normative base.

3. In the course of the study [4], a method for determination of the criterion of concrete bearing capacity $\varepsilon_{cu1,\theta}$ was developed for calculation of the fire resistance of reinforced concrete structures, on the basis of which data of characteristics of strength and deformation properties of concrete at higher temperatures given in Eurocodes EN 1992-1-2:2004 and ENV 1992-1-2:2005 Eurocode 2 were refined. The comparison of the diagram according to DBN B.2.6-98 [3] at the normal temperature (2) with a similar diagram of the standard [5] on the ascending branch at $\varepsilon \le \varepsilon_{c1,\theta}$ for the diagram under the formula (4) shows a significant overstatement of the value

$\varepsilon_{cu1,\theta=20°C} = 0.0200 = 20‰$ (on the [5] $\varepsilon_{cu1,ck} = 2.4...4.5‰$ depending on the concrete class).

4. The "stress-strain" diagram of concrete of class 30 under compression and temperature of 20 °C by formula (1) compared with formula (4) showed a large difference: at $\varepsilon_c = 0.0025$ by the formula (1) $\sigma_c = 47\text{MPa}$ by the formula (4) $\sigma_c = 28\text{MPa}$. This difference between diagrams occurs due that the concrete class is not taken into account. That is, there is no possibility for designers to continue the calculation of diagrams at different heating temperatures because they are offered a new algorithm with the absence of the concrete classes, criterion of the bearing capacity of concrete $\varepsilon_{cu1,0}$ of the elasticity of concrete E_{cm}, new characteristics of concrete based on a silicate and carbonate filler, and so on. The defects in the standard DSTU-N-P B V.2.6-159:2010 (EN 1994-1-2:2005, Eurocode 4. MOD) [7] harmonized with EN 1994-1-2:2005 were also identified.

5. The table of values of parameters of the "stress-strain" diagram for silicate-based concrete at higher temperatures was developed.

6. The correction in the indicated tables and the method for determination $\varepsilon_{cu1,\theta}$ on the basis of the energy approach [10] have made it possible to formulate the corrected dependence of the limit deformation on temperature (Fig. 2), the dependence of the maximum strain on temperature (Fig. 3) and Table 3 - Value of the parameters of the "stress-strain" diagram.

According to these data, using the formula (1), the "stress-strain" diagrams of concrete of class 30 were calculated under compression and heating according to EN 1992-1-2: 2004 (fig. 4).

References

1. S. Fomin, Y. Izbash, I. Plakhotnikova, S. Butenko and R. Shemet, MATEC Web of Conferences, **116,** 02013, 4 (2017)
2. EN 1992-1-1:2005 *Eurocode 2: Design of reinforced concrete structures - Part 1-1: General norms and rules for structures*, 136 (2005)
3. DBN V.2.6-98:2009 *Structures of buildings and facilities. Concrete and reinforced concrete structures. Basic provisions.* Ministry of regional development and construction of Ukraine, Kiev, 71 (2011)
4. S.L. Fomin, A.I. Davidenko, V.G. Poklonsky, News of Odessa State Academy of construction and architecture, **46**, 8 (2012)
5. DSTU-N-P B V.2.6-159:2010 *Structures of buildings and facilities. Design of composite steel and concrete structures. Part 1-2. General provisions. Structural fire design. (EN 1994-1-2:2005, Eurocode 4. MOD)*, 37 (2010)
6. ENV 1992-1-2:1995 *Eurocode 2: Design of Concrete Structures - Part 1-2: General Rules - Structural Fire Design*, 82 (1995)
7. DSTU-N B EN 1992-1-2:2012. *Eurocode 2: Design of concrete structures. Part 1-2: General rules. Structural fire design. (EN 1992-1-2:2004, IDT).* Minregionbud of Ukraine, Kiev, 131 (2012)
8. DBN V.2.6-160:2010 *Composite steel and concrete structures. Basic provisions.* Minregionbud of Ukraine, Kiev, 55 (2011)
9. EN 1992-1-2: 2004 *Eurocode 2: Design of concrete structures – Part 1-2: General rules - Structural fire design*, 75 (2004)
10. A.N. Bambura, First Ukr. Sc. Tech. Conf., *Scientific - practical problems of modern concrete*, **36**, 7 (1996)

Complete "stress-strain" diagrams of rolled steal beams

Stanislav Fomin[1,*], *Yuriy* Izbash[1], *Yuriy* Bondarenko[1], *Serhii* Butenko[1], and *Iryna* Plakhotnikova[1]

[1]Kharkiv National University of Construction and Architecture, 61002, Kharkiv, vul. Sumska, build. 40, Ukraine

Abstract. This paper covers features of rolled steel beams used in composite structures. Composite structures consist of reinforced concrete slabs and rolled steel beams. To calculate its fire resistance it is necessary to know their complete "stress-strain" diagrams. Steel girders, which work together with reinforced concrete slabs, are widely used in bridge building. The tests on samples cut from web and a rack of an I-beam were conducted. The results of experimental studies of the "stress-strain" diagrams of rolled beams indicate the difference in the physical-mechanical characteristics of steel in the web and flanges of the beams. The experimental studies of diagrams were carried out at temperatures of 20, 400, 500, 600, 700 and 800 °C. The properties of the steel were significantly affected by the temperature. The "stress-strain" diagrams of rolled beams and table of results of experimental studies were obtained. The complete "stress-strain" diagrams of the rolled beams were studied. Experimental studies have been conducted using a newly developed patented installation. The differences of diagrams in the web and flanges of the steel I-beams were found.

1 General provision

Composite structures consist of reinforced concrete slabs and rolled steel beams. To calculate its fire resistance it is necessary to know their complete "stress-strain" diagrams. Steel girders, which work together with reinforced concrete slabs, are widely used in bridge building.

The design of steel structures shall be carried out using the national standard, which corresponds to EN 1993-1-2:2005 Eurocode 3: Design of steel structures – Part 1-2: General rules – Structure fire design – Part 1-2: General rules - Calculation for fire resistance [1].

The mechanical properties of structural steels are determined by the National Institute of Standards and Technology (NIST). (2005) [2]. The behavior of high-strength structural steel at elevated temperatures is described in [3]. The behavior of steel framework in a fire condition, described in a candidate's dissertation [4] (University of Edinburgh, 2000); nonlinear analysis for a three-dimensional steel framework in a fire conditions is carried out

* Corresponding author: sfomin@ukr.net

in [5], mechanical properties of structural steel at elevated temperatures and after cooling are described in [6]. In [7] experimental studies of the critical temperature of compressed steel elements with limited thermal elongation are presented, and the fire resistance of steel columns with axial extensions and bending is investigated in [8]. The manufacture of I-beams by the hot rolling method is described in [9].

This paper considers features of rolled steel beam and tests on samples.

2 Technology for the production of steel twisted beam using the hot rolling method

Hot rolling of I-beams shall be carried out on rolling mills, which are long installations with a large number of rotating rolls of different diameters. The process involves the use of thick ingots or slabs that have a continuous section as a workpiece. When rolling, the ingots are moved along the roller table and processed using rollers, as a result of which the length of the workpieces is greatly increased, and the cross-section gradually decreases and becomes "I" shaped.

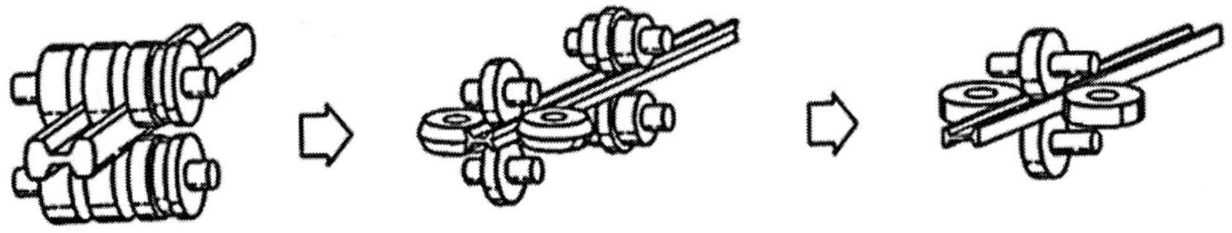

Fig. 1. The beam processing using rollers.

The method of hot rolling of I-beam is extremely effective. I-beam has unique mechanical properties. Its strength exceeds the strength of L-bar and square profile, and it has a high rigidity and a small specific density. In addition, one of the main structural properties of an I-beam is bending resistance, that is, to take over and transmit the vertical and horizontal load.

Beams have two types of flanges: with the sloping of internal faces of flanges; with parallel faces of flanges.

Fig. 2. The I-beam with sloping of faces of flanges.

Three types of I-beams are manufactured based on the ratio of sizes and purpose: (A) normal (B); column (K) wide-flanged (W); steel beams are the most demanded type of high-quality section-shaped rolled stock, thanks to a number of advantages: simplicity of design; ability to withheld increased loads; they allow large spans; they not burn (although they loss their bearing capacity when heated to the high temperature); they are resistant to the biological effects.

It can be assumed that different parts of beams have different stresses of compression at such rolling and their properties can differ from each other, all the more at influence of the high temperature. To check this assumption, the experimental studies were conducted using the new installation (testing stand), which was developed and patented by us [10].

3 Testing of samples

The stand for testing reinforcement that includes a metal frame, device for gripping the armature, devices for creation and measurement of the dynamic load, and measurement of deformations, system for the reinforcement heating, which is characterized in that the devices for gripping the reinforcement are connected with the system for creation of the dosed movements made of the worm movement gears with the steering wheels mounted on the pivot supports.

The operation principle of the reinforcement testing stand is as follows. The reinforcement sample is fixed in the collect clamps, heated to a given temperature by electric power and stretched for given displacements without accumulation of potential energy in the elements of the installation using the worm gears with steering wheels that create the system of the dosed movement. The value of the reinforcement deformation shall be measured using an extensometer with traction rods and indicators, the force value is indicated by the dynamometer. This allows us to obtain a complete “stress-strain” diagram of the reinforcement with a descending branch at elevated and high temperatures.

The tests were carried out on samples taken from the web and vertical member of the I-beam series 2062 with height $H = 200\text{mm}$, width of the upper and lower flanges - 100 mm, thickness $T_f = 8.6\text{mm}$, the thickness of web - $T_b = 5.2\text{mm}$, cross section area – 27.7 mm, used during experimental research of fragments of composite structures of bridges during heating.

Samples are shown in Fig. 3.

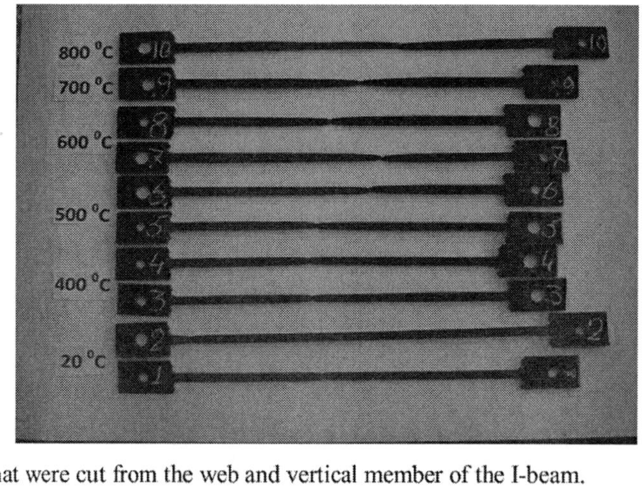

Fig. 3. The samples that were cut from the web and vertical member of the I-beam.

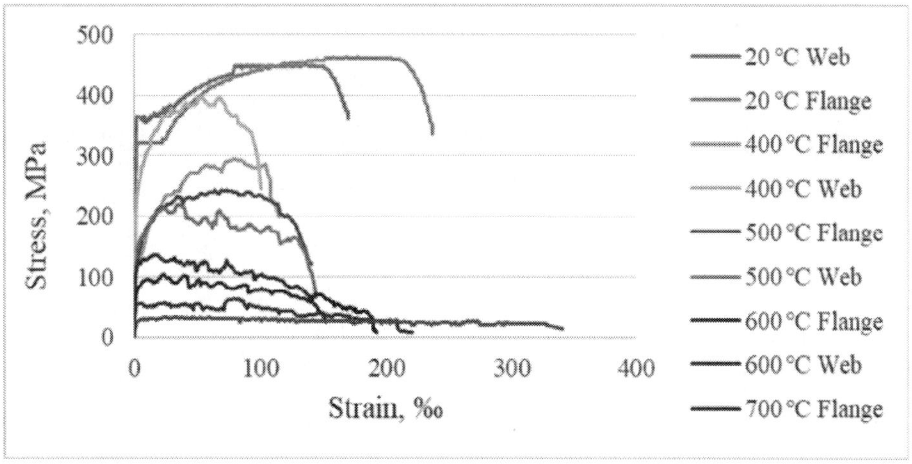

Fig. 4. General "stress-strain" diagrams (for web and flanges).

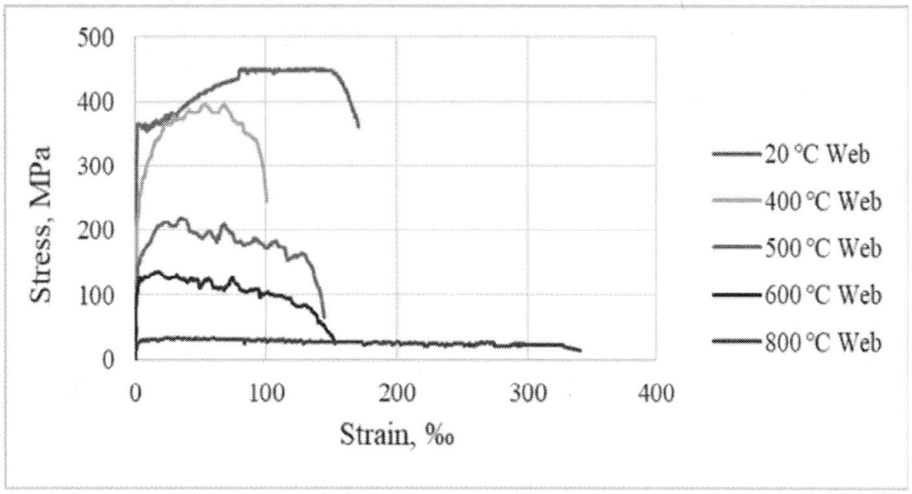

Fig. 5. "Stress-strain" diagrams for web of rolled beams.

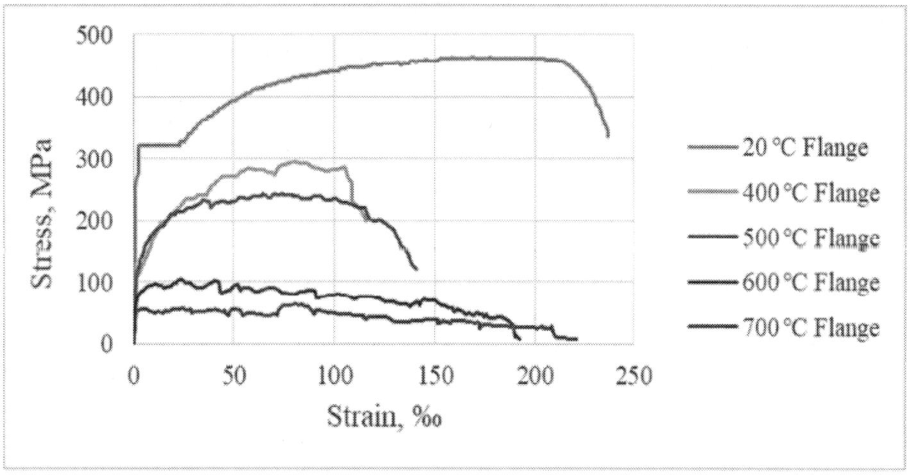

Fig. 6. "Stress-strain" diagrams for flanges of rolled beams.

Table 1. Results of study of "stress-strain" diagrams of rolled beams.

No. of a diagram	Tempe-rature °C	Part of beam	Deforma-tion, ‰	Stress, MPa	Straight line	Yield point	Maxi-mum stress	Destruction
1	20	wall	‰		2.2	15	79.5	170.3
				MPa	363.7	363.7	441.4	359.8
2	20	flange	‰		2.9	22.9	197.1	237
				MPa	320	320.8	469.5	396.5
3	400	wall	‰		1.4	-	56.6	99.4
				MPa	190		395	287
4	400	flange	‰		1.7	-	79.1	115.6
				MPa	106		294	208
5	500	wall	‰		1.9	-	35.5	144.7
				MPa	133		219	65.5
6	500	flange	‰		1.425	-	72.7	140
				MPa	119.6		243	122
7	600	wall	‰		0.475	-	18.3	152.5
				MPa	80.7		135.8	24.5
8	600	flange	‰		3.05	-	23.82	191
				MPa	82		105	12
9	700	flange	‰		0.7	-	79.5	217
				MPa	40.4		63.3	8.9
10	800	wall	‰		4.4	-	35.2	340
				MPa	28.6		33.6	13.7

4 Results of experimental studies

The results of experimental studies of the "stress-strain" diagrams of the rolled beams shown in Fig. 4 - 6 show the difference in physical and mechanical characteristics of steel in the web and flanges of beams.

So, at 20 °C (Table 1), the transition of the direct line to the yield point in the wall occurred at stress of 363.7 MPa, and in the flange - at stress of 320.7 MPa. The maximum stress at the yield point in the wall occurred at 441.4 MPa, and in the flange - at stress of 469.5 MPa, the destruction in the wall occurred at 359.8 MPa, in the flange - at stress of 396.5 MPa.

The properties of the steel were significantly affected by the temperature. Starting from 400 °C, the yield point has disappeared. The maximum stress in the wall has occurred at stress of 395 MPa and deformation 56.6 ‰, and in the flange - at stress of 294 MPa and a deformation of 79.1 ‰; the destruction has occurred at stress of 287 MPa and deformation of 99.4 ‰, and in the flange - at stress of 208 MPa and deformation of 115.5 ‰.

At 500 °C the maximum stress in the wall has occurred at the stress of 133 MPa and deformation of 1.9 ‰, in the flange - at stress of 119.6 MPa and deformation of 1.42 ‰; destruction in the wall - at stress of 219 MPa and deformation of 35.5 ‰, in the flange - at the stress of 243 MPa and deformation of 72.7 ‰.

At 600 °C the maximum stress in the wall has occurred at the stress of 135.8 MPa and deformation of 18.3 ‰, in the flange - at stress of 105 MPa and deformation of 23.82 ‰; destruction in the wall - at stress of 24.5 MPa and deformation of 152.5 ‰, in the flange - at the stress of 12 MPa and deformation of 191 ‰.

At 700 °C, the maximum stress in the flange has occurred at the stress of 63.3 MPa and deformation of 79.5 ‰, the destruction in the flange has occurred at the stress of 8.9 MPa and deformation of 217 ‰.

At 800 °C, the maximum stress in the flange has occurred at the stress of 33.6 MPa and deformation of 32.2 ‰, the destruction in the flange has occurred at the stress of 137.9 MPa and deformation of 340 ‰.

Conclusions

1. The complete "stress-strain" diagrams of the rolled beams were studied. Experimental studies have been conducted using a newly developed patented installation. The tests were carried out on samples taken from the web and vertical member of the I-beam series 2062 with height $H = 200\text{mm}$, width of the upper and lower flanges - 100 mm, thickness $T_f = 8.6\text{mm}$, the thickness of web - $T_b = 5.2\text{mm}$, cross section area – 27.7 mm, used during experimental research of fragments of composite structures of bridges during heating.

2. The "stress-strain" diagrams of rolled beams and table of results of experimental studies were obtained. The differences of diagrams in the web and flanges of the steel I-beams were found.

References

1. EN 1993-1-2:2005 *Eurocode 3: Design of steel structures – Part 1-2: General rules – Structure fire design (Eurocode 3: Design of steel structures - Part 1-2: General rules - Calculation for fire resistance) with a technical amendment EN 1993-1-2: 2005*/AC:2009
2. National Institute of Standards and Technology (NIST). Mechanical Properties of Structural Steels. (2005)
3. J. Chen, B. Young, and B. Uy, J. of Struc. Eng., **132**, 1948-1954, (2006)
4. M. Gillie, The behavior of steel-framed composite structures in fire conditions, PhD thesis, The University of Edinburgh, (2000)
5. S. Najjar, I. Burgess, Eng. Struc., **18**, 77-89, (1996)
6. J. Outinen, P. Makelainen, F. and Mat., **28**, 237-251, (2004)
7. J. Rodrigues, I. Neves, J. Valente, F. S. J., **35**, 77-98, (2000)
8. J. Valente, I. Neves, J. of Constr. S. R., **52**, 319-331, (1999)
9. A. P. Grudev, L. F. Mashkin, M. I. Khanin, *Technology of Rolling Production* (Metallurgy, 1994)
10. Patent for the utility model No. 120438. Testing stand for reinforcement. / Inventors: A. M. Kravchenko (UA), S. L. Fomin (UA), Y. M. Izbash (UA), S. V. Butenko (UA). Owner: State Enterprise "State Institute for design of the coke industry enterprises", SE "GIPROKOKS", vul. Sumska, building 60, m. Kharkiv, 61002 (UA), Bul. No. 20, 10/25/2017

Features of modeling of energy profile of construction

Lyudmyla Gaponova[1,*], *Serhii* Grebenchuk[1], *Nina* Psurtseva[1], *Oleg* Kalmykov[1] and *Ivan* Demyanenko[1]

O. M. Beketov National University of Urban Economy in Kharkiv, Department of Building Constructions, 61002, Kharkiv, Marshal Bazhanov str., 17 Ukraine

Abstract. Presentation and realization of the task of rationalization of structural parameters of reinforced concrete shells is presented. As a criterion for this task an energy principle is adopted, according to which it is assumed that from the whole set of possible values of the desired parameters of the system with a constant volume of material, the number of external and internal connections, the potential energy of deformation (DER) after the reconstruction will reach the lower limit on a rational combination of values of geometric parameters. The energy principle is taken as a criteria on for this task, according to which it is believed that out of numerous possible values of the searched parameters of the system with the constant amount of material, number of external and internal connections, the strain potential energy (SPE) after the reconstruction will reach the lower limit on the rational connection of the geometrical parameters describing the system. The determined interconnection of the rational parameters of the constructive anisotropic shell between SPE of the system of internal and external rational parameters: with ribs step value of about $l \approx 1000$ mm SPE reaches the lower limit.

1 Problem setting

The roofing of buildings and structures accepts essential loading which causes an essential stress-strain behavior (SSB) of the system. Therefore, it would be reasonable to consider such types of roofing which could accept, evenly distribute and transfer such loading to the supports. For this purpose, the most suitable would be the shells of different Gaussian curvature made of the reinforced concrete.

However, searching for the rational constructions of shells which accept the mentioned types of loading, is quite topical, but logically unsolved problem.

Besides, as the reinforced concrete is quite a heavy material, it would be essential to study the methods of its lighting. One of such methods is the embedment of the hole-forming insertions inside the construction, made of the foam polystyrene which at the same time form the shell angle cover. At the same time, giving some geometry to the angle fins

* Corresponding author: gaplyudmila@gmail.com

(due to the changing of the insertion forms), the most acceptable force distribution may be insured, which constitutes a given bearing capacity with the limited use of material.

To sum up, it may be stated that creating the methodology of forming the internal geometry with further experimental verification is the topical issue essential for the construction practice. Its solution will allow increasing the level of reliability of such systems, their durability and due to the rationalization of their construction parameters - realize the minimization of material consumption.

Key methods of calculating the reinforced concrete construction elements according to the main provisions of the deformation models were offered to the draft construction norm of Ukraine.

The shells made of the reinforced concrete and the stress-strain behavior of the reinforced concrete with the use of the calculation models were studied by the following scholars: A.Y. Barashykov, V.M. Bondarenko, O.I. Veinberg, H.M. Gasiy, Y.O. Klimov, H.A. Molodchenko, M.V. Savitskyi, L.I. Storozhenko, H.K. Khaidukov, A.A. Gvozdev, V.V. Shugaev, A.L. Shagin, V.S.Shmukler.

Despite numerous works in this sphere as of now unfortunately, there are no works of non-linear numerical study of two-zone anisotropic shells with the fins in different directions. The mentioned references provide the created theories and engineering methods of the shells calculation [1, 2].

We managed to somewhat specify and transform the notion of the "rational construction" with the help of interesting ideas and studies by H.V. Vasylkov and V.S. Shmukler [3, 4].

Modeling of the stress-strain behavior, geometrical parameters considering the energy principles of anisotropic shells are considered in the works [5-9].

The objective is the rationalization of the parameters of reinforced concrete constructive anisotropic shells. Research the revolutionary type of construction which has the specified external and calculated internal geometry; provide the mathematical modeling of the process of deforming the constructive anisotropic shell with different types of external and internal impacts.

2 Main material

V.S. Schmuckler in his work [10] stated that "for the regulated systems with the constant material volume, number of external and internal connections (external parameters) under the influence of the static external loading - dead weight, the strain potential energy (SPE) after the reconstruction will reach the lower limit on the rational combination of the geometrical parameters values.

$$U = \inf_{\alpha} U(\alpha^k), \quad k = 1,2,...\infty \quad , \tag{1}$$

where U – is SPE ; k - number of the comparison variant; $\alpha \in M$; M – variety of the acceptable values of the external geometrical parameters.

The approach realization presupposes the construction of interconnection between the system strain energy value and one or a group of the geometrical parameters. Setting the similar task in numerical form is possible when applying the calculation software complexes, however, the operation presupposes the formation of a big amount of models which complicates the process to some extent.

The work presents the setting and realization of the task of rationalization of the constructive parameters of suggested concrete shells. As a criterion the energy principle was taken, according to which out of the whole variety of possible values of the searched parameters of the system with the constant material volume, number of external and internal relations, strain potential energy (SPE) after the reconstruction will reach the lower limit on the rational combination of the geometrical parameters values describing the system.

The vector of control parameters of this system is introduced into the analysis:

$$\{x\}^T = \{H, B, V, R, L, l, q, \delta, \Delta, h\} \tag{2}$$

where H is a rising height, B is a shell length; V – material volume; R – curve radius, L is a shell passage, l – fin spacing, q – external loading, δ – plate thickness, Δ – fin thickness; h – height of the shell section.

Besides, the parameters H, B, R, L, q – are represented as the external ones, and the parameters l, δ, Δ, h – as internal ones.

The inner fins spacing of the shell l is taken as variable parameter (fig. 1).

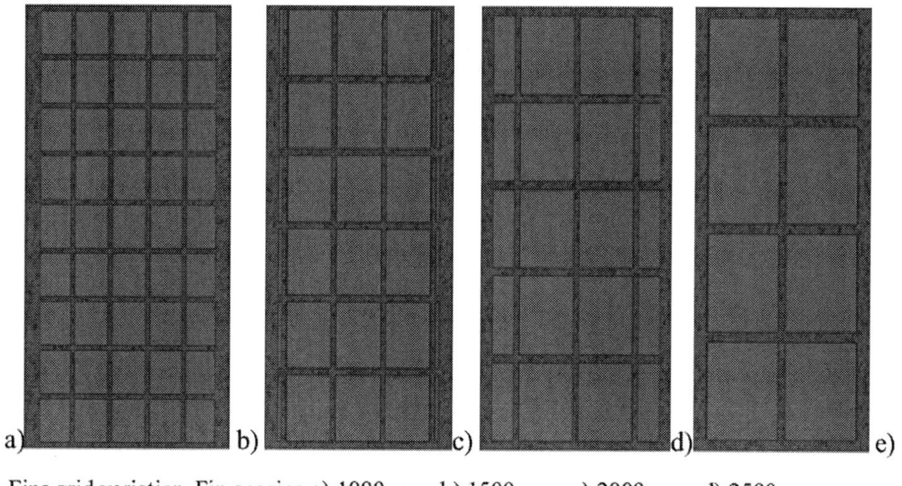

Fig. 1. Fins grid variation. Fin spacing a) 1000 mm; b) 1500 mm; c) 2000 mm; d) 2500 mm.

The remaining attributes of the calculation models are given similarly to the previously accepted, except for δ – plate thickness, and the rising height H, taken as 3.8 m. The stability of the material volume, in this case, is provided for the selection of the corresponding value Δ – fin thickness.

As it is seen from the diagram, if the fin spacing approaches $l \approx 1000$ mm SPE reaches the lower limit. Thus, the suggested approach allows determining the rational parameters of the reinforced concrete elements with the complex external and internal geometry.

The given analysis is extended due to studying the distribution of the strain energy density (internal parameter) in this case, the following dependency will constitute the criterion

$$e \rightarrow \text{const}, \tag{3}$$

where e is a density of the strain potential energy (DSPE). The field SPE (Lira PC) leveling was realized due to giving the complex form to the hole-forming insertions. In this case, the fins grid is not orthogonal. The area density of the potential energy depending upon the shell fin spacing with the chief beam 9000 mm are shown in fig. 2.

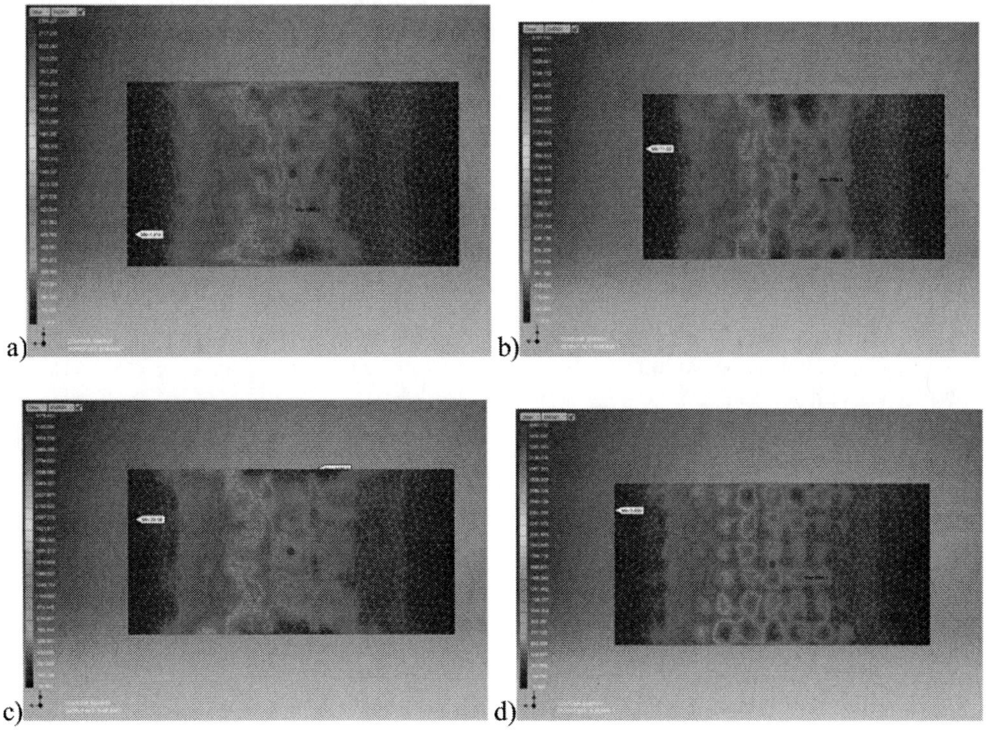

Fig. 2. The area density of the potential energy depending upon the shell fin spacing with the chief beam 9000mm with the rising height 2,250 mm: a) fin spacing 1000 mm, fins thickness 100mm; b) fin spacing 1500 mm, fins thickness 137.7 mm; c) fin spacing 2000 mm, fins thickness 169.9 mm; d) fin spacing 2500 mm and fins thickness 207.3 mm.

3 Conclusion

The work provides numerous studies of the two-zone anisotropic shells with the fins in different directions. The interconnection is determined of the rational parameters in the constructive anisotropic shell between SPE of the system of external and internal rational parameters: If the fin spacing approaches l≈1000 mm SPE reaches the lower limit.

The energy criterion is applied, as to the rationalization of the external constructive parameters of the reinforced concrete constructive anisotropic shells. According to the results of the study, it was determined that (for considered specific conditions) if H≈3.8m, the potential strain energy approaches the lower limit. The obtained results were verified through the analysis of natural vibration frequencies of the system for all *H* values and assessment of maximum bearing capacity (q_{max}) of the shell.

References

1. V. Bazhenov, N. Solovei, *Prikladnaya Mekhanika.* **45**, 3-40 (2009)

2. S. Lehnitsky, *Theoriia uprugosti anizotropnogo tela.* 416 (1977)

3. A. Gorodetsky, V. Shmukler, A. Bondarev, *Informacionnye tekhnologii rascheta I proektirovaniia stroitelnykh konstrukcii.* 889. (2003)

4. V. S. Shmuckler *Study of the work of gently sloping reinforced concrete shells.* diss Cand. tech. Sciences. Kyiv.188 (1977)

5. V. Babaev, V. Schmuckler, S. Feirushah, L. Gaponova, S. Grebenchuk, O. Kalmikov, *SU-ICEIT* 42-49. (2016)

6. O. Kalmykov, L. Gaponova, P. Reznik and S.Grebenchuk, *MATEC Web of Conf.* **116**, 02018 (2017)

7. O. Kalmykov, L. Gaponova, P. Reznik and S. Grebenchuk, *MATEC Web of Conf.* **116**, 02017 (2017)

8. L. V. Gaponova, O. A Kalmikov, S. S. Grebenchuk, Zb. nauk. Prats of UkrSURT **157**, 102-113. (2015)

9. L. Gaponova, O. Kalmikov, S. Grebenchuk *Communal economy of cities* **126,** 9-18 (2016)

10. V. Shmukler Zb. nauk. Prats of UkrSURT **167**, 54-69 (2017)

Methodological and computer-based support for choosing underground utility networks renovation method

Dmitriy Goncharenko, *Ihor* Shumakov, *Olha* Starkova*, *Alevtyna* Aleinikova, and *Revaz* Mikautadze [1]

Kharkiv National University of Civil Engineering and Architecture, Sumska st. 40, 61002 Kharkiv, Ukraine

Abstract. The article is devoted to the methodological and software tools for choosing organizational and technological solutions to maintain sewage collectors in an operable state to increase their service life. Such a tool must take into account a set of factors that affect the operation of sewer collectors and form a system, the implementation of which would improve the existing principles of monitoring the status of sewer collectors, identify effective organization and technological solutions of their renovation by taking into account factors that affect their safe operation, the facilities risk of accidents categories, conditions and parameters of works performance. For determining the optimal options for the renovation of the sewer collector using the developed mathematical and computer models, the results of problems solving were generalized: one-criterion optimization subject to the choice of one repair method; one-criterion optimization subject to the choice of different repair methods for the areas; multicriterion optimization.

1 Introduction

The sewage system of any human settlement is one of the most expensive and vulnerable parts of underground engineering infrastructure. The environment, the comfort of the population, and the efficient work of the city enterprises depend on its reliable and uninterrupted work [1, 2].

At the moment there is practically no planned approach in preparing a complex of organizational and technological solutions that would increase the service life of sewer collectors. In addition, such networks that are hidden underground, require attention only when there are emergency situations, and the complex of solutions for the support of these systems is reduced to emergency repair works at the collector site in order to eliminate the consequences of the accident [3-5]. This approach to the works management leads to a high cost of all types of utilities resources and, consequently, to adverse economic, environmental and social effects.

* Corresponding author: starkova.olha@gmail.com

Recent trends have shown that closed (trenchless) technologies have been used ever more commonly for the repair and renovation of sewage pipelines associated with the replacement of reinforced concrete, cast iron and other types of tubes with modern materials in the utilities of the European cities [4-13]. However, in contemporary practice of sewage pipeline repair there is no clear mechanism for selecting effective trenchless methods of underground communications repair.

In view of the above, developing methodological and software tools for choosing organizational and technological solutions is required to maintain sewage collectors in an operable state, which will increase their service life. Such a tool must take into account a set of factors that affect the operation of sewer collectors and form a system, the implementation of which would improve the existing principles of monitoring the status of sewer collectors, identify effective organization and technological solutions of their renovation by taking into account factors that affect their safe operation, the facilities risk of accidents categories, conditions and parameters of works performance [14].

2 Materials and methods

Today there are more than 20 contemporary trenchless methods of restoring water supply networks and sewage system of different diameters, with their advantages and disadvantages, each of them being unique and requiring justification for the effective use a particular method in accordance with the operational, structural and technical characteristics of the networks. Features of organizational and technological solutions do not allow to trace the direct connection between the choice of each particular method based on the degree of damage to the network area (its categorization). In addition, it is difficult to estimate the cost equivalent of each of the methods and choose a more acceptable method based, for instance, on cost parameters.

To solve this problem, a system has been proposed that allows to reasonably choose a restoration method or repair of a sewer collector focusing on most parameters of its operation, and also taking into account the features of organizational and technological repair solutions (fig. 1).

3 Results

To implement each of the steps given in fig. 1, the software toolkit has been developed. For example, in order to determine the categorization of sewer network areas, an expert system was implemented as a system of fuzzy logic output in a trial version of the MATLAB system using the Fuzzy Logic Toolbox, which contains a set of GUI modules providing the implementation structural identification stage in a dialog mode.

The connection between the input and output variables is carried out using the knowledge base. It contains over 200 rules that take into account all the combinations of input variables and the effect of these combinations on the categorization of the sewer collector area.

One of the results of the system based on the set values of input variables is presented in fig. 2. Thus, with a reduction of the cross section to 50%, the presence of cracks (their opening width) having size of 10 mm, deformation of about 25%, in the presence of breaks in pipes up to 3 cm, corrosion of reinforced concrete by 50%, pre-treated wastewater and lack the wetlands on the pipeline route, the system assigns the category 2 to an area – a critical condition.

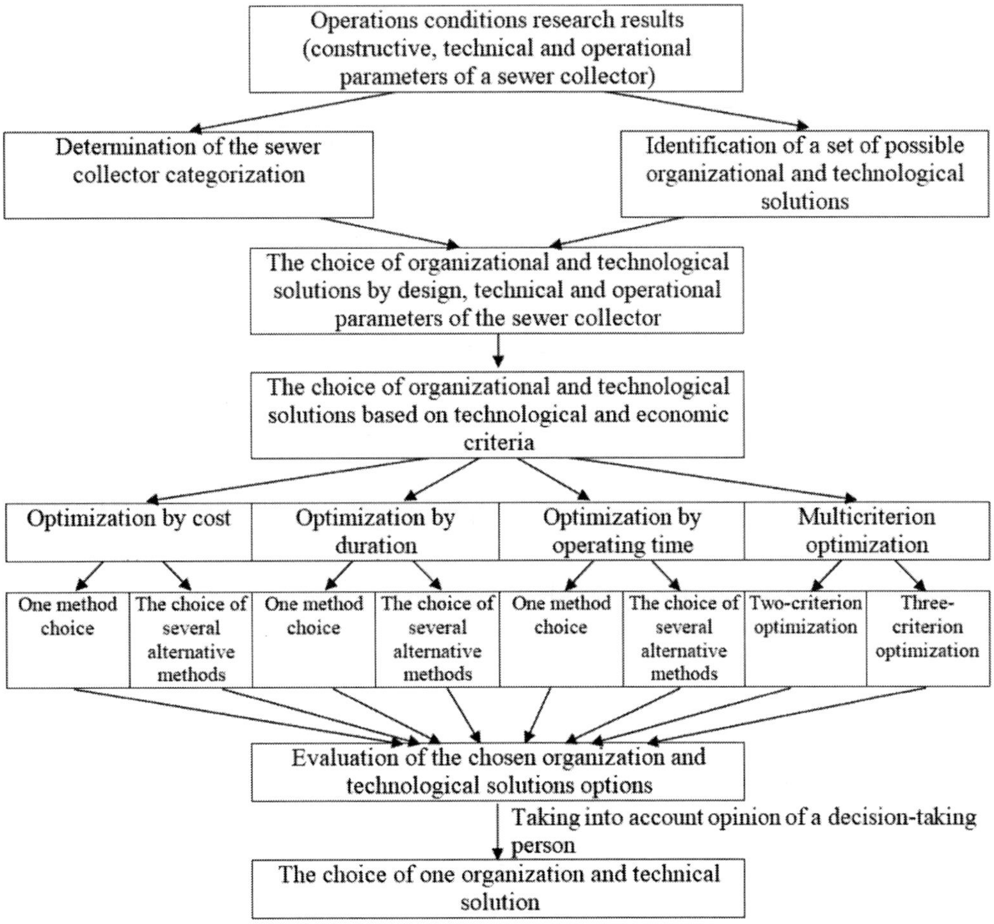

Fig. 1. The system of the solutions complex management for increasing the sewer collector service life.

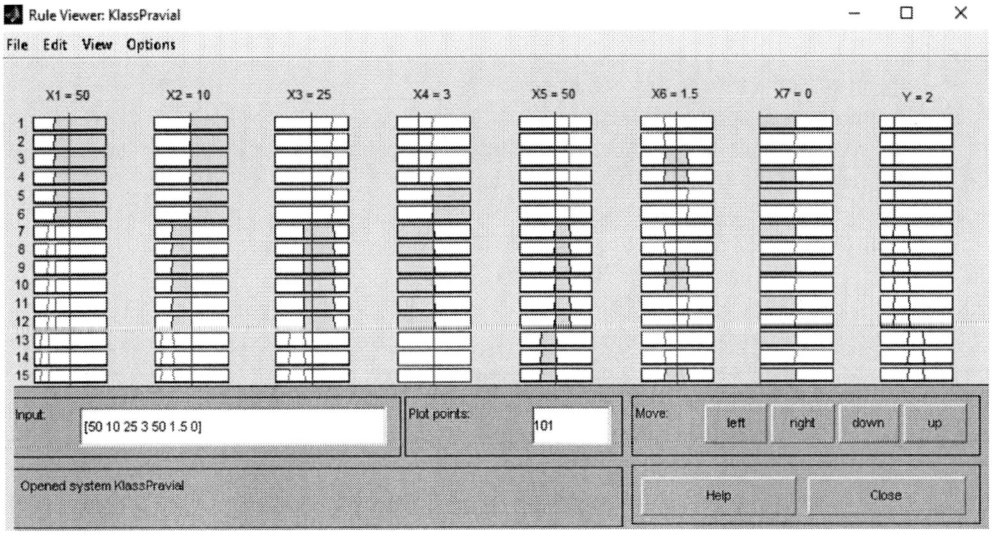

Fig. 2. The result of the system operation based on the set values of input variables.

A mechanism is developed which, based on the collector area parameters, allows choosing promising methods for its renovation. The solution of such a problem with the use of mathematical modeling is virtually impossible, since the diversity of the parameters of the area and the set of characteristics of each of the repair methods is very difficult to reduce to n-dimensional space of solutions, where n would equal to at least three. In addition, it is advisable to obtain more than one alternative method of repair of the sewer collector to choose from based on the design, technical and operational parameters of the network area. For this purpose, an algorithmic model of the reasoned choice of a repair method and restoration of the sewer collector area is proposed, which takes into account the design, operational and other characteristics of the site. The result of the model operation is a set of repair methods, which are selected taking into account the design, technical and operational parameters of the sewer collector area.

Mathematical models were developed to choose organizational and technological solutions based on technological and economic criteria: one-criterion optimization, subject to the choice of one method of the sewer collector area repair; one-criterion optimization, subject to the choice of several methods of the sewer collector area repair; multicriterion optimization subject to choosing the methods of the sewer collector area repair.

In a situation where the length of the area is insignificant, choosing one method of its renovation is the most expedient. For this purpose, mathematical models of one-criterion optimization were developed, subject to the choice of one method of sewer collector area repair:

$$\sum_{i=1}^{n} c_i \cdot l \cdot x_i \to min \qquad (1)$$

$$\begin{cases} x_i \in \{0;1\}, \quad i = \overline{1,n}; \\ \sum_{i=1}^{n_i} x_i = 1; \\ c_i \in C_{i_k}, \quad k = \overline{1,m}, \end{cases}$$

(1) – mathematical model in which optimization criterion is the cost of the repair works complex;

i – repair method and renovation number; $i=1,\dots, n$;

n – the number of repair methods;

c_i – the cost of repair works package (including materials and works performance costs) for 1 linear meter of the sewer collector by i-method;

l – the length of the repaired area, m;

x_i – logical variable reflecting the fact of choice of i-repair method:

$$x_i = \begin{cases} 1, \text{ in case } i\text{-repair method is chosen,} \\ 0, \text{ otherwise;} \end{cases}$$

C_{ik} – cost characteristics of i repair methods set which can be used in the area subject to its structural, technical, and operation k parameters;

m – the number of repair methods set which can be used in the area subject to its structural, technical, and operation k parameters.

Similarly formulated mathematical model in which optimization criterion is the duration of the repair works complex, mathematical model in which optimization criterion is the failure free operating period.

When operating entity has a sufficient time reserve, and the length of area is significant, or it is necessary to carry out the repair of a herring-bone pattern of sewer mains with tie-ins, the choice of different renovation methods is expedient. For this purpose, mathematical models of one-criterion optimization were developed, subject to the choice of several methods of sewer collector area repair:

$$\sum_{i=1}^{n}\sum_{j=1}^{m} c_{ij}\cdot l_j\cdot x_{ij}\rightarrow min \tag{2}$$

$$\begin{cases} x_{ij}\in\{0;1\}, \quad i=\overline{1,n}, \quad j=\overline{1,m}; \\ \sum_{j=1}^{n_i} x_{ij}=1, \quad i=\overline{1,n}; \\ c_{ij}\in C_{ij_k}, \quad k=\overline{1,p}, \end{cases}$$

(2) – mathematical model in which optimization criterion is the cost of the repair works package;

i – repair method and renovation number; i=1,..., n;

n – the number of repair methods;

j – the number of sewer collector area; j =1,..., m;

m – the number of sewer collector areas that are subject to repair;

c_{ij} – the cost of repair works package (including materials and works performance costs) for 1 linear meter of the j sewer collector area by i-method;

l_j – the length of the repaired j area, m;

x_{ij} – logical variable reflecting the fact of choice of i-repair method for j area:

$$x_{ij} = \begin{cases} 1, \text{ in case } i\text{-repair method for } j \text{ area is chosen,} \\ 0, \text{ in the opposite case;} \end{cases}$$

C_{ik} – cost characteristics of i repair methods set which can be used in the j area subject to its structural, technical, and operation k parameters;

p – the number of repair methods set which can be used in the area subject to its structural, technical, and operation parameters.

Similarly formulated mathematical model in which optimization criterion is the cost of the repair works package and mathematical model in which optimization criterion is the failure free operating period.

Besides, mathematical models of multicriterion optimization for solving the problem of choosing the method of the sewer collector area repair were developed.

With the use of modern specialized and standard computer technologies a software toolkit was developed for implementation of the mechanism for justified choice of the sewer collector area repair method and renovation.

The developed methodological principles of the choice and software tools were tested on the collector damaged section, which passes along the street Volohaevskaya (Kharkiv). This area includes seven pools. After a detailed study of the data on the areas technical condition and the definition of their categorization, the appropriate information is entered in the dialog box, based on which rational renovation methods are chosen for each individual case. The working area for introducing of the sewer collector area input parameters by the user and the result of an automated selection of promising repair methods on the basis of input data set by the user is provided in fig. 3. Here, the contrasting color automatically highlights those methods that can be applied based on the structural, technical and operational parameters of the area.

In order to analyze the optimal options for the renovation of the sewer collector using the developed mathematical and computer models, the results of problems solving were generalized:

– one-criterion optimization subject to the choice of one repair method;

– one-criterion optimization subject to the choice of different repair methods for the areas;

– multicriterion optimization.

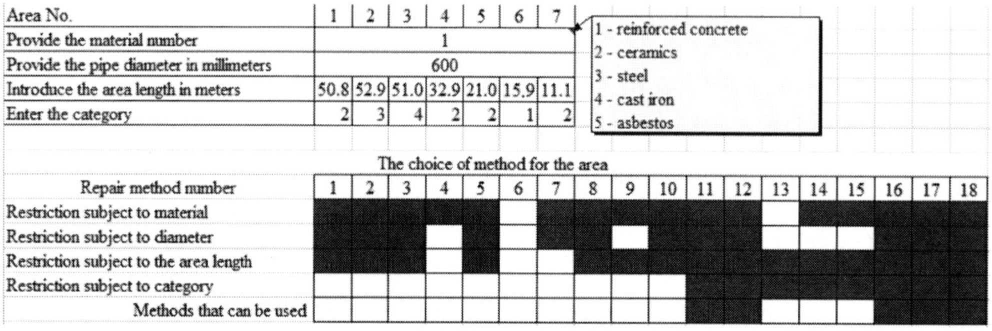

Fig. 3. Working area for introducing by the user input parameters of sewer collector area and automated selection of repair methods for the sewer collector area.

4 Conclusions

Thus, the introduction of methodological principles and software tools for choosing organizational and technological solutions to extend the service life of utility networks allows improvement of the efficiency of the decision-making process in identifying alternative methods of repair and renovation which is a reasoned argument in the context of insufficient financing of the plumbing and sewer industry.

References

1. A.I. Aleynikova, V.M. Volkov, D.F. Goncharenko, G.G. Zubko, O.V. Starkova. Rarities of Ukraine, *Methodological basis for underground utility networks service life extension: monograph: under the general editorship of Starkova O.V.*, **320** (2017)
2. A. Garmash, D. Bondarenko, G. Zubko, D. Goncharenko. World Journal of Engineering, *On renovation of the destroyed tunnel sewer collector in Kharkiv*, E **13**, 72 (2016)
3. D. Bondarenko, V. Bulhakov, A. Garmash, D. Goncharenko, S. Pilihram. Rarytety Ukrainy, *Sewage tunnels of Kharkiv: QUO VADIS: monograph: under the general editorship of D.F. Honcharenko*, **232** (2018)
4. I.A. Abramovich. Globus, *Wastewater Networks and Facilities. Calculations, Designing, Operation*, **288** (2005)
5. V.A. Orlov, V.A. Kharkin. ROST, *Development of a Strategy for Reconstructing Urban Drainage Networks*, E **3**, 20 (2001)
6. D. Stein. Ernst und Sohn, *Instandhaltung von Kanalisationen*, **941** (1999)
7. R. Stein, Sh. Ghaderi. Stein & Partner, *Wertemittung von Abwassernetzen*, **131** (2010)
8. K. Ochs. Bi UmweltBau, *Die Wahl einer Gesamtsanierungsstrategie*, E **5**, 19 (2013)
9. K. Körkemeyer, M. Schmitz. BauPortal ,*Werkstoffauswahl für Rohre und Schächte der Kanalisation*, E **7**, 22 (2014)

10. K. Körkemeyer, S. Kapitel. Springer-Verlag, *Zementgebundene Werkstoffe (Beton, Stahlbeton, Spannbeton), im Fachbuch „Rohrleitungen"*, E **1**, 25 (2015)
11. K. Körkemeyer. Betonwerk- und Fertigteiltechnik, *State-of-the-art sewer construction using precast elements*, E **2**, 59 (2015)
12. J. Kölbe. Wasser-Abwasser, *Benchmarking des Prozesser Wasserlustmanagement in Trinkwasserversorgungsystemen*, E **1**, 68 (2010)
13. O. Gutierre, J. Mohanakrishnan, K. Sharma, R. Meyer, J. Keller, Z. Yuan. Water Research, *Evaluation of oxygen injection as a means of controlling sulfide production in a sewer system*, E **42**, 17 (2010)
14. O.V. Starkova. KNUSEA, *System of scientifically grounded choice of organizational and technological solutions that provide increase in service life of sewer collectors: Thesis of a Dr.Sci.Tech,* **332** (2017)

Efficient method of eliminating accidental damage of sewer tunnel

Dmitriy Goncharenko[1], *Olga* Starkova[1], *Alevtyna* Aleinikova[1,*], *Yulyia* Kolomiiets[1], and *Oksana* Greenchuk[1]

[1]Kharkiv National University of Civil Engineering and Architecture, Sumska st. 40, 61002 Kharkiv, Ukraine

Abstract. The article is devoted to the development of technological and organizational solutions for elimination of local emergency damage to sewage tunnel collectors. In the article, conditions of operation and constructions of the sewage tunnel collector destroyed by corrosion are investigated on Grekovska street in Kharkiv, Ukraine; the methods of production of repair and restoration works are analyzed in this article; technological and organizational solutions have been developed for the liquidation of emergency damage to sewage tunnel collectors, taking into account technical, organizational and economic indicators. Thus, when comparing recovery options by different methods, it is noted that the proposed technology is more economically feasible in view of the relatively low cost of the material for this solutions. At the same time, the duration of the work by the proposed method of extrusion by introducing a secondary lining with anticorrosive properties will be reduced by 30%.

1 Introduction

The distribution system of sewerage utility systems is a sophisticated complex of facilities for sustainable sewage disposal. Its key elements include drainage pipelines and sewage tunnel collectors of various diameters characterizing the degree of development and redevelopment of a city. Under current conditions Ukrainian operating enterprises carry out their activities on the brink of their technological and organizational capabilities, as evidenced by high depreciation of fixed assets and emergency condition of many networks due to insufficient financing of the industry [1, 2]. As of 2015, the total length of drainage networks in Ukraine was 37.404 thousand km, including drainage networks in an emergency condition: 12.749 thousand km or 34% [3]. Compared with 2014, the total length of emergency drainage networks increased by 459 km or 3.6% for the whole country. Share of emergency drainage networks in Kharkiv in their whole length is 42 %. The trend of recent years indicates that the length of emergency networks increases annually by 0.67% [3].

Studies of many domestic and foreign scientists are devoted to operation, repair and reconstruction of sewage tunnel collectors [4-10]. D. Stein [11], M. Mahmoodian [12], N.

* Corresponding author: alevtynaal222@gmail.com

Rohem [13], W. Wang [14], M. Anbari [15] and other researchers of non-CIS countries studied various issues of construction of deep-laid sewage tunnel collectors, examination of their state, methods of repair and reconstruction. Meanwhile, the analysis of the publications of the above authors provides a basis to conclude that issues of research and development of economically viable private methods of reconstruction of sewage tunnel collectors is topical. This especially applies to tunnel collectors whose structures are subject to corrosion to a significant extent, which can lead to irreversible environmental consequences.

The improvement of operation of sewage pipelines and sewage tunnel collectors, the increase of their reliability is an important problem, which is still not fully resolved, so there is an urgent need to develop organizational and technological solutions for the sustainable operation of underground utility lines of the sewage system.

The purpose of this paper is to develop technological and organizational solutions for the elimination of accidental damages caused to sewage tunnel collectors.

The following tasks are formulated in the paper for achieving this purpose:

– to study conditions of operation and structures of the sewage tunnel collector on Grekovskaya street in Kharkiv destroyed by corrosion,

– to analyze methods of performance of repairing and reconstruction works,

– to develop technological and organizational solutions for the elimination of accidental damages caused to sewage tunnel collectors with the consideration of performance indicators.

2 Materials and methods

Discharges aggressive to concrete may likely appear in the urban sewerage system with its mixture of various discharges (industrial and household wastewater and storm water). At the same time, as numerous collapses of sewage tunnel collectors show, concrete and reinforced concrete structures cannot remain operable within their guaranteed service life (20-30 years) and often get out of order before its expiration [1]. With this regard, the main cause of collapses is the exposure of their arch surfaces to biogenic corrosion. When considering the possibility of reconstruction of structures of sewage tunnel collectors, special attention must be paid to corrosion of concrete and reinforced concrete structures: their widespread use resulted in a current emergency condition of almost tunnel collectors constructed and commissioned since the early 1950s.

For example, in January 2018 the asphalt-concrete surface on Grekovskaya street (Kharkiv, Ukraine) was initially damaged and sagged. As a result of the accident, a cavity with an approximate depth of 10 m and a diameter of 20 m appeared in a short time in the concrete sewage tunnel collector constructed in 1972, and the adjacent building collapsed. The tunnel was arch-type, its diameter was D=2540/2850 mm (diameter of the clear opening was D=2140/2450 mm), it was made of prefabricated reinforced concrete structures (M300 grade concrete) by shield tunneling with the use of the PShch-3,2 tunneling shield. It can be seen in the soil geological section that the tunnel was cut at a depth of 15 m in semi-solid clay and fine-grained watered sand at the groundwater level.

The accidental damage was caused to the deep-laid main collector tunnel in shafts 8 and 2 on the Grekovskaya street at the switching area of the main collector. The distance between the shaft No. 8 and the shaft No. 2 is 126 rm. The collapse area is in close proximity to the wastewater stilling chamber (shaft No. 8). The main cause of the local collapse is damage caused to the reinforced concrete arch of the tunnel collector by biogenic corrosion. Waste water in the stilling chamber falls from the mark 95.8 to the mark of the channel of the bypass sewage tunnel 83.9 (H=11.9 m). As a result of the stratification of the flow, gaseous sulfur trioxide is actively released, and it is converted into high-concentration sulfuric acid during subsequent reactions. In the absence of anticorrosive protection, concrete lining of the tunnel collector, which bears load of the soil column, becomes damaged. The collapse of the supporting arch structure resulted in moving soil inside the tunnel collector with further fall and collapse of the adjacent building.

The procedure for the elimination of accidents in sewage networks is stipulated by the technical operation rules for water supply and wastewater disposal systems of residential areas of Ukraine, approved by the Order No. 30 of July 05, 1995. Organizational and technological measures for the elimination of local damage to an operating tunnel collector with open access to soil inside it include pre-construction and basic works. At the first stage, it is necessary to stop the soil washing and to ensure the drainage of wastewater by switching to a reserve collector or by installing a bypass line. But there was no technical capability to switch the tunnel collector on the Grekovskaya street to another line, and a bypass line was installed in the shaft, that allowed to ensure the functioning of the water distribution and disposal system (Fig. 1). Main organizational and technological operations of the preparatory stage include:

– fencing of the emergency work site,

– arrangement of wastewater drainage via three lines of the PE pipeline (D=300 mm) with pumping stations,

– conducting the geological engineering survey in the emergency area,

– backfilling of the collapse cavern with soil to prevent further collapsing,

– conducting geophysical studies on the basis of the actual state of the soil mass with the indication of sheeting limits, karst caverns, existing ground water tables for the subsequent development of a practicable design solution for the method of reconstruction.

Fig. 1. Temporary disposal of wastewater.

The following was determined on the basis of the geophysical study report for the soil mass of the tunnel collector on Grekovskaya street from the shaft No. 8 to the shaft No. 2:

1. The absence of karst caverns in soil and the absence of loose soil over the arch part of the existing tunnel collector prove the integrity of the reinforced concrete lining of the sewage tunnel collector (its integrity was not established),

2. Small caverns and loose soil were found near the cavity over the existing tunnel line,

3. The survey was conducted in the subsoil waterlogging area. The ground water table is at the mark 98.5. The waterproof layer at the mark 94 consists of sandstones and marl. The aquifer is 4.5 m thick.

In the absence of a detailed study of the extent of the consequences of the accident and due to the unpredictability of conditions, it was resolved to fully fill the collapse cavern to enable subsequent development of technological solutions for the method of reconstruction of the tunnel collector on Grekovskaya street. To repair the damaged sewage tunnel collector, the following organizational and technological solutions may be used:

– full filling of the cavity, decommissioning of the damaged collector, construction of the new sewage tunnel collector (method 1);

– penetration into the cavity and reconstruction by inserting short polyethylene pipes (method 2);

– penetration into the cavity and reconstruction by the pipe-in-pipe method with the use of fiberglass pipes (method 3);

– penetration into the cavity by extrusion and reconstruction of the tunnel collector by introducing a secondary lining made of reinforced concrete rings with anti-corrosion coating (method 4).

The following technological scheme for repairing the sewage tunnel collector with penetration into the cavity by extrusion and further introducing a secondary lining made of reinforced concrete supporting rings with anticorrosive properties was developed. At the first stage, the tunnel shaft must be constructed on a curved section of the tunnel line to enable the performance of the works in both directions. Penetration into the cavity by extrusion must be made consistently from both sides by introducing reinforced concrete rings from the starting shaft No. 8 and the shaft No. 2. Reinforce concrete slabs in the stilling chamber No. 8 must be removed before such penetration to enable placing of working equipment and reinforced concrete rings inside the tunnel collector. Equipment for extrusion and reinforced concrete rings must also be installed in the reconstructed tunnel shaft. Then horizontal wells must be drilled through the soil in the cavity and steel rods with the detachable cutting edge must be pulled from both operating shafts. Pulling force applied to the cutting edge (ring-shaped) cuts the soil mass and introduces the rings in it. The supporting part of a reinforced concrete ring ensures the necessary rigidity and strength of the entire structure. Hydraulic jacks on the back side of the supporting ring embed the reinforced concrete rings with anti-corrosion coating in soil, and with this regard the precast lining is used as a jack pot for the jacks installed around the body. Then the jacks must be removed, and the area behind the supporting part must be reinforced with reinforced concrete components comprising the lining of the tunnel collector, if the cavity is being eliminated by extrusion. At the next stage, after the excavation of the next batch of soil, jacks push the cutting edge being supported by the components just installed.

The reconstruction is being performed step by step; each step depends on the maximum extension of rods of the hydraulic jacks. Soil is being excavated and removed manually with the use of hand-held power tools and small-size rock-removal wheeled machines.

After penetration into the cavity and installation of the secondary lining made of reinforced concrete rings, joints between reinforced concrete rings must be sealed with rapid setting expansive cement, and a cement and sand mortar must be injected in the interlining gap. The quality of the primary injection (i.e. filling the voids) must be checked by secondary control injection of the cement mortar.

When comparing the options of elimination of the accidental damage, it was taken into account that some operations to be performed in all four reconstruction options (methods 1-4) may be excluded, namely the preparation of the tunnel collector for reconstruction (in particular, switching waste water flows coming from the collector, preparation of starting and target shafts, etc. When reconstructing of the sewage tunnel collector with the use of trenchless reconstruction methods (methods 2, 3, 4), it was taken into account that the space between pipes has to be filled with concrete of the same volume and the same composition in both options: with polyethylene and fiberglass pipes. In the developed technology of reconstruction of the sewage tunnel collector by extrusion by introducing a secondary lining made of reinforced concrete rings with anticorrosive properties, the volume of concrete used for filling the space between pipes was less than the volumes of concrete used for methods 2 and 3.

The comparison of duration of reconstruction indicated for each option in the tables 1 and 2 gives us the possibility to conclude that the optimal duration of the work will be reached with the use of a method of penetration into the cavity by extrusion by introducing a secondary lining made of reinforced concrete rings with anticorrosive properties.

Table 1. Cost factors for materials used for elimination of the accidental damage caused to the sewage tunnel collector.

No. of the reconstruction method	Structures	Rated diameter DN	Unit	Price per unit*, UAH
1	Polyethylene pipe Spiro	1600	rm	35650.00
2	Polyethylene pipe Spiro	1600	rm	29300.00
3	Fiberglass pipe	1600	rm	28400.00
4	Reinforced concrete ring with anti-corrosion coating	1600	rm	9800.00

*prices are valid as of February 2018.

Table 2. Duration of elimination of the accidental damage caused to the tunnel collector on Grekovskaya street with the use of the mentioned methods (length of the site: 100 rm, diameter of the cavity: 20 rm).

No.	Reconstruction method	Duration in days, three-shift work
1	Full filling of the cavity, decommissioning of the damaged collector, construction of a new sewage tunnel collector (method 1)	15
2	Reconstruction by inserting short polyethylene pipes (method 2)	10
3	Reconstruction by the pipe-in-pipe method with the use of fiberglass pipes (method 3)	12
4	Reconstruction by extrusion by introducing a secondary lining (method 4)	9

4 Conclusions

The studies carried out gave the possibility to develop an economically and technologically effective technology of elimination of possible local collapses of sewage tunnel collectors. The method of extrusion for penetration into the cavity was proposed for reconstruction. When comparing different options of reconstruction (methods 1-4), it should be noted that the proposed technology is more economically feasible than the pipe-in-pipe method with the use of polyethylene or fiberglass pipes in view of high cost of the materials in the latter case. At the same time, the duration of the work by the proposed method of extrusion by introducing a secondary lining made of reinforced concrete rings with anticorrosive properties will be reduced by 30% in comparison with the methods 1-3.

References

1. D. Bondarenko, V. Bulhakov, A. Garmash, D. Goncharenko, S. Pilihram, *Kanalizatsiini tuneli Kharkova: QUO VADIS*, 232 (2018)
2. A. Aleynikova, V. Volkov, D. Goncharenko, G. Zubko, O. Starkova, *Methodological Foundations of Extension of the Operation Life of Underground Utility Networks,* 320 (2017)
3. O. Kravchenko, O. Yamko, Book of Reports of the International Congress and Engineering Exhibition "ETEWS-2017", **7**, 86 (2015)
4. I.A. Abramovich, Globus, *Wastewater Networks and Facilities. Calculations, Designing, Operation,* 288 (2005)
5. A. Garmash, D. Bondarenko, G. Zubko, D. Goncharenko, World Journal of Engineering, **13**, 72 (2016)
6. V. Vasilyev, Yu. Klementyev, Yu. Stolbikhin, Water Supply and Sanitary Engineering, **1**, 58 (2016)
7. V. Orlov, V. Kharkin. ROST, **3**, 20 (2001)
8. A. Kovalenko, *Development of Technological Solutions Used to Improve the Efficiency of Elimination of Accidents with Sewage Tunnel Collectors*, **138** (2002)
9. D. Goncharenko, A. Aleinikova, V. Volkov, S. Zabelin, Eastern-European Journal of Enterprise Technologies, **6**, 21 (2016)
10. A. Aleinikova, Actual Problems of Economics, **8**, 224 (2016)
11. D. Stein. Ernst und Sohn, *Instandhaltung von Kanalisationen*, 941 (1999)
12. M. Mahmoodian, A. Alani Mahmoodian, Journal of Materials in Civil Engineering, **29**, 1001 (2017)
13. N. Rohem et al., Composite Structures, **152**, 737 (2016)
14. W. Wang et al., Engineering Failure Analysis, **81**, 254 (2017)
15. M. Anbari, T. Massoud, R. Abbas, Journal of environmental management, **190**, 91 (2017)

 https://doi.org/10.1051/matecconf/201823002012

Bending deflection reinforced concrete elements determination

Dmitriy Kochkarev [1,*], *Taliat* Azizov[2], and *Tatyana* Galinska[3]

[1]National University of Water and Environmental Engineering, 33028 Rivne, Soborna St., 11, Ukraine
[2]Pavlo Tychyna Uman State Pedagogical University, 20300 Uman, Sadova St., 2, Ukraine
[3]Poltava National Technical Yuri Kondratyuk University, 36011 Poltava, Pershotravnevyi avenue, 24, Ukraine

Abstract. Longitudinal reinforced concrete elements stiffness exhaustion, often used in building practice, precedes obtaining bearing capacity, and therefore deflections determination becomes a determining factor in their design. In connection with it precise methods for determining such reinforced concrete elements deflections become especially relevant. The elastic-plastic properties of concrete and cracks in the stretched zone of reinforced concrete elements lead to a significant change in their bending stiffness. That is why the deflections determined by the materials classical resistance formulas differ significantly from the real ones. A large quantity of methods for determining deflections is based on the elastic characteristics correction of reinforced concrete elements consolidated section. Such methods, although providing calculation satisfactory results, are rather approximate and have empirical nature, due to it they have limited application. More precise calculation methods consist of curvature usage to determine deflections. The curvature of reinforced concrete elements cross sections is determined directly from the equilibrium equations, which are written taking into account nonlinear materials deformation diagrams. Calculation examples for bending reinforced concrete elements deflection are given.

1 Review of recent studies and publications

Bending deflection reinforced concrete elements determination is a very topical task, when designing large spans reinforced concrete elements. It is due to the connection with the fact that the elastic-plastic properties of concrete and cracks in the stretched zone of reinforced concrete elements lead to a significant change in their stiffness. That is why the deflections determined by the formulas of the classical materials resistance differ significantly from the real ones. A large number of methods for determining deflections are based on the correction of elastic characteristics of reinforced concrete elements consolidated sections [1, 2]. The most well-known is the Branson formula [3], which modified expression is included in the design norm AS3600-2001 [4]. Subsequently, it was corrected by such

* Corresponding author: dim7@ukr.net

researchers as Lutz [5], Al-Zaid, Al-Shaikh and Abu-Hussein [6], Fikry and S. Thomas [7]. The proposed methods, providing satisfactory calculation results, are rather approximate and have an empirical nature, due to their limited application.

More precise calculation methods are in curvature usage to determine deflections. Reinforced concrete elements cross sections curvature is determined directly from the equilibrium equations, which are written taking into account nonlinear materials deformation diagrams or piecewise linear diagrams [8, 9]. There are methods where a reinforced concrete beam is considered as a composite rod [10]. These methods are sufficiently precise, but not convenient in the complex repeatedly statically uncertain systems calculation.

Let's show curvature changes schedule in the bending reinforced concrete elements sections (Fig. 1). In Fig. 1 with points the stages of the cross sections stress-strain state are marked. Particular, p.1 corresponds to the moment of cracks formation; p.3 is fluidity obtaining moment in the stretched armature, p.4 and p.5 are destruction moment. At equation each stage, the equilibrium has its own form. Single dependence adoption on the entire gap necessarily leads to adjusting empirical coefficients usage. That is why the vast majority of scientists make curvature approximation only within certain areas. But all these methods are also approximate. It leads to averaged section concept usage. A more precise method for calculating deflections is to consider separate cross sections with cracks and between cracks, which allows getting a complete picture of a cracked formation. For most engineering calculations it is quite enough to use methods using the averaged section, which are proposed to be considered.

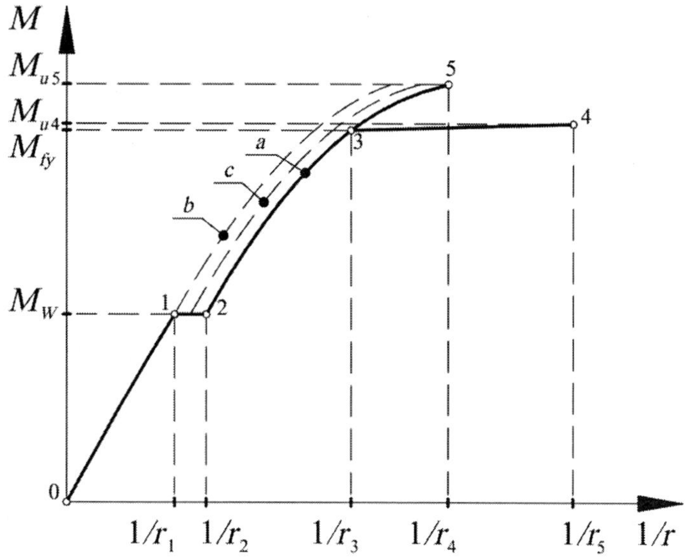

Fig. 1. "Moment-curvature" for bending elements dependence.

2 Basic material and results

Consider the methods for calculating bending deflection reinforced concrete elements, which are based on the method of calculation resistance of reinforced concrete [9] and on the use of cross-section curvature. In this case, deflections are determined by the known expression Mora

$$f = \int_0^l \overline{M}_x \left(\frac{1}{r}\right)_x dx. \quad (1)$$

Getting the general equation of the continuous curvature function $(1/r)_x$ along the length of the element is quite complicated due to the change in the deformation properties and cracks presence in the stretched zone. Therefore, for the practical calculations, the Simpson formula is used, using the discrete values of the curvature defined in the specific sections $(1/r)_x$

$$f = \frac{l^2}{3n}\left[\begin{array}{l}\alpha_0 \, 1/r_0 + \alpha_n \, 1/r_n + 4\left(\alpha_1 \, 1/r_1 + \alpha_3 \, 1/r_3 + \ldots + \alpha_{n-1} \, 1/r_{n-1}\right) \\ + 2\left(\alpha_2 \, 1/r_2 + \alpha_4 \, 1/r_4 + \ldots + \alpha_{n-2} \, 1/r_{n-2}\right)\end{array}\right], \quad (2)$$

where $1/r_i$ is beam curve i - section; α_i is unit force influence coefficient at the moment i -section; l is span element; n is section quantity along beam length in this case it should be paired.

A simplified equation is used for the preliminary approximation of deflections

$$f = \frac{1}{r_m} k_m l^2, \quad (3)$$

where $1/r_m$ is curved beam in a dangerous section; k_m is coefficient depending on the calculation scheme.

It is expedient to determine the curvature of an element directly from equations of equilibrium. To do it, take appropriate materials deformation diagram for concrete and reinforcement. The curvature values are determined by the formula

$$1/r = \frac{\sum \varepsilon}{d}, \quad (4)$$

where $\sum \varepsilon$ is a total relative deformation of the compressed concrete edge and stretched reinforced; d is working section height.

Total relative deformations value of compressed edge concrete and stretched reinforcement. $\sum \varepsilon$ is in a clear functional dependence on the percentage of reinforcement ρ_f, stresses in reinforced concrete σ_{zM}, concrete class C and reinforced class A

$$\sum \varepsilon = f(\rho_f, \sigma_{zM}, C, A). \quad (5)$$

For an example, let's look at a fragment of the table (see Table 1), which allows to determine the total relative deformation of the compressed concrete edge and the stretched reinforcement $\sum \varepsilon$.

The tensioning in reinforced concrete are determined by

$$\sigma_{z\grave{I}} = \frac{\grave{I}}{W_c}, \quad (6)$$

where M is bending moment in the section; W_c is elastic moment of working section concrete resistance, $W_c = bd^2/6$.

Another method for determining deflections is to approximate the dependence $M-1/r$. Such an approximation can be carried out only within the stages of the stress-strain state. Since the deflections of reinforced concrete elements, mainly, are within the range of 0.5-0.85 from the destructive load, therefore the approximation of dependence $M-1/r$ it is worth doing exactly within such limits. Dependence areas $M-1/r$ in such limits corresponds to the work of reinforced concrete elements until the yield strength in the reinforcement is reached. Approximation of the diagram section straight lines $M-1/r$ displayed sufficient convergence, where the correlation coefficient is within the range of 0.956-0.98 for various reinforcement proportion, reinforcement and concrete classes. For practical calculations, it is more expedient to perform approximation in coordinate's $\sigma_{zl} \; - \sum\varepsilon$.

As a result of the approximation by the least squares method, the dependence of the form is obtained

$$\sigma_{zM} = a_i + b_i \sum\varepsilon, \tag{7}$$

where a_i, b_i are power factors in MPa, which depend on the class of concrete and reinforcement.

Table 1. To determinate the stress-deformed state parameters of bending reinforced concrete elements for short-term load (reinforcement A400C).

Concrete class	Load level	Proportion of reinforcement, ρ_f								
		1			2			3		
		σ_{zM}, MPa	$\sum\varepsilon$ $\times10^4$	σ_s, MPa	σ_{zM}, MPa	$\sum\varepsilon$ $\times10^4$	σ_s, MPa	σ_{zM}, MPa	$\sum\varepsilon$ $\times10^4$	σ_s, MPa
C16/20	M_{W1}	3.7	3.0	31.0	5.0	3.3	30.6	6.2	3.5	30.3
	M_{W2}	3.7	4.3	57.2	5.0	4.2	40.1	6.2	4.1	34.1
	0.4	8.3	11.1	152.7	12.8	11.8	124.2	13.6	9.9	90.4
	0.6	12.4	17.6	235.4	19.1	18.8	192.6	20.4	15.9	140.8
	0.8	20.2	33.7	395.9	27.1	30.2	284.8	29.0	25.9	209.0
	1.0	20.6	65.7	400.0	31.9	46.0	380.3	34.1	40.5	281.6
C20/25	M_{W1}	4.1	3.0	31.8	5.4	3.3	31.4	6.7	3.5	31.1
	M_{W2}	4.1	4.4	63.0	5.4	4.4	43.3	6.7	4.2	36.6
	0.4	8.5	10.7	155.4	14.7	12.8	142.3	16.2	11.1	106.8
	0.6	12.8	16.9	239.4	22.1	20.3	219.7	24.3	17.7	165.7
	0.8	20.8	31.5	400.0	31.3	31.9	322.8	34.4	28.3	244.9
	1.0	21.3	72.4	400.0	36.8	49.3	400.0	40.5	43.1	323.9
C25/30	M_{W1}	4.7	3.2	33.7	6.1	3.4	33.4	7.5	3.6	33.1
	M_{W2}	4.7	4.7	72.2	6.1	4.7	48.7	7.5	4.6	40.6
	0.4	8.7	10.1	155.9	15.5	12.7	147.8	18.7	12.1	122.1
	0.6	13.0	16.2	241.4	23.3	20.0	227.9	28.0	19.3	189.1
	0.8	21.3	32.7	400.0	33.7	31.6	341.1	37.4	28.0	259.9
	1.0	21.7	77.7	400.0	38.8	50.6	400.0	46.7	45.0	362.3

Note. Intermediate values are determined by straight-line interpolation.

Change detailed analysis in coefficients has shown that the coefficient a_i depends on the change in the percentage of reinforcement in the section insignificantly. Without significant error, it can be taken as a class concrete function concrete, $a_i = a$. Coefficients value a and b_i are given in Table 2, which also contains the formulas for the calculation of intermediate values b_i and ρ_f at interpolation.

Using formulas (4), (6) and equation (5) we write

$$\frac{M_i}{W_c} = \grave{a} + b_i d \, {}^{1}\!/_{r_i}. \qquad (8)$$

Finally, from the equation (8) it will determine the curvature value

$${}^{1}\!/_{r_i} = \frac{M_i}{b_i W_c d} + \frac{a}{b_i d}. \qquad (9)$$

Equation (9) allows setting deflection formulas definition for the most common statically-defined calculation schemes. Table 3 displays the derived deflection formulas for some calculation schemes.

Analyzing them, it is clear that the second part of the formula does not depend on the type of load, but only on the type of beam and the place of the desired deflection. Consequently, if there are several power influences on the element, then the complete deflection will consist of the sum of deflections from each impact and will be determined by the formula

$$f = \sum_{i=1}^{n} f_{el,i} - \grave{A}_{\overline{I}} \grave{a} / b_i d, \qquad (11)$$

where $f_{el,i}$ is deflection of beams from power influences on formulas of classical materials resistance with replacement of flexural stiffness on equation $b_i W_c d$; $A_{\overline{M}}$ is distribution area of single strength moment; n is power influences quantity.

Table 2. Coefficients values for calculating reinforced concrete bending elements deflections, reinforced with reinforcement A400, A500, in operational stages.

Concrete class	a, MPa	Reinforced proportion ρ_f									
		0.50	0.5-1.00		1.00	1.00-2.00		2.00	2.00-3.00		3.00
			α	β		α	β		α	β	
		Coefficients b_i, 10^{-4} MPa									
C8/10	1.845	0.262	0.11	0.30	0.410	0.21	0.20	0.614	0.31	0.15	0.764
C12/15	2.116	0.301	0.12	0.37	0.484	0.23	0.26	0.741	0.37	0.19	0.929
C16/20	2.146	0.343	0.14	0.40	0.544	0.24	0.31	0.851	0.40	0.22	1.075
C20/25	2.256	0.355	0.11	0.49	0.598	0.27	0.33	0.929	0.44	0.25	1.175
C25/30	2.455	0.368	0.12	0.49	0.613	0.23	0.38	0.992	0.42	0.29	1.277
C30/35	2.701	0.374	0.10	0.54	0.644	0.26	0.38	1.024	0.44	0.29	1.315
C32/40	2.769	0.396	0.13	0.53	0.663	0.29	0.38	1.038	0.36	0.34	1.376
C35/45	2.803	0.398	0.11	0.57	0.684	0.27	0.42	1.102	0.50	0.30	1.404
C40/50	2.843	0.419	0.13	0.57	0.706	0.26	0.44	1.148	0.49	0.33	1.475
C45/55	2.849	0.437	0.15	0.57	0.723	0.26	0.46	1.183	0.50	0.34	1.526
C50/60	3.188	0.452	0.18	0.55	0.725	0.25	0.48	1.201	0.48	0.36	1.563
$b_i = \alpha + \beta \times \rho_f$, $\rho_f = (b_i - \alpha)/\beta$, $\frac{1}{r} = \frac{M_e}{b_i W_c d} - \frac{a}{b_i d}$.											

Table 3. Formulas for defining deflections for some calculation schemes.

№	Beam calculation scheme	Formula for determining deflection
1	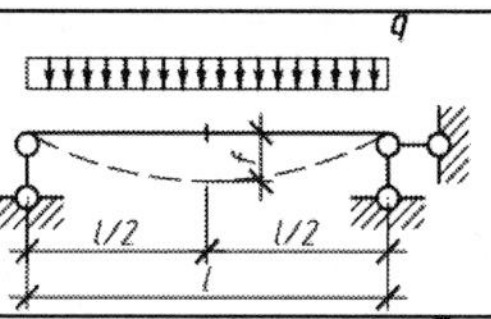	$f=\frac{5}{384}\frac{ql^4}{b_iW_cd}-\frac{l^2}{8}\frac{a}{b_id}.$
2	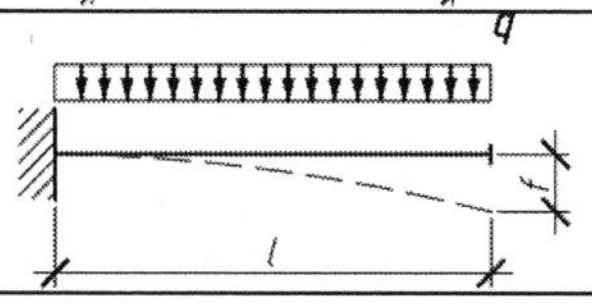	$f=\frac{1}{8}\frac{ql^4}{b_iW_cd}-\frac{l^2}{2}\frac{a}{b_id}.$

Let us compare the beam deflections on two supports with a uniformly distributed load in two methods: the Simpson formula, as the most accurate, and formula (9). We will accept reinforcement class A400C, concrete C16/20 and reinforcement proportion: 0.05; 0.25; 0.5; 1.0; 2.0; 3.0 for load levels 0.6; 0.8; 1.0. The calculation results are given in Table 4. It displays that the deflections, calculated by formula (9), are not significantly deviating from the most accurate result, even at a load level of 1.0. It allows recommending such a method for calculating deflections for practical applications.

Table 4. Reinforced concrete elements relative deflections comparison determined by different methods for the concrete class C16/20.

Reinforced Proportion ρ_f, %	Load level	$\left[\frac{f}{l}\frac{1}{\lambda}\right]_1$ for (2)	$\left[\frac{f}{l}\frac{1}{\lambda}\right]_2$ for (9)	$\left[\frac{f}{l}\frac{1}{\lambda}\right]_3$ for (3)	$\frac{\left[\frac{f}{l}\frac{1}{\lambda}\right]_2}{\left[\frac{f}{l}\frac{1}{\lambda}\right]_1}$	$\frac{\left[\frac{f}{l}\frac{1}{\lambda}\right]_3}{\left[\frac{f}{l}\frac{1}{\lambda}\right]_1}$
0.05	0.6	6.07	4.78	7.03	0.79	1.16
	0.8	10.62	10.70	11.89	1.01	1.12
	1.0	16.43	16.62	18.49	1.01	1.13
0.25	0.6	4.13	4.26	4.38	1.03	1.06
	0.8	6.22	6.71	6.59	1.08	1.06
	1.0	9.20	9.15	10.15	0.99	1.10
0.5	0.6	3.16	3.36	3.31	1.06	1.05
	0.8	4.67	5.06	4.91	1.08	1.05
	1.0	6.82	6.77	7.47	0.99	1.09
1.0	0.6	2.39	2.60	2.47	1.09	1.04
	0.8	3.49	3.82	3.65	1.09	1.05
	1.0	5.09	5.04	5.57	0.99	1.09
2.0	0.6	1.81	2.04	1.86	1.13	1.03
	0.8	2.64	2.95	2.76	1.12	1.05
	1.0	3.90	3.85	4.29	0.99	1.10
3.0	0.6	1.56	1.81	1.60	1.16	1.03
	0.8	2.28	2.60	2.39	1.14	1.05
	1.0	3.43	3.39	3.81	0.99	1.11
Average value x_m					1.04	1.08
Average quadratic value σ,%					12.54	2.48
Variation coefficient ν,%					12.04	2.31

3 Examples of application of the proposed calculation method

Calculate the one-pass reinforcement concrete beam working cross-section $b\times d$=40×36 cm, made of concrete class C50/60 and reinforcement A400C. Beam Cross section beam is reinforced 4∅18 A400 (A_s= 10.179 cm²) in the lower part. Estimated span l_0=6 m. It has an operational load q=25 kN/m.

Determine the deflection by close and precise methods. Solution.

1) The method of deflections linearization, by formula (2):

- Calculate the moment of working resistance and reinforcement portion section

$$W_c = \frac{bd^2}{6} = \frac{40\times 36^2}{6} = 8640 \text{ cm}^3;\ \rho_f = \frac{A_s}{bd}\times 100\% = \frac{10.179}{40\times 36}\times 100\% = 0.707\%,$$

- According to Table 1, determine linearization coefficients

a=3.188 MPa; $b_i=\alpha+\beta\times\rho_f$=0.18+0.55×0.707=0.57×10⁴ MPa;

- By formula (1, tab. 3), calculate the deflection of the beam

$$f_1 = \frac{5}{384}\frac{ql^4}{b_i W_c d} + \frac{l^2}{8}\frac{\grave{a}}{b_i d} = 1{,}695\,\text{cm}.$$

2) Determining one curvature deflections method, by the formula (3):

- Determine the tension $\sigma_{z\grave{i}} = \frac{ql^2}{8\times W_c} = \frac{25\times 6^2\times 10^3}{8\times 8640} = 13.02$ MPa.

- According to Table 2 where ρ_f= 1% determine $\Sigma\varepsilon$=18.719×10⁻⁴.

- Than $\frac{1}{r} = \frac{\Sigma\varepsilon}{d} = \frac{18.719\times 10^{-4}}{36} = 5.2\times 10^{-5}$ cm⁻¹.

$$f_2 = \frac{5}{48}\frac{1}{r_m} l^2 = 1.95\,\text{cm}.$$

3) Precise method for determining deflections using formula (4).

For calculation precise method, the element was broken in length by 10 sections, each having a defined curvature (see Table 5).

Table 5. Deflection definition by the precise method.

Distance from the supporting structure to the cross-section, m	α_i	σ_{zM}, MPa	ε_{zi} ×10⁴	$1/r_i$ ×10⁴, cm⁻¹	$\alpha_i\times 1/r_i$ ×10⁴, cm⁻¹
0	0	0	0	0	0
0.6	0.05	4.688	2.625	0.073	0.004
1.2	0.10	8.333	9.111	0.253	0.025
1.8	0.15	10.938	13.828	0.384	0.058
2.4	0.20	12.500	17.362	0.482	0.096
3.0	0.25	13.021	18.719	0.520	0.130

Deflection in terms of equation (4) was $f_3 = 1.485$ cm.

As can be seen from the calculations, the error of defining the deflection by the first method was 12.4%, by the second 23.85%.

To check the results can be used the following online resource http://sciencehunter.net/Services/Apps/Concrete.

4 Conclusions

The simplified and precise methods of deflection bending reinforced concrete elements determination according to curvature are offered. As shown by the calculations performed at operating load levels within the range of 0.5-0.85, approximate methods give satisfactory results. In some cases, simplified methods can exaggerate the actual deflections of reinforced concrete elements. In such cases it is worth using more accurate calculation methods.

References

1. J.K. Wight, J.G MacGregor. *Reinforced Concrete: mechanics and design* (New Jersey: Upper Saddle River, 2009)
2. J.C. McCormac, J.K. Nelson. *Design of Reinforced Concrete* (Hoboken, NJ:John Wiley & Sons, 2006)
3. D.E. Branson, HPR, **7**, 78 (1965)
4. AS3600-2001. *Australian Standard for Concrete Structures* (S.A, North Sydney, 2001)
5. L.A. Lutz, ACI Journal, **70**, 207-213 (1973)
6. R.Z. Al-Zaid, and A.H. Al-Shaikh, M. Abu-Hussein, ACI Journal, **88(2)**, 184-190 (1991)
7. A.M. Fikry, and C. Thomas, ACI Journal, **95(4)**, 444-455 (1998)
8. ENV 1992-1. Eurokode- 2. Design of concrete structure. Part 1, General rules and rules for buildings (GEN, 1993)
9. D. Kochkarev, T. Galinska, O. Tkachuk. UAE, **7(3)**, 176-182 (2018)
10. T. Azizov, A. Azizova, S. Al Ghadban. UAE, **7(3)**, 567-574 (2018)

Influence of loading level on the bearing capacity of RC columns strengthened by jacketing

Pavlo Krainskyi[1], *Yaroslav* Blikharskyy[2,*], *Roman* Khmil[1] and *Pavlo* Vegera[1]

[1]Lviv Polytechnic National University, Department of Building Constructions and Bridges, 79013, 6, Karpinskogo str., Lviv, Ukraine
[2]Lviv Polytechnic National University, Department of Highways and Bridges, 79013, 6, Karpinskogo str., Lviv, Ukraine

Abstract. Compressed reinforced concrete elements are quite common in construction. An example of such elements is the columns of industrial buildings, which are loaded with eccentric force, for example, from a bridge crane, columns of multistory buildings, structures subjected to compression and bending. When the reinforced concrete structure becomes unfit to future operation, it is often more economical and easy to retrofit it than to change it for a new one. Most of these structures are influenced by certain loads and unloading the structures before strengthening is not always possible. Therefore, in this work reinforced concrete structures strengthened by jacketing are investigated. To simulate the real life conditions, the loading level of 65-70% of the not strengthened column bearing capacity was maintained during jacketing. This load level simulates the actual service load on the structure. The bearing capacity load and serviceability limit state of not strengthened and jacketed reinforced concrete columns were investigated and the experimental results are presented in this paper.

Introduction

Reinforced concrete is a widespread material in construction. In structural elements, subjected to compression with bending, compressive stresses are taken by concrete while tensile stresses are transmitted to the rebar. At the same time, concrete serves as protection of structures from corrosion and various damages.

Many works are devoted to the use of improved high-strength concrete in reinforced concrete structures [1, 2]. In particular the investigations of special mineral additives like zeolite should be noted, which improve durability and corrosion resistance of cements [3].

Since most of the reinforced concrete structures in operation are made of traditional concrete, the issue of extending their serviceability time is relevant.

Many factors cause necessity for repairing or strengthening of buildings and constructions. Often it is corrosion or damages of construction [4, 5] during its lifetime

*Corresponding author: yaroslav.z.blikharskyy@lpnu.ua

from adverse environment influence. Also, it is new requirements for design, through changing of the functional purpose, installation of new equipment, upgrading engineering networks, improving architectural qualities of the building.

There are some new methods of strengthening RC constructions [6]. Reinforced concrete jacketing is one of the traditional and common methods of RC structures strengthening [7]. RC jacketing can be applied to both flexural and compressed members and has some significant advantages. One of the important in RC jacketing is the bond between repair materials and concrete substrate [8]. Jacketed member usually do not need additional corrosion or fire protection. But in jacketing need to provide shear strength or reinforce them too [9].

Many studies have been carried out to determine the stress-strain state of the jacketed RC members [10-13], bonding and surface preparation influence on the performance of jacketed structure [14] and seismic retrofitting using RC jacketing [15, 16].

Nevertheless, further study of this type of strengthening is relevant, since little is known about the performance of the structures strengthened under loads of different magnitude. The goal of this work to check the strengthening and serviceability effect of columns strengthened without and with initial load as 65-70% of C-1 and C-2 reference columns load-bearing capacity.

1 Research methodology

1.1 Test specimen description and material properties

All specimens were tested as pinned columns on a compression test stand (Fig. 1). This stand was designed to conduct compression tests in horizontal position which makes it more convenient to read data from mechanical gauges. Eccentrically loaded for every column eccentricity was 150mm. Load to the columns was applied incrementally by hydraulic jack. Loading level was determined from the readings of a ring dynamometer.

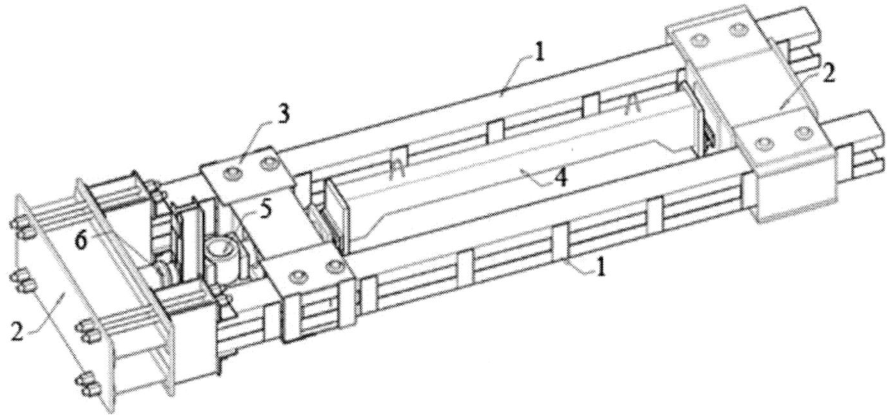

Fig. 1. Stand for columns compression test (1- steel constructions of stand; 2- rigid steel beam; 3- rigid steel beam with plunger; 4- tested column; 5- ring dynamometer; 6 - hydraulic jack).

The length of the columns was 2.2 m and a cross section dimensions before jacketing was 0.14 x 0.18 m. Columns were tested as compressed-bent elements and were manufactured with corbels on both ends for that purpose. Cross-section dimensions of the jacketed columns were increased to 0.2 x 0.26 m. All six columns were manufactured from the same materials. Material specimens were selected and tested to determine the mechanical properties. For cross section dimensions and reinforcement details see

Fig. 2 a, b. Reinforcement ratio of not strengthened column was 1.8%. Four 12 mm rebar were used as longitudinal reinforcement with 6mm rebar used for transverse reinforcement. The rupture stress of main rebar was 720 MPa; yield stress – 630 MPa; Young's modulus – 210 GPa. The test of rebar's were made as in [17]. Average concrete cube strength was 37 MPa, which corresponds to C25/30 concrete class.

Reinforced concrete jacketing length was 1.7 m, reinforcement ratio was 1.17%. Four 10 mm rebar were used as jacketing longitudinal reinforcement (Fig 2, 3) with 6mm rebar used for ties. Mechanical properties of concrete and rebar were the same as previous. The preparation of the concrete surface before jacketing consisted only from dust removal and sample hydration.

Fig.2. Reinforcement details of strengthened column specimen.

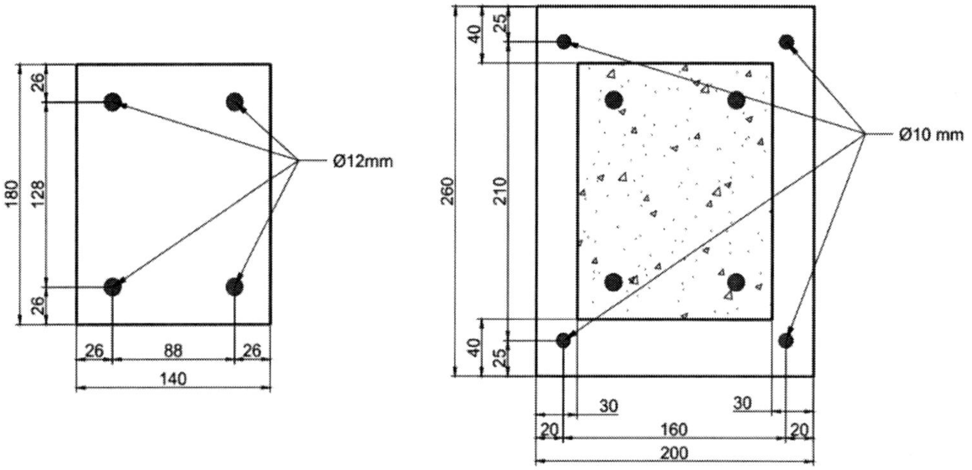

a) b)

Fig.3. Cross section dimensions and reinforcement details of unstrengthened (a) and strengthened (b) specimen's.

1.2 Experimental program

The first two columns C-1 and C-2 were reference columns. They were tested without strengthening to determine the bearing capacity of the specimen and to adopt the service load value for the next tests.

Columns C-3-0 and C-4-0 were also reference columns. They were tested after strengthening but without any initial load. The load-bearing capacity of these specimens was considered as maximum possible and was used for comparison of the strengthening effect of the following jacketed columns.

Columns C-5-0.7 and C-6-0.7 were initially loaded to a service load which was adopted as 65-70% of C-1 and C-2 reference columns load-bearing capacity. Afterwards, they were strengthened and tested to failure.

Columns were tested with pinned supports. Load was applied incrementally as a force on one of the supports. The following data was acquired during the tests of the columns: strain in the rebar; strain on the compressed face of the concrete; deflection of the mid-height of the column.

2 Results of the experimental study

All columns failed in a flexural manner at midsection. In case of columns C-01 and C-02 bearing-capacity was exceeded after column's tensioned reinforcement yielded, thus we consider the load when it happened a limit load. All strengthened specimens exceeded bearing-capacity after yielding of both column and jacketing reinforcement, which was accepted as a bearing-capacity load for the strengthened specimens.

Serviceability limit state allowed at load of maximum deflection. According to Ukrainian codes for serviceability limit state [18] the maximum deflection for test columns in this work 1/150 from length. This value 14.7 mm.

2.1 C-1, C-2 reference columns

C-1 and C-2 columns were tested to failure without strengthening. Specimens failed in flexural manner. Average load-bearing capacity of two specimens was 175 kN (Fig. 4).

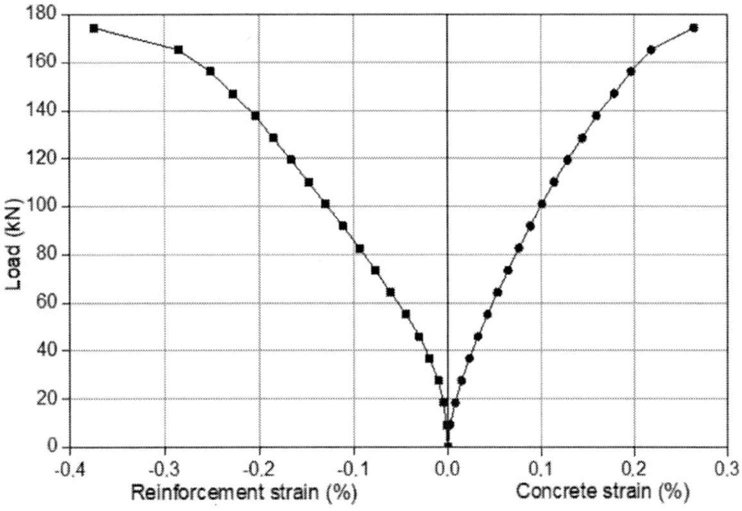

Fig. 4. Load vs reinforcement (concrete) strain for tested columns C-1 and C-2.

From Fig. 5 we can see that the loads that corresponded to a maximum allowed deflection of 14.7 mm were 122 kN and 125 kN.

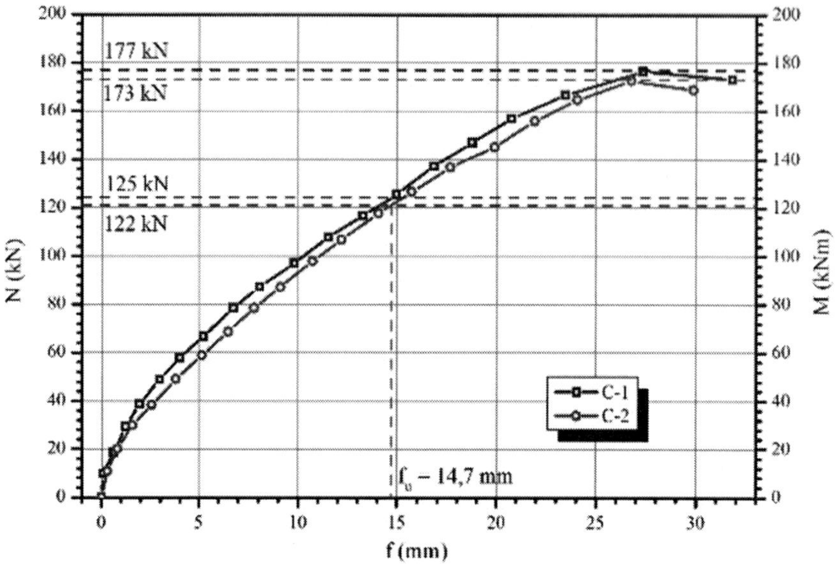

Fig. 5. C-1 and C-2 deflection at mid-height plot.

Considering these values we adopted a service loads for following tests of 123.5 kN.

2.2 C-3-0, C-4-0 reference strengthened columns

C-3-0 and C-4-0 columns were strengthened without initial loading and tested to failure. Specimens failed in flexural manner. Average load-bearing capacity of two specimens was 475 kN. It is 2.7 times bigger than the load-bearing capacity of C-1 and C-2 (see Fig. 6). It was consider it to be a maximum possible effect from strengthening since there were no initial stresses or deformations in the columns before strengthening.

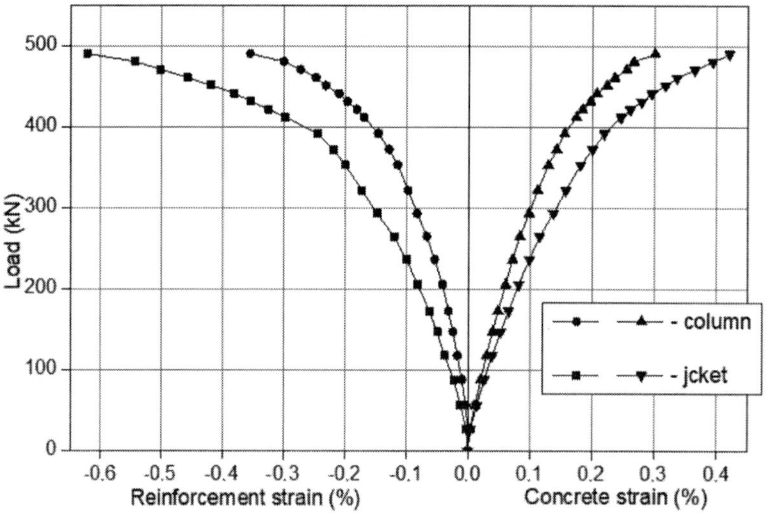

Fig. 6. Load vs reinforcement (concrete) strain for tested columns C-3-0 and C-4-0.

The load that corresponded to a maximum allowed deflection (Fig. 7) of 14.7 mm were 405 kN and 380 kN. Considering these values we adopted a service loads for following tests of 392.5 kN.

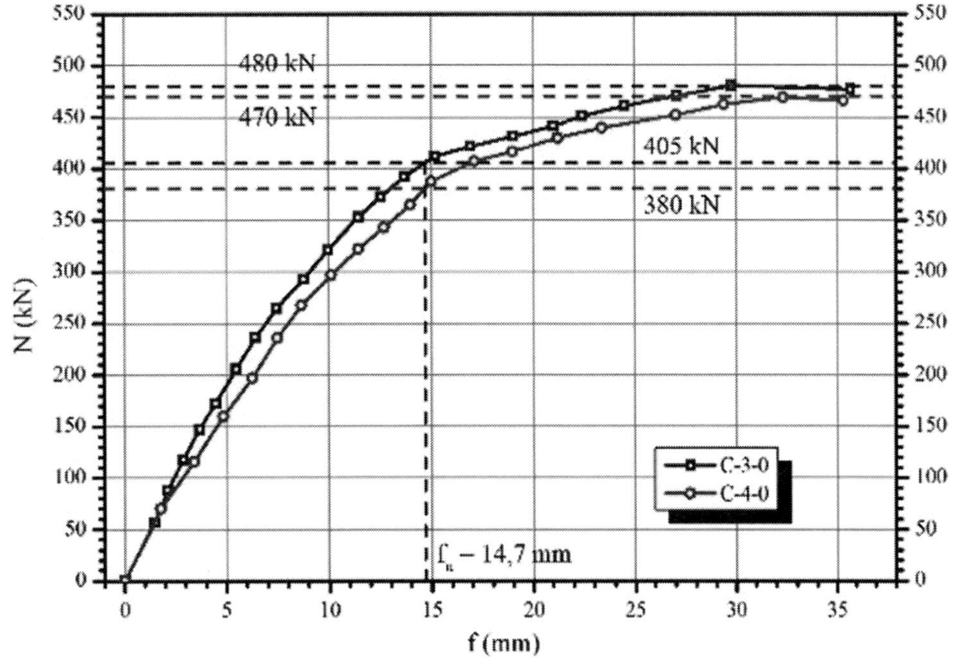

Fig.7. C-3-0 and C-4-0 deflection at mid-height plot.

2.3 C-5-0.7, C-6-0.7 main strengthened columns

C-5-0.7 and C-6-0.7 columns were initially loaded to a service load of 120kN, strengthened and then tested to failure. Specimens failed in flexural manner. Average load-bearing capacity of two specimens was 430 kN. It is 90% of a maximum possible effect and 2.6 times bigger than the load-bearing capacity of C-1 and C-2 (see Fig. 8).

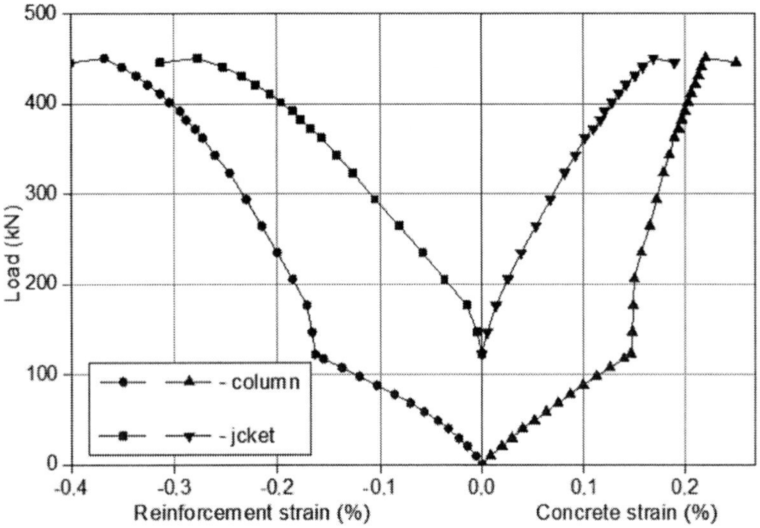

Fig. 8. Load vs reinforcement (concrete) strain for tested columns C-5-0.7 and C-6-0.7.

Although the application of the reinforced concrete jacketing was almost at the verge of exceeding the serviceability limit state requirements, some positive effect in deflection manner of the columns was still achieved. In Fig. 8 the increase in stiffness of the columns immediately after strengthening can be observed. This resulted in a very slow increase in deflections on first steps of loading after jacketing and 1.5 times increase in serviceability capacity.

The loads that corresponded to a maximum allowed deflection of 14.7 mm were 216 kN and 167 kN (Fig. 9). Considering these values we adopted a service loads for following tests of 191.5 kN.

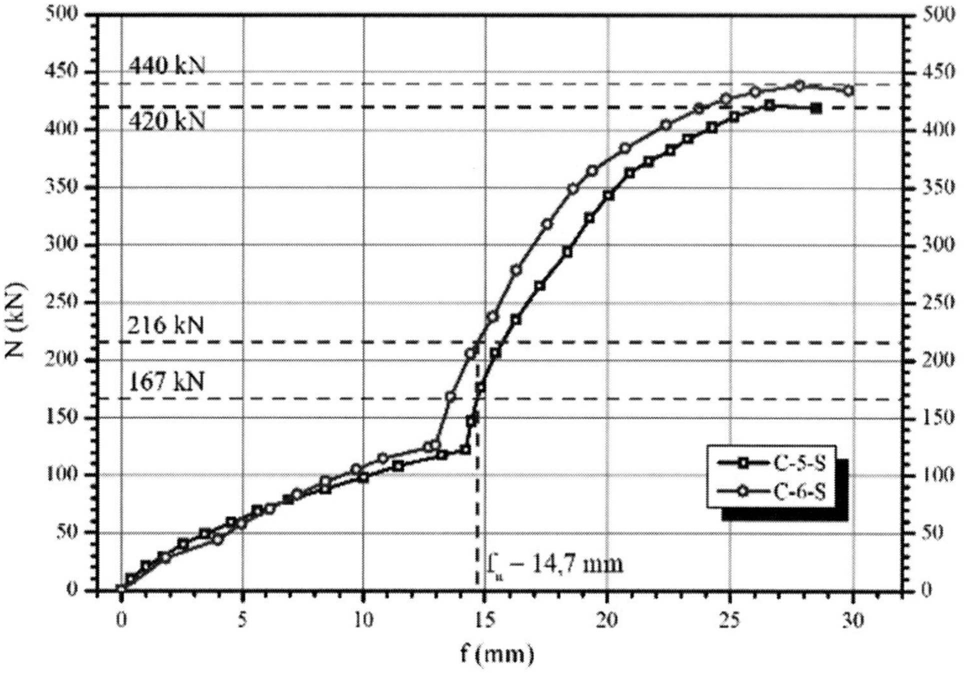

Fig.9. C-5-0.7 and C-6-0.7 deflection at mid-height plot.

For comparison of all samples the experimental data was summarized in Table1.

Table 1. Results of research.

Sample	**Load-bearing capacity, kN**		**Strengthening effect, %**		**Load of max. allowed deflection, kN**		**Strengthening effect, %**	
	sample	**average**	**sample**	**average**	**sample**	**average**	**sample**	**average**
C-1	177	175	-	-	122	123.5	-	-
C-2	173		-		125		-	
C-3-0	480	475	175	172	405	392.5	228	218
C-4-0	470		169		380		208	
C-5-0.7	420	430	142	146	216	191.5	75	55
C-6-0.7	440		151		167		35	

According to Table 1, the strengthening effect by load-bearing capacity for specimens C-3-0 and C-4-0 was 172%, for specimens C-5-S and C-6-S – 146%.

The strengthening effect by loading of maximum allowed deflection for specimens C-3-0 and C-4-0 was 218%, for specimens C-5-0.7 and C-6-0.7 – 55%.

The strengthening effect decreases when the structure is under loading during strengthening that can be seen from the obtained test results.

Conclusions

After experimental research of reinforced concrete columns strengthened by RC jacket there was determined that strengthening effect can be achieved and depends on the presence or absence of the load during the strengthening:

1. the strengthening effect by load-bearing capacity for specimens C-3-0 and C-4-0 was 172%, for specimens C-5-0.7 and C-6-0.7 – 146%.
2. the strengthening effect by loading of maximum allowed deflection for specimens C-3-0 and C-4-0 was 218%, for specimens C-5-0.7 and C-6-0.7 – 55%.
3. the strengthening effect has decreased from 172% to 146% by load-bearing capacity and from 218% to 55% by loading of maximum allowed deflection when existing loading was 65-70% from load-bearing capacity of unstrengthen one.

These obtained features should be taken into account when we design the system of strengthening by RC jacketing applied to real constructions.

References

1. Wu, Chengqing & Li, Jun & Su, Yu. *Development of Ultra-High Performance Concrete Against Blasts* (2018)
2. Ekkehard & Habil, Michael & Walraven, Joost & Leutbecher, Torsten & Fröhlich, Susanne. Ernst & Sohn GmbH & Co. KG (2014)
3. K. Sobol, Z. Blikharskyy, N. Petrovska, V. Terlyha. *Chemistry & Chemical Technology*. **8** (2014)
4. Ž. Kos, L. Gotal Dmitrović, E. Klimenko. Tehnički glasnik. **11**, 4 (2017)
5. J. Selejdak, R. Khmil, Z. Blikharskyy. *Matec Web of Conferences*. **183** (2018)
6. Y. Blikharskyy, R. Khmil, Z. Blikharskyy. *Matec Web of Conferences*. **174** (2018)
7. P. Krainskyi, Y. Blikharskyy, R. Khmil, Z. Blikharskyy. *Matec Web of Conferences*. **183** (2018)
8. Abu-Tair A., S. Rigden, E. Burley, ACI Mat. J. **93**, 6 (1996)
9. P. Vegera, R. Vashkevych, Z. Blikharskyy. *Matec Web of Conferences*. **174** (2018)
10. Sayed A. J. Eng. Sc., **37**, 3 (2009)
11. E. Julio, F. Branco, V. Silva. Prog Struct Eng Mat, **9** (2003)
12. A. Meda, G.A. Plizzari, Z. Rinaldi, G. Martinola, Con. Rep., Rehab. and Ret. **II**, 6 (2009)
13. Y. M. K. Vijaya, G. Siddappa, S. Chandra, IJRET **4** (2015)
14. C. Talbot, M. Pigeon, D. Beaupre, D. Morgan, ACI Mat. J. **91**, 7 (1994)
15. D. Stoppenhagen, J. Jirsa, L. Wyllie, ACI Str. J. **92**, 11 (1995)
16. S. Sugano, New Zeal. Nat. Soc. for Earthq. Eng. **14**,14 (1981)
17. Q. Zhang, Y.V. Mol'kov, Y.M. Sobko, Y.Z. Blikhars'kyi, R.E. Khmil. Materials Science, **50** (6) (2015)
18. National Standard of Ukraine, DBN B.2.6 - 98: 2009, 84 (2011)

Improved engineering method for calculating the strength of the supporting areas of reinforced concrete elements

Olena Krantovska[1], *Mykola* Petrov[1], *Liubov* Ksonshkevych[1], *Sergii* Synii[2,*], and *Pavlo* Sunak[2]

[1] Odesa State Academy of Construction and Architecture, 65029 Ditrikhson st. 4, Odesa, Ukraine
[2] Lutsk National Technical University, 43018 Lvivska st. 75, Lutsk, Ukraine

Abstract. According to the results of experimental studies, empirical dependencies were obtained (obtained from the appropriate mathematical models). The advanced (refined) method of engineering calculation of express-estimation of the strength of sloping cross sections of elements of flexible reinforced concrete structures on the basis of comparative analysis of the obtained empirical dependencies and standard accepted methods in the world was developed (nine basic norms are taken: DSTU B.V.2.6-156:2010 (Ukraine), previously operating SNiP 2.03.01-84*, SNB 5.03.01-02 (Belarus), SR 63.13330.2012 (Russia), Eurocode 2, 2004 (European Union), ACI 318-M14 (USA), AIJ Code, 2007 (Japan), CSA 23.3-04 (Canada), DIN 1045-1 (Germany). The coefficient of variation of bearing capacity and accident ratio according to the results of calculations of standard methods are determined. Their comparative analysis is carried out and informative graphic figures are presented.

1 Introduction

As research and design experience shows, due to the lack of a unified approach that would take into account the calculation of the inclined and normal sections, especially in complex stress-strain state, the problem of ensuring the strength of inclined sections of span reinforced concrete structures remains unsolved.

The construction of a common analysis model that would adequately simulate the stress-strain state and accurately describe the resistance of the structure the effect of the cut in combination with the bending, tensile, compression and torsion, is one of the most difficult tasks in the theory of reinforced concrete. In this regard, in practical calculations of the strength of inclined sections of reinforced concrete elements specialists uses conventional models, the so-called engineering techniques [1-4 etc.].

* Corresponding author: sergii.synii@gmail.com

2 Analysis of publications

The stress-strain state of complex loaded concrete elements was studied by T.N. Azizov, H.S. Aliev, Y.M. Babych, A.Y.Barashykov, L.V. Baykov, Z.Y. Blikharskii, O.O. Hvozdev, O.S. Horodetskii, B.H. Demchyna, S.A. Dmytriev, L.O. Doroshkevych, V.S. Dorofeev, O.S. Zaliesov, V.M. Karpiuk, P.M. Koval, R.L. Mailyan, A.M. Pavlikov, O.I. Storozhenko, V.V. Tur, Y.O. Shkola, O.F. Yaremenko etc. Among foreign works can be noted the work of R. Walter, A. Vista, L. Cany, M.F. Collins, F. Leongardt, J. Mitchell, K. Modi, J. Morrow, J. Regan, J. Taub, D. Hanson etc.

Study of the influence of longitudinal tensile forces on the strength of inclined sections of reinforced concrete elements described in the works of G. I. Kiriakadi, A. S. Zorich, B.A. Kalaturova, I. V. Volkova and others.

Issues of improvement of the standard calculation methods of durability of reinforced concrete designs are considered in works [5-18].

3 Purpose of the study

The purpose of the study is developing an improved (refined) engineering calculation method for express-estimation of the strength of sloping cross-sections of reinforced concrete elements bending on the basis of a comparative analysis of empirical dependencies and world-wide standard methods (nine standards are taken as the basis: Ukraine, previously acting SNiP2.03.01-84*, Belarus, Russia, the European Union, the USA, Japan, Canada, Germany).

4 Methods, equipment and main characteristics of the research samples

The set of equipment used in experimental research included standard equipment, provided by the relevant normative methods of conducting this kind of experiments. Also developed special equipment (in accordance with the developed method of planned experimental research). For example, designed and manufactured universal power unit. The principle of operation of these unit is due to the peculiarities of their constructional schemes (Fig. 1-2).

In accordance with the developed technique, everything was tested: 54 two-span continuous beams (fragments of the tests are shown in Fig. 1, b); 6 single-beams to bend; and 18 single-span beams for off-center stretching.

Investigated:

- single-span beams of a rectangular cross-section with dimensions of 100 × 200 mm with h_0=175 mm, span l=9·h_0=575 mm and shear sections a/h_0=1, 2, 3, made of heavy concrete of classes C12/15, C20/25, C30/35, reinforced by two plain frames with the lower rebar 2Ø12, 14, 16 A500S, with the top rebar 2Ø8, 10, 12 A500S and with transverse rebar 2Ø3, 4, 5 Vr-I;
- two-span continuous beams, respectively, b×h =100×180 mm with h_0 = 155 mm, L=8·h_0, a/h_0=1, 2, 3, C12/15, C20/25, C30/35, 2Ø10, 12, 14 500S, 2Ø10, 12, 14 A500S, 2Ø3, 4, 5 Vr-I.

Single-span beams were tested for non-centrally stretching at $N_{st}/(f_{ck}·b·h_0)$=0.05; 0.2; 0.35, non-centrally compression $N_{st}/(f_{ck}·b·h_0)$=0.2; 0.4; 0.6, in a numerical experiment, modeled using a deformation method based on the results of previous field experiments, on non-centrally compression with $N_{st}/(f_{ck}·b·h_0)$=0.05; 0.2; 0.35. The eccentricity of the application of the longitudinal force was changed: e/h_0= - 0.25; 0; +0.25.

The continuous same beams were tested on the bending by two concentrated forces.

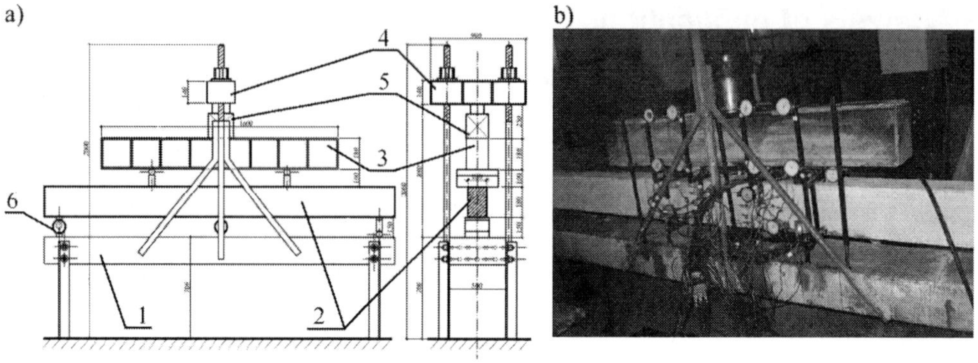

Fig. 1. The power unit developed for testing a continuous beam bending: a) - a constructional scheme; b) – photo fragment of conducting experiments; 1 - mounting bed; 2 – investigated beam specimen; 3, 4 - traverses; 5 - jack D 50; 6 – dynamometer.

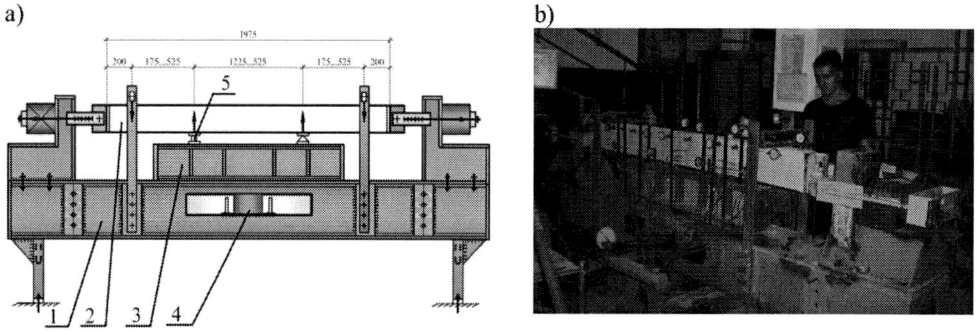

Fig. 2. The power unit for the testing of a single-span beam is developed for a non-center stretch: a) - a constructional scheme; b) – photo fragment of conducting experiments; 1 - mounting bed; 2 - investigated beam specimen; 3 - traverse; 4 - jack D 50; 5 – dynamometer.

The experimental samples were made according to five-factor three-level, close to the properties of the D-optimal, a plan of type Ha 5, [19], and non-centrally stretched samples - according to the Box 2 plan.

5 Research methods

According to the results of the experiment, new data on the stress-strain state of the supporting areas of non-centrally stretched, compressed and continuous reinforced concrete beams were obtained. The nature of the cracking and the destruction of the experimental elements is illustrated with the fragments of the photo recording experiments shown in Fig. 3, 4.

The express estimation of the strength of sloping cross sections of bending reinforced concrete elements is proposed to be carried out using an advanced engineering method using formulas (1)-(12). For this method, the dependences (7)-(12) are derived, which are based on adequate mathematical models of the main parameters of the workability of experimental beams. These models comprehensively take into account the effect of research factors and have sufficient information security. After removing the non-significant ones and recalculating remaining coefficients in the regression equations, replacing the coded variables with the natural values, these models turn into polynomials of the first or second degree.

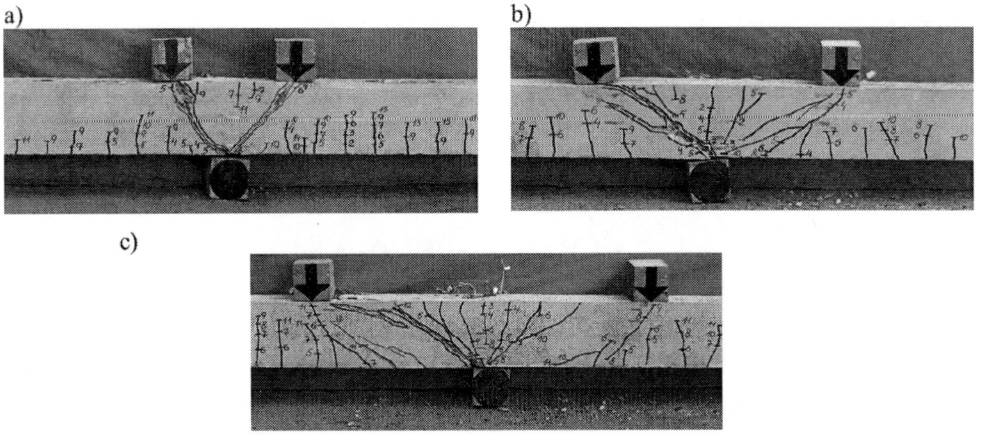

Fig. 3. The nature of cracking and destruction of experimental two-span continuous beams with small (a), middle (b) and large (c) shear span.

Experimental research has shown that part of the non-centrally stretched beams began to collapse due to the fluidity of the upper and lower longitudinal reinforcement (Fig. 4) due to their insufficient number.

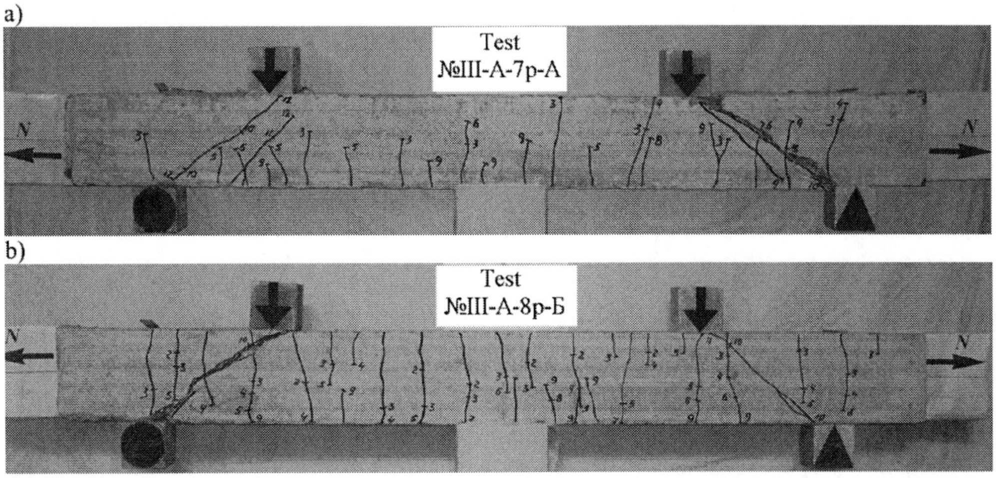

Fig. 4. The disclosure, the formation of new cracks and destruction during the test of a single-span reinforced concrete beam with a non-centrally stretch. The seventh test specimen (a), eighth test specimen (b).

The conditions of strength in this case have the appropriate appearance

$$N \le f'_{yd} A'_s (h_0 - a'_s) / e_s , \tag{1}$$

$$N \le f_{yd} A_s (h_0 - a'_s) / e'_s , \tag{2}$$

which allow you to pick up the required number of longitudinal reinforcement on a support or limit tensile force N.

In the case of a minimum number of longitudinal working reinforcement bars, the collapse of support area, usually a non-centrally extended element, in most cases is caused by the predominant action of the bending moment, which is accompanied by the fluidity of the longitudinal working reinforcements bars at the beginning of the sloping cracks, the minimum height or absence of a compressed concrete zone over a dangerous sloping cracks. The condition of the strength of the support area relative to the center of gravity of the upper (mounting) reinforcement bars has the form:

$$M \leq M_S + M_{SW} = M_{SN} + M_{SF} + M_{SW}^F . \tag{3}$$

Additional longitudinal force in the working reinforcement bars, at the beginning of the sloping crack:

$$N_{SF} = N_S - N_{SN} = (f_{yd} - \sigma_{SN})A_S, \tag{4}$$

where N_{SN}, σ_{SN} - respectively, the longitudinal force and normal stresses in the working reinforcement bars due to the longitudinal tensile force.

In general, the condition (3) can be represented:

$$M \leq V \cdot a + N \cdot e_s' - f_{yd} \cdot A_s \cdot Z_s - q_{sw} \cdot C_0^2 / 2 , \tag{5}$$

from

$$V \leq \left[f_{yd} \cdot A_s \left(h_0 - a_s' \right) + q_{sw} \cdot C_0^2 / 2 - N \cdot e_s' \right] / a , \tag{6}$$

where q_{sw} - the intensity of the transverse reinforcement, C_0 is the length of the horizontal projection of the dangerous sloping crack on the longitudinal axis of the beam.

According to the results of experimental studies, the C_0 index in practical calculations is recommended to be determined from the empirical dependences (7)-(10) obtained from the appropriate mathematical models:

- for ordinary single-span beams:

$$\begin{gathered} C_0 / h_0 = 1.114 + 0.24(a / h_0 - 2) - 0.05\left(\frac{\rho_W - 0.0030}{0.0014} \right) + 0.06\left(\frac{\rho_f - 0.0180}{0.0050} \right) - \\ -0.05(a / h_0 - 2)\left(\frac{\rho_W - 0.0030}{0.0014} \right) + 0.06(a / h_0 - 2)\left(\frac{\rho_f - 0.0180}{0.0050} \right) - 0.02\left(\frac{C - 25}{10} \right)\left(\frac{\rho_f - 0.0180}{0.0050} \right), \end{gathered} \tag{7}$$

where a / h_0 - the relative span of shear; ρ_W i ρ_f - coefficients of transverse and longitudinal beam reinforcement; C - class of concrete in MPa;

- for non-centrally stretched beams:

$$C_0 / h_0 = 1.246 - 0.007\left(\frac{N_p / f_{ck} b h_0 - 0.200}{0.148} \right) + 0.002(e / h_0) / 0.25, \tag{8}$$

where $N_p/(f_{ck} \cdot b \cdot h_0)$ - level of the stretching longitudinal force N_p, dimensions of the cross section bh_0, and characteristic resistance of the concrete to the compression is f_{ck} ; e / h_0 - relative eccentricity of applying longitudinal force;

- for non-centrally compressed beams:

$$C_0 / h_0 = 1.43 + 0.25(a / h_0 - 2) + 0.063\left(\frac{\rho_f - 0.0180}{0.0050} \right) + 0.063(a / h_0 - 2)\left(\frac{\rho_f - 0.0180}{0.0050} \right); \tag{9}$$

- for continuous two-span beams:

$$C_0/h_0 = 1.37 + 0.56(a/h_0 - 2) - 0.06\left(\frac{C-25}{10}\right) - 0.07\left(\frac{\rho_W - 0.0034}{0.0016}\right) - 0.21\left(\frac{\rho_{fn} - 0.0150}{0.0049}\right) +$$

$$+0.17\left(\frac{\rho_{fв} - 0.0150}{0.0049}\right) - 0.05\left(\frac{\rho_{fв} - 0.0150}{0.0049}\right)^2 - 0.05(a/h_0 - 2)\left(\frac{C-25}{10}\right) - 0.07(a/h_0 - 2)\times$$

$$\times\left(\frac{\rho_W - 0.0034}{0.0016}\right) - 0.15(a/h_0 - 2)\times\left(\frac{\rho_{fn} - 0.0150}{0.0049}\right) + 0.12(a/h_0 - 2)\times\left(\frac{\rho_{fв} - 0.0150}{0.0049}\right) - \quad (10)$$

$$-0.05(\frac{\rho_W - 0.0034}{0.0016})\times\left(\frac{\rho_{fn} - 0.0150}{0.0049}\right) + 0.05\left(\frac{\rho_W - 0.0034}{0.0016}\right)\left(\frac{\rho_{fв} - 0.0150}{0.0049}\right) -$$

$$-0.03(\frac{\rho_{fn} - 0.0150}{0.0049})\cdot(\frac{\rho_{fв} - 0.0150}{0.0049}),$$

where ρ_{fn} i $\rho_{fв}$ - the reinforcement coefficients, respectively, of the lower and upper zone of beam.

Also, there was a destruction of experimental reinforced concrete beams behind the sloping strip between the concentrated force and the support. According to the results of experiment, for such a case the destruction, the condition of the strength of the short strip in practical calculations is recommended to be determined by the empirical inequality:

$$V \le \varphi^*_{c_1} f_{cd} b h_0, \quad (11)$$

where $\varphi^*_{c_1}$ - a coefficient which integrally takes into account the influence of the concrete class, transverse reinforcement and the previous stresses on the working reinforcements. For ordinary and non-centrally compressed reinforced concrete beams with a small shear span experimentally ($\varphi^*_{c_1} = V / f_{cd} b h_0$) an adequate mathematical model of this coefficient has been obtained, which, after corresponding transformations, has become a form

$$\varphi^*_{c_1} = 0.3 - 0.09(C-25)/10 + 0.01(\rho_w - 0.0035)/0.00145, \quad (12)$$

where ρ_w - the coefficient of transverse reinforcement.

The following are the results of a comparative analysis of empirical dependencies and world-wide normative methodologies (the nine standards are taken as the basis: Ukraine, previously acting SNiP 2.03.01-84*, Belarus, Russia, the European Union, the USA, Japan, Canada, Germany).

For reliability and safety of the forecasting of the bearing capacity of the supporting areas of the span reinforced concrete elements, their bearing capacity according to the standards is determined taking into account their complex stress-strain state.

In the current Ukrainian [1], harmonized Belarusian [2], international European [3] and American [4] normative documents, the calculation of elements for the action of the transverse force is based on the "truss model", which consists of stretched and compressed rod elements. However, the use of the truss analogy in the normative documents of different countries has its own peculiarities [5]. For example, as noted in [6], despite the similarity of the methods used, the provisions of DBN and the corresponding Eurocode 2 differ significantlu, which should be taken into account when comparing the results of calculations on these norms among themselves.

The study of the strength of sloping sections is considered in [11-16]. Among the important criteria chosen for the comparative analysis was the coefficient of variation in strength (Fig. 5).

Among the important criteria chosen for the comparative analysis, the coefficient of variation in strength was selected (Fig. 5):

$$\upsilon = \frac{\sigma}{b_0} \cdot 100\% , \qquad (13)$$

where $\sigma = \sqrt{\sum \frac{\left(\hat{V}_{\exp} - V\right)^2}{n-1}}$; $\sum\left(\hat{V}_{\exp} - V\right)^2$ is average square deviation; b_0 is arithmetic average experimental value of bearing capacity $\hat{V}_{\exp}$; n is number of test specimens; and the accident ratio (Fig. 6):

$$k_a = \frac{n_a}{n} \qquad (14)$$

where n_a - number of test specimens, bearing capacity of the supporting areas of which more than 10 % exceeds the actual values.

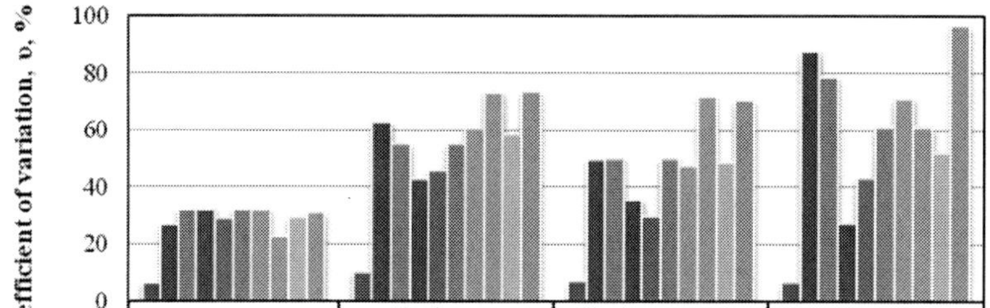

	Single-span non-centrally stretched beams (two-factor experiment)	Single-span non-centrally stretched beams (numerical five-factor experiment)	Single-span non-centrally compressed beam (numerical five-factor experiment)	Continuous two-span beams (five-factor experiment)
■ Advanced engineering methods	5.9	10.1	6.8	6.6
■ Ukrainian standard DSTU B V.2.6-156:2010	27	63.1	49.5	88
■ Belorussian standard SNB 5.03.01-02	32	55	50	79
■ Previously acting SNiP2.03.01- 84*	32.1	42.6	35.4	27.3
■ SR 63.13330.2012 (SET OF RULES, actualized) Russian standard	28.9	45.9	30	43.3
■ European standard EUROCODE-2	31.8	55.2	50.1	61.1
■ American standard ACI 318-14	32	60.7	47.3	71.3
■ Japanese standard AIJ Code, 2007	22.5	73.4	72.1	61.1
■ Canadian standard CSA 23.3-04	29.4	58.6	48.6	52.4
■ German standard DIN 1045-1	30.9	73.8	70.5	97

Fig. 5. Comparison of the results of calculations of the coefficient of variation.

Calculated formulas for the SNiP 2.03.01-84* previously operating in Ukraine and the new Russians formulas have the best convergence of calculation and research data of bearing capacity in the supporting areas of non-centrally compressed beams; according to other norms, the discrepancy is within the range of 53 to 73.8 %. Accident ratio $k_a = 0$, that is the most reliable. Similar results are obtained for non-centrally compressed beams. For continuous beams, the bearing capacity can be relatively satisfactorily determined according to the old standards (υ=27.3 %), while other norms show a higher coefficient (υ=43 to 97 %), with a rather low accident ratio.

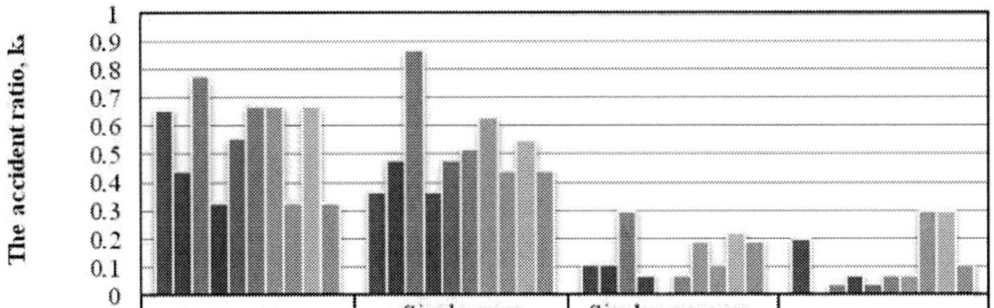

	Single-span non-centrally stretched beams (two-factor experiment)	Single-span non-centrally stretched beams (numerical five-factor experiment)	Single-span non-centrally compressed beam (numerical five-factor experiment)	Continuous two-span beams (five-factor experiment)
Advanced engineering methods	0.66	0.37	0.11	0.2
Ukrainian standard DSTU B V.2.6-156:2010	0.44	0.48	0.11	0
Belorussian standard SNB 5.03.01-02	0.78	0.87	0.3	0.04
Previously acting SNiP 2.03.01-84*	0.33	0.37	0.07	0.07
SR 63.13330.2012 (SET OF RULES, actualized) Russian standard	0.56	0.48	0	0.04
European standard EUROCODE-2	0.67	0.52	0.07	0.07
American standard ACI 318-14	0.67	0.63	0.19	0.07
Japanese standard AIJ Code, 2007	0.33	0.44	0.11	0.3
Canadian standard CSA 23.3-04	0.67	0.55	0.22	0.3
German standard DIN 1045-1	0.33	0.44	0.19	0.11

Fig. 6. Comparison of the results of calculations of the accident ratio.

Comparison of the experimental and calculated strength values of the supporting areas of non-centrally stretched and compressed beams, as well as continuous, determined by the recommendations and requirements of various standards, showed, on the one hand, their satisfactory convergence (coefficient of variation υ=20 to 74 %), and on the other hand, insufficient reliability of the calculated formulas of these norms, especially for samples with large shear spans.

6 Conclusions

An improved engineering methodology for calculating the strength of the supporting areas of the investigated reinforced concrete elements proposes, on the basis of the obtained empirical dependencies, a consistent consideration of all possible destructive schemes and adequately reflects the influence of the research factors on the bearing capacity of the beams. The comparison of the calculated values of the coefficient of variation of bearing capacity of the supporting areas of the experimental reinforcement concrete elements showed satisfactory convergence (υ<12%).

References

1. National standard of Ukraine, *DSTU B V.2.6-156:2010, Structures of buildings and erection. Concrete and reinforced concrete structures made of heavy concrete. Design rules,* Minregionbud Ukrainy, Kyiv, (2010).

2. National standard of Belarus, *SNB 5.03.01-02. Plain and reinforced concrete structures*, Minstroyarhitektury, Minsk, (2002).

3. European standard, *EN 1992-1-1 (2004). Eurocode 2: Design of Concrete structures – Part 1-1: General rules and rules for buildings*, CEN, Brussels, (2004).

4. National standard of the USA, *Building Code Requirements for Structural Concrete (ACI 318M-14) and Commentary (ACI 318RM-14)*, ACI, Mich., (2014).

5. V.M. Karpiuk, O.M. Krantovska, O.M. Kotsiurubenko, Gazette OSACEA, **57**, 182-188, (2015).

6. V.M. Babaiev, A.M. Bambura, O.M. Pustovoytova and others under general supervision by V.S. Shmukler, *Practical calculation of elements of reinforced concrete structures on DBN.2.6-98:2009 in comparison with calculations according to SNiP 2.03.01-84* and EN 1992-1-1 (Eurocode 2)*, Golden pages, Kharkiv, (2015).

7. V.M. Karpiuk, *Calculated force resistance models of span concrete structures in case of bending stress state (monograph)*, OSACEA, Odesa, (2014).

8. V.S. Dorofeev, V.M. Karpiuk, E.N. Krantovska, *Strength, crack resistance and deformability of continuous reinforced concrete beams (monograph)*, Even, Odesa, (2010).

9. V.S. Dorofeev, V.M. Karpiuk, N.N. Petrov, *Strength, deformability and crack resistance of the support of inclined sections of eccentrically stretched, compressed of reinforced concrete beams (monograph)*, Even, Odesa, (2011).

10. O.M. Kotsiurubenko, O.M. Krantovska, V.M. Karpiuk. Building constructions, SE NDIBK (NIISK), Kyiv, **82**, 166-175, (2015).

11. V.P. Mitrofanov. *Stress-strain state, strength and cracking of reinforced concrete elements in transverse bending*. - Auth. abstr. of Diss. PhD, speciality 05.23.01 "Building structures, buildings and structures", M., (1981).

12. O. Andriichuk, V. Babich, I. Yasyuk, S. Uzhehov, MATEC Web of Conferences, **116**, 02001, 1-9, (2017).

13. O.V. Andriichuk, E.M. Babych, Materials Science, New York, **52**, 4, 509-513 (2017).

14. Ahmed B. Shurain, J. of King Saud Univ. – Engineering Sciences, **25**, 65-74 (2013).

15. O. Mikulich, V. Shvabyuk, I. Pasternak, O. Andriichuk, Mechanics Research Communications, **91**, 107-111, (2018).

16. V.V. Tur, A.A. Kondratchik. *Calculation of reinforced concrete structures under the action of cutting forces: monograph*, Publ. BGTU, Brest, (2000).

17. Umer Farooq and K.S. Bedi, *Conf. Structural Engineering Convention 2014 (SEC2014)*, Delhi, 3477-3488, (2014).

18. F.J. Vecchio and M.P. Collins, ACI J., **83**, 2, 219-231, (1986).

19. V.A. Voznesensky, *Static methods of experiment planning in technical and economic research*, Finance and Statistics, (1981).

The search for sustainable parameters for steel-reinforced concrete section of a bridge superstructure

Yuriy Krul[1,*], and *Roman* Kaplin[2]

[1]Kharkiv National Automobile and Highway University, 61002 Yaroslava Mudrogo str, 25, Kharkiv, Ukraine
[2] Hon. Prof. M. S. Bokarius Kharkiv Research Institute of Forensic Examinations, 61117 Zolochevskaya str, 8a, Kharkiv, Ukraine

Abstract. The article contains the description of an efficient lightweight steel-reinforced concrete bridge superstructure, which includes metal blocks with a box-like section and reinforced concrete efficient roadway slab. Metal blocks are made of sheet perforated elements manufactured according to the no-waste technology. The blocks are interconnected as a single space bridge superstructure frame by means of high strength bolts. The reinforced concrete slab of the roadway is an efficient hollow slab. It includes upper and lower housing, inner ribs placed with a particular pitch, and all its remained part is filled with polystyrene foam extractable fillers along its height. At this, a rib height is equal to the full slab height. The metal and reinforced concrete parts are linked by means of a special system of shearing connections. The article describes an iteration algorithm of the construction enhancement, in course of which the geometrical parameters, such as bridge superstructure section full height, reinforced concrete slab thickness and a construction metal part height were defined. Theoretical models being the ground for the strain-stress state (SSS) of the sections studied analysis were developed.

1 Introduction

Steel-reinforced concrete bridge superstructures have become the up-to-date bridge structures widely applied worldwide [1-3]. The principle advantage of steel-reinforced concrete structures is the diminished dead weight in comparison with reinforced concrete systems as well as the dipped steel consumption and the resulted decrease in an initial cost comparing to metal bridge superstructures. These systems are the most sustainable at spans of 40-80 m [4]. One of the conditions to widen the range of use of steel-reinforced concrete bridge superstructures (firstly, according to the size of spans) is the diminished dead weight of a structure at a specified bearing capacity. This fact encourages searching for new efficient lightweight structures of steel-reinforced concrete bridge superstructures, whose

* Corresponding author: krul.yuriy@gmail.com

characteristics would be enhanced in comparison with fully reinforced concrete and steel systems.

2 Structural peculiarities

In connection with stated above, the modular two-component system of lightweight steel-reinforced concrete shell-type bridge superstructures was suggested [5]. The structure consists of metal blocks with box-like sections and an efficient lightweight reinforced concrete roadway slab put into operation by means of the shearing connections system (fig.1).

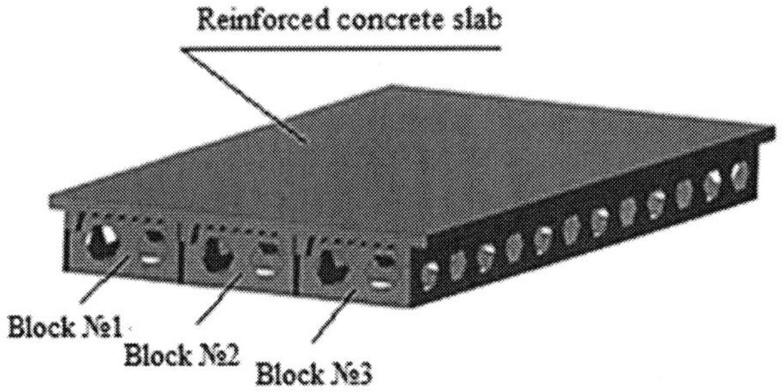

Fig. 1. The fragment of a suggested steel-reinforced concrete bridge superstructure.

Metal blocks are made of perforated sheet elements manufactured according to the no-waste technology. The block consists of two main bars, a lower metal plate connecting main bars and transverse membranes mounted with the calculated pitch according to the span length (Fig. 2).

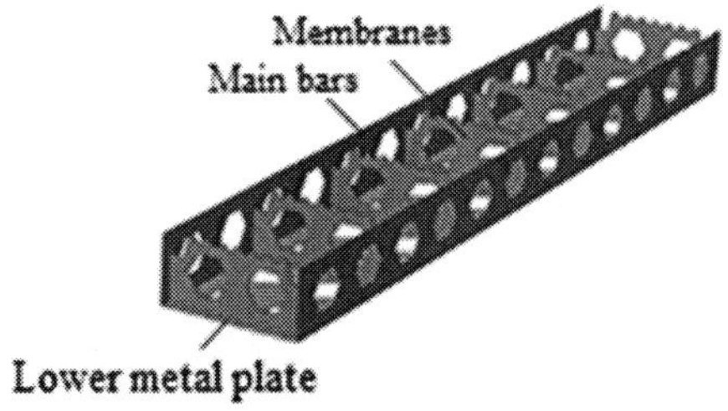

Fig. 2. Metal block of the steel-reinforced concrete bridge superstructure.

Metal bars, in turn, can be manufactured either with an upper horizontal shelf being a supporting element for the reinforced concrete slab, or without it. The blocks are interconnected as a single space bridge superstructure frame by means of high strength bolts (Fig. 3).

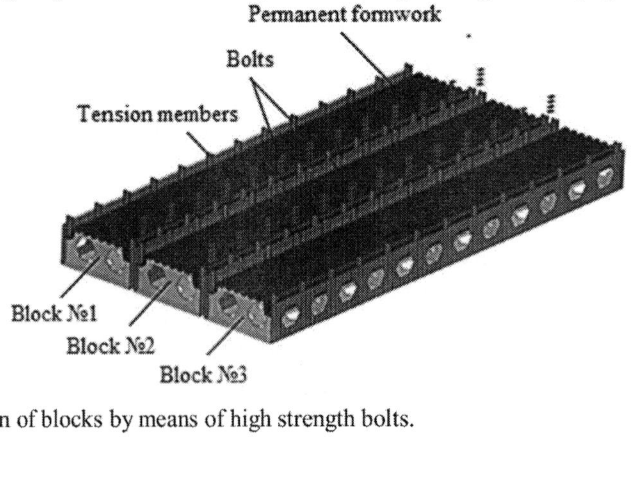

Fig. 3. Interconnection of blocks by means of high strength bolts.

A graded steel sheet is placed over the block membranes. It is a retained form for the reinforced concrete slab. The reinforced concrete slab of the roadway is an efficient hollow slab. It includes upper and lower housing, inner ribs placed with a particular pitch, and its entire remained part is filled with polystyrene foam extractable fillers along its height. Thus, a rib height is equal to the full slab height (Fig. 4).

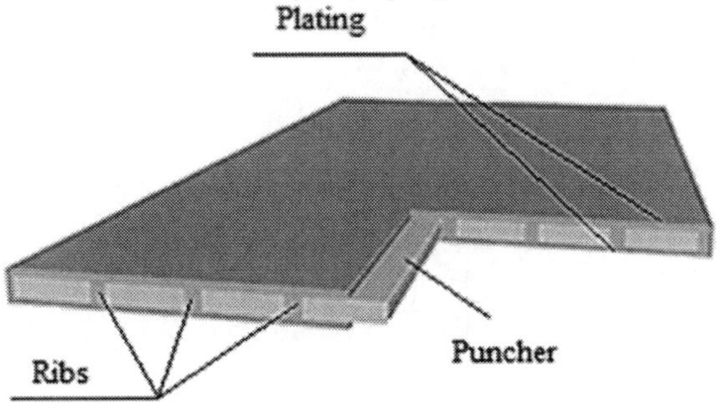

Fig. 4. The roadway slab construction.

To provide joint operation of the metal part with the reinforced concrete slab the system of discrete-and-continual shear connections in the form of dead stops jointed at the transverse and longitudinal directions with deformed section steel bars jointed with the slab reinforcement cage was suggested (fig. 5).

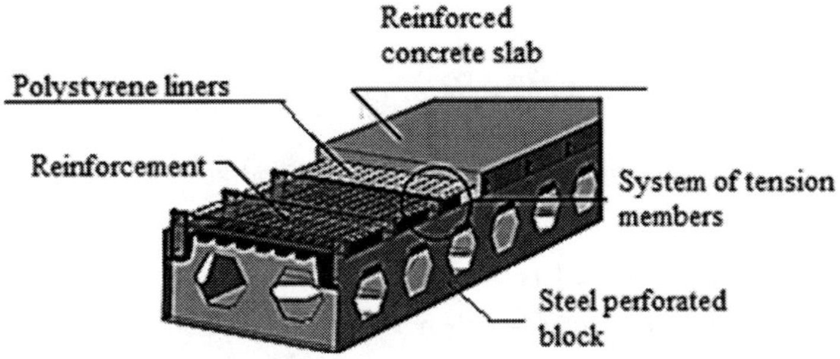

Fig. 5. The system of discrete-and-continual shear connections. General view.

3 Problem statement

The article [6] suggests determining a minimum height of the metal perforated bar on the ground of necessary toughness conditions by the following formula:

$$h = 2 \cdot \beta \cdot \frac{\sigma}{E} \cdot \left[\frac{l}{f}\right] \cdot l \tag{1}$$

where σ – calculated steel resistance; E – steel deflection module; f – allowed standard sag; l – bar span.

It is recommended to adopt the T-bar belt height in compliance with [4] as follows: $h_f = h/4$

The guaranteed inclusion of a reinforced concrete slab into joint operation allows decreasing the metal part height, which results in shrinking of the studied section full height. The following solution to the given problem is suggested to determine the reinforced concrete slab application efficiency: to compare characteristics of metal perforated bar and steel-reinforced concrete section of the bridge superstructure fragment.

4 Hypothesis and assumptions

An algorithm to determine the optimal section height of the bridge superstructure and the impact of the reinforced concrete slab on the height of metal part of the structure was suggested within the stated design solution. For this purpose, a fragment is extracted from the shell. It is a metal perforated two-T-bar with the efficient reinforced concrete slab placed on the upper shelf of the bar (Fig. 6).

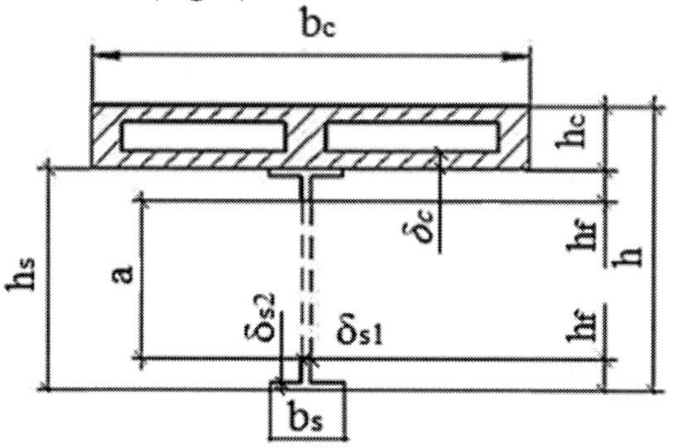

Fig. 6. Transverse section of the bridge superstructure fragment.

We introduce parameters vector [7,8,9], which describe the suggested steel-reinforced concrete section:

$$\{X\}^T = \{h, h_s, h_f, \delta_{s1}, \delta_{s2}, \delta_c, h_c, a, b_c, b_s\} \tag{2}$$

where h – full section height; h_s – metal perforated part section height; h_f – height of T-bar belt of the metal perforated part section; δ_{s1} – wall thickness of metal perforated part section; δ_{s2} – metal perforated part section shelf thickness; δ_c – housing thickness of reinforced concrete part section; a – hole height of metal perforated part section; b_s – shelf width of metal perforated part section; h_c – thickness of reinforced concrete part section; b_c – width of reinforced concrete part section (*bar pitch*).

We adopt several assumptions during calculations:
–the suggested system of shearing connections is assumed to provide joint operation of metal and reinforced concrete parts section that allows considering the given fragment integrally;
– at this, at limit state it is adopted that the reinforced concrete part section is in the compressive side, while metal part is in the tensile one. This allows assuming that the neutral axis comes along the boarder of the metal bar connection with the reinforced concrete slab.

5 Calculations procedure

Studying the metal two-T-bar by formula (1), we determine a minimum bar height on the basis of necessary toughness conditions. We further determine several geometric characteristics for the metal bar:

$$\begin{cases} A_1 = b\cdot\delta_1\cdot 2 + (h_f-\delta_1)\cdot\delta_2\cdot 2 \\ S_1 = b\cdot\delta_1\cdot\left(h-\frac{\delta_1}{2}\right) + (h_f-\delta_1)\cdot\delta_2\cdot(h-\frac{h_f-\delta_1}{2}) + (h_f-\delta_1)\cdot\delta_2\cdot(\frac{h_f-\delta_1}{2}+\delta_1) + b\cdot\delta_1\cdot\left(\frac{\delta_1}{2}\right) \\ y_1^* = \frac{S_z}{A} \\ J_1 = 2\cdot\left(\frac{b\cdot\delta_1^3}{12} + \frac{\delta_2\cdot(h_f-\delta_1)^3}{12} + b\cdot\delta_1\cdot\left(h-\frac{\delta_1}{2}-y^*\right)^2 + \delta_2\cdot h_f\cdot\left(h-\frac{h_f-\delta_1}{2}-y^*\right)^2\right) \\ W_1 = J_{z1}\cdot 2/h \end{cases}$$

where A_1 – metal cross-sectional area; S_1 – first moment of the area of metal section; y_1^* – metal section gravity center; J_1 – second moment of the metal section; W_1 - sideway moment of metal section resistance.

We determine the metal bar bearing capacity by the following formula:

$$M_{US} = R_S\cdot W_{PL} \quad (3)$$

where R_S – calculated steel resistance; W_{PL} – elastic-plastic moment of section resistance.

It can be determined as follows:

$$W_{PL} = \gamma\cdot W \quad (4)$$

where $\gamma = 1,25...2,0$ (depends on the section shape and is to be determined by means of manual tables).

We further study the steel-reinforced concrete fragment section. To simplify calculations, we substitute the efficient reinforced concrete slab with the reinforced concrete slab with an effective thickness (Fig. 7).

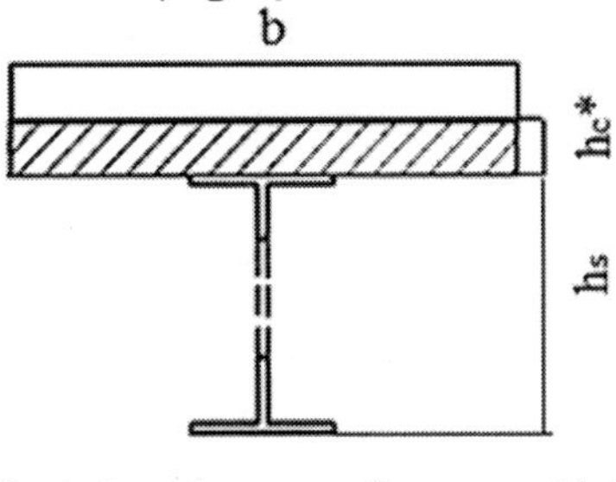

Fig. 7. Transverse section of the steel-reinforced concrete fragment with the effective reinforced concrete slab.

Determination of the effective slab thickness is made on the ground of section toughness equality. Since reinforced concrete consumption for slabs is equal, second moment of the section must coincide:

$$J_2 = J_2^* \quad (5)$$

where J_2 – second moment of the section with the efficient reinforced concrete slab; J_2^* – section second moment with the effective slab.

Based on the above, an expected diagram of tensions is demonstrated in Fig. 8.

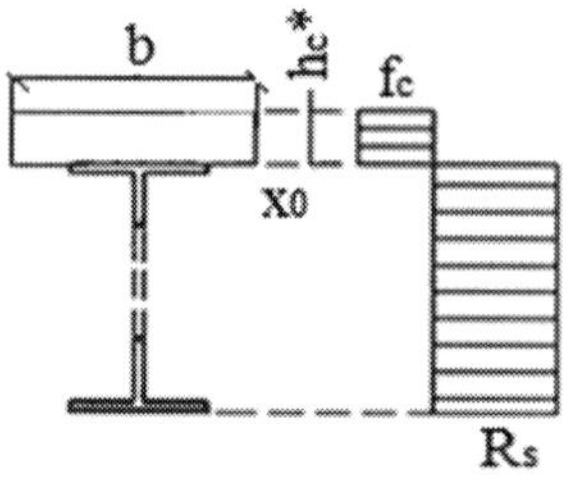

Fig. 8. Diagram of tensions of a steel-reinforced concrete element (limit state).

The hypothesis introduced above permits the following equation for limit balance:

$$b_c^{red} \cdot h_c^* \cdot f_c - A_s \sigma_{ys} = 0 \tag{6}$$

Based on (6), effective equivalent constituent thickness of the reinforced concrete slab is equal to:

$$h_c^* = \frac{A_s \cdot R_s}{b_c^{red} \cdot f_c} \tag{7}$$

The following stage of calculation is to perform constructive reduction to a single material. For this we apply reduction factor method. After this operation, the metal bar section with the equivalent constituent reflecting reinforced concrete slab at the upper belt will become equal in strength with metal two-T-bar with the efficient hollow reinforced concrete slab (Fig. 9).

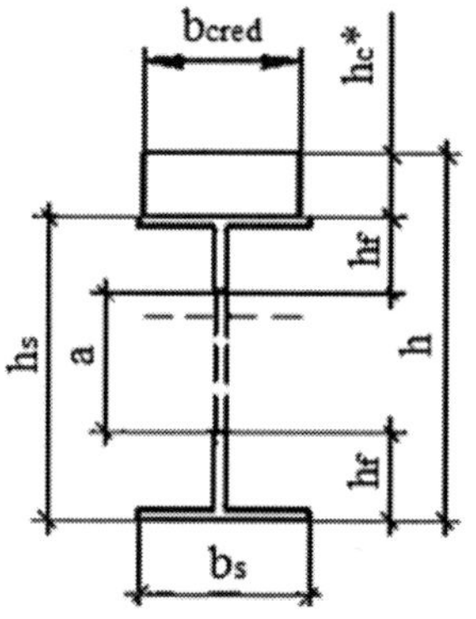

Fig. 9. Section with the equivalent constituent, reflecting reinforced concrete slab.

Considering that the metal bar and section with the equivalent constituent, reflecting reinforced concrete slab, we take them as equivalent, and they must have equal toughness, i.e. the condition must be performed (8):

$$J_2^{*red} = J_1 \tag{8}$$

where J_1 – metal section second moment; J_2^{*red} – section second moment with the equivalent constituent, reflecting reinforced concrete slab.

Equation for section second moment with the equivalent constituent, reflecting reinforced concrete slab:

$$J_2^{red} = \frac{b_c^{red} \cdot h_c^{*3}}{12} + \left(\frac{b_s \cdot \delta_1^3}{12}\right) \cdot 2 + \frac{\delta_1 \cdot \left(h_f - 2 \cdot \delta_2\right)^3}{12} + b_c^{red} \cdot h_c^* \cdot \left(h_s - \frac{h_c}{2} - y_2^*\right)^2 +$$
$$+ b_1 \cdot \delta_2 \cdot \left(h_s - \frac{\delta_2}{2} - y_2^*\right)^2 + \delta_1 \cdot \left(h_s - 2 \cdot \delta_2\right) \cdot \left(\frac{h_s}{2} - y_2^*\right)^2 \quad (9)$$

Putting expression (7) into equation (9) as well as into condition (8) we obtain the opportunity to determine height of the metal bar h_s. By obtaining the height of the metal part section and putting it into expression (7), we determine the reinforced concrete slab equivalent constituent thickness.

Condition for bearing capacity of the studied section must be filfilled in the same way:

$$\frac{M_U}{M_{US}} \geq 1 \quad (10)$$

where M_U – bearing capacity of the section with the reinforced concrete slab equivalent constituent; M_{US} – bearing capacity of the reinforced concrete slab.

Bearing capacity of the section with the reinforced concrete slab equivalent constituent is equal to:

$$M_U = b \cdot h_c \cdot f_c \cdot \frac{h_c}{2} + A_s \cdot \sigma_{ys} \cdot \frac{h_s}{2} \quad (11)$$

In case of non-fulfillment of one of conditions (8) and (10), gradual rise in reinforced concrete slab thickness will be taking place until each condition is fulfilled.

We would like to give an example of calculation of a fragment of the steel-reinforced concrete bridge superstructure. The studied construction fragment is considered to have preliminary agreed parameters, namely:

- span length $L = 24$ m;
- allowed structural deflection $f = 1/400L = 60$ mm;
- reinforced concrete part section made of concrete C30/35;
- metal part section made of steel 15HSND;
- slab roughness ribs placed with the 1 m pitch. Accepted rib thickness is 100 mm;
- upper and lower housing with 50 mm thickness;
- two-T-bar shelves with 300 mm width and 14 mm thickness;
- two-T-bar wall with 12 mm thickness.

By putting above parameters into expression (1) we obtain minimum height of the metal perforated bar $h = 1960$ mm. T-bar belt height is accepted as $h_f = h_0/4 = 490$

In the result of calculations the two-T-bar section becomes like in Fig. 10.

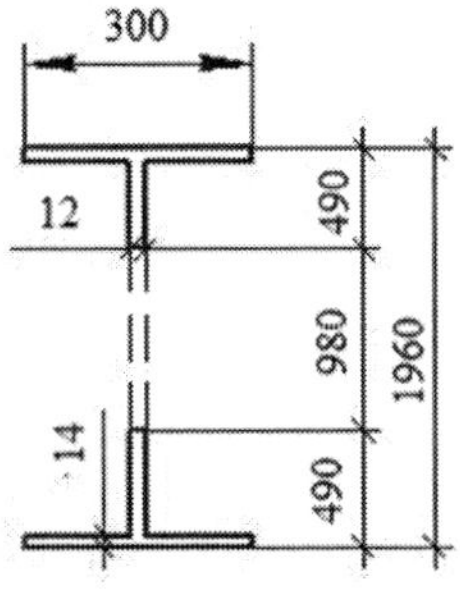

Fig. 10. Metal perforated two-T-bar.

We determine geometric characteristics of the metal bar:

$$A_1=19820\text{mm}^2$$
$$S_1=1.951{\cdot}10^7\text{mm}^3$$
$$J_1=8.177{\cdot}10^9\text{mm}^4$$
$$W_1=8.344{\cdot}10^6\text{mm3}$$

Further, we determine the metal bar height and the reinforced concrete slab equivalent constituent thickness in accordance with (7) and (8).

$$h_s=1100\text{mm}$$
$$h_c^*=140\text{mm}$$

And by means of condition (5) we find the efficient hollow reinforced concrete slab thickness:

$$h_c=300\text{mm}$$

The obtained transverse section of the bridge superstructure is demonstrated in Fig. 11.

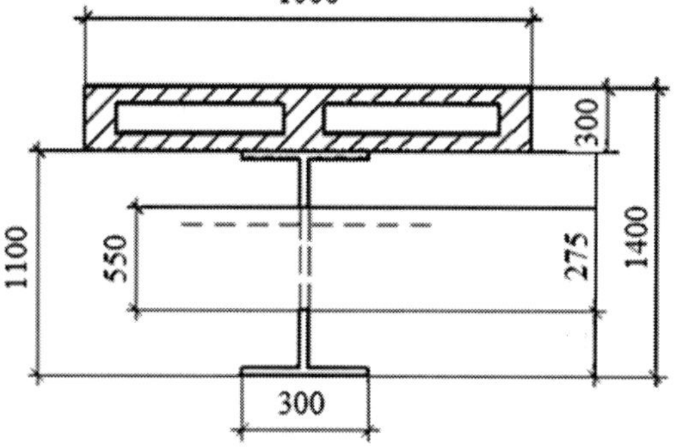

Fig. 11. Final fragment of the steel-reinforced concrete bridge superstructure.

The demonstrated formation algorithm of the efficient steel-reinforced concrete structure allows shrinking the metal bar height 1.8 times in comparison with all-metal bar. At this, the steel-reinforced concrete structure height (line meter) is 35% less than of a standard metal perforated two-T-bar. Economy is based on the reinforced concrete (nearly 1.300 UAH/m^3) and steel (nearly 2.500 UAH/t) price difference. Considering the substantially lower price of the reinforced concrete, the difference in structure cost constitutes nearly 40%. Due to inclusion into operation the efficient reinforced concrete slab the opportunity to dip tension in the metal part section occurs, which results in somewhat unloading the metal bar. All listed allows concluding that the suggested structure is really efficient.

Aiming at checking the suggested solution to the problem, we developed theoretical models (Fig. 12) that became the basis for conducting strain-stress state (SSS) structure analysis with determination of potential deformation energy (PDE) values.

a) b)

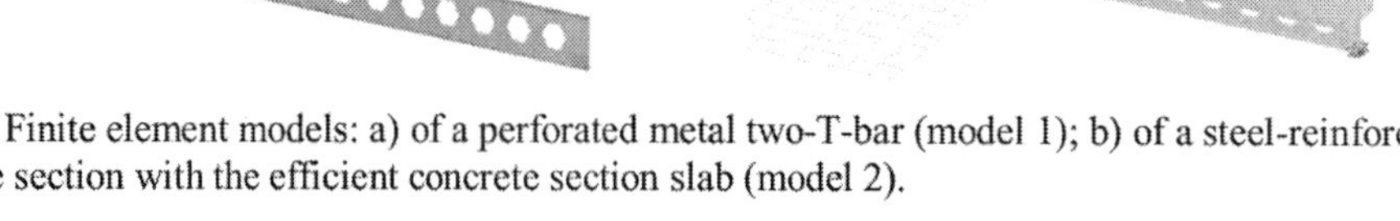

Fig. 12. Finite element models: a) of a perforated metal two-T-bar (model 1); b) of a steel-reinforced concrete section with the efficient concrete section slab (model 2).

The values obtained by the calculation are given in Table 1.

Table 1 Calculation results.

Parameter	Model 1	Model 2
Potential deformation energy (PDE), U	2.505	2.402
Sag Δ, mm	5.736	5.730
Tension N_x, MPa	1.750	1.610
Tension N_y, MPa	4.720	4.630
Tension τ_{xy}, MPa	2.810	2.708
Free frequency υ, Hz	10	12

On the basis of the mentioned theoretical analysis we confirmed the studied constructive topology and its compliance with solidity, deformations, stability and free frequency limits.

Conclusions:

• The modular two-component system of the lightweight steel-reinforced concrete shell-type bridge superstructure was suggested.
• As a result of conducted calculations, we obtained geometrical characteristics of the steel-reinforced concrete section of the suggested structure.
• We evaluated the impact of the efficient reinforced concrete slab on the height of the metal part and the height of the entire structure. This approach allows shrinking the metal bar height 1.8 times, decreasing the finite structure line meter weight by 35% and cutting the cost of used materials by 40%.
• We created and analyzed two calculation models of the studied sections. At this, the discrepancy in values obtained in the course of calculation did not exceed 5%.

References

1. E.E. Gibshman, *Mostyi so stalnyimi balkami, ob'edinennyimi s zhelezobetonnoy plitoy* (1952)
2. E.E. Gibshman, *Proektirovanie stalnyih konstruktsiy, ob'edinennyih s zhelezobetonom, v avtodorozhnyih mostah* (1956)
3. N.N. Streletskiy, *Stalezhelezobetonnyie proletnyie stroeniya mostov* (1981)
4. M.M. Korneev, *Stalezhelezobetonnyie mostyi: teoreticheskoe i prakticheskoe posobie po proektirovaniyu* (2015)
5. Y.N. Krul, *O formirovanii ratsionalnoy konstruktsii proletnogo stroeniya avtomobilno-dorozhnogo mosta*, Kom. Hoz. Gor. **101**, 31 (2011)
6. V.V. Biryulyov, *Proektirovanie metallicheskih konstruktsiy* (1990)
7. G.V. Vasilkov, *Evolyutsionnaya teoriya zhiznennogo tsikla mehanicheskih sistem. Teoriya sooruzheniy* (2008)
8. V.S. Shmukler, Y.A. Klimov, N.P Burak, *Karkasnyie sistemyi oblegchennogo tipa* (2008)
9. V.M. Babaev, S.O. Bugaevskiy,S.M. Evel, I.D. Evzerov, A.I. Lantuh-Lyaschenko, V.V. Shevetovskiy, O.V. Shimanovskiy, V.S. Shmukler, *Chiselni ta eksperimentalni metodi ratsionalnogo proektuvannya ta zvedennya konstruktivnih sistem* (2017)

Analytic formulas for the natural frequencies of hinged structures with taking into account the dead weight

Yurii Krutii[1,*], *Mykola* Suriyaninov[1], and *Victor* Vandynskyi[1]

[1]Odessa State Academy of Civil Engineering and Architecture, Didrikson str., 4, Odessa, 65029, Ukraine

Abstract. Free bending vibrations of hinged vertical uniform rod with taking into account the dead weight are investigated. The research is based on the exact solution of the partial differential vibration equation with variable coefficients. This approach allows to get more reliable picture of rod's vibration because only the exact solution carries information of a qualitative nature and forms the most complete picture of the physical phenomenon under consideration. The frequencies equation of problem was written in dimensionless form and the way of its root finding is shown. It has been shown that the problem of determination the nature frequencies of structures is reduced to finding corresponding dimensionless vibration coefficients from equation. The formulas for the first three vibration frequencies of structures were obtained in analytical form. An analytic relationship between the frequencies with and without taking into account the dead weight of the structures was established. The nature of the dependence of frequencies on the value of the longitudinal load was revealed. The presence of conclusive analytical formulas for determining the vibration frequencies of hinged vertical structures with taking into account the dead weight is a real alternative for using the approximate methods for solving this class of problems of solid mechanics.

1 Introduction

When designing vertical structures, it is necessary to consider all possible impacts they may be exposed during construction and operation. Taking into account the fact, that recently the structures of high-rise building is intensively developing, including at the earthquake-prone areas, the relevance of investigation the different types of structures with dynamic load is undisputed.

The qualitative evaluation of the dynamic parameters of the relevant structures is required for the quality structural design. This evaluation can be obtained by analyzing such important characteristics as free vibration frequencies and mode shapes, obtained by investigation of the structure in the unloaded state. Based on these characteristics, we can

[*] Corresponding author: yurii.krutii@gmail.com

make the conclusion about the rigidity and operability of the structure, correctness of design solutions.

Many materials and publications are concerned with research of different type of construction for bending vibrations with taking into account effect of longitudinal force.

In article [1] bending vibrations of uniform beams under various boundary conditions and axial loads have been studied. The paper [2] deals with the case when the axial force takes different constant values in different parts of the beam. Here, the first three vibration frequencies for various combinations of boundary conditions have been calculated. The authors of [3] study vibrations of frame structures taking into account the influence of axial load. The load is calculated based on the total weight and number of floors of the building. Natural frequencies of frame structures are obtained by solving the differential equation written for an equivalent rod whose rigidity and weight must be uniformly distributed along the length. In the paper [4] the necessity of taking into account the influence of an axial force on the free vibration frequencies of marine steel platform support blocks' grid elements during the design and reliability prediction of welded joints is proved. In the investigations of influences of load on bending vibration the axial load is often [5–7] accepted constant along the height for simplification. However, in the real structures the axial forces in different cross sections have different values.

One of the most widely used design schemes in structural design is the hinged vertical uniform rod under the variable longitudinal force, represented with dead weight of the structure.

The mathematical model of such a physical phenomenon is a differential variable-coefficient equation [8, 9]. Its solution had been unknown until the work [10] published. Therefore, the studies were conducted by approximate, as a rule, variational methods [9]. However, the presence of an exact solution provides for qualitative evaluation of mechanical system. Based on the previously obtained results [10, 11], the objective of the work is to determinate in analytical form the formulas for the free vibration frequencies of vertical hinged structures with taking into account the dead weight; to establish in an analytical form the relationship between frequencies with taking into account the dead weight of the structures and without it.

2 Results

2.1 General symbols and equations

Let us analyze the free bending vibrations of the rod with uniform flexural rigidity and intensity of the distributed mass taking into account the longitudinal distribution force (Fig.1, Fig.2). Fig. 1 shows the general scheme of vibrations of the vertical hinged rod, and Fig. 2 shows the external and internal forces acting on its element.

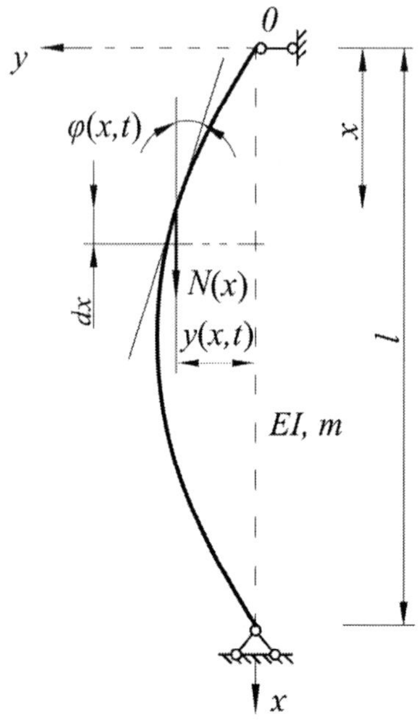

Fig. 1. The scheme of free bending vibrations of a straight rod

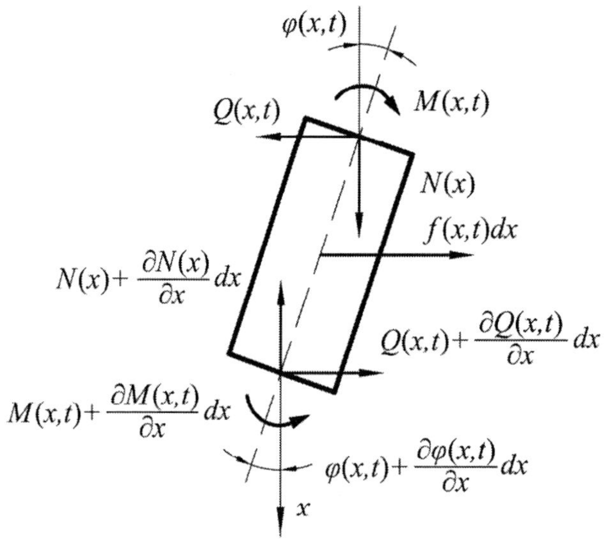

Fig. 2. The scheme of the forces acting upon the element of a rod

List of symbols:

$EI-$ flexural rigidity of a rod;

$m-$ intensity of the distributed mass (dead weight) of the rod;

$N(x)=qx-$ variable axial (compressive) force, where $q-$ weight per unit length of beam;

$y(x,t)-$ cross motion of the axis point of the rod with coordinate x at time t (dynamic deflection);

$\varphi(x,t)-$ dynamic angular displacement;

$M(x,t)-$ dynamic bending moment;

$Q(x,t)$ – the dynamic transverse force;

$f(x,t)$ – the intensity of inertial forces that appear during vibration (D'Alembert force).

It is known [8, 9], that the equation of free transverse vibrations account of own weight is written as

$$EI\frac{\partial^4 y}{\partial x^4}+q\frac{\partial}{\partial x}\left(x\frac{\partial y}{\partial x}\right)+m\frac{\partial^2 y}{\partial t^2}=0\,. \tag{1}$$

In [10], the authors first found the exact solution of this differential equation. As a consequence, formulas were also obtained for the dynamic parameters of the state of the rod (deflection, angular displacement, bending moment and transverse force), which are expressed in terms of dimensionless fundamental solutions (1). The presence of these formulas makes it possible to investigate the bending vibrations of rods with dead weight taken into account under any boundary conditions. For example, in the work [11] the authors proposed new method of cantilever rod structures calculation with taking into account the dead weight.

Dynamic boundary conditions for the hinged rod have the form:

$$y(0,t)=0;\ M(0,t)=0;\ y(l,t)=0;\ M(l,t)=0\,.$$

Realization of these conditions by exact formulas from [10] leads to the frequency equation, on the left side of which there appears a convergent series in powers of the dimensionless parameter K

$$\eta_0+\eta_1 K^2+\eta_2 K^4+\eta_3 K^6+...=0\,. \tag{2}$$

The dimensionless coefficients of the equation $\eta_k\ (k=1,2,3,...)$ are defined by the formulas

$$\eta_0=1,\ \eta_k=\sum_{j=0}^{k}(\beta_{2,j}(l)\widehat{\beta}_{4,k-j}(l)-\widehat{\beta}_{2,j}(l)\beta_{4,k-j}(l))\,, \tag{3}$$

where

$$\beta_{2,0}(l)=c_{2,0,0}+\sum_{j=1}^{\infty}(-1)^j\alpha^j\,c_{2,0,j}\,, \tag{4}$$

$$\beta_{4,0}(l)=c_{4,0,0}+\sum_{j=1}^{\infty}(-1)^j\alpha^j\,c_{4,0,j}\,, \tag{5}$$

$$\widehat{\beta}_{2,0}(l)=(f_{2,0,0}-1)\,f_{2,0,0}\,c_{2,0,0}+\sum_{j=1}^{\infty}(-1)^j\alpha^j(f_{2,0,j}-1)f_{2,0,j}\,c_{2,0,j}\,, \tag{6}$$

$$\widehat{\beta}_{4,0}(l)=(f_{4,0,0}-1)f_{4,0,0}\,c_{4,0,0}+\sum_{j=1}^{\infty}(-1)^j\alpha^j(f_{4,0,j}-1)f_{4,0,j}\,c_{4,0,j}\,, \tag{7}$$

$$\beta_{n,k}(l)=c_{n,k,0}+\sum_{j=1}^{\infty}(-1)^j\alpha^j\,c_{n,k,j}\,, \tag{8}$$

$$\widehat{\beta}_{n,k}(l)=(f_{n,k,0}-1)f_{n,k,0}\,c_{n,k,0}+\sum_{j=1}^{\infty}(-1)^{j}\alpha^{j}(f_{n,k,j}-1)f_{n,k,j}\,c_{n,k,j}\,, \tag{9}$$

$$\alpha=\frac{ql^{3}}{EI}\,. \tag{10}$$

To determine the dimensionless coefficients of series (4)-(9), we use the set of recurrence formulas [10]:

$$c_{n,0,0}=\frac{1}{(n-1)!}\,;$$

$$c_{n,k,0}=\frac{1}{(n+4k-1)!}\,;$$

$$c_{n,0,j}=\frac{(n-1)(n+2)...(n+3j-4)}{(n+3j-1)!}\,;$$

$$c_{n,k,j}=\frac{c_{n,k-1,j}+(f_{n,k,j}-3)^{2}c_{n,k,j-1}}{(f_{n,k,j}-3)(f_{n,k,j}-2)(f_{n,k,j}-1)f_{n,k,j}}\,,$$

where

$$f_{n,k,j}=n+4k+3j-1\ \ (n=2,4)(k=1,2,3,...)(j=1,2,3,...)\,.$$

The dimensionless parameter K will further be called the coefficient of vibrations.

Equations of the form (2) are often encountered in mechanics. To find solutions of such equations, the method of comparing the roots calculated for different numbers of retained terms of the series is used [12]. Such an approach makes it possible to calculate the roots of an equation with any given accuracy.

2.3. Analytical formulas for the vibration frequencies

In the publication [10] the formula was obtained for frequencies of free vibration of a rod with arbitrary boundary conditions:

$$p_{j}=\frac{K_{j}}{l^{2}}\sqrt{\frac{EI}{m}}\ \ (j=1,2,3,...)\,, \tag{11}$$

where $K_1, K_2, K_3,...$ –roots of the corresponding frequency equation. In our case, the frequency equation has the form (2). Consequently, the problem of determining the frequencies for a vertical hinged rod is reduced to finding dimensionless vibration coefficients from equation (2).

As can be seen from (2)-(9), the values of the roots $K_1, K_2, K_3,...$ will depend on the dimensionless parameter α . Let's set the actual range of this parameter change.

From the theory of stability, it is known [13, 14] that a hinged column under the influence of longitudinal distribution load with a value

$$q_{cr} = 18.5687 \frac{EI}{l^3} \tag{12}$$

will lose stability. Therefore, from the practical point of view it is meaningless to consider the problem of rod's vibrations for values of the parameter q greater than q_{cr}. Assuming in formula (10) $q = q_{cr}$ and using (12), we find the maximum possible value of α :

$$\alpha_{\max} = \frac{q_{cr} l^3}{EI} = 18.5687 .$$

Consequently, the values α relevant for the study lie in the interval

$$0 \le \alpha \le 18.5687 . \tag{13}$$

Table 1 shows the results of the calculation of the first three vibration coefficients for eleven values of the parameter α from the specified interval, equidistant from each other in steps 1.85687.

Table 1. Vibration coefficients

α	K_1	K_2	K_3
0.0000	9.8696	39.4784	88.8264
1.8569	10.2490	38.7144	88.4026
3.7137	10.6561	37.9218	87.9758
5.5706	11.0950	37.0967	87.5461
7.4275	11.5708	36.2339	87.1133
9.2844	12.0901	35.3270	86.6776
11.1412	12.6617	34.3670	86.2388
12.9981	13.2982	33.3416	85.7970
14.8550	14.0173	32.2330	85.3520
16.7118	14.8477	31.0126	84.9040
18.5687	15.8392	29.6305	84.4528

Using the data from Table 1, we represent the found correspondence between parameters α and K in the analytical form. Considering the coefficient K as a function of variable α, and also having the set of values of this function, we approximate K by a polynomial. To achieve high accuracy of approximation, the degree of the polynomial will be chosen from the condition that the coefficient of determination is not less than 0.9999.

As calculations show, for the first and the second vibration coefficients the value of the determination coefficient will practically equal one for forth power approximation, and for the third – for a quadratic approximation. The final formulas have the form:

$$K_1 = \pi^2 + 0.189\alpha + 0.00659\alpha^2 - 0.000291\alpha^3 + 1.72 \cdot 10^{-5} \alpha^4 ; \tag{14}$$

$$K_2 = 4\pi^2 - 0.3959\alpha - 0.00676\alpha^2 + 0.00029\alpha^3 - 1.7\cdot10^{-5}\alpha^4\,; \tag{15}$$

$$K_3 = 9\pi^2 - 0.22734\alpha - 0.00044\alpha^2\,. \tag{16}$$

As a result, according to formulas (11), (14)-(16), to determine the first three frequencies of free vibrations of vertical hinged structures, we have the following analytical formulas:

$$p_1 = \frac{\pi^2 + 0.189\alpha + 0.00659\alpha^2 - 0.000291\alpha^3 + 1.72\cdot10^{-5}\alpha^4}{l^2}\sqrt{\frac{EI}{m}}\,; \tag{17}$$

$$p_2 = \frac{4\pi^2 - 0.3959\alpha - 0.00676\alpha^2 + 0.00029\alpha^3 - 1.7\cdot10^{-5}\alpha^4}{l^2}\sqrt{\frac{EI}{m}}\,; \tag{18}$$

$$p_3 = \frac{9\pi^2 - 0.22734\alpha - 0.00044\alpha^2}{l^2}\sqrt{\frac{EI}{m}}\,. \tag{19}$$

Using formulas (10), (17)-(19), knowing the characteristics of $q,\ l,\ EI$, it is possible in the future to calculate the vibration frequencies of hinged structures with dead weight taken into account without using approximate methods and program complexes.

The vibrations frequencies for a weightless hinged rod are given by the known formulas [15, 16]:

$$\omega_1 = \frac{\pi^2}{l^2}\sqrt{\frac{EI}{m}}\,; \tag{20}$$

$$\omega_2 = \frac{4\pi^2}{l^2}\sqrt{\frac{EI}{m}}\,; \tag{21}$$

$$\omega_3 = \frac{9\pi^2}{l^2}\sqrt{\frac{EI}{m}}\,. \tag{22}$$

It is obvious that formulas (20)-(22) are a special case of formulas (17)-(19).

With the help of (17)-(22), it is easy to establish a connection between the frequencies taking into account the dead weight and frequencies without taking into account the dead weight of the rod:

$$\frac{p_1}{\omega_1} = 1 + 0.0191\alpha + 0.000668\alpha^2 - 2.95\cdot10^{-5}\alpha^3 + 1.74\cdot10^{-6}\alpha^4\,; \tag{23}$$

$$\frac{p_2}{\omega_2} = 1 - 0.010\alpha - 0.000171\alpha^2 + 7.35\cdot10^{-6}\alpha^3 - 4.31\cdot10^{-7}\alpha^4\,; \tag{24}$$

$$\frac{p_3}{\omega_3} = 1 - 0.00256\alpha - 4.95\cdot10^{-6}\alpha^2\,. \tag{25}$$

Based on (13), (23)-(25), we can also establish the change limits for the frequency ratios:

$$1 \le \frac{p_1}{\omega_1} < 1.61; \tag{26}$$

$$0.75 < \frac{p_2}{\omega_2} \le 1; \tag{27}$$

$$0.95 < \frac{p_3}{\omega_3} \le 1. \tag{28}$$

Formula (10) with taking into account (12) can be rewritten in the form

$$\alpha = 18.5687 \frac{q}{q_{cr}}. \tag{29}$$

Substituting the value (29) instead of parameter α in formulas (23)-(25), we express the frequency ratios in terms of the distributed longitudinal load q:

$$\frac{p_1}{\omega_1} = 1 + 0.3556 \frac{q}{q_{cr}} + 0.2302 \left(\frac{q}{q_{cr}} \right)^2 - 0.1887 \left(\frac{q}{q_{cr}} \right)^3 + 0.2071 \left(\frac{q}{q_{cr}} \right)^4; \tag{30}$$

$$\frac{p_2}{\omega_2} = 1 - 0.1862 \frac{q}{q_{cr}} - 0.059 \left(\frac{q}{q_{cr}} \right)^2 + 0.0470 \left(\frac{q}{q_{cr}} \right)^3 - 0.0512 \left(\frac{q}{q_{cr}} \right)^4; \tag{31}$$

$$\frac{p_3}{\omega_3} = 1 - 0.04752 \frac{q}{q_{cr}} - 0.0017 \left(\frac{q}{q_{cr}} \right)^2. \tag{32}$$

Fig. 3-5 show the graphical dependence of the frequency ratios on parameter q.

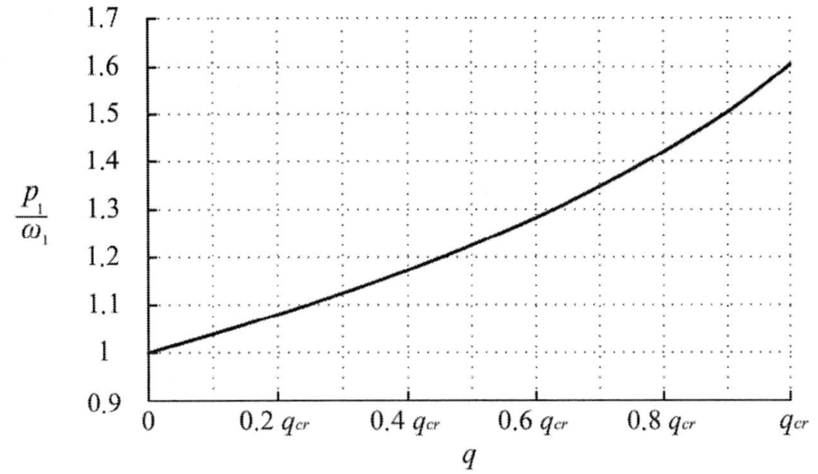

Fig. 3. Graph of the dependence of p_1/ω_1 on the longitudinal load q.

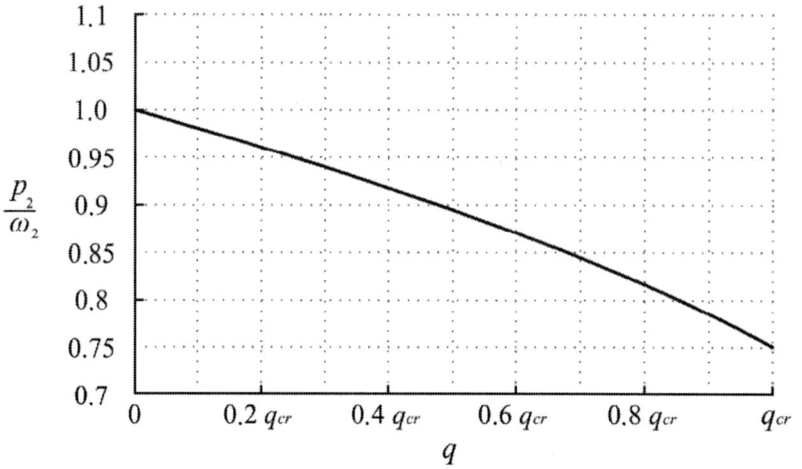

Fig. 4. Graph of the dependence of p_2/ω_2 on the longitudinal load q.

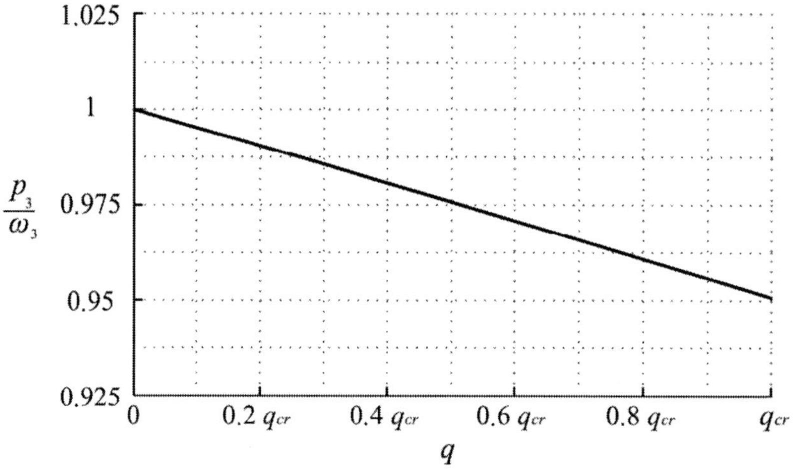

Fig. 5. Graph of the dependence of p_3/ω_3 on the longitudinal load q.

As can be seen from the graphs, taking into account the rod's dead weight leads to an increase the first frequency and a decrease the second and the third one. The greater the value of the longitudinal load, the greater the difference between the frequencies of a weighty rod and the frequencies of a weightless rod.

3 Conclusions

Based on the exact solution of the differential equation of vibrations of a weighted uniform rod, the following results for hinged vertical structures with dead weight taken into account are obtained:

1. Formulas for the first three frequencies of vibrations (17)-(19) are found in an analytical form. Using these formulas, it is possible to calculate frequencies without using approximate methods and program complexes.

2. The dependence between the frequencies with taking into account the dead weight of structure and without it, was established analytically (23)-(25). It was established that taking into account dead weight leads to an increase in the first frequency and a decrease in

the second and the third frequency, in comparison with the corresponding frequencies for the weightless rod.

3. The greater the value of the longitudinal load, the greater the difference between the frequencies of the weighty rod and the frequencies for the weightless rod. From the formulas (26)-(28) we come to the following conclusions:

– the difference between p_1 and ω_1 does not exceed 61%;

– the difference between p_2 and ω_2 does not exceed 25%;

– the difference between p_3 and ω_3 does not exceed 5%.

The maximum difference will be achieved at a value of $q = q_{cr}$.

4. Formulas (30)-(32) established an analytical relationship $p_j / \omega_j \ (j = 1,2,3)$ between the frequencies when the longitudinal load value changes.

In the future, based on the results of the publication [10], similar studies can be carried out for structures under the other boundary conditions.

References

1. R. Luo, Journal of Mechanical Engineering Science, **228 (1)** (2013)
2. S. Naguleswaran, International Journal of Mechanical Engineering Education, **33 (1)** (2005)
3. E. Ghandi, B. Rafezy, Journal of Structural and Construction engineering, **84** (2016).
4. S. Azhermachev, Construction and industrial safety, 33–34 (2010)
5. S. S. Rao, *Mechanical vibrations* (Pearson, Prentice Hall, 2004)
6. F. J. Shaker, Lewis Research Center Report NASA TN D-8109 (December 1975)
7. A. Bokaian, Journal of Sound and Vibration, **126 (1)** (1998)
8. M. Vasilenko, A. Alekseychuk, *Teoriya kolyvan' ta stiykosti rukhu* (Kyiv, High School, 2004)
9. E. Khachiyan, *Seysmicheskoye vozdeystvie na vysotnyie zdaniya i sooruzheniya* (Ayastan, 1973)
10. Yu. Krutii, M. Suriyaninov, V. Vandynskyi, *MATEC Web of Conferences*, **116** (2017)
11. Yu. Krutii, M. Suriyaninov, V. Vandynskyi, Eastern-European Journal of Enterprise Technologies, **3/7 (93)** (2018)
12. V. P. Ilyin, A. M. Karpov, A. M. Maslennikova, *Chislennye metody resheniya zadach stroitelnoy mekhaniki* (Vysheyshaya shkola, 1990)
13. C. M. Wang, C. Y. Wang, J. N. Reddy, *Exact solutions for buckling of structural members* (CRC Press, 2005)
14. N. S. Volmir, *Ustoychivost uprugih sistem* (Fizmatgiz, 1963)
15. V. A. Bazhenov, A. V. Perelmuter, O. V. Shishov, *Budivel'na mekhanika. Kompyuterni tekhnolohii* (Kyiv, Karavela, 2009)
16. V. A. Kiseliov, *Stroitelnaya mekhanika* (Moscow, Stroyizdat, 1980)

Peculiarities of calculation and design of slabs on elastic cushion reinforced with non-metallic composite reinforcement

Olena Luhchenko[1*], *Ali* Nazhem[1], and *Dmitriy* Oreshkin[2]

[1]O.M. Beketov National University of Urban Economy in Kharkiv, 17 Marshala Bazhanova str., Kharkiv, Ukraine
[2] Ekipazh Technological group, Kharkiv, Ukraine

Abstract. Fiber-glass composite reinforcement becomes increasingly popular in different spheres of industry and public economy. Unique technology of the reinforcement production allows receiving the high-quality construction material which complies with modern requirements of liability, quality and safety. Pavlivska Square in Kharkiv may serve as an example of efficient application of non-metallic composite reinforcement for the reinforcement of slabs on elastic cushion during the construction of which the composite grids SKS-100/100 manufactured by Ekipage Technological Group were used as the reinforcement units. Pavement slab reinforcement scheme (including the working draft) was elaborated by the specialists of the Building Constructions Department of O.M. Beketov National University of Urban Economy. The arrangement of the pavement slab of Pavlivska Square was conducted within the short period of time. Due to this, for the purpose of mounting the two-layer reinforcing cage of the slab, the workers of Ekipage Technological Group elaborated a new mounting method which allows the essential reduction of the performance time.

1 Introduction

Today the fiber-glass and basalt-plastic composite reinforcement becomes increasingly popular in different spheres of industry and public economy. Unique technology of the reinforcement production allows receiving the high-quality construction material which complies with modern liability, quality and safety requirements. Noteworthy, the composite reinforcement is undemanding as to the application environment [5, 7, 8, 12]. Thus, reinforcement may be used at different temperatures, from -700°C up to +1000°C. However, the demand for this material is conditioned, first of all, by the fact that the fiber-glass reinforcement is resistant against corrosion and has longer service life [8-10, 12]. The practicability if replacement the metal reinforcement by composite is constituted by a number of advantages, mainly: tear resistance is two times higher than the strength properties of the steel reinforcement A400C; stainless material; the reinforcement is acid-resistant, works well in the sea water; the reinforcement does not bend (higher elasticity); is

* Corresponding author: elenalugchenko@gmail.com

non-conductive, is dielectric; is not heat-transmitting; is radio-transparent; is magnetically inert (the change of the strength properties of the composite reinforcement under the impact of electro-magnetic fields is excluded); does not lose its strength properties under the influence of very low temperatures; is 5 times lighter than the metal reinforcement, and at the replacement equal in strength is 9 times lighter; any construction length; economic efficiency (reduction of transportation costs) [11]. The comparative characteristics of the metal and fiber-glass reinforcement are shown in Table 1. Based on the data received after many years of studies, a set of regulatory documents was elaborated regulating the application of non-metallic composite reinforcement and items there of for the construction purposes in different countries [2, 3].

Table 1. Replacement equal in strength according to physical and chemical properties.

Metal class A400C	Composite polymeric fiber-glass reinforcement AKS
6 A400C	4 AKS
8 A400C	5.5 AKS
10 A400C	6 AKS
12 A400C	8 AKS
14 A400C	10 AKS
16 A400C	12 AKS
18 A400C	14 AKS
20 A400C	16 AKS

2 Project solution

Pavlivska Square in Kharkiv may serve as the example of efficient application of non-metallic composite reinforcement of slabs on elastic cushion (Fig. 1) during the construction of which the composite grids SKS-100/100 manufactured by Ekipage Technological Group were used as the reinforcement units [5-7].

(a) (b)

Fig. 1 a, b. Standard form of the pavement slab in Pavlivska sq., Kharkiv.

The square is located in the central part of Kharkiv. The location is unfavorable, primarily, due to a high location and essential fluctuations of the ground water level (GWL). During the floods GWL may reach the land surface. Besides, the soils occurring on such territories, as a rule, have a poor load bearing capacity; they are non-uniform, poorly consolidated, and saturated with water. The sands are quick. The quaternary deposits represented by modern filling layer take part in the geological construction. The filling layer is represented by the humous clay loam and sand with the construction waste, broken brick and break stone. It is 0.6 – 2.1 m strong; occurring everywhere. The ground waters of the underground reservoir have the following properties with regard to the material: concretes – slightly aggressive for all makes of concrete; reinforced concrete – medium-aggressive during the periodic immersion; metal constructions – highly aggressive. The factors above constitute the reason for the composite reinforcement application at this object.

3 Static calculations of slabs

The slab calculation shall be performed in compliance with [1, 11]. Slab loading according is 25 tons on axis (loading NK-100). In case of infinitely great slab bearing the localized loading P. Load deflection reaches the maximum value

$$\omega_{\max} = \frac{Pl^2}{8D} \quad (1)$$

where D is the flexural rigidity, calculated as

$$D = \frac{E_{cm}\delta^3}{12(1-v)} \quad (2)$$

l – characteristic length,

$$l = \sqrt[4]{\frac{D}{k}} \quad (3)$$

E_{cm} – slab concrete stress-strain modulus; k – coefficient of the soil reaction, depends upon the type of the subsoil, according to [1] $k = 2760$ t/m^3; δ – slab thickness.
Subsoil reaction in the place of the greatest slab deflection

$$\rho_{\max} = \frac{P}{8l^2} \quad (4)$$

Bending moment values may be easily calculated using the formulas [1]:

$$M_r = D\left(\frac{d\varphi}{dr} + \frac{v}{r}\varphi\right) \quad (5)$$

$$M_t = D\left(\frac{\varphi}{r} + v\frac{d\varphi}{dr}\right) \quad (6)$$

Pavement slab reinforcement scheme (including the working draft) was elaborated by the specialists of the Building Constructions Department of O.M. Beketov National University of Urban Economy under the supervision of prof. V.S. Shmukler and is shown in Fig. 2. The reinforcement cage has two layers, is made of glass composite reinforcement ∅10 mm (Ekibar 10 AKSP TU U B.2.7-25.2-21191464-024:2011) housing parameters 150×150 mm, elastic hydraulic seal. The area of longitudinal extended non-metallic composite reinforcement for the slab was taken on the basis of provisions of section 9 DBN B.2.6-98, section 8 DSTU B B.2.6-156, and corresponding provisions of section 3 DBN B.1.1-12:

$$A_{f,\min} = 0.26\frac{f_{ctm}}{f_{fk}}\ b_t d \tag{7}$$

where b_t is a medium width of the extended area; f_{ctm} – medium concrete tension capacity determined for the corresonding strength category, according to DBN B.2.6-98, but not less than 0.0013 $b_t d$.

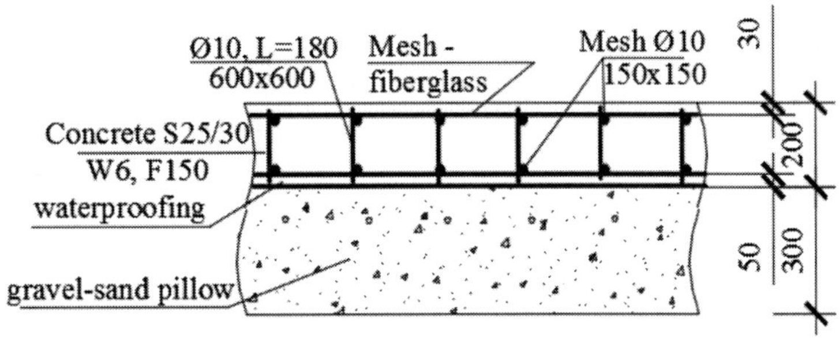

Fig. 2. General pavement slab reinforcement scheme.

Physical and mechanical characteristics of the glass composite reinforcement Ø10 mm (Ekibar 10 AKSP TU U B.2.7-25.2-21191464-024:2011) are shown in Table 2 [2].

Table 2. Physical and mechanical characteristics of the glass composite reinforcement.

Reinforcement class	Mechanical characteristics					
	f_{fk}, H/mm^2	f_{fd}, H/mm^2	f_{fcd}, H/mm^2	$E_f \cdot 10^{-3}$, H/mm^2	ε_{fu}, %	ε_{f_0} ,%
AKS 600	600	400	100	50	1.2	0.80

The shank spacing of the main longitudinal non-metallic composite reinforcement does not exceed 3*h* and is smaller than 400 mm (*h* –slab length). In the localized load or maximum torque areas, the shank spacing of the main reinforcement does not exceed 2h and is less than 250 mm.

The determination of slab bearing capacity at the suggested variant of reinforcement is conducted with the help of calculation software elaborated (based on the library DSTU_156.dll) by the specialists of the Building Constructions Department of O.M. Beketov National University of Urban Economy [4]. The calculation results are shown in graphic form (fig. 3-5).

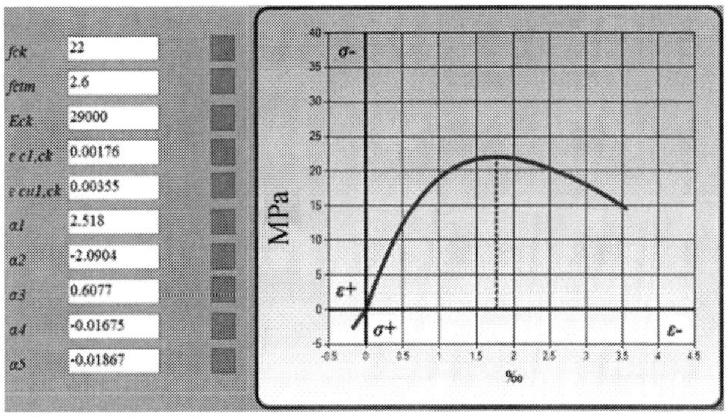

Fig. 3. The objective of physical and mechanical characteristics of concrete. Screenshot of the application window.

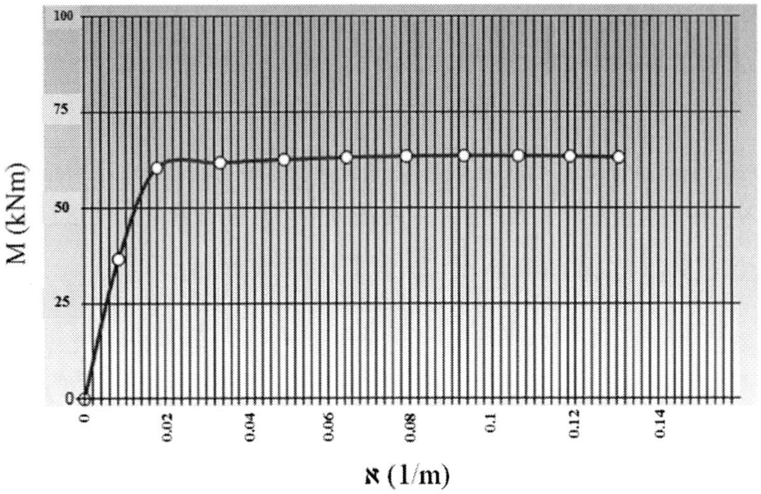

Fig. 4. The sectional diagram «torque-curve». Screenshot of the application window.

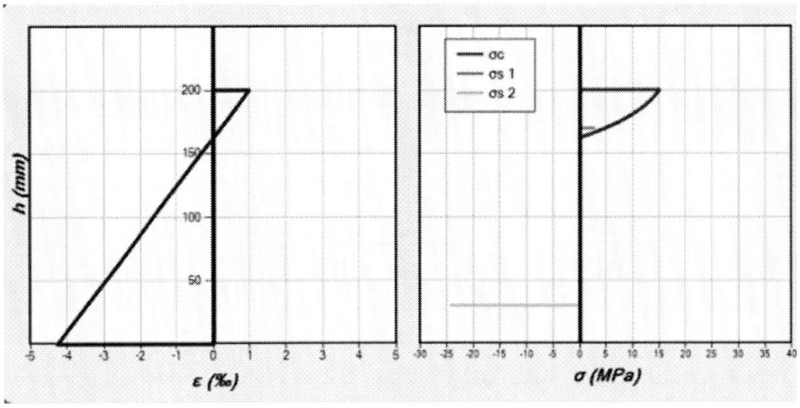

Fig. 5. The strain ε and stress σ diagram. Screenshot of the application window.

The table of results contains the strain E_{ps} (‰) on both section planes, torque M (kNm), height of the compressive zone of the concrete x_1 (mm), section curve ℵ (1/m).

Table 3. The table of results.

№	E_{ps} c(1) (‰)	E_{ps} c(2) (‰)	M (kNm)	x_1 (mm)	ℵ (1/m)
1	2	3	4	5	6
1	0.0656	-1.0222	7.1444	24.1211	0.00272
2	0.1312	-2.0183	14.0967	24.4152	0.00537
3	0.1968	-2.9887	20.8612	24.712	0.00796
4	0.2624	-3.9341	27.4425	25.0113	0.01049
5	0.328	-4.8551	33.8447	25.313	0.01296
6	0.3936	-5.7523	40.0722	25.6172	0.01536
7	0.4592	-6.6262	46.1291	25.9337	0.01771
8	0.5248	-7.4774	52.0197	26.2326	0.02001
9	0.5904	-8.3066	57.7481	26.5438	0.02224
10	0.656	-9.1142	63.3183	26.8572	0.02443

Bearing capacity of the slab section turned out to be enough to bear the operational load.

4 Technology of slab reinforcement

The arrangement of the pavement slab of Pavlivska Square was conducted within the short period of time. Due to this, for the purpose of mounting the two-layer reinforcing cage of the slab, the workers of Ekipage Technological Group elaborated a new mounting method which allows the essential reduction of the performance time. According to the reinforcement scheme - the sickness of the concrete slab is 200 mm; the distance from the lower grid of the reinforcing cage to the hydraulic seal layer is 50 mm; the thickness of the concrete wall protection layer is 30 mm. The distance of the lower grid of the reinforcing cage from the hydraulic seal layer is provided for by the supporting elements, distributed around the lower grid (fig.6).

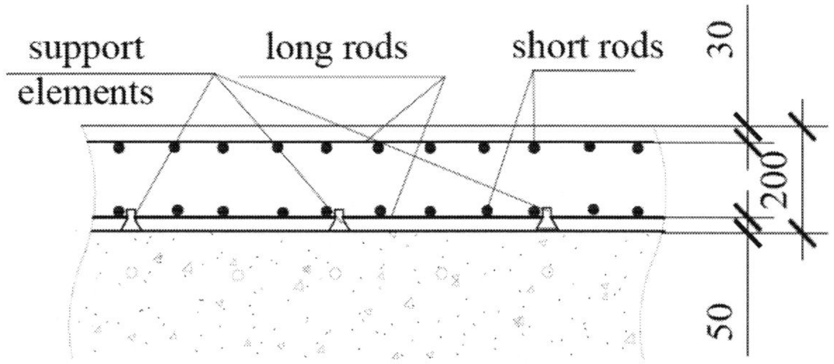

Fig. 6. General reinforcement scheme through the thickness of the concrete slab.

The sequence of mounting the supporting elements (SE), h = 50 mm, is performed according to 5-point envelope scheme, with their installation with the spacing of 1 m. The scheme of the SE location is shown in fig.7. SE are installed directly on the fitting shanks of the lower grid chart, which, in their turn, are located in the future lower slab area, which rests on the soil (elastic cushion). In this case, necessary thickness of the protective layer of concrete is ensured due to the SE installation.

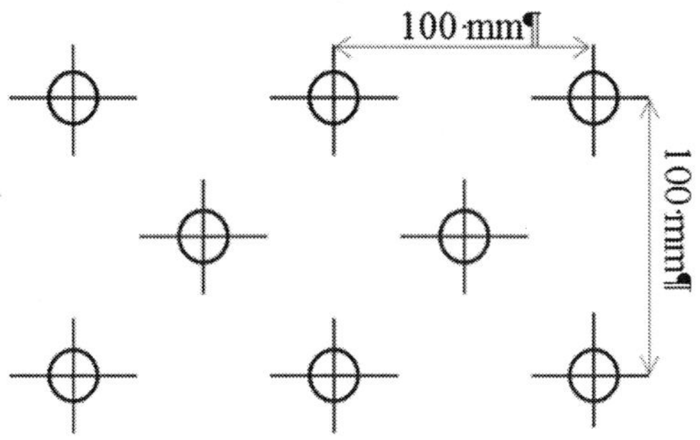

(a)

(b)

Fig. 7. The scheme of the SE location (a) and standard form of SE at the building site (b).

The peculiarity of formation of the spatial reinforcing cage is the installation of mounting elements inside it. And, in particular, the vertical pillar (VP), L = 96 mm, designed for the installation as a bridging, perpendicular grid plane, to ensure the design spacing between the lower and upper grids of the reinforcing cage. Short Rod (ShR), L = 191 mm, is used for the installation as diagonal bridging between the internal parallel reinforcing shanks of upper and lower grids of the reinforcing cage. Long Rod (LR), L = 196 mm, (marked with the paint) necessary for the installation as diagonal bridging between the external parallel reinforcing shanks. General pattern of the installation of VP, ShR and LR is shown in fig. 8-9. VP are installed on the internal reinforcement shanks of the reinforcing cage, right against the grid nodes, near the SE fixing (fig.8). ShR are installed on the internal parallel reinforcing shanks of the reinforcing cage grids, in the housing adjacent to the earlier installed SE. ShR are installed between the parallel internal reinforcing shanks (fig. 8) LR are installed on the external parallel reinforcing shanks of the reinforcing cage grids, in the housing adjacent to the earlier installed SE. ShR are installed between the parallel internal reinforcing shanks (fig. 8-9).

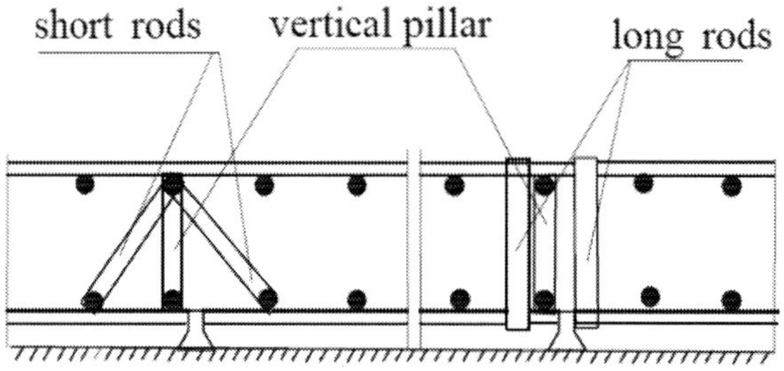

Fig. 8. General pattern of the installation of vertical pillar (VP), Short Rod (ShR) and Long Rod (LR).

(a) (b)

Fig. 9 a, b. Installation of vertical pillar (VP), Short Rod (ShR) and Long Rod (LR).

To prevent the shifting of upper grid against the lower, during the mounting operations of reinforcing cage and its filling with the concrete, the pairs of ShR and LR shall be mounted in chessboard order, according to fig. 10.

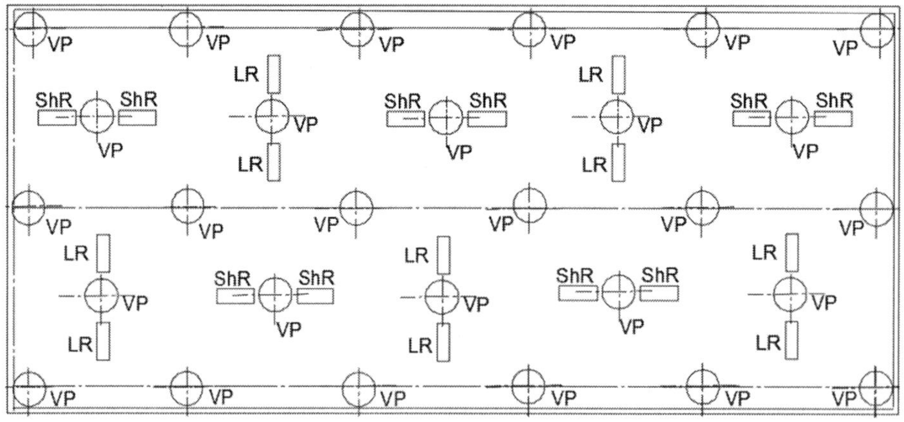

Fig. 10. General pattern of the bracing elements location (in the plan) around the reinforcing cage.

The general order of the reinforcing cage mounting is as follows: the composite grids of the lower reinforcement cage band are installed and interconnected (according to the design documentation); composite grids of the upper reinforcement cage band are installed and interconnected directly on the lower band (according to the design documentation), as a rule, with the half-section flooring; between the upper and lower bands reinforcing the pillars (VP) and brackets (ShR and LR) are installed, besides, in this case the mounting bunks with the supports are used, the construction of which does not damage the below hydraulic seal layer; final operation of mounting the two-layer reinforcing cage is the procedure of lifting the reinforcement cage on the supporting elements (SE) (fig. 11).

(a)

(b)

Fig. 11. The reinforcement cage lifted on the supporting elements (SE).

Final stage of the slab building is filling the reinforcing cage with the concrete according to the design documentation (fig.12).

(a)

(b)

Fig. 12 a, b. Filling the reinforcing cage with the concrete.

Conclusions

General technical and statistical data on the realized project of reinforcing composite cage of the pavement slab in Pavlivska sq., Kharkiv, are as follows:

1. Total area of the concrete slab with the composite two-layer reinforcing cage is 8.7 ths. sq.m.
2. The tear resistance of the two-layer reinforcing cage of the composite reinforcement manufactured by Ekipage Technological Group is 420 kN/m.
3. The amount of meters of the composite reinforcement manufactured by Ekipage Technological Group ("Ekibar" AKSP-10-800 TU U B.2.7-25.2-21191464-024:2011), used for the reinforcement cage arrangement is 260 km.
4. Total weight of the composite reinforcement used for the reinforcement cage arrangement is 41 tons.
5. Total amount of the binding points for the reinforcing cage construction is 879040 pieces.

6. Total amount of the binding points for the standard grids-charts 6x2 m under the plant conditions – 852800 pieces.

7. Total amount of the binding points for mounting the standard two-layer building blocks (sandwiches) to the mounted slab reinforcement cage at the building site is 26240 pieces.

8. The comparative amount of binding points in case of steel reinforcement, (mounting according to the traditional technology with the application of Z- or U-shaped inter-layer insertions-spacers) – 1036480 pieces.

As a result of applying the new quick mounting technology of the two-layer composite reinforcement cage the following was achieved:

1. Reduction of the total amount of binding points of the reinforcement cage by 15.2 %.

2. Reduction of the total labor intensity of the reinforcement cage mounting by 35 %.

3. Reduction of the labor intensity of the reinforcement cage mounting directly at the building site by more than 10 times.

References

1. S.P. Timoshenko, *Plastinki i obolochki*.469 (1948)
2. DSTU-N B V.2.6-185:2012: *Nastanova z proektuvannia ta vyhotovlennia betonnykh konstruktsii z nemetalevoi kompozytnoi armatury na osnovi bazalto- i sklorovinhu* Kiev, Minrehion Ukrainy (2012)
3. DBN V.2.6-98:2009: *Betonni ta zalizobetonni konstruktsii. Osnovni polozhennia* Kiev, Minrehion Ukrainy (2011)
4. V.M. Babaiev, A.M. Bambura, O.M. Pustovoitova, V.S. Shmukler *Praktychnyi rozrakhunok elementiv zalizobetonnykh konstruktsii za DBN V.2.6-98:2009 u porivnianni z rozrakhunkamy za SNiP 2.03.01-84* i EN 1992-1-1 (Eurocode 2).* 208 (2015)
5. Yu.A. Klimov, D.O. Orieshkin, O.S Soldatchenko, *Eksperymentalni doslidzhennia zcheplennia kompozytnoi nemetalevoi armatury z betonom, Teoriia i praktyka budivnytstva*. 71. **3**, 237–241 (2010)
6. D.O. Orieshkin, *Kompozitnye geosetki dlya armirovaniya nezhestkikh dorozhnyh odezhd.* Proceedings of the Suchasni tekhnologhii budivnytstva j ekspluatasii avtomobiljnykh dorih. 220–225 (2013)
7. O.S. Chapiuk, A.V. Hryshkova *Porivnialnyi analiz zcheplennia stalevoi armatury serpovydnoho profiliu z vazhkym betonom za pryzovym ta balkovym metodom vyprobuvan.* Resursoekonomni materialy, konstruktsii, budivli ta sporudy, **29**, 391-399 (2014)
8. J.J. Luchko, I.I. Ghlaghola, B.L. Nazarevych *Metody pidvyshchennia koroziinoi stiikosti ta dovhovichnosti betonnykh ta zalizobetonnykh konstruktsii i sporud.* 229 (1999)
9. L.C. Bank, M. *Puterman Microscopic study of surface degradation of glass fiber – reinforced polymer rods embedded in concrete castings subjected to environmental conditioning.* ASTM International, **2**, 191-205 (1997)
10. E. Cosenza, G. Manfredi, A. Nanni *Composites in Construction: A Reality.* Proceedings of the International Workshop. Capri. 277 (2001)
11. Dzh. Sendetski (ed.) *Mekhanika kompozitsionnyh materialov*. 2. **8,** 564 (1978)
12. L.Yu. Ogrel *Povyshenie effektivnosti stroitelnyh polimernyh kompozitov, ekspluatiruemyh v agressivnyh sredah*. PhD Thesis. Belgorod. 40 (2006)

Stress-strain state of the storage silos under the action of the asymmetric load

*Nataliia*Makhinko[1,*]

[1]National Aviation University, Kosmonavt Komarov avenue, 1, 03058, Kyiv, Ukraine

Abstract. This paper deals with the investigation of work of the vertical steel cylindrical storage silos for grain with the different thickness of casing sheets and stiffeners at different height sunder the asymmetric wind load. Construction is designed using the analytical method involving decomposition of load into the completed trigonometric series and adding separate stress-strain states of k influence. The construction design model is a thin-walled cylindrical shell reinforced with vertical stiffeners. According to the shell theory, all unknown elements of the stress-strain states from the wind load components are expressed by unknown function $\xi_k(x)$. In the analysis, the introduced hypotheses were used, which provides for using exponential dependence to describe the thickness changing law of the casing sheets and stiffeners and the equality of the change rate of these characteristics. The obtained results for silos with corrugated wall showed that the desired function which characterizes the stress-strain state of the construction is determined by ratio of the total cross-section area of all stiffeners to the cross-sectional area of the silo provided this ratio is constant at all heights. Represented calculation formulas for radial, circular and longitudinal displacements, and for strains in cross-sections are simple enough for using in the engineering practice.

1 Introduction

The main type of load at the vertical cylindrical silos with flexible external frame in the form of a system of vertical stiffeners is vertical and horizontal pressure of a bulk material, which causes axially symmetric loading of the silo, and the pressure of the wind flow, which is an unsymmetrical load and causes the greatest difficulties in determining internal efforts and displacements. Generally, the presence of vertical stiffeners in the silo is considered approximately, by bringing it to the scheme of orthotropic shell, which has different reduced meridional and circular wall thicknesses. For the most of real constructions of this type, the casing sheets and vertical stiffeners are made with different thickness at different heights in order to spare the metal. Considering this feature is causing mathematical difficulties when studying stress-strain state and using the analytical technique. According to the outlined problem, the necessity of proposals development is arising about simplifying analytical procedure of construction calculation.

*Correspondingauthor: pasargada1985@gmail.com

The analysis of behavior of the vertical steel silos is a relevant problem for researchers, which is demonstrated by a number of scientific works [1-6], including studies of the authors [7, 8, 9]. It is quite logical that the obtained results of investigation of the stress-strain state of silo constructions are based on works in the shells theory [10, 11, 12] and verified by calculation of real silo models in the environment of the finite-element analysis. However, because of constantly growing demand to the construction of this type and not always sufficient qualification of the software users, the resulting assessment of work of the structure is often wrong that eventually results in failures, the statistics of which is quite not optimistic now [13, 14]. Therefore analytical (engineering) methods of solving must be definitely used in the practice of calculation of the steel storage silos.

In general, vertical cylindrical storage silo is a thin-walled cylindrical shell reinforced with vertical stiffeners. The finite-element method is most accurate method for defining the full range of internal efforts and displacements of the shell construction. However, investigation of silos with different thicknesses of elements in the finite-element analysis is complicated, since the model formation is multivariable. Accordingly, the results not always correctly reflect the true stress-strain state. This problem especially worsens, when the user doesn't understand the principle of calculation, which leads to the unreasonable choice of the design model and use of the finite-element library. Consequently, authors' proposals are primarily oriented on the development of engineering calculation, and the obtained conclusions will help reach reasonable compromise – to solve the problem with the accuracy sufficient for practical purposes without complicating it with extra calculations.

2Research results

The analytical calculation technique for the vertical cylindrical grain storage silos under the unsymmetrical wind load is based on the possibility of expansion of the aerodynamic coefficient C_{aer}, and correspondingly wind load, into the finite trigonometric series (Fig. 1)

$$C_{aer}(\varphi) = a_0 + a_1\cos(\varphi) + a_2\cos(2\varphi) + \ldots + a_k\cos(k\varphi) + \ldots + a_m\cos(m\varphi), \quad (1)$$

where a_k are coefficients of expansion.

This, accordingly, allows making a calculation to each load element separately. Unknown elements of the stress-strain state under load $W_k = A_k\cos(k\varphi)$ are defined using the differential equation

$$\frac{d^4\xi_k(x)}{dx^4} + 4\eta_{w,k}^4\xi_k(x) = 4\eta_{w,k}^4\xi_{k,0}, \quad (2)$$

where A_k is an amplitude value of the k_{th} element; φ is an incidence angle which consider the wind direction; $\eta_{w,k}^4$ is the value which consider rigid characteristics of shell in the predetermined conditions of the k_{th} loading harmonic; $\xi_{k,0}$ is the proportionality coefficient which has dimension of the function $\xi_k(y)$ and considers the character of the harmonious load W_k.

The calculation problem of the stress-strain state of the silo with different thickness of the casing sheets $t_w(x)$ and vertical stiffeners $t_p(x)$ from the mathematical point of view is complicated by the dependence of value $\eta_{w,k}$ from longitudinal coordinate x in equation (2). Consequently, this differential equation is considered in the form of

$$\frac{d^2}{dx^2}\left[\frac{1}{4\eta_{\eta,k}^4(x)}\frac{d^2\xi_k(x)}{dx^2}\right]+\xi_k(x)=\xi_{k,0}\,. \tag{3}$$

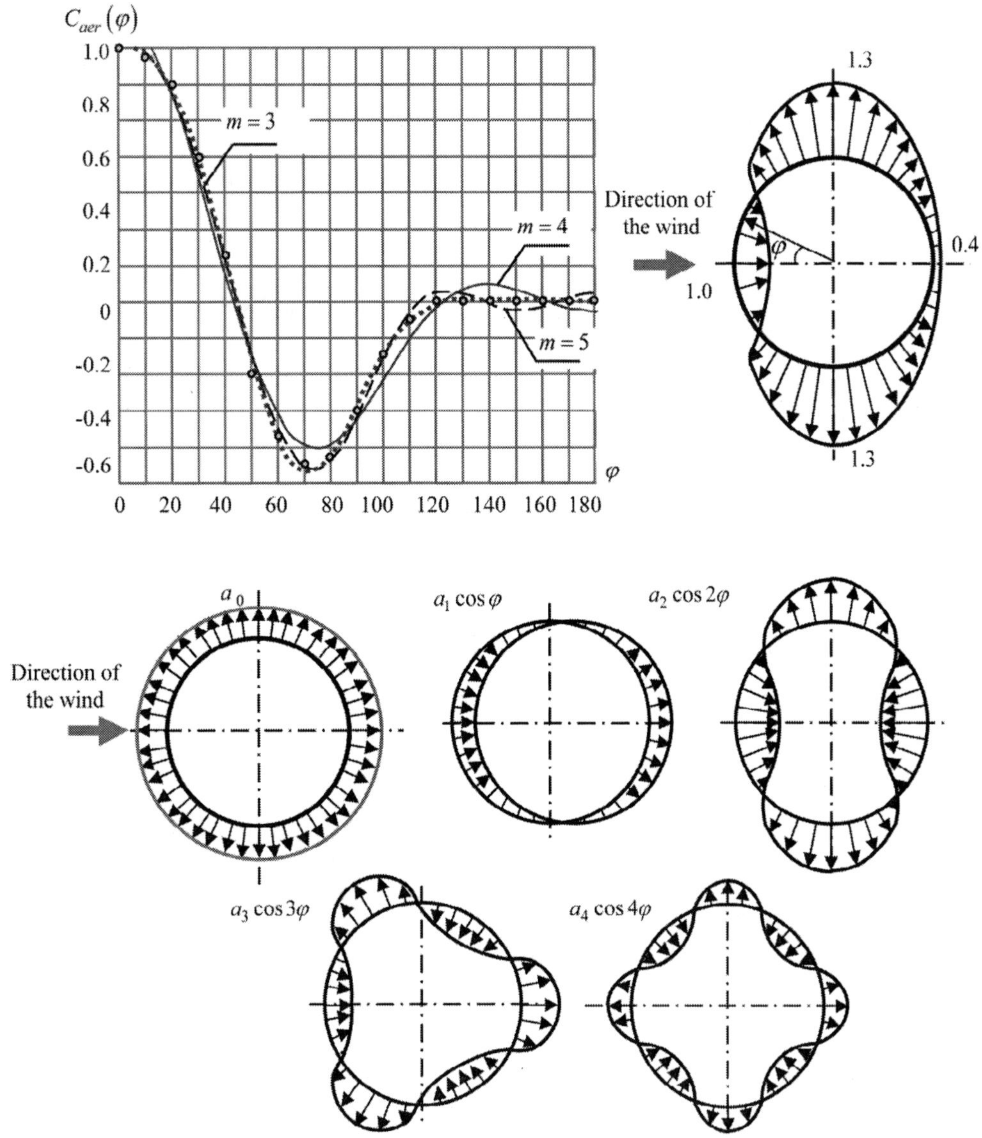

Fig. 1. Series expansion of the aerodynamic coefficient of the wind load around the perimeter of a cylindrical silo.

However, when two hypotheses are formulated which do not contradict the experience of designing storage silos, the problem could be solved in a very simple way. The first hypothesis involves using of exponential dependence for describing the thickness changing law for the casing and stiffeners along the height [9]

$$t_{ef}(x)=t_{ef,r}g_w(x)=t_{ef,r}\exp\left[-\varepsilon_w\left(\frac{x}{H_w}\right)\right], \tag{4}$$

where H_w is the height of the storage silo; $t_{ef,r}=t_w\ell_{w,1}/\ell_w$ is the thickness of sheets (with the rigidity which is equivalent to the one of the corrugated sheet with the thickness t_{ef}) which perceive circular efforts; $\ell_{w,1}$ is the length of the corrugation development with the wave pitch of corrugated sheet of the shell ℓ_w; ε_w is a non-dimensional parameter which is responsible for the curve shape $g_w(x)$.

According to the second hypothesis, it is supposed that the changing rate of the casing sheets and stiffeners is equal. It means that the same parameters ε_w were used in both dependences. Then after additional expanding of the formula of cylindrical rigidity $D_r(x)$, for function $\eta_{w,k}$ we can write

$$\eta_{w,k}^4(x)=\frac{16k^4(k^2-1)^2}{(1+\alpha_{wp})D_w^6}t_w^2(x)\Delta J_w(x). \tag{5}$$

where $\alpha_{wp}=\dfrac{A_pn_p}{\pi D_wt_w}$ is the ratio of the corresponding areas of the stiffeners and the shell; D_w is the diameter of the silo; A_p is the cross-sectional area; n_p is the number of vertical stiffeners; ΔJ_w is the ratio of inertia moments of cross-sections of corrugated and flat wall sheets of the silo casing.

Product of $t_w^2(x)\Delta J_w(x)$ is the element which defines dependences of value $\eta_{w,k}$ on x. For flat wall silos, ratio ΔJ_w is 1, and for the corrugated wall silo it is expressed through the formula [15]

$$\Delta J_w=\frac{h_w^2\psi_2(\delta_w)}{t_w^2(x)\delta_w};$$
$$\psi_2(\delta_w)=\sqrt{32+\delta_w^2}+2\sqrt{12+\delta_w^2}+2.78\sqrt{1.5+\delta_w^2}, \tag{6}$$

where $\delta_w=\ell_w/h_w$ is the ratio of the period of the sheet profile of the silo wall to its height.

We can see that thicknesses reduce and value $\eta_{w,k}$ becomes independent not only from coordinate x, but also from thickness of the casing t_w. Consequently, there is an important conclusion: for silos with corrugated casing reinforced with vertical stiffeners, the desired function $\xi_k(x)$ is not dependent on the absolute thickness value and the law of their change in height, but determined only by the ratio of the total cross-sectional area of all stiffeners to the cross-sectional area of the silo provided this ratio is constant at all heights.

Consequently, according to the general dependences of the shell theory and results, for elements of the stress-strain state of the cylindrical silo with the corrugated wall and different thickness of the casing and stiffeners in different heights, we can write down calculation formulas for radial $w_k(y,\varphi)$, circular $v_k(y,\varphi)$, longitudinal $u_k(y,\varphi)$ displacements and also for strains in cross sections $\sigma_{p,k}(y,\varphi)$ and $\tau_{p,k}(y,\varphi)$

$$w_k(y,\varphi) = 2D_w\Lambda_w \frac{[1-\psi_{m,k}(y)]}{(k^2-1)^2} a_k \cos(k\varphi). \quad (7)$$

$$v_k(y,\varphi) = 2D_w\Lambda_w \frac{[1-\psi_{m,k}(y)]}{k(k^2-1)^2} a_k \sin(k\varphi). \quad (8)$$

$$u_k(y,\varphi) = \frac{D_w}{\Delta_w}\Lambda_w \frac{\psi_{u,k}(y)}{k^2(k^2-1)^2} \lambda_{\mu,k} a_k \cos(k\varphi). \quad (9)$$

$$\sigma_{p,k}(y,\varphi) = \frac{E}{\Delta_w^2}\Lambda_w \frac{\psi_{\sigma,k}(y)}{k^2(k^2-1)^2} \lambda_{\eta,k}^2 a_k \cos(k\varphi). \quad (10)$$

$$\tau_{p,k}(y,\varphi) = \frac{E}{\Delta_w^3}\Lambda_w \frac{\psi_{\tau,k}(y)}{k^3(k^2-1)^2} \lambda_{\eta,k}^3 a_k \sin(k\varphi), \quad (11)$$

where $\psi_{m,k}(y)$ is calculated according to the formula

$$\psi_{m,k}(y) = M_{1,k}\cosh(y\lambda_{\eta,k})\sin(y\lambda_{\eta,k}) + \cos(y\lambda_{\eta,k})\left[M_{2,k}\exp(-y\lambda_{\eta,k}) + M_{3,k}\exp(y\lambda_{\eta,k})\right]. \quad (12)$$

$M_{1,k}$, $M_{2,k}$ and $M_{3,k}$ are non-dimensional functions of response of the silo; functions $\psi_{u,k}(y)$, $\psi_{\sigma,k}(y)$ and $\psi_{\tau,k}(y)$ are similar according to the mathematic nature of $\psi_{m,k}(y)$ and determined by expressions of the first, second and third derivatives; E is the modulus of the material resilience.

Special attention must be paid to compliance coefficient Λ_w of the silo which is the proportionality coefficient in formulas (7)-(11) and is the function of a non-dimension height $y = x / H_w$. Due to the received dependence (4) for thicknesses of the casing $t_w(x)$ and stiffeners $t_p(x)$, we will have for the confluence (in this case the origin is considered to be at the top of the silo)

$$\Lambda_w(y) = \frac{3\delta_{Dt}^6(1-\mu^2)}{8\Delta J_w}\frac{w_p}{E}\exp[\varepsilon_w(1-y)], \quad (13)$$

where all parameters which include wall thickness t_w and stiffener area A_p, refer to the confinement area of the silo.

For flat wall silos, there is no such a simplifying procedure and the solution of equation (3) must be searched according to the rules of differential calculation. But if the flat sheet considers as the borderline case of the corrugated one, then all formulated above conclusions will remain in force. In fact, corrugating parameters can always be selected so "bad" to provide equality of bending rigidity to the flat sheet.

Accordingly, formula (13) could be extended to the flat wall silos without considering lengthy solution of the differential equality. Acceptability of such idealization was verified at some finite-element models, which confirm the assumption. Results are also confirmed by conclusions of study [15], where the main differential equation of the stress-strain state of the silo with different thickness under the axially symmetric load was solved.

3Conclusions

1. All unknown elements of the stress-strain state from the wind load elements can be expressed through function ξ_k which satisfies the linear differential equation (1).

2. Introduced hypotheses to describe the thickness changing law for the casing sheets and vertical stiffeners along the height and equality of changing rate of these thicknesses allow simplifying the problem of location of strains in cross-sections and the radial, circular and longitudinal displacements, and also allow using easy calculation formulas.

3. For silos with corrugated casing reinforced with vertical stiffeners, the desired function is not dependent on the absolute thickness value and the change law along the height, and is determined only by the ratio of the total cross-section area of all stiffeners to the cross-sectional area of the silo provided this ratio is constant at all heights.

4. For flat wall containers the obtained results are true, only if the flat sheet is considered as the borderline case of the corrugated one.

References

1. M. Wojcik, P.Iwicki, J. Tejchman, Thin-Walled Str.**49**, 8 947-963 (2011).
2. A. Sadowski, M. Rotter,Thin-Walled Str.**61** 169-179 (2012).
3. Y. Zhao, Q. Cao, L. Su,Thin-Walled Str.**73** 337-349 (2013).
4. R. Stepanov, D. Romensky, V. Mushchanov, Science**9**, 6152-160 (2017).
5. C. Maraveas, G. Balokas, K. Tsavdaridis, Thin-Walled Str.**95**152-160 (2015).
6. S. Maleki, A. Mehretehran, Engineering Failure Analysis**90** 156-167 (2018).
7. A. Makhinko, N. Makhinko,Pidvodni tekhnolohii**06** 16-27 (2017).
8. N. Makhinko,Visnyk ODABA**70**68-74, (2018).
9. A. Makhinko, N. Makhinko, Industrial Machine Building69-78 (2018).
10. V.Vlasov,*Obshchaia teoriia obolochek*(Akademiia nauk, Moskva, 1962).
11. S. Kahn,*Stroitelnaia mekhanika obolochek*(Mashinostroenie, Moskva, 1966).
12. A. Goldenveizer,*Teoriia uprugikh tonkikh obolochek* (Nauka, Moskva, 1976).
13. M. Kazakevich, D. Bannikov,Metallicheskie konstruktsii**1** 59-66 (2002).
14. A. Raeesi, H. Ghaednia, J. Zohrehheydariha, S. Das, Engineering Failure Analysis**79** 749-761 (2017).
15. O. Lapenko, N. Makhinko,Nauka i budivnytstvo,**2(16)** 40-45 (2018).

MATEC Web of Conferences **230**, 02019 (2018) https://doi.org/10.1051/matecconf/201823002019
Transbud-2018

Probable reliability prediction of the dam constructed with ground materials of the Nam Chien hydraulic power system in Vietnam using the Monte Carlo method

Andrii Mozgovuy[1,*], and *Svitlana* Butnik[1]

[1] Kharkiv National University of Construction and Architecture, 61002, Sumskaya Str., 40, Kharkiv, Ukraine

Abstract. The current method of probabilistic safety assesses dams with ground material by the criterion of water overflow through the dam crest. Due to the stochastic nature of the material properties, loads and impacts method involves the use of the Monte Carlo method (statistical tests). The algorithm was tested on the example of the ground material dams of Nam Chien hydropower in the Vietnam.

Calculation and design of modern hydraulic engineering works require assessment of their reliability and safety on the basis of probabilistic methods. Quite a common type of water power systems backwater front structures are dams constructed of ground materials. In this paper, a modern method of probabilistic assessment of the reliability of dams constructed of ground materials is applied. The criterion of ensuring the edge state is water overflow over the dam crest. During the study of the dam reliability, the probabilistic nature of the loads and influences perceived by the dam is taken into account. Reliability assessment of the dam is fulfilled in the software complex Math Cad with the use of modern numerical Monte-Carlo method (statistical simulations method). The algorithm was tested on the example of the earth-and-rockfill dam Nam Chien in Vietnam. The earth-and-rockfill dam is designed homogeneous with a clay soil core with upstream slope pavement using reinforced concrete 0.5 m thick slabs on a 0.4 m thick gravel layer. The slope base of the upstream slope is 1:1.1. From the downstream slope, topsoil ramping is carried. The slope base of the downstream slope is 1:1.2. (Fig. 1).

The analysis of literary sources [1-10] shows that the statistical simulations method is the most convenient to perform calculations to assess the probability of water overflow over the dam crest.

According to the Monte Carlo method, N simulations are performed. The sequence of calculations for each simulation is accepted as follows.

1. Water level marks in front of the dam P_Q are given by evenly distributed random probability in the range from 0 to 1, caused by the maximum flood flow.

2. The quantile is determined by the value of P_Q, - the mark of the static water level in front of the dam Z_Q caused by the maximum flood flow, according to the relation Z_Q

* Corresponding author: mozgovoyandrey@i.ua

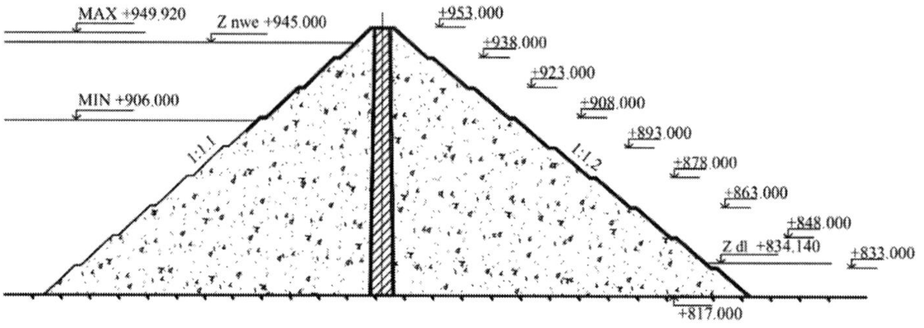

Fig. 1. The section of the earth-and-rockfill dam of the hydraulic power system Nam Chien

3. Due to the well-known mark Z_Q the values of the fetch L and the average water depth in the reservoir H are found, using well-known relations $L = L\ (Z_Q)$, $H = H\ (Z_Q)$.

4. Wind speed P_V is given by evenly distributed random probability in the range from 0 to 1.

5. The quantile is determined by the value of P_V - wind speed V.

6. Using the normative technique based on the known values of fetch L, the average water depth in the reservoir H and wind speed V, the wave parameters are calculated.

6.1. The average wave height is found.

6.2. Average wave periods are calculated.

6.3. The average wave length is determined.

6.4. The wave length 1% of exceedance probability in the wave system is calculated.

6.5. The height of wave run-up on the slope is determined.

6.6. The wind setup height is found.

7. The water level mark in front of the dam Z is determined.

8. The fulfillment of the condition $Z < Z_{dc}$ is checked.

According to the numerical experiment program, the planned number of statistical simulations was carried out. The number of statistical simulations, in which the condition that the water level elevation in front of the dam would be less than the earth dam crest mark was not fulfilled, was attributed to the total number of simulations and determined the probability of the risk of water overflow over the dam crest.

After performing all N simulations, the probability of water overflow over the dam crest P_{OVT} is calculated as the ratio of the number of simulations N_1, with $Z_{ul} < Z_{dc}$, to the number of all simulations N.

It should be noted that due to the smallness of the values R_{OVT}, the number of statistical simulations should be relatively larger in order to provide sufficient probability of the derived values.

To determine the risk of water overflow over the earth dam crest, random probability of wind speed pV is set distributed from 0 to 1 (see Fig. 2). According to the outcoming data, the probabilities of annual maximum wind speed can be represented by a normal distribution.

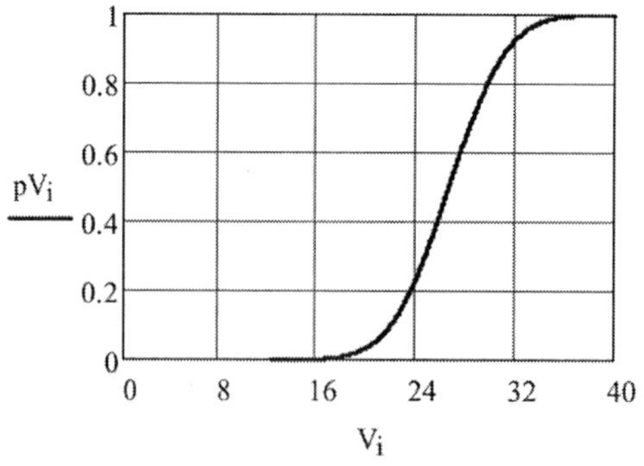

Fig. 2. The maximum wind speed V distribution probabilistic curve, m/s, in the geographical location of the Nam Chien dam

The quantile is determined by the value of P_V - wind speed V.

Random probability pZ of water level mark Z in front of the dam constructed of ground materials is set, distributed from 0 to 1, m (see Fig. 3).

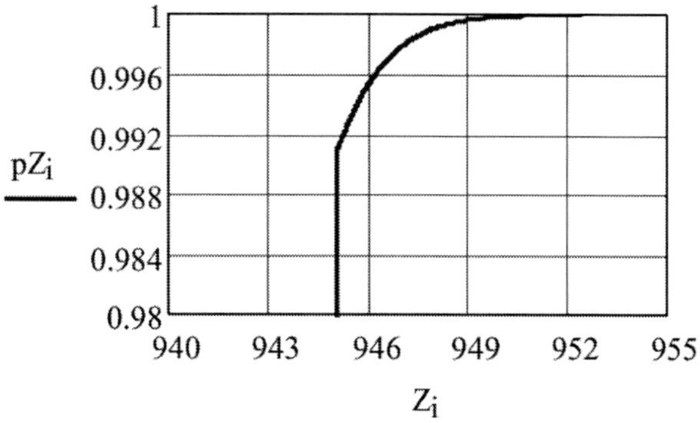

Fig. 3. Probabilistic curve of the static water level distribution in front the dam constructed of ground materials Z, m Nam Chien dam

At the same time, the static water level is caused by the maximum flood flow.

The quantile is determined by the value of pZ - the water level mark Z, m in front of the dam constructed of ground materials.

Depending on static water level Z, m, average value of the water depth in the reservoir H, m and water depth in the reservoir Hs, m, in front of the earth dam are determined (see Fig. 4).

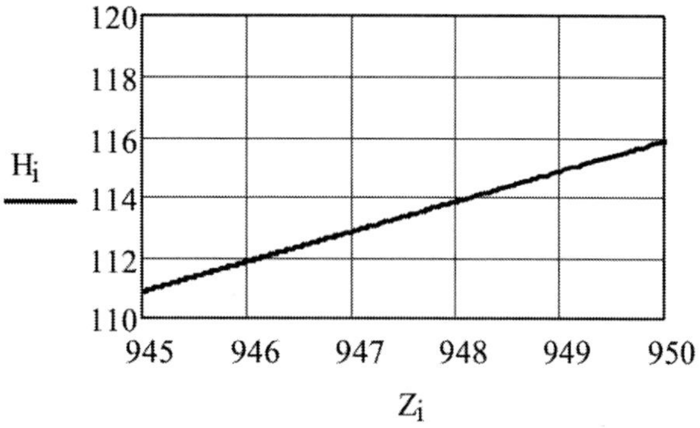

Fig. 4. Relation between the average value of water depth in the reservoir H, m and the static water level mark Z, m

Depending on the average water depth in the reservoir H, m the fetch L is determined, m.

Depending on fetch L, m, water depth in the reservoir Ns, m in front of the dam and wind speed V, m/s, wind setup height dh, m (see Fig. 5) is determined according to the known formula 1:

$$dh = (0.3 + 0.09 \times V) \times 10^{-6} \frac{V^2 \times L}{g \times H_C} \times \cos(\alpha_W), \tag{1}$$

where 0.3 and 0.09 – empirical coefficients, g – gravitational acceleration, m/s^2, α_w – the angle between the longitudinal axis of the reservoir and the airflow direction.

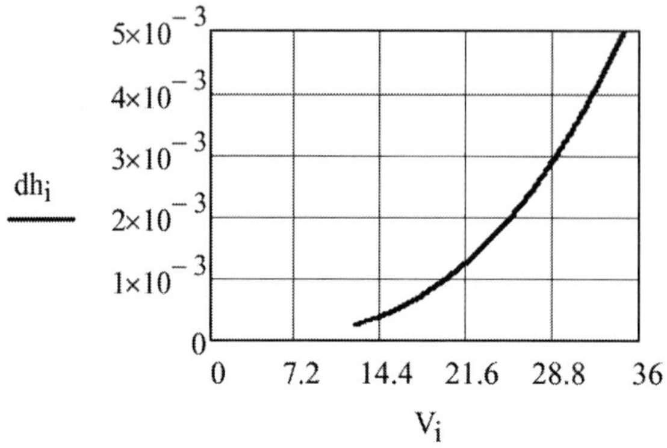

Fig. 5. Relation between the wind setup height dh, m in the reservoir *H*, m and the wind speed *V*, m/s

Depending on fetch *L*, m, water depth in the reservoir *N*, m in front of the dam and wind speed *V*, m/s, average wave height h_{av}, m (see Fig. 6) is determined according to the formula 2:

$$h_{av} = 0.16\frac{V^2}{g}\left[1-\left[1+0.006\sqrt{g\frac{L}{V^2}}\right]^{-2}\right]\times th\left[\frac{0.625\left[g\frac{H}{V^2}\right]^{0.8}}{1-\left[1+0.006\sqrt{g\frac{L}{V^2}}\right]^{-2}}\right], \quad (2)$$

where 0.006, 0.16, 0.625 - empirical coefficients.

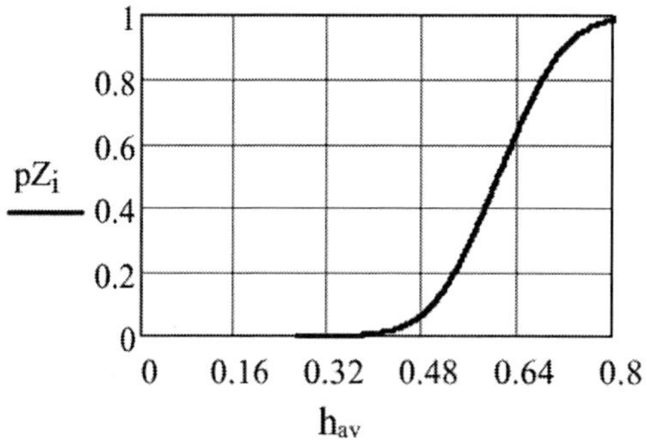

Fig. 6. Relation between the average wave height h_{av},m in the reservoir H, m and the random value - the static water level mark Z, m

Depending on average wave height h_{av}, m and wind speed *V*, m/s, average wave period T_{av} is determined using formula 3:

$$T_{av} = 3.1\times 2\pi\frac{V}{g}\left[g\frac{h_{av}}{V^2}\right]^{0.625}, \quad (3)$$

where 0.625, 3.1 - empirical coefficients.

Depending on average wave period T_{av}, average wave length λ_{av}, m is determined using formula 4:

$$\lambda_{av} = \frac{g\times T_{av}^2}{2\pi}, \quad (4)$$

Depending on fetch L, m, wind speed V, m/s, and average wave length h_{av}, m, the wave length 1% of exceedance probability in the wave system is determined using formula 5:

$$h_{1\%} = h_{av} \frac{(-\ln(0.01)) \times 0.001\left[\sqrt{g\frac{L}{V^2}} + 390\right]}{G\left[0.001\left[\sqrt{g\frac{L}{V^2}} + 390\right] + 1\right]}, \tag{5}$$

where 0.001, 1, 390 - empirical coefficients; 0.01 - wave height exceedance probability; G - value of the gamma-function.

Depending on the wave height 1% of exceedance probability in the wave system $h_{1\%}$, m the wave run-up on the slope $h_{run1\%}$, m is determined according to the formula 6:

$$h_{run1\%} = k_r \times k_p \times k_{sp} \times k_{run} \times h_{1\%}, \tag{6}$$

where k_r - coefficient of slope paving roughness; k_p - coefficient of slope paving penetration; k_{sp} - coefficient, which depends on wind speed and slope ratio; k_{run} - coefficient, which is determined using formula 7:

$$k_{run} = 1,2 \times \cos\alpha \times \sqrt{\frac{\lambda}{h_{1\%} \times (1 + ctg^2\phi)}}, \tag{7}$$

where α is the angle between the wave line and the normal line to the bank line; φ - canting angle of the slope.

Depending on the water level mark Z, m, the height of wave run-up on the slope $h_{run1\%}$, m and the wind setup height dh, m the water level mark in front of the dam Z_f, m is determined using formula 8 (see Fig. 7):

$$Z_f = Z + h_{run1\%} + dh. \tag{8}$$

The fulfillment of the condition 9 is checked

$$Z_f < Z_{DC}. \tag{9}$$

where Z_{DC} is the mark of the earth dam crest.

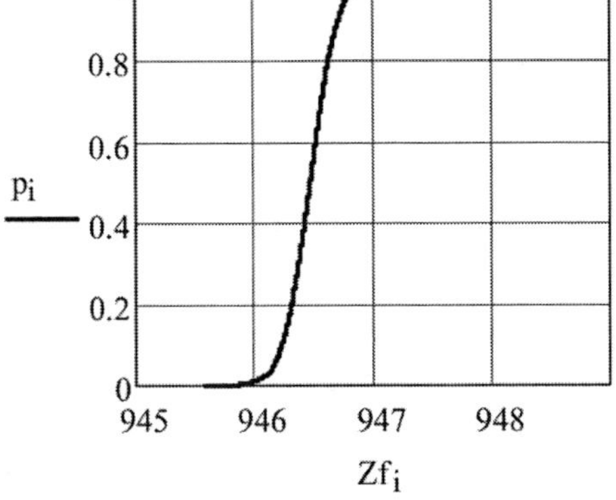

Fig. 7. Probabilistic curve of the water level mark distribution in front the dam constructed of ground materials Z_f, m of the hydraulic power system Nam Chien

Calculation of the earth-and-rockfill dam of the hydraulic power system Nam Chien under the condition of preventing the water overflow over the crest was carried out according to the normative method for the first and for the second settlement cases.

The results of the calculations of the earth-and-rockfill dam of the hydraulic power system Nam Chien under the condition of preventing the water overflow over the crest carried out according to the normative method for the first settlement case are given in Table 1, for the second settlement case - in Table 2, the results of the probabilistic calculations of the earth-and-rockfill dam - in Table 3.

Table 1. The results of calculations of the earth-and-rockfill dam of the hydraulic power system Nam Chien under the condition of preventing the water overflow over the crest was carried out according to the normative method for the first settlement case

No.	Name of the estimated grandeur in	Value
1	Normal water surface elevation Z_{NWE}, m	945.0
2	Maximum rated wind speed V, m/s 2% of exceedance probability	35.19
3	Average value of water depth in the reservoir H, m	110.86
4	Water depth in the reservoir Hc, m in front of the dam	127.0
5	The fetch L, m	1500
6	Wind tide height dh, m	$4.757 \cdot 10^{-3}$
7	Average wave height h_{av}, m	0.787
8	Average wave period T_{av}, s	2.944
9	Average wave length λ_{av}, m	13.533
10	Wave height 1% of exceedance probability in the wave system $h_{1\%}$ m	1.617
11	Coefficients: $k_r = 1.0$; $k_p = 0.9$; $k_{sp} = 1.5$; $k_{run} = 1.10$	
12	Swash height on the slope $h_{run1\%}$, m	2.396
13	Mark of the dam crest required for the calculation Z_{dcr}, m	947.9
14	Actual mark of the dam crest Z_{dcaF}, m	953.0

Table 2. Results of calculations of the earth-and-rockfill dam of the hydraulic power system Nam Chien under the condition of preventing the water overflow over the crest was carried out according to the normative method for the second settlement case

No.	Name of the calculated values	Value
1	Maximum water surface elevation Z_{MWE}, m	949.92
2	Maximum rated wind speed V, m /s 20% of exceedance probability	29.77
3	Average value of water depth in the reservoir H, m	115.78
4	Water depth in the reservoir Hc, m before the dam	131.92
5	The fetch L, m	1510
6	Wind tide height dh, m	$3.082 \cdot 10^{-3}$
7	Average wave height h_{av}, m	0.684
8	Average wave period T_{av}, s	2.792
9	Average wave length λ_{av}, m	12.176
10	Wave height 1% of exceedance probability in the wave system $h_{1\%}$, m	1.407
11	Coefficients: $k_r = 1.0$; $k_p = 0.9$; $k_{sp} = 1.5$; $k_{run} = 1.12$	
12	Swash height on the slope $h_{run1\%}$, m	2.120
13	Mark of the dam crest required for the calculation Z_{dcr}, m	952.54
14	Actual mark of the dam crest Z_{dcaF}, m	953.0

Table 3. The results of probabilistic calculations of the earth-and-rockfill dam

CC (consequence class)	Number of statistical tests	Estimated value of the probability of water overflow over the crest, $year^{-1}$	Confidence interval, $year^{-1}$	Accepted value of the probability of water overflow over the crest, $year^{-1}$
The earth-and-rockfill dam of the hydraulic power system Nam Chien				
CC3	$1\cdot10^6$	$2\cdot10^{-5}$	$1.66\cdot10^{-5}$ – $1.83\cdot10^{-5}$	$5\cdot10^{-5}$

Conclusions

According to the probabilistic assessment of the dam of the hydraulic power system Nam Chien, it may be concluded that the probability of water overflow over the crest of the dam $p_{est.} = 2\cdot10^{-5}$ and it does not exceed the accepted value of $p_{acc.} = 5\cdot10^{-5}$, and the reliability of the earth-and-rockfill dam is provided.

References

1. N.S. Arunraj, S. Mandal, J. Maiti, Acc. An. & Prev. **55**, 242–255 (2013)
2. L. Altarejos-García, I. Escuder-Bueno, A. Serrano-Lombillo, M.G. Membrillera-Ortuño, Struc. Saf. **36–37**, 1–13 (2012)
3. O. Morales-Nápoles, D.J. Delgado-Hernández, D. De-León-Escobedo, Arteaga-Arcos, Struc. Infr. Eng. **10**, 589–603 (2014)
4. L. Peyras, C. Carvajal, H. Felix, C. Bacconnet, P. Royet, J.P. Becue, D. Boissier, Eur. J Env. Civ. Eng. **16**, 795–817 (2012)
5. Z.Y. Wu, Y.L. Li, J.K. Chen, H. Zhang, L. Pei, Comp. Geot. **51**, 42–49 (2013)
6. Y. Li, Y. Sun, B. Li, Z. Xu, Comp. Geot. **81**, 19–25 (2016)
7. A.O. Mozgovuy, Naukovii visnyk budivnitstva, **54**, 272–277 (2009)
8. A.O. Mozgovuy, Naukovii visnyk budivnitstva, **58**, 264–270 (2010)
9. A.O. Mozgovuy, Naukovii visnyk budivnitstva, **65**, 364–370 (2011)
10. S.V. Butnik, K.S. Belousova, Naukovii visnyk budivnitstva, **68**, 132–137 (2012)

Influence of long-term compressive stresses on strength of concrete and steel-fiber concrete prismatic element

Stepan Neutov[1], *Maryna* Sydorchuk[1], and *Oleksii* Shyliaiev[1,*]

[1]Odessa State Academy of Civil Engineering and Architecture, Didrikson str., 4, Odessa, 65029, Ukraine

Abstract. The strength and deformation characteristics of concrete and steel-fiber concrete on prisms and cubes are determined. Short-term and long-term tests of the samples were carried out taking into account the recommendations of the regulatory documents for the testing of concrete. Loading of samples with a long load was carried out in force installations (stands) consisting of four metal rods to which rigid loading plates were fixed at certain levels with threaded connections. To cover the entire operational spectrum of the stress state of real reinforced concrete elements, levels of 0.3; 0.4; 0.5; 0.67 and 0.8 from the short-term destructive load were adopted as long-term loading levels. All samples, which were more than a year under the action of a long-acting load, increased their bearing capacity. Prisms from steel-fiber concrete, which were under the influence of a long-acting load during 370 days, increased the bearing capacity, depending on the level of load, by 30-50%. The higher load level, the higher the creep rupture strength. When the loading was repeated before the destruction, the deformations of steel-fiber-concrete prisms changed linearly.

1 Introduction

Among materials used in construction, concrete and reinforced concrete are the most common ones. According to some reports, the production of concrete per year is one ton per inhabitant of our planet. Increasing use in construction is found in high-strength powdered concrete obtained by the use of modern modifiers and finely dispersed mineral aggregates. However, along with obvious advantages, they have a significant disadvantage compared with traditional concrete - brittleness. This is one of the reasons limiting the wide use of high-strength powdered concrete in construction. An effective solution of this problem is the addition of steel fiber to the concrete matrix. Thus, a spatial framework was created that increases the resistance to cracking and the fracture toughness of concrete, frost resistance, tensile strength, bending strength, torsion strength, etc. [1-4]. Despite the fact that, both in our country and abroad, studies of steel-fiber-concrete are conducted quite a

*Corresponding author: shyliaiev@gmail.com

long time, information on the physicomechanical and rheological properties of steel-fiber concrete is relatively small, and the available data vary in a considerable interval.

For solving multiple problems in the design of concrete and reinforced concrete structures, it is very important to take into account the long processes taking place in concrete [5-8], since practically all structures in real conditions are loaded with long-term loads [9]. In determining the strength, rigidity and crack resistance it is necessary to know the parameters characterizing the long processes occurring in concrete, which correspond to the concrete types and compositions of the used concrete [10].

2 Results

In our studies, the strength and deformation characteristics of concrete and steel-fiber concrete were determined on concrete prisms with dimensions 100 × 100 × 400 mm and cubes with dimensions 100 × 100 × 100 mm. Depending on the tasks to be performed during the experiments, the twin samples were combined into groups and series. Samples of each group were made in one step. Prisms and cubes were concreted in metal cassette formwork. The concrete mixture was made in a free-fall concrete mixer. Before laying in the formwork, the mixture for further homogeneity was additionally mixed by hand. Concrete consolidation in the formwork was carried out on a laboratory vibration table.
Before the beginning of the experimental studies, the samples were stored at a temperature of 18-24°C. Before the test, the attachments for the installation of dial indicators were fastened to the side edges of the prism.

The method of short-term and long-term testing of samples was made taking into account the recommendations of the regulatory documents for the testing of concrete [11].

The physical and mechanical characteristics of concrete and steel-fiber concrete elements under short-term loading were determined on a universal hydraulic press (500 kN).

Determination of the modulus of elasticity and prismatic strength was preceded by the alignment of the samples along the physical center by applying test loads, causing a stress of up to 0.2 from the prismatic strength. The centering was performed to ensure that the deformations along the four faces of the sample were approximately the same and almost completely reversible. The loading was carried out in steps of 0.1 from the prismatic strength. Deformations were measured at each loading stage. Maximum force, perceived by the sample before destruction, was taken as the magnitude of the destructive load.

Loading of the samples with a long-term load was carried out in force installations (stands) consisting of 4 metal rods with a diameter of 46 mm, to which rigid loading plates (4 pcs) were attached to certain levels using threaded connections (Fig. 1). To maintain the load during long-term tests at a given level, the transfer of forces from the hydraulic jack to the prisms studied was carried out through a force installation consisting of 4 springs and 2 load plates. The strength of each spring is 100 kN.

All forces generated by the force device, are balanced within the upper and lower loading plates. From the installation on the foundation, only its own weight and possible dynamic impact during the fragile destruction of the prisms are transmitted.

Before loading, each spring cassette was calibrated using a standard dynamometer (500 kN), located at the location of the experimental sample. The load during the experiment was maintained at the required level and controlled by the deformations of the spring unit and the pressure gauge of the pumping station.

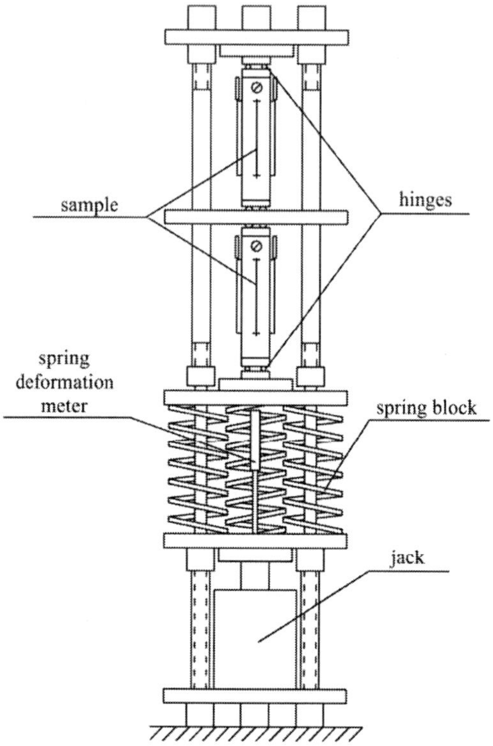

Fig. 1. Force installation scheme.

The height of the installation in one power line, i.e. in two floors, there were two samples: one of ordinary concrete, and the second - from the steel-fiber concrete. Thus, during the entire experiment (370 days), the mode and load level for both samples were exactly the same.

To determine the bearing capacity, as well as to compare the results of short-term and long-term loading, in each group of samples, three prisms were brought to destruction by a short-time load on the hydraulic press. Five prisms of each composition of concrete were loaded with a long-term load in force stands and four prisms of each group were stored until the end of the experiment to determine the corresponding prism strength. As a result of short-term tests it was established that the prismatic strength of ordinary concrete is 235 kN, and that of steel-fiber concrete is 252 kN.

To cover the entire operational spectrum of the stress state of real reinforced concrete elements, levels of 0.3; 0.4; 0.5; 0.67 and 0.8 from the short-term destructive load were adopted as long-term loading levels. Fig. 2 shows the results of long-term tests for three levels - 0.3; 0.5 and 0.8. The solid lines show the deformation of concrete, and the dashed lines show the steel-fiber concrete. Creep deformations of steel-fiber reinforced concrete are on 20% lower than concrete in average.

From the results presented in Fig. 2, it can be seen that the deformation development under long-term loading can be conditionally divided into 3 stages. At the first stage there is an accelerated deformation. The deformations appeared during this period are almost 75% of the total deformations for the entire time of observations. At the second stage, deformation occurs at a conditionally constant rate, i.e. the increasing of deformations is practically linear. At a load level of 0.3 R, the linear part is stable after 35-40 days, for level 0.8 - after 65-70 days. The deformations appeared at this stage are 20-23% of their total value. And, finally, at the third stage the rate of increasing of deformations practically tends to zero. The graph of the deformations is almost parallel to the horizontal axis.

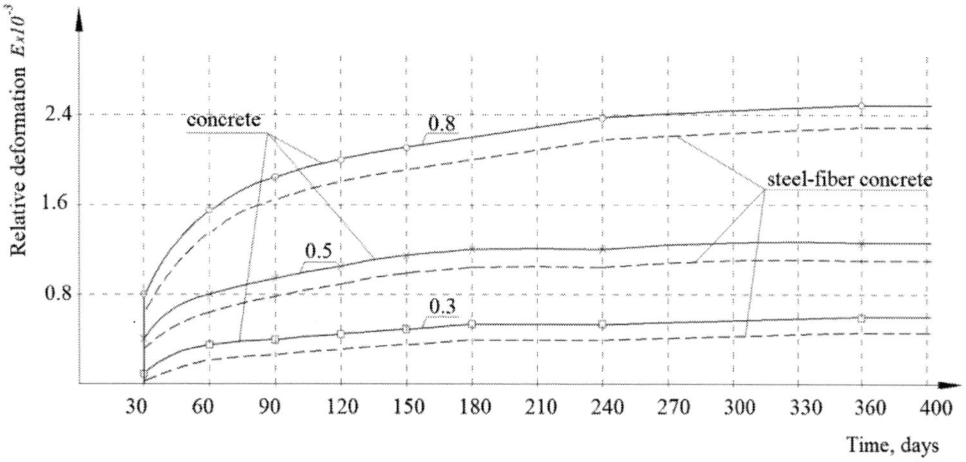

Fig. 2. Results of long-term tests (creep curves).

Fig. 3 shows the results of three stages of loading of concrete prisms in force stands. At the first stage, as already noted, the prisms were loaded with a short-term load to the planned level of 0.3; 0.5 and 0.8. The deformations corresponding to this stage of loading are given in the first part of Table. 1. From the obtained results it is clear that at this stage of loading the strain values are practically proportional to the level of the load, namely, at a level of 0.8 the strain is 3 times higher than at a level of 0.3, etc.

Table 1. Deformations $\varepsilon \, x \, 10^3$

Level	I stage		II stage			III stage			Limit of defomati on
	Value	Difference, %	Value	Increment to I stage	Difference, %	Value	Increment to II stage	Difference, %	
0.3	0.25	100	0.68	0.43	100	1.4	0.72	170	1.4
0.5	0.37	148	1.44	1.07	149	2.1	0.66	157	2.1
0.8	0.75	300	2.48	1.73	302	2.9	0.42	100	2.9

At the second stage, a constant load was applied to the prisms during 370 days, corresponding to the selected loading levels. In Fig. 3 this stage is represented by horizontal lines. Load is constant, concrete "creeps". The deformations corresponding to this stage of loading are presented in the second part of Table 1. Differences in deformations accumulated at this stage of loading are more significant than in the previous stage. At a level of 0.8, the strain accumulated due to the creep of concrete is 3 times higher than at a level of 0.3 R.

And, finally, at the 3rd stage, the prisms were loaded to destruction in force stands without preliminary unloading. The deformations appeared in the samples at this stage are listed in part 3 of Table 1. The obtained data indicate that the deformation pattern here is opposite to the first two stages. With repeated loading of the deformation of samples preloaded to a level of 0.3 R are 1.7 times greater than similar deformations at a level of 0.8 R. The limiting deformations accumulated in the loaded samples at the time of destruction for the entire three-stage load period are the following:

- at level 0.3 – $\varepsilon = 1.4 \times 10^3$;
- at level 0.5 – $\varepsilon = 2.1 \times 10^3$;
- at level 0.8 – $\varepsilon = 2.9 \times 10^3$.

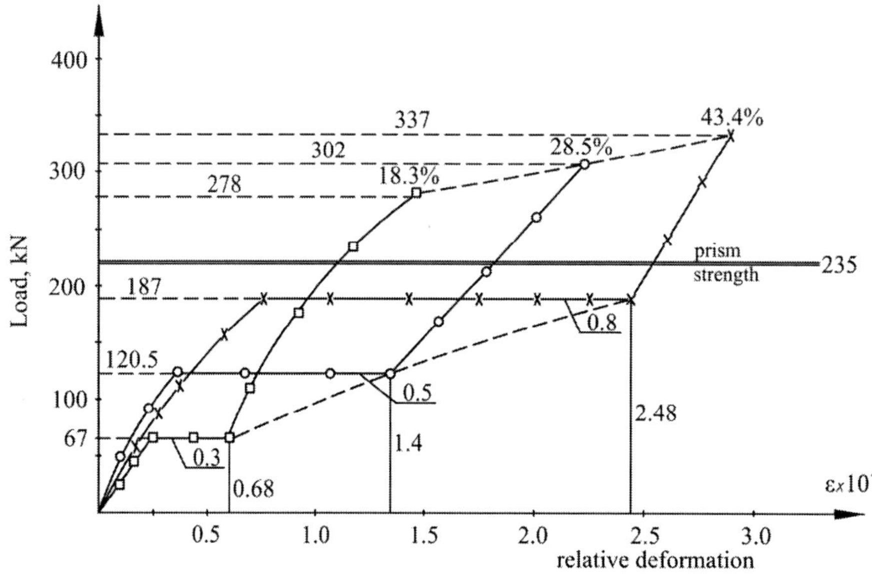

Fig. 3. Three-stage loading of concrete prisms in force stands.

From the results presented in Fig. 3, it follows that all samples that have been more than a year under the influence of a long-acting load have increased their carrying capacity. The higher the level of the continuous load, the more compacted concrete (see ultimate deformations ε) and, naturally, its stress is higher. Samples loaded to the level of 0.8 R, in 370 days increase their bearing capacity by 40.5%. For the same time the limit stress of the sample, loaded to a level of 0.3 R, increased by 18.3%.

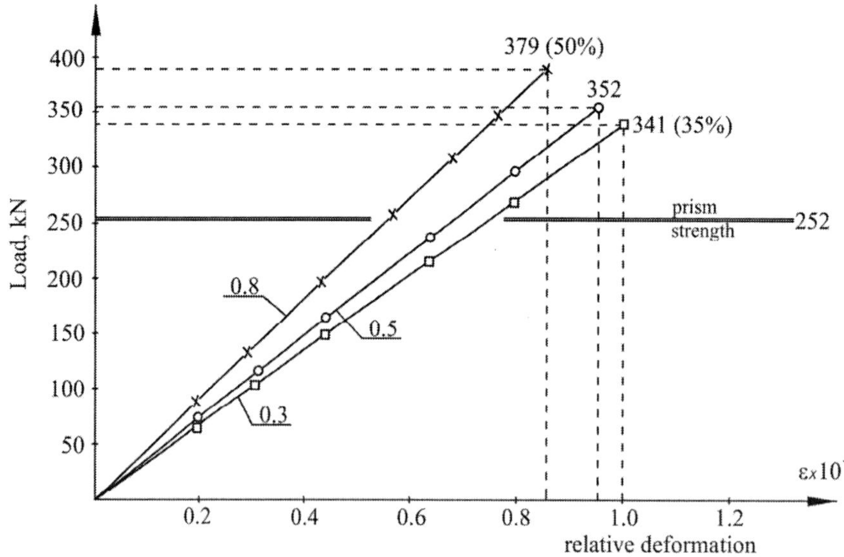

Fig. 4. The bearing capacity of steel-fiber concrete prisms previously loaded with a long-acting load.

In each stand two samples were simultaneously loaded. In the process of loading, of course, only one was destroyed, the bearing capacity of which was lower. In all three installations, where simultaneously one under the other was installed concrete and steel-fiber concrete samples, the prism from ordinary concrete was broken first. The results of the experiment are shown in Fig. 3.

The unbroken and simultaneously unloaded steel-fiber concrete prism after the destruction of the concrete sample was moved to a hydraulic press and subjected to short-term loading until failure.

The loading speed on the press was about the same as at the stand. The results of the tests are shown in Fig. 4.

3 Conclusions

From the results obtained, it can be seen that the prisms of steel-fiber concrete, which were during 370 days under the influence of a prolonged load, increased the bearing capacity, depending on the level of load, by 30-50%. The higher the load level, the higher the limiting stress. When the loading was repeated again, the deformations of the steel-fiber concrete prisms changed linearly according to a linear law (Fig. 4). This is explained by the fact that in the process of three-stage loading, fast-flowing creep of concrete was chosen.

References

1. K.V. Talantova, Arch. i stroit. Tez. dokl. konf., Tomsk, 16-17 (2002)
2. V.B. Aronchik, *Study of the work of reinforcing fiber in fiber concrete* (Riga, 1983)
3. S.H. Kwon, R.P. Ferron, Y. Akkaya, S.P. Shah, Int. Journ. Of Conc. Str. And Mat., **54**, 205-221, (2007)
4. Y. Yuguang, J.C. Walraven, J.A. D. Uijl, Heron, **54**, 205-221, (2009)
5. A.V. Sakvarelidze, Bet. I Zhelez., **8**, 12-13, (1986)
6. A.V. Sakvarelidze, Bet. I Zhelez., **3**, 8-9, (1987)
7. Yu.V. Puharenko, I.U. Aubakirova, V.Yu. Golubev, Tehn. Bet., **03-04**, 44-45 (2010)
8. A.V. Mishina, Nauch.-tehn. Journ. "Str. I rekonctr.", **6(38)**, 70-74, (2011)
9. Yu.V. Veryuzhskiy, A.B. Golyshev, V.I. Kolchunov, *Reference manual for structural mechanics* (ASV, Moscow, 2014)
10. D.A. Smirnov, Vestn. Grazhdan. Inzhen., **3(28)**, 56-60, (2010)
11. GOST 24544-81 *Concretes. Methods of shrinkage and creep flow determination* (1981)

A method of experimental studies of heat transfer processes between industrial constructions

Vadym Nizhnyk[1,*], *Oksana* Kyrychenko, *Olexandr* Tarasenko[3], *Andrii* Shvydenko, and *Serhii* Hovalenkov[3]

[1]The Ukrainian Civil Protection Research Institute, 18, Rybalska St., Kyiv, 01011, Ukraine
[2]Cherkasy Institute of Fire Safety named after Chornobyl Heroes of National University of Civil Defense of Ukraine, 8, Onoprienko St.,Cherkasy, 18034, Ukraine
[3]National University of Civil Defense of Ukraine, 94, Chernyshevska St., Kharkiv, 61023, Ukraine

Abstract. A method of experimental study of heat transfer processes between industrial constructions during a fire was developed. Types of equipment necessary for the conduction of the experimental studies were determined. A new type of specimen to be used as the one to be studied when conducting experimental studies was developed. Installation sites of thermocouples and heat flux detector on the specimen under study are shown as layout. Installation sites of the specimens under study relative to heat radiation source were substantiated experimentally. Succession of the conduction of the experimental studies of heat transfer processes between industrial constructions during a fire was developed which consists of the following procedures: production of specimens for the studies, measurement and recording on ambient conditions, measuring of temperature and heat flux on the surface of the specimen under study, and filling 55B test fire with water and diesel fuel. It was established that average temperatures and heat flux density values as well as flame geometry should be measured when performing experimental studies.

1 Introduction

Specific features of fires in industrial constructions, unlike buildings, lies in open burning without any flame restriction with non-combustible building constructions having some fire resistance class. This stipulates for more powerful effect of heat irradiation caused by fire due to larger area of flame on adjacent buildings and constructions.

Full-scale fire tests should be conducted to study the heat transfer processes between adjacent industrial constructions (one of which is burning and another one is being irradiated) in order to implement calculation method of the determination of fire breaks between manufacturing and industrial designation constructions. Essence of the method lies in the determination of heat flux density q_2 (kW/m^2) and temperature T_2 (°C) at the surface of the object being irradiated (2) by flame (1) produced at the burning object (see Figure 1).

* Corresponding author: nignyk@ukr.net

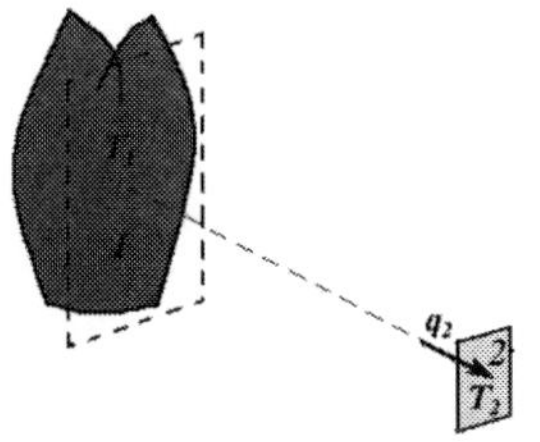

Fig. 1. Layout of item 2 irradiated by flame 1: 1 - surface of the flame reduced to rectangle; 2 – surface of the object being irradiated; r – distance between the planes of the rectangles.

1.1 Literature review

Analysis of relevant scientific papers showed that heat influence of fire was realized by heat transfer on account of radiative heat exchange [1 – 10]. This phenomenon shall be taken into account when substantiating fire breaks between buildings, too [9 – 12].

Classical theory of radiative heat exchange is laid down to the basis of substantiation of fire breaks between buildings. Essence of the tasks is reduced to the comparison of heat flux density irradiated by the facility being on fire q_{irr} with maximum allowable heat flux density of the materials used at the irradiated facility q_{all}, equation (1) [11].:

$$(q_{irr} \leq 0.8q_{all})\mathrm{U}(\theta_{irr} \leq 0.8(\theta_{all}) \qquad (1)$$

where q_{irr} or θ_{all} - stands for heat flux density irradiated by the facility on fire;
q_{all} or θ_{all} stands for maximum allowable heat flux density for appropriate materials

1.2 Purpose, objectives and results

Object of this paper is the development of a method of experimental studies of heat transfer processes between adjacent facilities when simulating fire. The following tasks were to be solved in order to reach the object preset:

– to determine types of test equipment necessary for the experimental studies;

– to develop configuration of the specimen for studies;

– to substantiate points of placement of the specimens for studies relative to the heat radiation source;

– to develop sequence of conducting experimental studies.

Equipment used for the conduction of the full-scale fire experimental studies includes the following: class B test fire, specimen for studies, measuring instruments and equipment for photo and video shooting.

55B is used as test fire when performing studies in an industrial construction [13], this test fire represents metal pan (1.480 ± 0.5) mm in diameter, height of its side board equals to (150 ± 5) mm and thickness of the latter is (2.5 ± 0.5) mm, 18 l of water and 37 l of diesel fuel are poured into it.

The following measuring instruments were used in the course of the study:

– thermocouples;

– "Thermocount" measuring and calculating system;

– RAP-12M.2 thermal radiation receiver;

– heat flux meter;

– control unit;

– moisture gauge for wood;

– R10 UZK measuring tape;

– MV-4M aspiration hygrometer;

– M67 aneroid barometer;
– ASO-3 winged anemometer;
– still and video cameras;
– Testo 885 infrared imager.

6 test specimens are used in the course of the studies.

We chose the following items in order to fabricate a specimen: 6 solid pine wood bars measuring 50 mm × 50 mm × 4,000 mm, 30 solid pine wood boards measuring 20 mm × 100 mm × 700 mm, 12 bars measuring 50 mm × 50 mm × 1.000 mm, 120 nails measuring 4 mm × 100 mm, 72 nails measuring 2.5 mm × 60 mm, rotational processing machine and drying chamber.

Wood drying is done inside a drying chamber at the first stage of preparation until its moisture content decreases to 12 % [12]. Its moisture content is measured by electric moisture gauge.

Interconnection of wood parts is performed at the second stage of preparation. Figure 2 shows layout of the specimen used for the study after interconnection of its constituent parts.

Some gaps are drilled at the specimen's surface in order to measure its surface temperature, this is done with a drill bit being 4 mm in diameter for the purpose of installation of the thermocouples (T1, T2, T3) according to the layout shown on Figure 2. Gap drilling is done in such a manner that thickness of the undrilled distance to the surface of the specimen is within 1 to 2 mm; this ensures determination of the most exact data as to warming of the surface of the specimen under study.

Thermal radiation receiver is installed on the surface of the specimen (see Figure 2).

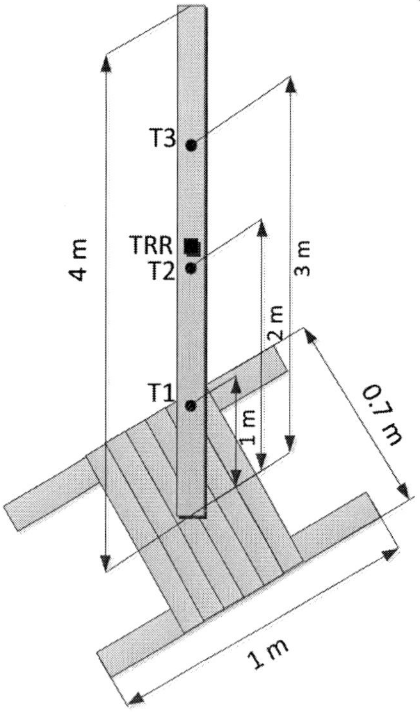

Fig. 2. Layout of the specimen under study (T1, T2, T3 stand for the places for the thermocouples installation, TRR stands for the place of the thermal radiation receiver installation).

Some trial experimental studies for the determination of temperature distribution during 55B test fire burning were conducted preliminarily for the purpose of substantiation of the places where the specimens under study shall be placed relative to heat radiation source.

The test specimens are placed as per layout shown on Figure 3.

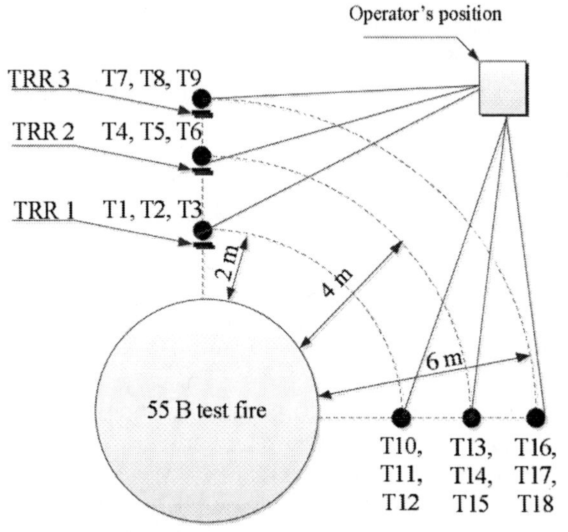

Fig. 3. Layout of the test specimens and measuring instruments installation (T1 to T18 stand for thermocouples, TRR1 to TRR 3 stand for thermal radiation receivers).

Temperatures versus duration of heat irradiation curves at various places relative to 55B test fire are shown on Figure 4.

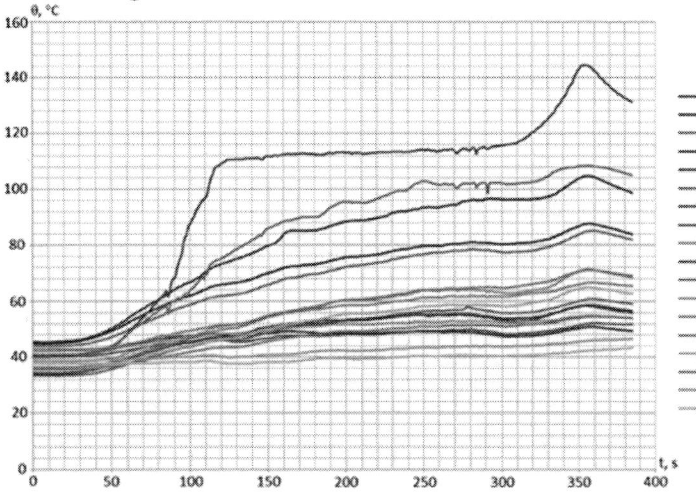

Fig. 4. Temperature versus duration of heat irradiation curves.

Thus, one can conclude that it is parts of the samples under study located 1 to 2 m above ground level that experience the most intensive heat irradiation. Temperature of the specimens' surface lowers as distance to 55B test fire increases.

Therefore when performing experimental studies of the heat transfer processes between industrial constructions thermocouples installation at the specimens under study shall be taken as 1, 2 and 3 m; this case corresponds to maximum intensity of heat irradiation of the specimens under study caused by 55B test fire. At that we propose to place the specimens under study 2 m, 4 m and 6 m away from 55B test fire.

Experimental studies are conducted using at least three specimens under the following weather conditions: no atmospheric precipitations, air temperature +5°C to +40°C, maximum wind velocity 3 m/s.

We conduct experimental studies with the following sequence:

1. Preparation:

1.1 Fabrication of the specimens for studies.

1.2 Placement of 55B test fire at fire testing ground.

1.3 Positioning of the specimens and measuring instruments as per layout shown on Figure 3.

1.4 Installation of measuring instruments (thermocouples, heat radiation receivers) in accordance with the layout shown on Figure 2. The thermocouples are connected with the "Thermocount" information and measuring system, and heat radiation receivers as well as heat flux meter are checked for their proper operation.

1.5 Preparation of the equipment necessary for photo and video shooting.

1.6 Installation of Testo 885 thermal imager on the stand.

1.7 Preparation of appropriate first aid fire-fighting means.

2. Sequence:

2.1 Measuring and recording of the environmental conditions (air temperature, wind velocity, humidity and atmospheric pressure) are performed:

– air temperature and humidity are measured with MV-4M aspiration hygrometer;

– atmospheric pressure is measured with M 67 aneroid barometer;

– wind velocity is measured with ASO-3 winged anemometer

2.2 Measurement and recording of the initial values of temperatures and heat radiation are conducted.

2.3 55B test fire is filled with water and fuel no later than 3 minutes before starting the studies.

2.4 Information and measuring devices are switched to record mode and video shooting and thermal imager are triggered; 55B test fire is ignited.

2.5 Operations using thermal imager:

– accumulator battery is connected;

– objective lens cap is removed;

– memory card is installed$

– thermal imager is switched on;

– basic adjustments are performed. Irradiation factor is set as per recommendations on the device operation in order to record temperatures more qualitatively (irradiation factor for wood equals to 0.74);

– measurement results recording is started;

– device adjustment as well as changing its position relative to the specimens under study as well as heat radiation source are possible if necessary when conducting measurements;

– recorded data are saved upon cessation of the study and equipment being constituent parts of the device are packed.

2.6 When conducting studies using information and measuring system we measure and record temperature values at the exposed surface of the specimen under study as well as heat flux intensity at appropriate distances from the test fire. Temperature and heat flux recording is done at intervals not exceeding 1 min. Video shooting is performed at the moment of the most intensive burning of the fire load. Photo and video shooting are done in order to determine flame geometrical parameters.

2.7 Studies are conducted until complete burning-out of the diesel fuel poured into 55B test fire.

3. Estimation:

3.1 Using results of the studies for each time moment tj when the measurement was taken we determine surface temperature of the specimen under study θj and heat flux density qj as well as geometrical parameters of the flame (hj, δj). Experimental results are put down to the log of experimental studies.

3.2 Using results of the studies we calculate measuring error using the following formula:

$$\Delta A = \pm k\sqrt{(\Delta A_1)^2 + (\Delta A_2)^2} \qquad (2)$$

where ΔA – absolute error of the figures being measured;
ΔA_1 – error of detectors (thermocouples, heat flux meter, and instrumental one);
ΔA_2 – error of measuring devices ("Thermocount" information and measuring system, heat flux meter, and error of reading of the results when measuring dimensions);
κ is a factor which depends on probability (k = 1.1 at P = 0.95) [11].
When performing experimental studies we derive average temperature and heat flux values as well as flame geometrical parameters. Temperatures, heat flux and flame geometrical parameters versus time are plotted.

3 Conclusions

We proposed a method of experimental studies of heat transfer processes between adjacent constructions in this paper, and at that:

points of placement of the specimens for studies relative to heat flux source were substantiated experimentally and it was revealed that when conducting experimental studies of heat transfer processes between 55B test fire and specimens under study the highest heat irradiation of the specimen takes place at 1 to 2 m height;

sequence of conduction of experimental studies of heat transfer processes between adjacent facilities during fire was developed.

References

1. S. Pozdieiev, O. Nuianzin, S. Sidnei, S. Shchipets, MATEC Web of Conferences, **116**, 02027 (2017).
2. O.Nekora, V. Slovynsky, S. Pozdieiev, MATEC Web of Conferences, **116**, 02024 (2017).
3. K. Mygalenko, V. Nuyanzin, A. Zemlianskyi, A. Dominik, S. Pozdieiev, EasternEuropean Journal of Enterprise Technologies **1**, 31-37 (2018).
4. Y. Danchenko, V. Andronov, M. Teslenko, V. Permiakov, E. Rybka, R. Meleshchenko, A. Kosse, Eastern European Journal of Enterprise Technologies **1**, 9-17 (2018).
5. B. Pospelov, V. Andronov, E. Rybka, V. Popov, A. Romin, EasternEuropean Journal of Enterprise Technologies, **1**, 50-55. (2018).
6. B. Pospelov, V. Andronov, E. Rybka, V. Popov, O. Semkiv, EasternEuropean Journal of Enterprise Technologies,**10,** 2 (2018).
7. B.V. Hrushevskiy, A.I. Yakovlev, I.A. Krivosheyev, *Fire prevention in building industry*, Moscow, 451, (1985).
8. M.Ya. Roitman., Ye.P. Komissarov., V.A.. Pchelintsev, *Thermal dynamics and heat transfer in fire-fighting procedures*, Moscow, 415, (1977)
9. P.N. Romanenko, Yu.A. Koshmarov, M.P. Bashkirtsev, *Fire prevention in building industry*, Moscow, 363, (1978).
10. Morgan J. Hurley, *SFPE Handbook of fire protection engineering*, Greenbelt, 351, (2016).

11. GOST 12.1.004-91 *Fire and explosion safety of substances and materials and methods of their determination*, (1996).
12. DSTU-N B EN 1995-1-2:2012 *Eurocode 5. Designing of wood constructions. Part 1-2. General regulations. Calculation of fire resistance of the constructions (EN 1995-1-2:2004, IDT)*, (2013).
13. DSTU 3675-98 *Fire-fighting equipment. Portable fire extinguishers. General technical requirements and test methods*, (1999).

Investigation of the regularities of temperature regime of fire in cable tunnels depending on its parameters

Oleksandr Nuianzin[1*], *Mykola* Kryshtal[1], *Artem* Nesterenko[1], *Dmytro* Kryshtal[1], and *Taras* Samchenko[2]

[1]Cherkassy Institute of Fire Safety named after Chernobyl Heroes National University of Civil Defence of Ukraine, Onoprienka, 8, Cherkassy, 18000, Ukraine
[2]Ukrainian Civil Protection Research Institute, Rybalska, 18, Kyiv, 01001, Ukraine

Abstract. Simulation, as a method of scientific research, makes it possible, without performing costly and labor-intensive field experiments on models, to carry out all necessary experiments to determine the temperature modes of fire in cable tunnels. The purpose of the research of this work was to determine the temperature regime of fire in a cable tunnel depending on its shape, size and fire load. Mathematical models of cable tunnels were created in one of the CFD software systems. Cable products are constantly evolving and improving. For tests on the fire resistance of building structures of cable tunnels, a standard temperature mode of fire is used which may not correspond to fire mode in a real cable tunnel. The computational experiments were carried out and the temperature regimes of fires in tunnels with different parameters were determined. The obtained results showed the parameters of cable tunnels, which influence the temperature regime of fire in tunnels most. In this paper the use of computational experiments for the study of heat and mass transfer processes in fires in cable tunnels was examined further. CFD Fire Dynamics Simulator 6.2 was used.

1 Problem statement

Despite significant advances in solving problems to improve fire safety of cables now there are also many issues concerning both cables and cable lines, including cable tunnels of a rectangular cross section [1].

Cable products are constantly evolving and improving. For tests on the fire resistance of building structures of cable tunnels, a standard temperature mode of fire is used which may not correspond to fire mode in a real cable tunnel.

Simulation, as a method of scientific research, makes it possible to carry out all necessary experiments to determine the temperature modes of fire in cable tunnels without performing costly and labor-intensive field experiments on models. The computational experiments were carried out and the temperature regimes of fires in tunnels with different

* Corresponding author: nuyanzin@i.ua

parameters were determined. The obtained results showed the parameters of cable tunnels which influence the temperature regime of fire in tunnels most.

2 Analysis of recent achievements and publications

In the works [2-4], studies on the development and spread of dangerous fire factors in cable structures were carried out. The behavior of building structures in case of fire was also examined. The conclusions were made on the basis of real and computational experiments.

In work [5] it was established that the rate of burnout of a fire load almost does not depend on the width of the tunnel. The natural experiments carried out in work [6] describe the distribution of temperature in tunnels with different aerodynamic parameters. The causes of fires in tunnels are analyzed in work [7]. In work [8], it was analyzed which parameters influence the rate of burnout of cable isolation. Analyzed The affect of the method of laying cables on the linear speed of the spread of fire. Despite the practical significance of such results, it is necessary to remember that cable products and construction constructions of cable tunnels are constantly being improved, therefore, additional ones need to be made. In previous studies [1, 9], we carried out natural experiments and verified established mathematical models. The obtained data made it possible to confirm the adequacy of computer models of fires in cable tunnels and to conduct mathematical modeling to determine the temperature regime of a fire for testing fire resistance of building structures of cable tunnels.

3 Purpose

The purpose is to study the patterns of temperature regime of fire in cable tunnels, depending on its geometrical, aerodynamic parameters and fire load as a scientific basis for calculating the limit of fire resistance of building structures of cable tunnels.

To achieve the goal, the following tasks must be solved:

1. To create mathematical models of cable tunnels with different geometric, aerodynamic parameters and fire load.
2. To carry out computational experiments of heat and mass transfer during a fire in the created mathematical models and to allocate the parameters of the tunnel which most significantly affect the temperature regime of the fire.
3. To construct the dependence of temperature modes of fire on time in cable tunnels depending on its geometric, aerodynamic parameters and fire load.

4 Method

Theoretical studies were carried out on the basis of systems of differential equations of continuous media of the type of Navier-Stokes equations and Fourier thermal conductivity equations. To solve the equations in the work, the method of finite or boundary elements, non-relation methods, Galerkin method, and optimization methods are used. Computational experiments for the study of heat and mass transfer processes during fire in cable tunnels were carried out in the CFD Fire Dynamics Simulator 6.2.

5 Consideration on methods and results

For the computing experiments 8 mathematical models of cable tunnels were created. For the tests, the following sequence of settlement procedures is used.

1. The CAD software creates a geometric configuration of the cable tunnel of the required size. Transverse dimensions of the tunnel varied. The height is 1.8 m to 2.2 m, the width is 1.6 m to 2.0 m. The length of the tunnel is always 10 m. The models of cables, steel angel bars, openings for the output of combustion products are created inside, and the horizontal component of the speed of movement of air is set. The geometric model is imported into the environment of the FDS settlement system.

2. The initial parameters of the simulation are introduced, which are impossible to change during the calculation: the initial ambient temperature, the air support on one side of the tunnel, the required time of fire (35 minutes).

3. A combustion process in the middle of the tunnel with a focus on the bottom line of the cable is initiated.

4. During the calculation, the temperature monitoring of the corresponding points in the tunnel and the temperature gradient is monitored online..

In order to control the temperature regime, by means of the FDS computer system 45 monitoring points were created (Figures 1-3). The places of temperature control were created according to the following principle:

1) in the plane of the cross-section of the inflammation zone there is 1 place of control in the geometric center and 4 more places in the geometric centers of the formed quaternion (Fig. 1).

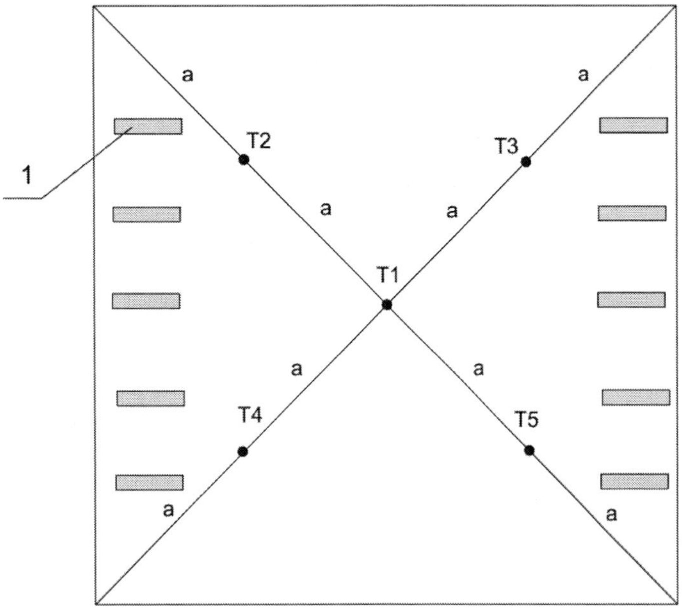

Fig. 1: The scheme of location of control points of temperature in the geometric section of the cable tunnel: 1 - cable line, T1 - T5 – the places of temperature control, a - geometrical dimension parameter depending on the cross section of the tunnel.

2) 10 planes were created, 5 control points in each: 2 planes at a distance of 0.5 m from the fire cell in different directions. Also, at a distance of 2, 4, 6 m. The general layout of the planes is shown in Fig. 2.

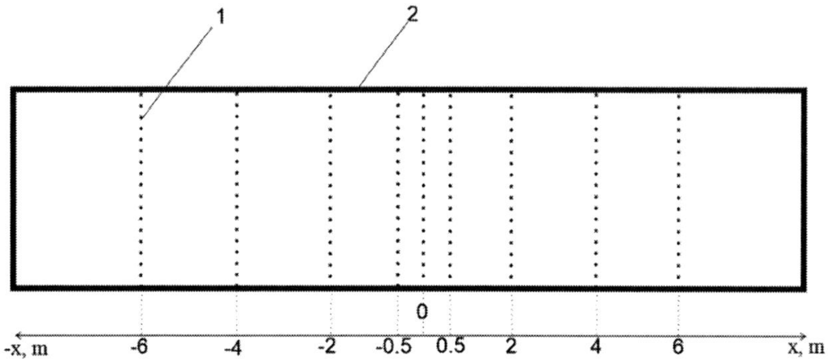

Fig. 2: The layout of the planes in which there were places of temperature control along the length of the cable tunnel: 1 - a plane in which there are 5 temperature control points in the manner shown in Fig. 4, 2 - fencing of a cable tunnel.

The general view of a cable tunnel with places for temperature control is shown in Fig. 3.

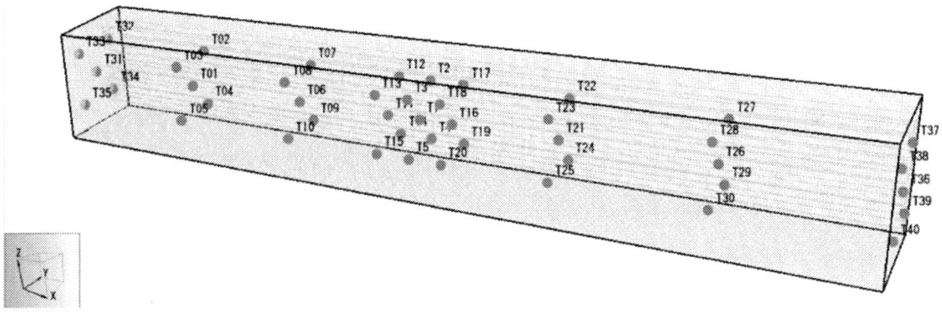

Fig. 3: General View of a Cable Tunnel with Places of Temperature Control: T01 - T40, T1 - T5 - places of control.

The scheme of the tunnel with geometric parameters which were changed is shown in Fig. 4.

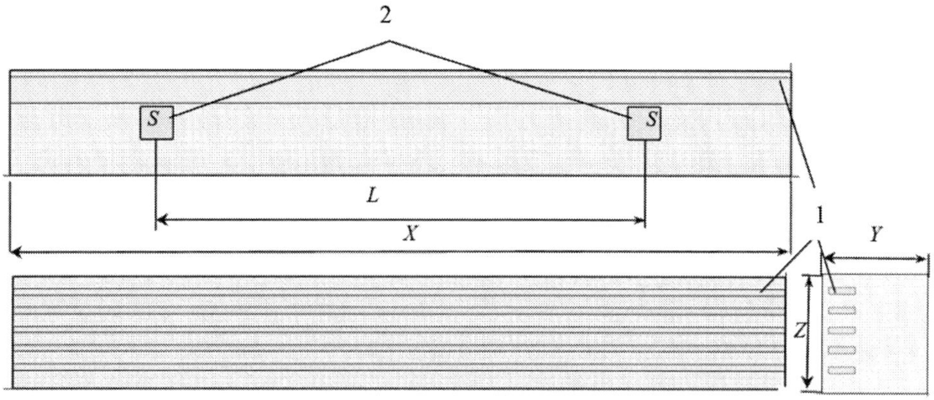

Fig. 4: The scheme of the tunnel with changed geometric parameters: 1 - model of cable lines; 2 - model of ventilation and inspection hatches with the area *S*, m^2; *X*, *Y*, *Z*, *L* –distance between the holes.

In accordance with the calculation scheme of the mathematical model of the tunnel (Fig. 3) for the purpose of conducting a complete analysis of the parameters influencing the temperature regime of the fire the following parameters varied:

1. Fire load:

- cable lines were modeled on two and one side;
- the number of cable lines is from 1 to 10;
- the material of isolation of cables and conductors has changed. Due to this, the speed of thermal energy output from 1 m^2 of cable lines changed (Table 1).

2. Geometric parameters of the cable tunnel:

- varied as a cross-sectional area, using a combination of parameters *Y* and *Z*, the general parameter is the cross-sectional area of the cable tunnel (Table 1).

3. Aerodynamic parameters:

- the horizontal component of the speed of air was set as a parameter that characterizes the excess or lack of oxidants and influences the intensity of the development and spread of the fire.

To construct a mathematical model of the temperature regime of a fire in a cable tunnel, it is necessary to conduct a complete factor computing experiment. In this case, there are three independent factors - the cross-sectional area of the cable tunnel, the fire load, as well as the horizontal component of the air flow inside the tunnel. In table. 1 there are intervals of the parameters in the experiment, which are selected as factors.

Table 1. Intervals of variation of factors in a computational experiment.

Parameter	Minimum value	Average value	Maximum value
Fire load (speed of thermal energy output from 1 m^2 of cable lines), kW/m^2	749	4120	7490
Cross-sectional area of the cable tunnel, m^2	2.88	3.64	4.4
The horizontal component of air velocity, m/s	0	2.5	5

8 computational experiments were conducted to determine the parameters that have the most significant effect on the temperature regime of the fire. A typical matrix for planning a complete factor experiment for determining the temperature regime of a fire in a tunnel is shown in Table 2.

Table 2. A typical matrix for planning a complete factor experiment for determining the temperature regime of a fire in a tunnel.

№	x1	x2	x3	x1x2	x1x3	x2x3	x1x2x3
1	+	+	+	+	+	+	+
2	-	+	+	-	-	+	-
3	+	-	+	-	+	-	-
4	-	-	+	+	-	-	+
5	+	+	-	+	-	-	-
6	-	+	-	-	+	-	+
7	+	-	-	-	-	+	+
8	-	-	-	-	-	-	-

The results of full factorial experimental determination of temperature control of fire in the tunnel are presented further. The order number of the computing experiment meets the conditions of the line from Table 2 and the conditions that were scheduled in Table 1.

After the completion of computational experiments, the temperature-time dependence data were obtained and processed.

According to the results of the mathematical modeling of heat and mass transfer during a fire in cable tunnels of various geometric sizes, fire loads and aerodynamic characteristics, corresponding to [10], the temperature regimes of the fire were calculated and shown in Fig. 5.

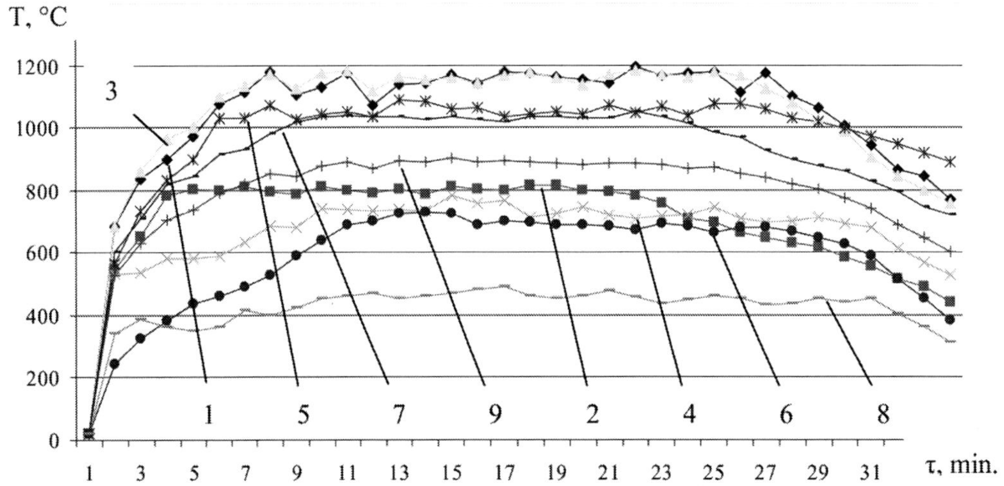

Fig. 5: The results of a full factor experiment for determining the temperature regime of a fire in a tunnel: 1-8 correspond to the matrix of planning (Table 2); 9 is the average value of computational experiments 1-8.

Analyzing the obtained graphs (Fig. 5) it can be stated that the fire load of the tunnel is influenced most by the temperature regime of the fire. At its maximum, the temperature reached 1200 °C, with a minimum of 500 °C. With an average level of fire load and variation of the remaining parameters, the maximum temperature inside the tunnel reaches 700-800 °C.

At the first stage (5-9 minutes), the temperature in the tunnel section rise to the maximum. At the second stage, within 25-30 minutes, the maximum temperature in the tunnel section is maintained. At the third stage there is a gradual cooling in the selected section of the cable tunnel.

6. Conclusions

As a result of researches of this work the regularities of the temperature regime of a fire in cable tunnels depending on its geometrical, aerodynamic parameters and fire loading as a scientific basis of calculation of the limit of fire resistance of building structures of cable tunnels are studied.

Thus, to test the building designs of cable tunnels for fire resistance, it is necessary to choose the most severe temperature regime, which differs from the standard [1]. According to mathematical modeling, the highest temperature is observed in the area of combustion. It is within 1200 °C with the maximum possible fire load in accordance with [10] (Fig. 5). However, if the parameters and the fire load of the tunnel are known, it may be possible to use a less intense test mode. However, in this case, it will be impossible to add cable lines or to change the design of the tunnel, as this can increase the fire load and change the temperature mode of the fire.

References

1. O. Nuianzin, S Pozdieiev, V. Hora, A. Shvydenko, T. Samchenko, Cable tunnels temperature fire mode experimental study, Eastern European Journal of Enterprise Technologies, **3**, 21 (2018).
2. Y. Niu, & W. Li, *Simulation Study on Value of Cable Fire in the Cable Tunnel*, Procedia Engineering, **43**, 569 (2012).
3. S. Pozdieiev, O. Nuianzin, S.Sidnei, S. Shchipets, *Computational study of bearing walls fire resistance tests efficiency using different combustion furnaces configurations*, MATEC Web of Conferences, **116**, 02027 (2017).
4. Y. Zhao, G. Zhu, Y. Gao, Experimental Study on Smoke Temperature Distribution under Different Power Conditions in Utility Tunnel, Case Studies in Thermal Engineering, **12**, 69 (2018).
5. J. Ji et al., *Influence of aspect ratio of tunnel on smoke temperature distribution under ceiling in near field of fire source*, Applied Thermal Engineering, **106**, 1094 (2016).
6. X. Tian et al., *Full–scale tunnel fire experimental study of fire–induced smoke temperature profiles with methanol–gasoline blends*, Applied Thermal Engineering, **116**, 223 (2017).
7. J. Modic, *Fire simulation in road tunnels*, Tunnelling and underground space technology, **18/5**, 525 (2003).
8. J. Vaari et al., Numerical simulations on the performance of water–based fire suppression systems - VTT Technol, 54, (2012).
9. O. Nuianzin, T. Samchenko, S. Kasiarum, K. Hryhorenko, M. Kryshtal, *The Heat Exchange Mathematical Model of Fire in Cable Tunnel Adequacy Research,* International Journal of Engineering & Technology, **7**, 303 (2018).
10. HBN V. 2.2-34620942-002:2015. *Liniino-kabelni sporudy telekomunikatsiy. Proektuvannia*. 135 (2015).

Methodology remaining lifetime determination of the building structures

Yuriy Otrosh[1,*], *Andrii* Kovalov[2], *Oleg* Semkiv[1], *Irina* Rudeshko[2], and *Valentin* Diven[2]

[1]National University of Civil Defence of Ukraine, Chernishevska str., 94, 61023 Kharkov, Ukraine
[2]Cherkassy Institute of Fire Safety of National University of Civil Defence of Ukraine, Onoprienko str., 8, 18034 Cherkassy, Ukraine

Abstract. The paper proposes an approach to determining the increase of the normative lifetime of building structures on the basis of the algorithm of numerical calculation of the residual life of structures, which is performed on the basis of the results of defining parameters obtained during the technical examination and evaluation of defects, failures and structural damage. To calculate the structures, the simulation of the stress-strain state of the building structures with the involvement of OK LIRA was used. Simulation allowed performing optimization of technical solutions and strengthening of structures for buildings. The methodology is used to design elements for strengthening the structures of buildings and structures. Reliable estimation and forecasting of the technical condition enables to prevent the occurrence of accidents of structures and their associated losses, to use rational use of funds for current and capital repairs, and to regulate the technical condition in such a way as to achieve the most effective use of fixed assets.

1 Introduction

The growth of the risk for technogenic emergencies in Ukraine is due to the fact that in the recent years, in most important industries, design service life of the high-risk facilities and potentially dangerous objects have expired by 50-70%, sometimes reaching the pre-emergency level. If structural measures on protection against differential settlement of the foundation are not available (which is typical of buildings built in the postwar period), the cracks and other defects appear caused by differential deformations of the foundation [1-3].

The security means the absence of unacceptable risk associated with the possibility of causing any harm to people's life, health and property, as well as for the environment [4-5]. The unsatisfactory technical state of many construction facilities is a consequence of their dilapidation and needs a rapid assessment of the technical state of the structures in order to prevent the occurrence of emergencies [6].

Based on the above, an important task is to create an appropriate, available and simple, but effective approach for determining the increase in the normative useful life of building structures or their reassigned lifetime based on their remaining lifetime calculation, since in the country with a difficult economic situation, the task of extending the service life of

* Corresponding author: yuriyotrosh@gmail.com

building structures is possible up to two or three warranty operation lives or service lives, specified in the documentation. This is important and requires clarification of the known methods.

2 Unsolved issues

In Ukraine, an emphasis is given to the problem of determining the remaining lifetime of building structures, especially within the comprehensive program "Problems of Lifetime and Safe Operation of Structures, Constructions and Machines" of the National Academy of Sciences in Ukraine, which coordinates the works of both as the high-risk facilities, so the ordinary buildings and structures. The program was launched after the publication of the relevant resolution of the Cabinet of Ministers of Ukraine.

Within the framework of this program, approaches are known for determining the remaining lifetime and increase in the normative operation life of high-risk facilities, for example [7-9]. But this approach does not provide objective information on the use of such methods in the construction industry.

There are also many methods for determining the remaining lifetime of steel structures, one of which, for example, is described in a scientific article of the following authors [10]. This approach for determining the remaining lifetime of construction facilities includes an algorithm for numerical calculation of the remaining lifetime of construction facilities based on the results of the constitutive parameters obtained in the course of technical inspection and evaluation of defects, failures and damages of construction facilities. On the basis of this approach, the physical and mathematical models of facilities are constructed, the constitutive parameters of which have temporal development t, or on the basis of measurements of constitutive parameters in-situ in different time intervals. It is proposed to take into account the development in time of the identified defects and to take into account the defects and damages of steel structures, which were not identified during the technical inspection. The differentiation of threats is performed in the models, that make up a certain types of hazards (local risks), as well as the threats of a complex hazards according to the appropriate combinations and types of hazard (global risk). In these approaches for calculating the remaining lifetime, the occurrence and development of unfavourable events are used as assigned at different time moments, and the calculated and postulated dangerous processes having a development in time t. With this approach, the time scales of risks $R(t)$ are used. But such calculations algorithms are designed to be used for calculating steel structures, such as pipelines, and they do not meet the requirements of the analysis of the remaining lifetime of building structures operation. Thus, their use in the construction industry is not appropriate.

3 The purpose of the work

The work suggests an approach for determining an increase in the standard term of building structures operation based on the algorithm for numerical calculation of the remaining lifetime of construction facilities, which is performed on the basis of the results of constitutive parameters obtained in the course of technical inspection and evaluation of defects, failures and damages of building structures. The calculation is performed on the basis of constructed physical and mathematical models of facilities, the constitutive parameters of which have temporal development, or on the basis of measurements of constitutive parameters in-situ in different time intervals.

4 Main part

The methodology of the element lifetime reassignment should:

- assess adequately the changes in assessment parameters of the technical state in time (reaching by the dilapidation parameter of the value corresponding to the limiting one is considered as a reaching by the element of the limiting state);
- assess the probability of reaching the limiting state for the period of lifetime reassignment.

The methodology of lifetime reassignment should be substantiated with:

- the accuracy and reliability of the initial data and the results being obtained;
- the methodology of experimental data collection and processing;
- measurement accuracy of dilapidation parameter;
- reliability requirements to the value of the element lifetime;
- availability of an appropriate diagnostic and control system.

In accordance with the requirements of the valid normative documents, the element (construction) is considered in operating condition, and its technical state is normal or satisfactory, if it is not met:

– the failure condition of the structures (the reaching of the limiting states of the first group)

$$F \geq F_u, \tag{1}$$

– the condition when the structure reaches the limiting states of the second group

$$f \geq f_u, \qquad a_{crc} \geq a_{crc,u}. \tag{2}$$

The functions of the load-bearing capacity can be accepted in accordance with the applicable regulatory documents.

In as much as the changes in time happen in the junction nodes, concrete and reinforcement, the base metal, welded joints, the nodes of the elements connection due to the corrosion processes, the effects of radiation and elevated temperatures, and also mechanical stresses, some (or all) of the above inequality components are functions of time. Each of the inequalities (1) and (2) can be transformed to:

$$\Phi_u[x_1(t), x_2(t), \dots, x_m(t), y_1, y_2, \dots, y_n] < \Phi(t), \tag{3}$$

where $\Phi_u[x_1(t), x_2(t), \dots, x_m(t), y_1, y_2, \dots, y_n]$ is the function of load-bearing capacity (deformability) of elements (structures), which is determined according to recommendations valid normative documents, with account of changes in time of the input parameters;

$\Phi(t)$ is the acting maximum force (deformation) in the element (structure). The form of the functional dependence should be determined after algebraic transformations of inequalities of the form (1), (2) with integrating, if it is possible, of all time-dependent variables into the set of arguments.

Accordingly, in the inequality (3) the function arguments $\Phi(t)$ are divided into two groups. The arguments of the first group x_1, x_2 ,.., x_m depend on time and are called constitutive parameters. The arguments of the second group $y_1, y_2, \dots, y_n$ do not depend on time and are constants.

The assessment of the possibility of lifetime reassignment is carried out in accordance with the physical (deterministic) model developed on the basis of data on dilapidation parameters and reaching the limiting values by the elements, analysis of failures and damages of the studied elements, states and operating conditions that obtained in the course of inspection and evaluation of the technical state.

The model for assessing the possibility of lifetime reassignment should provide:

adequate assessment of changes in time of the technical state parameters;

possibility of reaching the limiting state.

When assessing the possibility of increase in the service life by the condition of the load-bearing capacity preserving, use the representation of inequalities in formula (3). The dependence of the constitutive parameters on time is recommended to accept linear form:

$$x_i(t)=x_i^0+s_i t \quad (i = 1, 2,..,m), \tag{4}$$

where x_i^0 - value of the constitutive parameter with the number i at the moment of examination; s_i - rate of the constitutive parameter change with the number i; t - time, calculated from the examination moment.

For the constitutive parameters, take up the rates of their change (linear or nonlinear) on the basis of an analysis of behaviour and operating conditions of the structures.

The rate of the constitutive parameter change should be determined on the basis of measurements of this parameter in-situ at two different moments of time t_1 and t_2. To obtain reliable results, the time base:

$$\Delta t=t_2 - t_1, \tag{5}$$

should be significant (measured in years).

The rate of the constitutive parameter change x_i should be calculated by the formula

$$s_i = \frac{x_i^2 - x_i^1}{\Delta t} \quad (i = 1, 2,..,m), \tag{6}$$

where x_i^1 - the value of the constitutive parameter x_i at the time moment $t = t_1$, x_i^2 - the value of the constitutive parameter x_i at the time moment $t = t_2$.

It is recommended to perform as many independent measurements of the same constitutive parameter as possible. In this case x_i^1 - is the arithmetic average of the measurement results of the parameter x_i at the time moment $t = t_1$, x_i^2 - is the arithmetic average of the measurement results of the parameter x_i at the time moment $t = t_2$.

If it is impossible to make the measurements, the rate of the constitutive parameter change can be determined on the basis of theoretical or laboratory studies.

The assessment of the possibility to increase the service life according to the load-bearing capacity should be made in the following sequence:

- estimate the load-bearing capacity of the element (structure) according to the designing data F_{pr};
- based on the inspection, determine the parameters of the load-bearing capacity function $\Phi_u[x_1(t),x_2(t),...,x_m(t),y_1,y_2,...,y_n]$ for a particular element (structure) and determine the loadbearing capacity of the structure $F_{cr}=\Phi_u[x_1(t),x_2(t),...,x_m(t),y_1,y_2,...,y_n]$;
- from the calculation results, estimate maximum forces F in the element (structure). Compare

$$F_{cr} \geq F, \tag{7}$$

if the inequality is met, the service life has not run off;

- the lifetime reassignment t_R should be determined using the supposition with respect to a linear dependence of the controlled parameters change in time

$$t_R = \Delta t \frac{F_{cr} - F}{F_{pr} - F_{cr}}, \tag{8}$$

If a failure is possible for two or more elements of a statically indeterminate system, the calculation of the system should be made with account of the redistribution of forces after the elimination of the ruined elements from the computational scheme.

For reinforced concrete structures, along with the strength, deformability and crack resistance calculations, it is allowed to use the methods of mechanics of concrete deterioration, which consider the structural concrete changes in time [3, 11].

The need for inspection of the building structure of Imeni Gagarina Cinema at 5 Shchusev St. in the Shevchenkivsky district in Kiev, arose as a result of the termination of the building use for its intended purpose without the development and implementation in-situ of measures for conservation of the building. The most probable cause of damage in the walls is considered the unsatisfactory performance of drainage from the roof of the building, the termination of the building operation without the development and implementation of conservation measures, the unsatisfactory state of the blind area, and the lack of drainage of precipitation from the site.

To determine the minimum required reinforcement of structures, it is necessary to determine the stress-strain state of the building as a whole (forces that arise in the building structures, as well as the structures movement). To achieve this purpose, a finite element model of the building in one-dimensional problem setting has been developed (fig.1). The calculations of the building were made with account of yielding of a foundation and the stiffness coefficient of the monolithic foundation plate and the frame structures. The purpose of the complex calculation was to obtain the stress-strain state of the building and foundation structures, as an integrally working system.

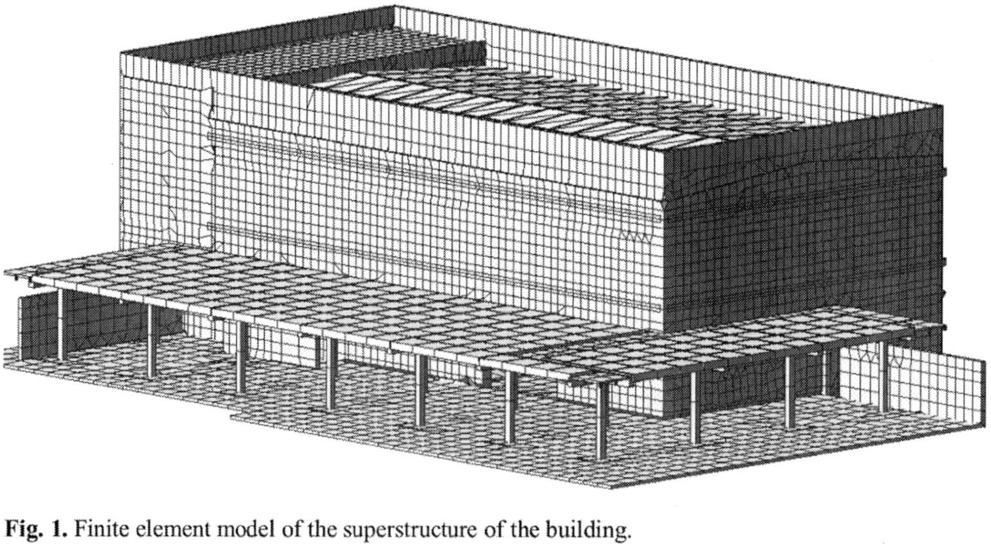

Fig. 1. Finite element model of the superstructure of the building.

Based on the results of the performed calculations of the system, the greatest deformation of the foundation soils is 22.19 mm under the walls of the hall part of the building. According to the recommended normative document, for multistoried frameless buildings with load-bearing walls of brickwork with the armor-clad belts, the average yieldings of the building foundation should not exceed 18 cm, that is, the calculated yieldings of the building do not exceed the limiting values.

Based on the results of the performed calculations of the system, it was established that the main stress N_1 and N_3 have the highest values (both compression and tension) in the wall of the first floor in the local zones of the areas of window apertures and door apertures location. The greatest value of the main stresses N_1 and N_3 is 4.03 MPa (compression) and 2.51 MPa (tension). In other parts of the walls of the first, second and third floors, the main stresses have significantly lower values of compression against the strength of the brick and mortar solution.

In the development of thereconstruction project, it is necessary to make such decisions which will contribute to a uniform loadingon the foundation structures.When working on the serviceability restoration of the building structure of Imeni Gagarina Cinema, it is necessary to perform the reinforcement of structures that have significant dilapidation (floor slab, partition walls, external staircases on steel stringboards, etc.), as well as the wall insulation and roof coating replacement. You need to install the windows and doors assembly, to perform the replacement of the floor and interior decoration in the premises.

5 Conclusions

A methodology has been developed for diagnosis of the technical state and determination of the remaining lifetime of the building structures, which is used when designing reinforcement elements of buildings and structures. For the calculation of structures, the modelling of the stress-strain state of the building structures with the use of LIRA software has been used. This allowed by way of calculation to perform optimization of technical solutions and to ruggedize structures of the houses, which are in operation, and to perform the further strengthening with minimal costs. An approach for determining the remaining lifetime of structures depending on operating conditions has been developed. The reliable assessment and forecasting of the technical state makes it possible to prevent the occurrence of structural failures and related losses, to use rationally the funds for performing current and capital repairs, as well as to regulate the technical state in such a way as to achieve the maximum efficiency in the use of basic funds. Since the service life of most residential, public and industrial buildings exceeded the duration of effective operation prior to capital repair, the methodology has been developed for remaining lifetime assessment of such buildings on the basis of modern concepts of the building materials and structures performance under loading with the aim of deciding on possible further operation or setting the buildings for major repairs (reconstruction) under the combined action of force, deformation and high-temperature effects.

References

1. M.I. Vasiliev, I.O. Movchan, O.M. Koval, Naukovyi Visnyk Natsionalnoho Hirnychoho Universytetu, **5**, 106-113 (2014)
2. Y. Abramov, V. Borisenko, V. Krivtsova, Eastern European Journal of Enterprise Technologies, **5**, 16-21 (2017)
3. O. Nekora, V. Slovynsky, S. Pozdieiev, MATEC Web of Conferences, **116**, 02024 (2017)
4. S. Vambol, V. Vambol, O. Kondratenko, Y. Suchikova, O. Hurenko, Eastern European Journal of Enterprise Technologies, **3**, 63-73 (2017)
5. S. Vambol, V. Vambol, Y. Suchikova, N. Deyneko, EasternEuropean Journal of Enterprise Technologies, **1**, 27-36 (2017)
6. V.V. Tiutiunyk, H.V. Ivanets, I.A. Tolkunov, E.I. Stetsyuk, Naukovyi Visnyk Natsionalnoho Hirnychoho Universytetu, **1**, 99-105 (2018)
7. Patent Ukrainy na KM # 18066, MPK G01N 3/00 *Protses vyznachennia zalyshkovoho*

resursu naftohazoprovodiv ta zbilshennia normatyvnoho terminu yikh ekspluatatsii

8. Patent Ukrainy na KM # 9473, MPK G01 N3/00, G01 M3/22 *Protses vyznachennia zalyshkovoho resursu obiekta pidvyshchenoi nebezpeky ta zbilshennia normatyvnoho terminu yoho ekspluatatsii*
9. V. Andronov, B. Pospelov, E. Rybka, Eastern European Journal of Enterprise Technologies, **4**, 38-44 (2016)
10. S. V. Kolesnichenko, Yu. V. Seliutin, I. Yu. Chernykh, K. B. Mnatsakanian, Scientific Journal «ScienceRise», **11**, 37-42 (2017)
11. S. Pozdieiev, O. Nuianzin, S. Sidnei, S. Shchipets, MATEC Web of Conferences, **116**, 02027 (2017)

Heat-insulating materials in modern building industry

Kyrylo Plakhotnikov[1], *Dmitry* Bondarenko[1], *Elena* Dedenyova[1], *Medeia* Saliia[1,*], and Tatiana Kostuk[1]

[1]Kharkiv National University of Civil Engineering and Architecture, 61002 Sumska st, 40, Kharkiv, Ukraine

Abstract. The characteristic feature of the modern construction industry is the broad appliance of contemporary design and technology solutions, which require new constructive, water-proof, finishing and heat-insulating materials. Implementation of new energy saving norms for the residential sector and increasing requirements for thermal protection of buildings have sharpened an issue how to develop new thermal insulation materials. Although the thick-layered heat insulation is known for a while, the heat insulation made of thin layers requires additional studies. This group of constructive materials is relatively new; there is no equipment and techniques for studying that. The structure of these materials is composite, comprising hollow ceramic and glass microspheres and a binder substance. This combination lends the lightness to the material and makes it highly adhesive to a surface where it should be applied. After application to a surface, the microspheres create a multilayered combination, reflecting up to 80% of heat stream. It makes the premises more comfortable, cutting down the expenditures on heating in winter, and summer air conditioning as well.

1 Introduction

The demand for the thermal insulating materials is growing routinely. Ever more manufacturers offer various materials, such as mineral cotton, foam plastic, liquid heat insulation [1-4]. One of the reasons of high interest to the liquid heat insulation is low labor input of ventilated front mounting, and the problem of heat insulation in old houses which have complex configurations of facades and require restoration. After application of the heat insulating composition to outer or inner walls and ceilings, the values of heat egress and heat ingress by a frame filling lowered. The methods for determination of its properties are to be elaborated [5,7,8].

2 Materials and methods

As an outcome of the studies, the heat insulation composition was obtained. It comprises white portland cement, hollow microspheres of glass and aluminosilicate with diameters

* Corresponding author: medeasalia@gmail.com

ranging from 5 to 30 μm, and the chemical complex additive (CCA) for regulation of hydration processes and strength enhancement. Furthermore, an additive of methylcellulose (MC) is merged into the composition for correction of its rheological properties, to enhance adhesion to the concrete surface and other stone-like bases. The concrete base was chosen for studies of the adhesive strength of the heat insulating thin-layered coating. The application of the thin-layered coating was performed by a paint brush. In order to provide the compression strength studies, the samples 4x4x8 cm were produced and cut into cubes before the tests. The samples are shown in Fig.1.

a) b)

Fig. 1. Pictures of the samples for experimental determination of: a – adhesive strength, b – compressive strength.

The adhesive strength of this coating was defined by the pull-off method, as a function of the concentration of methylcellulose [6,9,10]. The results of the studies of adhesive strength in the coating based on aluminosilicate and glass hollow microspheres are presented in fig.2. The diagram shows, that the maximal adhesive strength of the coating based on hollow microspheres corresponds to the concentration of MC 0.25-0.3% and comprises 1.2-1.4 MPa. It meets the normative requirements to the cement-based coatings, DSTU B.V.2.7-126:2011 "Mixtures constructive, dry, and modified. General technical requirements".

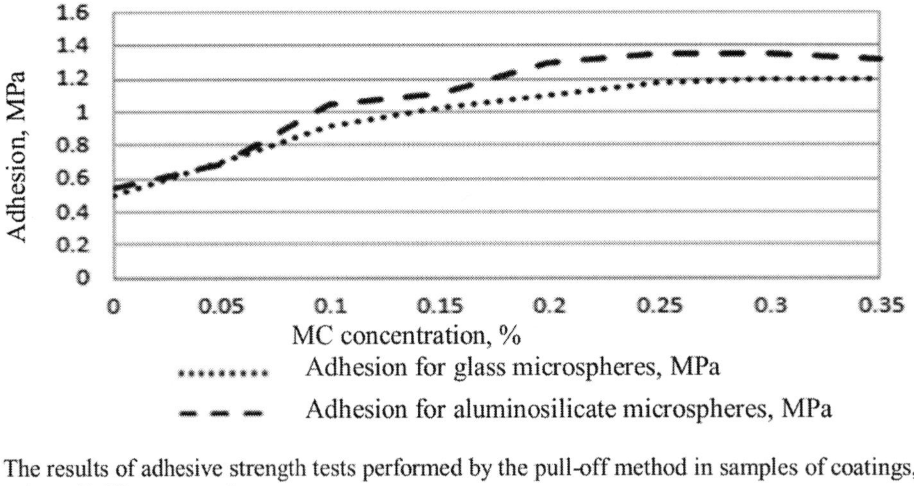

Fig. 2. The results of adhesive strength tests performed by the pull-off method in samples of coatings, as a function of MC concentration.

Also, the strength in compression for the samples of compositions, containing glass and aluminosilicate microspheres, was determined, see fig. 3.

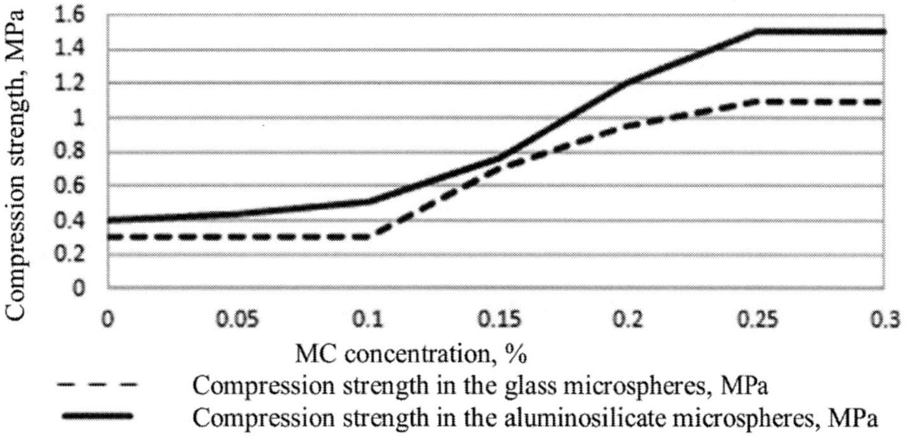

Fig. 3. The test results for the compression strength limit on the 28th day in the samples of the coating composition as a function of MC concentration.

Fig.3 shows, that the highest values of strength are found in the composites where the values of MC concentration comprise 0.25-0.30%. Among them, the maximal value is shown for the composites based on the aluminosilicate hollow microspheres. On the grounds of the results, we obtained the samples of the compositions of heat insulating coatings with aluminosilicate and glass microspheres and fixed 0.25% MC concentration. Meanwhile, the concentration of CCA added to the composition for intensification of the hydration processes was variable. The results of the experiment are presented in fig.4.

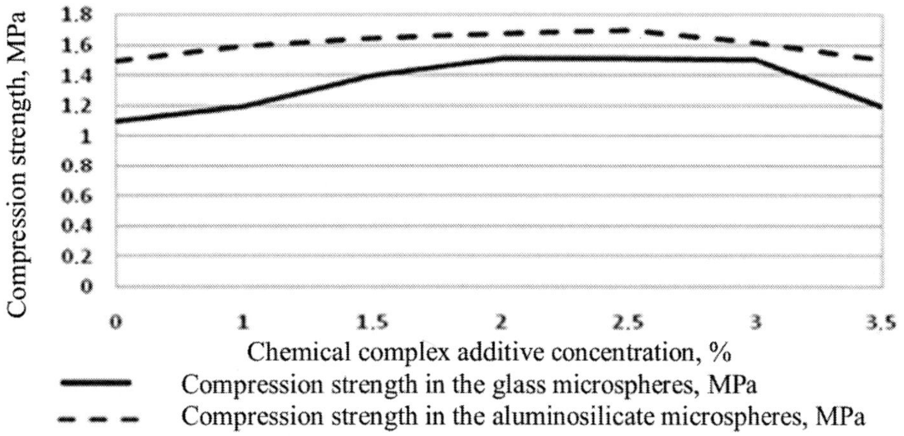

Fig. 4. The results of the compression strength limit test carried out on the 28th day in the samples of the coating composition as a function of CCA concentration.

Fig. 4 shows, that the composites containing the aluminosilicate microspheres have the highest strength, and the optimal amount of the additive is 2-3%.

Out of the composites, which were optimal in terms of the methylcellulose concentration and the CCA concentration, the heat insulating coatings were produced and then applied to 1 mm thick concrete slabs. The slabs were placed on stands open-air for testing in natural conditions (Fig.5). The coatings were checked every 3 days by a 4x magnifier.

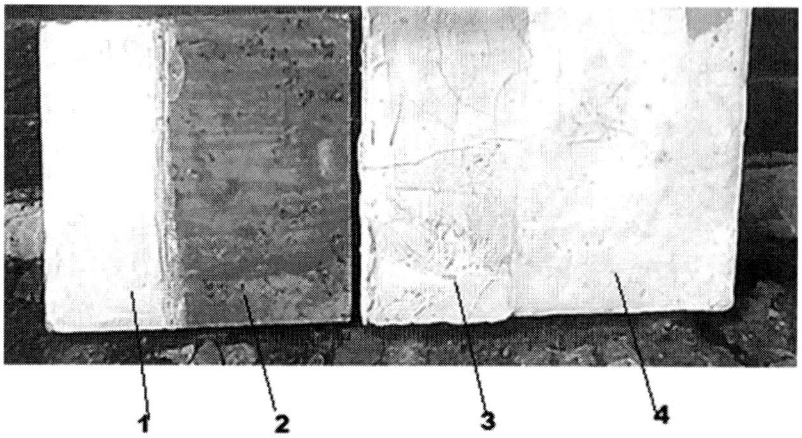

Fig. 5. Testing the heat insulating covers in natural conditions: 1 – aluminosilicate microspheres based coating; 2 – a concrete slab; 3 – a concrete slab; 4 – glass microspheres based cover.

Observation of the heat insulating coatings in natural conditions during 12 months has shown that the surfaces remain without any changes; no cracks, delaminations, or distentions have been found.

3 Conclusions

The cement compositions for thin layered heat insulation based on the hollow microspheres of aluminosilicate and glass, which have been obtained in result of the studies, have the good adhesive strength to concrete bases, and the compression strength, that allows the coating to preserve its integrity and weather resistance.

References

1. B.M. Shoykhet. The constructive materials. *New quality and nomenclature of the heat insulating materials Saint Gobain Isover*, **3**, 19-21 (2005)
2. Yu.G. Skrypnyk, V.V. Petrovych, T.V. Skrypnyk. Building Industry of Ukraine. *Foam polystyrene as a reality of today*, **6**, 37-39 (2002)
3. R.I. Burangulov, G.V. Tenenbaum, D.M. Khabirov. The constructive materials. *The technology of obtaining and application of superlight and light polystyrene foam concrete mixtures*, **12**, 16-17 (2003)
4. The constructive materials. *Heat insulating items URSA GLASSWOOL*, **5**, 38-39 (2005)
5. M. Tchaika. *Heat insulating paints as an effective way of saving energy sources. http://naukarus.com/teploizolyatsionnye-kraski-effektivnyy-sposob-ekonomii-energoresursov*
6. A.S. Inozemtsev, E.V. Korolev. Industrial and civil construction: a journal. *Hollow microspheres - an effective filler for high-strength lightweight concrete*, **10**, 80-83 (2013)
7. Liquid thermal insulation ThermoRon facade. *https://www.microspheres.com.ua/en/ produktsiya/stroitelnaya-teploizolyatsiya/140-zhidkaya-teploizolyatsiya-termoron-fasad*
8. Aluminosilicate microsphere. *https://en.wikipedia.org/wiki*

9. ISO 4624:2013. *Paint materials. Pull-off test for adhesion*
10. K.V. Plakhotnikov, O.I Dyomina O.I Bondarenko, O.B Dedenova. Scientific Bulletin of Construction. *Effect of water-retaining additives on the adhesion strength of thin-layer heat-insulating coatings based on mineral binders*, **1,** 152-155 (Kharkiv, 2018)

The improved technique for calculating the concrete elements strength under local compression

Volodymyr Pohribnyi[1], *Oksana* Dovzhenko[1*], *Iryna* Kuznietsova[1], and *Dmytro* Usenko[1]

[1] Poltava National Technical Yuri Kondratyuk University, Pershotravneviy sq., 24, Poltava, 36011, Ukraine

Abstract. The central one-sided concrete element crushing is considered when the loading platform width is less than the element width (in three-dimensional stress state conditions). A variational method in the concrete plasticity theory, which is developed in the Poltava National Technical Yuri Kondratyuk University, is proposed to calculate the strength. The formula for determining the ultimate load is based on the destruction kinematics, which is observed in the experiments, and takes into account the loading platform size ratio and the element, both concrete strength characteristics, the frictional forces between the support surface and the element. The calculation dependence is rather precise during the cubes central crushing when the element height ratio and the loading platform greater than 4. With other geometric parameters the kinematic scheme changes, which must be taken into account when deriving the formulas for calculating the strength by the variational method. The proposed method allows considering all possible calculation schemes and determining its application limits.

1 Introduction

In the construction there are large concentrated loads transfer cases to concrete elements. As an example, it is possible to support the inclined colonnade reinforced concrete columns on the shell and the hyperbolic cooling tower pool wall, multi-story buildings reinforced concrete columns or suspended walls supports on concrete elements of different height and configuration in the plan that are basement part or other buildings structural parts [1]. In this case, there is a concrete local compression in the structural elements and its destruction probability from punching shear or crushing depending on the geometric dimensions that fall under the relevant design schemes adopted in the standards.

One of the reserves for improving the concrete and reinforced concrete elements operational reliability is to improve its calculation methods. It is necessary to switch from empirical dependences to other alternatives dependences, which take into account the stress-strain state and the element kinematic destruction characteristic under crushing. Thus, the strength problem solution under local compression remains an urgent task and requires further theoretical and experimental studies.

[*] Corresponding author: o.o.dovzhenko@gmail.com

The work aim is to improve the concrete elements strength calculating method under local compression.

According to experimental studies, the ratio between loading platform length and the element height [2-5] or length, the element sides ratio [6], size effect [7], the eccentricity applied load [3, 6], the friction on the contact between the load (support) platform and the element surface, the concrete class and type [3, 8], loading prehistory [9] are main factors influencing the concrete crushing strength.

Most of the calculation methods [10] are based on the Bauschinger's formula and its variations, which operate at most two or three significant factors, in particular, the calculated area A_{c1} and the compressive concrete strength f_{cd}, then at least eight of them. This often leads to significant differences between experimental and theoretical results.

2 The analysis of proposed design dependencies

The normative in different periods (Fig. 1) and author's dependencies are considered and analyzed (Fig. 2, Table 1) for the local central one-sided compression. The bearing capacity was determined for the column support variant on the basement wall (Fig. 3) from concrete with a class on the compressive strength C20/25.

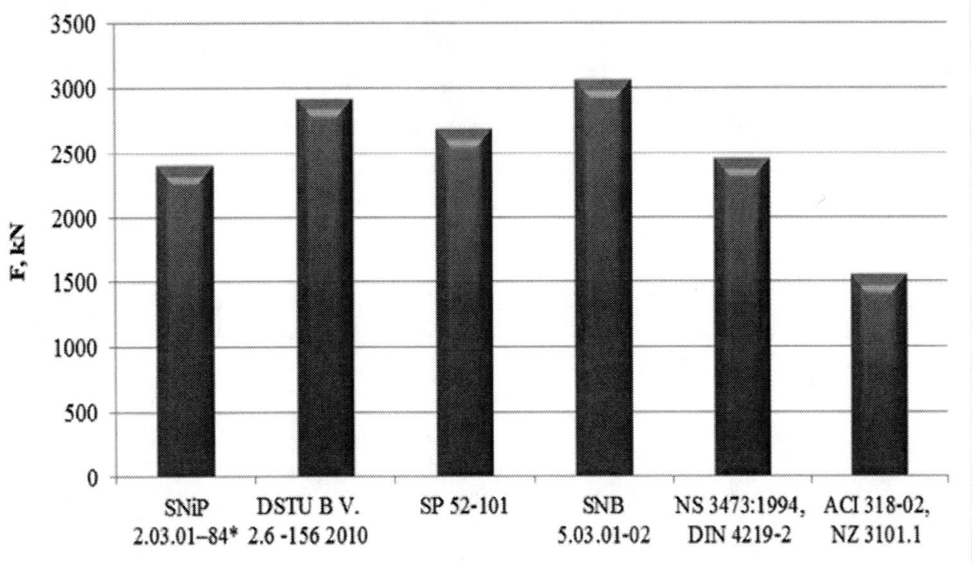

Fig. 1. Concrete elements strength under local central one-sided compression obtained according to normative documents.

As a calculations result, we can conclude that the largest bearing capacity for the considered case was obtained according to the method presented in Belarusian norms. The smallest bearing capacity was obtained according to the method presented in American and New Zealand norms (the difference between the bearing capacity values is 50%). The theoretical strength convergence analysis with experimental for 35 samples [11, 12] was performed. The best statistical indicators are the normative methodology of Ukraine and Eurocode 2: for its, the theoretical strength ratio average arithmetic value to the experimental is $m = 0.99$, the variation coefficient is $\nu = 29.42\%$.

The results most closely related to the experiments among authors has the V. Jensen's dependence with the following statistical characteristics: $m = 0.84$, $\nu = 28.52\%$. At the same time, the maximal values for the considered case are obtained by the

V. Chervonobaba's formula, the minimal values – V. Gladyshev's, and they differ more than twice.

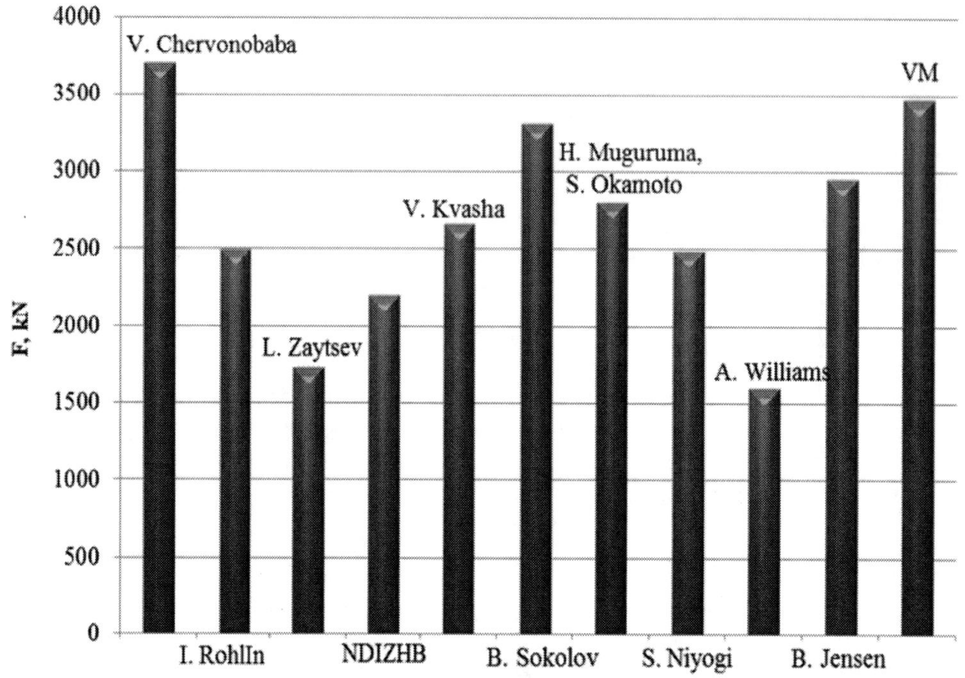

Fig. 2. Concrete elements strength under local central one-sided compression obtained according to author's methods.

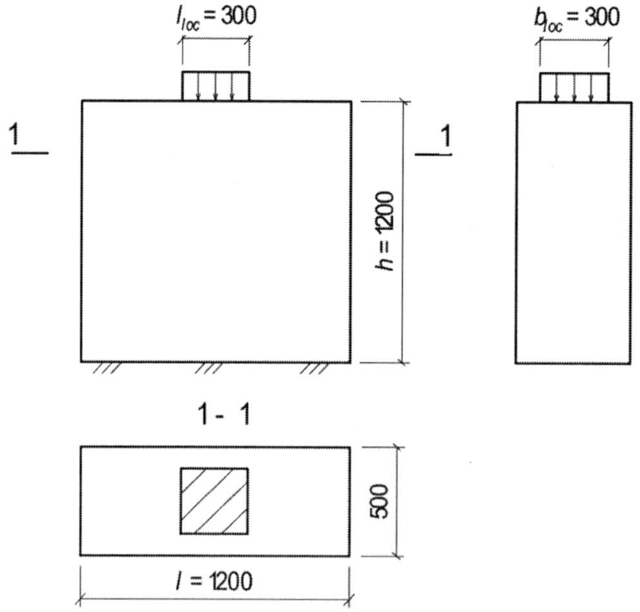

Fig. 3. Local central one-sided compression.

A large number of proposed dependencies and the results widespread scatter, necessitates the need to improve the calculation method under crushing.

Table 1. Design dependencies proposed for concrete strength under local central one-sided compression.

№	The author	Estimated dependencies
1	Chervonobaba V.A.	$f_{c,loc}=f_c\left[4-3\left(\frac{A_{c1}}{A_{c0}}\right)^{\frac{1}{2}}\right]$
2	Rohlin I.A.	$f_{c,loc}=f_c\left(\frac{A_{c0}}{A_{c1}}\right)^{-0,3424}$
3	Hladyshev B.M.	$f_{c,loc}=0.4f_{ct}\frac{h}{l_{loc}}\left(2.3+5\frac{l_{loc}}{l}\right)\left(2.5-\frac{h}{l}\right)$
4	Zaytsev L.N.	$f_{c,loc}=\frac{f_c h/l_{loc}}{\left(\frac{2}{\pi k_3 m_2}\right)\left(1+\frac{f_c}{f_{ct}}\right)\left[1-1.41(\delta-0.1)+0.53(\delta-0.1)^2\right]+t}$, $k_3=1;\quad m_2=1.6;\quad \delta=\frac{l_{loc}}{h};\quad t=\frac{0.75\delta+0.275}{1.4\delta}$
5	NDIZHB	$f_{c,loc}=m_1 f_{ct}+m_2 f_{ct}\left(h/l_{loc}-1\right)$, $m_1=8;\quad m_2=2$
6	Kvasha V.H.	$f_{c,loc}=f_c\left\{1+\frac{k}{f_{ct}}\left[\left(\left(\frac{A_{c1}}{A_{c0}}\right)^{\frac{1}{2}}-1\right)\right]\left[f-f_{ct}-2\left(f^2-ff_{ct}+\frac{f_{ct}^2}{3}\right)^{\frac{1}{2}}\right]\right\}$
7	Sokolov B.S.	$f_{c,loc}=f_c\frac{ctg\alpha f_{ct}}{f_c}\left[2mk\left(\frac{1}{\delta}-\cos\alpha\sin\alpha\right)+6\left(1+\sin^2\alpha\right)+\sin^4\alpha\right]$, $tg\alpha=\frac{0.25f_c}{f_{ct}}-1.56;\quad m=1...0.6;\quad k=\frac{b}{l_{loc}};\quad \delta=\frac{l_{loc}}{h}$
8	Muguruma H., Okamoto S.	$f_{c,loc}=f_c\left(\frac{A_{c1}}{A_{c0}}\right)^k\left(0.71+\frac{l}{(6.67h-2.91l)}\right)$, $k=\frac{7.61h/l-3.54}{16.44h/l-6.65}$
9	Niyogi S.K.	$f_{c,loc}=f\left\{0.42\left(\frac{l}{l_{loc}}+2\right)-0.29\left[\left(\frac{l}{l_{loc}}-1\right)^2+5.06\right]^{\frac{1}{2}}\right\}$
10	Williams A.	$f_{c,loc}=6.92f_{ct}\left(\frac{A_{c0}}{A_{c1}}\right)^{-0,47}$
11	Jensen B.C.	$f_{c,loc}=f_c\left[0.2+0.8\left(\frac{A_{c1}}{A_{c0}}\right)^{\frac{1}{2}}\right]$

3 Strength theoretical studies under crushing by the variational method

In PoltNTU a variational method (VM) in the concrete plasticity theory [13-15] is developed. The method is confirmed by experimental studies under shear [16, 17] and is proposed for strength calculating under crushing.

The concrete element central one-sided crushing when the loading platform width is less than the element width is considered.

Presented in Fig. 4 the sample destruction kinematic scheme corresponds to the destruction experimental nature.

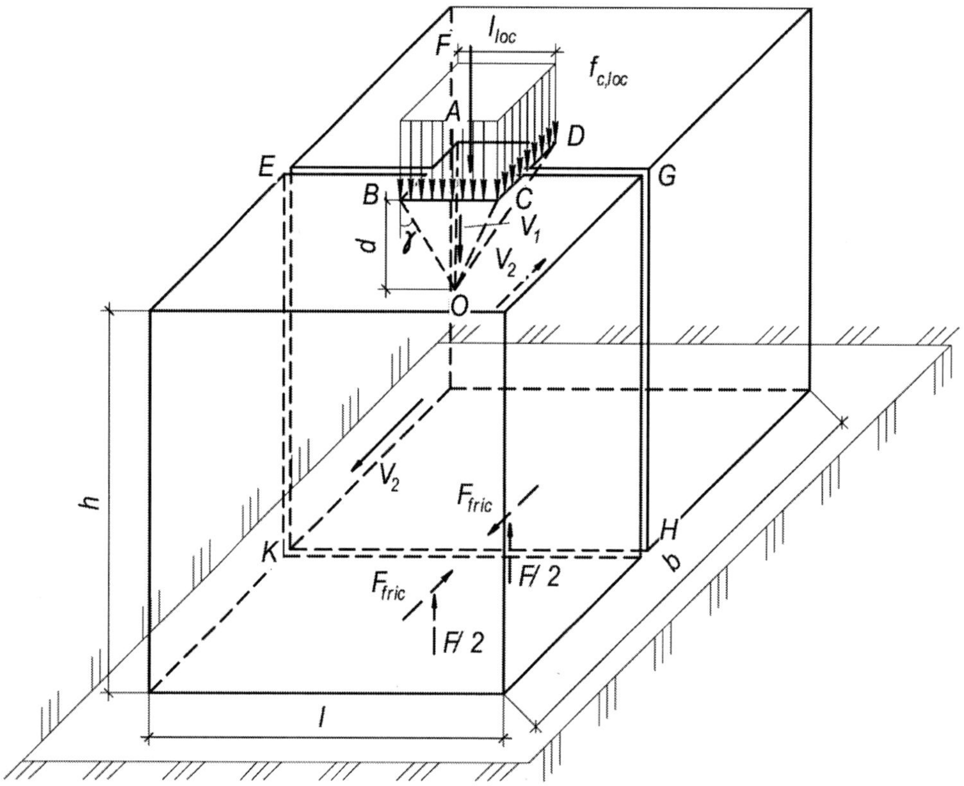

Fig. 4. The concrete element destruction kinematic scheme under central one-sided crushing.

In the limit state, a solid concrete element is divided into three rigid bodies: a pyramid with a base equal to the loading area, and two halves, separated by a splitting plane and a sliding surface. The pyramid moves down with the speed V_1, and the sample halves are removed horizontally at the same speeds V_2 and $-V_2$. The splitting plane passes through the sample gravity center in parallel to the short lateral face. On it, there are stresses that are taken equal to the concrete tensile strength f_{ctd}. The velocities ratio $k = V_1 / V_2$ and the pyramid relative height $D = 2d / l_{loc}$ are unknown for this task. On the destruction surface sections, the normal and the tangent speed component jumps are determined.

The ABO pyramid surface

$$\begin{cases} \Delta V_n = V_2 \cos\gamma - V_1 \sin\gamma; \\ \Delta V_t = V_2 \sin\gamma + V_1 \cos\gamma. \end{cases} \quad (1)$$

Its square

$$A_{\mathrm{ABO}} = \frac{l_{loc}^2}{4\sin\gamma}. \tag{2}$$

The BCO pyramid surface

$$\begin{cases} \Delta V_n = V_1 \sin\gamma; \\ \Delta V_t = \sqrt{(V_1 \cos\gamma)^2 + V_2^2}. \end{cases} \tag{3}$$

Its square

$$A_{\mathrm{BCO}} = \frac{l_{loc}^2}{4\sin\gamma}. \tag{4}$$

Splitting surface EGHK

$$\Delta V_t = V_2. \tag{5}$$

Its square

$$A_{\mathrm{EGHK}} = lh - \frac{l_{loc}^2}{4tg\gamma}, \tag{6}$$

where $tg\gamma = \dfrac{l_{loc}}{2d}$.

The virtual velocities principle functional for a three-dimensional stress state is recorded and equated to 0

$$\begin{aligned} I = 2m\left[B^2 + 0.25\frac{(V_2\sin\gamma + V_1\cos\gamma)^2}{(V_2\cos\gamma - V_1\sin\gamma)^2}\right](V_2\cos\gamma - V_1\sin\gamma)\frac{l_{loc}^2}{4\sin\gamma} + \\ + 2m\left[B^2 + 0.25\frac{(V_1\cos\gamma)^2 + V_2^2}{(V_1\sin\gamma)^2}\right]V_1\sin\gamma\frac{l_{loc}^2}{4\sin\gamma} + 2f_{ctd}V_2\left(lh - \frac{l_{loc}^2}{4tg\gamma}\right) - f_{c,loc}l_{loc}^2V_1 = 0. \end{aligned} \tag{7}$$

Limit load is defined as

$$\frac{f_{c,loc}}{m} = 0.5\left[B^2 + 0.25\left(k^2\left(1+D^2\right)+D^2\right)\right] + \frac{0.5\left[B^2(Dk-1)^2 + 0.25(k+D)^2\right]}{Dk-1} + \frac{f_{ctd}k\left(\alpha'^2 - D\right)}{2m}, \tag{8}$$

where $B^2 = \left(1+\chi/(1-\chi)^2\right)/3$, $\chi = f_{ct}/f_{c,prizm}$, $m = f_{c,prizm} - f_{ct}$, $\alpha' = 2h/l_{loc}$.

From the condition $f_{c,loc} = \min$ unknown values k and D are determined, after which the limit value is calculated.

Dependences for determining the bearing capacity under crushing from the strength determinants are constructed (Fig. 5).

The proposed kinematic scheme is also implemented at the cubes central one-sided local loading. The theoretical strength comparison with the experimental results indicates that the formula (8) with experiment is in good agreement ($m = 1.02$, $v = 9.5\%$), however, it should be noted that the accepted kinematic scheme is implemented in relation to the element height to the loading platform $h / l_{coc} \geq 4$.

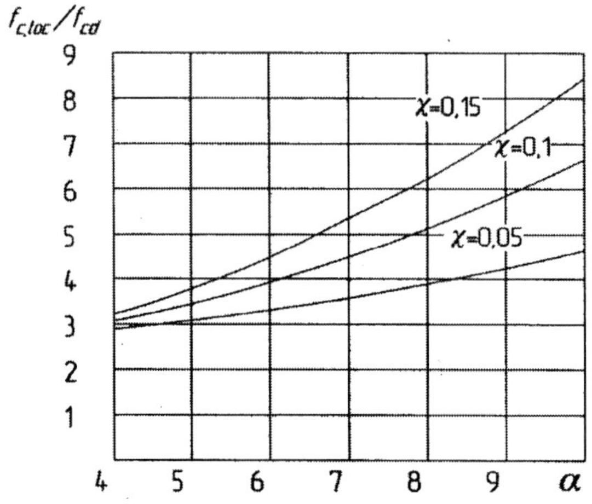

Fig. 5. The dependence for crushing strength determination for definite $\alpha = h/l_{loc}$ and χ .

In the larger loading areas case the stress uneven distribution on it may lead to the other destruction schemes implementation. In particular, the load is concentrated on the cargo area part (for example, on ¼ its area) and a possible split option along the planes parallel to the conditional (reduced) decomposition site diagonals. Scattering along diagonal sites may also be due to the experimental sample concrete structure heterogeneity.

If the load is transmitted through a platform which width is equal to the width of the element, the strength problem in the two-dimensional stress state case should be considered. Its solution is provided in [18]. The proposed calculation technique allows us to solve the strength problem for various central one-sided crushing cases, as well as to find its possible implementation limits.

Conclusions

During calculating of concrete elements strength under local compression, empirical dependencies are widely used, which have an application narrow scope, limited by the experimental conditions. The concrete plasticity theory is promising to improve the calculation methodology. The proposed formula for determining the concrete elements strength under central one-sided local compression in three-dimensional stress conditions is quite precise. Necessary accuracy is achieved by considering the destruction kinematics, which is observed in the experiments, and the strength determinants set.

References

1. D. H. Hladyshev, Resursoekonomni materialy, konstruktsiyi, budivli ta sporudy, Rovno, **29**, 94-101, (2014)
2. S. Anjali, P. Lakshmi, IJIRT, **3(4)**, 173-179, (2016)
3. R. Breitenbücher, G. Meschke, F. Song, M. Hofmann, & Y. Zhan, Proceedings of Joint ACI-fib International Workshop: *Fibre Reinforced Concrete: from Design to Structural Applications* (FRC 2014), Montreal, 417-429, (2014)

4. V. Venckevičius, Journal of Civil Engineering and Management, **11(3)**, 243-248, (2005)
5. S. K. Niyogi, ASCE Journal of Structural Division, **114(3)**, 932-940, (1993)
6. Krishna Arathi, S Anjali, International Journal of Civil Engineering and Technology(UAE), **9(1)**, 41-47, (2018)
7. Fanbing Song, *Steel Fiber Reinforced Concrete Under Concentrated Load,* Ph.D thesis in Engineering science, Bochum, 169, (2017)
8. Zhenpeng Yu, Qiao Huang, Yushi Shan, & Zheng Ren, Journal of Materials in Civil Engineering, **30(10)**, (2018)
9. V. V. Tur, N. A. Rak, *Prochnost i deformatsii betona v raschetah konstruktsiy*, Monografiya, Brest, 252, (2003)
10. B. M. Gladyishev *Mehanicheskoe vzaimodeystvie elementov strukturyi i prochnost betona*, Harkov, 168, (1987)
11. A. B. Piradov, *Konstruktivnyie svoystva legkogo betona i zhelezobetona,* Moskow, 135, (1973)
12. V. P. Mitrofanov, Stroitelnaya mehanika i raschet sooruzheniy, Moskow, **6**, 23-28, (1990)
13. O. Dovzhenko, V. Pogrebnyi, I. Yurko, News NAS RK. Series of geology and technical science, **2 (428)**, 212-219, (2018)
14. V. Pohribnyi, O. Dovzhenko, O. Maliovana, International Journal of Engineering and Technology(UAE), **7 (3.2)**, 19-26, (2018)
15. O. Dovzhenko, V. Pohribnyi, I. Yurko, I. Shostak, Web of Conferences, **116**, 02011, (2017)
16. O. Dovzhenko, V. Pohribnyi, L. Karabash, Journal of Engineering & Technology (UAE), **7 (3.2)**, 354-359, (2018)
17. V. Pohribnyi, O. Dovzhenko, L. Karabash, I. Usenko, Web of Conferences, **116**, 02026, (2017)

Method of the calculated estimation of the possibility of progressive destruction of buildings in result of fire

Serhii Pozdieiev[1,*], *Olha* Nekora[2], *Tetiana* Kryshtal[2], *Vitalii* Zazhoma[2], and *Stanislav* Sidnei[2]

[1]Cherkasy State Technological University (CSTU), 18006, Cherkasy, Shevchenko blvd. 460, Ukraine
[2]Cherkasy Institute of Fire Safety named after Chornobyl Heroes of National University of Civil Defense of Ukraine, 18034 Cherkassy, Onoprienko Str., 8, Ukraine

Abstract. When analyzing the fire safety of building structures, in addition to considering their fire resistance, the limits of flame expansion and other fire-safety engineering features, there should be considered also the probability of their progressive collapse in case of a damage caused to some their elements. In case of progressive collapse of the building structures, socio-economic losses become the worst possible. The article proposes a method of calculation for the estimation of the possibility of progressive destruction. The method is based on the assumption that one or several compressed elements are damaged and must be removed from the system, which ensures the rigidity and geometric immutability. The basic point of the method is the hypothesis of formation of the line of plastic hinge joints in the ceiling plate. The estimation of the possibility of progressive destruction is made according to the energy criterion on the basis of a comparison of the work of internal and external forces on the possible movements of the system, which, under these conditions, is geometrically variable. The proposed method is productive and economical comparing to the existing methods which involve complex mathematical models and software complexes.

1 Introduction

The statistics of fires and fire-related accidents reveals that one of the most dangerous factors is the destruction of load-carrying reinforced concrete structures, as shown by the works [1-6]. In this case, the destruction reaches its highest level as the damage caused to one or more elements of structures leads to a series of collapses of other elements, resulting in collapse of a part or entire building. In case of progressive destruction, socio-economic losses are the highest possible. One of the effective measures to ensure the survivability of buildings and their structures is the introduction of modern calculation methods for predicting the possibility of progressive destruction and the use of appropriate engineering measures to strengthen the building structures.

[*] Corresponding author: svp_chipbbk@ukr.net

1.1 Literature review

According to the researches [7], one of the most efficient and economical methods of engineering estimation of the possibility of progressive destruction of buildings and structures is the kinematic method. The method, used in the researches, determines the work of internal forces (W) and external loading (U) for each of the previously adopted mechanisms of progressive destruction on the possible movements of the considered mechanism, into which the static system is transformed. The condition for preserving the static system unchanged is a solution to the inequality:

$$W \geq U. \quad (1)$$

The model of a building frame, which is in critical condition, assumes a complete removal of the destroyed column from the structure of the rigidity of the building. The removed column is not considered anymore to be a part of the mechanism, into which the building is transformed, along with the existing plastic hinge joints in it. At the same time, this method has a certain shortcoming: the lines of plastic hinge joints are straight, while the separated zones between the lines of plastic hinge joints have a form of triangular facets. This approach may lead then to the fact that unreasonable overcharging of the model of plastic hinge joints lines with additional lines may lead to significant distortion of the obtained results and imposes the need to introduce additional coefficients of safety margins for the investigated structures.

Another approach to the calculated forecasting of the progressive destruction is described and efficiently applied in the researches [8]. This approach is based on the application of the finite element method in the combination with the method of penalty functions in separating and co-inflicting of the parts with finite elements, formed again in the process of calculating the emergency system. This approach requires the use of labor-intensive and long-term calculations involving complex software and skilled engineers.

1.2 Purpose, objectives and results

The purpose of the research (its main results are provided by the article) is developing a mathematical description of the work of external and internal forces in the kinematic model of the system, into which the structure of the building is transformed in case of inserting the plastic hinge joint lines in the ceilings, limiting the curvilinear portions of this geometrically variable system as the basis of the improved kinematic calculation method for predicting the progressive destruction in buildings with reinforced concrete structures.

2 Basic theoretical provisions for the calculation

To considerate the progressive destruction, based on the assumptions, there was used a universal calculation model (shown in Fig. 1). This model corresponds to the kinematic calculation of a part of a building with one element, which is assumed to be destroyed in case of fire. It is also taken into account that the destruction of the column occurs after the destruction of the ceiling that may happen in result of a possible fire.

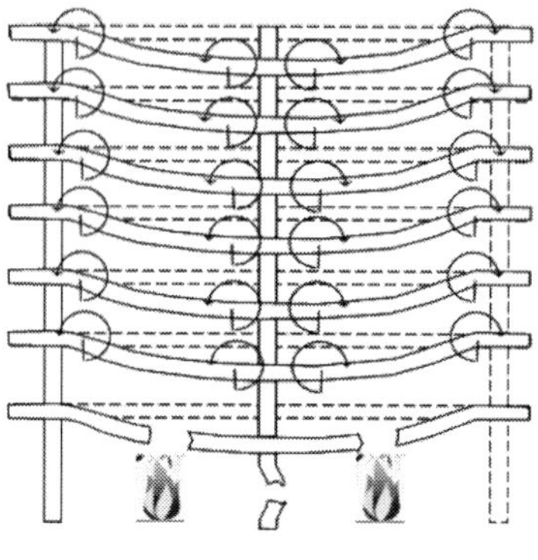

Fig. 1. The calculation model used for calculating the progressive destruction of a building with reinforced concrete structures.

Fig. 2 shows a model of reproduction of the plastic hinge joint lines, according to the Bezier curves. The hypothesis that plastic hinge joint lines can be recreated using the Bezier curves was adopted by the assumption on the location of plastic hinge joints on the tangents, which divide the parallel lines to series of undamaged columns and walls in the percentage correlated with the variable function of the Bezier curves.

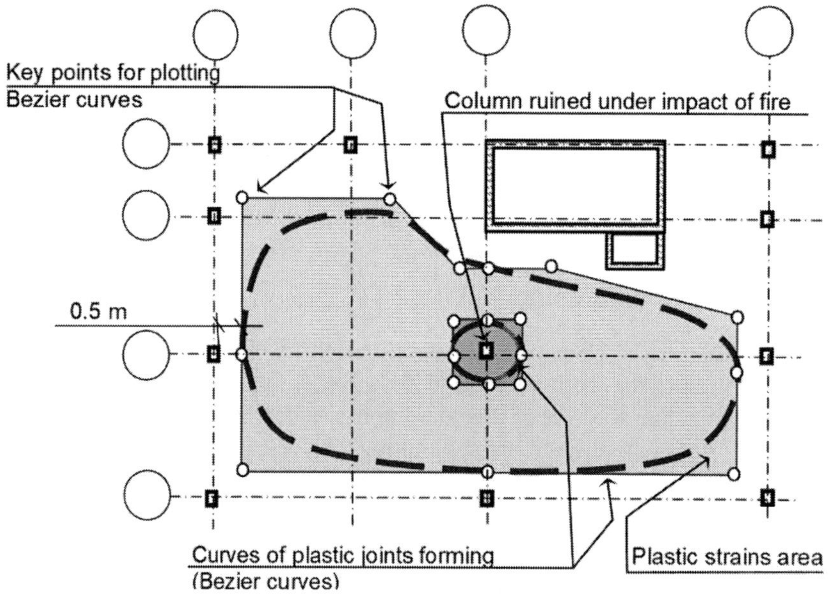

Fig. 2. Reproduction of lines of formation the plastic hinge joint using Bezier curves.

According to Fig. 2, two lines of the formation of plastic hinge joints should be considered: a line around the column (rigidity diaphragm) and a line that envelops the compressed load-bearing elements of the selected part of the building which is subjected to progressive destruction. The application of the approximation is due to the fact that the Bezier curves are continuous to the derivatives of the first and second order, therefore, they allow taking into account the specific features of the zone of plastic deformation and define the current values of the corresponding geometric parameters used for the determination of both internal and external force factors.

The equation of the function describing the Bezier curve is written in an implicit form and represents a system of two parametric equations [9]:

$$x(u)=\sum_{k=1}^{n} x_k B_{k,n}(u),\ y(u)=\sum_{k=1}^{n} y_k B_{k,n}(u), \tag{2}$$

where x_k and y_k are the coordinates of the nodal points for construction of the Bezier curve. In formula (2) the Bernstein polynomials are determined by the formula [9]:

$$B_{k,n}(u)=\frac{n!}{k!(n-k)!}u^k(1-u)^{n-k}. \tag{3}$$

where n - number of control points; k - current number of the nodal point.

3 Mathematical apparatus to calculate the work of internal forces

Using expressions (2) and (3), the virtual work of internal forces can be determined according to the models shown in Fig. 3 and Fig. 4.

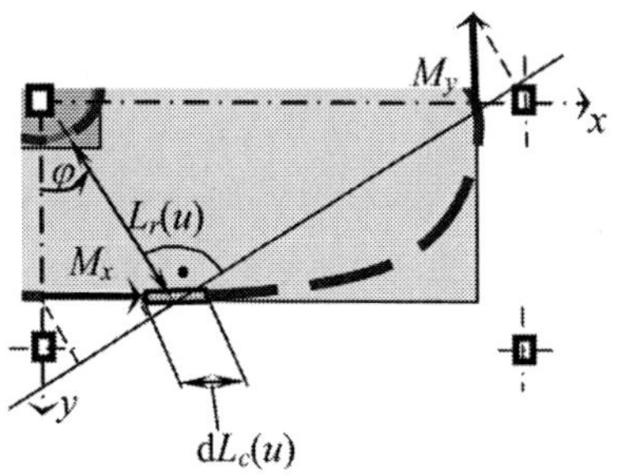

Fig. 3 Model for determining the work of the boundary internal forces in plastic hinge joints of ceiling.

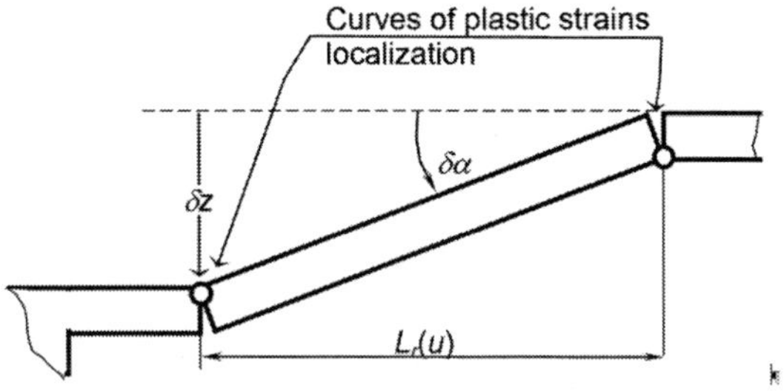

Fig. 4 Model for determining possible displacements in case of determining the work of internal forces in ceilings.

The following calculation models determine the work of the internal forces on the possible displacement δz, using the following expressions:

$$W_{sl1}=\int_0^1 (M_{x1}\cos\varphi+M_{y1}\sin\varphi)L_{c1}(u)\frac{\delta z}{L_r(u)}du,\ W_{sl2}=\int_0^1 (M_{x2}\cos\varphi+M_{y2}\sin\varphi)L_{c2}(u)\frac{\delta z}{L_r(u)}du, \tag{4}$$

$$\cos\varphi=\frac{tg\varphi}{\sqrt{1+tg^2\varphi}},\ \sin\varphi=\frac{1}{\sqrt{1+tg^2\varphi}},\ tg\varphi=y(u)/x(u), \tag{5}$$

$$L_{c1}(u)=\sqrt{\left(\frac{dx_1(u)}{du}\right)^2+\left(\frac{dy_1(u)}{du}\right)^2},\ L_{c2}(u)=\sqrt{\left(\frac{dx_2(u)}{du}\right)^2+\left(\frac{dy_2(u)}{du}\right)^2} \tag{6}$$

$$L_r(u)=\sqrt{(x_1(u))^2+(y_1(u))^2}-\sqrt{(x_2(u))^2+(y_2(u))^2}\ ,\ W=W_{sl1}+W_{sl2}, \tag{7}$$

where M_x, M_y are the lateral boundary moments during the plastic boundary deformation of ceiling slabs, determined by the method of boundary deformations.

The first derivatives of the functions that describe Bezier curves are determined by the following expressions [6]:

$$\frac{dx(u)}{du}=\sum_{k=1}^{n}x_kB'_{k,n}(u),\ \frac{dy(u)}{du}=\sum_{k=1}^{n}y_kB'_{k,n}(u),\ B'_{k,n}(u)=B_{k,n}(u)\frac{k-nu}{u(1-u)} \tag{8}$$

When implementing this calculation method, a value of the variable should be obtained to determine the function of the Bezier curve around the vertical supporting element, corresponding to the current angle φ. For this purpose, the following formulas can be applied:

$$x_2(u)=\sum_{k=1}^{n}x_kB_{k,n}(p(u)),\ y_2(u)=\sum_{k=1}^{n}y_kB_{k,n}(p(u)), \tag{9}$$

$$p(u)=\frac{p_1(u)-p_2(u)+\sqrt{p_2^2-p_1(u)p_3(u)}}{p_1(u)-2p_2(u)+p_3(u)},\ p_k(u)=y_k-x_k\frac{y_1(u)}{x_1(u)}. \tag{10}$$

4 Mathematical apparatus to calculating external forces work

The following calculation models determine the work of internal forces in one quadrant on the possible displacement δz using the following expressions:

$$U_{sl}=U_{sl1}+U_{sl2},\ U_{sl1}=\frac{1}{6}\eta_{fi}Q_{sl}\delta z\int_0^1 L_{c1}(u)L_{r1}(u)du-\frac{1}{2}\eta_{fi}Q_{sl}\delta z\int_0^1\frac{L_{c2}(u)L_{r2}(u)}{L_r(u)}\left(\frac{1}{3}L_{r2}(u)+L_r(u)\right)du, \tag{11}$$

$$U_{sl2}=\frac{1}{2}\eta_{fi}Q_{sl}\delta z\int_0^1 L_{c2}(u)L_{r2}(u)du-0.25\eta_{fi}Q_{sl}\delta z\cdot S_c, \tag{12}$$

$$L_{r1}(u)=\sqrt{(x_1(u))^2+(y_1(u))^2}\ ,\ L_{r2}(u)=\sqrt{(x_2(u))^2+(y_2(u))^2}\ , \tag{13}$$

We also need to determine the work of external forces due to the own weight of the columns. For this purpose the following formula is applied:

$$U_c=(\rho_cA_c+\rho_sA_s)L_c(1+1/N)\delta z. \tag{14}$$

where ρ_c - density of concrete; ρ_s - density of reinforcement steel; N - number of ceilings that were not destroyed by fire; S_c - cross-sectional area of a column or a diaphragm of stiffness; S_s - cross-sectional area of the column reinforcement or a diaphragm of stiffness; L_p – length of the column within the framework of the floor.

The general work of external forces in possible displacements is as follows:

$$U=U_{sl}+U_c. \tag{15}$$

Integration of the expressions (4), (11) and (12) is performed by one of the numerical methods. In this particular case, it is recommended to apply the Romberg's method [10].

5 Calculation method of estimation the progressive destruction possibility

For the calculation aimed at determining the possibility of progressive destruction of a building in result of fire, the following procedures should be performed.

1. One or a group of columns (diaphragms) is determined to be removed from the structure of the rigidity of the building as destroyed in result of fire. 2. The points of the boundary of the zone of plastic deformations for the first and second lines of plastic hinge joints are to be determined. The obtained zone should have boundaries passing through the points located 0.5 m away from groups of undamaged columns and rigidity diaphragms. The obtained zone is divided into several parts by the axial lines drawn through the center of gravity of the column section or the sections of a group of columns. 3. The boundary moments in the ceiling slabs are determined on conditions of normal temperatures. 4. The location of the points, through which the curves of local plastic deformation should pass (the Bezier line), are to be determined. The coordinate vector of these points is written for the parametric functions, which are described in accordance with the formula (2). 5. Applying the formulas (4)-(10), there is determined the possible work of the internal forces in each of the parts, into which the zone of plastic deformations, located around the removed columns, was previously divided. The total amount of possible work is defined as the sum of all obtained components. 6. Applying the formulas (11) and (15), the summed amount of the possible work of external forces is defined. 7. A performance of the condition (1) is checked and a conclusion on the possibility of progressive destruction of the building in case of fire is drawn.

6 Conclusions

As a result of the research, the following results were obtained:

- there was developed a mathematical apparatus for the determination of virtual work for the external and internal forces of kinematic systems, into which the structural systems of buildings, including the ones in disrepair, are transformed as a result of a fire;
- on the basis of the proposed mathematical apparatus there was improved an advanced kinematic calculation method for predicting the possibility of progressive destruction.

References

1. S. Pozdieiev, O. Nuianzin, S. Sidnei, S. Shchipets *Computational study of bearing walls fire resistance tests efficiency using different combustion furnaces configurations* MATEC Web of Conferences, **116**, 02027, (2017).
2. O. Nekora, V. Slovynsky, S. Pozdieiev *The research of bearing capacity of reinforced concrete beam with use combined experimental-computational method* MATEC Web of Conferences, **116**, 02024, (2017).
3. Y. Danchenko, V. Andronov, M. Teslenko, V. Permiakov, E. Rybka, R. Meleshchenko, A. Kosse Study of the free surface energy of epoxy composites using an automated measurement system EasternEuropean Journal of Enterprise Technologies, **1,** (12-91), (2018).

4. B. Pospelov, V. Andronov, E. Rybka, V. Popov, A. Romin Experimental study of the fluctuations of gas medium parameters as early signs of fire. EasternEuropean Journal of Enterprise Technologies, **1,** (10-91), (2018).
5. B. Pospelov, V. Andronov, E. Rybka, Vadym Popov, O. Semkiv Development of the method of frequency-temporal representation of fluctuations of gaseous medium parameters at fire **2**, 10, (2018).
6. DBN V.1.1-7-2016: *State Building of Ukraine, Fire protection. Building structures fire safety* (2017).
7. G.I. Shapiro, Yu.A. Eisman, A.S. Zalesov, *Rekomendacii po zachite monolitnyh zdanii ot progressiruyuchego obrushenia. Moskomarchitectury*. Moscow. 28, (2005).
8. J.O. Hallquist: *LS-DYNA Theory Manual, Livermore Software Technology Corporation: California*, USA, 680, (2005).
9. D Hearn., M.P. Baker, W.R. Carithers *Computer graphics with Open GL. 4-th edition. Harlow*: Pearson Education Limited, 812, (2014).
10. N. Piskunov *Differential and integral calculus for technical institutes*. 13-th edition - Moscow: Science, Main redaction of technical literature. 560, (1985)

Evaluation of bond between reinforcement and concrete

Olena Romashko[1], and *Vasyl* Romashko[1,*]

[1]National University of Water Management and Nature Recourses Use, Architectural Designing Bases, Constructing and Graphics Department, Soborna str. 11, 33028 Rivne, Ukraine

Abstract. The scheme of stressed state changes of the concrete contact layer and reinforcing rod during its pulling out of concrete is considered. The most important drawbacks of the known methods for evaluating the bond strength of reinforcement to concrete have been analyzed. Particular attention is focused on bond estimating by average tangential stresses. It is substantiated why the linear dependence of these stresses on normal stresses in the reinforcement cannot cover the processes of reinforcement bonding with concrete in the stage of use of reinforced concrete elements. A universal dependence of the average tangential stresses for estimating the reinforcement bond to concrete at any stage of elements deformation is proposed. Statistical evaluation of the obtained dependence was made on the basis of theoretical and experimental values of the average tangential bonding stresses comparison. The area of use of reinforcement bond with concrete average stresses dependence is defined.

1 Introduction

Concerning the reinforcement bond with concrete has always been one of the most important issues in the theory of reinforced concrete. It is well known that, together with the design measures of anchoring, the bond provides a consistent operation of reinforcement and concrete in structures, reflecting the essence of the reinforced concrete itself.

Of particular importance are the issues of reinforcement bond with concrete in the calculation of reinforced concrete elements for crack resistance, as well as in their designing for determining the length of the rods anchoring. It should be emphasized separately that the calculation of pre-stressed elements and elements with mixed reinforcement, even with simplified methods, as well as their designing, can not be realized without consideration of reinforcement bond with concrete.

In general, today numerous models, theories and methods for estimating the reinforcement bond with concrete have been proposed. However, despite the results of extensive experimental and theoretical research, the overwhelming majority of mentioned models, theories and methods [1-5] remain rather complicated and unsuitable for use in engineering calculations.

* Corresponding author: romashkovasyl@gmail.com

2 Analysis of recent studies and publications

Early research was mainly aimed at estimating the reinforcement bond strength with concrete that was extremely important for calculating the anchoring of reinforcing rods. The criteria for the bond strength are usually taken:

• average values of tangential stresses along the conditional cylindrical-drill surface of the reinforcing rod in the limiting deformation stage τ_{bmu} [6];

• the forces corresponding to the beginning of the displacement of the unloaded end of the rod $\sigma_{\mathrm{sb}}A_{\mathrm{s}}$ or the general destruction of the sample when the concrete reinforcement is pulled out $\sigma_{\mathrm{su}}A_{\mathrm{s}}$ [1];

• coefficient of completeness of the diagram of stresses arising in the rod along the length of its laying in concrete from the external forces action at the limiting deformation stage $\alpha = \int_0^l \sigma_{\mathrm{s}}(x) \cdot dx / (\sigma_{\mathrm{y}} \cdot l)$ [7];

• maximum (peak) tangential bond strain at critical displacement of reinforcement relative to concrete s_1 on the diagram $\tau_{\mathrm{b}} - s$ [2, 8].

Although each of the above methods has its own disadvantages, one disadvantage is common for all of them – the limited use of the above criteria for coupling only for the limiting stage of deformation of reinforced concrete elements.

A noticeable extension of modern studies of reinforcement bond with concrete is largely associated with the crack resistance of reinforced concrete elements in general and with the cracks formation levels in them in particular [4, 9]. In such circumstances, there is a need to estimate the reinforcement bond with concrete not only in the limit, but also in the operational stages of reinforced concrete elements deformation. The overwhelming majority of researchers propose to solve a similar problem with the help of the diagram "tangential bond stresses – displacement of reinforcement relative to concrete" [4, 5]. However, the description of this diagram as the only analytical function remains rather problematic.

The attempts by individual researchers to estimate the reinforcement bond with concrete by average tangential bond stresses along the surface of the reinforcing rod at any stage of reinforced concrete element deformation are also known [9]. However, the linear function proposed for the calculation of these stresses causes a series of warnings:

• in the absence of normal stresses in the reinforcing rod ($\sigma_{\mathrm{si}} = 0$), the average tangential bond stresses does not equal zero ($\tau_{\mathrm{bmi}} \neq 0$);

• dependence cannot describe the bonding processes to the crack formation and after the shift of the reinforcing rod unloaded end;

• the function is purely empirical and cannot reflect the essence of the physical patterns of reinforcement bond with concrete at intermediate stages of reinforced concrete element deformation.

3 Issues not solved within the common problem

In the theory and practice of calculation of the reinforced concrete elements the reinforcement bond with concrete is proposed to be evaluated by complex mathematical dependencies $\tau_{\mathrm{b}} - s$, the determining parameters of which, however, need to be installed experimentally. However, in most cases, neither type of fittings, nor geometric characteristics of its section, nor the type of concrete, etc., are taken into account. In addition, the models, theories and laws, methods and techniques for assessing the reinforcement bond with concrete proposed by the researchers, differ from each other, not only functionally, but formally. This is due to both subjective and objective reasons, including the influence of a large number of powers, deformation, structural and technological factors on the patterns of reinforcement bond with concrete. In other words, the evaluation of reinforcement bond with concrete due to the inevitable integration of dependence $\tau_b - s$ along the laying area of reinforcing rod remains a rather difficult task even with the use of special software.

The linear function of the reinforcement bond with concrete average stresses, proposed in [9], due to the above-mentioned disadvantages, can not be recommended for engineering calculations. It is obvious that only a nonlinear function of the reinforcement bond with concrete average stresses can cover all the stages of reinforced concrete element deformation.

4 Purpose and objectives of the study

The purpose of the study is developing an engineering method for estimating the reinforcement bond with concrete at any stage of reinforced concrete elements deformation. Of all the above-mentioned methods for estimating the reinforcement bond with concrete, the advantage is given by the average tangential bond stresses along the conditional cylindrical surface of the reinforcing rod. The main task of the study is obtaining a universal nonlinear dependence of these stresses from normal stresses in the reinforcement when squeezing it out of concrete.

5 Results and discussion

Each of the aforementioned methods for estimating the reinforcement bond with concrete has its own disadvantages, by which they mainly differ. However, the common to these methods is that they are all one way or another aimed at determining the effort of reinforcement bond with concrete at a certain stage of reinforced concrete element deformation.

Undoubtedly, the most simple bond effort can be determined at any stage of deformation, if the function of the average tangential stresses of reinforcement bond with concrete is known. To establish its character we consider the change scheme and the features of the stressed state of the concrete contact layer and reinforcing rod when it is removed from concrete (Fig. 1).

I – conditionally elastic stage. With a little effort, the concrete and reinforcement are deformed almost elastically: the diagram of normal stresses in the rod is close to the triangular, the maximum tangential bond stresses τ_{bmax^2} arise near the loaded end of the concrete sample, and their average values can be assumed to be approximately equal $\tau_{\mathrm{bm}^2} = \tau_{\mathrm{bmax}^2} / 2$.

II – elastic-plastic stage. The appearance of noticeable plastic deformations in concrete leads to distortion the reinforcement normal stresses pattern, the maximum tangential bond stresses τ_{bmaxI^2} remain close to the loaded end of the concrete sample, but the completeness coefficient of the diagram of the specified stresses $\tau_{\mathrm{bm}^2} / \tau_{\mathrm{bmax}^2\mathrm{I}}$ begins to grow.

III – offset stage. At significant displacements of the rod loaded end may occur not only plastic concrete deformation under its projections, but also partial concrete destruction in the contact layer. Maximum tangential bond stresses τ_{bmaxII^2} gradually shifted to the concrete sample unloaded end with constant growth of the diagram completeness coefficient of the specified stresses $\tau_{bm^2} / \tau_{bmax^2II}$.

IV – destruction stage. The slightest increase in effort is accompanied by the slipping of the reinforcing rod throughout its length.

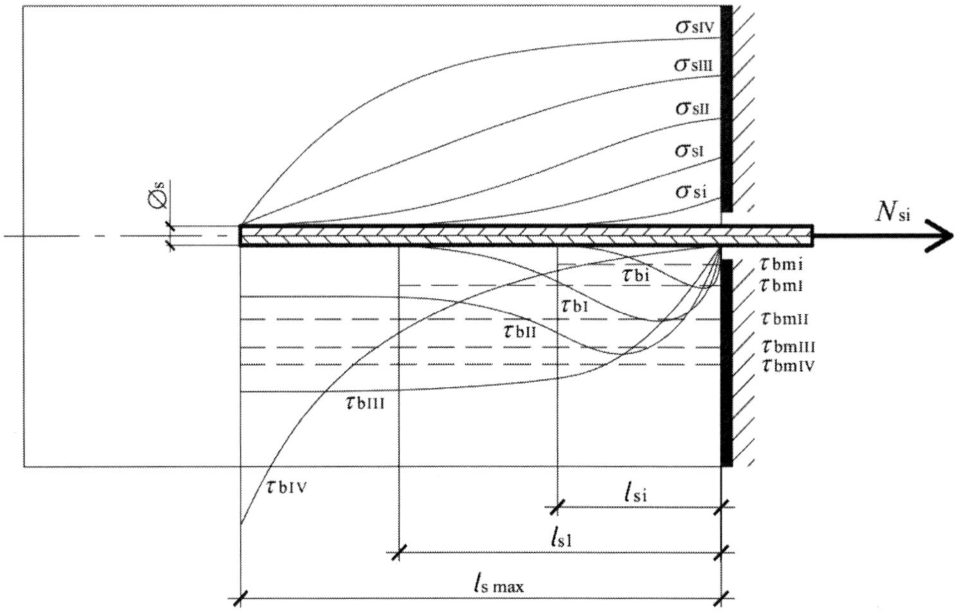

Fig. 1. Scheme of stressed state change of the concrete contact layer and reinforcing rod when it is removed from concrete.

The considered scheme is typical for a case where the reinforcement deformations remain elastic to the destruction of the concrete sample without the appearance of cracks in it. Deformation of reinforced concrete elements with cracks (with stretched concrete) takes place under a slightly different scheme. For them, the main is the second multilevel stage of cracks formation with a certain approach to the third stage only at the last level of their development.

Taking into account the above, the average tangential stresses of reinforcement bond with concrete at intermediate stages τ_{bm} will be described similarly to the limit stage f_{bd} according to the known formula

$$\tau_{bm} = \sigma_s \cdot \varnothing_s / (4 \cdot l_i) , \tag{1}$$

where σ_s – the greatest stresses in the reinforcement on the area of its bond to concrete; $\varnothing_s$ – diameter of the reinforcing rod; l_i – area of reinforcement active bond with concrete.

It is easy to notice that the level of the average tangential bond stresses

$$\tau_{bm} / f_{bd} = \sigma_s / \sigma_y \cdot l_s / l_i , \tag{2}$$

depends not only on the level of the boundary stresses in the reinforcement, but also on the relative area of active bond, which, by the results of processing of experimental data [5, 7, 8, 10-14], is recommended to calculate the expression

$$l_i / l_s \approx (\sigma_s / \sigma_y)^{1/\eta_s} , \tag{3}$$

where σ_y – extreme stresses in reinforcement ($\sigma_y = f_y$); l_s – area of reinforcement active bond with concrete at the limit stage; $1/\eta_s$ – bond intensity parameter.

According to the results of processing the experimental data, shown in Fig. 2 and Fig. 3, coefficient η_s is recommended to take equal: η_1 – for fittings of periodic profile and $6 \cdot \eta_1$ – for fittings of smooth profile.

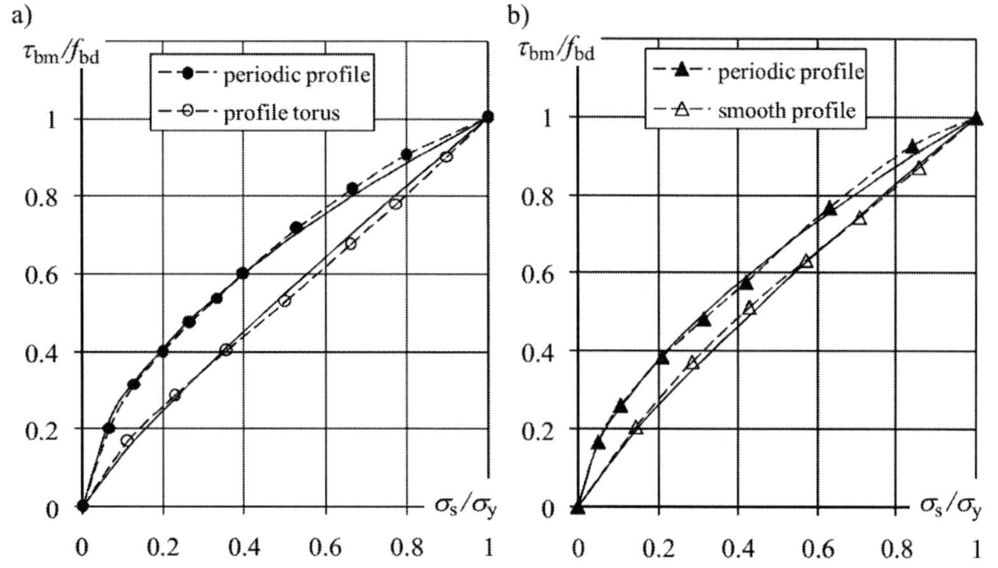

Fig. 2. Theoretical (——) and experimental (– – –) dependencies of average tangential bond stresses from normal stresses in reinforcement for experiments: (a) – Amstutz E. [11]; (b) – Bernander K. [12].

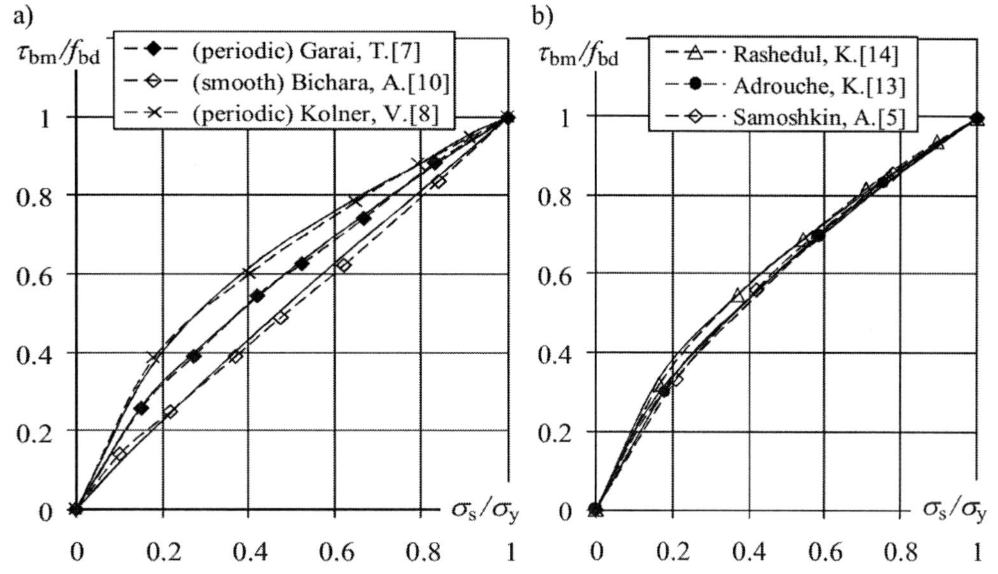

Fig. 3. Theoretical (——) and experimental (– – –) dependencies of average tangential bond stresses from normal stresses in the reinforcement: (a) - different profiles; (b) - periodic profile.

Taking into account the limit values of average bond stresses

$$f_{bd} = \eta_1 \cdot \eta_2 \cdot f_{\text{ñtk}} \tag{4}$$

the universal dependence of the average tangential stresses of the reinforcement bond with concrete will take the final form

$$\tau_{\text{bm}} = \eta_1 \cdot \eta_2 \cdot f_{\text{ñtk}} \cdot (\sigma_s / \sigma_y)^{1-1/\eta_s}, \qquad (5)$$

where η_2 – coefficient taking into account the influence of the reinforcement diameter [3].

The statistical estimation of the obtained dependence (5) on the evaluation of the reinforcement bond with concrete, shown in Table1, confirms the lawfulness of its use in practical calculations.

Table 1. Comparison of theoretical and experimental values of the average tangential stresses of reinforcement bond with concrete

The authors of the research	Year	Profile and diameter of fittings, mm	Deviation from experimental data		
			Δ	σ	v, %
Bichara, A. [10]	1951	smooth, 39	1.01	5.97	5.94
Amstutz, E. [11]	1955	torus, 30	1.00	4.71	4.71
		periodic, 30	0.99	3.22	3.25
Bernander, K. [12]	1957	smooth, 15	1.02	2.59	2.55
		periodic, 16	1.01	2.26	2.25
Garai, T. [7]	1959	periodic, 20	0.99	1.34	1.35
Kolner, V. [8]	1965	periodic, 20	1.00	1.73	1.73
Adrouche, K. [13]	1987	periodic, 16	0.995	2.00	2.01
Rashedul, K . [14]	2014	periodic, 20	0.994	2.29	2.30
Samoshkin, A. [5]	2017	periodic, 16	0.998	2.28	2.29
Notes: Δ and σ - arithmetic mean and standard deviations between experimental and calculated values, v - variance coefficient of deviations					

6 Conclusions

Proceeding from the foregoing, the estimating of the reinforcement bond with concrete in practical calculations is recommended to be performed with the help of a nonlinear function of the average tangential bond stresses. This will allow not only to clarify, but also to significantly simplify the calculation of the reinforced concrete elements crack strength [15, 16], even taking into account the levels of crack formation.

References

1. T.I. Astrova, *Treschinostoykost i deformativnost obyichnyih i predvaritelno napryazhennyih zhelezobetonnyih konstruktsiy*, 223-270 (Stroyizdat, Moscow, 1965)
2. M.M. Holmyanskiy, *Kontakt armaturyi s betonom* (Stroyizdat, Moscow, 1981)
3. *EN 1992-1-1, Eurocode 2: Design of Concrete Structures* (CEN, Brussels, 2004)
4. V.I. Kolchunov and etc., Resursoekonomni materialy, konstruktsii, budivli ta sporudy: zb. nauk. prats, **32**, 183-196 (2016)
5. A.S. Samoshkin, V.M. Tihomirov, Vyichislitelnyie tehnologii, **22(1)**, 75-86 (2017)
6. D.A. Abrams, PROC ASTM, **25**, 2-10 (1925)

7. T. Garai, Issledovanie prochnosti elementov zhelezobetonnyih konstruktsiy: sbornik trudov NIIZhB, **5**, 78-109 (1959)
8. V.M. Kolner and etc., Beton i zhelezobeton, **11**, 15–19 (1965)
9. D.V. Kochkarov, Visnyk NUVHP (tekh. nauky), **1(65)**, 176-185 (2014)
10. A. Bichara, *Cahier du Centre Scientifique et Technique du Batiment*, 117-127 (C.S.T.B, Paris, 1951)
11. E. Amstutz, BAUINGENIEUR-GERMANY, **30(10)**, 353-359 (1955)
12. K.G. Bernander, RILEM Symposium on Bond and Crack Formation in Reinforced Concrete, **1**, 203-210 (1957)
13. K. Adrouche, M. Lorrain, *MATER STRUCT, RILEM*, **20**, 315-320 (1987)
14. K. Rashedul, I. Mashfiqul, INT J CIV STRUCT ENG, **5(1)**, 80-90 (2014)
15. V.M. Romashko, *Deformatsiino-sylova model oporu betonu ta zalizobetonu* (O. Zen, Rivne, 2016)
16. V. Romashko, O. Romashko, *MATEC WEB CONF*, **116**, 02028 (2017)

Calculation of the crack resistance of reinforced concrete elements with allowance for the levels of normal crack formation

Vasyl Romashko[1,*], and *Olena* Romashko[1]

[1]National University of Water Management and Nature Recourses Use, Architectural Designing Bases, Constructing and Graphics Department, Soborna str. 11, 33028 Rivne, Ukraine

Abstract. The separate experimental investigations results and the major drawbacks of existing methods of calculating the reinforced concrete elements crack resistance are considered in detail, taking into account the process of multilevel formation and crack opening. Based on the basic provisions of the deformation and strength model of concrete and reinforced concrete resistance, an engineering method for calculating the formation and disclosure of normal cracks in the reinforced concrete elements and structures at the operational stages of their deformation is proposed. It is implemented on the system of generally accepted static, geometric and physical relations of a deformed solid mechanics. The simplicity of the proposed calculation method is due to the use of the nonlinear function of the average stresses of the reinforcement bond with concrete on the sections of the reinforced concrete element between adjacent cracks. The main advantages of the developed calculation method in comparison with others are outlined.

1 Introduction

The issues of crack resistance of reinforced concrete elements in the theory of concrete and reinforced concrete have always been and still remain one of the defining. It is connected, first of all, with the need to assess the actual stress-strain state of these elements in the operational stages. But describing this state is extremely difficult, since the length of the individual blocks, into which the reinforced concrete element is successively divided during the cracks formation, is variable and depends on many factors.

Since the process of crack formation and detection in reinforced concrete leads to violation of its continuity, the evaluation of the stress strain state of reinforced concrete elements tends to be associated with the basic provisions of the destruction mechanics. At the same time, the fact that the task of calculating distances between cracks in principle cannot have an exact analytical solution due to initial shrinkage micro-breaks of concrete continuity on the surface of its stretched zone is often neglected. Therefore, engineering methods for calculating the crack resistance of reinforced concrete elements, based on the general laws of changing the distances between cracks and reinforcement deformations and

* Corresponding author: romashkovasyl@gmail.com

the resistance of the concrete displacement in the areas between cracks, will remain relevant in the future.

2 Analysis of recent studies and publications

A rather detailed classification and a thorough analysis of numerous investigations of the crack resistance of reinforced concrete elements are given in [1]. It should be noted that the vast majority of calculation methods based on the results of the above studies were developed for the stage of stabilized cracking of reinforced concrete elements. As for the methods of calculating the crack resistance of elements, taking into account the multilevel formation of cracks, then, despite the accumulation of experimental research results of [2-9], they are very limited and practically are only beginning to form. By their orientation, these techniques can be classified as based on:

• models of a "two-console" element in a section with a crack [4];
• solutions strictly within the framework of brittle fracture mechanics [9];
• models of stretched concrete resistance to reinforcement displacement in sections between adjacent cracks [7, 10, 11].

In general, a "two-console" element in a section with a crack is used to translate rather complicated dependencies of fracture mechanics into less complicated dependencies of a mechanics of deformed solid [4]. But the main disadvantage of this method is that the correct allocation of this "two-console" element is a rather complex task, which, in turn, should be carefully connected not only with the stress and strain state of the reinforced concrete element cross-section, but also with the laws of reinforcement bond with concrete.

Solutions proposed strictly within the fracture mechanics framework [9] are so complex, that today they can be implemented only software, in particular, using the finite element method. At the same time, the physical essence of the actual deformation process of reinforced concrete elements is often lost or lowed.

With regard to the calculation methods based on the models of the stretched concrete resistance to reinforcement displacement in the areas between adjacent cracks, then they can be divided into three groups.

In the methods of the first group, the concrete resistance to the reinforcement displacement is considered from the standpoint of the elastic-plastic material with strengthening [7] and is based on the use of a number of significant simplifications and empirical parameters and coefficients. Because of this, these techniques very closely reflect the processes of reinforced concrete elements crack resistance in the real conditions of their deformation.

The methods of the second group include the stepwise numerical integration of the differential equation of the composite transfer relative to the reinforcing rod [10] using a number of empirical parameters and coefficients. This is due to the direct integration impossibility of the spline function of the concrete and reinforcement mutual displacement over the entire range of reinforced concrete element deformation due to its lump approximation.

One of the simplest methods is the third group, by which the concrete resistance to reinforcement displacement is determined by the average stresses of reinforcement bond with concrete in the area between adjacent cracks [11]. However, there are a number of reservations about their methodological justification in the following issues:

• determination of the true curvature and stiffness of the reinforced concrete element in cross sections with cracks;
• evaluation of the element state according to the materials state in separate sections, on the one hand, and the average stresses of reinforcement bond with concrete in the area between adjacent cracks, on the other side;

• the linearity of the medium stresses of the reinforcement bond with concrete over the entire range of reinforced concrete element deformation.

3 Issues not solved within the common problem

The results of the above analysis allow to note: none of the above-mentioned techniques can be implemented in the engineering non-program version. The universal method for calculating the reinforced concrete elements strength, including those taking into account the levels of normal cracks formation, must be equally successful in both engineering and program execution.

4 Purpose and objectives of research

The purpose of the study is developing an engineering method for calculating the formation and disclosure of discrete cracks at any stage reinforced concrete elements of deformation. In this article two major tasks are solved:

• ensuring the universality of the method with the help of the basic provisions of the deformation and strength model of reinforced concrete resistance [1, 12];

• to ensure maximum simplicity of the method due to the use of nonlinear function of medium bond stresses of reinforcement with concrete on element sections between adjacent cracks.

5 Results and discussion

According to the deformation and force model [1], the appearance of normal first-level cracks in a reinforced concrete element is calculated according with two generally accepted equilibrium equations with boundary deformations of stretched concrete $\varepsilon_{\tilde{n}tu}$, which are fixed by the extreme criterion of its strength $dN_{ct} / d\varepsilon_{ct} = 0$ or are accepted equal $\varepsilon_{\tilde{n}tu} = 2 f_{ctk} / E_{\tilde{n}o}$. In this case, the distance between cracks is calculated from the equilibrium condition of maximum efforts in stretched concrete $N_{ct,cr} = f(\varepsilon_{\tilde{n}tu})$ and the efforts of active bond on the area between cracks $N_{bd,cr}$.

Strength of reinforced concrete active bond with stretched concrete, due to the known function of the average bond stresses τ_{bmi}, can be determined in the cracks formation of the corresponding level by the formula

$$N_{bd,cr} = \int_{s_r} u \cdot \tau_b(z) \cdot dz \cong u \cdot \tau_{bmi} \cdot s_{ri}, \tag{1}$$

where u – cross-section perimeter of the reinforcing rod in diameter $\varnothing_s$, which has an area A_s, equal $u = 4 \cdot A_s / \varnothing_s$; s_{ri} – the area of the reinforcing rod active bond with concrete (distance) between the cracks of the corresponding level; $\tau_{bm\,i}$ – the value of the average stresses of reinforcement bond with concrete in the area between cracks of the corresponding level.

By the processing results of numerous experimental data [13-20] they are proposed to calculate by the following degree expression

$$\tau_{bmi} = \eta_1 \cdot \eta_2 \cdot f_{\tilde{n}tk} \cdot (\sigma_{si} / \sigma_y)^{1-1/\eta_s}, \tag{2}$$

where η_1 –coefficient taking into account the profile of the reinforcing rod [1] by the Rehm's criterion [21]; η_2 – coefficient taking into account the influence of the rod diameter [22]; $f_{\tilde{n}tk}$ – characteristic value of concrete strength on stretching; σ_{si} – tension in the rod in the section with a crack of the corresponding level; σ_y – extreme stresses in rods ($\sigma_y = f_y$); $1/\eta_s$ – bond intensity parameter (for periodic profile rods – $\eta_s = \eta_1$, for smooth profile rods – $\eta_s = 6 \cdot \eta_1$).

Thus, taking into account all of the foregoing provisions, the distance between adjacent first-level cracks at the stretched rod gravity center, with stresses in it $\sigma_{si} = \varepsilon_{ctu} \cdot E_s$, can be determined by the following dependence

$$s_{r1} = \frac{\varnothing_s}{4 \cdot \eta_1 \cdot \eta_2 \cdot f_{ctk} \cdot (\varepsilon_{ctu} \cdot E_s / \sigma_y)^{1-1/\eta_s}} \cdot \frac{N_{ct,cr}}{A_s}. \qquad (3)$$

Since the average bond stresses in the cross sections between the cracks are different due to the direct connection with the size of the normal tensions in the reinforcement, then the distance between the normal cracks of the new and previous levels in the bending elements (Fig.1) will always be differentiated among themselves. They can be found in terms of expressions

$$s_{ri+1,j} = \frac{s_{ri,j} \cdot \tau_{bmi+1,j+1}}{\tau_{bmi+1,j} + \tau_{bmi+1,j+1}}; \; s_{ri+1,j+1} = \frac{s_{ri,j} \cdot \tau_{bmi+1,j}}{\tau_{bmi+1,j} + \tau_{bmi+1,j+1}}, \qquad (4)$$

where j – the crack number of the corresponding level of their formation – i .

If, however, neglected the difference in the rods stresses in the cross sections with adjacent cracks, then further calculations, including the cracks opening width, can be substantially simplified by taking $s_{ri+1} = s_{ri} / 2$.

The normal cracks opening calculation is expedient to perform in accordance with Thomas's hypothesis [23] from the positions of successive accumulation of reinforcement and concrete mutual displacements $w_k = 2 \int_0^{0.5 s_r} \varepsilon_\tau(z)\, dz$ on active bond areas located on both sides of the crack. And because the dependence of mutual reinforcement and concrete displacement $\varepsilon_\tau(z) = \varepsilon_s(z) - \varepsilon_{ct}(z)$ is rather complex and can not be described by a single function, then the cracks opening width, taking into account the foregoing, is proposed to be calculated according to the following formula

$$w_k = s_{r1} \cdot (\sum_{i=1}^{n} \varepsilon_{sm,i} - \varepsilon_{ctm}) - s_{ri,j} \cdot (\sum_{i=2}^{n} \varepsilon_{sm,i} - \varepsilon_{ctm}) - \ldots - s_{ri,j} \cdot (\sum_{i=n}^{n} \varepsilon_{sm,i} - \varepsilon_{ctm}), \qquad (5)$$

where $\varepsilon_{sm,i}$ – mean values of relative strains of stretched armature in the area between cracks of a certain level; ε_{ctm} – mean values of relative deformations of stretched concrete on the same site.

In general, the average reinforcement deformation in the area between cracks of a certain level should be determined directly from the diagram of the element state (by its curvature) in the averaged section of the block between the cracks $\varepsilon_{sm,i} = f(1/r)$. As for the average deformations of stretched concrete in the corresponding section, it is recommended to calculate them by the formula

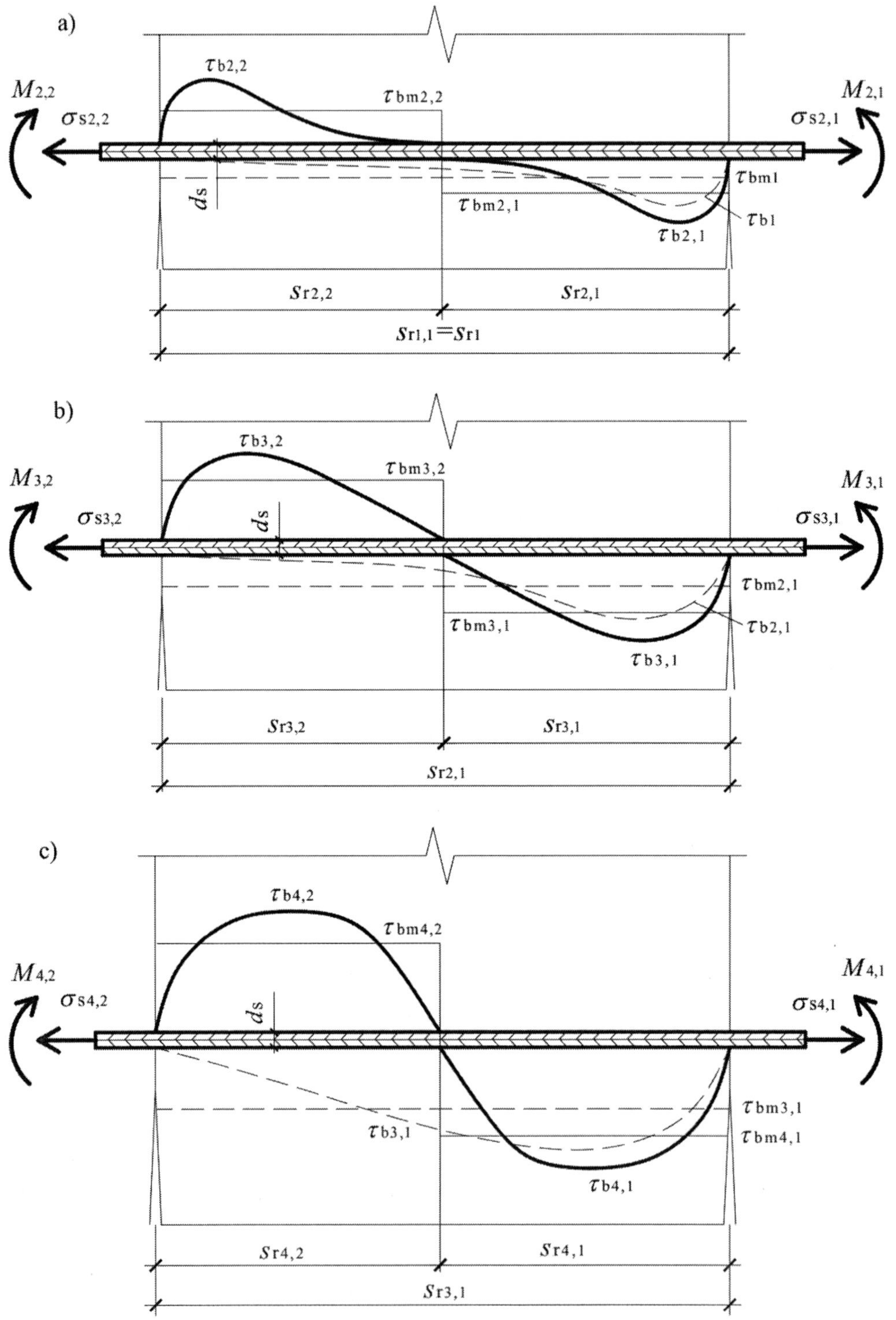

Fig. 1. Scheme of tangential stresses of reinforcement bond with concrete after the formation of cracks of the first (a), second (b) and third (c) levels.

$$\varepsilon_{\text{ctm}} = 0.5 \cdot \varepsilon_{\text{ctu}} \cdot (d - x) / (h - x), \tag{6}$$

where d – working height of the element cross-section; h – section height of the element of a rectangular profile; x – compressed zone height in the cross section of the element.

Since in the deformation-power model [1, 12], the value of the average reinforcement deformations in the sections between cracks is controlled by the diagrams of the reinforced concrete elements state throughout the process of their deformation, this allows to link the normal cracks opening width with the determining parameters of the stress and strain state of these elements by a system of generally accepted static, geometric and physical correlations, or equations of mechanics of a deformed solid state:

- $w_{\text{k}} = s_{\text{r1}} \cdot (\sum_{\text{i=1}}^{\text{n}} \varepsilon_{\text{sm,i}} - \varepsilon_{\text{ctm}}) - s_{\text{ri,j}} \cdot (\sum_{\text{i=2}}^{\text{n}} \varepsilon_{\text{sm,i}} - \varepsilon_{\text{ctm}}) - \ldots - s_{\text{ri,j}} \cdot (\sum_{\text{i=n}}^{\text{n}} \varepsilon_{\text{sm,i}} - \varepsilon_{\text{ctm}})$;
- static relations $\grave{I}_{з} = f(\varepsilon_{\tilde{n},з}, \varepsilon_{\tilde{n}t,з}, \varepsilon_{s,з})$, $N = f(\varepsilon_{\tilde{n},з}, \varepsilon_{\tilde{n}t,з}, \varepsilon_{s,з})$;
- geometrical relations with the hypothesis of plane cross sections $1/r_{з} = f(\varepsilon_{\tilde{n},з}, \varepsilon_{s,з})$ or $\varepsilon_{s,з} = f(1/r_{з})$;
- physical relations: the elements state $1/r_{з} = f(\grave{I}_{з})$ and materials state $\sigma_{\tilde{n},з} = f(\varepsilon_{\tilde{n},з})$, $\sigma_{\tilde{n}t,з} = f(\varepsilon_{\tilde{n}t,з})$, $\sigma_{s,з} = f(\varepsilon_{s,з})$.

(7)

6 Conclusions

Thus, on the basis of the above, it should be noted that the developed model of the step formation and the disclosure of normal cracks:

- complements the generalized model of reinforced concrete elements deformation;
- is universal, because its basis is as much as possible deprived of the empiricism influence and is acceptable for any reinforced concrete elements undergoing heterogeneous deformation;
- takes into account the qualitative and quantitative difference between the parameters of reinforcement bond with concrete and examines the distance between adjacent cracks as a value that varies discretely throughout the deformation process;
- reflects the dependence of the average tangential stresses of reinforcement with expanded concrete on a number of factors, including the type and profile of the reinforcement itself;
- directly correlates the reinforced concrete elements crack strength with the main parameters of their stress and strain state (curvilinear values $1/r_{\text{i}}$, bending moments values M_{i}, stresses in the reinforcement $\sigma_{\text{s,i}}$, its deformations $\varepsilon_{\text{s,i}}$ etc.);
- allows control the process of the formation and disclosure of normal cracks in reinforced concrete elements at any stage of their deformation from the positions of successive accumulation of mutual displacements of the reinforcement relative to the concrete in the active bond areas.

References

1. V.M. Romashko, *Deformatsiino-sylova model oporu betonu ta zalizobetonu* (O. Zen, Rivne, 2016)
2. G.A. Molodchenko, Stroitelnyie konstruktsii, **XIX**, 80-84 (1972)

3. Sh.A. Hakimov, *Vozdeystvie staticheskih, dinamicheskih i mnogokratno povtoryayuschihsya nagruzok na beton i elementyi zhelezobetonnyih konstruktsiy*, 65-86 (Stroyizdat, Moscow, 1972)
4. V.M. Bondarenko, V.I. Kolchunov, *Raschetnyie modeli silovogo soprotivleniya zhelezobetona* (ASV, St. Petersburg , Moscow, 2004)
5. V.I. Kolchunov, N.E. Masud, Budivnytstvo Ukrainy, **3**, 36-38 (2006)
6. C.B. Peña, *Serviceability behaviour of fibre reinforced polymer reinforced concrete beams* (Thesis of Doct. PhD, Girona, 2010)
7. V.I. Mayorov, P.K. Kuzmin, Stroitelnaya mehanika inzhenernyih konstruktsiy i sooruzheniy, **2**, 22-28 (2011)
8. H.Z. Bashirov, *Zhelezobetonnyie sostavnyie konstruktsii transportnyih zdaniy i sooruzheniy* (Avtoref. Dis. Dokt. Tehn. Nauk, Moscow, 2013)
9. I.N. Shardakov, Frattura ed Integrità Strutturale, **3**, 339-350, (2016)
10. M. Alvarez, *Einfluss des Verbundverhaltens auf das Verformungsvermögen von Stahlbeton* (Abhandlung des Titels Dokt. der Tech. Wissenschaften, Zürich, 1998)
11. D.V. Kochkarov, Resursoekonomni materialy, konstruktsii, budivli ta sporudy, **28**, 228-236 (2014)
12. V. Romashko, O. Romashko, *MATEC WEB CONF*, **116**, 02028, (2017)
13. A. Bichara, *Cahiere du Centre Scientifique et Technique du Batiment*, 117-127 (C.S.T.B, Paris, 1951)
14. E. Amstutz, BAUINGENIEUR-GERMANY, **30(10)**, 353-359 (1955)
15. K.G. Bernander, RILEM Symposium on Bond and Crack Formation in Reinforced Concrete, **1**, 203-210 (1957)
16. T. Garai, Issledovanie prochnosti elementov zhelezobetonnyih konstruktsiy: sb. tr. NIIZhB, **5**, 78-109 (1959)
17. V.M. Kolner and etc., Beton i zhelezobeton, **11**, 15–19 (1965)
18. K. Adrouche, M. Lorrain, MATER STRUCT, RILEM, **20**, 315-320 (1987)
19. K. Rashedul, I. Mashfiqul, INT J CIV STRUCT ENG, **5(1)**, 80-90 (2014)
20. A.S. Samoshkin, V.M. Tihomirov, Vyichislitelnyie tehnologii, **22(1)**, 75-86 (2017)
21. G. Rehm, Deutscher Ausschusses für Stahlbeton, **138,** 1-59 (1961)
22. *EN 1992-1-1, Eurocode 2: Design of Concrete Structures (*CEN, Brussels, 2004)
23. F.G. Thomas, STRUC ENG, **14(7)**, 298-320 (1936)

Influence of external and internal cooling at solidification on strength of brittle duralumin in compression

Oleksandr Semko[1], *Oleksii* Fenko[1], *Anton* Hasenko[1], *Olha* Harkava[1,*], and *Volodymyr* Kyrychenko[1]

[1]Poltava National Technical Yuri Kondratyuk University, Department of Reinforced Concrete and Masonry Structures and Strength of Materials, 24, Pershotravneva Avenue, Poltava, 36011, Ukraine

Abstract. The causes of the occurrence and influence of inherent stresses, unevenly distributed across the cross section, on the strength of materials are analyzed. Methods of creating artificial distribution of inherent stresses with opposite signs in comparison with the usual distribution obtained with natural solidification of the material are given. The sequence of an experiment on cylindrical specimens made of epoxy resin and brittle silicate aluminum alloy in creating distributions of inherent stresses with opposite signs is described. The results of experimental tests of the influence of inherent stresses unevenly distributed across the cross section, on the strength of the epoxy resin and of the fragile silicate aluminum alloy with compression are given. Quantitative characteristics of the influence of unevenly distributed inherent stresses on the strength of materials under compression are obtained. The reasons of the influence of inherent stresses on the strength of materials under compression are analyzed and grounded.

1 Introduction

The detection of places with the highest stresses in elements of the structures has always been a topical issue in optimizing the design schemes of buildings. The sequence of the destruction of samples under the action of longitudinal force is well known: destruction begins with the surface layers of the sample in compressive tests, and the inside layers break up first in the case of tensile tests [1]. Such a sequence of destruction can be explained by the presence of irregularly distributed inherent stresses in the cross-section.

1.1 Review of well-known publications and researches

Details of machines, mechanisms or other structural elements can be made of metal castings. Cooling of such castings starts from the outside during solidification under normal conditions. The temperature of the inside layers exceeds the temperature of the surface layers in this process [2]. After reaching the ambient temperature, the surface layers stop

* Corresponding author: author@email.org

shortening due to a decrease in temperature. Thus, they prevent the reduction of the inside layers, the temperature of which is higher. When the temperature is aligned in the cross section due to the final cooling of the casting, the surface layers will be compressed, and the inside will be tensioned. The stresses fluctuate smoothly from the surface of the sample to the core (the inside part), trying to deplan the section (they are called deplanation stresses). The above-mentioned deplanation stresses, in addition to the case during the cooling of metallic castings in their solidification, may occur in the concrete cross section in uneven shrinkage [3 – 5], which influences its work under biaxial bending [6, 7] and in reduction [8, 9]. The same picture is observed in hammered or rolled elements: the surface layers undergo an intense deformation, which causes the described distribution of inherent stresses after cooling. The deplanation stresses will increase with increasing sample sizes, (due to a greater difference in the temperature of the inside and surface layers of the casting), which must also affect the scale effect [10, 11].

1.2 Highlight a previously unresolved issue

During the loading of the sample along the axis by the compressive force, the stress field caused by mechanical loading will be imposed on the distribution of its inherent deplanation stresses obtained during solidification. Thus, the stresses in the surface layers and the stress caused by the mechanical load are added (since they are like-sign) when testing samples on compression. Therefore, the destruction in compression begins with the surface layers. At the same time, the stresses in the inside layers are smaller than in the surface (in the inside layers the stresses are opposite sign). The surface layers begin to collapse, after which the destruction spreads to the middle of the sample.

During the tensile test, the opposite effect occurs, which is also due to imposition of stress distributions (inherent and caused by external load). As with compression, the stresses will be added with only the difference, that now the inherent and the mechanical stresses in the core of the cross section are like-sign. The stresses in the surface layers will be subtracted, because the inherent and the mechanical stresses are opposite sign. Thus, the destruction begins with the surface layers of the sample and extends outward.

1.3 Research objective

The purpose of the study is to confirm experimentally the theoretical conclusions about the influence of the distributions of the inherent deplanation stresses caused by the conditions of solidification during cooling, on the strength of the samples under compression.

Achievement of the set goal is possible by creating an artificial distribution of the inherent (deplanation) stresses with opposite signs compared to the usual distribution obtained with the natural solidification of the material. When artificially creating a distribution of the deplanation stresses, reversed to the natural distribution, dangerous stresses will first appear in the core of the cross-section in compressing tests. However, the sample will better resist fracture due to the casing effect created by the less stressed surface layers. Thus, the effect of increasing the strength can be achieved in compression tests through the created field of inherent stresses. It should be noted that this effect could not be achieved in tensile tests. When creating an inverse natural distribution of the inherent stresses, the sequence of destruction of the sample will be from the surface layers to the core, and strength increase cannot be obtained.

2 Main material and results

2.1 Experiments on samples of epoxy resin

Experiments on samples of epoxy resin [1] were carried out for the preliminary confirmation of the conclusions made regarding the imposition of stress fields. Two series of samples were manufactured and tested for the purpose of the work (see Table 1). Each series of samples was consisted of two groups.

Table 1. Nomenclature of epoxy resin experimental samples.

<table>
<tr><th>Series sample number</th><th>Series name (sample material)</th><th>Purpose of research of series of samples</th><th>Sample group code</th><th colspan="2">Group name (the essence of the manufacturing)</th></tr>
<tr><td rowspan="4">1</td><td rowspan="4">Three-layer cylinders made of epoxy resin</td><td rowspan="4">Preliminary detection of the investigated effect</td><td>LISE</td><td rowspan="2">Samples with previous chain loading of layers during manufacturing:</td><td>from inside to outside</td></tr>
<tr><td>LSIE</td><td>from outside to inside</td></tr>
<tr><td>ISE</td><td rowspan="2">Control unloaded samples:</td><td>made from inside to outside</td></tr>
<tr><td>SIE</td><td>made from outside to inside</td></tr>
<tr><td rowspan="2">2</td><td rowspan="2">Castings from duralumin (brittle silicate aluminum alloy A-12)</td><td rowspan="2">Detection of the influence of internal and external cooling on compressive strength</td><td>ISD</td><td colspan="2">Samples of artificial solidification (cooling from the inside layers to the outside)</td></tr>
<tr><td>SID</td><td colspan="2">Samples of natural solidification (cooling of the outside layers to the inside)</td></tr>
</table>

Cylindrical specimens with a diameter of 30 mm and a height of 60 mm were made of epoxy resin in three stages. The sequence of the production of samples of LISE group (see Table 1, series 1) is from the inside layers to the outside ones, with the interval between the casting of the layers of 10 days (after the set of sufficient strength of layers). The first inside layer, which cured for 10 days, was loaded with the compressive force q_1 and covered around by the next layer (Fig. 1, a). After 10 days, both layers (inside and middle) were loaded with more force q_2. After that, the last surface layer was poured. Then the samples gained strength for 10 days, after which they were tested in compression.

Samples of the LSIE group (see Table 1, series 1) were made in the reverse order (Fig. 1, b). The surface layer was made the first. It was loaded by compressive force q_1 after 10 days and the middle layer was poured. Both layers were loaded with force q_2 after hardening and the last inside layer was poured.

The field of inherent stresses was created after unloading of the LISE samples, in which the inside part of the sample was compressed and the outside was tensioned. There was the field of inherent stresses with opposite signs in the samples of the LSIE: the surface layers were compressed and the inside were tensioned. Thus, with the same strength of the material of different layers, the effect of their inherent stresses on samples of LSIE is close to the distribution of stresses with natural hardening of casting, and in samples of LISE the effect is opposite to the natural (artificially created distribution). Therefore, a greater strength of LISE samples was expected. It should be noted that the strength of the layers was different due to different periods of their hardening in the manufacturing.

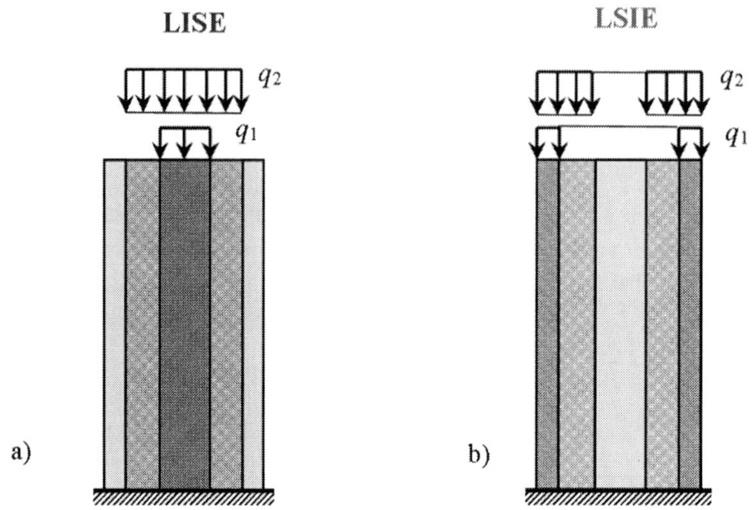

Fig. 1. Sequence of solidification and loading the epoxy resin specimens during manufacturing: a) LISE group (from inside to outside); b) LSIE group (from outside to inside).

In addition, for assessing the effect of the resin strength difference in layers ISE and SIE samples were made at the same time with LISE and LSIE, which were not loaded in the manufacture (see Table 1, Series 1). The results of the experiment are shown in Figure 2.

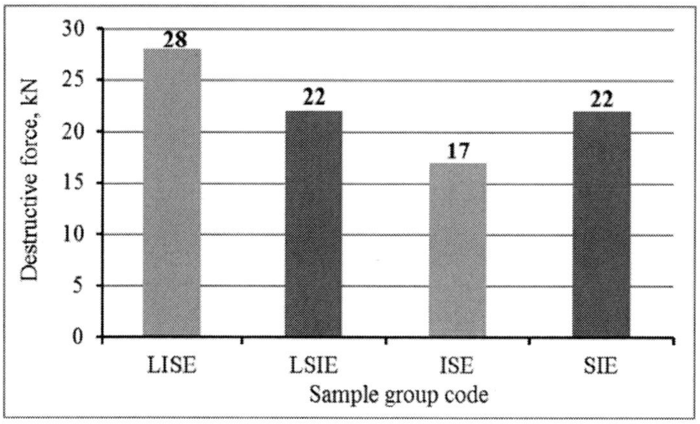

Fig. 2. Results of the first series samples tests.

As it can be seen from Fig. 2, the results of the experiment confirm the previous conclusions. LISE examples were stronger than LSIE, whereas ISE samples had less strength than SIE. This suggests that the effect of the inherent (deploating) stresses blocked the effect of the material strength difference in the layers of samples.

The disadvantage of the experiments with samples of the first series, made of epoxy resin, was that their own stresses were distributed gradually across the cross section. Some errors in the results of experiments were due to the epoxy resin was an aging material, and in different layers there were different terms of hardening of the resin due to the technology of making the samples. It is logical to conduct experiments on non-aging material to eliminate these shortcomings. At the same time, the samples should be manufactured at one casting.

2.2 Experiments on samples of duralumin

The planned second series of samples eliminates the disadvantages of the first series. The experiments on the determination of the influence on the strength of the unevenly distributed inherent stresses due to different cooling sequences were carried out on duralumin specimens (fragile silicate aluminium alloy A-12) under compression. The height of the samples is 60 mm, the diameter is 30 mm. Samples should be made with the artificial distribution of the inherent stresses (Fig. 3, b) of the opposite sign to the field of the inherent stresses that occurs in natural hardening (Fig. 3, a) to prove the influence of the inherent stresses on the strength of material. That is, the surface layers should be tensioned, and the internal ones should be compressed. This distribution of stresses can be achieved by a reciprocal cooling of the castings during their hardening, that is, starting from the inside.

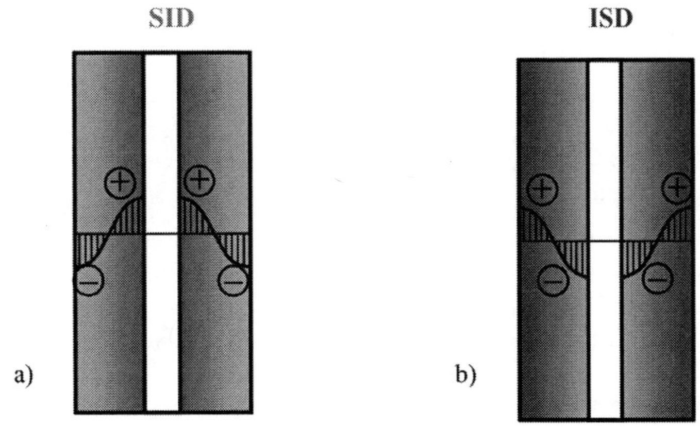

Fig. 3. Distribution of the inherent stresses when cooling the second series samples: a) starting from the surface layers (SID); b) from the inside layers (ISD).

The samples in each test lot were cast out of one fusion to prevent inaccuracies in the obtained results and increase the reliability. SID samples of "natural" hardening were cast in the steel moulding boxes and cooled from the surface layers. In the steel moulding boxes (of the same size as the previous ones), designed for the production of ISD samples, a steel tube with a diameter of 6 mm was mounted in the centre (see Fig. 4). The moulding boxes were in cylindrical asbestos thermal insulators. An electric heating helix was mounted inside of the heat insulators.

In order to ensure the objectivity of the results of the experiment, the succession of duralumin casting in each test lot was consistently changed: in some test lots, samples of natural hardening were casted first, in other test lots, on the contrary: the moulding boxes of the "opposite" stress field were poured first. The height of the castings was one and a half times bigger than the planned height of the samples to ensure the prevention of the possible ingress of oxidized metal into samples.

The order of samples production with the opposite artificial field of inherent stresses was as follows. An electric spiral and water supply were included before the filling of the duralumin. The moulding box was heated to the melting point of duralumin. When the duralumin was casted, the lid in the thermal insulation was opened for several seconds and the electric spiral was switched off. Water was fed until complete hardening and considerable cooling of the sample.

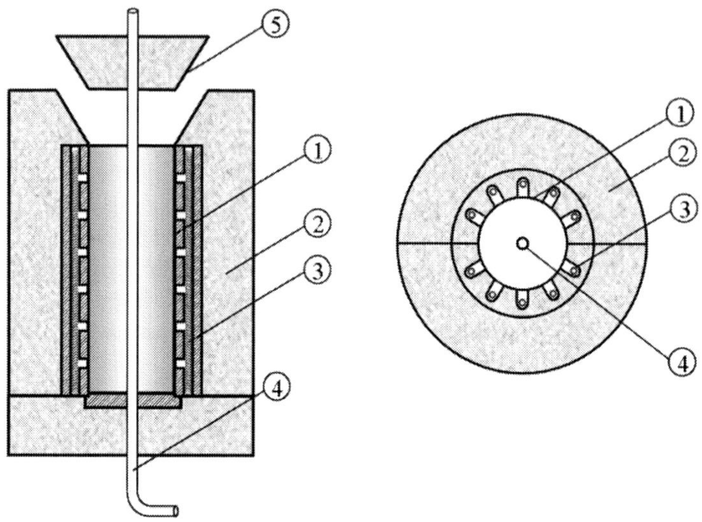

Fig. 4. Thermo-insulated moulding box for samples with "opposite" distribution of the inherent stresses: 1 - steel form; 2 - asbestos thermal insulation; 3 - electric spiral; 4 - water supply pipe; 5 - thermal insulation cover.

In order to ensure the objectivity of the results of the experiment, the succession of duralumin casting in each test lot was consistently changed: in some test lots, samples of natural hardening were casted first, in other test lots, on the contrary: the moulding boxes of the "opposite" stress field were poured first. The height of the castings was one and a half times bigger than the planned height of the samples to ensure the prevention of the possible ingress of oxidized metal into samples. The order of samples production with the opposite artificial field of inherent stresses was as follows. An electric spiral and water supply were included before the filling of the duralumin. The form was heated to the melting point of duralumin. When the duralumin was casted, the lid in the thermal insulation was opened for several seconds and the electric spiral was switched off. Water was fed until complete hardening and considerable cooling of the sample.

The effect of versatile cooling (internal and external) was detected before experimental testing: in the SID samples, the inside part of the specimen was concave inside the sample, and in the ISD samples it was convex outward (see Fig. 5). This indicates the accumulation of the inherent stresses in the casting material during its cooling.

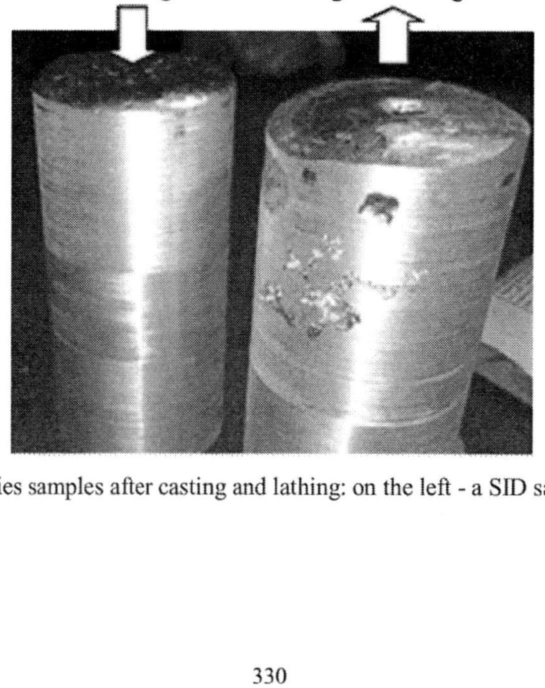

Fig. 5. The second series samples after casting and lathing: on the left - a SID sample, on the right – an ISD sample.

The samples were lathed to the planned sizes after cooling. The upper part (30 mm high) of the work piece was cut off. The opening diameter 6 mm was turned along the axis in the SID samples. Thus, all geometric characteristics were the same in the samples.

The nature of the destruction of samples completely confirmed the fragility of the tested duralumin during the tests (see Fig. 6). The results of the tests are shown in Figure 7. Comparison of the strength of the tested samples of the two groups are given in Table 2.

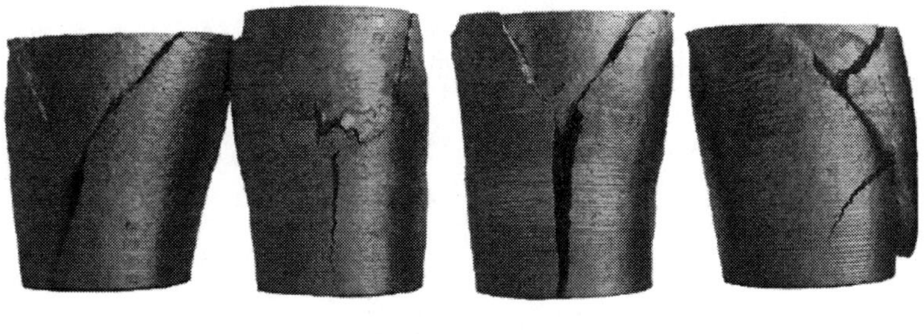

Fig. 6. General view of the second series samples after the test.

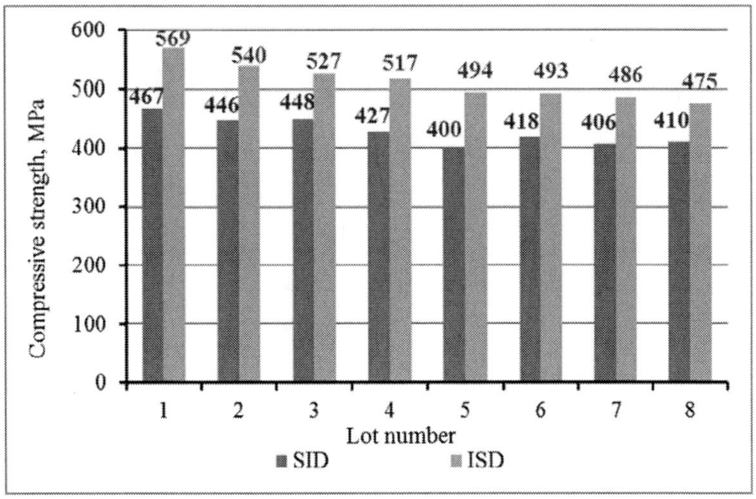

Fig. 7. Results of the second series samples tests.

Table 2. Comparison of duralumin specimens' strength with opposite signs of the inherent stresses distribution.

Casting lot number	Strength, MPa		Strength difference	
	SID	ISD	MPa	%
1	467	569	102	21.8
2	446	540	94	21.1
3	448	527	79	17.6
4	427	517	90	21.1
5	400	494	94	23.5
6	418	493	75	17.9
7	406	486	80	19.7
8	410	475	65	15.9

3 Conclusions

Experiments confirmed preliminary conclusions about the influence of the unevenly distributed across the cross section inherent stresses on the strength of the material. The difference in strength caused by opposite distributions of the inherent stresses due to different cooling sequences is about 20%. Thus, each field of the inherent stress leads to a change in strength by an average of 10%.

Some differences in the strength of samples in different castings can be explained by the fact that the samples after the tests were melting and used for the production of new samples, and the repeated melting led to some changes in the properties of duralumin.

References

1. YA.B. Fridman, *Mehanicheskie svoystva metallov. CHast 2. Mehanicheskie ispyitaniya. Konstruktsionnaya prochnost* (1972)
2. A.N. Krutilin, P.E. Luschik, A.A. Andrits, Lite i metallurgiya, **4**, 10 (2014)
3. O.G. Fenko, O.A. Krupchenko, I.A. Yurko, Acad. journal industrial machine building, civil eng., **2(47)**, 155 (2016)
4. X. Chen, W. Huang, J. Zhou, Indian Journal of Eng. and Materials Sc., **19**, 427 (2012)
5. H. Hotta, K. Takiguchi, Nucl. Eng. Des., **156(6)**, 218 (1995)
6. A.M. Pavlikov *Neliniina model napruzheno-deformovanoho stanu kosozavantazhenykh zalizobetonnykh elementiv u zakrytychnii stadii* (2007)
7. A.M. Pavlikov, D.V. Kochkarov, Naukovo-tekhnichnyi, vyrobnychyi ta informatsiino-analitychnyi zhurnal "Nauka ta budivnytstvo", **3**, 22 (2017)
8. D.A. Yermolenko, Stroytelstvo, materyalovedenye, mashynostroenye: Sb. nauch. Trudov, **61**, 172 (2011)
9. S.A. Hudz, Stalezalizobetonni konstruktsii: doslidzhennia, proektuvannia, budivnytstvo, ekspluatatsiia: Zb. nauk. Statei, **12**, 76 (2016)
10. O.G. Fenko, Zb. nauk. prats (haluzeve mashynobuduvannia, budivnytstvo), **2**(44), 86 (2015)
11. Z.P. Bazant, Jia-Liang Le, Sadhana (Indian Acad. of Sc.), **37**(1), 1731 (2012)

Estimation of the composition of the forced and free vibrations of the casing of the steel rope roof of Ukraine Cinema & Concert Hall located in Kharkov over the area of main line tunnel of working metropolitan railway

Valeriy Shmukler[1], [1], *Valeriy* Zhylyakov[1], and *Oleksandr* Shapovalov[1]

[1]O.M. Beketov National University of Urban Economy in Kharkiv, 17, Marshal Bazhanov Street, Khariv, 61002, Ukraine

Abstract. The paper discusses issues related to the dynamic parameters of the metro train influence on the building construction of Ukraine Cinema & Concert Hall complicated in terms of construction located in Kharkiv City. The hall roof is of peculiar complexity made in the shape of hyperbolic paraboloid hung to the cable roof system and supported by the inclining reinforced concrete arches. Using the equipment by DIAMEX 2000 LLC we have measured vertical and horizontal forced vibrations in the most relevant points of the casing. Natural vibrations were defined by creating dynamic impact of the falling load weighing 10 kg from a height og 2 m. The vibration parameter measurement was carried out by means of Ahat-M vibration analyzer. As a result we have obtained amplitude-frequency characteristics of the forced vibrations caused by metro train movements. The frequency of base and obertones is within 30-50 Hz. At the same time, vibration amplitudes is up 200-350 μm. Specific range of vibrations leads to the occurrence of the for framing roof framing (concrete and steel strand guy ropes), noise penetrating into the hall significantly worsens operating characteristics for the spectators. A special calculation of guy rope element taking into account fatigue factors has been made.

1 Structural Design

Determination of the dynamic load impact on the building constructions is of great importance for estimation of their reliable and safe service during the defined design period. At the same time, it is essential to define nature of the dynamic impact and main parameters of this load (natural and forced vibrations, frequency and amplitude of vibrations, risk of resonance etc.). There is a large number of studies in this direction and methods to calculate dynamic impacts on the building constructions [1-4].

Among these studies the works, a special place stands for the studies that are dedicated to analyse dynamic parameters of the movement of railway transport [5, 6], but there are a

[1] Corresponding author: bk@kname.edu.ua

few studies devoted to the impact of metro train traffic on buildings located over and near the tunnels of metro railway. In this case the dynamic load transfer depends on many parameters, in particular, on the thickness of soil layer over the metropolitan, composition and physical and mechanical properties of ground, running tunnel construction arrangement, railing construction etc.

In some cases, the impact of metro train traffic on the building above-ground significantly worsens the operational parameters of these buildings, especially, if this building is classified as a public building. Ukraine Cinema & Concert Hall located in Kharkiv (Ukraine) built in 1963 belongs precisely to such a construction. At that time there were no analogues of construct of such building in the world's practice. The authors of this project were V.S. Vasyliev, Yu.A. Plaksiiev, V.A. Reusov, design engineer L.B. Fridgan.

The building of Ukraine Cinema & Concert Hall is distinguished by the originality of construction solution – for the first time in the world's practice the asymmetric scheme of saddle-shaped roof was performed, the casing of 90x48 m in size was constructed which was closed off with reinforced concrete slabs that was the major achievement of the urban architecture of the second half of the 20th century.

Initially the building design provided the work of hall as a pavilion, however, later building envelope and all required attributes were added to ensure holding concerts and events all year through. In 1980 the building was placed under the protection of the government as a monument of architecture and city development of local significance, protection number 50.

The façade of Ukraine Cinema & Concert Hall is shown in Figure 1, the roof plan is shown in Figure 2.

Fig. 1. The front façade of Ukraine Cinema and Concert Hall.

The audience space is designed in the form of one whole amphitheatre with the overall dimensions in plan view of 48.0×45.0 m.

The roof is made in the form of casing outlined in regards to the surface of the hyperbolic paraboloid. The roof is fenced with two parabolic arches with arrows of lift of 31.275 m and 20.365 m. The angles of inclination of arches to the horizon is 12 (canopy arch) and 45 (steep arch) (Figure 3).

The scheme of the drawing in longitudinal section of the building is shown in Figure 4.

Structural scheme of the hall roof is a casing consisting of reinforced concrete slabs (28 mm in thickness) suspended to the system of stressed support and shrinking (stabilizing) strands and connected to these slabs of the reinforced concrete plate-casing of 30 mm in thickness. Combined action of the reinforced concrete slabs (initially serving as a timbering for the reinforced concrete plate) and the plate-casing itself is carried out by means of simultaneous embedding of strands using shotcreting process both in the deepening of reinforced concrete slabs and in the body of the reinforced concrete encasement.

Load due to the roof is taken up by the ferrocement contour of two parabolic arches and transmitted to two foundations. The foundations are located in the arches' crossing point.

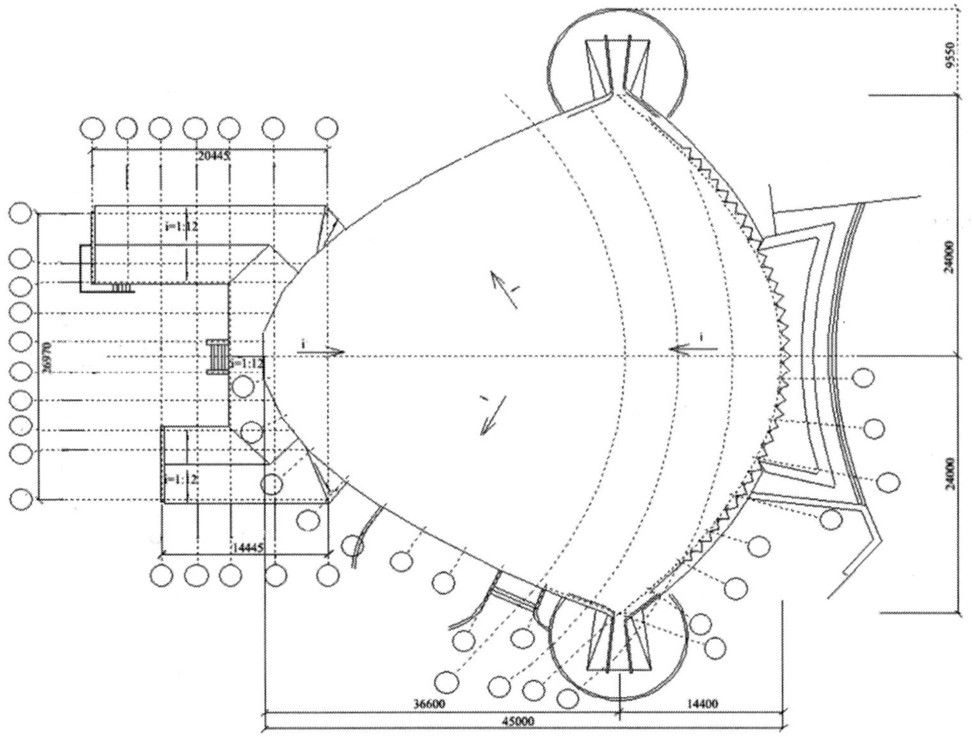

Fig. 2. The roof scheme of Ukraine Cinema and Concert Hall.

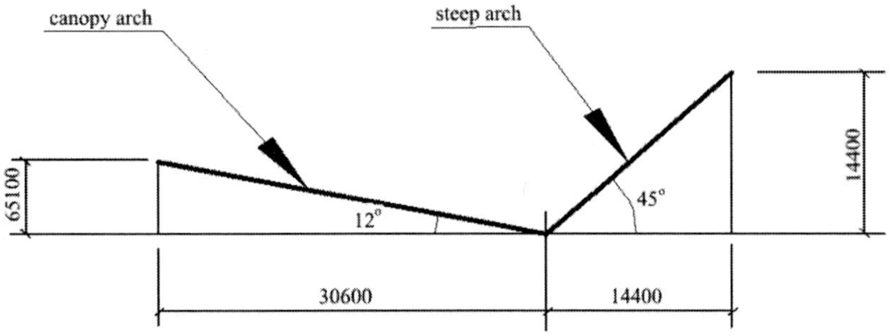

Fig. 3. Geometrical arrangement of the roof arches' incline.

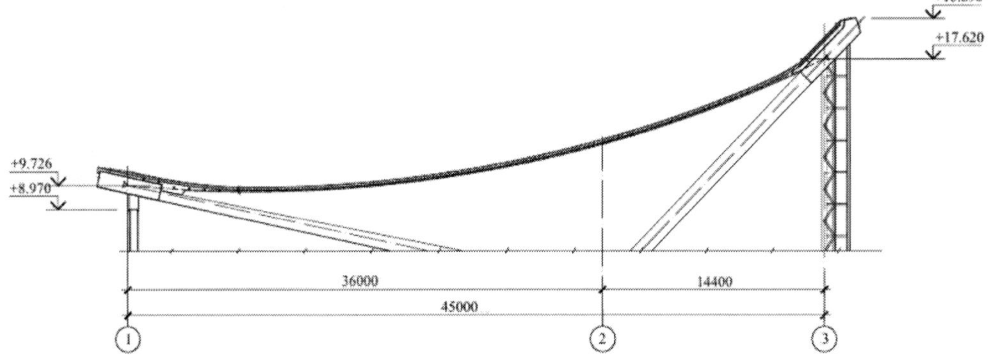

Fig. 4. Scheme of drawing in longitudinal section of the building.

2 Dynamical Characteristics

To define parameters of dynamic impacts on the building construction of Ukraine Cinema and Concert Hall we used equipment manufactured by DIAMEX 2000 and Ahat-M vibration analyser, serial number 1535.

The main purpose of these measures was to estimate amplitude and frequency components of the forced and free vibrations (absolute vibrations) of supporting elements of the structures of Ukraine Cinema and Concert Hall in the predetermined points. All works were performed in accordance with the standards of Ukraine [6-9].

Analysis of these measurements was carried out using DIAMANT 2 software, version 2.082.13. General layout of the points is shown in the Figure 5.

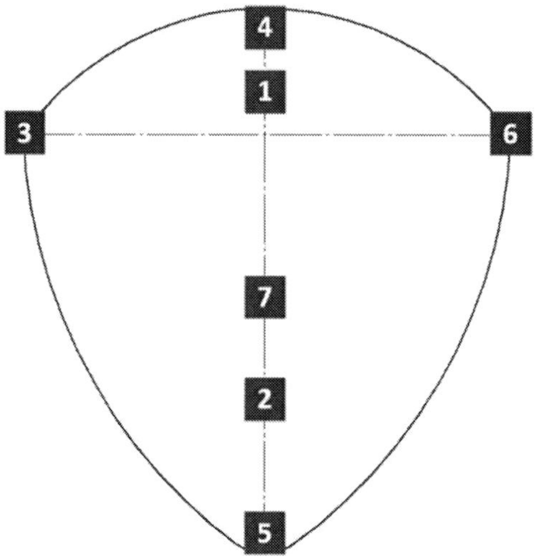

Fig. 5. Location of measurement points to define dynamic characteristics.

Where 1st and 2nd points mean location of control detectors on the supporting floor elements, in the middle part of the hall for spectators; the 3rd, 4th, 5th, 6th, and 7th are control points on the support elements of the building roof.

Vibration properties of the constructions were defined in horizontal and vertical directions. The Figure 6a, 6b shows frequencies and amplitudes of vibrations for the point 1, the Figures 7 and 8 show frequencies and amplitudes of vibrations for the points 4 and 6 correspondingly.

The analysis of the diagrams showed points to the fact that the frequencies of the forced vibrations in the horizontal and vertical directions in peak points differ by 20-40%, at the same time the frequencies of vibrations in horizontal direction exceed the frequency rate in the vertical direction.

The peak amplitude of the floor vibrations in Ukraine Cinema and Concert Hall is 2.52 μm, and the peak amplitude of the vibrations of roof is 0.65 μm.

The peak frequency of the floor vibration in Ukraine Cinema and Concert Hall is 71.69 Hz, and the peak frequency of the vibrations of roof is 59.76 Hz.

Dot №: 1 Direction of reconciliation: Vertical

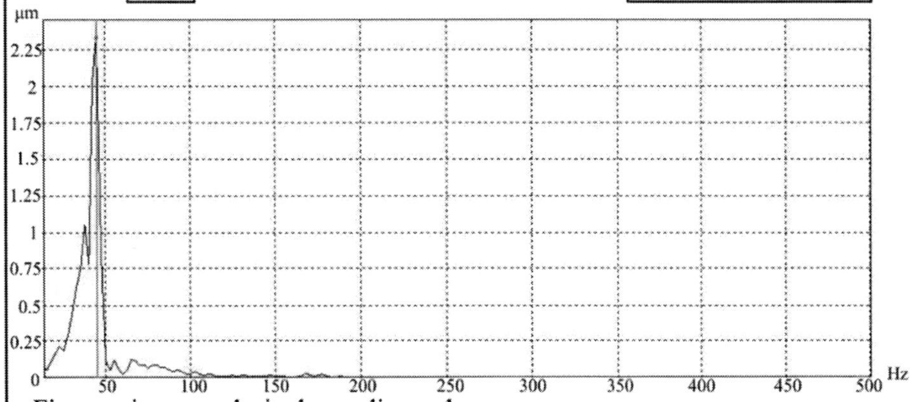

Five maximum peaks in descending order:

№ peak	Frequency, Hz	Amplitude, μm
1.	44.072	2.5221
2.	37.584	1.0544
3.	22.899	0.2165
4.	66.071	0.1355
5.	55.053	0.1188

Fig. 6a. Frequencies and Amplitudes of Vibrations for the Point 1

Dot №: 1 Direction of reconciliation: Horizontal

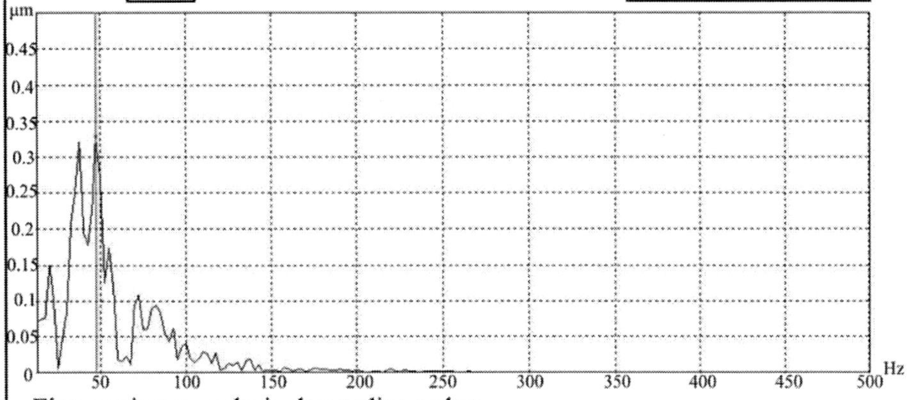

Five maximum peaks in descending order:

№ peak	Frequency, Hz	Amplitude, μm
1.	47.943	0.3328
2.	37.034	0.3260
3.	54.889	0.1739
4.	20.362	0.1524
5.	71.688	0.1130

Fig. 6b. Frequencies and Amplitudes of Vibrations for the Point 1.

Dot №: 4 Direction of reconciliation: Vertical

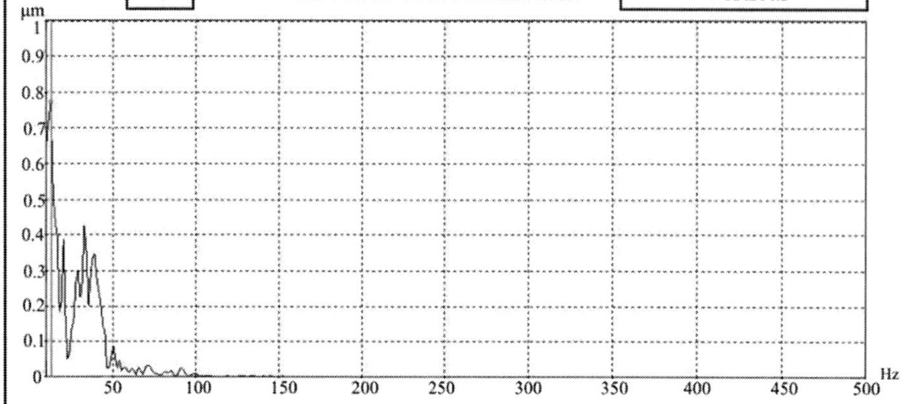

Five maximum peaks in descending order:

№ peak	Frequency, Hz	Amplitude, µm
1.	11.992	0.8147
2.	32.896	0.4444
3.	19.909	0.3888
4.	38.325	0.3547
5.	28.444	0.3045

Fig. 7. Frequencies and Amplitudes of Vibrations for the Point 4.

Dot №: 6 Direction of reconciliation: Vertical

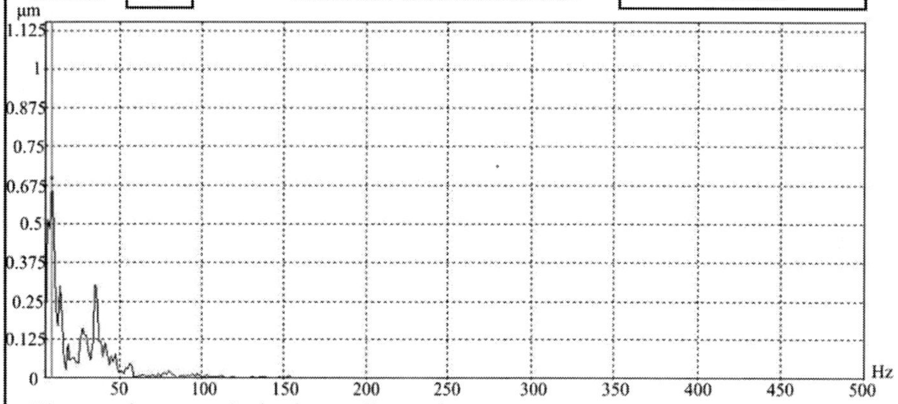

Five maximum peaks in descending order:

№ peak	Frequency, Hz	Amplitude, µm
1.	8.750	0.6495
2.	35.584	0.3399
3.	13.750	0.2998
4.	27.744	0.1620
5.	41.360	0.1135

Fig. 8. Frequencies and Amplitudes of Vibrations for the Point 6.

3 Stress Calculation in the Roof Guy Ropes

Approved:
- bay – 48.0 m;
- bending deflection – 5.0 m;
- loads – 0.5 t/m^2;
- guy rope (beam) intervals – 1.0 m.

Forcing in guy rope [1, 3] is:

$$N = \sqrt{H^2 + Q^2}, \tag{1}$$

where H – thrust; Q – maximum shearing force.

In this case,

$$Q = \frac{qL}{2}; \tag{2}$$

$$H = \frac{qL}{2\sqrt{6}} \cdot \frac{1}{\sqrt{\frac{L}{l} - 1}}, \tag{3}$$

where q – shearing load along guy rope q=*const*; L – length of guy rope; l – bay.

The length of guy rope is defined approximately:

$$L = l + \frac{D}{2H^2}, \tag{4}$$

where

$$D = \int_0^L q(x) Mb dx,$$

Mb – girder moment.

Making square area, taking into cons $Q = \frac{qL}{2}$;

$M(x) = \frac{q}{2}(Lx - x^2)$ we will obtain:

$$L = l + \frac{8 \cdot f^2}{3L}, \tag{5}$$

where f – guy rope bending deflection, when substituting with numeric values, we have:

$$L \approx 49.4\text{m}; \ \frac{L}{l} \approx \frac{49.4}{48.0} = 1.03$$

$$H = \frac{0.5 \cdot 48.0}{2\sqrt{6}} \cdot \frac{1}{\sqrt{0.03}} = 283 \text{ kN}$$

$$Q = \frac{0.5 \cdot 48.0}{2} = 120\,\text{kN}$$

$$N = \sqrt{2.3^2 + 12.0^2} = 307.4\text{kN}$$

Taking into account the previous stringing (3.3 kN under the project) $N \approx 34.0$ kN.

3.1 Determination of Wearing Process of Steel Beams

Using diagram at the page 398 [2] we will obtain that as a result of 55-year operation, the corrosive wear makes up 0.3 mm. Then, the relative wear per one wire (F=5 mm) is equal to $\frac{4.7}{5.0} = 0.94$. Taking into account of operation during 50 years as well as considering that the defined relation of the destructive force of wire with corrosion and without it is $\frac{31.93}{34.25} = 0.93$. It is taken for the further calculations

$$\Phi_{\kappa} = 0.9 \cdot 5 = 4.5\text{mm}$$

where Φ_{κ} – diameter of corroded element.

The mentioned above determines definition of stress in guy ropes, taking into account corrosive wear as a middle tension element, according to the formula [2]:

$$\sigma = \frac{N}{A \cdot K_A} \le R_y \cdot \gamma_c,$$

$$\text{where } K_A = \frac{t_k}{t_o} = \frac{4.5}{5.0} = 0.9;$$

$$A = 16 \cdot \frac{\pi \cdot \Phi_{\kappa}^2}{4} = \frac{\pi \cdot 0.45^2}{4} = 2.545\,\text{cm}^2,$$

n=16 – number of wires in beam.

Conclusion

The specified range of frequencies determines:

1. Possible accelerated fatigue of materials of roof (concrete and steel guy ropes).
2. Significant noise occurring because of metro train traffic, essentially worsens the set of parameters, determining normal and comfort operation for viewing audience.

Recommendations Related to the Further Building Operation

1. To ensure substitution of rigid supports of rail tracks with rubber one (dampers) at the area under Ukraine Cinema and Concert Hall building with dimensions mentioned on the scheme equal to the dimensions of building on the scheme. This requirement is caused by a necessity of increase of fatigue failure level of construction (first of all, cable wires) and creation of comfortable atmosphere in the premises during its operation under necessary conditions.
2. To install a roof deformation monitoring system on Ukraine Cinema and Concert Hall building, operating in the on-line mode.

References

1. I.A. Birger, Ya.G. Panovko, *Prochnost'. Ustoychivost'. Kolebaniya.* Spravochnik v trekh tomakh. M. Mashinostroenie, 567, (1968).
2. V.V. Biriulev and others. *Proyektirovaniye metallicheskikh konstruktsiy. Spetsyukurs. Ucheb.posobiye dlya vuzov.* L.: Stroiizdat, 432, (1990).
3. V.M. Babaiev, S.O. Bugaievskyi, S.M. Evel, I.D. Ievzerov, A.I. Lantukh-Lyashchenko, V.V. Shevetoskyi, O.V. Shymanovskyi, V.S. Shmukler *Chysel'ni ta eksperymental'ni metody ratsional'noho proektuvannya ta zvedennya konstruktyvnykh system.* K.: Publishing office "Stal", 404, (2017).
4. V.I. Feodosev *Soprotivleniye materialov.* Publishing Office. "Nauka", M.: 552, (1967).
5. L. Li, M.-X. Zhang, H.-M. Wu. Influence of metro train loading calculation methods on dynamic responses of shield tunnel. Article. July. (2015).
6. I.L. Nikiforova Dynamic Study of Land Surface Movement When Constructing Tunnels. Mining Research and Information Bulletin (scientific and technical journal) **6**, (2014).
7. DSTU-NBV.1.2-18:2016 *Nastonova shchodo obstezhennya budivel' i sporud dlya vyznachennya ta otsinky yikh tekhnichnoho stanu*, Kyiv.: Minenergbud of Ukraine, (2017)
8. DBN B.1.2-14:2009 *Zahal'ni pryntsypy zabezpechennya nadiynosti ta konstruktyvnoyi bezpeky budivel', sporud, budivel'nykh konstruktsiy ta osnov*, Kyiv. Minenergbud of Ukraine, (2009)
9. DBN B.2.6-98:2009 *Konstruktsiyi budynkiv i sporud. Betonni ta zalizobetonni konstruktsiyi. Osnovni polozhennya*, Kyiv. Minenergbud of Ukraine, (2011)

Rationalization of the parameters of the cylindrical bridge support (theoretical basis)

Valery Shmukler[1*], *Olena* Petrova[1], and *Hammoud* Mohammad[1]

[1]O.M.Beketov National University of Urban Economy in Kharkiv, 17 Marshala Bazhanova Str., Kharkiv, 61002, Ukraine

Abstract. Paper deals with a new constructive solution of a massive concrete pillar intended, mostly, for construction of railway and road bridge supports and with procedure of rationalization of its parameters. Statement and solution of the problem of strength and stability of a concrete support, which has an external mesh cage, reinforcing rings (frames) and an internal cavity is considered. The contact stresses between the concrete core and the metal cage are determined, which, in turn, depend on the attributes of the cage and the frames. The rationalization criterion based on minimizing the potential energy of deformation is introduced. For the indicated task, the variables are the thickness of the mesh cage, the size of its cell, the step and the cross-section of the frames. Support with rational parameters which depend on characteristics of cage and frames gives the lowest material consumption and the highest bearing capacity that is the aim of the approach of investigation.

Introduction

Developing the idea of a concrete support with external reinforcement, presented in [1], this paper proposes the detail approach to the rationalization of its parameters. The effectiveness of external reinforcement made of expanded metal sheet can be extremely increased in case of using the additional transverse steel rings. One can see the interest in the task of finding the rational dependency between parameters of the mesh cage and reinforcing rings. However, together with mentioned above, the task of this work is to establish the bond between the value of the potential energy of deformations (PED) and characteristics of external reinforcement (the cage and the frames at one time) and, using the energy approach [2], to find out the decision which gives minimum value of PED.

1 Problem Statement

The problem of rationalization of the structure of a hollow cylindrical, reinforced concrete pillar is considered. Structurally, this pillar represents a concrete element of a tubular type, enclosed in a mesh steel cage. The thickness of the cylinder is such that local buckling is

* Corresponding author: petrova.bcd@gmail.com

excluded. In turn, mesh cage is supported by transverse steel rings (frames), set with a_c step (Fig. 1).

For high massive supports, the cross-sectional area $A(z)$ should be formulated in accordance with the expression [3]:

$$A(z) = \frac{\pi}{4}(D^2 - d_0^2 \exp\frac{2\gamma_c}{f_c} \cdot z) \tag{1}$$

where d_0 - diameter of the internal cavity at the origin of the coordinates of the cylindrical system (Fig. 2);
D – external diameter of the support;
γ_c – specific weight of concrete;
f_c – compression tensile strength of concrete.

To the ends of the element (concrete part), axially applied constant stresses of q intensity. Equilibrium of these stresses is equal to F.

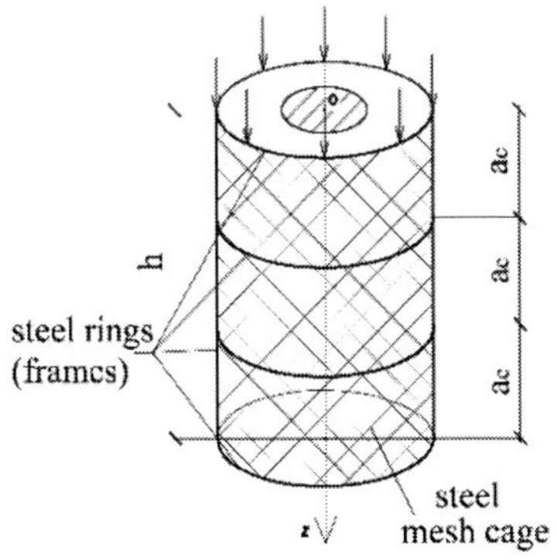

Fig. 1. The support – concrete element in a mesh cage.

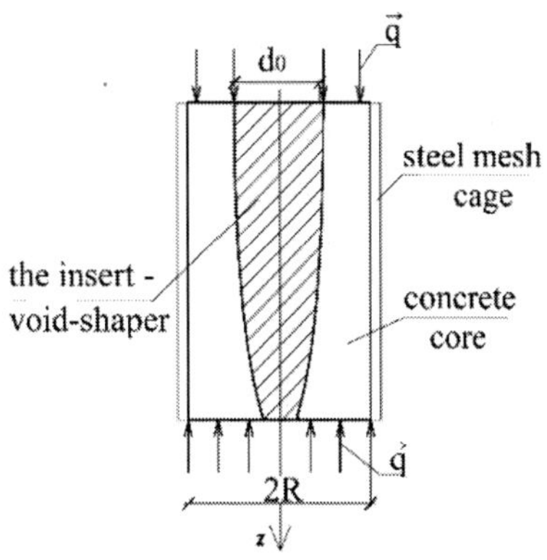

Fig. 2. The longitudinal cross section of the support.

2 Strength Problem

Let us denote by p yet unknown contact voltages emerging between the mesh cage and the concrete core. Here it is assumed that radial deformations of the core and circular mesh cage are joint throughout the whole load range (life-cycle).

The transverse cross sections of the considered element (core + cage) under the designated central compression will have the following stress strain state (Fig. 3):

- the core - axially symmetric transverse compression;
- mesh cage - axially symmetric radial tension.

Let's set the following designations:
R – external radius of a concrete core;
D – internal radius of a concrete core;
r – current radius (coordinate of cylindrical system);
E_c, G_c – deformation modules of the 1st and 2nd kind of concrete core, $G_c = \frac{E_c}{2(1+\mu_c)}$;

μ_c – Poisson ratio of a concrete core.

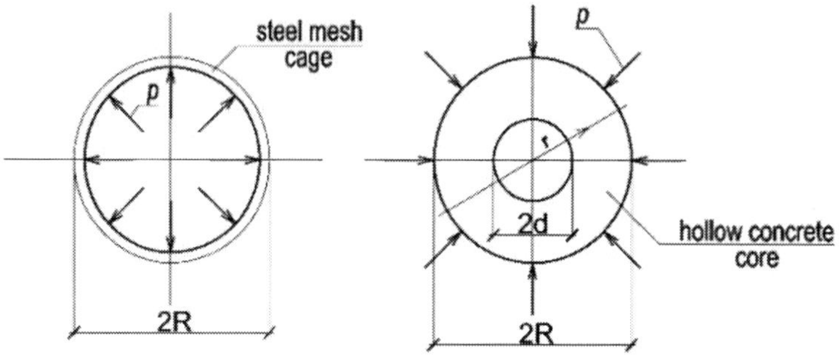

Fig. 3. Cross section of the core and the cage.

2.1 Concrete core

To determine the radial deformations of the concrete core, we use the solution of the Lame problem [4].

Radial displacement of the core under compression:

$$u_r = \frac{1}{2G_c(R^2-d^2)}\left[\frac{p\cdot d^2\cdot R^2}{r}+\frac{(1-\mu_c)pR^2}{1+\mu_c}\cdot r\right] \tag{2}$$

At the point of contact between the core and the cage $r = R$.

Deformation of the concrete in the transverse direction:

$$\varepsilon_r = \frac{\partial u_r}{\partial r} = \frac{pR^2}{2G_c(R^2-d^2)}\left[d^2(-\frac{1}{r^2})+\frac{1-\mu_c}{1+\mu_c}\right] \tag{3}$$

At the point of contact between the core and the cage $r = R$ and

$$\bar{\varepsilon}_r = \frac{(1+\mu)pR^2}{(R^2-d^2)\cdot E_c}\cdot\left(-\frac{d^2}{R^2}+\frac{(1-\mu)}{(1+\mu)}\right) \tag{4}$$

Expansion of the core under longitudinal forces in the case of axial symmetry:

$$\bar{\bar{\varepsilon}}_r = \frac{1}{2G_c}\cdot\left(-\frac{\mu_c}{1+\mu}q\right) = -\frac{(1+\mu_c)\mu}{(1+\mu_c)E_c}q \tag{5}$$

or

$$\bar{\bar{\varepsilon}}_r = \frac{\mu_c}{E_c} \tag{6}$$

In case of simultaneous action of q and p (solid cylinder) [5]:

$$\varepsilon_r = \bar{\varepsilon}_r + \bar{\bar{\varepsilon}}_r = \frac{p}{E_c} - p\mu_c\frac{1}{E_c} - \frac{\mu_c q}{E_c} = \frac{p}{E_c} - \frac{\mu_c}{E_c}(p+q) \tag{7}$$

Considering the direction of compression forces

$$\varepsilon_r = \frac{p}{E_c} - \frac{\mu_c}{E_c}(p+q) \tag{8}$$

In case of an empty cylinder, after transformations, and considering the results [6]:

$$\varepsilon_r = \frac{(1+\mu_c)R^2}{(R^2-d^2)E_c}\left[\left(\frac{d^2}{R^2}-\frac{1-\mu_c}{1+\mu_c}\right)p+\mu_c q\right] \tag{9}$$

For verification when $d=0$ we have:

$$\varepsilon_r = \frac{(1+\mu_c)}{E_c}\left[-\frac{1-\mu_c}{1+\mu_c}p+\mu_c q\right]=\frac{1}{E_c}\left[-p+\mu_c p+\mu_c q+\mu_c^2 q\right] \tag{10}$$

And assuming the validity of $\mu^2 \ll \varepsilon$ ($\varepsilon>0$ – is a small value), we get

$$\varepsilon_r = \frac{p}{E_c}+\frac{\mu_c}{E_c}(p+q)$$

which gives a complete coincidence with [5].

2.2 Steel Mesh Cage

Let's assume that the size of the cells of mesh cage are small in comparison with radius and height of the support. Then, on the basis of [7], the mesh cage can be roughly replaced by a continuous orthotropic shell. Introducing further a three-orthogonal coordinate system, in the spirit of cited work, we will have:

$$\begin{gathered} N_\alpha = B_{11}\cdot\varepsilon_\alpha + B_{12}\cdot\varepsilon_\beta,\; N_\beta = B_{21}\cdot\varepsilon_\alpha + B_{22}\cdot\varepsilon_\beta,\; N_{\alpha\beta}=N_{\alpha\beta}=B_{33}\cdot\gamma_{\alpha\beta}, \\ M_\alpha = D_{11}\cdot K_\alpha + D_{12}\cdot K_\beta,\; M_\beta = D_{21}\cdot K_\alpha + D_{22}\cdot K_\beta,\; M_{\alpha\beta}=M_{\alpha\beta}=D_{33}\cdot\gamma_{\alpha\beta}, \\ Q_\alpha = C_{11}\cdot\Psi_\alpha,\; Q_\beta = C_{22}\cdot\Psi_\beta \end{gathered} \tag{11}$$

where ε_α, ε_β, $\varepsilon_{\alpha\beta}$ – deformations of the median surface;
K_α, K_β, $K_{\alpha\beta}$ – bending deformations of the shell (curvature);
Ψ_α, Ψ_β – transverse shear deformation.

$$\begin{gathered} B_{11}=2\frac{B_h}{\alpha_h}\cdot c^4,\; B_{12}=B_{21}=B_{33}=2\frac{B_h}{\alpha_h}\cdot s^2\cdot c^2,\; B_{22}=2\frac{B_h}{\alpha_h}\cdot s^4+\frac{B_c}{\alpha_c}, \\ D_{mn}=\frac{h^2}{12}\cdot B_{mn}\,(mn = 11,\,12,\,22,\,33),\; C_{11}=2\frac{C_h}{\alpha_h}\cdot c^2,\; C_{22}=2\frac{C_h}{\alpha_h}\cdot s^4+\frac{C_c}{\alpha_c}, \\ B_h=\delta_h\cdot h\cdot E_h,\; B_c=\delta_c\cdot h\cdot E_c,\; C_h=\delta_h\cdot h\cdot G_h,\; C_c=\delta_c\cdot h\cdot G_c \end{gathered} \tag{12}$$

In expressions of stiffness (12) for spiral and circular ribs under tension-compression and shear conditions it is indicated (Fig. 4):

$$s=\sin\varphi,\; c=\cos\varphi \tag{13}$$

φ is the angle of the inclination of the spiral ribs to the creature;
h – thickness of the mesh structure (height of ribs);
δ_h, δ_c – the width of the spiral and circular ribs, respectively;
a_h, a_c – the distance between spiral and circular ribs;
E_s, G_s – elastic and shear modules of ribs (steel);
n_h – number of pairs of spiral ribs;

D is the diameter of the shell.

$$\alpha = \frac{\pi D}{n_h}; \quad \alpha_h = a\cos\varphi; \quad \alpha_c = \frac{a}{2} ctg\varphi; \quad l_h = \frac{\alpha}{2\sin\varphi},$$

where a is the distance between the points of intersection of the symmetrical spiral ribs over the circumference of the cross-section of the shell.

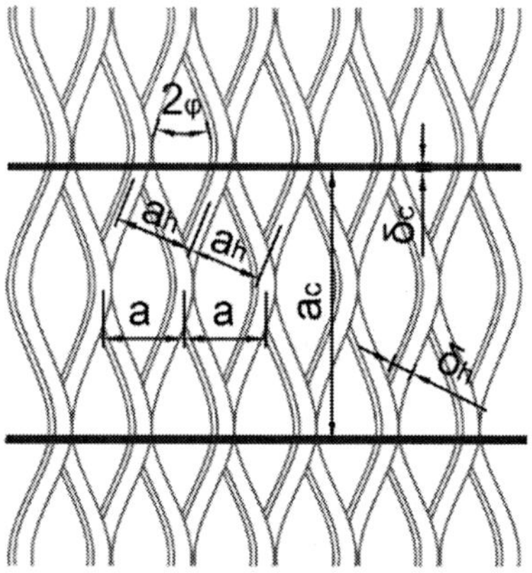

Fig. 4. The main attributes of the cage.

Main attributes of the mesh cage are [8]:

$$\varepsilon_\alpha^e = -\frac{1}{B_e} B_{12}^e N_\beta \tag{14}$$

$$\varepsilon_\beta^e = \frac{1}{B_e} B_{11}^e N_\beta; \quad \varepsilon_\beta^e = \frac{B_{11}}{B_e} pR \tag{15}$$

$$B_e = B_{11}^e B_{22}^e - (B_{12}^e)^2 = B_{11}^e B_{22}^e - B_{12}^e \cdot B_{21}^e \tag{16}$$

$$\text{and} \quad N_\beta = pR, \quad N_\alpha = 0.0.$$

Equating the radial deformation of the concrete core to the annular deformation of steel cage, we get:

$$\frac{(1+\mu_c)\cdot R^2}{E_c \cdot (R^2 - d^2)} \left\{ \left[\frac{d^2}{R^2} - \frac{1-\mu_c}{1+\mu_c} \right] \cdot p + \mu_c \cdot q \right\} = \frac{B_{11}^s}{B_e^s} \cdot pR \tag{17}$$

Let's set the designation:

$$\gamma_1 = \frac{(1+\mu_c)\cdot d^2}{E_c \cdot (R^2 - d^2)}, \quad \gamma_2 = \frac{(1-\mu_c)\cdot R^2}{E_c \cdot (R^2 - d^2)}, \quad \gamma_3 = \frac{\mu_c(1+\mu_c)\cdot R^2}{E_c \cdot (R^2 - d^2)}, \quad \gamma_4 = \frac{B_{11}^s}{B_e^s} \cdot R \tag{18}$$

Then

$$(-\gamma_1 + \gamma_2 + \gamma_4)\cdot p = \gamma_3 \cdot q \tag{19}$$

Or

$$p = \frac{\gamma_3}{(\gamma_2 + \gamma_4 - \gamma_1)} \cdot q \tag{20}$$

Futher, considering the relationship between the contact tensions p and the longitudinal (given) stresses q, written for the isotropic shell, we obtain [5]:

$$p = \frac{\mu_c}{\frac{R}{h_{red}} \cdot n + 1 - \mu_c} \cdot q \tag{21}$$

where $n = \frac{E_c}{E_s}$ – reduction factor;

h_{red} – reduced thickness of the mesh cage.

Having a sort of cut sheets, it is possible on the basis of (20) to determine the contact tension between the cage and the core, depending on the main attributes of the sheet and the step of the steel rings (frames).

2.3 Stress State of the Core

To estimate the stress state of the core we define the main stresses, considering that the components of the stress tensor, in this case, are the following [2, 4]:

$$\sigma_r = \frac{-R^2}{R^2 - d^2} \cdot \left(\frac{d^2}{r^2} - 1 \right) \cdot p; \tag{22}$$

$$\sigma_\beta = \frac{-R^2}{R^2 - d^2} \cdot \left(\frac{d^2}{r^2} + 1 \right) \cdot p; \tag{23}$$

$$\sigma_z = -q; \tag{24}$$

$$\sigma_{rz} = \sigma_{\beta z} = \sigma_{r\beta} = 0. \tag{25}$$

As a consequence, invariants of the stress tensor will be:

$$I_1(T_\sigma) = \sigma_r + \sigma_\beta + \sigma_z = -\frac{2R^2}{R^2 - d^2} \cdot p - q; \tag{26}$$

$$I_1(T_\sigma) = -\sigma_r \cdot \sigma_\beta - \sigma_r \cdot \sigma_z - \sigma_\beta \cdot \sigma_z = \frac{-R^2 \cdot p}{R^2 - d^2}\left(\frac{d^2}{r^2} - 1 \right)\left[-\frac{R^2 \cdot p}{R^2 - d^2}\left(\frac{d^2}{r^2} + 1 \right) \right] +$$

$$+ \frac{R^2 \cdot p \cdot q}{R^2 - d^2}\left(\frac{d^2}{r^2} - 1 \right) - \frac{R^2 \cdot p \cdot q}{R^2 - d^2}\left(\frac{d^2}{r^2} + 1 \right) = \frac{R^4 \cdot p^2}{\left(R^2 - d^2\right)^2}\left(\frac{d^4}{r^2} - 1 \right) - \frac{2R^2 \cdot p \cdot q}{R^2 - d^2} = \tag{27}$$

$$= \frac{R^2 \cdot p}{R^2 - d^2}\left[\frac{p\left(d^4 - r^4\right)R^2}{\left(R^2 - d^2\right)r^4} - 2q \right]$$

$$I_3(T_\sigma) = \sigma_r \cdot \sigma_\beta \cdot \sigma_z = \frac{R^4}{\left(R^2 - d^2\right)^2}\left(\frac{d^4}{r^4} - 1\right) \cdot p^2 q; \tag{28}$$

Decisive equation concerning main stresses becomes:

$$\sigma^3 - I_1(T_\sigma) \cdot \sigma^2 - I_2(T_\sigma) \cdot \sigma - I_3(T_\sigma) = 0$$

Or with respect of (26-28):

$$\sigma^3 + \left[\frac{2R^2}{R^2 - d^2} \cdot p + q\right] \cdot \sigma^2 - \frac{R^2 \cdot p}{R^2 - d^2}\left[\frac{(d^4 - r^4) \cdot R^2 p}{(R^2 - d^2) \cdot r^4} - 2q\right] \cdot \sigma - \frac{R^4}{(R^2 - d^2)^2}\left(\frac{d^4}{r^4} - 1\right) = 0 \tag{29}$$

At the point where $r = d$, (29) is converted to

$$\sigma^2 + \left[\frac{2R^2}{R^2 - d^2} \cdot p + q\right] \cdot \sigma + \frac{2R^2 \cdot p \cdot q}{R^2 - d^2} \tag{30}$$

Herefrom

$$\sigma_{1,2} = -\frac{1}{2}\left(\frac{2R^2}{R^2 - d^2} \cdot p + q\right) \pm \sqrt{\frac{1}{4}\left(\frac{2R^2}{R^2 - d^2} \cdot p + q\right)^2 - \frac{2R^2 \cdot p \cdot q}{R^2 - d^2}} \tag{31}$$

At the point where $r = R$, the equation (29) will have a look of:

$$\sigma^3 + \left[\frac{2R^2}{R^2 - d^2} \cdot p + q\right] \cdot \sigma^2 + \frac{R^2 \cdot p}{R^2 - d^2} \cdot \left(\frac{R^2 + d^2}{R^2} \cdot p - 2q\right)\sigma + \frac{R^2 + d^4}{R^2 - d^2} \cdot p^2 q = 0 \tag{32}$$

As a result of the solution (29), (30), (32), we obtain the main stresses:

$$\sigma_1 \geq \sigma_2 \geq \sigma_3, \tag{33}$$

Knowing the main stresses, one can determine the density of the potential energy of deformations:

$$e = \frac{1}{2E_c}\left[\sigma_1^2 + \sigma_2^2 + \sigma_3^2 - 2\mu_c\left(\sigma_1 \cdot \sigma_2 + \sigma_1 \cdot \sigma_3 + \sigma_2 \cdot \sigma_3\right)\right], \tag{34}$$

and, finally, the potential energy (PED)

$$U = \int_V e dV, \tag{35}$$

where U - potential energy of deformation;
V - the volume of the support.

Apparently that due to presented algorithm, the main stresses, the density of the potential energy of deformation and PED also depend on the parameters of the mesh cage.

2.3 Criterion and procedure of rationalization

As a criterion for rationalization of the considered design, it is accepted that:

$$U \to \inf,$$

with limitation

$$e_i \le e_{nc}, \tag{36}$$

where e_{nc} is the normalized value of the density of deformation energy (DPED);
e_i - DPED in the i-th point of the rack.

$$e_n \le \gamma \cdot e_u, \tag{37}$$

where e_u is the limiting value of the density of deformation energy (DPED);
γ - statistically substantiated safety ratio, in the first approximation it is acceptable to take $\gamma = 0,77$.

In turn, the DPED is given in accordance with the Vasil'kov - Shmukler criterion [9]:

$$e_u = 0,5\chi_\sigma^2\left[(\chi_\sigma + 1)e_{cu} - (\chi_\sigma - 1)e_{tu}\right] + (1 - \chi_\sigma^2)e_{shu} \tag{38}$$

where $\chi_\sigma = \dfrac{2\sigma_2 - \sigma_1 - \sigma_3}{\sigma_1 - \sigma_3}$ – the Lode-Nadai parameter of the stress state (39)

e_{su}, e_{tu}, e_{shu} - partial values of the density of deformation energy for compression, tension, and shear, respectively.

Let's denote that for uniaxial tension:
$\sigma_1 \ne 0; \sigma_2 = 0; \sigma_3 = 0; \chi_\sigma = -1.0,$
for uniaxial compression:
$\sigma_1 = 0; \sigma_2 = 0; \sigma_3 \ne 0; \chi_\sigma = -1.0,$
for uniaxial shear:
$\sigma_1 = \sigma_3; \sigma_2 = 0; \chi_\sigma = 0.0.$

Each partial value of the limiting density of the deformation's energy is determined by:

$$e_{iu} = \int_0^{\varepsilon_u} \sigma(\varepsilon)d\varepsilon, \quad i = c, t, sh \tag{40}$$

where $\sigma_i(\varepsilon)$ is the known dependence between stresses and deformations, which is obtained experimentally for compression (c), tensile (t) and shear (sh);
ε_u is the limiting deformation.

An important point here is the fact (and (40) expression reflects it) that, in general case, the limiting deformation does not correspond to the maximum tension (due to physical nonlinearity).

And since the connection between stresses and deformations for concrete is nonlinear, the polynomial of the fifth degree is accepted for its description [9,10].

In this case, deformation diagrams of concrete for compression, tension and shear are taken on the basis of:

$$\sigma_L = \alpha_1 \cdot \varepsilon_L + \alpha_2 \cdot \varepsilon_L^2 + \alpha_3 \cdot \varepsilon_L^3 + \alpha_4 \cdot \varepsilon_L^4 + \alpha_5 \cdot \varepsilon_L^5 \tag{41}$$

Conclusions

Listed, in aggregate, predetermines the following conclusion:

The developed approach of assessment of the effect of a mesh cage reinforced by frames on the stress-strain state of a cylindrical concrete support allows one to determine the rational parameters of the reinforcement elements (characteristics of the mesh sheet and steel frames) on the basis of the energy portrait of a structure based on new energy methods [9].

References

1. O. Petrova, O.Lugchenko, A. Nazhem. MATEC Web of Conferences **116**, 02025 (2017).
2. G.V. Vasilkov. *Evolutionary theory of the life cycle of mechanical systems. Theory of constructions.* Moscow: LKI (2008).
3. V.S. Shmukler, O.O. Petrova, O.V. Sinkovska. Scientific Bulletin of Construction. - Kh.: KhNUBA, KhOTV ABU, **74**, 148-156 (2013).
4. V.G. Rekach. *Kerivnytstvo do vyrishennia zadach z teorii pruzhnosti.* - Moscow: Higher School (1966).
5. B.N. Zhemochkin. *Teoria pruzhnosti.* - M .: State Publishing House of Literature on Construction and Architecture (1957).
6. Tymoshenko, S.P. *Opir materialiv.* - Moscow: Nauka Volume I (1965).
7. A.V. Azarov. Composites and nanostructures, Moscow, **7(3)**, 151-161 (2015).
8. A.V. Egorov, A.V. Azarov. Electronic Magazine of Labor "MAI", **73**, 1-16 (2014).
9. V.S. Shmukler. Collection of scientific works of the Ukrainian State University of Railway Transport. - Kh., **167**, 54-70 (2017).
10. M. Hammoud, V. Shmukler, O. Petrova. Proc. of the 12th fib International PhD Symposium in Civil Engineering. CTU, Prague, Czech Republic, 593-599 (2018).

To the solution of the problem of bending of a cylindrical shell by the boundary elements method

Mykola Surianinov[1,*], and *Yurii* Krutii[1]

[1]Odessa State Academy of Civil Engineering and Architecture, Didrikson str. 4, Odessa, 65029, Ukraine

Abstract. The solution of the problem of the long cylindrical shell bending by a numerical and analytical boundary elements method is considered. The method is based on the analytical construction of a fundamental system of solutions and Green's functions for the differential equation of the problem under consideration. This paper is devoted to the determination of these expressions. The semi-moment theory of the cylindrical shell calculation, proposed by V.Z. Vlasov, which for the problem under consideration leads to one eighth-order partial differential equation is used. The problem of the bending of a cylindrical shell is two-dimensional, and in the numerical and analytical boundary elements method, plates and shells are considered as generalized one-dimensional modules, so the variational method of Kantorovich-Vlasov was applied to this equation to obtain an ordinary differential equation of the eighth order. Sixty-four expressions of all the fundamental functions of the problem are constructed, as well as an analytic expression for the Green's function, which makes it possible to construct a load vector (without any restrictions on the nature of its application), and then proceed to the solution of boundary-value problems for the bending of long cylindrical shells under various boundary conditions.

1 Introduction

The stress state of a cylindrical shell and the corresponding theory of its calculation depend substantially on the length of the shell. In the case of a long shell supported along curvilinear edges and loaded with an arbitrary smoothly varying load, its stress state is close to the one of the beam. With an uneven distribution of the load along the shell and along the cross-section, this approach gives incorrect results, and we must take into account the deformation of the contour. This also applies to the case of securing the longitudinal edges of the shell.

V.Z. Vlasov proposed a theory of long cylindrical shells calculation, which was called a semi-moment theory [1, 2]. According to this model the shell consists of an infinite number of transversal curved elementary strips, joined connected by a system of rods with hinged

* Corresponding author: sng@ogasa.org.ua

joints (Fig. 1). Each such rod can transmit only normal N_1 and shear S forces. The bending moment in the longitudinal direction, the transverse force and the torque can be neglected, so the state vector $\overline{\mathbf{P}}$ of such a shell is represented in the form

$$\overline{\mathbf{P}}^{\mathrm{T}} = \left\| W, W', M, Q, S, N, U, V \right\|. \tag{1}$$

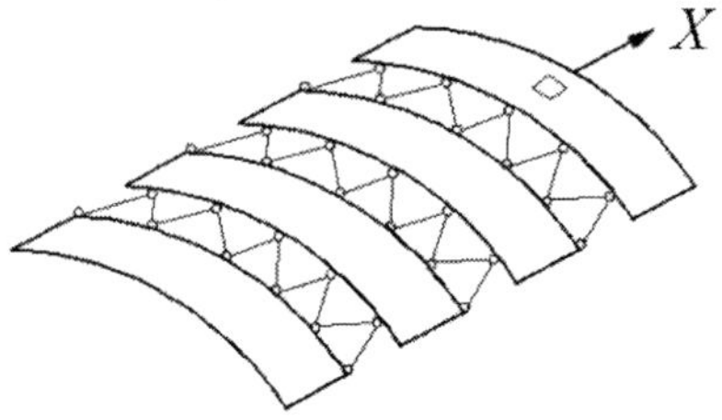

Fig. 1. Model of V.Z. Vlasov.

In total, in the state vector (1) there are 4 forces and 4 displacements (Fig. 2).

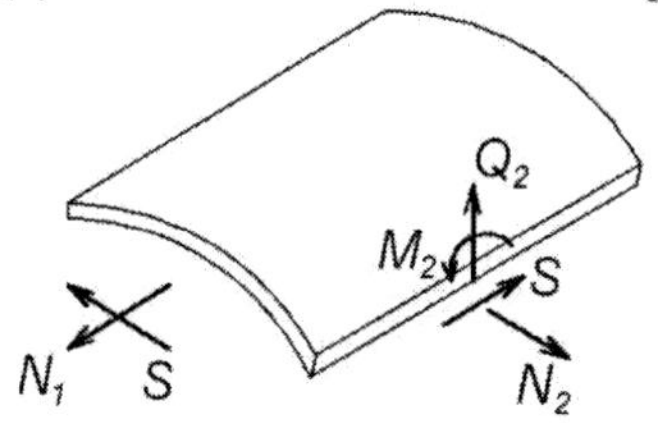

Fig. 2. Forces and displacements in shell.

In view of what has been said, all the equations of the general theory of shells are simplified [3-5], and as a result we get the basic resolving equation of the problem

$$\left(\frac{\partial^8}{\partial \theta^8} + 2\frac{\partial^6}{\partial \theta^6} + \frac{\partial^4}{\partial \theta^4} \right) F + \frac{12R^2}{h^2} \frac{\partial^4 F}{\partial \beta^4} = 0, \tag{2}$$

where F — new function (analog of stress function in the theory of plates), through which all the parameters of the state of the shell are expressed, θ — dimensionless coordinate (Fig. 3):

$$\beta = x / R; \quad \theta = s / R\,.$$

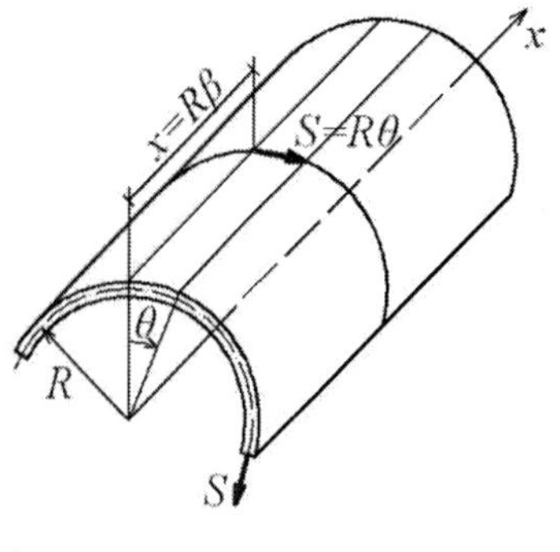

Fig. 3. The coordinate system used.

2 Results

2.1 Construction of the system of fundamental functions

To solve the problem, we will use the numerical and analytical boundary elements method (NA BEM) [6–9]. The method is based on the analytical construction of a fundamental system of solutions and Green's functions for the differential equation of the problem under consideration. The boundary conditions, or the conditions of contact between modules (individual elements of the system) are taken into account by numerically solving a small system of linear algebraic equations.

The algorithm of the method can be determined by transformation according to the scheme:

$$\vec{\mathbf{Y}}(l)=\overline{\mathbf{A}}(l)\vec{X}(0)+\vec{\mathbf{B}}(l)\rightarrow\overline{\mathbf{A}}(l)\vec{\mathbf{X}}(0)-\vec{\mathbf{Y}}(l)=-\vec{\mathbf{B}}(l)\rightarrow\overline{\mathbf{A}}_*(l)\vec{\mathbf{X}}_*(0,l)=-\vec{\mathbf{B}}(l),$$

where the final boundary parameters of the matrix $\vec{\mathbf{Y}}$ are transferred to the place of zero vector parameters $\vec{\mathbf{X}}$. Moreover, these vectors are supplemented by equations of equilibrium and compatibility of displacements of nodal points and boundary conditions.

The purpose of this paper is to construct analytical expressions for the matrix $\overline{\mathbf{A}}$ and the Green's function, which is used in the formation of the load vector $\vec{\mathbf{B}}$.

The problem of a cylindrical shell bending is two-dimensional, but plates and shells are considered as generalized one-dimensional modules in the NA BEM, therefore, we applied Kantorovich-Vlasov's variational method to equation (2), resulting in a differential equation with constant coefficients:

$$Y^{VIII}(\theta)+2Y^{VI}(\theta)+Y^{IV}(\theta)+\frac{12R^2\lambda^4}{h^2}Y(\theta)=0. \quad (3)$$

Here λ — a certain parameter related to the frequency of natural oscillations; in the shell problem $\lambda=\frac{Rm}{l}$, where h and l — thickness and length of shell; R — shell radius; m — the characteristic number determined when setting the boundary conditions.

Characteristic equation for (3) has a form [10]

$$t^8+2t^6+t^4+k=0, \quad (4)$$

where $k=\frac{12R^2\lambda^4}{h^2}$.

Characteristic equation (4) has 8 roots:

$$t_{1-4}=\pm\alpha_1\pm i\beta_1;\ \ t_{5-8}=\pm\alpha_2\pm i\beta_2. \quad (5)$$

We write the function $Y(\theta)$ in the form

$$Y(\theta)=C_1\Phi_1+C_2\Phi_2+C_3\Phi_3+C_4\Phi_4+C_5\Phi_5+C_6\Phi_6+C_7\Phi_7+C_8\Phi_8, \quad (6)$$

where $\Phi_1-\Phi_8$ — hyperbolic-trigonometric functions:

$$\begin{cases} \Phi_1 = ch\alpha_1 y \sin\beta_1 y; \quad \Phi_2 = ch\alpha_1 y \cos\beta_1 y; \\ \Phi_3 = sh\alpha_1 y \cos\beta_1 y; \quad \Phi_4 = sh\alpha_1 y \sin\beta_1 y; \\ \Phi_5 = ch\alpha_2 y \sin\beta_2 y; \quad \Phi_6 = ch\alpha_2 y \cos\beta_2 y; \\ \Phi_7 = sh\alpha_2 y \cos\beta_2 y; \quad \Phi_8 = sh\alpha_2 y \sin\beta_2 y. \end{cases} \tag{7}$$

Constants $C_1 - C_8$ we determine from relations

$$\overline{\Phi_i}(0)\vec{C}_i = \vec{P}_i(0), \quad i = 1, 2, ..., 8. \tag{8}$$

Expanding (8), we have eight algebraic equations that form two independent systems with respect to sets of constants with even and odd indices. In the coefficients of systems, functions $\Phi_1 - \Phi_8$ and their derivatives up to the seventh order inclusive are used, therefore we introduce additional notation, with allowance for which the systems of equations are written in the form

$$\begin{cases} H_{21}C_1 + H_{23}C_3 + H_{25}C_5 + H_{27}C_7 = W_0'; \\ H_{41}C_1 + H_{43}C_3 + H_{45}C_5 + H_{47}C_7 = Q_0; \\ H_{51}C_1 + H_{53}C_3 + H_{55}C_5 + H_{57}C_7 = S_0; \\ H_{81}C_1 + H_{83}C_3 + H_{85}C_5 + H_{87}C_7 = V_0. \end{cases} \tag{9}$$

$$\begin{cases} H_{12}C_2 + H_{14}C_4 + H_{16}C_6 + H_{18}C_8 = W_0; \\ H_{32}C_2 + H_{34}C_4 + H_{36}C_6 + H_{38}C_8 = M_0; \\ H_{62}C_2 + H_{64}C_4 + H_{66}C_6 + H_{68}C_8 = N_0; \\ H_{72}C_2 + H_{74}C_4 + H_{76}C_6 + H_{78}C_8 = U_0, \end{cases} \tag{10}$$

where

$$H_{12} = (\alpha_1^4 - 6\alpha_1^2\beta_1^2 + \beta_1^4); \; H_{14} = 4\alpha_1\beta_1(\alpha_1^2 - \beta_1^2);$$

$$H_{16} = (\alpha_2^4 - 6\alpha_2^2\beta_2^2 + \beta_2^4); \; H_{18} = 4\alpha_2\beta_2(\alpha_2^2 - \beta_2^2);$$

$$H_{21} = \beta_1(5\alpha_1^4 - 10\alpha_1^2\beta_1^2 + \beta_1^4); \; H_{23} = \alpha_1(\alpha_1^4 - 10\alpha_1^2\beta_1^2 + 5\beta_1^4);$$

$$H_{25} = \beta_2(5\alpha_2^4 - 10\alpha_2^2\beta_2^2 + \beta_2^4); \; H_{27} = \alpha_2(\alpha_2^4 - 10\alpha_2^2\beta_2^2 + 5\beta_2^4);$$

$$H_{32} = -\frac{D}{R^2}(\alpha_1^6 + \alpha_1^4 - 15\alpha_1^4\beta_1^2 - 6\alpha_1^2\beta_1^2 + 15\alpha_1^2\beta_1^4 + \beta_1^4 - \beta_1^6);$$

$$H_{34} = -\frac{D}{R^2} \cdot 2\alpha_1\beta_1(3\alpha_1^4 - 10\alpha_1^2\beta_1^2 + 3\beta_1^4 + 2\alpha_1^2 - 2\beta_1^2);$$

$$H_{36} = -\frac{D}{R^2}(\alpha_2^6 + \alpha_2^4 - 15\alpha_2^4\beta_2^2 - 6\alpha_2^2\beta_2^2 + 15\alpha_2^2\beta_2^4 + \beta_2^4 - \beta_2^6);$$

$$H_{38} = -\frac{D}{R^2} 2\alpha_2\beta_2(3\alpha_2^4 - 10\alpha_2^2\beta_2^2 + 3\beta_2^4 + 2\alpha_2^2 - 2\beta_2^2);$$

$$H_{41} = \frac{D}{R^3} \cdot \beta_1(7\alpha_1^6 - 35\alpha_1^4\beta_1^2 + 21\alpha_1^2\beta_1^4 - \beta_1^6 + 5\alpha_1^4 - 10\alpha_1^2\beta_1^2 + \beta_1^4);$$

$$H_{43} = \frac{D}{R^3} \cdot \alpha_1(\alpha_1^6 - 21\alpha_1^4\beta_1^2 + 35\alpha_1^2\beta_1^4 - 7\beta_1^6 + \alpha_1^4 - 10\alpha_1^2\beta_1^2 + 5\beta_1^4);$$

$$H_{45} = \frac{D}{R^3} \cdot \beta_2(7\alpha_2^6 - 35\alpha_2^4\beta_2^2 + 21\alpha_2^2\beta_2^4 - \beta_2^6 + 5\alpha_2^4 - 10\alpha_2^2\beta_2^2 + \beta_2^4);$$

$$H_{47} = \frac{D}{R^3} \cdot \alpha_2(\alpha_2^6 - 21\alpha_2^4\beta_2^2 + 35\alpha_2^2\beta_2^4 - 7\beta_2^6 + \alpha_2^4 - 10\alpha_2^2\beta_2^2 + 5\beta_2^4);$$

$$H_{51} = \beta_1\frac{Eh}{R};\; H_{53} = \alpha_1\frac{Eh}{R};\; H_{55} = \beta_2\frac{Eh}{R};\; H_{57} = \alpha_2\frac{Eh}{R};$$

$$H_{62} = -\frac{Eh}{R}[a + \alpha_1^4 - 6\alpha_1^2\beta_1^2 + \beta_1^4 + b(\alpha_1^6 - 15\alpha_1^4\beta_1^2 + 15\alpha_1^2\beta_1^4 - \beta_1^6)];$$

$$H_{64} = -\frac{Eh}{R}[4\alpha_1\beta_1(\alpha_1^2 - \beta_1^2) + 2b\alpha_1\beta_1(3\alpha_1^4 - 10\alpha_1^2\beta_1^2 + 3\beta_1^4)];$$

$$H_{66} = -\frac{Eh}{R}[a + \alpha_2^4 - 6\alpha_2^2\beta_2^2 + \beta_2^4 + b(\alpha_2^6 - 15\alpha_2^4\beta_2^2 + 15\alpha_2^2\beta_2^4 - \beta_2^6)];$$

$$H_{68} = -\frac{Eh}{R}[4\alpha_2\beta_2(\alpha_2^2 - \beta_2^2) + 2b\alpha_2\beta_2(3\alpha_2^4 - 10\alpha_2^2\beta_2^2 + 3\beta_2^4)];$$

$$H_{72} = -(\alpha_1^2 - \beta_1^2);\; H_{74} = -2\alpha_1\beta_1;\; H_{76} = -(\alpha_2^2 - \beta_2^2);\; H_{78} = -2\alpha_2\beta_2;$$

$$H_{81} = \beta_1(3\alpha_1^2 - \beta_1^2);\; H_{83} = \alpha_1(\alpha_1^2 - 3\beta_1^2);\; H_{85} = \beta_2(3\alpha_2^2 - \beta_2^2);$$
$$H_{87} = \alpha_2(\alpha_2^2 - 3\beta_2^2).$$

Coefficients H_{ij} of systems (9), (10) are arithmetic operators that are easily programmed and calculated, but the right-hand parts of these systems are parameters of the shell state vector. In accordance with the NA BEM algorithm [3], the constants must be expressed in a general form as a function of the parameters of the state vector, which will allow us to obtain analytic expressions for the fundamental functions.

This is achieved by solving the systems (9) and (10) by the determinants method. Consider, for example, the solution of system (9).

We assume that the principal determinant of this system is computed (this can be done in any corresponding mathematical program) i.e. is a known number, which we denote as Δ_{odd}.

The determinants Δ_{C_1}, Δ_{C_3}, Δ_{C_5}, Δ_{C_7}, can also be calculated by expanding them along the parameter column of the right-hand side of (9).

The solution of the system (9) takes the form

$$\begin{cases} C_1 = \dfrac{1}{\Delta_{odd}}(\Delta^{(1)}_{W_0'} W_0' - \Delta^{(1)}_{Q_0} Q_0 + \Delta^{(1)}_{S_0} S_0 - \Delta^{(1)}_{V_0} V_0); \\ C_3 = \dfrac{-1}{\Delta_{odd}}(\Delta^{(3)}_{W_0'} W_0' - \Delta^{(3)}_{Q_0} Q_0 + \Delta^{(3)}_{S_0} S_0 - \Delta^{(3)}_{V_0} V_0); \\ C_5 = \dfrac{1}{\Delta_{odd}}(\Delta^{(5)}_{W_0'} W_0' - \Delta^{(5)}_{Q_0} Q_0 + \Delta^{(5)}_{S_0} S_0 - \Delta^{(5)}_{V_0} V_0); \\ C_7 = \dfrac{-1}{\Delta_{odd}}(\Delta^{(7)}_{W_0'} W_0' - \Delta^{(7)}_{Q_0} Q_0 + \Delta^{(7)}_{S_0} S_0 - \Delta^{(7)}_{V_0} V_0). \end{cases} \tag{11}$$

Similarly, we approach the definition of a set of constants C_2, C_4, C_6, C_8:

$$\begin{cases} C_2 = \dfrac{1}{\Delta_{even}}(\Delta^{(2)}_{W_0} W_0 - \Delta^{(2)}_{M_0} M_0 + \Delta^{(2)}_{N_0} N_0 - \Delta^{(2)}_{U_0} U_0); \\ C_4 = \dfrac{-1}{\Delta_{even}}(\Delta^{(4)}_{W_0} W_0 - \Delta^{(4)}_{M_0} M_0 + \Delta^{(4)}_{N_0} N_0 - \Delta^{(4)}_{U_0} U_0); \\ C_6 = \dfrac{1}{\Delta_{even}}(\Delta^{(6)}_{W_0} W_0 - \Delta^{(6)}_{M_0} M_0 + \Delta^{(6)}_{N_0} N_0 - \Delta^{(6)}_{U_0} U_0); \\ C_8 = \dfrac{-1}{\Delta_{even}}(\Delta^{(8)}_{W_0} W_0 - \Delta^{(8)}_{M_0} M_0 + \Delta^{(8)}_{N_0} N_0 - \Delta^{(8)}_{U_0} U_0). \end{cases} \tag{12}$$

We proceed to the definition of fundamental orthonormal functions.

$$W = A_{11}W_0 + A_{12}W_0' + A_{13}M_0 + A_{14}Q_0 + A_{15}S_0 + A_{16}N_0 + A_{17}U_0 + A_{18}V_0. \tag{13}$$

On the other hand

$$\begin{aligned} W = Y^{IV} &= C_1\Phi_1^{IV} + C_2\Phi_2^{IV} + C_3\Phi_3^{IV} + C_4\Phi_4^{IV} + C_5\Phi_5^{IV} + \\ &+ C_6\Phi_6^{IV} + C_7\Phi_7^{IV} + C_8\Phi_8^{IV}. \end{aligned} \tag{14}$$

We substitute the values of the constants (11), (12) into expression (14) and regroup it with respect to the parameters of the state vector. Taking into account (13), the coefficients at these parameters will form the first eight fundamental functions:

$$\begin{aligned} A_{11} = \frac{1}{\Delta_{even}}&[(H_{12}\Phi_2 - H_{14}\Phi_4)\Delta^{(2)}_{W_0} - (H_{12}\Phi_4 + H_{14}\Phi_2)\Delta^{(4)}_{W_0} + \\ &+ (H_{16}\Phi_6 - H_{18}\Phi_8)\Delta^{(6)}_{W_0} - (H_{16}\Phi_8 + H_{18}\Phi_6)\Delta^{(8)}_{W_0}]; \end{aligned}$$

$$A_{12} = \frac{1}{\Delta_{odd}}[(H_{12}\Phi_1 + H_{14}\Phi_3)\Delta^{(1)}_{W'_0} - (H_{12}\Phi_3 - H_{14}\Phi_1)\Delta^{(3)}_{W'_0} +$$
$$+ (H_{16}\Phi_5 + H_{18}\Phi_7)\Delta^{(5)}_{W'_0} - (H_{16}\Phi_7 - H_{18}\Phi_5)\Delta^{(7)}_{W'_0}];$$

$$A_{13} = \frac{1}{\Delta_{even}}[-(H_{12}\Phi_2 - H_{14}\Phi_4)\Delta^{(2)}_{M_0} + (H_{12}\Phi_4 + H_{14}\Phi_2)\Delta^{(4)}_{M_0} -$$
$$- (H_{16}\Phi_6 - H_{18}\Phi_8)\Delta^{(6)}_{M_0} + (H_{16}\Phi_8 + H_{18}\Phi_6)\Delta^{(8)}_{M_0}];$$

$$A_{14} = \frac{1}{\Delta_{odd}}[-(H_{12}\Phi_1 + H_{14}\Phi_3)\Delta^{(1)}_{Q_0} + (H_{12}\Phi_3 - H_{14}\Phi_1)\Delta^{(3)}_{Q_0} -$$
$$- (H_{16}\Phi_5 + H_{18}\Phi_7)\Delta^{(5)}_{Q_0} + (H_{16}\Phi_7 - H_{18}\Phi_5)\Delta^{(7)}_{Q_0}];$$

$$A_{15} = \frac{1}{\Delta_{odd}}[(H_{12}\Phi_1 + H_{14}\Phi_3)\Delta^{(1)}_{S_0} - (H_{12}\Phi_3 - H_{14}\Phi_1)\Delta^{(3)}_{S_0} +$$
$$+ (H_{16}\Phi_5 + H_{18}\Phi_7)\Delta^{(5)}_{S_0} - (H_{16}\Phi_7 - H_{18}\Phi_5)\Delta^{(7)}_{S_0}];$$

$$A_{16} = \frac{1}{\Delta_{even}}[(H_{12}\Phi_2 - H_{14}\Phi_4)\Delta^{(2)}_{N_0} - (H_{12}\Phi_4 + H_{14}\Phi_2)\Delta^{(4)}_{N_0} +$$
$$+ (H_{16}\Phi_6 - H_{18}\Phi_8)\Delta^{(6)}_{N_0} - (H_{16}\Phi_8 + H_{18}\Phi_6)\Delta^{(8)}_{N_0}];$$

$$A_{17} = \frac{1}{\Delta_{even}}[-(H_{12}\Phi_2 - H_{14}\Phi_4)\Delta^{(2)}_{U_0} + (H_{12}\Phi_4 + H_{14}\Phi_2)\Delta^{(4)}_{U_0} -$$
$$- (H_{16}\Phi_6 - H_{18}\Phi_8)\Delta^{(6)}_{U_0} + (H_{16}\Phi_8 + H_{18}\Phi_6)\Delta^{(8)}_{U_0}];$$

$$A_{18} = \frac{1}{\Delta_{odd}}[-(H_{12}\Phi_1 + H_{14}\Phi_3)\Delta^{(1)}_{V_0} + (H_{12}\Phi_3 - H_{14}\Phi_1)\Delta^{(3)}_{V_0} -$$
$$- (H_{16}\Phi_5 + H_{18}\Phi_7)\Delta^{(5)}_{V_0} + (H_{16}\Phi_7 - H_{18}\Phi_5)\Delta^{(7)}_{V_0}].$$

The expressions for the remaining 56 fundamental functions are defined in a similar way.

2.2 The construction of the Green's function

One of the main operations for the implementation of the algorithm of NA BEM with respect to the problem of a long cylindrical shell bending is the formation of the vector of the right-hand sides (the load vector) of equation

$$\overline{\mathbf{A}}_*(l) \cdot \vec{\mathbf{X}}(0, l) = -\vec{\mathbf{B}}(l), \tag{15}$$

where $\overline{\mathbf{A}}_*(l)$ — a transformed matrix of orthonormal fundamental functions; $\vec{\mathbf{X}}(0, l)$ — vector of unknown initial and final parameters; $\vec{\mathbf{B}}(l)$ — vector of load elements [8, 9].

The construction of the vector of cargo elements involves the definition of the Green's function and its seven derivatives, since the stress-strain state of a long cylindrical shell is characterized by eight parameters when it is bent.

As is known [10], the mechanical mea`ning of the Green's function is that this function determines the magnitude of the displacement from the unit concentrated impact. Taking into account the vector of the shell state parameters, this means that under the action of normal and tangential loads, the Green's function has the form

$$G(y-\xi) = A_{14}(y-\xi) + A_{18}(y-\xi), \tag{16}$$

where expressions $A_{14}(y-\xi)$ и $A_{18}(y-\xi)$ coincide with the expressions for the fundamental functions A_{14} and A_{18} obtained earlier, but from argument $(y-\xi)$.

Under real operating conditions, tangential loads on the shell are rare, and then will be equality

$$G(y-\xi) = A_{14}(y-\xi)\,. \tag{17}$$

This assumption is confirmed by all our previous studies [8, 9], but we need to verify its validity with respect to a long cylindrical shell.

In accordance with the well-known algorithm for constructing the Green's function [8, 10], it can be represented in the form

$$G(y,\xi) = Y(y,\xi)H(y-\xi), \tag{18}$$

where $Y(y,\xi)$ — a particular solution of the differential equation of problem; $H(y-\xi)$ — Heaviside function.

For the Green's function, inequality $y > \xi$ performs, therefore $H(y-\xi)$ in (18) can be omitted.

A particular solution is formed in the form

$$Y(y,\xi) = C_1(\xi)y_1(y) + C_2(\xi)y_2(y) + \ldots + C_8(\xi)y_8(y). \tag{19}$$

Constants $C_i(\xi)\,(i = 1, 2, \ldots, 8)$ are determined from condition

$$\begin{vmatrix} y_1(\xi) & y_2(\xi) & \ldots\ldots & y_8(\xi) \\ y_1'(\xi) & y_2'(\xi) & \ldots\ldots & y_8'(\xi) \\ \ldots & \ldots & \ldots & \ldots \\ y_1^{VII}(\xi) & y_2^{VII}(\xi) & \ldots\ldots & y_8^{VII}(\xi) \end{vmatrix} \begin{vmatrix} C_1 \\ C_2 \\ \vdots \\ C_8 \end{vmatrix} = \begin{vmatrix} 0 \\ 0 \\ \vdots \\ 1/a_0 \end{vmatrix}, \tag{20}$$

where

$$\begin{cases} y_1 = ch\alpha_1\xi \sin\beta_1\xi; \;\; y_2 = ch\alpha_1\xi \cos\beta_1\xi; \;\; y_3 = sh\alpha_1\xi \cos\beta_1\xi; \\ y_4 = sh\alpha_1\xi \sin\beta_1\xi; \;\; y_5 = ch\alpha_2\xi \sin\beta_2\xi; \;\; y_6 = ch\alpha_2\xi \cos\beta_2\xi; \\ y_7 = sh\alpha_2\xi \cos\beta_2\xi; \;\; y_8 = sh\alpha_2\xi \sin\beta_2\xi. \end{cases} \tag{21}$$

Here $\alpha_1, \alpha_2, \beta_1, \beta_2$ — the coefficients of the roots of the characteristic equation corresponding to the differential equation of the shell bending [1].

The determinant of the system (20) is the Wronskian determinant, and is not zero for $y \in (0, y_{boundary})$, therefore (20) has a unique solution. Coefficient a_0 (coefficient at the highest degree of the derivative in the differential equation of the shell bending) is equal to one, so we obtain a system of linear algebraic equations:

$$\begin{cases} y_1 C_1 + y_2 C_2 + \ldots + y_8 C_8 = 0; \\ y_1' C_1 + y_2' C_2 + \ldots + y_8' C_8 = 0; \\ \ldots\ldots\ldots\ldots\ldots\ldots\ldots\ldots \\ y_1^{VII} C_1 + y_2^{VII} C_2 + \ldots + y_8^{VII} C_8 = 1. \end{cases} \quad (22)$$

The derivatives of the functions (21) that enter into equations (22) can be expressed in terms of the functions y_i. The corresponding expressions are given in [8]. Denoting this coefficients by H_{ij} $(i, j = 1,2,\ldots 8)$, we obtain a system of algebraic equations of the form

$$\begin{cases} y_1 C_1 + y_2 C_2 + \ldots + y_8 C_8 = 0; \\ (H_{53} y_4 + H_{51} y_2) C_1 + (H_{53} y_3 + H_{51} y_1) C_2 + \ldots + \\ + (H_{57} y_5 + H_{55} y_7) C_8 = 0; \\ (-H_{72} y_1 - H_{74} y_3) C_1 + (-H_{72} y_2 + H_{74} y_4) C_2 + \ldots + \\ + (-H_{76} y_8 - H_{78} y_6) C_8 = 0; \\ \ldots\ldots\ldots\ldots\ldots\ldots\ldots\ldots\ldots\ldots\ldots\ldots \\ \left[\left(\frac{R^3}{D} H_{43} - H_{23} \right) y_4 + \left(\frac{R^3}{D} H_{41} - H_{21} \right) y_2 \right] C_1 + \\ + \left[\left(\frac{R^3}{D} H_{43} - H_{23} \right) y_3 - \left(\frac{R^3}{D} H_{41} - H_{21} \right) y_1 \right] C_2 + \ldots + \\ + \left[\left(\frac{R^3}{D} H_{47} - H_{27} \right) y_5 + \left(\frac{R^3}{D} H_{45} - H_{25} \right) y_7 \right] C_8 = 1, \end{cases} \quad (23)$$

where $D = \dfrac{Eh^3}{12}$ — cylindrical stiffness; h and R — thickness and radius of the shell.

Solving the system (23) by the Gauss method (note that in this case this process leads to rather cumbersome intermediate calculations), we determine the constants $C_i(\xi)$ $(i = 1,2,\ldots 8)$.

Finally, for the Green's function we obtain expression

$$G(y,\xi)=Y(y,\xi)H(y-\xi)=C_1(\xi)y_1(y)+C_2(\xi)y_2(y)+C_8(\xi)y_8(y)=$$
$$=\frac{1}{\Delta_{odd}}[-(H_{12}y_1+H_{14}y_3)\Delta_{Q_0}^{(1)}+(H_{12}y_3-H_{14}y_1)\Delta_{Q_0}^{(3)}- \quad (24)$$
$$-(H_{16}y_5+H_{18}y_7)\Delta_{Q_0}^{(5)}+(H_{16}y_7-H_{18}y_5)\Delta_{Q_0}^{(7)}],$$

where through $\Delta_{Q_0}^{(1)}, \Delta_{Q_0}^{(3)}, \Delta_{Q_0}^{(5)}, \Delta_{Q_0}^{(7)}$ the determinants are denoted:

$$\Delta_{Q_0}^{(1)}=\begin{vmatrix}H_{23} & H_{25} & H_{27}\\ H_{53} & H_{55} & H_{57}\\ H_{83} & H_{85} & H_{87}\end{vmatrix};\quad \Delta_{Q_0}^{(3)}=\begin{vmatrix}H_{21} & H_{25} & H_{27}\\ H_{51} & H_{55} & H_{57}\\ H_{81} & H_{85} & H_{87}\end{vmatrix};$$

$$\Delta_{Q_0}^{(5)}=\begin{vmatrix}H_{21} & H_{23} & H_{27}\\ H_{51} & H_{53} & H_{57}\\ H_{81} & H_{83} & H_{87}\end{vmatrix};\quad \Delta_{Q_0}^{(7)}=\begin{vmatrix}H_{21} & H_{23} & H_{25}\\ H_{51} & H_{53} & H_{55}\\ H_{81} & H_{83} & H_{85}\end{vmatrix}.$$

A comparison of expression (24) with the analytical expressions obtained earlier for orthonormal fundamental functions confirms the assumption made at first, namely:

$$G(y-\xi)=A_{14}(y-\xi). \quad (25)$$

It can be verified that the constructed function $G(y,\xi)$ has all the properties characteristic of the Green's function:

a) $G(y,\xi)=0$ at $y<\xi$;

b) $G(y,\xi)$ as function of y at fixed ξ in the interval $(0, y_{boundary})$, excluding point $y=\xi$, satisfy the differential equation of problem;

c) $G(y,\xi)$ and its derivatives by y up to n-th order for $y\in(0, y_{boundary})$, excluding point $y=\xi$, in which the derivatives with respect to y are continuous only up to $(n-2)$ order, and $(n-1)$ the derivative has a discontinuity of the first kind with a jump

$$\left.\frac{d^{(n-1)}G(y,\xi)}{dy^{(n-1)}}\right|_{y=\xi+0}-\left.\frac{d^{(n-1)}G(y,\xi)}{dy^{(n-1)}}\right|_{y=\xi-0}=\frac{1}{a_0};$$

d) at $y=\xi$,

$$G(\xi,\xi)=G'(\xi,\xi)=...=G^{(n-2)}(\xi,\xi)=0;\ G^{(n-1)}(\xi,\xi)=\frac{1}{a_0}=1;$$

e) $G(y,\xi)$ for the differential equation of the problem (and this is a differential equation with constant coefficients) depends only on the difference $(y-\xi)$.

3 Conclusions

Thus, the basic analytical apparatus for solving the problem of the bending of a long cylindrical shell by a numerical and analytical boundary elements method is constructed. 64 expressions for all the fundamental functions of the problem are obtained, as well as an analytic expression for the Green's function, which makes it possible to construct a load vector (without any restrictions on the nature of its application), and then proceed to the solution of boundary-value problems for the bending of long cylindrical shells under various boundary conditions.

References

1. V.Z. Vlasov, *General theory of shells and its application in technics* (State Publishing House of Technical and Theoretical Literature, Moscow, 1948)
2. N.V. Kolkunov, *Basics of elastic shells calculations* (High School, Moskow, 1972)
3. S.P. Timoshenko, S. Woinowsky-Krieger, *Plastiny i obolochki* (Science, Moscow, 1966)
4. E. Reissner, *On the foundations of Generalized linear shell theory* (Copenhagen, 1967)
5. P.M. Naghdi, Fl. Han. Der Ph. Bd., **VI**, 425-640, (1972)
6. S.L. Crouch, A. M. Starfield, *Boundary Element Methods in Solid Mechanics: With Applications in Rock Mechanics and Geological Engineering* (George Allen & Unwin, Crows Nest, 1983)
7. C.A. Brebbia, *Recent Advances in Boundary Element Methods* (Pentech Press, London, 1978)
8. A.F. Daschenko, L.V. Kolomiets, V.F. Orobey, M.H. Surianinov, *Numerical and analytical method of boundary elements* (VMV, Odesa, 2010)
9. V.F. Orobey, M.H. Surianinov, *Praktworkshop on solving boundary value problems of mechanics* (Astroprint, Odesa, 2011)
10. A.F. Bermant, *Mathematical Analysis Course* (State Publishing House of Technical and Theoretical Literature t, Moscow, 1953)

Abrasion of cement-concrete and his contents investigation

Sergii Tolmachov[1,*], *Olena* Belichenko[1], and *Dmitrii* Tolmachov[2]

[1]Kharkov National Automobile und Highway University, ul. Yaroslava Mudrogo, 25, 61002 Kharkov, Ukraine
[2]LLC "Elavtopas", ul. Gvardijciv-Shy`roninciv, 33, 61170 Kharkov, Ukraine

Abstract. The paper presents studies of abrasion of concretes made on various rubble: granite, syenite and limestone. The purpose of the research was to assess the wear of each element of the concrete structure and its contribution to the overall wear resistance of concrete. For this purpose, experimental studies were carried out on the wearability of coarse aggregate, cured cement paste, cement-sand mortar and concrete on different aggregates. Studies of abrasion were carried out on concretes hardened in various conditions: normal, air-dry and moisture-proof moisture (under the film). The hardening conditions of the samples play a major role in the change in abrasion and can lead to an increase in abrasion by several times. It is shown that the greatest abrasion is observed in the hardened cement test, as well as in cement-sand mortar. It is established that the wearability of a coarse aggregate from dense rocks increases with an increase in the attrition path (the number of cycles of abrasion). The abrasion of the moisture-proof moisture of the lower part of the samples from the hardened cement stone, mortar or concrete may be lower than the abrasion of the upper, not protected from evaporation of moisture.

1 State of the issue

1.1 Analysis of studies and publications

Aggressive factors that influence the road cement-concrete result in its destruction [1-6]. These factors are aqueous solutions of salts, gases, heating and cooling, freezing and thawing, etc. Therefore it is important to quickly determine how external conditions affect the change, that takes place in the structure of concrete. Most of researches accept change in concrete strength as a parameter that allows for conducting such estimation.

They compare concrete strength before the beginning of effect of aggressive factors and after the fixed exposure time. However, for road concrete, strength is not always a parameter that allows for defining just how much its durability goes down. It is necessary to take into account that aggressive actions affect primarily the concrete topping that is often in a wet state, which leads to its destruction.

* Corresponding author: Tolmachov.serg@gmail.com

The action of aggressive factors takes place jointly with the mechanical motor transport loading [7]. Under such conditions, first of all, wear-resistance decreases rather than strength. Its criterion is wearability. Combination of physicochemical and mechanical factors leads to rapid destruction of road concrete pavements [7-10].

1.2 Relevance

According to ideas of Prof. I.M. Grushko and other researchers, cement concrete can be viewed as a structure within another structure. Concrete structure contains cement mortar. Cement mortar structure contains hardened cement paste. Furthermore, concrete is composite material that consists of crushed stone, sand, cement and water. Wearability of each element is different, but there is very little research on this matter. For this reason, assessment of wear of each element of structure and its contribution to general wear-resistance of concrete is relevant.

2 Materials and methods of research

In research the following typed of crushed stone were used: granite, cyenit and limestone. Fraction of crushed stone is 5-10 mm. Crushed stone strength grade was M 1200 (120 MPa). Fine aggregate was quartz and lime sand with fineness modulus = 2.1...2.4. Binding cement was of PC I 500 (CEM I 52,5) grade.

Wearability was determined on cubic samples with dimensions of 7x7x7 cm. They were made of rocks as well as hardened cement paste, cement-sand mortar and concrete. Cement paste, mortars and concrete had been hardening for 28 days in different conditions. After that samples were subjected to abrasion on laboratory abrasion table (LKI-3) in accordance with standard DSTU B.2.7-212 (fig. 1, 2).

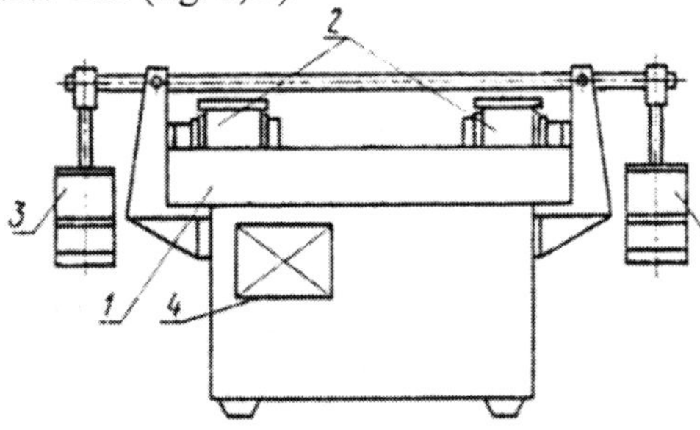

Fig. 1. Diagrammatical representation of the abrasion disk of LKI-3 type: 1 - abrasion disk; 2 - the tested samples; 3 - loading device; 4 - number of turns of disk counter

Fig. 2. The exterior of abrasion disk of LKI-3 type

According to the requirements of DSTU B.2.7-212, cubic samples were placed in two metal holders secured above the surface of the disk, so that their lower edge was on the surface of the disk. Vertical loading of 300±5 N, which corresponds to a pressure of 60 ± 1 kPa, was applied to each sample. 20 g of the special sand were evenly poured on the surface of the disk made from cast-iron. During tests of the water-saturated (moist) samples sand was moisturized by 15 cm^3 of water. After that, the device was switched on. After 28 turns of disk, which corresponds to 30 m of the abrasion track, the device was switched off. Then sand and pieces of the destroyed samples were removed from the surface of disk. A new portion of sand was poured and device was switched on once again. This operation was repeated 5 times, which corresponded to 1 cycle or 150 m of the abrasion track.

After that samples were taken out from a holder, turned 90° around vertical axis and then cycle was restarted. According to the DSTU B.2.7-212, a sample must pass 4 cycles of tests. In our research, 4 and 6 cycles of tests were conducted. According to DSTU B.2.7-212, a test must be applied only to the bottom part of sample. But in our research we tested bottom and top sides of samples.

Samples were weighed before the beginning of tests and after every cycle. Also the height of sample was measured before the beginning of tests and after every cycle.

Abrasion was determined by an Eq. (1):

$$G = \frac{m_1 - m_2}{S}, \tag{1}$$

where: m_1 is the mass of sample before the beginning of tests, g; m_2 is the mass of sample after the cycles of tests, g; S - an area of the side of sample, that is subjected to abrasion, cm^2.

During testing abrasion was applied not only to bottom but also to top side of mortar and concrete. This is due to the fact that in case of violation of conditions of hardening and absence of curing of hardening mortar or concrete, the surface of sample becomes too flimsy due to the increase of porosity in the top part of the sample. Wearability must increase in this case.

3 Experimental research

The results of tests of wearability of coarse aggregate, hardened cement paste, cement-sand mortar and concrete are shown in Table 1.

It is evident that magmatic rock aggregates, for example, granite or syenite, wear off slowly. The maximal value of wearability of granite does not exceed 0.028, and syenite – 0.036 g/cm^2 after 4 test cycles and 0.036 and 0.053 g/cm^2 after 6 test cycles. A metamorphosed rock, such as a marble limestone, is worn down by 1.5…2.2 times more than magmatic rocks. It is possible to see a conformity: with the increase in a number of test cycles wearability of all rocks increases approximately by the same value. This shows high homogeneity of textures of these rocks, their high density and low porosity.

In comparison with wearability of concrete, it is possible to see that the wear of the samples made from these rocks is insignificant. Therefore, wearability of the crushed stone made from these rocks is 9…19 % comparing to the general wearability of concrete, at the wear of its friable top part. As compared with wearability of bottom side of concrete, wearability of rocks is even less and makes 4…8 % (Table. 1). Then wearability of hardened cement paste was studied.

Will abrasion investigation of hardening cement paste. Cement paste samples were hardening in a laboratory at the temperature of air 22...25 °C and relative humidity 50…60 %. No special actions on hardening cement curing were conducted. A top side of a sample was open and moisture evaporated through it. The bottom side of a sample was protected from evaporation of moisture by polyethylene.

Investigations of the effect of various additives on the strength of concrete were carried out on equally mobile concrete mix in the Table 1. Studies have shown that when the mineral additive is added into the concrete, strength increases, but for compressive strength and flexural strength this increase is different. The compressive strength of concrete with a mineral additive increases by only 3...6 % at different times of hardening. At the same time, the flexural strength of concrete with a mineral additive increases by 18, 10 and 8 % on 3, 7 and 28 days, respectively.

Table 1. Wearability of elements of concrete structure

Tape of material	Wearability, face	Wearability, *G*, g/cm^2 before wear cycles	
		4	6
Granite	-	0.028	0.036
Cyenit	-	0.036	0.053
Marble limestone	-	0.057	0.081
Cement stone	bottom / top	0.173/1.000	0.258/1.080
Cement-sand mortar of the composition:			
C/Q = 1:1	bottom / top	0.176/0.372	0.230/0.417
C/Q = 1:2	bottom / top	0.201/0.552	0.265/0.623
C/Q = 1:3	bottom / top	0.205/0.617	0.306/0.705
C/Q = 1:4	bottom / top	0.372/1.110	0.447/1.260
Concrete composition C:S:K = 1:1.6:3.1	bottom / top	0.292/0.714	0.380/0.825

A distinction is obvious between wearability of the exposed top side of the sample from hardened paste and its protected top side (table. 1). After 4 cycles of tests wearability of top side was 5.6 times more. This confirms that top layer of a sample of hardened paste, from which moisture has evaporated, becomes friable and wears down easily.

After 6 cycles of wear, the difference in wearability of top and bottom sides decreased to 4.1 times. It was also found that while wearability of bottom side of sample continues to increase with increasing of number of cycles, wearability of top side decreased considerably after 4 cycles. This shows that there is a limit of thickness of a layer of heavy wear and it is possible to determine its value.

Table 2. Wearability of hardened cement paste

Bottom side	cycles						
	0	1	2	3	4	5	6
Sample height, *h*, mm	70.1	69.8	69.6	69.4	69.2	69.1	69.0
Wearability, *G*, g/cm^2	0	0.065	0.107	0.145	0.180	0.223	0.264
Top side	cycles						
	0	1	2	3	4	5	6
Sample height, *h*, mm	70.0	69.0	68.4	68.0	67.6	67.3	67.1
Wearability, *G*, g/cm^2	0	0.587	0.765	0.886	1.00	1.04	1.08

As can be seen from the Table. 2, the bottom side of a sample is worn down evenly, which is confirmed by data of change of wearability and height of sample. The depth of the layer of wear of bottom side of a sample does not exceed 0.9 mm. Top layer of the sample became quit friable and is subjected to rapid wear, which is confirmed by results of the research. Experimentally determined thickness of friable zone of cement stone at 4 cycles of wear was 2.4 mm from the surface of sample (table 2). At the further increase of amount of test cycles from 4 to 6, the wear of top side increases insignificantly, both in value and depth, reaching 2.9 mm.

It may be noted that wearability of cement stone in relation to general wearability of concrete is quite big. For example, when comparing wearability of protected sides, wearability of concrete stands at 0.292 g/cm^2, and wearability of hardened cement paste reaches 0.18 g/cm^2, which makes 62 % of general wear concrete.

When comparing wearability of unprotected sides of samples of concrete and hardened cement paste, wearability of cement sample is 3.4 times more than wear of concrete sample. This proves that even under normal conditions of hardening wearability of concrete heavily depends on wearability of hardened cement paste. Violation of the conditions of normal hardening leads to the situation when wearability of concrete primarily depends on extent of wearability of hardened cement paste.

Research has shown that wearability of protected bottom side of samples of cement mortars of different composition is significantly less than that of unprotected top side (table 1). There is the same conformity as for estimations of wearability of hardened cement paste.

Obviously, with decreasing of content of cement in the composition of mortar wearability increases for both top and bottom sides. In the mortars of compositions C:P = 1:1…1:3, containing 900…450 kg of cement per 1 m^3, wearability of bottom sides differs insignificantly (0.176…0.205 g/cm^2). When switching to mortars of compositions C:P = 1:4 with cement consumption 350…400 kg/m^3, wearability of bottom side of the sample increases significantly, 1.85 times as compared to other compositions. This can be explained by decreasing of density of solution due to the increase of porosity, and open porosity in the first place. There is even bigger difference in wearability for top sides of the samples.

Table 3. Water absorption of mortars

Proofness of sides of samples during hardening	Water absorption mortar, W, %, by 24 h			
	C:S=1:1	C:S =1:2	C:S =1:3	C:S =1:4
Both sides are protected	3.8	5.1	6.2	8.5
Top side is open	5.9	7.7	9.3	12.8

In our research applied the standard method of determination of water absorption of simples was applied, in accordance with DSTU B.2.7-170:2008. According to this method, in order to determine water absorption, samples were dried to constant mass at the temperature of 105±10C°. Then they were cooled at the temperature of 18±2C° and submerged in water at the same temperature. Samples were periodically taken out from water and weighed. According to the last weighing, in 24 hours after water-logging of samples, water absorption of mortar (concrete) on mass was determined. We used an Eq. (2):

$$W = \frac{m_2 - m_1}{m_1} \cdot 100\%, \tag{2}$$

where m_1 is the mass of the sample dried to constant mass; m_2 is mass of sample that had been in water for 24 hours.

Research of water absorption of mortars of different composition showed that while going from mortars of C:S = 1:3 to mortars of C:S = 1:4 there is a sharp increase in this index (Table. 3). In the case when both sides of samples of mortars of composition of C:S = 1:4 are protected from water evaporation during period of hardening, water absorption increases by 2.3% or on 37 %. Approximately the same increase of water absorption takes place in going from mortars of composition of C:S = 1:1 to C:S = 1:3 (increase cement content from 900 to 450 kg/m^3). A similar increase of water absorption takes place for mortars where bottom side wasn't protected from evaporation of moisture during period of hardening.

It can be noted that for the samples of mortars where the top side was open water absorption is 1.5 time more than for samples where both sides were protected from evaporation of moisture in the period of hardening. It confirms that the increase of wearability of mortars is proportional to the increase of open porosity of mortars. The method of water absorption allows for accurate enough estimation of a number of open pores mainly (first of all, capillar). Therefore the index of water absorption depends on a number of open pores. With the increase of an amount of evaporated moisture, the volume of open pores increases steadily. Accordingly, water absorption of samples must increase. In parallel, due to the increase of porosity, especially of the top layer of a sample, wearability will increase. Results of our research which are presented in a Table 4, confirm this.

Table 4. Wearability of cement mortars of composition of C:S = 1:3

Conditions of manufacturing	Average density	Water absorption		Wearability	
		15 min	24 h	wearing depth	G, g/cm^2
Vibroflotation	2200 kg/m^3	6.6%	8.0%	2.4 mm	0.523
Pressing with 20 MPa	2020 kg/m^3	4.9%	5.8%	2.7 mm	0.538
Pressing with 90 MPa	2280 kg/m^3	3.5%	3.9%	1.9 mm	0.418

Cement mortars of composition of C:S = 1:3 were compacted at different pressure and the indexes of water absorption and readability were estimated. For this purpose vibration was applied, normal pressing at pressure 20 MPa and hyper pressing at pressure 90 MPa. In the first 15 minutes, there is a dampness penetration of the largest open pores and capillaries. The biggest macroporosity occurs in samples that were compacted by vibration – 6.6 %. The least macroporosity occurs in samples that were compacted at high external pressure – 3.5 % (so-called hyper pressing). Accordingly, the least wearability, both in the depth of the layer of wear (h = 1.9 mm) and in general size (G = 0.418 g/cm^2) was discovered in the hyper pressed samples. The density of these samples is higher than that of samples compacted in another ways.

The results of experiments showed that there is correlation between wearability of mortars and number of open pores in them, first of all in top side of the sample. Assuming that density of mortars with decrease in the number of binder in their composition goes down, in this case conditions for the increase of an amount of open pores are automatically created. Most of open pores are in the top part of a sample, from the surface of which moisture evaporates. Then friable top layer of a mortar wears down easier and quicker. Taking into account that most of open pores are in mortar phase, wearability of concretes with the cement consumption of more than 350…400 kg/m^3 will depend on wearability of cement mortar.

The estimation of wearability of concrete was conducted on the samples of composition that is most often used for manufacturing (cement consumption is 350.400 kg/m^3). The data show (Table. 1) that wearability of such concrete is comparable to wearability of mortars with cement consumption of 400…500 kg/m^3. This is true for top and bottom sides of samples and proves the assumption form above. It may also be noted that wearability of top side of concrete sample is 2.2…2.4 times higher, than that of bottom side, which correlates with similar data for a cement stone and mortars.

Taking into account that the conditions of hardening strongly affects wearability of concretes, research was conducted on wearability of samples made from concretes that were hardened in different humidity conditions (Table. 5).

Table 5. Wearability of concretes that were hardening in various conditions

Conditions of concrete hardening	Side under wear	Wearability, G, g/cm^2	
		C-350 C-1250 S-750 kg/m^3 (Granite crushed stone, quartz sand)	C-350 C-1250 S-750 kg/m^3 (Limestone crushed stone and sand)
Normal	top	0.496	1.682
	bottom	0.271	0.968
Air-dry	top	0.601	2.025
	bottom	0.365	1.368
Under the tape	top	0.368	1.092
	bottom	0.243	0.912

The results of experiments show the following: when hardening of samples in air-dry terms (the temperature of 25...30 C° and relative humidity of 40…50 %) wearability of top side of the samples of concretes increases by 21…63 % comparing to the wearability of samples, that were hardening in normal conditions (the temperature of 18…22 °C, and relative humidity of 95…100 %) and were covered by polyethylene tape. This applies to the samples made on granite crushed stone and quartz sand. For the samples made on a limestone crushed stone and sand, this difference is 21…85 %.

The conditions of hardening change the wearability of bottom side of samples of concrete too. The difference in wearability of samples that were hardening in air-dry conditions is 35 % compared to samples that were hardening in normal conditions and 50 % compared to the samples that were hardening under tape. For samples on limestone aggregates, the difference in wearability of bottom side is 41% and 50 % respectively.

Analysis of data of Tables 1-5 shows that there is a critical cement consumption in the composition of concrete, and it equals 350…450 kg/m^3. With reducing of expense of cement below the level of 350 kg/m^3 or increasing it for more than 450 kg/m^3 wearability increases sharply, which results in reduction of life of concrete.

With the increase in cement consumption in concretes from 250 to 450 kg/m^3 wearability of samples also changes. For example, for samples that were hardening in laboratory conditions, wearability of bottom side with the increase of consumption cement within the specific limits goes down from 0.332to 0.221 g/cm^2, and for the top side, vice versa increases from 0.425 to 1.521 g/cm^2. These data are consistent with results of the above-mentioned research (Table. 1).

Conclusion

1. It was extablished that wearability of coarse aggregate from dense rocks increases with the increase of path of abrasion (number of cycles of wear) by the same value. This value is very small in comparison with the general wearability of concrete.

2. It is shown that the conditions of hardening of samples play a key role in the change of wearability and can lead to increasing of in the height of wearability by 5…6 times.

3. It has been proven that wearability of the protected from evaporation of moisture bottom parts of the samples, made of hardened cement stone, mortar or concrete, can be up to 6 times lower than wearability of top side, which was not protected.

4. The greatest wearability was detected for hardening cement paste, and also cement-sand mortar, especially in case when hardening took place in air-dry terms.

5. Wearability of cement mortars and concretes depends on their content of binder and increases when cement consumption is less than 350 kg/m^3 or more than 450 kg/m^3.

References

1. Luiz Fernandez, Malhotra V. Mohan, Cement, Concrete and Aggregates, **2**, 87-100 (1990)

2. M. Moukwa, Cement and Concrete Research, **3**, 439-446, (1990)

3. A.N. Plugin, I.G. Prokopova, S. Wild, A.A. Plugin, Goteborg: Inform Trycket AB, **4 (4iv075)**, (1997)

4. S.N. Tolmachev, Scientific Herald of DonGGAA, **2004-1(43), 2**, 133-137 (2004)

5. I. Stark, B. Whitch, *The durability of concrete* (Kiev: Oranta, 2004)

6. Gheorghe Tudorie, Alexendru Pasnicu, Rev.transp.: Auto, drumuri, navig, **3-4**, 65-69 (1991)

7. S.V. Shestoperov *The durability of concrete and reinforced concrete constructions* (Moscow: Stroyizdat, 1966)

8. E. Koskinen, Tienpaaallysteiden ongelmat. "Tie ja liikenne", **3 (51)**, 116 (1981)

9. S.N. Tolmachev, I.G. Kondratyeva, A.N. Chugueenko, Scientific Bulletin of Construction, **43**, 300-304 (2007)

10. S.N. Tolmachev, R.A. Grinchenko, L.D. Marakina, Scientific Bulletin of Construction, **39**, 198-204 (2006)

Increase of fire resistance of coating wood with adding mineral fillers

Yuriy Tsapko[1,2*], *Volodymyr* Kyrycyok[1], *Aleksii* Tsapko[2], *Olga* Bondarenko[1], and *Sergii* Guzii[1]

[1]Kyiv National University of construction and Architecture, Scientific research Institute for binders and materials named after V. Glukhovskii, Povitroflotsky Avenue 31, 03037, Kiyv, Ukraine
[2]National University of Life and Environmental Sciences of Ukraine, Heroes of Defense st. 15, 03041, Kiyv, Ukraine

Abstract. The article covers all the stages of the process of fire protection: as the separation of gases, heat transfer, the swelling of the coating, which is formed during by the action of fire. To determine the fire protection efficiency in the coatings developed, studies have been carried out on determining the combustibility of wood in terms of mass loss and increase in the temperature of flue gases, and it has been established that mass loss of coated wood does not exceed 3% and the temperature does not exceed 160°C. The results of the determination of the protection efficiency for the system have shown that under the influence of the high-temperature flame of burning of the material and the loss of the mass of the coating is reduced due to the formation of high-temperature compounds, while the time to reach the limiting temperature increases. Studies on determining the quality of fire protection wood treatment by coating showed that the rate of cooking for a rough sample of wood was more than 2 mm/min, for fire-proof samples - did not exceed 0.5 mm/min, which makes it possible to conclude that it is advisable to use them.

1 Introduction

Since cellulosic materials and products (wood, plywood, paper and fiber materials, and others) are the main conductors of flame propagation, they require adequate protection. Therefore, fire safety is placing increasingly high demands on the effectiveness of fire protection means, as well as on the quality of fire protection materials.

The main requirements for fire protection of cellulose-containing materials are to provide them with the ability to withstand the effects of fire and not to spread the flames on the surface. The change in the decomposition of the material during fire protection is directed toward the formation of non-combustible gases and heavy-duty coke overshoot, as well as inhibition of oxidation in the gas and condensed phase. Fire protection of coatings based on organic binders is effective under the condition of their operation without the action of water and ultraviolet radiation. Therefore, the creation of organo-mineral coatings is a topical issue of creating a modern fire protection of building structures.

* Corresponding author: juriyts@ukr.net

2 Analysis of recent research and publications

The peculiarity of fire protection of building structures is to create on the surface elements of constructions of heat-insulating, which withstand high temperatures and direct effect of fire. Their presence allows to slow heating of material and keep functions of construction at a fire during the set period of time and makes wood hard burning materials [1].

Modern methods of fire protection of building structures are based on the use of expanding coatings, which are complex systems of organic and inorganic components. But with prolonged action of the flame, they are capable of gradually burning and, accordingly, reducing the fire-resistance of the structure, which requires the addition of mineral fibers capable of forming a more stable foam coke layer [2].

In the recent years, from the proposed direction of research, works are being known to the synthesis of coatings using organic lacquers, refractory oxides and silicates, which in the process of heating form thermo- and heat-resistant ceramic phases [3]. The most common enamel and glass-crystalline coatings [4-6], however, can not provide reliable protection of constructions at temperatures above 1000 °C, since the higher temperatures of exploitation results in the destruction of the organic component, and the inflicted coating becomes porous, which greatly impairs its operational properties.

The influence of inorganic fillers on the flame retardant coating has shown its effectiveness, however, which coating work is not specified and no changes in heat conductivity are observed during the transition of coatings to coke [7-9]. The authors showed only an analytical model for calculating the thermal conductivity of the protective layer of foam coke during the coating process, which takes into account the forms of the pore, but which transformations in the material occur at combustion temperatures is unknown.

Therefore, the installation of the influence of the components included in them on the thermal conductivity and the integrated determination of the effectiveness of fire protection is an unresolved component of ensuring the fire resistance of building structures. This led to the need for research in this direction.

3 Determine the purpose and objectives of the study

The conducted researches were aimed at determining the features of fire protection of structures from the effect of high temperature when added to the coating of mineral fillers.

To achieve the goal, the following tasks were solved:

– to determine the features of combustion of of wood with a protective coating on the basis of organo-mineral substances;

– to establish the efficiency of application of mineral fillers of coatings after the thermal action of combustible mixtures, which are characterized by high-temperature effect.

4 Materials and methods of research

Experimental samples of wood coatings were prepared on the basis of an organo-inorganic system containing 18÷20% PFA, 12÷14% melamine, 10÷12% PE and astringent, consisting of 16% PVA dispersion and water. For the coating of wood, the given organo-inorganic mass was mixed, filled with fillers (minerals - aluminosilicate microspheres, perlite, basaltic scales, metallurgical sludge, ash) in the amount of 10% and applied to wooden samples made from wood of pine wood thickness of 30 mm average size of 150×60 mm.

For investigations of the effectiveness of fire protection under the influence of high-temperature flame during combustion of metal, model samples of wooden structures made of wood with a thickness of 19 mm, average size of 190×155 mm and a height of 140 mm, were used, and treated with coatings.

The study on determining the combustibility group of wood treated with the proposed coating was carried out in accordance with the current normative base. The essence of the test method for the experimental determination of the group of heavy and combustible solids and materials according to the influence of the sample is located in the ceramic tube of the OTM installation, the flame of the burner with the given parameters (the temperature of the gaseous combustion products at the exit from the ceramic tube is 200oC ± 5oC) In the course of experimental studies, the maximum increase in the temperature of the gaseous combustion products (Δt) and the sample mass loss (Δm) are recorded. According to test results, materials are classified as:

– heavyweight – $\Delta t < 60$°C and $\Delta m < 60\%$;

– combustible – $\Delta t \geq 60$°C or $\Delta m \geq 60\%$.

Determination of fire-fighting efficiency of wood during field tests was carried out using a working methodology, the essence of which was to experimentally determine the effectiveness of fire protection of a building structure made of wood and treated with protective means in the effect of a sample of magnesium flame (model solid ignition) with specified parameters and registration mass loss and degree of charring of the sample after the test.

5 The main part of the study

The results of studies to determine the mass loss of samples (Δm,%) and the increase in the maximum temperature of gaseous combustion products (Δt, °C) of samples of fire retardant wood carried out in laboratory conditions are shown in Fig. 1, 2.

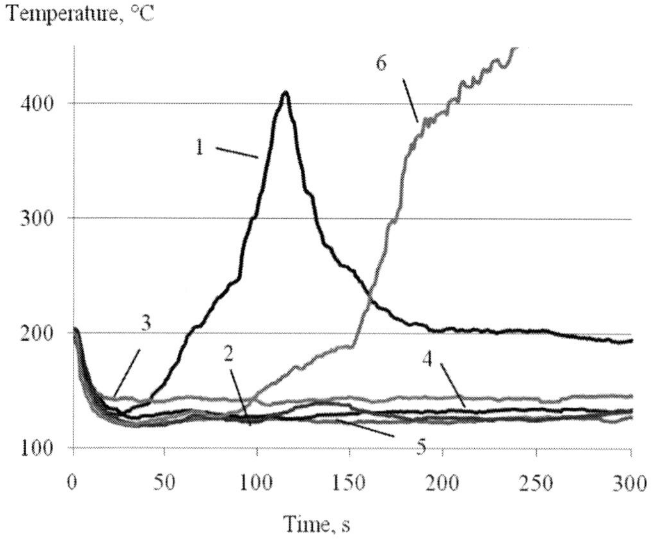

Fig. 1. Dynamics of increase of temperature of flue gases in tests of fire-proof wood on an organic basis (1) with the addition of: 2 – aluminosilicate microspheres; 3 – perlite; 4 – basalt scales; 5 – ash; 6 – metallurgical sludge.

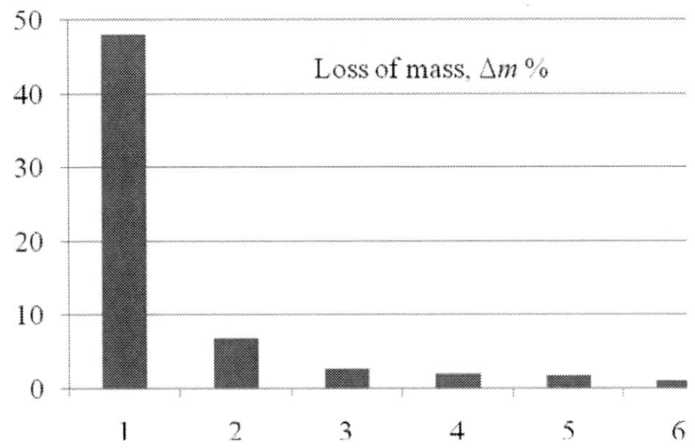

Fig. 2. Results of mass loss of samples Dm,% of fireproof wood with organic coating, with addition: 1 – untilled; 2 – metallurgical sludge; 3 – perlite; 4 – basalt scales; 5 – ash; 6 – aluminosilicate microspheres.

Studies have shown that rough wood is a combustible material, fire-retardant wood that has been treated with an organic coating with the addition of aluminosilicate microspheres, perlite, basaltic scales, and ash has withstood the temperature effect and relates to heavy materials on the basis of weight loss and flue gas temperature rise. However, the addition to the coating of metallurgical sludge initially countered the thermal effects, but over the years led to an increase in temperature and, consequently, greater loss of sample mass.

Therefore, to determine the effectiveness of fire protection under the influence of mineral fillers on the coating, under the action of a higher temperature, conducted appropriate physical tests.

Table 1 shows the results of mass loss and combustion time of the model center of the wooden construction, Table 2 wood charcoal depth.

Table 1. Results of tests of wooden construction.

Protective coating with addition	Sample weight, kg		Burning time of sample, s	Area of sample damage S, m^2	Speed of burning down of standard υ, $kg/(m^2\cdot s)$
	To	After			
	tests				
untilled	1.620	0.972	1301	0.0342	0.0145
aluminosilicate microspheres	1.874	1.785	567	0.0340	0.0046
perlite	1.861	1.754	783	0.0342	0.0040
basalt scales	1.569	1.498	447	0.0341	0.0047
ash	1.761	1.671	552	0.0342	0.0048
metallurgical sludge	1.621	1.439	501	0.0342	0.0106

During the tests of combustion of magnesium shavings the formation of a slag layer on the surface of the material was established under which the process of thermal decomposition of wood continued and depending on the properties of the coating, its suspension was carried out in different ways and affected the depth of carbonation. The depth of roasting of rough wood reached 16÷20 mm. For a model sample treated with a protective coating with the addition of mineral substances. the outline of the protective coating was observed during the interaction with the flame and there was charcoal of the wood under the coating layer at a depth of only 5÷6 mm, respectively.

Table 2. Results of tests of wooden construction.

Protective coating with addition	Depth of charring. mm	Speed of charring. mm/min	The time of the flame of magnesium on the surface of the material *t*. min
untilled	16÷20	1.93	20.63999
aluminosilicate microspheres	4÷5	0.61	9.019946
perlite	5÷5.2	0.53	13.66021
basalt scales	3÷4	0.62	7.530295
ash	4÷5	0.64	9.937369
metallurgical sludge	5.5÷6.0	1.41	6.097788

So the time of exposure of the surface of fire-proof wood to magnesium flame can be estimated if its charcoal depth is known by the equation [10, 11]:

$$t = \frac{x_e^{1,23}}{c}, \qquad (1)$$

where t - time of fire influence, min;

c – charcoal speed, mm/min;

x_e – charcoal depth, mm.

The prediction of the average rate of charcoal was carried out according to the dependence:

$$c = \frac{\upsilon}{\rho_w}, \qquad (2)$$

where m – burning rate of wood (speed of mass loss), kg / (m^2·s);

ρ_w – density of wood in dry condition (humidity 10÷12%), 450 kg/m^3.

The rate of loss of mass of timber samples was calculated by the equation:

$$\upsilon = \frac{\Delta m}{\tau \cdot S}, \qquad (3)$$

where Δm – sample mass loss after testing;

τ – test time;

S – area of sample damage.

As can be seen from Table 2, the samples is treated with a protective coating with the addition of mineral fillers which leads to the creation of an expanded layer of coke and prevents the influence of high temperature and the rate of charcoal wood, and the addition of metallurgical sludge somewhat reduces resistance to high temperature.

6 Conclusions from research and prospects. further development in this direction

These studies have established the effectiveness of mixtures of organo-mineral mixtures as fire retardant coatings for wood. in particular:

– in the case of the effect of heat from standard burner burning on samples of treated coatings on the basis of organo-mineral mixtures. the weight loss and the temperature of flue gases did not exceed the specified value. however. a protective layer which has a substantial thickness to the building structure to ensure its protection should be used. but the case of a sample treated with a coating on the basis of organic-mineral compounds. the fire resistance of wood is significantly increased due to the formation of a coke layer with minimal consumption of coating;

– full-scale tests using model designs of wooden construction constructions under the influence of high-temperature flames have shown that coatings based on mixtures with the addition of minerals can withstand high temperatures. effectively prevent the penetration of heat to the material due to the formation of expanded layer of coke. which affects the rate and depth of absorption.

Further research will focus on the processes of forming the structure of the protective layer and establishing a connection between the nature of the components and the properties of the coatings. such as the formation of foam coke. as well as resistance to heat and weather.

References

1. V. Artemenko, Collection of scientific works of LDU BZD. Fire safety, **25**, 6-11, (2014)
2. A. Kravchenko, Yu. Tsapko, O. Tsapko, Building materials. products and sanitary equipment NDIBMV, **57**, 61-67, (2016)
3. M. Givlyud, O. Bashinsky, S. Vovk, Collection of scientific works of Lviv State University, **18**, 40-45, (2011)
4. S. Guzii, R. Hela, V. Kirichok, Advanced Materials Research, **688**, 107-112, (2013)
5. B. Ciripi, Y. Wang, B. Rogers, Fire Safety Journal, **81**, 74-84, (2016)
6. P. Kryvenko, V. Kyrychok, S. Guzii, Eastern-european Journal of Enterprise Technologies, **5 (83)**, 40-48, (2016)
7. Yu. Tsapko, A. Tsapko, East European Journal Enterprise Technologies, **5/10 (89)**, 55–60, (2017)
8. O. Bondarenko, S. Guzii, E. Zacharchenko, East European Journal Enterprise Technologies, **6/11 (78)**, 41–47, (2015)
9. F. Fan, Z. Xia, Q. Li, Z. Li, Progress in Organic Coatings, **76 (5)**, 844-851, (2013)
10. T. Shnal, *Fire resistance and fire protection of wooden structures* (Lviv, Lviv Polytechnic, 2006)
11. Yu. Tsapko, A. Tsapko, East European Journal Enterprise Technologies, **3/10(87)**, 50–55, (2017)

Durability of beams with hybrid reinforcement from metal and basalt fiber reinforced polymer (BFRP) armature

Alexander Valovoi[1,*], *Peter* Koval[2], *Alexander* Eremenko[1], *Maksim* Valovoi[1], and *Sergei* Volkov[1]

[1] Kryvyi Rih National University, Vitaly Matusevich Str. 11, Kryvyi Rih, Ukraine, 50036.
[2] National academy of fine arts and architecture, Voznesensk Privoz Str.20, Kyiv, Ukraine, 04053.

Abstract. The article presents the program and the results of experimental tests of beam samples with metal, basalt fiber reinforced polymer (BFRP) and hybrid (metal and BFRP) reinforcement. The samples were made of standard concrete with quartz sand as the fine aggregate and concrete with fine fraction wastes of a mining and beneficiation complex (MBC) used instead of the sand. Short-run tests of the beams under monotonous static loading until destruction enabled the conclusion that durability of the BFRP reinforced beams increased by 37-44% as compared to the metal reinforced beams. When hybrid reinforcing, reduction of the (BFRP) content did not produce an effect on decrease of durability indices; durability gains compared to the beams reinforced by metal made 38-41%. In the BFRP reinforced beams, due to the absence of plastic deformations in this reinforcement, there were no residual deformations after cessation of loading despite significant damage and deterioration of the concrete. Samples of beams made of concrete on fine fraction wastes of MBC, showed 1-8% higher strengths in comparison with similar beams made of concrete with quartz sand as the fine aggregate.

1 Introduction

Composite reinforcement is known to possess high tensile strength, but a lower modulus of elasticity as compared with steel [1-3]. It was found [2-8] that the experimental samples showed a significant increase in durability, by 25-90%, vs. the samples reinforced by metal armature. Due to the low modulus of elasticity, the deformability of the samples reinforced by BFRP was higher. Thus, in [8], it is stated that the deflections and width of cracks in bended samples, reinforced by composite armature, were 60% higher than those reinforced by metal reinforcement, at a stress level of 70% of the destructive one. Similar results were obtained in the works [2-10, 11].

It should be noted that the increased deformability of structures is specific both for (BFRP), and for glass fiber reinforced polymer (GFRP) [9, 10].

One of the ways to reduce the deformability of bending structures, reinforced with BFRP armature, is to use mixed (hybrid) reinforcement, in which both the BFRP armature and the metal one are placed in the stretched zone at the same time. In this case, the metal armature should provide the required level of deflection and crack resistance, and the BFRP

* Corresponding author: valovojua@gmail.com

armature should increase the cross-section strength and reduce the destructive impact of the aggressive environment on the overall bearing ability of the structure.

At present, there is a very limited base of experimental data on the work of such hybrid reinforced structures. In particular, existing research works do not clarify the question of how the percentage reduction of BFRP armature in hybrid reinforcement impacts the strength values of prototype samples.

2 Test program

The experimental tests were carried out in the laboratory of the Civil Engineering faculty at Kryvyi Rih National University, Ukraine.

2.1 Examples

The program of experimental tests provided production of 6 series of concrete reinforced beams samples, with three beams in each series. Samples by the series differed in the following features:

- the BM series - beams reinforced by metal armature (control series);
- the BB series - beams reinforced by BFRP armature;
- the BMB series - beams with hybrid reinforcement with metal and BFRP armature;
-the BMD series - beams reinforced by metal armature using concrete on fine wastes of mining and beneficiation complex (MBC);
- the BBD series - beams reinforced by BFRP using concrete with fine wastes of MBC;
- the BMBD series - beams with hybrid reinforcement by metal and BFRP armature using concrete with fine wastes of MBC.

Reinforcement of cross-sections of beams of the series is given in Table 1.

To provide correct comparison of the test results of beams of different series, the sizes of the cross sections of the beams and the percentage of their reinforcement were assumed to be unchanged for the samples of all series. The pattern of reinforcing samples of beams is shown in Fig. 1.

Manufacturing of armature grating, concrete mixture and concreting of experimental samples of beams took place at the PJSC "Kryvorizhindustrbud", Ukraine. Concreting of experimental samples of beams and cubes was carried out in metallic forms.

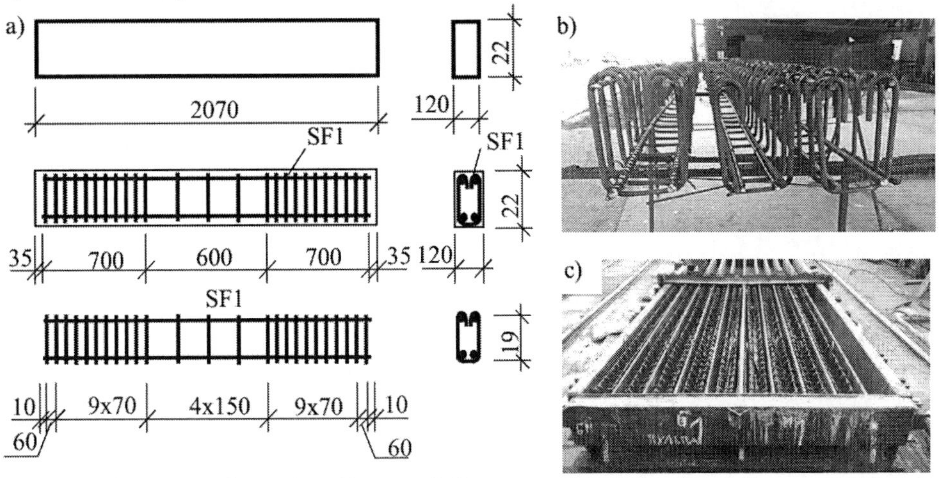

Fig. 1. Reinforcement of samples of beams: a) Framing armature drawings of beams of BM series; b) Armature grating of beams; c) Framing forms of beams; SF – space framework.

Table1. Reinforcement pattern of cross sections of beams by series.

Series No.	Beam mark	Cross-section of beams	Series No.	Beam mark	Cross section of beams
1	2	3	1	2	3
1	BM-1 BM-2 BM-3	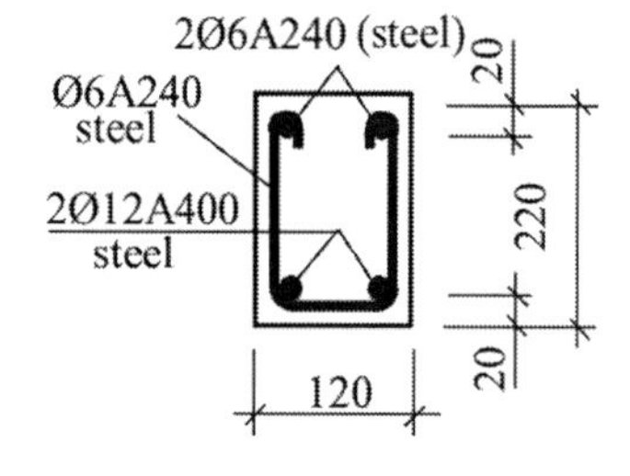	4	BMD-1 BMD-2 BMD-3	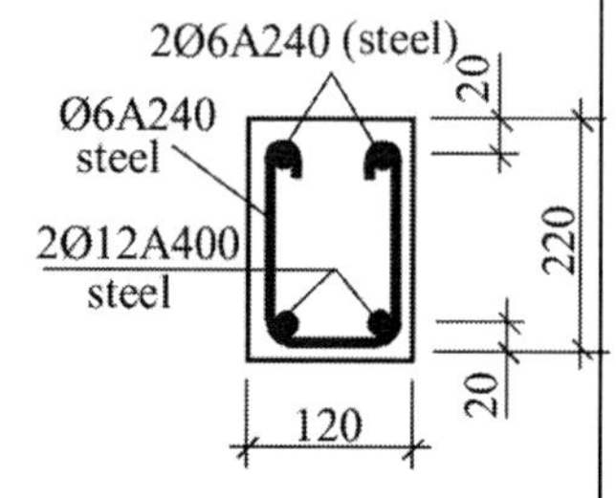
2	BB-1 BB-2 BB-3	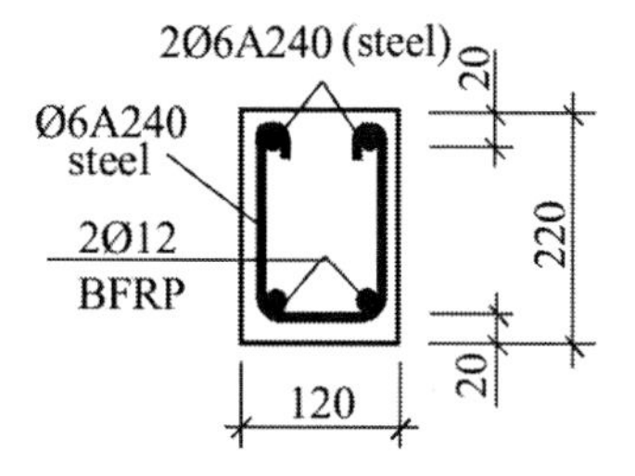	5	BBD-1 BBD-2 BBD-3	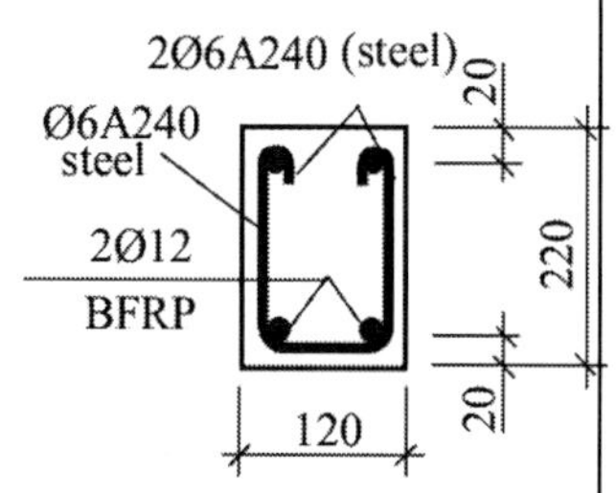
3	BMB-1 BMB-2 BMB-3	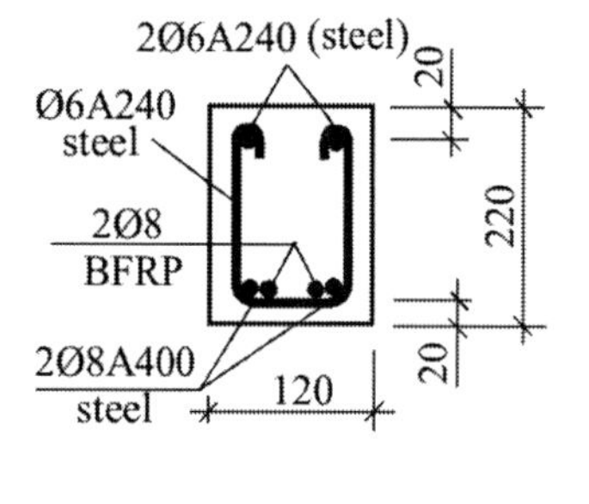	6	BMBD-1 BMBD-2 BMBD-3	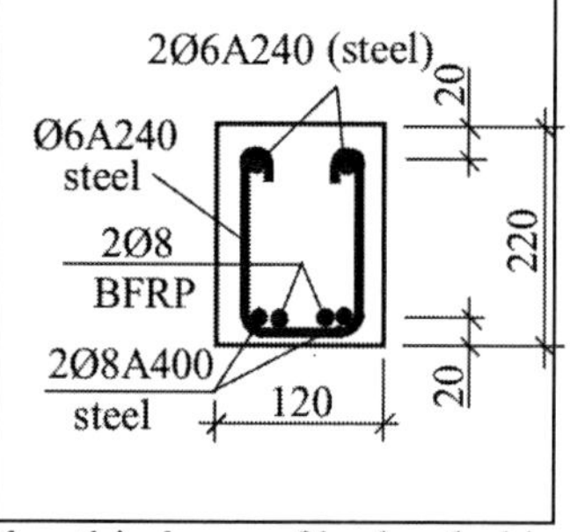

Before concreting the surface of the forms was greased with a thin layer of hydrophobic lubricant. Joined spatial gratings were installed in the forms. The designed position of the gratings was provided with the help of concrete linings to ensure the formation of the protective layer of concrete. The preparation of a concrete mixture was carried out at a concrete-soluble knot. Placing the mixture into the formwork was carried out using a bucket with subsequent seal on the vibration table (Fig. 2). After forming, the samples were placed in a heating chamber. The form was removed 28 days after concreting.

a) b)

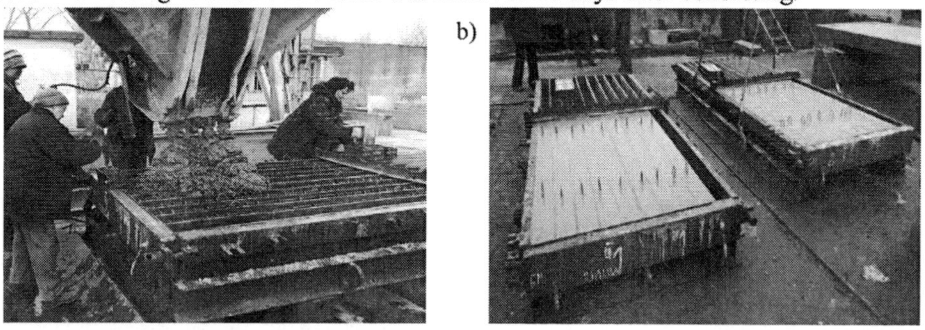

Fig. 2. Samples of beams manufacturing: a) laying of concrete mixture; b) general view of concreted samples.

2.2 Materials

For reinforcing samples, metal and BFRP armature were used in accordance with the reinforcement schemes shown in Table. 1. BFRP armature was provided by the manufacturer of composite armature LLC "Technobazalt-Invest", Ukraine.

Mechanical characteristics of the armature are given in the Table 2.

Table 2. Mechanical characteristics of armature for samples of beams.

Armature class	The boundary of fluidity, N/mm^2	Temporary resistance to breaking, N/mm^2	Modulus of elasticity, N/mm^2	Relative elongation after breaking, %
A400C (steel)	400	500	200×10^3	16
A240 (steel)	240	370	200×10^3	25
BFRP	-	800	45×10^3	1.78

Concrete of class C25/30 made at a concrete-soluble knot was used for concreting. As mentioned above, for concreting beams of series BM, BB, BMB concrete with quartz sand as a fine aggregate was used. In the beams of the series BMD, BBD, BMBD we used concrete on fine MBC wastes which are a fractionated quartz-glandular mineral mixture with the density of 1500-1600 kg/m^3. The actual use of MBC wastes as a fine aggregate for concrete mixture is economically justified (MBC wastes cost 8-10 times less than quartz sand) and in part, it allows solving the issues of industrial wastes recycling.

The numerous studies [12] of the past decades indicate the possibility of equivalent replacement of quartz sand by fine wastes of MBC when manufacturing a wide range of reinforced concrete products.

In order to determine the strength and deformation characteristics of the concrete types used, there was carried out concreting of samples of cubes (100x100x100 mm) of each concrete type simultaneously with concreting beams. Tests of concrete samples were performed both at the age of 28 days and at the age of 151 days after their manufacturing, immediately before testing the samples of the beams. At the age of 28 days the compression strength of both types of concrete corresponded to the accepted designed class of concrete (C25/30). At the age of 151 days it was 46.16 N/mm^2 for concrete on quartz sand and 51.72 N/mm^2 for concrete on fine MBC wastes.

2.3 Tests of the samples

Testing of beams was carried out on a hydraulic press P-125 according to the scheme of a single-run free beam loaded with two lumped forces in the thirds of gear. During the testing the following devices were used: the Maximov deflection indicator; engineer's dial gauge with a price of 0.01 mm readability for determining the deformations of beam concrete and sitting beams bearing supports; the MPB-2 microscope with a 24x magnification and 0.05 mm readability to determine the width of the opening of normal and inclined cracks.

The loading of tested beams was carried out by 0.1 of the ultimate breaking load. The program of experimental tests provided for determination of strength, rigidity and crack resistance of test samples which conditioned use of the above-mentioned devices. The stand for testing the beams for bending and its scheme are shown in Fig. 3.

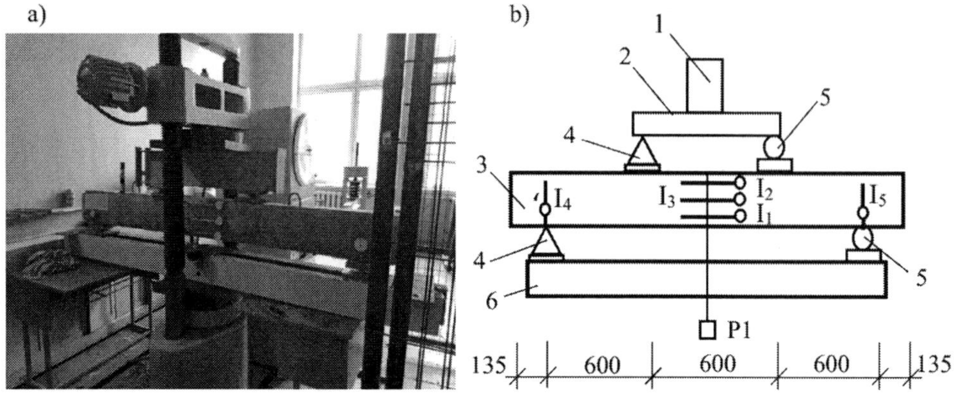

Fig. 3. a) Stand for testing of beams; b) Plant for testing of beams 1 - press plate; 2 - traverse upper distributive; 3 - experimental beam; 4 - fixed bearing support; 5 - moving traverse; 6 - lower traverse; I_1, I_2, I_3, I_4, I_5 – engineer's dial gauge of 0.01 mm readability; P1 - the Maximov deflection indicator of 0.1 mm readability.

3 Results and discussion

The work of experimental samples of beams under load and their destruction were in direct dependence on the applied type of reinforcement.

Beams of the series BM, BMD showed the work and nature of destruction typical for beams reinforced by metal armature. With 0.3-0.4 of the destructive load, the first normal cracks began to form in the concrete in the zone of pure bending. The further load increase resulted in the formation of new normal cracks, increase of the width opening and the growth of deflections. With 0.9 of the destructive load, the beginning of flowability of metal armature was registered, which led to a rapid increase in the width of the cracks opening, significant increase in deflections and the destruction of samples with the cracking of concrete in the compressed zone. The moment of the beginning of metal armature flowability was considered as starting point of the destruction of the tested beams.

Due to the reduced modulus of BFRP, the series BB, BBD proved to be more deformative than the beams of the BM, BMD series. The appearance of the first normal cracks was recorded at a loading level of 0.1-0.2 of the destructive one. The further increase of the load resulted in the appearance of new normal cracks, and at the loading of 0.5-0.6 of the destructive one the sloping cracks began to appear. The location of the cracks was characterized by a uniform distribution with beam spacing of 100-150 mm. The width of the cracks opening and deflections linearly increased with load increasing. It is typical for the structures reinforced by composite armature. The destruction of these samples was caused by fragmentation of the concrete in the compressed zone. The breakage or slipping of the BFRP rods was not registered.

Work of the beams of the series BMB, BMBD under load can be divided into two stages - before the beginning of flowability of metal armature and after it. The first normal cracks arose at a loading level of 0.2-0.3 of a destructive one. To the level of loads 0.4-0.5 of the destructive one the beam worked similarly to the beams of the series BM, BMD. The latter is due to the prevailing influence of metal armature. The further increase in load resulted in the appearance of the armature flowability. At this stage, the main resistance to the load is made by BFRP armature. The character of the work of the beam is similar to the work of the beams of the series BB, DBD - the number and width of the opening of cracks increases and the deflections grow. The destruction of the samples occurred due to the crushing of the concrete of the compressed zone with the simultaneous complete or partial breaking of the fibers of the BFRP armature. In Fig. 4 it is shown the general view of the ruined beam sample of the BMB series and the complete and partial breaking of the extended BFRP armature of this sample.

Durability indicators of tested beams by series are given in Table 3.

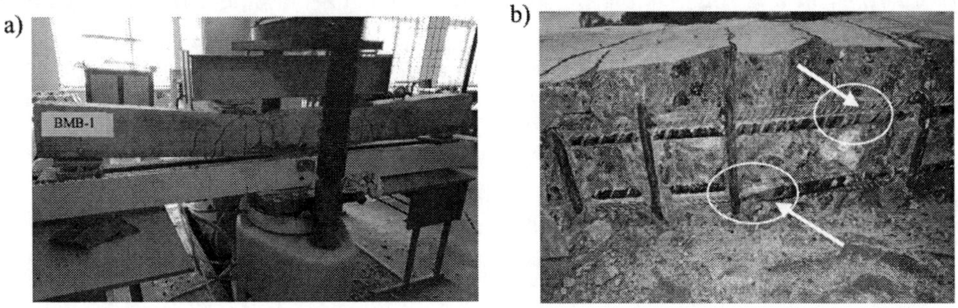

Fig. 4. Ruined beam sample of BMB series: a) General view; b) Complete or partial breaking of fibers of BFRP armature.

Table 3. Parameters of durability of samples of beams.

Series of beams	Destructive effort, kN	Average destructive effort, F_{ui}, kN	Relative durability value, F_{ui}/F_u
BM-1	69.16*		
BM-2	69.9*	70.22*	1
BM-3	71.61*		
BMD-1	73.58*		
BMD-2	74.56*	75.87*	1.08
BMD-3	79.46*		
BB-1	98.1		
BB-2	95.65	100.88	1.43
BB-3	108.89		
BBD-1	104.23		
BBD-2	98.1	101.17	1.44
BBD-3	85.84**		
BMD-1	93.2(44.15*)		
BMB-2	98.1(49.05*)	96.1 (45.13*)	1.38
BMB-3	96.87(42.18*)		
BMBD-1	98.1(47.09*)		
BMBD-2	100.55(51.99*)	99.24 (49.7*)	1.41
BMBD-3	99.08(50.03*)		

Notes:

1. F_u – destructive force of the beams of the series BM;

2.* - an effort that corresponds to the beginning of flowability of metal armature.

3.** - the datum is not considered when calculating the average destructive load.

The analysis of the data presented in Table 3 indicates that replacement of metal armature by BFRP armature has led to increase in the bearing power of the beam samples by 37-44% depending on the series. In the beams of BMB, BMBD series with the hybrid reinforcement, the part of BFRP armature was replaced by metal armature that had no impact on their bearing power. The durability increase compared with the beams of the BMW series is 38-41%. This indicates that there are no reasons against the using of hybrid reinforcement of the bending structures from the perspective the performance of the bearing power.

Use of the fine aggregate of MBC for producing concrete for the beams of the BMD, BBD, BMBD series did not significantly impact the marginal durability. The beams of the BMD, BMBD, BBD series showed an increase in the bearing power from 1% to 8%. This can be explained by higher roughness of the surface of grains of MBC wastes compared with quartz sand and suggests the possibility of efficient replacement of quartz sand by cheaper industrial wastes.

In Fig. 5 a general view of broken samples of beams is shown. Careful examination reveals that samples of the BM series (Fig.5, a) have significant residual deformation, samples of the BMB series (Fig. 5, b) are smaller. In the samples of the BB series the deformations are absent (Fig. 5, c). Flowability of the metal armature results in the significant residual structural deformations, which makes them unrebuildable. Due to the elastic performance of BFRP armature, the samples of the BB series did not have residual deformations, even after the destruction of the compressed zone concrete. This property of BFRP reinforcement can provide greater reliability of structures under the influence of overloads (earthquakes, accidents, etc.) due to the restoration of the design position after the cessation of power influences, even with significant damage. The latter statement needs further study by conducting appropriate experimental tests.

Fig. 5. General view of ruined samples of beams: a) Series BM; b) Series BMB; c) Series BB.

4 Conclusions and recommendations

4.1 Conclusions

The carried out experimental research of concrete beams with different types of reinforcement allows drawing the following conclusions:

- replacement of metal armature by BFRP armature allowed increasing the bearing ability of the samples of beams by 37-44%;

-use of hybrid reinforcement allowed to improvement of the deformation characteristics of beams in comparison with the reinforcement of the BFRP armature, and did not cause a decrease in their bearing ability (durability increase was made 38-41%);

-elastic work of BFRP results in the absence of residual structural deformations of the structure after the load decreasing despite the insignificant damage and concrete destruction;

- fine fractionated MBC wastes are an effective replacement of quartz sand in the manufacture of concrete mixture.

4.2 Recommendations

The given results can be used for the development of the program of additional experimental tests of bending structures with hybrid reinforcement using metal and BFRP armature to determine the optimal ratios of different types of armature, as well as to developing recommendations for the design of such structures.

References

1. M. Wang, Z. Zhang, Y. Li, M. Li, Z. Sun, Chemical Durability and Mechanical Properties of Alkali-proof Basalt Fiber and its Reinforced Epoxy Composites. Journal of Reinforced Plastics and Composites **27(4)**, 393-407, (2008).
2. B. Pouya, *"Experimental investigation of the mechanical and creep rupture properties of basalt fiber reinforced polymer (bfrp)"*, PhD thesis, University of Akron, the USA, 216, (2011).
3. O. S. Soldatchenko, *"Durability, rigidity and crack resistance of bending structures with glass fiber reinforced polymer and basalt fiber reinforced polymer"*, PhD thesis, Kyiv, Ukraine, 196, (2012).
4. T. Ovitigala and M. Issa, *Flexural behavior of concrete beams reinforced with basalt fiber reinforcement polymer (BFRP) bars.* Paper presented at the **11**th International Symposium on Fiber Reinforced Polymer for Reinforced Concrete Structures, Guimarães, Portugal (2013).
5. D. Pawłowski and M. Szumigała, *Flexural behavior of full-scale basalt FRPRC beams–experimental and numerical studies.* Procedia Engineering, **108**, 518-525 (2015).
6. K. L. Kudyakov, V. S. Plevkov and A. V. Nevskii, *Strength and deformability of concrete beams reinforced by non-metallic fiber and composite rebar*, IOP Conf. Series: Materials Science and Engineering **71** /1/012030, (2015).
7. A. A. Suzan Mustafa, A. Hilal Hassan, Behavior of concrete beams reinforced with hybrid steel and FRP composites, HBRC Journal, Housing and Building National Research Center [Electronic resource] (2017). – Mode of access: https://doi.org/10.1016/j.hbrcj.2017.01.001.
8. Akiel Mohammad, El-Maaddawy Tamer, El Refai Ahmed, *Flexural tests of continuous concrete slabs reinforced with basalt fiber-reinforced polymer bars*, CSCE 2016 Resilient Infrastructure, London, Ontario, 1-7, (2016).
9. G. Naveen Kumar, Karthik Sundaravadivelu, "Experimental Study On Flexural Behaviour Of Beams Reinforced With GFRP Rebars". IOP Conference Series: Earth and Environmental Science, **80**, 1-11, (2017).
10. S. Vivekanand, A. Sumathi, *"Static Behaviour of Concrete Beams Reinforced in Shear with GFRP Bars"*. International Journal of ChemTech Research, **8**, 2, 635-642, (2015).
11. E. Mohamed, *"Behaviour of continuous concrete slabs reinforced with frp bars"*, PhD thesis, University of Bradford, UK, 177, (2013).
12. L. I. Storozhenko, B. N. Shevchenko, V. M. Ilienko, *Zhelezobetonnyie konstruktsii iz betona na othodah gornorudnoy i metallurgicheskoy promyishlennosti.* – Kyev: Budivelnyk, – 72, (1982).

Feature of fire resistance calculation of steel structures with intumescent coating

Alexey Vasilchenko[1,*], *Yuriy* Otrosh[1], *Nikolay* Adamenko[2], *Evgeny* Doronin[2], and *Andrey* Kovalov[3]

[1]National University of Civil Defence of Ukraine, Chernishevska str. 94, 61023, Kharkov, Ukraine
[2]V. N. Karazin Kharkov National University, Svobody Sq. 4, 61022, Kharkov, Ukraine
[3]Cherkassy Institute of Fire Safety of National University of Civil Defence of Ukraine, Onoprienko str. 8, 18034, Cherkassy, Ukraine

Abstract. The problem of estimation of fire resistance of steel frame structures with intumescent coatings is considered. It implies that both physical properties of a covering (its thickness and structure) and mechanical properties of a metal structure change critically at heating. All above changes should be considered to maintain the standard values of fire resistance of a construction at calculation. Usually, known technical characteristics of fire resistance of intumescent coverings are used for estimation of fire resistance of steel structures with intumescent coverings. Importance of taking into account the influence of strength loss time at heating of a steel structure on calculation of fire resistance limit of system "intumescent fireproof coating – steel structure" is shown in the article. On an example of calculation of heating time to the critical temperature of steel columns and beams protected by intumescent coating, it is shown that own heating time of steel structures before they lose strength makes 10 to 16 % from a settlement limit of fire resistance. This fact should be considered at the forecast of fire resistance of steel frame structures with intumescent coatings.

1 Introduction

Common use of steel structures in industrial frame buildings causes special requirements to their fire safety. The basic danger at heating of metal structures is their considerable deformations and fast loss of strength. This process can start with temperature of 350 °C. An effective method of fire resistance of steel structures is use intumescent coatings. These coatings at contact with fire changes their structure, considerably increase in volume and forming a heat-protection layer [1, 2, 3].

For maintenance of standard values of fire resistance of the steel frame structures protected by intumescent coatings, it is necessary to predict their characteristics taking into account power loadings. Here two approaches are possible:

* Corresponding author: avas2006@ukr.net

– To estimate limits of fire resistance of the existing steel structures protected by intumescent coatings;
– To offer type and a thickness of a layer intumescent coatings for maintenance of demanded fire resistance of steel structures.

However, to define experimentally fire resistance of the large steel structures protected by intumescent coatings, being in construction structure it is almost impossible. Therefore the settlement estimation of their fire resistance limits is necessary.

The problem of settlement estimation consists that by working out of a method of problem solving it is necessary to consider such circumstances simultaneously: cardinal change at heating of properties of covering, its thickness and structure, and also change of properties of a metal structure [4].

Feature of intumescent fireproof coatings is the fast increase in their volume at heating and rough facing formation. This porous coating has low heat conductivity and protects the basic material from heating [1, 2]. Existing calculation methods allow to count efficiency of such covering authentically. They consider time of warming up of an initial layer prior to the beginning of it intumescent and, then, time during which the intumescent layer is capable to protect a structure [2, 3, 5]. For steel structures in connection with high heat conductivity of metal the time of achievement of fire resistance limit consider the moment when temperature on border "covering – metal" becomes to equal critical temperature of a structure [6]. Values of fire resistance limits, calculated thus, do not consider time of warming up of actually metal structure which is in a stress state. Therefore they can be a little underestimated. It can lead to the over-expenditure of materials or the inexact forecast of behaviour of steel structures at a fire.

2 Statement of a problem and its decision

A work problem is calculation of warming up time of the steel structure protected by a intumescent coating, to critical temperature depending on an initial thickness of a covering.

It is possible to present a limit of fire resistance of a steel structure τ_{kp} as the sum of warming up times:
– Of a protective coat τ_{Fb} to its temperature of intumescent t_{Fb};
– Of a intumescent layer τ_{Fs} to critical temperature of a steel structure t_{kS};
– Of a steel structure before loss of strength τ_{kS}:

$$\tau_{kp} = \tau_{Fb} + \tau_{Fs} + \tau_{kS} \,. \tag{1}$$

Calculation of fire resistance of a steel structure protected by a intumescent coating is based on the equation of heat conductivity of Fourier for the description of a one-dimensional temperature field

$$\rho c \frac{\partial t}{\partial \tau} = \frac{\partial}{\partial x}\left(\lambda \frac{\partial t}{\partial x} \right), \tag{2}$$

where ρ – density of a material taking into account porosity; c – specific thermal capacity of a material; t – temperature; λ – factor of heat conductivity of a material; x – the co-ordinate directed on a thickness of a material; τ – time.

The equation (2) numerically solves by finite element method for a covering layer, both to it swelling, and after swelling, considering following boundary conditions:

a) For an external (heated up) surface of a fireproof covering boundary conditions of 3rd sort are accepted:
– Temperature of environment (t_F) changes according to a curve of a "standard" fire:

$$t_F = 345\lg(8\tau + 1) + t_0\,; \tag{3}$$

– Heat transfer factor (α) from the gas environment to a structure surface equals:

$$\alpha = 29 + 5.75\,\varepsilon_\gamma \frac{\left(\frac{t_F}{100}\right)^4 - \left(\frac{t_{Fb}}{100}\right)^4}{t_F - t_{Fb}}, \tag{4}$$

where ε_γ – the reduced emissivity factor of an irradiated surface;

b) For border between a covering and a steel structure conditions are accepted:

– Heat transfer factor $\alpha = 0$ W/(m^2·K) (adiabatic wall);

– Boundary conditions:

$$t\big|_{x=x_\gamma-0} = t\big|_{x=x_\gamma+0}\,; \tag{5}$$

$$\lambda\left.\frac{\partial t}{\partial x}\right|_{x=x_\gamma-0} = \lambda\left.\frac{\partial t}{\partial x}\right|_{x=x_\gamma+0}. \tag{6}$$

For an example of calculation such steel structures warmed from 4 parties have been chosen: single-member axially-loaded columns and compound welded I-beams.

As a coating the fireproof intumescent structure "Terma" with such characteristics have been chosen: density $\rho = 0.8$ kg/m^3; effective factor of heat conductivity $\lambda_{ef} = 0.05$ to 0.32 W/(m·K); intumescent factor $k_{sw} = 40$; critical value of temperature $t_{Fb} = 177$ °C.

At calculations following assumptions have been made:

– Warming up process for a local site as for a semi-infinite body with boundary conditions of 3 sorts was considered;

– Intumescent time of a protective film was not considered;

– It was supposed that thermal contact between a layer of covering both to swelling, and after swelling and a steel structure is ideal;

To calculate a column applies such method. Critical temperature find for each column, having set constant loading and having calculated factor of decrease in bearing ability at heating γ_T:

$$\gamma_T = \frac{N}{\varphi A_K R_S \gamma_c}, \tag{7}$$

where N – loading on a column; φ – factor of a longitudinal bend; A_K – area of section of a column; R_S – limit of resistance of a column material; γ_c – working conditions factor.

Calculation of reduced thickness of a column δ_k, does possible to define time of loss of strength [6-8].

Calculations for steel columns of various sections were spent. For each calculation loading on a column which provided critical temperature $t_{kS} = 500$ °C was accepted.

To calculate a beam applies such method. Factor of decrease in bearing ability at heating γ_T (on which critical temperatures in the chosen sections were defined) was calculated under the formula:

$$\gamma_T = \frac{M_x}{W_x R_s}, \tag{8}$$

where M_x – the bending moment in settlement section at a cross-section bend, kN·cm; W_x – modulus of section, cm^3; R_s – ultimate strength of steel, kN/cm^2.

Parameters of section of steel composite welded I-beams in length of 18 m, 21 m and 24 m calculated on [9]. Stiffening ribs are located with pitch of 1.5 m. Beams are made of C345 steel, category 1, with limiting resistance $R_s = 45$ kN/cm^2. For comparability of results the distributed linear load on beams was accepted such to provide critical temperature $t_{kS} = 500$ °C.

Calculations were made in SCAD software. Received values were applied to an estimation of fire resistance limits on a method [8].

Results of calculations are shown in the table 1.

Table 1. Settlement values of a fire resistance limit of the steel structures covered with fireproof intumescent structure "Terma".

Structure profile	Warmed perimeter, mm	Section area, mm^2	Resulted thickness, mm	Thickness of a coating, mm	Time of warming up of a intumescent coating, $\tau_{Fb} + \tau_{Fs}$, minutes	Time of warming up of a structure, τ_{kS}, minutes	Fire resistance limit, τ_{kp}, minutes
Column Pipe 200×200	800	3790	4.74	1,0	58	8	66
				2,0	68		76
Column I-beam 200×100×5.6×8.5	788	2849	3.61	1.0	44	8	52
				2.0	62		70
Column I-beam 300×140×5.8×8.5	1140	4192	3.67	1.0	46	8	54
				2.0	62		70
Column I-beam 395×165×7×9.5	1430	6125	4.28	1.0	56	8	64
				2.0	66		74
Column I-beam 492×200×8.8×12	1766	9298	5.26	1.0	59	10	69
				2.0	70		80
Column H-beam 196×193×6×9	1152	3895	3.38	1.0	30	7	37
				2.0	60		67
Column H-beam 291×200×8×11	1366	6831	5	1.0	58	9	67
				2.0	70		79
Column H-beam 388×300×9.5×14	1957	12240	11	1.0	60	11	71
				2.0	71		82
Column H-beam 484×300×11×15	2146	14570	12	1.0	60	12	72
				2.0	72		84
Structural I-Beam 18 m: 2000×446×36×18	5784	88056	15.2	1.0	62	17	79
				2.0	80		97
Structural I-Beam 21 m: 2100×495×45×20	6170	114300	18.5	1.0	65	19	84
				2.0	85		104
Structural I-Beam 24 m: 2400×655×55×28	7422	168680	22.7	1.0	67	22	89
				2.0	95		117

The resulted example shows that at estimation of fire resistance of steel structures protected by a intumescent coating "Terma", the time of warming up of actually steel structures before they will lose strength, makes 10 to 16 % of a settlement limit of fire resistance. And this contribution will increase at increase in the resulted thickness of a structure. Also it is necessary to take into consideration that as the critical temperature of steel structures depends on loading rate on them and ultimate strength of steel these criteria should be considered at designing of fire protection.

3 Conclusion

Thus, it is shown on an example that at calculations of fire resistance limit of steel structure protected by intumescent coating, it is necessary to consider as time of warming up of this coating to critical temperature, and time of strength loss of actually steel structure at the reached temperature. This time depends also on parameters of the structure profile, of loading rate on it and of ultimate strength of steel.

References

1. G. Camino, S. Lomakin, *Fire retardant materials*/ Ed. by Horrocks A. and Price D. (CRC Press and Woodhead Publishing Ltd., Cambridge, 2001)
2. I.G. Romanenkov, V.N. Zigern-Korn, *Ognestojkost' stroitel'nyh konstrukcij iz effektivnyh materialov* (Strojizdat, Moskow, 1984)
3. S. V. Bazhenov, Pozharnaya bezopasnost', **5**, 97–102 (2005)
4. O. Nekora, V. Slovynsky, S. Pozdieiev, MATEC Web of Conferences, **116**, 02024, (2017).
5. D.E. Cagliostro, S.K. Riccitielio, K.J. Clare, J. Fire and Flammabie, **2**, 205–291 (1975)
6. V.I. Golovanov, A.V. Ruzhinskij, Pozharo vzryvo bezopasnost', **2**, 37-39 (1994)
7. N.M. Bessonov, T.Yu. Eremina, Yu.N. Dmitrieva, M.V. Krasheninnikova, Pozharnaya bezopasnost', **1**, 22-28 (2007)
8. V.M. Rojtman, *Inzhenernye resheniya po ocenke ognestojkosti proektiruemyh i rekonstruiruemyh zdanij* (Associaciya "Pozharnaya bezopasnost' i nauka", Moskow, 2001)
9. E.I. Belenya, V.A. Baldin, *Metallicheskie konstrukcii. Obshchij kurs: Uchebnik dlya vuzov* (Strojizdat, Moskow, 1986)

Optimization of the truss beam. Verification of the calculation results

Glib Vatulia[1], *Sophia* Komagorova[1*], *and Mykhailo* Pavliuchenkov[1]

[1]Ukrainian State University of Railway Transport, 61050 Feierbakh Sq. 7, Kharkiv, Ukraine

Abstract. The article presents a comparison of results of optimized calculation of a truss beam which was chosen as a combined construction. The results of calculation of a beam are compared using the method based on the properties of spacer systems and the calculation of the construction designed in LIRA software complex. The article is dedicated to verification of adequacy of the results of theoretical calculations of construction optimization. Values of longitudinal forces and bending moments appearing in a truss beam are chosen as convergence criteria. Two variants of construction loading are considered: a truss beam exposed to constant load only and a truss beam exposed to constant and temporary load. In the case under consideration, the minimum value (weight) of construction is an optimality criteria, variable parameters include beam panel length and camber height of a trussing rod. As a result, the construction will be considered optimal, if bearing and maximal (between the pillars) bending moments are equal in it. The result of verification of the obtained data is the value of error.

1 Introduction

In solving optimal design problems, regularities are found which can be generalized to formulate axiomatical statements – postulates [1, 2]. Using them in creating optimization algorithms allows clearer understanding of ways of achieving goals, i.e. simplifying problem solving.

1. Increase in the degree of equal strength generally reduces the objective function value. The lowest value of the objective function is reached in the system with fully equal strength.

2. Increase in the number of project variables with the unchanged system of stress-strain state restrictions either results in decreasing of the objective function of the optimal solution or leaves it unchanged.

3. With the permanent number of project variables, adding restrictions narrows the range of feasible solutions and cannot reduce the value of the objective function of the optimal solution – it either remains unchanged or increases.

* Corresponding author: sinchuk.sophia@gmail.com

1.1 Optimization technique

On the basis of the above postulates, an optimization method for a truss beam is proposed, which is based on following aspects:

1. In combined beam structures, the material consumption is determined by elements that experience a stressed state in the form of bending compression.

2. In the elements exposed to bending compression, the reduction in the cross-section is determined rather by the reduction of the bending moment.

3. Bending moments increase from the pillar to the middle of the span.

4. Truss beams belong to spacer systems, which also include arches. Thus, both have some common properties, for example, in tree-hinged statically determinate arches, the cross-sections are affected by eccentric compression i.e. a bending moment and a longitudinal force appear in them. Naturally, the lower the values of bending moments are, the smaller the cross-sectional dimensions are, based on the strength condition, which means less material consumption for making the arch and less costs. Therefore, a rational or an optimal arch, affected, for example, only by constant load, it generally understood as an arch the axis of which is described as a funicular curve determined by the type of load. In all cross-sections of such an arch, bending moments are equal to zero and its manufacture requires less material then that of arches described by other curves.

It can be assumed that truss beams also have this property under certain conditions. Particularly, if the lower truss boom is determined along a curve corresponding to the acting constant load and the braces of the beam element form a continuum with the lower truss boom, then bending moments in the beam elements are equal to zero.

In fact, the truss beam design (Fig. 1) assumes that the braces of the beam element and lower truss boom are made in the form of a limited number of rods (pillars). As a result, bending moments still appear in the beam – negative ones over the pillars and mostly positive between them. Their values depend on the number of pillars and their location in the span. The values of some moments increase, while others decrease when the position of the pillars changes. Rational is such an arrangement in which the absolute values of the moments are equalized and are the lowest [3].

5. Bending moments do not depend on the truss boom if the truss rod is delineated by the funicular curve for the three-hinged arch [4-6]. This will be seen from formulas (1) and (2) below

$$M_j^{\sup} = \frac{q(\ell - x_j)x_j}{2}\left[1 - \frac{\ell^2}{4(\ell - x_{n/2})x_{n/2}}\right], \; j = 1, 2, \ldots, n/2, \tag{1}$$

$$M_j^{\max} = \frac{q\ell}{2}x_{j,\max} - \frac{q(x_{j,\max})^2}{2} - \frac{q\ell^2}{8(\ell - x_{n/2})x_{n/2}}\Big[(\ell - x_{j-1})x_{j-1} +$$

$$+ \frac{(\ell - x_j)x_j - (\ell - x_{j-1})x_{j-1}}{z_j}(x_{j,\max} - x_{j-1})\Big], \; j = 1, 2, \ldots, n/2, \tag{2}$$

where $x_{j,\max}$ – distance from the beam beginning to the extreme in the j th panel,

$$x_{j,\max} = \frac{\ell}{2}\left\{1 - \frac{\ell\left[(\ell - x_j)x_j - (\ell - x_{j-1})x_{j-1}\right]}{4z_j(\ell - x_{n/2})x_{n/2}}\right\}. \tag{3}$$

Based on the analysis of the above positions, the method for truss beam optimization is proposed, which is based on the minimization not of volume, but of the bending moments in the beam element of construction, i.e.

$$M_{\text{calc}} = \max_{j=1,2,\ldots,n/2}\left(\left|M_j^{\text{sup}}\right|, \left|M_j^{\max}\right|\right) \rightarrow \min, \tag{4}$$

where M_j^{sup} – bending moment in the beam element cross-section above the j th pillar; $M_j^{\max}$ – extreme bending moment in the cross-section between ($j-1$)th and the j th pillars (within the limits of the j th panel).

Two optimality conditions must be satisfied herein. The first condition

$$\left|M_{n/2}^{\text{sup}}\right| = \left|M_{n/2}^{\max}\right| \tag{5}$$

must be satisfied every time.

The second condition will be different, it is chosen depending on the type of optimized beam element design:
– with constant cross-section of a beam element along the entire span length

$$M_j^{\max} = M_{n/2}^{\max}, \quad j = 1, 2, \ldots, n/2-1; \tag{6}$$

– with piecewise constant cross-section of a beam element (constant in the intervals between the pillars)

$$M_j^{\max} = \left|M_j^{\text{sup}}\right|, \quad j = 1, 2, \ldots, n/2-1. \tag{7}$$

2 Calculation results

2.1 Calculation results for a truss beam under action of constant load

Span beam is accepted $\ell = 24\,\text{m}$; boom $f = 3.6\,\text{m}$; distributed load intensity $q = 60\,\text{kN/m}$; number of pillars connecting the beam element with the lower truss boom $n = 6$; beam material – steel (design resistance $mR_y = 200\,\text{MPa}$). The beam element has a rectangular cross-section with the height to width ratio $h/b = 5$; pillars and lower truss beam elements have annular cross-section with a diameter D and wall thickness $0.025D$ [7].

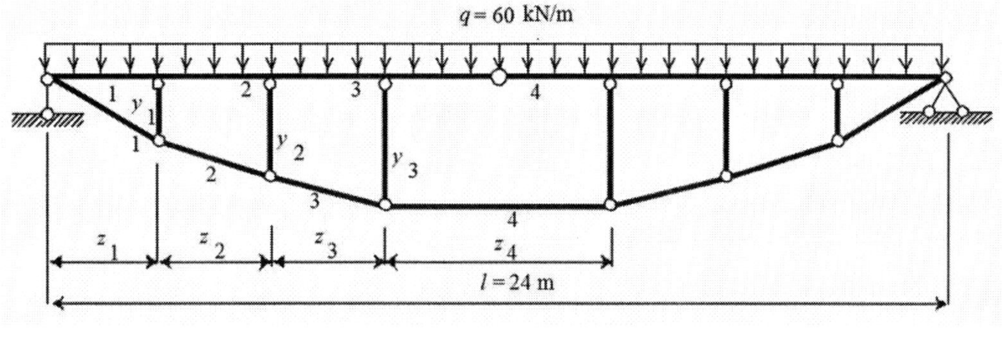

Fig. 1. Truss beam diagram and numbering of its elements.

Next, we will change each value z_j, $j = 1, \ldots, n/2$ with the earlier chosen increment, for example, $\Delta z = 0.01\,\text{m}$ until a condition specified below is achieved. The rules of change z_j are given in the description of the optimization process [3].

Table 1. Calculation results.

Approximation	Value z, m			Moments value, kNm					
	z_1	z_2	z_3	M_1^{max}	M_1^{sup}	M_2^{max}	M_2^{sup}	M_3^{max}	M_3^{sup}
0	3	3	3	19.2	-126	-96	-216	-172.8	-270
Optimal	3.11	3.60	3.88	59.5	-27.3	59.5	-48.7	58.8	-59.6

The construction is designed and calculated in LIRA software complex to verify the obtained data.

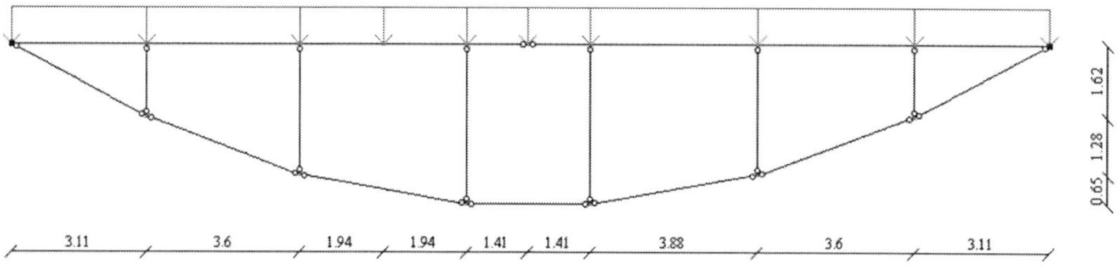

Fig. 2. The optimal truss beam design under action of constant load.

Values of forces, cross-sectional areas, and volumes of lower elements on all optimization approximations are given in Table 2.

Table 2. Results of the truss beam optimization.

	Pillars			Beam element				Lower belt			
0 approximation	1	2	3	1	2	3	4	1	2	3	4
N, kN	-192	-192	-288	1280				1446	1367	1312	1280
M_{max}, kNm				270							
A, cm^2	9.6	11.34	17.1	280.13				72.28	68.35	65.6	64
V, cm^3	20689.39			673014.50				171528.00			
	V=865231.9 cm^3										
Optimal approximation	1	2	3	1	2	3	4	1	2	3	4
N, kN	-204.1	-227.5	-203.8	1217				1373	1291	1234	1217
M_{max}, kNm				59.5							
A, cm^2	10.21	13.40	13.18	130.68				68.64	64.55	61.69	60.84
V, cm^3	20444.0			314014.4				163166.8			
	V=497625.2 cm^3 (decreased by 45.6 %)										

As a result of calculation in LIRA software complex, a diagram of longitudinal forces (Fig. 3) and of bended moments (Fig. 4) were obtained.

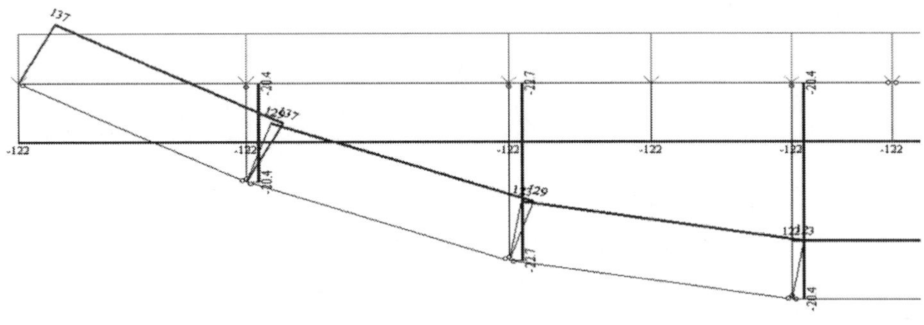

Fig. 3. The diagram of longitudinal forces arising in the truss beam under action of constant load, t.

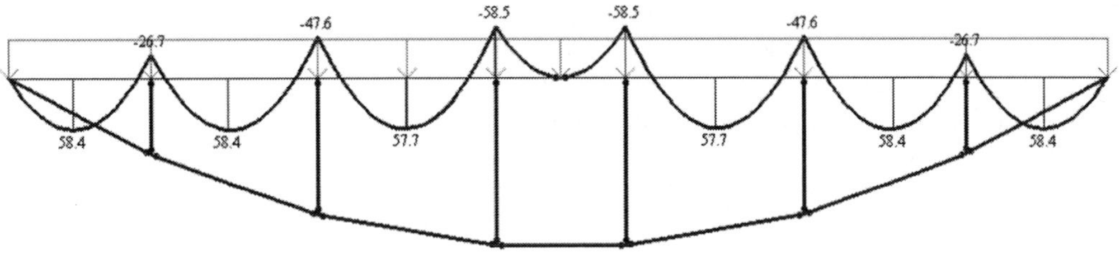

Fig. 4. The bending moments diagram arising in the truss beam under action of constant load, kNm.

Comparing the results of the design calculation with the action of constant load with the calculated ones, obtained an infelicity of 0.06% and 0.05%, respectively.

2.2 Calculation results for a truss beam under action of constant and temporary load

To obtain the optimal beam design, we will assume the precondition substantiated above [8]. In addition, we assume here that under the combined action of loads – uniform constant load $q = \text{const}$ and pedestrian load $p = \text{const}$ it would be rational to position the nodes of the lower boom on a square parabola, as under action of only constant load. The expression for determining the ordinates of such a curve is

$$y = \frac{4fx(\ell - x)}{\ell^2}, \tag{8}$$

where f –theoretical sagitta of parabola, on which the nodes of the lower boom are positioned.

Considering (8), influence line ordinates (9)–(13) are:

$$a_j = \frac{x_j(\ell - x_j)}{\ell} - \frac{x_j^2(\ell - x_j)}{2x_{n/2}(\ell - x_{n/2})} = \frac{x_j(\ell - x_j)}{\ell}\left[1 - \frac{x_j \ell}{2x_{n/2}(\ell - x_{n/2})}\right], \tag{9}$$

$$b_j = \frac{x_j}{2}\left[1 - \frac{\ell(\ell - x_j)}{2x_{n/2}(\ell - x_{n/2})}\right], \tag{10}$$

$$M_j^q = \frac{q}{2}\left[a_j u_j + b_j(\ell - u_j)\right], \tag{11}$$

$$M_j^{-p} = \frac{p}{2} b_j (\ell - u_j), \tag{12}$$

$$M_j^{q-p} = \frac{q}{2} a_j u_j + \frac{q+p}{2} b_j (\ell - u_j). \tag{13}$$

On the basis of the properties of truss beams described above, we can formulate the following algorithm for choosing the optimal panel dimensions z_j, $j = 1, 2, \ldots, n/2$.

The arrangement of nodes is optimal if all values of the maximum span moments $M_j^{\max} = M_j^{q+p}(x_{j,\max})$ and all values of the moments at support $M_j^{\sup} = M_j^{q-p}(x_j)$ have minimum values. The experience of optimizing the truss beam under action of constant load shows that in such case, these moments become approximately equal, which means equal strength of the system.

The following actions are performed to achieve these conditions.

The initial values are chosen z_j, for example,

$$z_j^{(0)} = \frac{\ell}{2(n/2+1)}, \; j = 1, \ldots, n/2,$$

$$x_j^{(0)} = \sum_{k=1}^{j} z_k^{(0)}, \; j = 1, \ldots, n/2.$$

Values $M_j^{\sup} = M_j^{q-p}(x_j^{(0)})$ are calculated according to (14).

$$M_j^{q-p} = \frac{q}{2} a_j u_j + \frac{q+p}{2} b_j (\ell - u_j) \tag{14}$$

Values $M_j^{\max} = M_j^{q+p}(x_{j,\max})$ are calculated, where the extremum point $x_{j,\max}$ is determined from equation

$$\frac{q+p}{2}\left(\frac{\partial a_{jx}}{\partial x} u_{jx} + a_{jx}\frac{\partial u_{jx}}{\partial x}\right) + \frac{q}{2}\left[\frac{\partial b_{jx}}{\partial x}(\ell - u_{jx}) - b_{jx}\frac{\partial u_{jx}}{\partial x}\right] = 0 \tag{15}$$

The increments are specified $z_j^{(1)} = z_j^{(0)} + \Delta z_j$.

If $\max\limits_{j=1,n/2}\left(M_j^{q+p}\left(x_{j,\max}^{(1)}\right), \left|M_j^{q-p}\left(x_j^{(1)}\right)\right|\right) < \max\limits_{j=1,n/2}\left(M_j^{q+p}\left(x_{j,\max}^{(0)}\right), \left|M_j^{q-p}\left(x_j^{(0)}\right)\right|\right)$, then $x_j^{(1)}$ are accepted as new values x_j (and respectively z_j).

Similarly, increments are specified in an arbitrary order $z_j^{(m)} = z_j^{(m-1)} + \Delta z_j$, $j = 1, \ldots, n_j$.

Values $M_j^{\sup} = M_j^{q-p}(x_j^{(m)})$ are calculating according to (14).

Values $M_j^{\max} = M_j^{q+p}(x_{j,\max}^{(m)})$ are calculated, where the extremum point $x_{j,\max}^{(m)}$ is determined from equation (15).

If $\max\limits_{j=1,n/2}\left(M_j^{q+p}\left(x_{j,\max}^{(m)}\right), \left|M_j^{q-p}\left(x_j^{(m)}\right)\right|\right) < \max\limits_{j=1,n/2}\left(M_j^{q+p}\left(x_{j,\max}^{(m-1)}\right), \left|M_j^{q-p}\left(x_j^{(m-1)}\right)\right|\right)$, then $x_j^{(m)}$ are accepted as new values x_j (and respectively z_j).

The process continues until x_j reaches some values limited by the design features of the system, for example, $x_{n/2} < \ell/2$. Values z_j, $j = 1,...,n/2$ obtained at the last approximation are accepted as optimal z_j^{opt}.

The proposed algorithm is used to optimize the truss beam given above with the calculation of the structure under action of constant load (Fig. 1) with the same dimensions and initial data. Moving load is accepted $p_1 = p_2 = p = q = 6tf/m$. Initial values $z_1^{(0)} = z_2^{(0)} = z_3^{(0)} = 3m$. The obtained optimization results are given in Table 3.

Table 3. Calculation results.

z \ M	M^{max}, kNm	M^{sup}, kNm
$z_1 = 4.50m$	671.4	-673.8
$z_2 = 2.58m$	673.7	-673.6
$z_3 = 3.70m$	600.1	-276.0

The construction is designed and calculated in LIRA software complex to verify the obtained data.

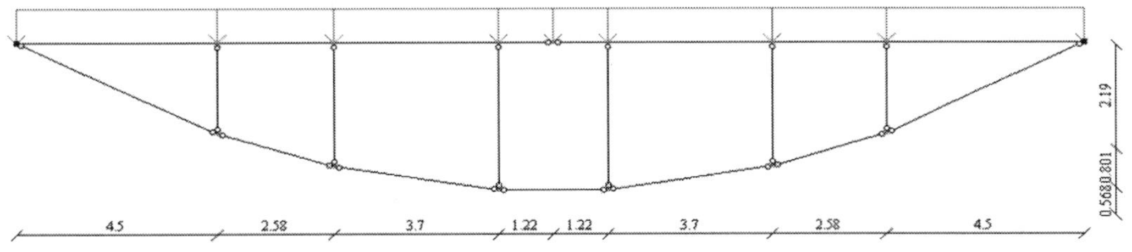

Fig. 5. The optimal truss beam design under action of constant and temporary load.

The calculation results on the initial (at $z_j^{(0)}$) and the final (at z_j^{opt}) optimization approximation are shown in Table 4.

Table 4. Results of optimization of the truss beam.

	Pillars			Beam element				Lower belt			
0 approximation	1	2	3	1	2	3	4	1	2	3	4
N, kN	-384	-384	-576	2560				2891	2734	2324	2560
M_{max}, kNm				1014							
A, cm^2	19.2	19.2	28.8	657.8				144.6	136.7	131.2	128
V, cm^3	35856			1578656				343056			
	V=1957568 cm^3										
Optimal approximation	1	2	3	1	2	3	4	1	2	3	4
N, kN	-429.2	-380.7	-372.2	2425				2698	2539	2453	2425
M_{max}, kNm				673.8							
A, cm^2	21.46	19.04	21.73	518.23				134.9	127.0	122.7	121.3
V, cm^3	36305.6			1243751.0				325089.1			
	V=1605146 cm^3 (decreased by 18 %)										

Conclusion

Analysis of the possibilities of structure designing in LIRA software complex suggested that in this software complex, only the pedestrian load cannot be applied during calculating the mobile load. The calculation will be carried out only when the structure is loaded with vehicle or rolling stock mobile load, and the pedestrian load will be considered as an additional one within the sidewalks [9, 10].

The proposed optimization algorithm for combined structures allows calculating the design when any kinds of loads and their combinations are applied. The results are verified for compliance with the conditions of strength limitations, stability and structural flexibility limitations of the rods.

References

1. A.I. Vinogradov, *Problema optimalnogo proektirovaniia v stroitelnoi mekhanike* (Vyscha shkola, Kharkiv, 1973).
2. E.D. Chikhladze, G.L. Vatulia, Yu.P. Kitov, *Osnovy rascheta i proektirovaniia kombinirovannykh i stalebetonnykh konstruktsii* (Transport, Kiev, 2006).
3. Yu.P. Kitov, G.L. Vatulia, M.A. Verevicheva, Zb. nauk. prats / CSP, UkrDAZT **143**, 124 – 131 (2014).
4. S.P. Fesyk, *Spravochnik po soprotivleniyu materialov* (Budivelnyk, Kyiv, 1982).
5. E. Hog, *Prikladnoe optimalnoe proektirovanie* (Mir, Moskva, 1983).
6. P.M. Salamakhin, Tr. Izd. **4**, 20-23 (2005).
7. SSN V.2. 3-14:2006. *Mosty ta truby. Pravyla proektuvannia.* (MBAZKH, Kyiv, 2006).
8. Yu. Kitov, M. Verevicheva, G. Vatulia, Ye. Orel, S. Deryzemlia *Matec Web of Conferences,* **133,** 03001 (2017).
9. A.S. Gorodetsky, Yu. D. Geraimovych, D.V. Marchenko, SADC and G. **1,** 70-74, (2010).
10. D.A. Gorodetsky, M.S. Barabash, R.Yu. Vodopyanov, V.P. Titok, A.E. Artamanova, *Programmnyi kompleks LIRA-SAPR* (Kyiv-Moscow, 2013).

The heat loss calculating methods of external walls in the buildings reconstruction

Varvara Vinnichenko[1,*], *Azat* Gabitov[2], *Aleksandr* Salov[2], *Askar* Gaisin[2], and *Dmitriy* Kuznetsov[2]

[1]Kharkiv National University of Civil Engineering and Architecture, 61002, Sumska st. 40, Kharkiv, Ukraine
[2]Ufa State Petroleum Technological University, 450000, Mendeleeva St. 195, Ufa, Russia

Abstract. Heat loss analysis in cladding of brick buildings under reconstruction is presented. Thermograms obtained under thermovision inspection and window systems operating experience in conditions of the Republic of Bashkortostan are thoroughly studied. Live issue of increasing buildings energy performance in public utilities sector may largely be solved by replacement of existing window units made of wood to new PVC profile windows equipped with multi-glazed glass units both in brick and frame-panel old buildings. Significant heat loss occurs in junctions between the window frames and the wall in jamb areas. Therefore for the heat conduction matrix for the finite element is used the heat balance equation. Use of the software application to choose certain measures to eliminate the thermal bridges enables to get the thermotechnical calculations in the junction between the window and the exterior wall in the form of temperature fields. Practical recommendations for arrangement of heat insulation in junctions between the window frames and the wall to eliminate actual defects and for normal room conditions are made under examinations. To get the heat conduction matrix for the finite element we will use the heat balance equation.

1 Introduction

The majority of residential buildings using wooden windows was built in the 20th century in the Soviet Union. Window constructions using PVC profile were being developed for the last 15 years due to strengthening of requirements to buildings heat protection and new technologies for making windows, with two-casement glass windows being replaced by single-casement equipped with glass units [1]. New domestic standards, regulating thermotechnical parameters for enclosing structures, that were applied in engineering and construction of new and reconstruction and renovation of available housing and public facilities were consistently toughening the requirements to thermal resistance all over the enclosing structures of buildings (walls, coating, windows, etc.).

* Corresponding author: vvinnichenko@ukr.net

2 Bibliography analysis

Procedure to define the reduced heat-transfer resistance was improved by new SNiP (construction rules and regulations), thereby changing understanding of this value itself. Now the reduced heat-transfer resistance is to be understood as complex value characterizing the whole selected segment of envelope of the building considering all current non-homogeneities. So far, however, there is such a variety of engineering solutions for enclosing structures together with non-homogeneities peculiar thereto that the simplified geometric layout of heat-transfer inclusions specified in previous heat insulation regulatory documents cannot be used in making due calculations and engineering design [2]. Inconsistency of residential buildings, designed by SNiP 23-02-2003 before updating, with present standards by the reduced resistance values of external walls and by specific heat performance [3, 4] found by some scientists under analysis of cladding structural solutions is absolutely explained by this fact.

Various software applications [5, 6] are currently applied for calculation of 2D and 3D temperature fields when defining the reduced heat-transfer resistance values of external walls. In 2015 SP 230.1325800.2015 "Enclosing structures of buildings. Thermal homogeneity characteristics" regulation containing reference tables of specific loss of heat through some standard heat-transfer inclusions for the most commonly used engineering solutions to external walls came into force. Good repeatability was demonstrated by the study in calculating results of the reduced heat-transfer resistance of the wall by temperature fields using both software applications [7, 8] and SP 230.1325800.2015 methods with reference tables.

3 Materials and methods

General problem to increase energy performance of buildings in public utilities sector may largely be solved by replacement of existing window units made of wood to new PVC profile windows equipped with multi-glazed glass units both in brick and frame-panel old buildings [9, 10].

However, replacement of window units is stipulated by some principal peculiarities:

- significant increase of high-shielding performance of a window unit occurs but due to reduction in frame thickness from 120mm to 60-80mm there is a risk of jamb freezing;
- PVC window unit exhibits minimum air permeability (i.e. sealed) thereby significantly (and not for the better) changing room air conditions;
- if the technology of window installation is violated, the area connecting translucent unit with the wall will be the thermal bridge.

Furthermore, increase of high-shielding performance of window units and sealing of junctions thereof together with current ventilation system lead to high humidity and if combined with window installation defects often result in freezing of joints between the window unit and the wall thereby leaving moisture and even frost on the jamb surface [11, 12]. Experimental evaluation of thermal and physical characteristics changes in exterior walls after installation of new windows with glass units was made through thermovision inspection (Fluke Ti20 thermal imager).

According to the thermograms (Fig.1.) temperature in the corners of the window area from the inside is 13°C, revealing incorrect installation and freezing of junction between the window and exterior brick wall [12, 13]. Similar defects are found under examination of residential and municipal facilities in the Republic of Bashkortostan. Junction between the window (new unit with narrow frame) and single-layer walls is found to be the weakest point in heat shielding.

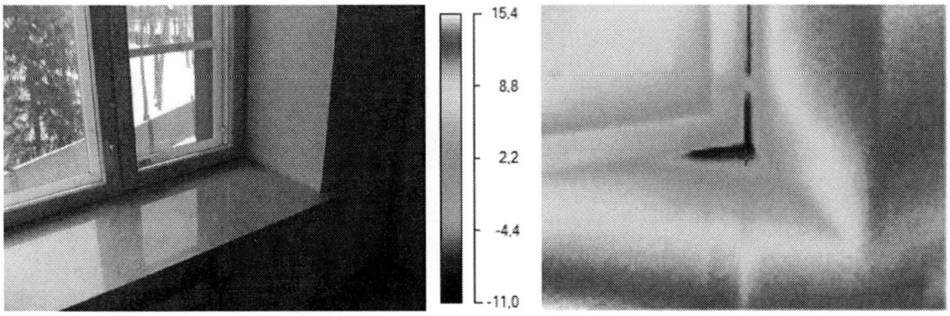

Fig.1. Window area thermogram.

The reason for this is heat loss by the window frame through the jambs. Window unit is in the negative temperature area in single-layer walls and heat is lost not only along the window frame but through the wall itself. This is not happened when the exterior wall is efficiently insulated and if combined with window frame insulation [14, 15].

Junction between the window unit and exterior wall of the building is quite complex and multi-element therefore its manual thermotechnical calculation by SP 50.133330.2012 is difficult enough. Thermotechnical calculations made as per this standard may be done using the finite elements method (FEM) [16, 17]. In our case the finite element (junction between the wall and the window) is taken as a rectangle for making thermotechnical calculations. Mathematical tools are presented in [18] by the authors.

To get the heat conduction matrix for the finite element we will use the heat balance equation:

$$\chi = \int_V \frac{1}{2}[\{g\}^T[D]\{g\}]dV + \int_S qTds + \int_s \frac{1}{2}h(T - T_\infty)^2 dS \quad (1)$$

As function T is not continuous in the whole area instead of it we will introduce function Ti defined on separate elements. Integral (1) will be broken into integrals by separate elements

$$\chi = \sum_{i=1}^{n}(\int_{V^i} \frac{1}{2}[\{g^i\}^T[D^i]\{g^i\}]dV + \int_{S^i} q^i T^i ds + \int_{s^i} \frac{1}{2}h^i(T^i - T_\infty^i)^2 dS) \quad (2)$$

where, n – total number of elements.

Substituting (1) to (2) we get the following:

$$\chi = \sum_{i=1}^{n}(\int_{V^i} \frac{1}{2}\{T\}^T[B^e]^T[D^i][B^e]\{T\}]dV + \int_{S^i} q[B^i]\{T\}ds + \int_{S^i} \frac{1}{2}h^i(T^i - T_\infty^i)^2 dS) \quad (3)$$

If there is no convective heat exchange in the boundary and heat flow q is equal to zero then (3) will be as follows

$$\chi = \sum_{i=1}^{n}(\int_{V^i} \frac{1}{2}\{T\}^T[B^e]^T[D^i][B^e]\{T\}]dV \quad (4)$$

Heat conduction matrix for rectangular finite element will be defined from (4) by

$$R = \int_{-\frac{b}{2}}^{\frac{b}{2}} \int_{-\frac{a}{2}}^{\frac{a}{2}} B^T DBdxdy \tag{5}$$

on integrating (5) the heat conduction matrix in tabular form will be as follows:

$R = \frac{k_t}{3}$

$\frac{b}{a}+\frac{a}{b}$	$-\frac{b}{a}+\frac{a}{2b}$	$\frac{b}{2a}-\frac{a}{b}$	$-\frac{b}{2a}-\frac{a}{2b}$
	$\frac{b}{a}+\frac{a}{b}$	$-\frac{b}{2a}-\frac{a}{2b}$	$\frac{b}{2a}-\frac{a}{b}$
		$\frac{b}{a}+\frac{a}{b}$	$-\frac{b}{a}+\frac{a}{2b}$
			$\frac{b}{a}+\frac{a}{b}$

(6)

The heat conduction matrix was run through the test case considering convective heat exchange, by making the heating pattern of the wall under the results thereof.

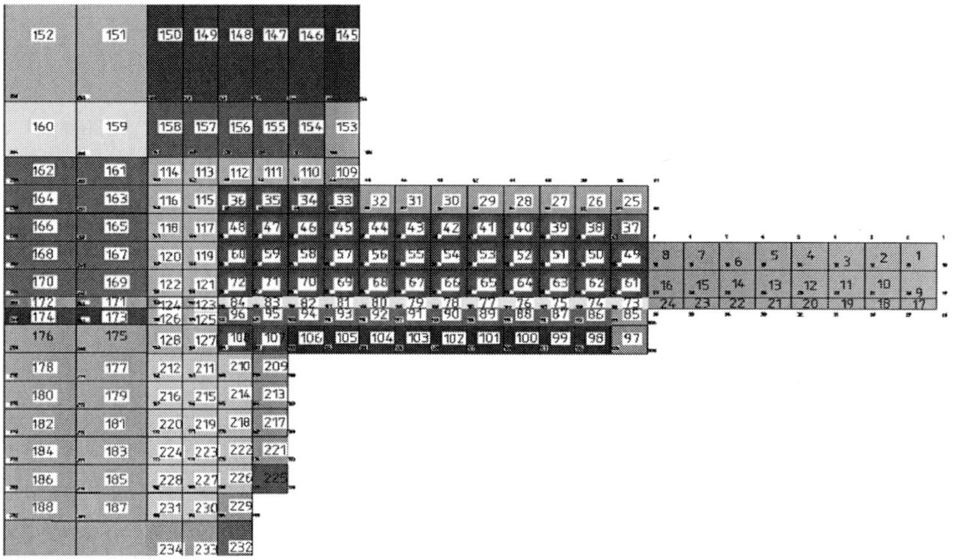

Fig.2. Computational model of a part of the wall with the window attached thereto.

According to analysis of the obtained data the dew point is inside the wall [19] (for the coldest five-day period of –35°C). Above mentioned method (Fig.2) was realized by "ASPECT" software application set up by authors of this research.

Thermotechnical calculations results for the junction between the window unit and the exterior wall of the building may be likely to be obtained using this software application as the temperature fields [20]. This application enables to define the coldest area for choosing certain measures to eliminate the thermal bridges thereby confirming the research results of various authors. The jamb appears to be in the negative temperature area according to the calculations made.

4 Conclusions

In a number of cases significant heat loss is found to take place in jamb areas when analyzing heat loss in exterior walls of brick buildings. Experimental evaluation of thermal and physical characteristics changes in exterior walls after installation of new windows with glass units was made through thermovision inspection. Use of the software application [21] to choose certain measures to eliminate the thermal bridges enables to get the thermotechnical calculations in the junction between the window and the exterior wall in the form of temperature fields.

According to the study made the heat insulation placed in the junction between the window unit and the wall is the important element of the window frame to eliminate thermal bridges and for keeping normal room conditions.

References

1. A.I. Bedov, A.M. Gaysin, A.I. Gabitov, News of higher educational institutions. Textile industry technology, **3**, 231-236 (2017)
2. A.M. Gaysin, M.V. Sharafutdinova, A.I. Gabitov, E.A. Udalova, History of science and technology, **10**, 21-28 (2014)
3. A.M. Gaisin, V.V. Babkov, Construction materials, **1**, 55-58 (2016)
4. A.S. Salov, L.I. Khabibullina, A.I. Gabitov, E.A. Udalova, V.A. Timofeev, A.A. Timofeev, History of science and technology, **11**, 37-43 (2017)
5. A.I. Bedov, A.I. Gabitov, Bulletin of MSUCE, 328 (2017)
6. A.A. Semenov, A.I. Gabitov, *Part 1 Static analysis* M: Association of Construction Institutions Publishers, 152 (2005)
7. A.S. Salov, I.G. Terekhov, I.V. Nedoseko, Ufa USPTU Publishers, 138 (2016)
8. A.I. Bedov, A.S. Salov, A.I. Gabitov, D.V. Kuznetsov, E.A. Sadykova, International Journal for Computational Civil and Structural Engineering, **4**, 37-46 (2017)
9. S.I. Kryshov, I.S. Kurilyuk, Housing construction, **7**, 3-5 (2016)
10. A.M. Gaisin, R.R. Gareev, V.V. Babkov, I.V. Nedoseko, S.Y. Samokhodova, Construction materials, **4**, 82-86 (2015)
11. A.I. Gabitov, E.A. Udalova, A.S. Salov, A.R. Chernova, D.V. Pyzhjanova, V.V. Yamilova, History of science and technology, **6**, 58-65 (2017)
12. A.I. Bedov, V.V. Znamensky, A.I. Gabitov, *Structural Assessment, Renovation and Strengthening of Frames of Operating Buildings and Structures. Part 1. Structural Survey and Assessment of Frames of Operating Buildings and Structures* (M: ASV Press, 2014)
13. A.I. Bedov, A.I. Gabitov, V.V. Znamensky, *Structural Assessment, Renovation and Strengthening of Frames of Operating Buildings and Structures. In two parts. Part 2. Structural Renovation and Strengthening of Frames of Operating Buildings and Structures* (M: ASV Press, 2017)
14. A.I. Bedov, A.M. Gaisin, A.I. Gabitov, A.S. Salov, A.A. Gallyamov, Industrial and Civil Construction, **3**, 58-63 (2017)
15. A.M. Gaisin, S.Y. Samokhodova, I.V. Nedoseko, A.Y. Paymetkina, Residential construction, **5**, 36-40 (2016)
16. A.I. Bedov, A.I. Gabitov, A.S. Salov, A.M. Gaisin, L.I. Khabibullina, Building and reconstruction, **6(74)**, 3-12 (2017)

17. A.I. Bedov, A.S. Salov, A.I. Gabitov, XXI International Scientific Conference on Advanced in Civil Engineering "Construction - The Formation of Living Environment", **365**, 1-8 (2018)
18. A.I. Bedov, A.M. Gaisin, A.I. Gabitov, R.G. Galeev, A.S. Salov, M.S. Shibirkina, Bulletin of MSUCE, **11**, 46-57 (2015)
19. A.I. Bedov, A.I. Gabitov, *Engineering, Renovation and Strengthening of Masonry and Reinforced Masonry Structures* (M.: ASV Press, 2006)
20. D.N. Sursanov, A.B. Ponomaryov, Bulletin of National Research Perm Polytechnic University Construction and Architecture, **4**, 144-165 (2015)
21. R.G. Galeev, A.M. Gaisin, M.S. Shibirkina, D.A. Sinitsyn, A.S. Salov, *Application for making thermotechnical calculation of junction between window unit and exterior wall of the building: certificate of computer software registration*, patent 2016612591 (2016)

Study of heat shielding qualities of a brick wall corner with additional insulation on the brick

Oleg Yurin [1], *Anna* Azizova[1], and *Tatyana* Galinska[1*]

[1]Poltava National Technical Yuri Kondratyuk University, 36011 Poltava, Pershotravnevyi avenue, 24, Ukraine

Abstract. This work presents the theoretical investigations results for the brick wall thermal protection properties exterior corner with the insulated layer SUPERROCK mineral wool. As the outer corner of the enclosing structures has lower thermal protection properties than the other part of the wall, it needs additional insulation. Two variants of additional insulation were considered in the work, in the deepened outer surface and in the middle of the brick wall. In this case, the additional heater was not continuous, but divided into two parts and shifted from the corner. The length and volume of additional insulation for different variants were determined. The optimal variant of the proposed schemes was determined by the minimum volume of the heater and compared with the existing ones. The volume of additional insulation is set at 1 meter of the fence height. For the proposed variant of the location for the heater in the brick wall middle, its thickness in the brick cavity, the length and offset from the corner magnitude were determined. For a location-variant of the heater in the deepened on the outer surface of the fence bricklayer, the length of the additional insulation and its displacement from the corner were determined. The thickness of additional insulation was taken 130 mm (width of the brick and seam between the bricks). Studies were carried out for the first temperature zone.

1 Introduction

One of the way of saving the thermal energy going to the heating of buildings and structures is improvement the thermal units of enclosure constructions. One of such thermal units is the outer corner of the external enclosures. Typically, heat nodes are places with a lower heat transfer resistance than the main fence, and therefore contribute to increased heat loss. Improving the insulation of the outer corner of the brick wall is an urgent task. The optimal location and dimensions of additional insulation are determined. By the criterion of optimality, the minimum amount of additional insulation is adopted, at which the requirements of thermal protection are fulfilled.

[*] Corresponding author: galinska@i.ua

1.1 Review of recent research and publications sources

Many authors' works are dedicated to corners increasing heat shielding properties issue according to the norms requirements. Recently, the variants for insulation of the external brick walls corner were considered in the works [1-3]. In article [1] there is proposition of constructive solution in which additional insulation is located in the deepening of the brick wall from the outside. Such an arrangement of additional insulation, according to the authors, increases the temperature in the corner at 0.5 to 1.7 °C, which provides more comfortable indoor conditions. The authors obtained mathematical dependence, which allows to calculate the value of temperature in the corner without using numerical simulation and experimental methods, which simplifies the ad variant of engineering solutions.

The article [2] discussed variants to deepen in the external surface sections of window and door openings, exterior walls in the corners and at the level of overlap which placed an additional layer of insulation. According to the authors, the proposed constructive solutions can be used in the new construction and reconstruction of buildings. The considered constructive decisions do not complicate the process of execution of work and can be fulfilled with the appropriate qualifications of the workers. Additional insulation increases the value of the given resistance of the heat transfer of the wall and thereby reduces the heat loss. The economic efficiency of the proposed solutions for a two-story building with a capacity of about 1000 m^3 is 20,000 UAH, which indicates the availability of its use in building practice.

In article [3], a method for locating thermoclines in places of special depressions of the walls from the outside of the corners, openings and the level of inter-floor overlap is proposed. The work limits the use of this method and provides constructive recommendations for the practical arrangement of nodes. The decrease in linear coefficients in the corners, in the openings and also in the joints of inter-floor overlap at 1.3 – 9.4 times is fixed. The minimum lengths of the thermocline, which is equal to the thickness of the wall, are substantiated. As a variant, an analysis of the location of the heater in the brick masonry in the area of the external corner is given.

In work [4], variants of additional insulation of the corner of frame-monolithic buildings were considered. [5-6] The influence of the mutual arrangement of the external corner of the house and the window opening on the temperature of the internal surface of the corner was considered in article [7-10]. Definition the parameters of thermal insulation in the foundation zone [11].

1.2 Previously unsolved assignment aspects of the general problem

In publications devoted to research ways to improve the heat shielding properties of corner to the heat transfer norms proposed use continuous additional layer of insulation located near external corner of the enclosure construction. The variant of dividing additional insulation into two parts and displacement it from the corner within the design scheme is not considered.

1.3 Setting objectives

Purpose of this work was to determine the optimal location and the possible minimum size of additional insulation, divided into two parts and shifted from the corner of the external brick wall within the design scheme. For optimality criterion was adopted minimum required capacity of additional insulation that provides thermal performance standards.

2 Basic material and results

In the works [1-3], proposed variants of additional insulation of the external corner of brick walls that are shown in Figure 1.

Studies were carried out for the 1st temperature zone. Enclosing structure of a wall of a residential building with a thickness of 510 mm. Heater insulation made of basalt mineral wool SUPERROCK with a density 135 kg/m^3. The thickness of the main layer of the heater insulation is 90 mm, which ensures the implementation of the rules of thermal protection in the areas of walls without heat conductors.

The proposed additional insulation variants were compared with those given in [1-3] (Fig.°1).

For variants given in [1-3] the minimum possible length of additional insulation was determined at which the requirements of thermal protection are fulfilled.

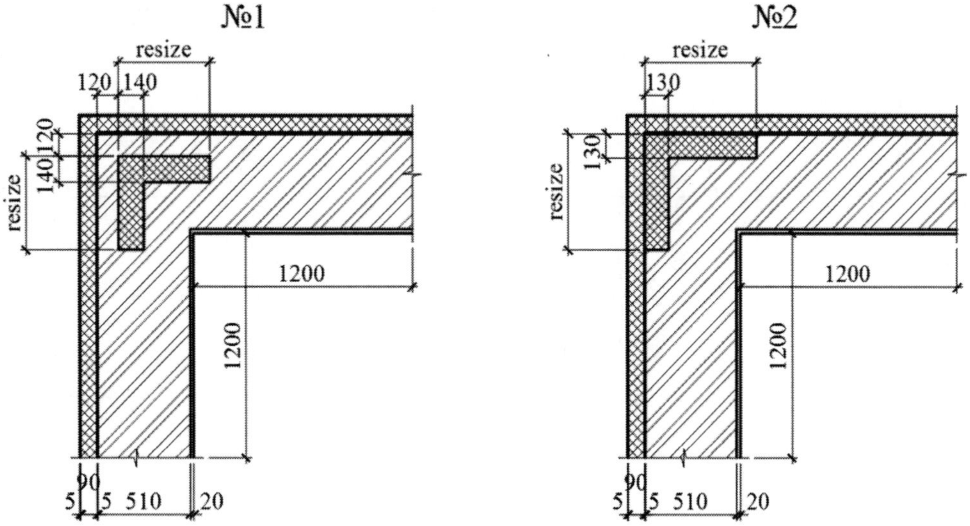

Fig. 1. Additional insulation of the external corner of brick wall by [1-3]. 1 – deep in the brickwork; 2 – in the deepening from the outside.

Figure 2 shows the variants for additional insulation that are offered.

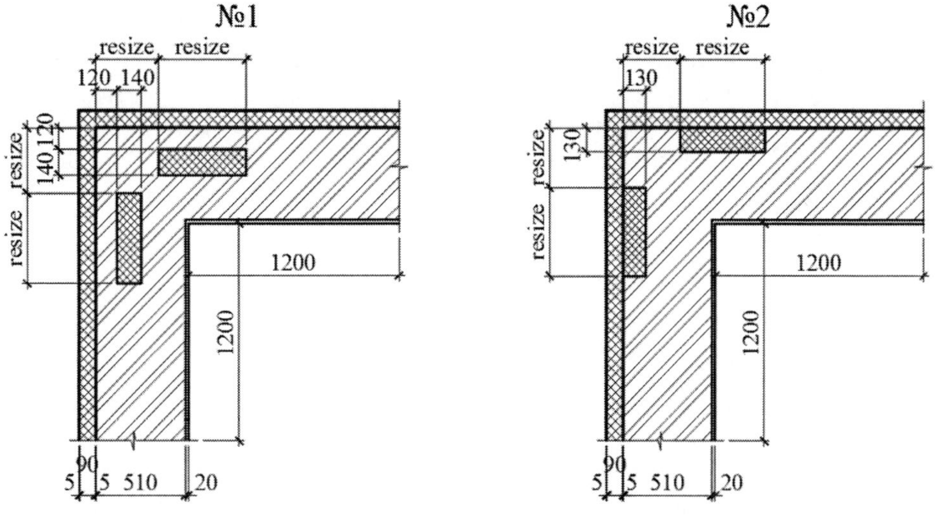

Fig. 2. Additional insulation of the external corner of brick wall that are offered. 1 – deep in the brickwork; 2 – in the deepening from the outside.

Under the scheme number 1, the length of additional insulation was taken (multiple sizes of the brick, taking into account the seams between the bricks), from 270 mm to 920 mm, with a graduation of 130 mm. The reduced resistance of the heat transfer of the calculated wall section was determined on the basis of the calculation of temperature fields. The results of the studies are presented in Table 1.

According to [6], for the external walls of residential buildings in the 1st temperature zone, the following requirements of thermal protection must be fulfilled: $\tau_{(в.min)} > t_p = 10.7$ °C.

As can be seen from Table 1, the second and third heat protection requirements are fulfilled at all lengths of additional insulation. The graph shown in Figure 3 allows you to determine the length of additional insulation in which the first requirement is fulfilled.

Table 1. Heat protection properties of the calculated area of the external corner according to the scheme of additional insulation N°1.

Length of additional insulation, mm	$R_{\Sigma\ red}$, m^2K/W	Δt_{red} °C	$\tau_{(в.min)}$ °C
270	2.636	1.8	16.4
400	2.719	1.8	16.8
530	2.832	1.7	17.4
660	2.971	1.6	17.8
790	3.133	1.5	18.1
920	3.317	1.5	18.2

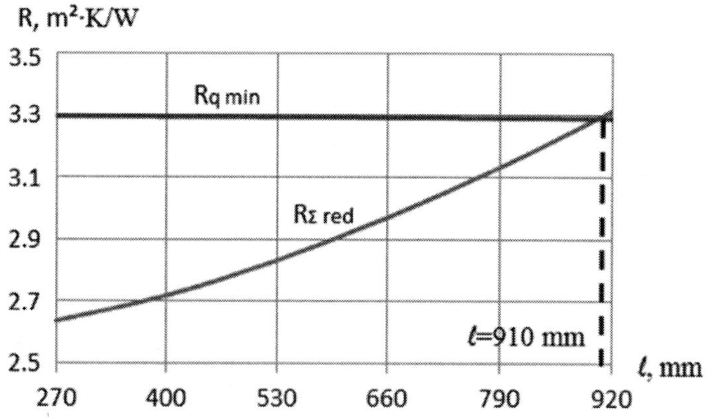

Fig. 3. Determination of the minimum length of additional insulation in which the first heat protection requirement is fulfilled.

As can be seen from Figure 3, the minimum length of additional insulation at which the first heat protection requirement is fulfilled is 910 mm. At this length, the additional heat insulation was 0.2352 m^3 per 1 m of height of the fence.

Similar studies have been performed for scheme 2. In considering this scheme, the length of additional insulation was taken (multiple sizes of the brick, taking into account the seams between the bricks), from 390 mm to 910 mm, with a gradation of 130 mm. The results of the research are presented in Table 2.

As can be seen from Table 2, the second and third heat protection requirements are also fulfilled at all lengths of additional insulation.

Table 2. Heat protection properties of the calculated area of the external corner according to the scheme of additional insulation N°2.

Length of additional insulation, mm	$R_{\Sigma\ red}$,m^{2}K/W	Δt_{red} ,°C	$\tau_{(в.min)}$,°C
390	2.803	1.7	16.7
520	2.909	1.7	17
650	3.032	1.6	17.3
780	3.175	1.5	17.5
910	3.337	1.4	17.6

The graph shown in Figure 4 allows to determine the length of additional insulation in which the first requirement is fulfilled

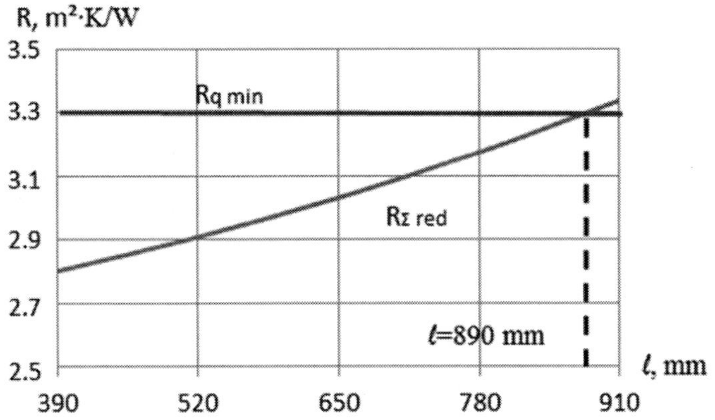

Fig. 4. Determination of the minimum length of additional insulation in which the first heat protection requirement is fulfilled.

As can be seen from Figure 3, the minimum length of additional insulation at which the first heat protection requirement is fulfilled is 890 mm. At this length, the capacity of additional insulation was 0.2145 m^3 per 1 m of height of the fence.

The capacity of additional insulation at its location in the deepness of the brickwork is less than when located in the deepening of the outside. That is, it is more expedient to approach it to the internal surface of the enclosure. But this worsens the humidity of the heater insulation. In addition, it is inappropriate to located the heater insulation in the zone of slabs peeling on the walls.

The variant of proposed additional insulation of the corner of the brick wall involves separating the heater insulation into two parts and shifting it from the corner (Figure 2). For a more accurate determination of the optimal value of the shifting of additional insulation, the value of the shifting was taken a multiple of 10 mm (not a multiple size of the brick). The optimal result was the result of which the capacity of additional insulation was minimal.

According to a variant N°.1 (Figure 2) additional insulation thickness taken from 140 mm (total cavity filling) to 30 mm. At the same time, the width of the cavity in which the additional insulation was arranged remained constant and equaled 140 mm. That is, with a decrease in the thickness of additional insulation, a closed air layer appeared. Length of insulation taken so that provides thermal performance requirements.

The results of the study of heat-shielding properties of the corner at a thickness of additional insulation of 140 mm are given in Table 3. As can be seen from Table 2, the optimal shifting value of additional insulation from the wall corner is 100 mm.

Table 3. The research results of heat-shielding properties of the corner of the proposed scheme with additional insulation thickness of 140 mm.

Scheme	The value of the shifting of additional insulation from the corner, mm	Length of additional insulation, mm	The reduced resistance of heat transfer, m^2K/W	Capacity of additional insulation, m^3
Present in [1-3] (without separation heater insulation)	0	910	3.302	0.2352
Proposed (heater insulation consists of two parts)	50	860	3.301	0.2327
	90	830	3.305	0.2299
	100	820	3.301	0.228
	110	820	3.308	0.2287
	130	820	3.309	0.2295
	150	900	3.311	0.252

The fulfillment of the requirements of thermal protection provides the length of the heater insulation of 820 mm. The capacity of insulation at 1 m of the height of the fence is 0.228°m^3. This is 3% less than the additional insulation scheme without dividing it into two parts and shifting from the corner that given in [1-3].

The research results of the heat-shielding properties of the corner at a thickness of additional insulation from 140 mm to 30 mm are shown in Table 4. The capacity of additional insulation is set at 1 m of the height of the fence.

Table 4. The research results of heat-shielding properties of the corner of the proposed scheme with additional insulation thickness from 140 mm to 30 mm.

Thickness of additional insulation, mm	The optimum value of the shifting of additional insulation from the corner, mm	Length of additional insulation, mm	The reduced resistance of heat transfer, m^2K/W	Capacity of additional insulation, m^3
140	100	820	3.301	0.228
130	100	820	3.3	0.2116
120	100	840	3.301	0.2
110	110	860	3.306	0.1883
100	110	890	3.304	0.1771
90	120	920	3.301	0.1652
80	120	970	3.305	0.1548
70	120	1030	3.307	0.1438
60	120	1110	3.307	0.1328
50	120	1210	3.301	0.1206
40	130	1360	3.303	0.1087
30	0	1500	3.301	0.0891

From Table 4 it is seen that there is the smallest amount of additional heater with its thickness of 30 mm without shifting from the corner. The length of the heater insulation, which ensures the implementation of the conditions of thermal protection is 1500 mm (throughout the length of the calculation scheme). Figure 5 shows the scheme of the optimal location of the heater insulation in the corner according to the proposed scheme of additional insulation N°.1.

Table 4 shows that it is more appropriate to reduce the thickness of the additional insulation with a corresponding increase in its length. That is, the best result is the uniform distribution of the heater insulation within the limits of the calculation scheme.

To shift a heater insulation from an corner at its thickness of 30 mm is impossible, because it is not enough length of the calculation scheme for fulfillment of the first requirement of thermal protection. The scheme shown in Figure 5 allows reducing the capacity of additional insulation at 62% compared with the scheme with the full filling of the cavity between the layers of the brick that given in [1-3].

Similar studies were performed on the proposed variant of additional insulation No.2 (Figure 2). The thickness of additional insulation was taken 130 mm (width of the brick and seam between the bricks). The distance from the additional insulation from the corner was taken multiple times the size of the brick. For more accurate results, the length of the additional insulation, which ensures the implementation of the standards of thermal insulation, was determined with a multiplicity of 10 mm.

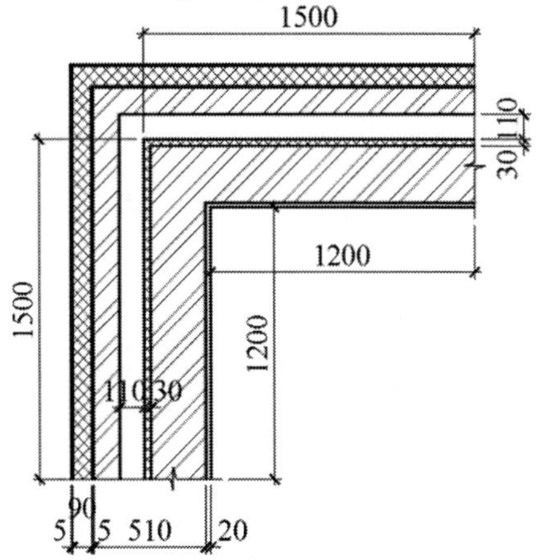

Fig.5. The scheme of the optimal arrangement of the heater insulation in the corner according to the proposed scheme of additional insulation N°.1.

The research results of heat-shielding properties of the corner of the scheme additional insulation N°.2 are given in Table 5. The capacity of additional insulation is set at 1 m of the height of the fence.

Table 5. The research results of heat-shielding properties of the corner of the proposed scheme N°.2.

The value of the shifting of additional insulation from the corner, mm	Length of additional insulation, mm	The reduced resistance of heat transfer, m^2K/W	Capacity of additional insulation, m^3
0	890	3.31	0.2145
260	870	3.3	0.2262
390	850	3.301	0.221
520	840	3.311	0.2184
650	820	3.302	0.2132
780	810	3.301	0.2106
910	800	3.307	0.208
On the edge of the calculation scheme	760	3.312	0.1976

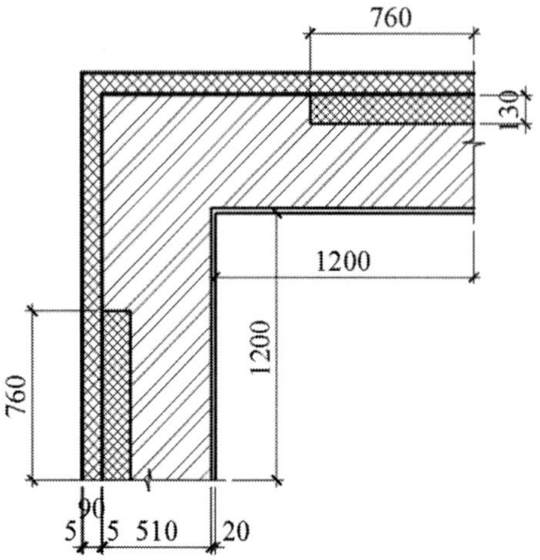

Fig.6. Scheme of the optimal location of the heater insulation in the corner according to the proposed scheme of additional insulation N°.2.

The performed studies have shown that the smallest capacity of additional insulation is observed at its location along the edge of the calculation scheme. The length of the additional insulation is 760 mm.

Figure 6 shows the scheme of the optimal location of the heater insulation in the corner according to the proposed scheme of additional insulation N°.2.

The scheme shown in Figure 6 allows to reduce the capacity of additional insulation at 8% compared with the scheme with full filling of the cavity between the layers of the brick that given in [1-3].

3 Conclusions

1. With the application of additional insulation of the corner of the scheme N°.1 with a decrease in its thickness and corresponding increase in length, the capacity of insulation needed to meet the requirements of thermal insulation decreases. The best result gives an even distribution of the heater insulation within the limits of the calculation scheme.
2. The smallest capacity of additional insulation is observed at its thickness of 30 mm, if it is located throughout the length of the calculation scheme without shifting from the corner.
3. Such an arrangement of an additional insulation, and the use of a closed air layer, can reduce its capacity at 62% compared with the scheme with the full filling of the cavity insulation between the layers of brick that given in [1-3].
4. The smallest capacity of additional insulation when using additional insulation of corner scheme No. 2 observed when it is on the edge of the settlement scheme.
5. Such an arrangement of an additional heater allows to reduce its capacity at 8% compared with its arrangement according to the scheme that given in [1-3].
6. From the two schemes considered, the scheme N°.1 is more optimal where the capacity of additional insulation is less.

References

1. *DBN V.2.6-31:2006. Konstruktsiyi budinkiv i sporud. Teplova izolyatsiya budivel* / Minbud Ukrayini. Kiyiv, (2006).
2. A.E Piir., O.A Kozak., V.B Kuntysh. *Science & Technique*. **16(2)** 931(2017).

3. A.E.Zakharevich . *Science & Technique*. **15(6)** 940(2016).
4. A. M. Prishchenko, T. V. Zhmykhova, M. V. Tymofieiev. Budivelni konstruktsii. **79** 16(2013).
5. A.N. Pryshchenko Visnyk Donbaskoi natsionalnoi akademii budivnytstva i arkhitektury. **2** 13(2014).
6. G.Vasilyev, V.Leskov, V.Gornov, V. Lichman, M. Kolesova, A.Dmitriev. *MATEC Web of Conf.* **40** 05004 (2016).
7. N. Bhikhoo, A. Hashemi. Sustainability (Switzerland), **9(8)** (2017)
8. O. V. Semko, O. I. Yurin Academic journal Series: Industrial Machine Building, Civil Engineering **4(2)** 182 (2013).
9. O.Yurin, T.Galinska. *MATEC Web of Conf.* **116** 02039 (2017).
10. X.Zhou, J.Carmeliet, D.Derome. *Building and Environment*, **135**, 246-256. (2018).
11. O. Filonenko. *Energy Efficiency*, **11**(3), 603-626. (2018).

Ground base deformation by circular plate peculiarities

Mykola Zotsenko[1,*], *Yuriy* Vynnykov[1], *Iryna* Lartseva[1], and *Svitlana* Sivitska[1]

[1]Poltava National Technical Yuri Kondratyuk University, Pershotravnevyi Avenue, 24, Poltava, 36011, Ukraine

Abstract. It is established that at the moment of the onset of the boundary equilibrium of the base under a central loaded plate, an elastic core is formed in the form of a cone, all particles move along with the plate progressively down. In this case, within the limits of the "area of influence" the boundary condition of the base is violated by protrusion or consolidation of soils. The volume of the "area of influence" is determined by the height of the elastic core, from which its diameter is calculated. Experiments using conical tips were made and the height of the elastic core was determined and the diameter of the "area of influence" was determined. For creation at penetration and probing friction of soil on a ground the conic tip is executed rough with cylindrical stages. Under the steps of the tip, formed areas of compacted soil, moving along with it, creating a so-called "ground shirt" in the form of a cone with a creature at an angle α to the vertical. The obtained data allowed specifying coefficients of bearing capacity of soils, which are used in construction norms.

1 Introduction

Modern methods of mathematical modeling of geotechnical processes [1] provide the model construction of the structure surrounding soil foundation interaction [2-6].

With the gradual loading of a rigid circular plate from the surface the deformation of its base includes three phases [7]:

1 – a compaction phase, during which the elastic core of compacted soil is formed;

2 – a local sliding phase, where with further increase in the load on outside the core shifts begin to develop changes that occur first in the individual points of the array, and the growth of tensions occupy a larger area;

3 – a phase of destruction, where zones of local shifts are gradually guided, creating so-called sliding surfaces, with the complete plate formation significantly precipitates when bulging the soil on the surface and in the sides. Under a central loaded circular plate, an elastic core is formed in the form of a cone, all particles, along with the plate, move progressively downward.

* Corresponding author: vynnykov@hotmail.com

2 Analysis of resent research and publications

Theoretical and experimental studies have established that at the time of the onset of the basis limit state of the plate correspond to [8, 9]:

– in case of the flow is the general shift of the soil around the plate behind the surfaces of the slide;

– in the absence of flow is soil consolidation around the plate.

Immersion of the plate occurs due to the displacement of the soil in the zone of the least resistance. Such zones, depending on the relative penetration of the die and the density of the sand, may be zones above and below the stem of the plate.

According to tests [10] the interaction of soil zones shift with parallel immersion of plates is felt in much wider boundaries than is predicted by the theory [11, 12]. Therefore, the determination of the geometric boundaries of the soil flow zone by immersing a plate is especially important for constructing a design destruction soil model.

3 Establishing the objective and tasks of the study

As mentioned above, under a rigid circular plate in the soil, an elastic core is formed, which after further plate loading disperses and densifies the soil around. Thus, a protrusion zone forms soil compaction. The effort that is spent on the formation of the zone is the limiting load on the base.

In the classical solution of Professor Berezantsev V., which is used in building codes, an elastic cone with an angle at the apex of 90° is taken. Professor Malyshev M. took an elastic cone with an angle at the top of 60° [8].

Of course, the height of the hard core determines the size of the calculation model scheme for the base destruction under a circular plate and, accordingly, the marginal load magnitude on the plate. Consequently, determination of the actual size of this zone is the main task when compiling the calculation base destruction model scheme under a circular plate.

Therefore, the purpose of the work is to clarify the known solutions for determining the boundary load on the plate. To do this, it is necessary to determine experimentally the height of the elastic core, which is formed during the circular plate loading.

4 The main part of the study

Soils penetration and sounding are usually carried out by immersion in metal of conical polished tips of different sizes and opening angles 2α. At penetration, the conical tip is immersed in the soil only at its height; with further immersion of sounding soils [13, 14].

In the process of penetration (sensing), the process of soil destruction is modeled in accordance with the theory of boundary equilibrium of a disperse medium [15]. As stated above, in accordance with this theory, the soil destroys the elastic core of the compacted soil, and the use of polished metal tips is a certain deviation from the usage theory.

To create cone-shaped tips of soil friction, they were made rough with cylindrical steps, as shown in Fig. 1 [16]. The rough tip consists of a cone and a shank. Cone, made with cylindrical steps, which have a correlation

$$\delta/h_c = tg\alpha/2, \quad (1)$$

where δ – ledge; h_c – rise; α – angle between vertical and a cone generator.

Under the conical tip steps there are formed compacted soil area, which move along with the tip and form so-called "ground shirt" in the form of a cone with a creature at an angle α to the vertical.

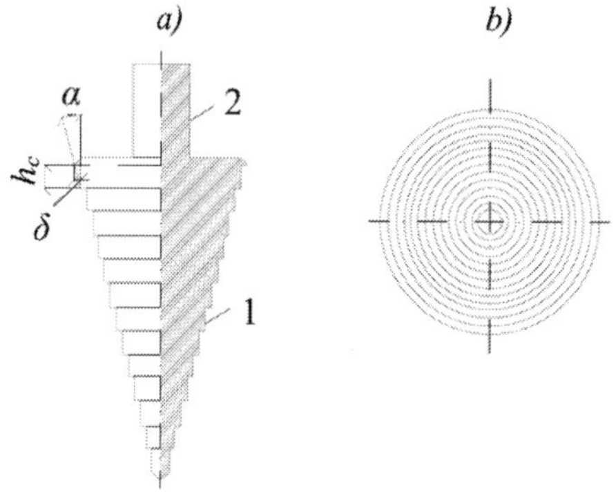

Fig. 1. Staggered tapered tip for penetration and sensing: a – section; b – plan;
1 – conical section; 2 – shank, h_c – rise; δ – ledge; α – angle between vertical and a cone generator.

In Fig. 2, a the results of small-grained, quartz sand with a conical tip with an angle at the top tests $2\alpha = 30°$ are given.

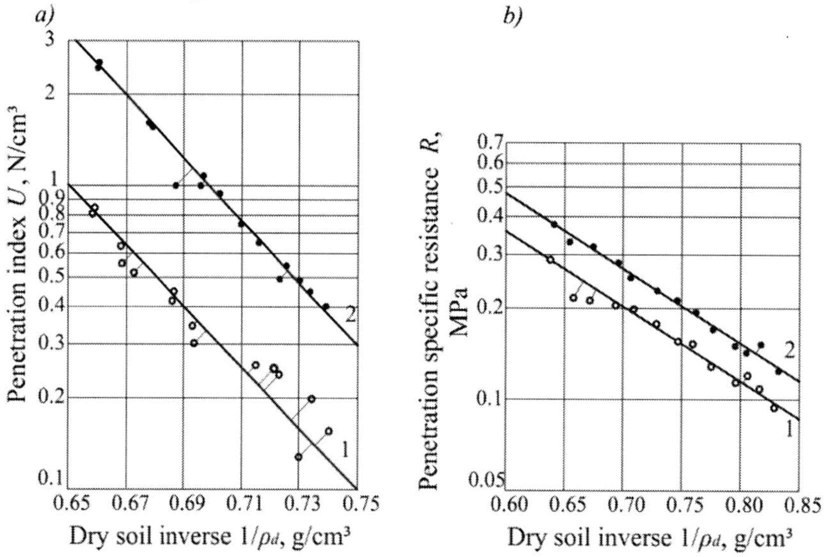

Fig. 2. Results of soils penetration tests with conical tips with an angle at the vertex $2\alpha = 30°$ (with surface: 1 – smooth; 2 – rough: *a* – sand; *b* – loess soil.

The tests are performed using a laboratory penetrometer LP-1. Soil skeleton density varied within ρ_d = 1.37-1.52 g/cm^3. In this case, the angle of internal friction of sand fluctuated within $\varphi = 32\text{-}36^0$. The averaging line 1 of the penetration index U, N/cm^3, of the dry soil inverse $1/\rho_d$, cm^3/g, obtained using a tip with a smooth surface. The equation of this dependence is represented by formula (2) based on the results of 16 individual tests

$$lgU = 5.329 - 7.058\cdot 1/\rho_d. \quad (2)$$

Equation correlation coefficient is r = 0.934, and variation coefficient υ = 0.195, accordingly, straight line 2 is obtained using a rough tip on the results of 14 individual tests which indicates a close connection between research data and empirical dependence.

$$lgU = 18.11 - 24.279\cdot 1/\rho_d. \quad (3)$$

The equation correlation coefficient is r = 0.946 and variation coefficient υ = 0.109, which indicates the close connection between the experimental data and empirical dependence.

There is a significant difference in large sand penetration values. In the skeleton sand density range ρ_d = 1.37-1.52 g/cm^3 penetration index *U*, obtained by a rough tip, in 2.18-3.37 times higher.

In Fig. 2, b shows similar data obtained for the pulled loam of the broken structure. The angle of internal friction varied within φ = 14-16°, and the skeleton sand density ρ_d = 1.20-1.56 g/cm^3.

Straight line 1 dependence of the penetration specific resistance *R*, MPa, on the inverse dry soil density $1/\rho_d$, cm^3/g, obtained using a tip with a smooth surface. The equation of this dependence is represented by formula (4) based on the results of 13 individual tests

$$lgR = 0.721 - 0.732 \cdot 1/\rho_d. \quad (4)$$

Equation correlation coefficient is r = 0.943, and variation coefficient υ = 0.136, which indicates a close connection between research data and empirical dependence.

Straight line 2 dependences of the penetration specific resistance *R*, MPa, on the inverse dry soil density value $1/\rho_d$, cm^3/g, obtained with a rough surface tip. This dependence is represented by formula (5) according to the 13 individual tests

$$lgR = 1.003 - 1.036 \cdot 1/\rho_d. \quad (5)$$

Equation correlation coefficient is r = 0.944, and variation coefficient υ = 0.122, which indicates a close connection between research data and empirical dependence.

The pattern established for the sand remains, but the difference between the data obtained by tips with a smooth surface and rough, is much smaller.

According to the study [17], the friction angle of the soil beyond the surface of a smooth metal tip does not exceed 10°, which corresponds to the coefficient of friction f = 0.177.

With the same surface area of the tips, the difference in penetration indicators values corresponds to the tangents ratio to the friction angles. For the sand $tg34° / tg10°$ = = 0.67 / 0.177 = 3.8, and for the pulled loam of the broken structure $tg15° / tg10°$ = = 0.268 / 0.177 = 1.514. The results of the calculations are close enough to the experiment.

A series of experiments was conducted to verify the impact of the angle at the apex of the conical tips 2α on the friction value of their surface during the static probing in a laboratory environment. The tests were carried out using a laboratory probing device LZU-1 of PoltNTU design. As soil, sand, quartz, large, slightly wet, which, at different density values, was placed in a tray with a cross section of 100x100 cm and a height of 70 cm. Conical metal tips with angles at the top 2α = 20, 30, 45, 60, 90° were made for sensing. For each of the angles there were produced tips with smooth and rough surfaces which are cylindrical stages. The diameter of the all the tips base was equal to 50 mm. additionally a flat plate of the same diameter was made.

For each of the tips, sensing was carried out under seven indicators of dry soil. Further sensing with flat plate was conducted. The value of a sensing index *V*, N/cm^3 was agreed on as probing characteristic, defined according to formula

$$V = (P - P_0) / (H \cdot A), \quad (6)$$

where *P* – sensing force, P_0 – penetration force; *H* – sensing depth; *A* – base of the cone (plate).

In total there were 77 sensing tests performed, with sensing indices *V* defined for them. Generalization of sensing results of sand of quartz, large, slightly wet, of different density with stage and smooth conical tips with different angles 2α are given in Fig. 3.

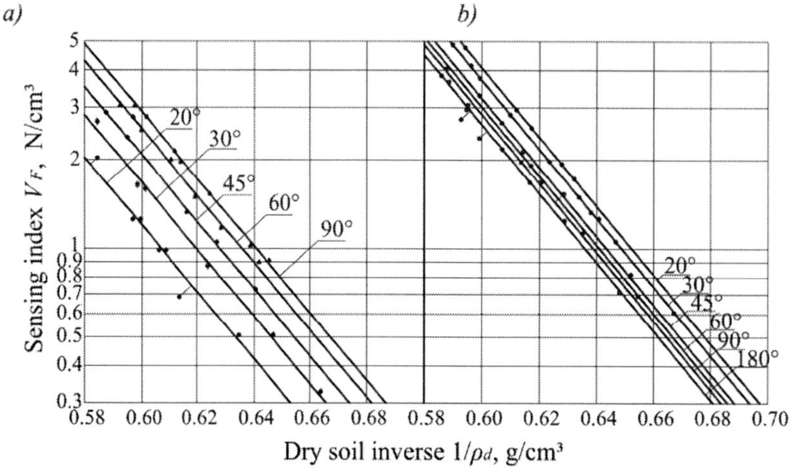

Fig. 3. Results of sensing tests of soils with conical tips with different angles at the vertex: *a* – with a smooth outer surface; *b* – with rough outer surface (stage).

For tips with a smooth surface received A_0 = 15.738; A_2 = 0.0398; A_3 = -9.753; α_0 = 1; ρ_{d0} = 1. Coefficient of correlation of the equation is r = 0.997, and Fisher's test F = 178.35.

For tips with rough surface received A_0 = -3.573; A_2 = 0.03; A_3 = -2.753; α_0 = 1; ρ_{d0} = 1. The equation correlation index makes r = 0.993, and Fisher criterion makes F = 75.29.

While sensing sand with smooth external surface tips, the least sensing index was obtained for cone with $2\alpha = 20^\circ$ angle and the biggest for cone with $2\alpha = 90^\circ$ angle. This testifies that along the whole height of the elastic core there is friction of the polished metal with large sand that means observing the lowest coefficient of friction of the cone with the soil. For a cone with $2\alpha = 20^\circ$, area of the metal polished surface within elastic core is the smallest. That is why the biggest resistance is observed. It means that the height of the elastic core is less than that of the cone $2\alpha = 20^\circ$ and larger than the cone $2\alpha = 90^\circ$.

When sensing sand with rough cones and a stamp of friction within the height of an elastic core, only the soil was ground. Therefore, the least resistance to sensing demonstrate a stamp and a cone with an angle of $2\alpha = 90^\circ$. At smaller angles 2α we have an artificially enlarged elastic core and, correspondingly, a larger surface of soil friction on the ground.

To determine the height of elastic core, an experiment was conducted, based on sandy soils probing with halves of rough metal cones with different angles at 2α vertex, diameter of which is equal to the diameter of the plate. The immersion in such cones was carried out along a transparent screen that limited the sandbox. The sand in the tray was put in layers of a given density, which in height were separated by indicative layers of ground chalk.

When immersing the halves of the cones along the transparent screen, the soil near the plate deforms. This phenomenon was recorded by certain distortions of the indicative layers in the thickness of the sand. In Fig. 4 photos of submerged halves of rough (stepped) cones in large sand with indicative layers are shown. The experience of tray experiments [10, 12] shows that interaction of zones of influence altogether with parallel immersion of stamps is observed in broader boundaries than in theory. Therefore, determining the boundaries of the zone of influence on the surrounding soil while immersing plates, is especially important for constructing a calculation scheme of the soil destruction model. Under the "zone of influence" we mean the area of the soil base around the plate, within which it is possible to fix the stress from the load of the plate. Typically, the problem of fixing stresses in soils is solved by installing different types of pressure sensors (string, strain gauge, hydraulic, electromagnetic). These works require high costs and special equipment for performing them. However, such experiments do not always produce unambiguous results. The results of these tests are significantly influenced by the sensors contact conditions of the soil.

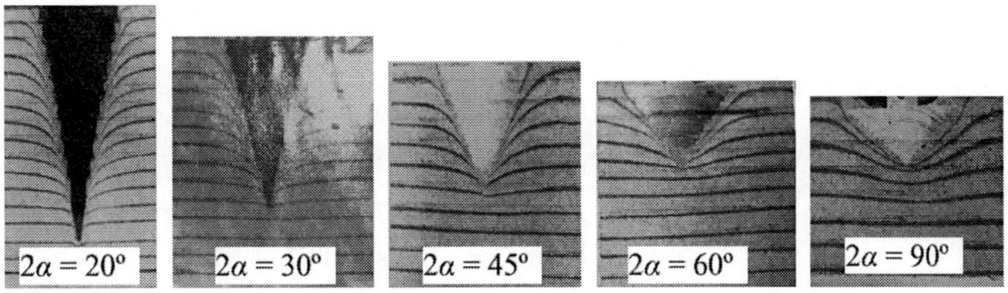

Fig. 4. Photo of the rough halves (stage) cones with angles 2α, what are immersed in coarse sand.

We used a direct way to determine the zone of stresses distribution on the base model with conical foundation. It is based on the effect of a ring walls influence with a soil sample while conducting penetration tests. The effect is that under immersion of a conical tip with diameter d and angle of opening $2\alpha = 30°$ in the soil placed in a metal ring at the certain depth of immersion of the tip (the critical depth of immersion h_k) there is a violation of invariance of specific penetration resistance, R, for clay soils, or Penetrance, U, for sands. This effect definitely determines the achievement of the "zone of influence" boundary the tip of the ring walls. The method of a sample preparation in advance guaranteed perfect homogeneity of the soil in the ring for each of clay and sandy soils sample in the given density ranges.

Examples of such violation of the penetration indicator invariance are given in Fig. 5. Here should be noted the high sensitivity of the method to the slightest violations of the conditions of experiment conducting, which in turn testifies of its sufficient accuracy. According to the graphs in Fig. 5 a similar violation of the invariance of the penetration specific resistance (penetration indicator) is characterized by a marked deviation of experimental points to the right of the averaged line.

a) *b)*

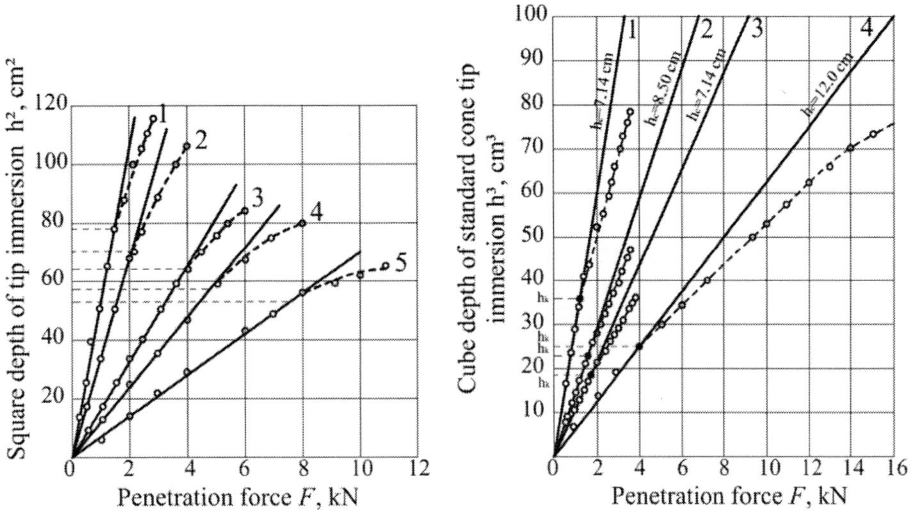

Fig. 5. Graphs of soil penetration with a conical tip: a – loams and sandy loam in the range of the angle of internal friction $\varphi = 12 - 26°$; b – quartz sand, shallow, air-dried in the range of $\varphi = 22 - 38°$.

Thus, for each penetration test of clay and sandy soils of different densities and correspondingly, the angle φ, the diameter of the "zone of influence" D_i is determined by the critical depth of the conical tip h_k immersion. In Fig. 6, curve 1 corresponds to the results of clay and sandy soils sensing in the range of the angle $\varphi = 12 - 38°$. Each point on the graph is set according to the results of five parallel tests. The equation of the curve, established by correlation analysis, has the following form (at $r = 0.996$ and $\upsilon = 0.048$)

$$D_i/d = 1.79 + 4.76 \cdot 10^{-7} \cdot \varphi^{4,5}. \quad (8)$$

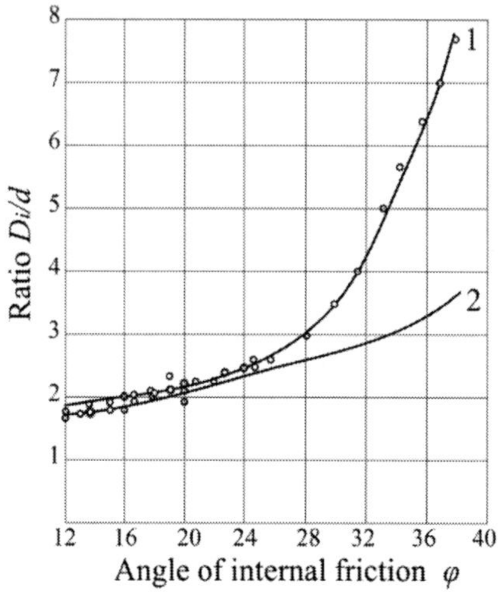

Fig. 6. Curves for determination of diameters 1 – "zone of influence" and 2 – "zone of compaction", depending on the angle of soil internal friction φ and diameter of plate d_p.

For comparison in Fig. 6 shows curve 2, which is set for the pyramidal piles based on the results of field tests of the diameter of the "zone of compaction" of soil D_c by methods of penetration and sensing. The corresponding equation has the form

$$D_c/d_p = 1.39 + exp(0{,}024 \cdot \varphi), \quad (9)$$

where d_p – diameter of pile; at $r = 0.99$ and $\upsilon = 0.024$.

Stipulates have shown that for clay soils with an angle $\varphi \le 24^0$ the dimensions of the "zone of influence" practically coincide with the limits of the "zone of compaction". But for sandy soils at $\varphi > 24^0$, the "zone of influence" in size exceeds the "zone of compaction". This difference is the more, the more the density of the sand, that is, φ.

5 Conclusions

Thus, on the basis of the conducted researches, a method for compilation of the calculation scheme of a solid model for modelling the soil destruction process with axial-symmetric loading of a flat plate is developed (Fig. 7).

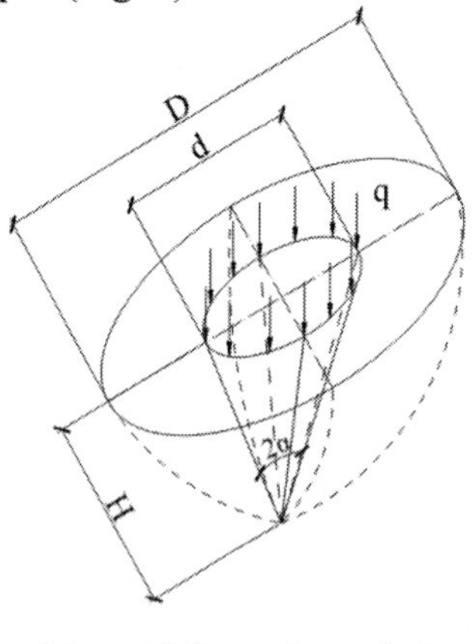

Fig. 7. Schematic diagram of the spatial model for mathematical modelling of the soil destruction process with axially loaded flat plate.

On the surface of the half-space (the surface of the soil) we apply the area of the plate. Displays the elastic core in the form of a cone with the area of the plate in the base and the angle of opening $2\alpha = 44^{\circ}$. The diameter of the "zone of influence" (the distribution of stresses on the soil foundation) is determined by the graph of Fig. 6 by the magnitude of the angle of the soil internal friction. At a certain diameter of the "zone of influence" we take a circle on the surface of the half-space. The surface of the "zone of influence" is constructed by connecting a circle line with the top of an elastic core. To simulate the process at a certain depth of the base, we place the half-space surface evenly distributed load from the weight of the soil layer in a height equal to the depth of the plate immersion.

References

1. K. Chau, *Numerical Methods*, Proc. of the 18th Intern. Conf. on Soil Mechanics and Geotechnical Engineering, Paris (2013).
2. M. Kelm, J. Grabe, *FE-simulation of soil compaction*, Proc. XIIIth European Conf. on Soil Mechanics and Geotechnical Engineering, Prague, **1** (2003).
3. N. Zotsenko, Y. Vynnykov, *Designing the compacted subsoil's using mathematical simulation method*, Active Geotechnical Design in Infrastructure Development, Proc. of XIIIth Danube-European Conf. on Geotechnical Engineering, Ljubljana, **2** (2006).
4. S. Henke, J. Grabe, *Numerical modeling of pile installation*, Proc. of the 17th Intern. Conf. on Soil Mechanics and Geotechnical Engineering, Amsterdam: IOS Press (2009).
5. K. Edip, M. Garevski, V. Sheshov, J. Bojadjeva, *Numerical simulation of consolidation problem*, Proc. of the XVI ECSMGE Geotechnical Engineering for Infrastructure and Development, Edinburg (2015).
6. P. Kryvosheiev, G. Farenyuk, V. Tytarenko, I. Boyko, M. Kornienko, M. Zotsenko, Yu. Vynnykov, V. Siedin, V. Shokarev, *Innovative projects in difficult soil conditions using artificial foundation and base, arranged without soil excavation*, Proc. of the 19th Intern. Conf. on Soil Mechanics and Geotechnical Engineering, Seoul (2017).
7. A. Manjriker, I. Gunarante, *Foundation Engineering*, New York: Taylor and Francis (2006).
8. V. Berezancev, *Raschet osnovanij sooruzhenij*, Leningrad (1970).
9. M. Malyshev, *Prochnost' gruntov i ustojchivost' osnovanij sooruzhenij*, Moskva (1994).
10. V. Rasorenov, *Penetracionnye ispytaniya gruntov*, Moskva (1980).
11. E. Schultze, K. Melzer, *The Determination of the Density and the Modulus Compressibility of Noun-cohesive Soils by Soundings*, Proc. of the VI Intern.Conf. on Soil Mechanics and Foundation Engineering, Montreal, vol. **I** (1966).
12. M. Iskander, R. Bathurst, M. Omidvar, *Past, present, and future of transparent soils*, Geotechnical Testing Journal, **38 (5)**, (2015).
13. M. Omidvar, M. Iskander, *Modeling of projectile penetration using transparent soils*, Proc. of the 19th Intern. Conf. on Soil Mechanics and Geotechnical Engineering, Seoul (2017).
14. H. Sturm, *The tip resistance in layered soils during static penetration*, Proc. of the 18th Intern. Conf. on Soil Mechanics and Geotechnical Engineering, Paris (2013).
15. M. Zotsenko, Y. Vynnykov, A. Yakovlev, *Modern practice of determination of strength characteristics of cohesive soils by penetration methods*, Proc. of XIVth Danube – European Conf. on Geotechnical Eng., Bratislava: University of Technology (2010).
16. Borvik Tore, Dey Sumita, L. Olovsson, *Penetration of granular materials by small-arms bullets*, Intern. Journal of Impact Engineering, **75** (2015).
17. N. Zotsenko, Y. Vynnykov, *Rapid Investigation Methods of Soil Properties and Interpretation of their Results for Bridge Foundations Design*, IABSE Colloquium reports on "Foundations for Major Bridges: Design and Construction", N. Delhi (1999).

Predicting the strength and maturity of hardening concrete

Irina Belykh, *Viktor* Sopov*, *Larisa* Butska, *Lidiya* Pershina, and *Olga* Makarenko

Kharkiv National University of Civil Engineering and Architecture, Sumskast. 40, 61002 Kharkiv, Ukraine

Abstract. When predicting the growth of strength and maturity of hardening concrete, the role of the temperature factorshould be taken into account. Taking into account the temperature factor requires at the initial stage of hardening of concrete low positive temperatures and low speed of heating concrete. The level of maximum heating of concrete will depend not only on the exothermy of the cement, but also on the amplitude of the fluctuations in the ambient temperature. The contribution to the cracking of concrete of various types of cements from the point of view of heat release during their hydration is analyzed. The existing models for predicting the character of the set of strength and maturity of concrete on the basis of data on heat dissipation during hydration of cement in concrete structures are analyzed. It is shown that laboratory tests are insufficient to evaluate the nature of heat generation. To effectively predict the physical and mechanical properties, it is necessary to conduct temperature-time monitoring during the concrete hardening in the design to take into account the influence of external factors.

1 Introduction

One of the main challenges of monolithic concreting is the problem of ensuring the temperature crack resistance of concrete. The formation of temperature cracks is associated with heating and cooling of concrete when it hardens in the structure. Heating is caused by heat release in concrete due to exothermic reactions of hydration of cement during its interaction with water. Since heat cannot be transferred instantaneously, the temperature gradients will be observed in the volume of the laid concrete, which will lead to the development of significant tensile stresses at the outer edges of the concrete massif. Also, crack formation can be caused by the effect of fluctuating outdoor air temperatures and the heat exchange conditions of the structural elements with the environment.

Temperature control in a concrete structure is one of the main measures to ensure the fracture toughness of concrete [1-4].

When predicting the growth of strength and maturity of hardening concrete, it is necessary to take into account the role of the temperature factor.

Taking into account the temperature factor requires low positive temperatures and low speed of heating concreteat the initial stage of hardening of concrete. In this case, the level

*Corresponding author: vsopov@ukr.net

of maximum heating of concrete will depend not only on the exothermy of the cement, but also on the amplitude of the variations in the ambient temperature.

The prevention of crack formation from various types of temperature influences, caused by the peculiarities of external and internal heat exchange in the structures at the stage of their erection, in order to ensure high consumer properties, today puts in monolithic construction an important and urgent task that predetermines the purpose of this work and the direction of research.

2 Materials and methods of research

Various types of cements were investigated in accordance with the classification of DSTU B.V 2.7-46: 2010 [5], whose compositions are given in Table 1.

Table 1. Types and mineral compositions of cements.

Type of cement	Mineral content, %				Specific surface, cm^2/g	Activity, MPa	Total heat evolution for 7 days, kJ/kg
	C_3S	C_2S	C_3A	C_4AF			
I (CEM I)	57	18	10	8	3650	42.3	
II (CEM II/B-S)	56	19	7	10	3730	38.6	
III (CEM III/A)	55	17	10	9	3530	37.3	
IV (CEM IV/A)	54	22	5	11	3820	38.2	

To estimate the magnitude of heat release during hydration of cements and concretes, the pseudo-adiabatic calorimetry method was used [6-8].

Determination of compressive strength of concrete samples 10 × 10 × 10 cm was carried out in accordance with DSTU B.V.2.7-214: 2009 [9].

3 Main part

It is known that with increasing temperature the rate of cement hydration processes in concrete increases in accordance with the Arrhenius equation. This contributes to a faster growth of concrete strength. Varying the conditions of hardening, relationship between the strength of the concrete and the temperature of its hardeningcan be established.

However, the exothermic nature of cement hydration in concrete contributes to an additional contribution to the temperature regime of hardening. The presence of resistance to heat transfer leads to the formation of an uneven temperature distribution in the volume of concrete [10-11]. As a consequence, there are temperature gradients between different parts of the structure, which contributes to the occurrence of cracks. Thus, cracking in hardening concrete can be caused by both hardening conditions and the nature of heat release.

Analysis of the literature sources [12-14] allowed to determine the correlation between the temperature gradient (temperature difference at the surface and in the core of the concrete structure) and the strength of the concrete (Fig. 1).

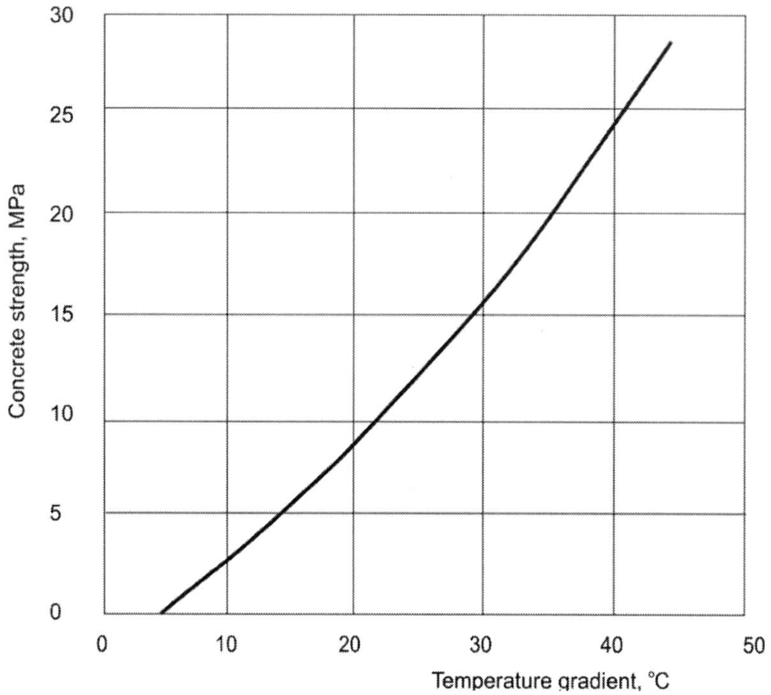

Fig. 1. Dependence of the compressive strength of concrete on the magnitude of the temperature gradient.

Depending on the conditions of hardening, the time of the year, the massiveness of the concrete structure, the time of aging in the formwork, it is possible to determine the optimum compositions of concrete of specified properties by the value of heat evolution.

When predicting the growth of strength and maturity of hardening concrete, it is necessary to take into account the effect of the temperature factor on:

1) magnitude of the temperature gradient;

2) magnitude of the temperature stresses between different parts of the structure.

Calorimetric analysis of hardening cement pastes presented in Table 1 is shown in Fig. 2.

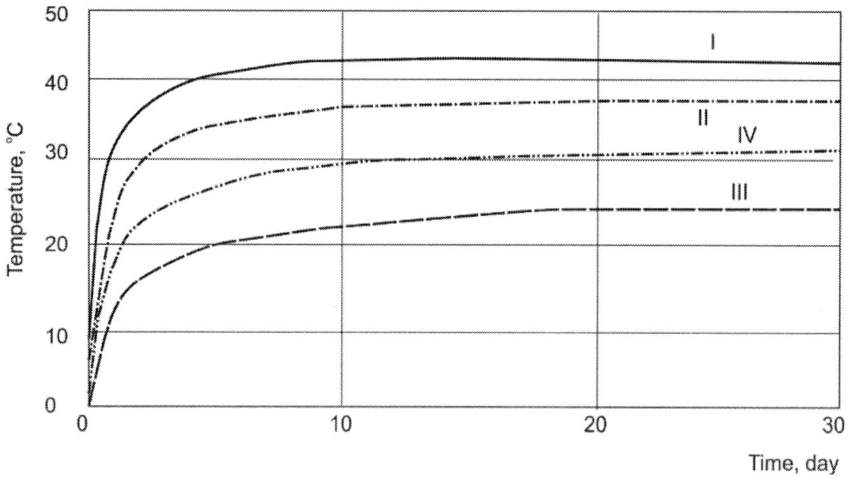

Fig. 2. Temperature value of hardening cement pastes: I, II, III, IV - types of cements according to Table 1.

The change in the temperature of hardening concrete based on the cement under study is shown in Fig. 3.

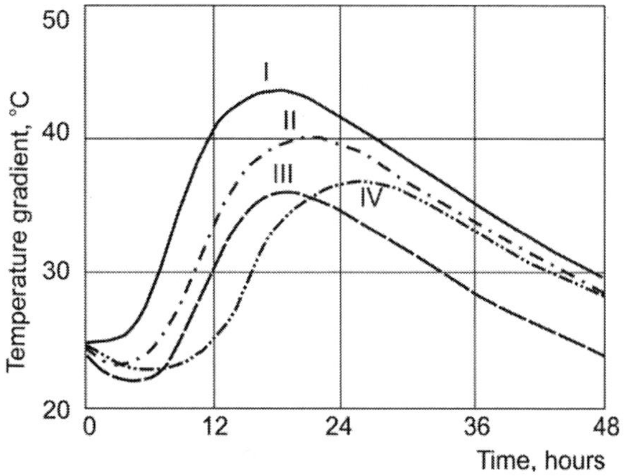

Fig. 3. Change in temperature gradient during hardening of concrete: I, II, III, IV - types of cements according to Table 1.

As can be seen from Fig. 3, the magnitude of the temperature gradient and the time of the onset of the maximum depend on the mineral composition of the cements, the type and amount of mineral additives in their composition. Temperature measurements were carried out in concrete samples measuring 1.0 × 1.0 × 1.0 m on the surface of the concrete block and in the center. An increase in the block size by a factor of 2 leads to an increase in the temperature gradient by a factor of 1.2.

The magnitude of the gradient is also affected by fluctuations in the ambient temperature, the presence and force of the wind, the form of the formwork and many other factors that can not be cited within the framework of this article.

Accounting for all factors is very difficult due to not always predictable weather changes. Therefore, to estimate the maturity of concrete M (τ, t), the averaged approach is used[15-16]. In this paper we used the following relation:

$$M(\tau, t) = \sum(t_\tau - t_0)\Delta\tau, \quad (1)$$

Wheret_τ is the temperature of the concrete at time τ; t_0 is the temperature of the concrete mixture at the time of laying; $\Delta\tau$ is the curing time.

The maturity value of concrete makes it possible to estimate the average strength of concrete at a given time R_τ:

$$R_\tau = R_{28} exp^{-\left[\frac{\tau}{M}\right]^a} \quad (2)$$

Where R_{28} - given strength of concrete;$\alpha = \frac{t_\tau - t_0}{t_n - t_0}$; t_n = 25°C.

Dependence of strength of concrete class B30 on maturity for the types of cement studied is shown in Fig. 4.

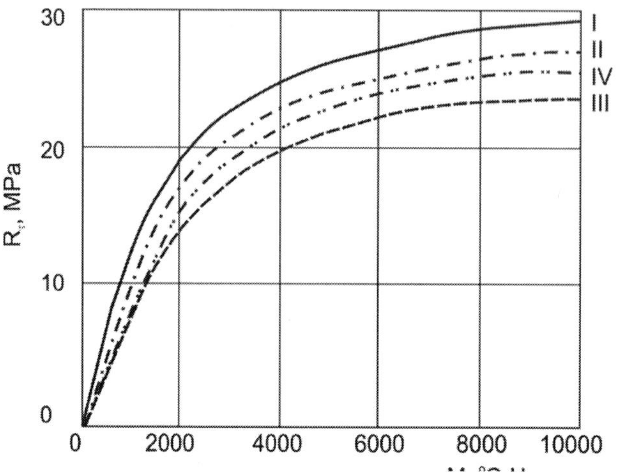

Fig. 4. Dependence of strength of concrete from maturity for the types of cement studied.

Thus, without rigidity tests, it is possible to obtain sufficiently accurate strength values of concrete at any time of hardening, based on temperature monitoring data.

4 Conclusion

The increase in temperature as a result of the release of heat of hydration, together with shrinkage deformations, is one of the possible load-independent causes of the appearance of cracks in young concrete.

The conducted studies enable establishinga correlation between the composition of cement and the nature of the strength set of concrete on its basis, based on the calorimetric determination of the temperature change during cement hydration in concrete.

The obtained dependences of strength of concrete from its maturity make it possible to determine the time of removal of the formwork from a concrete construction using different types of cements.

Observing the temperature changes of concrete in a certain area of the structure, you can:

- rather accurately determine the current strength of concrete in this zone;
- to know that the concrete is not frozen and not overheated;
- to control and observe the temperature parameters of the rates of heating-cooling of concrete normalized for monolithic structures;
- evaluate the air-concrete temperature gradient and correlate it with the allowed normalized values when the formwork is removed.

References

1. G. Wischers, Beton,**14,** H. 1 und H. 2 (1964)
2. R. Springenschmid, P. Nischer, Beton- und Stahlbetonbau, **68**, 9 (1973)
3. G. Thielen, H. Grube, Beton-und Stahlbetonbau, **85**, 6(1990)
4. *Prediction of creep, shrinkage, and temperature effects in concrete structures*, ACI 209R-92, Report of Committee 209, Manual of Concrete Practice, American Concrete Institute,Farmington Hills, MI (Reapproved, 1997)

5. *DSTU B.V 2.7-46: 2010.Cements for general construction purposes.* Technical specifications (State standard of Ukraine, Kyiv, 2010)
6. I.M. Belykh, V.P. Sopov, Building materials and products,**1-2** (**97**)(2018)
7. M. Azenha, R. Faria, D. Ferreira, Cement and Concrete Composites. Elsevier: Banthia N., **31** (2009)
8. I.M. Belykh, V.P. Sopov, L. Butskaya, P. Dolya, Motrol. Commission of motorization and energetics in agriculture, **18**, 6(2016)
9. *DSTU B.V.2.7-214: 2009. Concretes. Methods of determining the validity for control models* (State standard of Ukraine, Kyiv, 2010)
10. Y. Ballim, Cement & Concrete Composites, **26** (2004)
11. S.W. Phillips, W. Aquino, W.M. Chirdon, Journal of Engineering Mechanics, **133** (2007)
12. W. Hintzen, Schriftenreihe der Zementindustrie, **59** (1998)
13. M.L. Goral, J. Am. Conc. Inst., **53(2)**, 215 (1956)
14. F.K. Chin, J. Am. Conc. Inst., **68(3)**, 196 (1971)
15. J.M. Plowman, Mag. Concr. Res., **8(22)**, 13 (1956)
16. V.M. Malhotra, *Maturity Concepts and the Estimation of Concrete Strength - a Review* (Department of Energy, Mines and Resources, Mines Branch, Ottawa, IC277, November 1971)

Surface active properties of silicate and aluminosilicate surfaces

Dmitry Bondarenko, *Iryna* Plakhotnikova, *Medeia* Saliia*, *Olga* Demina, and *Alexander* Bondarenko

Kharkiv National University of Civil Engineering and Architecture, 61002 Sumska st, 40, Kharkiv, Ukraine

Abstract. Liquid heat-insulation based on hollow microspheres and polymer compounds become more and more popular in the construction industry. This material enjoys indisputable technological advantages – easiness of application, low mass, flexibility, good adhesion to different materials. Traditional non-organic heat insulating materials, widely presented on modern construction market, have some disadvantages. Foam concrete and gas silicate have relatively high average density, which enables obtaining items and materials with a fairly low thermal conductivity coefficient. These materials have high saturation coefficient and low values of cold endurance. Inside the mineral-wool plates, which are quite popular on the modern market, the hyperfine fibers transform into dust particles by time, and agglutinative compounds destruct, releasing toxic substances. The aluminosilicate microspheres have a high potential as a commercial product, but their market has just started forming. According to rough calculations, the cost of the product is next lower order to the hollow microspheres, obtained by industrial methods.

1 Introduction

The requirements for ecological friendliness and durability of the construction materials are increasing rapidly. As a result, effective and ecologically safe materials are getting increasing demand and actuality [1 - 3].

One of these materials can be thin-layered coating based on a mineral binder or a polymeric compound, filled with glass or aluminosilicate hollow microspheres. Functional properties of such materials spark off a lot of questions among the experts. Firstly, it is about the thermal conduction coefficient, which is varying among manufacturers between 0.022 – 0.057 V/mK [2].

The second issue is the principles of heat transfer. For traditional heat-insulating materials, the mechanism is connected with the photons. The function of the materials, filled with hollow microspheres made of glass or aluminosilicate up to 90% and more, lies in the fact of a threefold mechanism of heat transmission: by the photons inside the material itself; by air convection; and by reflection of the main types of radiation, including heat.

* Corresponding author: medeasalia@gmail.com

The reflection effect increases by the presence of inner partitions inside the microspheres, which slows down air convection. There is a great amount of light ash and slugger, occurring at the heat power plants, which is currently accumulating around the world. An upcoming trend of this kind of waste utilization is obtaining and application of the hollow microspheres. They are glass-like hollow spherical particles, occurring by combustion of coal from certain deposits in the caldrons of the heat power stations. The value of the microspheres is caused by their low bulk density, low heat conductivity and spherical form of the particles. In average, the concentration of microspheres in fly ash can fluctuate from 0.3 to 3-5%. Significantly, it is relatively easy to extract the hollow microspheres out of the ash. Because of low density, they surface in waterworks of a heat power plant and it is easy to collect them. The aluminosilicate microspheres (ASM) have a high potential as a commercial product, but their market has just started forming. According to rough calculations, the cost of the product is next lower order to the hollow microspheres, obtained by industrial methods. Application of ASM has a positive ecological effect because it enables the utilization of the wastes of production.

2 Materials and methods of research

It is known, that effective heat insulating materials are obtained from the expanded perlite (slabs, plasters), the grains of which have septate structure, and this structure allows decreasing heat transferring inside the material by means of convective heat. Microscopic researches have been carried out during the execution of the work; it has been shown that the structure of the glass and aluminosilicate microspheres also belong to a developed septate type.

This work investigates the possibilities of glass and aluminosilicate microspheres application in cement compounds. Exposition in alkaline condition by pH12 within 28 days has shown that the surfaces of microspheres remain intact, Fig.1-2.

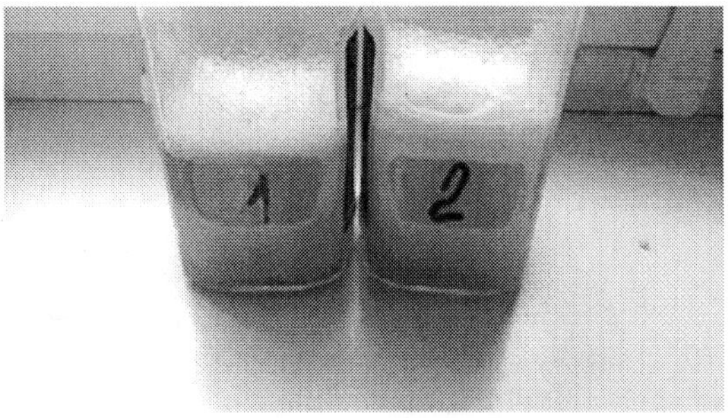

Fig. 1. Exposition of glass and aluminosilicate microspheres in water condition by pH=12.

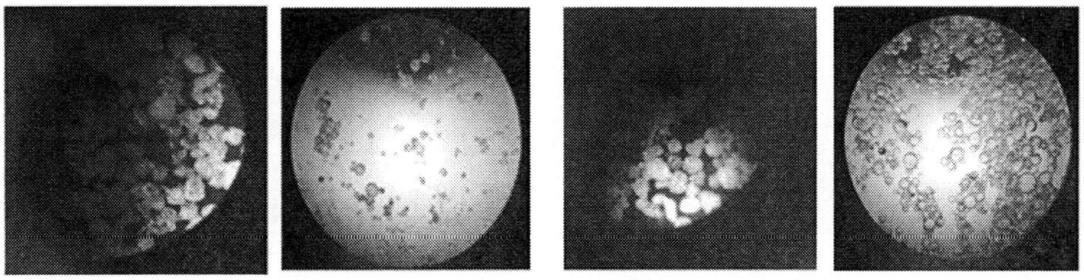

(a) (b) (c) (d)

Fig. 2. The research of surface condition in glass and aliminosilicate microspheres before exposition (a, b), and afterwards (c, d).

Surface properties of the microspheres have been detected quantitatively and qualitatively by means of the technique, described in works by A.P. Nechyporenko, V.A. Matviienko, V.I. Babushkin, A.A. Plugin and others [4-8]. Changes in the electronic system of the microspheres surface under the influence of various factors bring about a shift of the acid-alkali balance in the respective direction, which express in the change of the dissociation constant, Kd. The surface of a solid substance is a combination of established acids and alkali, which amount and strength can be fixed by means of molecular indicators adsorption and expressed by their Kd, i.e. in pH scale units.

The research, carried out in this work, namely detection of adsorption centers on the surface of the aluminosilicate microspheres, has shown the following. The concentration of the Bronsted base centers (of negative surface charge) comprises 23.1 × 105 mg eq/g. In cement stone, this value comprises 54 × 105 mg eq/g, that is 2.3 times more. The concentration of the Lewis base centers (of positive surface charge) comprises in aluminosilicate microspheres 4.5 × 105 mg eq/g, against 0.7 × 105 mg eq/g in cement stone, which is 6 times more. These balance of active adsorption centers in cement particles and aluminosilicate microspheres allowed strengthening of intergranular space in cement composite by forming silicate gel over their surfaces that has been proven by electronic microscopy researches, Fig. 3.

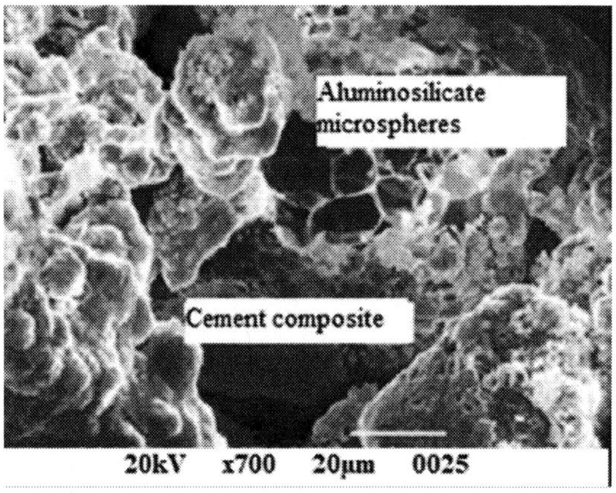

Fig. 3. Electronic microscopy pictures of cement composite with aluminosilicate microspheres on the seventh day.

In order to provide the highest structural strength and to narrow down the intergranular space, it is necessary to create the balance between the particles of the opposite surface charge. It was shown by previous studies that on the silicate base layers of glass microfibers in cement stone, the formation of the crystalline hydrates takes place, and the compounds of positive surface charge are the hydroaluminates and calcium carbonate, Fig. 4.

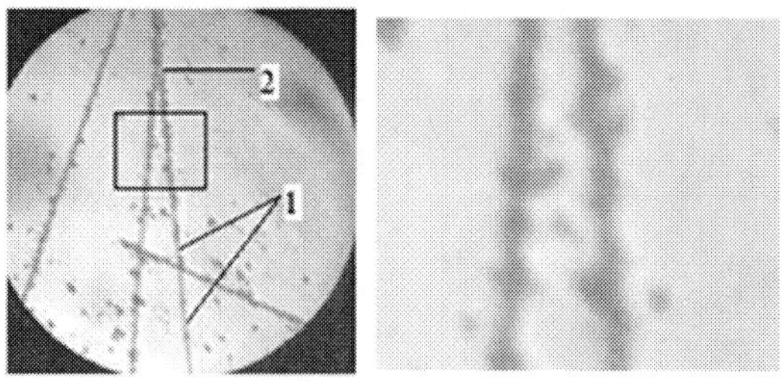

Fig. 4. Formation of the crystalline hydrates with a positive surface charge on silicate surface of a glass microfiber: 1 - a glass microfiber (enlargement 96x), 2 - the crystalline hydrates fouling of a glass microfiber on the 28th day in cement extract.

Equally spread inside the mass, the process of formation of crystals with positive and negative surface charge, allows obtaining more solid and hard structure, Fig.5.

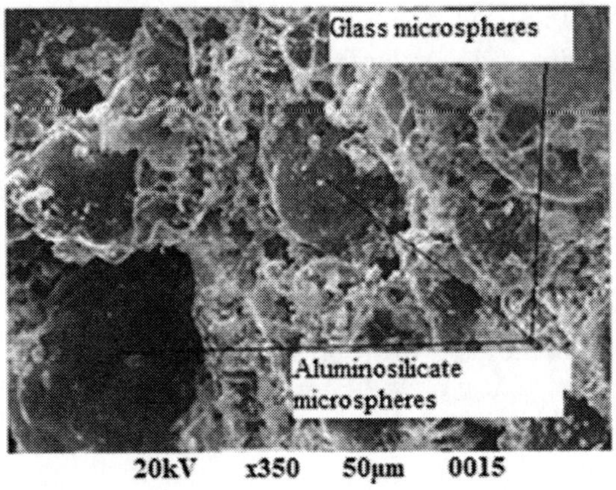

Fig. 5. Electronic microscopy pictures of cement composite, containing silicate and aluminosilicate hollow microspheres on the 28th day.

Electrical surface properties of the components, comprising the system of dry construction mixtures, which we research, have been checked under the impact of an electrostatic field of high voltage by the *Razryad* device following the technique [8, 10-12]. Fig. 5 shows a device for quantitative evaluation of surface polarity by separation of particles in high voltage field [13].

3 Conclusions

In order to provide the highest durability in cement composites, obtaining the balance between positively and negatively charged surfaces of all dispersive particles and crystalline hydrates is necessary. Adding 5-50 µm glass and aluminosilicate microspheres to cement structure shows the positive impact on the formation of the more dense structure as on early stages, as in later terms of solidification.

References

1. V. Demchenko, V. Sviders'kyi. *Continuous development, its condition, and perspectives*, International Scientific Symposium SDEV'2018 Materials. Heat insulating building mixtures with modified ash microspheres, 205-208 (2018)
2. *The myths of cosmic technologies or marketers fantasies* https://www.microspheres.com.ua
3. K.V. Plakhotnikov, O.I. Bondarenko, O.B. Dedeniova. Research journal of building industry, **3 (89)**, 226-229 (2017)
4. A.P. Netchyporenko, A.I. Kudriashova. Journal of Applied Chemistry, **9 (60)** (1957-1961)
5. V.I. Babushkin, A.A. Plugin, T.A. Kostyuk, V.A. Matviyenko, Research journal of building industry, **5**, 85-88 (1998)
6. V.M. Vorobiov, D.R. Kadyrova, G.Sh. Talipov. Kinetics, and catalysis, **1(15)**, 170-175 (1974)
7. V.A. Matviienko, *Electrical activation in the technology of concrete production.* PhD sci. diss. (Makiyivka, 1993)

8. A.A. Plugin, Research Journal of NTU KhPI, **15** (2003)
9. Saliia M.G. *Waterproofing coating on a cement base with increased crack resistance*. PhD sci. diss. (Kharkiv, 2012)
10. A.A. Plugin, *Longevity of concrete and reinforced concrete in drown constructions. Colloid chemical foundations*. Doctoral dissertation in technical sciences (Kharkiv, 2005)
11. A.N. Plugin, *Electrical heterogenous interaction when hardening of cement binders*. Doctoral dissertation in chemistry, **282** (Kyiv, 1989)
12. O.I. Demina, A.A. Plugin, E.B. Dedenyova, D.O. Bondarenko, T.A. Kostuk, Functional Materials, **3(24)**, 415-419 (2017)
13. A.A. Plugin, Journal of the NTU «KhPI», **33** (2004)

Use of a highly dispersed chalk additive for the production of concrete for transport structures

Olga Borziak[1], *Svitlana* Chepurna[2,*], *Tatyana* Zidkova[2], *Anna* Zhyhlo[2], and *Andrii* Ismagilov[1]

[1]Ukrainian State University of Railway Transport, Feuerbach sq. 7, 61050 Kharkiv, Ukraine
[2]O.M. Beketov National University of Urban Economy in Kharkiv, Kulikovska descent 12, 61002 Kharkiv, Ukraine

Abstract. The concretes used for transport infrastructure facilities are exposed to the active impact of corrosive environment; therefore, such concretes face raised demands as for water repellency, frost resistance and corrosion resistance. These properties can be improved due to the use of additives that have a different kind of effect. This research paper gives consideration to the influence of added highly-dispersed chalk on the physical and mechanical properties of concrete, in particular, water repellency and frost resistance. Experimental data show that the doping of highly-dispersed chalk results in a reduced water repellency and an increased frost-resistance of concrete. The basic physical and mechanical properties of concretes depend on the structure with a special particle packing of the matrix in the interlayers between the structure –forming elements of each level. Assumingly, the influence of added highly-dispersedchalk on the water-repellency and frost-resistance of pavement concrete is conditioned by the optimal arrangement of chalk particles (matrix particles) in the interlayers between the cement particles and the filler (structure-forming elements).

1 Introduction

Recently, much attention has been paid to the production of new composite binders, which have improved physical and mechanical characteristics and durability indicators, which is achieved through the use of mineral additives [1-3]. The use of local mineral resources and industrial wastes [4], in particular, sedimentary gypsum rocks and chalk [5], became especially urgent. This will solve environmental issues and get low-price local raw materials for obtaining new materials. It is known [6], that the use of various types of limestone and chalk powders as fillers improves the strength properties of concrete, including road concrete as well.

Concrete used at transport constructions is exposed to active influence of aggressive environment, therefore its durability depends on filtration properties [7]. The introduction of fine-dispersed mineral powders, which are used as active mineral additives or inert fillers for binders, leads to a change in the phase composition of Portland cement hydration

* Corresponding author: s.chepurna0274@gmail.com

products [8] and the structure of concrete, leading to the formation of a more dense structure of the composite and, consequently, will increase the durability of the structures as a whole [9-11]. The effectiveness of the mineral additive depends on the uniformity of its distribution over the volume of the composite, so it is advisable to add the admixture to the concrete mixture in the form of a suspension.

Proceeding from the foregoing, it was suggested that replacing part of the astringent additive with highly disperse chalk would improve the water resistance, frost resistance and corrosion resistance of concrete, without impairing its strength characteristics.

The aim of the paper is to study the effect of the addition of highly disperse chalk on the physical and mechanical characteristics of concrete, in particular water and frost resistance.

2 Research methods and materials

2.1 Materials

To evaluate the effect of the addition of highly disperse chalk on the water and frost resistance of concrete, a series of samples were manufactured, the consumption of components for the manufacture of which is given in the Table 1.

As a binder, Portland cement was used without additives (CEM I 42.5 N). The fine aggregate used was 0/4 fraction sand. The coarse aggregate used was 5/10 and 10/20 fraction crushed stone. The water-cement ratio was limited by the indices of workability of the concrete mixture.

Table 1. Consumption of components for making concrete samples.

Batch number	Consumption of components, kg per 1 m^3 concrete mixture					W/C
	Cement	Sand 0/4	Crushed stone 5/10	Crushed stone 10/20	Chalk	
1	377	580	420	780	0	0.466
2	340	580	420	780	37	0.460
3	302	580	420	780	75	0.462
4	275	580	420	780	105	0.468
5	277	580	420	780	150	0.472

The specific surface of particles of chalk is 10,000 cm^2/g, which excludes the influence of the strength of the rock itself. The chalk was pre-grounded in a dispersant. After dispersion, the size of the chalk particles in the suspension did not exceed 5 microns. For uniform distribution, the addition of highly dispersed chalk was introduced into a concrete slurry mixture. Formulations based on Portland cement were prepared with the replacement of 10, 20, 30 and 40% of binder to highly-dispersed chalk to determine water resistance and frost resistance.

2.2 Method for determining the water resistance of concrete

The waterproof test was carried out in accordance with [12] by the "wet spots" method. For the test, sample cylinders, with the height and diameter of 150 mm were made. Samples were placed in a metal cage and exposed to unilateral action of water under pressure. The pressure was increased by steps of 0.2 MPa for 2 minutes and retained for 16 hours. The method is based on measuring the maximum pressure at which water is not filtered through the sample. at the same time with the definition of water resistance, the compressive strength of concrete samples of each series was determined.

2.3 Method for determining the frost resistance of concrete.

The frost resistance was determined in accordance with [13]. To determine the frost-resistance of the concrete mixture, sample-cubes were made, in the size 100×100×100 mm. The samples were frozen at a temperature of -18 (± 2)°C for 4 hours, and then thawed at a temperature of +20 (± 5)°C for at least 2 hours. Frost resistance was determined in two ways: the first method - samples of concrete subjected to freezing and thawing in water, the second - freezing and thawing is carried out in a 5% solution of sodium chloride (NaCl). After a certain number of freezing and thawing cycles, the strength of the samples was determined. The frost resistance was estimated by the coefficient of frost resistance C_f according to the formula:

$$C_f = \frac{R^f}{R}, \quad (1)$$

where R^f is the strength of the concrete samples after alternating freezing and thawing; R - strength of samples of concrete that have not been frozen. The concrete grade is determined by the number of cycles of freezing and thawing, after which concrete samples lose strength by no more than 15%.

3 Results of experimental studies

The experiments showed that the water resistance index of a series of samples modified with 10% highly-dispersed chalk increased by 20%, compressive strength by 8.4% compared to control samples without an additive (Fig 1).

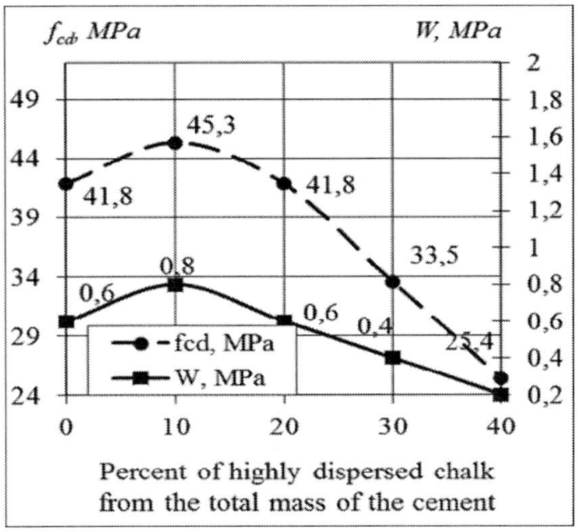

Fig. 1. Dependence of strength at contraction *(f_{cd})* and waterproofness *(W)* of concrete samples according to the chalk inclusion.

Waterproofness grade of concretes modified with 20% chalk corresponds to the brand of the control sample. It should be noted that concretes modified with 30% and 40% chalk have high water resistance grades (W4, W2, respectively), despite the low compressive strength.

According to the results of the study of frost resistance the maximum frost resistance grade is 400, the samples modified by 10 and 20% chalk correspond. The loss of strength in these samples after 400 cycles of freezing and thawing in the 1st method of determining frost resistance, and 150 cycles – in the second method, was no more than 5%. In this case, control samples and samples modified with 30 and 40% chalk after 400 (150) cycles were completely destroyed. The concrete grade for frost resistance of these samples was 300.

Table 2. The value of the coefficient of frost resistance.

Batch number	Number of freezing and thawing cycles							
	1st method				2nd method			
	100	200	300	400	45	75	110	150
1	1.16	1.2	1.18	-	1.21	1.18	1.0	-
2	1.18	1.25	1.2	0.98	1.28	1.26	1.15	0.95
3	1.14	1.18	1.12	0.95	1.25	1.21	1.1	0.8
4	1.09	1.12	1.05	–	1.15	1.11	0.89	–
5	1.02	0.95	0.75	–	1.11	1.04	0.82	–

4 Discussion

It is known [14] that the basic physical and mechanical properties of concrete depend on the structure, which is represented by the nature of the packing of the matrix particles in the inter-layers between the structure-forming elements of each level (Fig 2).

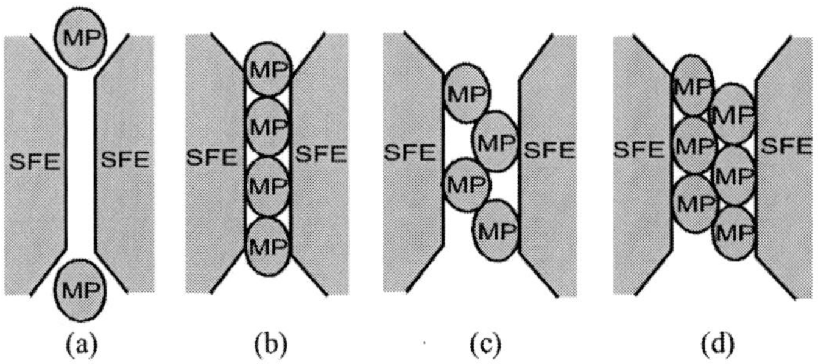

Fig. 2. Packing of particles of a matrix (MP) in a layer between the structure-forming elements (SFE): (a), (b) - the non-optimal structure of the interlayer; (c), (d) is the optimal structure of the interlayer [14].

The fine chalk used as an additive is a mixture of calcite grains (5 to 1 microns) has an angled or coiled form. The size of the particles of highly dispersed chalk can be compared with the size of the capillaries of the cement stone, so the replacement of chalk with cement parts contributes to the formation of a more dense structure. At the micro level, an optimal package of matrix particles is formed in the layers between the structure-forming elements: carbonate particles, which dissolve poorly in water, fill the layers between the particles of cement and fine aggregate, which increases the density and reduces the permeability of the concrete (Fig. 3). Since the volume of capillary pores does not usually exceed 40%, therefore, the total volume of the binding agent can be increased by 40% due to the highly dispersed chalk to fill them.

It is known that carbonate additives affect the hydration process of Portland cement [8, 15], in particular, such stable compounds are formed as calcium hydrobromoaluminates $CaO \times Al_2O_3 \times CaCO_3 \times nH_2O$.

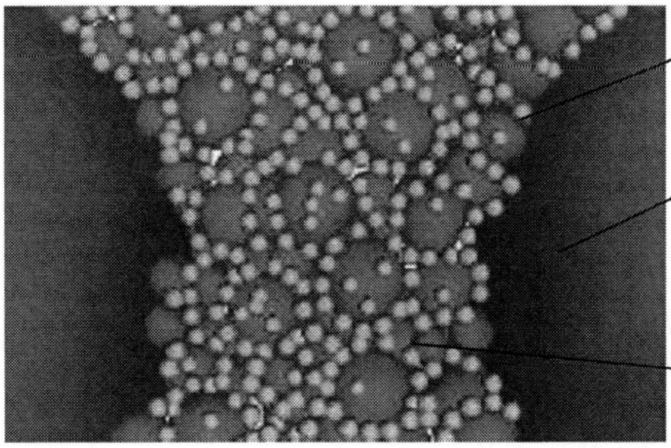

Fig. 3. Spatial-structural scheme of the placement of particles of cement and highly dispersed chalk.

According to [16], the surface of particles of organogenic carbonate rocks, in particular chalk, is covered with amorphous silica. When calcium hydroxide reacts with silica, low-basic calcium hydrosilicates are formed. Thus, the use of highly disperse chalk leads to a decrease in the content and binding of free calcium hydroxide, as well as the aluminate phase, to stable hydration products. This contributes to the increase of the corrosion resistance of cement stone under the influence of aggressive media containing sulfate and chloride ions, as well as magnesium ions [17].

5 Conclusions

According to the studies, the use of a highly dispersed chalk additive reduces the capillary porosity and compacts the structure of the cement stone, which helps reduce the effective diffusion coefficient of aggressive ions, increase the water resistance and frost resistance of concrete. The change in the phase composition of the products of Portland cement hydration in the presence of a fine chalk additive and the formation of such hardy compounds as calcium hydrogencarbonate and low-basic calcium hydrosilicates also increases the corrosion resistance of cement stone and concrete. This will increase the stability of structures to the influence of aggressive media and, therefore, increase their durability. The developed compositions can be used in the production of road concrete.

References

1. U. Marushchak, M. Sanytsky, T. Mazurak, Y. Olevych, Eastern-European Journal of Enterprise Technologies, **6/6 (84)**, 50-57 (2016)
2. M. Sanytsky, T. Kropyvnytska, R. Kotiv, Advanced Materials Research, **923**, 42-47 (2014)
3. U. Maruchchak, M. Sanytsky, S. Korolko, Y. Shabatura, N. Sydor, Eastern-European Journal of Enterprise Technologies, **2/6 (92)**, 34-41 (2018)
4. O. Shynkevych, Y. Lutskin, A. Aniskin, Technical Journal, **11 (4-2017)**, 160-165 (2017)
5. Mridul Garg, A.K. Minocha, Neeraj Jain, Construction and Building Materials, **25 (2)**, 944-949 (2011)

6. Wenzhong Zhu, John C. Gibbs, Cement and Concrete Research, **35 (8)**, 1457 (2005)
7. Peter A. Claisse, *Transport Properties of Concrete* (Woodhead Publishing, 2014)
8. A.A. Plugin, R.F. Runova, International Journal of Engineering Research in Africa, **36**, 69-73 (2018)
9. O.Yu. Bazhenova, ISAEECE, AER, **124** (2017)
10. E. Lutskin, E. Shinkevich, Technical Journal, **9 (1-2015)**, 27-33 (2015)
11. T. Kropyvnytska, R. Semeniv, H. Ivashchyshyn, MATEC Web of Conferences, **116**, 01007 (2017)
12. *DSTU B V.2.7-170:2008. Construction Materials. Concretes. Methods for determining average density, humidity, water absorption, porosity and water repellency* (State standard of Ukraine, Kyiv, 2008)
13. *DSTU B V.2.7-42-97. Building materials. Methods of determination of water absorption, density and frost resistance of building materials and products* (State standard of Ukraine, Kyiv, 1997)
14. A.N. Plugin, A.A. Plugin et al. *Osnovy teorii tverdeniya, prochnosti, razrusheniya i dolgovechnosti portlandtsementa, betona i konstruktsiy iz nikh* (Naukova dumka, Kyiv, 2012)
15. O.S. Borzyak, S.M. Chepurna, Collected scientific works of Ukrainian State University of Railway Transport, **175**, 110-117 (2018)
16. I.M. Gorkova et al., *Priroda prochnosti i deformatsionnyie osobennosti mela i nekotoryih melopodobnyih porod* (Akademiya nauk, 1962)
17. S.N. Chepurna, A.A. Plugin, O.S. Borzyak, Naukovii visnyk budivnitstva, **1 (91)**, 292-298 (2018)

Acid-basic surface properties of clay disperse fillers

Yuliya Danchenko[1,*], *Vladimir* Andronov[2] , *Viktor* Sopov[1], *Igor* Khmyrov[2], and *Anton* Khryapynskyy[2]

[1]Kharkiv National University of Civil Engineering and Architecture, Sumska str., 40, 61002, Kharkiv, Ukraine
[2]National University of Civil Protection of Ukraine, Chernyshevska str., 94, 61023, Kharkiv, Ukraine

Abstract. The mineral composition, surface morphology, dispersion and specific surface of dispersed clay fillers of various chemical nature have been experimentally determined. As materials for research used clay mineral disperse fillers of various chemical and mineral composition: bentonite clay, kaolinite clay and diabase flour. Using the thermogravimetric (TG) analysis and the differential thermal analysis (DTA), the structure of the hydroxyl-hydrated surface layer of the filler surface has been studied. The amount and acidity of the surface active centers of clay fillers have been investigated using pK-metric and pH-metric methods. It was established that the mineral composition is the determining factor in the acid-base surface properties of clay fillers. The thickness of the hydroxyl-hydrated surface layer affects the acidity of the surface, at certain large sizes. It has been shown that clay fillers, which include montmorillonite as the dominant minerals, are characterized by higher acidity than the fillers that consist of kaolinite. In the mixed clay fillers, the acid-base properties of the surface depend on the chemical nature of the impurities and related minerals.

1 Introduction

Natural clay disperse fillers are widely used in the development of new polymer composites and other composite materials of various applications. Cheap clay fillers are added to composites in order to increase the economic and environmental attractiveness of building and restoration polymer materials. Addition of clay fillers allows adjusting the technological properties of compositions and improving the physical, mechanical and performance characteristics of materials [1-7]. Disperse clay fillers of various chemical and mineral nature can provide special properties to materials (for example, resistance to biologically aggressive media, fire resistance, vibration absorbing properties, etc.) [2-4]. In this case, the materials become multifunctional, that is, they can simultaneously provide several necessary operational or technological requirements. Moreover, adding clay fillers ensures the production of polymer materials with certain decorative properties, which is important when used in architecture, restoration and design [8, 9]. Materials of different

* Corresponding author: u_danchenko@ukr.net

colours, either matte or glossy, may be produced, with imitation of the appearance of natural materials, etc.

2 Literature review

Investigation of the patterns of the influence of disperse clay fillers on the properties of dispersed filled materials solves the problem of obtaining new multifunctional composites with adjustable characteristics for construction, architecture, restoration, transport and other industries. Interest in this area is primarily due to the study of the physic-chemistry of surface phenomena in composite materials. The urgency of this direction arises in the process of creating new polymer composites, which are heterogeneous systems with highly developed phase separation surfaces. Properties of the surface of the dispersed phase determine the initial stage of adhesion interaction between the components - adsorption, wetting, spreading of the polymer on the surface of the filler, forming the boundary of phase separation, impregnation of filled and reinforced systems. Of all the existing theories of phase interaction in polymer composite materials, the least studied acid-base, the existence of which is confirmed by studies of recent years [10, 11]. When choosing components of composite material, it is important to predict interactions on the interphase surface, which are simultaneously determined by the acid-base properties of the polymer and the acid strength of adsorption centers on the surface of the fillers.

Due to the natural origin of clay fillers, the chemical and mineralogical composition, and, consequently, the surface properties, are extremely unstable. In this case, the particles have superfluous surface energy and easily adsorb molecules of water from the air [12, 13]. The effect of chemically and physically adsorbed water on the acid-base properties of the surface of inorganic materials is confirmed experimentally. It has been proved that the number and acid-base parameters of surface active centers change during burning or machining (milling) [6]. At adsorption of water on the surface formed hydroxyl-hydrated layer, which is characterized by a wide spectrum of active adsorption centers, the acid-basic properties of which are of great scientific interest [14], but not sufficiently studied.

The structure of the adsorption surface layer consists of chemically bound water molecules that form the first hydroxyl layer of OH-groups (Brensted's centers) and the following hydrated layers from water molecules bound to a hydroxyl layer due to hydrogen bonds. On the surface, the aprotonic acid and the base centers of Lewis, which are apparently small on air-dry surfaces, may be present. The determination of the acid-base characteristics of the active centers of the surface layer of inorganic materials is a complex but very important task. For this purpose, methods are used to determine the quantity and acid-base parameters of surface active centers, mainly experimentally [15, 16]. For example, the methods of IR spectroscopy, adsorption of gases in a gaseous medium or color indicators in liquids, by the method of potentiometric titration of suspensions. Experimental methods are informative but time-consuming and do not allow to determine the chemical nature of active centers. For the determination of the chemical nature of the active centers and the investigation of the protolytic equilibrium on the surface of inorganic materials, a method of quantum-chemical modeling is proposed [14].

Taking into account the fact that the surface properties of clay materials are directly related to the chemical and mineral composition, it is necessary to study the correlation between them. Thus, the task of this work involves installing the above-mentioned relationships. This will create opportunities for predicting the surface acid-base properties of clay disperse fillers, and also help to optimize the processes of developing new dispersed polymer composite materials with certain technological and operational characteristics.

3 Materials and methods of the study

As materials for research, clay disperses fillers of various chemical and mineral compositions: bentonite clay (BC), kaolinite clay (KC) and diabase flour (DF) have been selected. For selected materials, experimental studies were carried out on the establishment of mineral composition, dispersion, surface morphology and surface area of particles. The surface area of the fillers was determined by the method of BET (Brunauer, Emmett, Teller). X-ray diffraction analysis was carried out on a X-ray diffractometer DRON-2 (Russia). Surface morphology and dispersion studies were performed using scanning electron microscope of the JSM-6390LV mark (Japan). Thermogravimetric (TG) analysis and Differential Scanning Calorimetry / Differential Thermal Analysis (DTA) with the use of the thermal analyzer of the brand SDT Q600 manufactured by TA Indstruments (USA) were performed for the investigation of the hydroxyl-hydrated surface layer of fillers.

Two methods were used to characterize the surface acid-base properties of clay fillers.

1) pKa-metric experimental method for determining the number of surface active centers by adsorption of Gamet color indicators from aqueous solutions and photocolorimetric measurements of optical density. As a result, the calculated total acid centers Σq_A with $pK_a <$ 7, the base centers Σq_B with $pK_B > 7$, the total number of surface centers $\Sigma q = \Sigma q_A + \Sigma q_B$ and the acid-basic index $Q = \Sigma q_A / \Sigma q_B$. The obtained index makes it possible to estimate the surface as a whole as a neutral ($Q \approx 1$), acidic ($Q > 1$) or basic ($Q < 1$). It is assumed that the higher the value of Q, the more acidic properties are characterized by the surface. The integral surface acidity index Q has some advantages and disadvantages. The main disadvantage is that the value of Q does not allow estimating the acid-base properties of the surface by Lewis, and this requires additional experimental studies (for example, the method of adsorption of gas molecules). Advantages are the possibility of a more complete understanding of the acid-base nature of the surface of any disperse materials or fillers within a small number of experimental and settlement operations [5, 7].

2) pH-metric experimental method with potentiometric measurements of pH^{eq}_{susp} of aqueous suspensions of fillers in the process of establishing equilibrium state. Taking into account the pH_0 of pure water, the values pH_{IIP} (isoadsorption point) of aqueous suspensions in the equilibrium state were calculated, which correspond to the arithmetic mean of pK_a of the surface active centers of fillers:

$$pH_{IIP} = 1/n \cdot \sum^{n}_{i=1}(pK_{a1} + pK_{a2} + \ldots + pK_{ai}), \quad (1)$$

$$pH_{IIP} = pH_0 + (pH^{eq}_{susp} - pH_0) \cdot (pH_0 - 6). \quad (2)$$

For the approximate estimation of the spectrum of active surface centers of fillers, the value of $\Delta pH = pH_{IIP} - pH^{eq}_{susp}$, which indicates the width of the range of pK_a centers on the surface, was calculated. An increase in the value of ΔpH shows the expansion of the spectrum of active centers on the surface of the fillers. According to the potentiometric studies, the duration of the establishment of the equilibrium state in the suspension was established for an approximate quantitative assessment of the active surface centers. The duration of the establishment of the equilibrium state in the suspension can characterize the total number of acid-base surface active centers that take part in interactions with water molecules.

4 Discussion of the results of the study

The results of the study of mineral composition, dispersion and surface area of clay fillers are presented in Table 1 and in Fig. 1-3.

Table 1. Properties of clay disperse fillers.

Filler	Mineralogical composition	Surface area, m^2/g	Average particle size *d*, μm
Bentonite clay (BC)	Montmorillonite $(Al,Mg)_2(OH)_2[Si_4O_{10}]\cdot nH_2O^{\bullet}$ β-quartz SiO_2^{+} Kaolinite $Al_2Si_2O_5(OH)_4^{+}$	15.62	10-20
Diabase flour (DF)	β-quartz $SiO_2^{\bullet}$ Montmorillonite $(Al,Mg)_2(OH)_2[Si_4O_{10}]\cdot nH_2O^{\bullet}$ Orthoclase $K_2O\cdot Al_2O_3\cdot 6SiO_2^{\bullet}$ Leucite $K_2O\cdot Al_2O_3\cdot 4SiO_2^{\bullet}$ Lymonit $Fe_2O_3\cdot H_2O^{+}$	1.90	2-4
Kaoline clay (KC)	Kaolinite $Al_2Si_2O_5(OH)_4^{\bullet}$ β-quartz SiO_2^{+} Biotite $(Fe,Mg)_3[OH]_2(Al,Fe)Si_3O_{10}^{+}$ Hydrous micas $(K,Na)Al_2(Si,Al)_4O_{10}(OH)_2\cdot nH_2O^{+}$	8.78	4-5

Remarks: • – basic mineral phase; + - side mineral phase; * - traces.

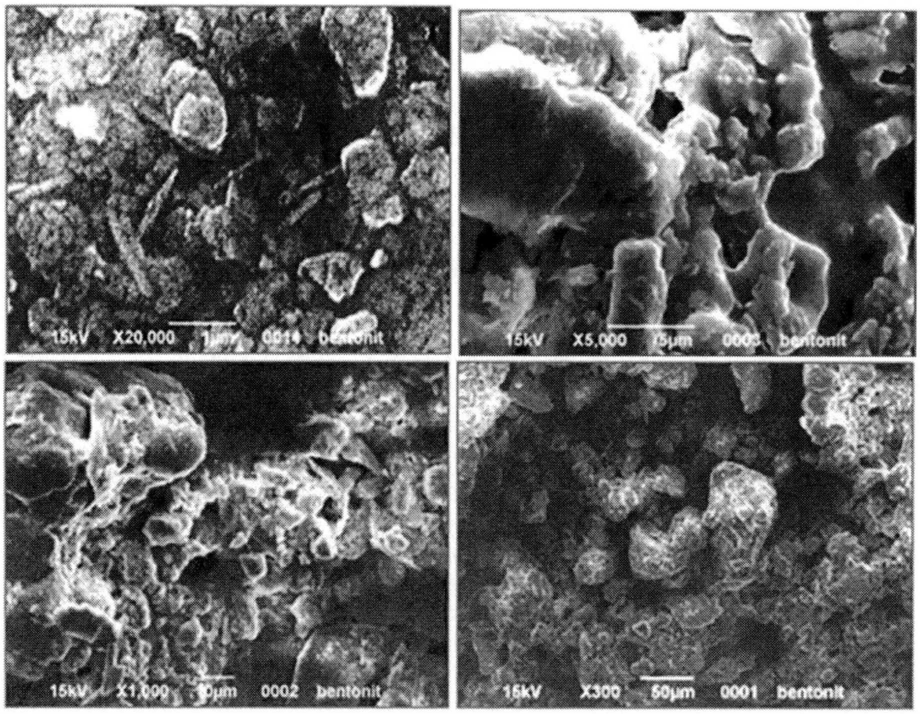

Fig. 1. Microphotographs for bentonite clay.

Fig. 2. Microphotographs for diabase flour.

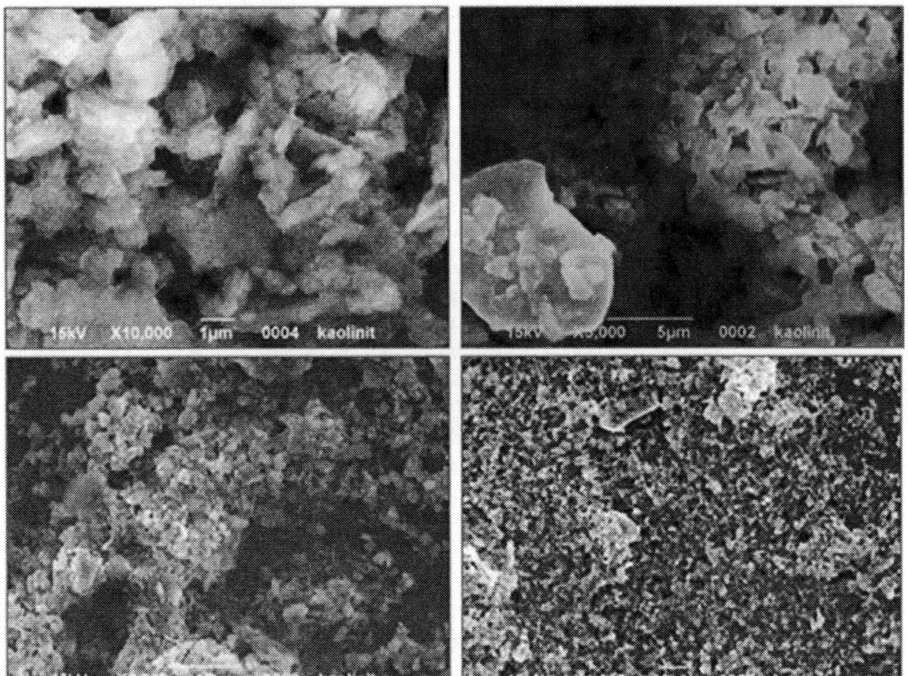

Fig. 3. Microphotographs for kaolinite clay.

Microphotographs of bentonite clay are presented in Fig. 1. The roentgenogram of bentonite clay is represented by a set of diffuse diffraction maxima of weak intensity corresponding to minerals of the montmorillonite group (montmorillonite and beidelite).

Accompanying are kaolinite and β-quartz. Microphotographs show scaly grains of clay minerals and individual rhomboidric grains of quartz with a size of 1-1.5 μm. Fine grains of clay minerals smaller than 0.5 μm are assembled in aggregates of 10-20 μm, and some large (50-60 μm) particles occur. Microphotographs of diabase flour are presented in Fig. 2. Diabase flour is a product of fine grinding in ball mills of acid-resistant magmatic rocks (diabase). Diabase flour is an agglomerated powder containing the leucite $K_2O{\cdot}Al_2O_3{\cdot}4SiO_2$, orthoclase $K_2O{\cdot}Al_2O_3{\cdot}6SiO_2$, SiO_2 quartz, montmorillonite $(Al,Mg)_2(OH)_2[Si_4O_{10}]{\cdot}nH_2O$ and lemonite $Fe_2O_3{\cdot}H_2O$ as the main phases) (Table 1). Isometric particles, collected in agglomerates with a size of 2-4 μm. Large particles with a size of 40-50 μm have a shell fragmentation, covered with scaly formations, apparently montmorillonite and limonite. Microphotographs of kaolinite clay are presented in Fig. 3. Kaolinite clay is a dispersed mineral material of yellowish color. Hydrous micas, biotite $(Fe,Mg)_3[OH]_2(Al,Fe)Si_3O_{10}$ and β-quartz are present as accompanying minerals in powder composition (Table 1). The powder has a homogeneous disperse composition. Microphotographs show agglomerates of 4-5 μm in size, which are formed from scalable kaolinite crystals of hydrous micas crystals with a size of 0.5-1 μm. Less common are large dense isometric particles of quartz size 10-12 μm.

In Table 2 show the dynamics of mass loss of clay fillers in the process of increasing temperature.

Table 2. Loss of mass of clay disperse fillers in the process of temperature increase.

Filler	Weight loss of the filler in the temperature range, %			
	0-150°C	150-300°C	300-600°C	600-900°C
Bentonite clay (BC)	0.7	0	8.8	0.5
Diabase flour (DF)	1.0	0.8	1.1	0.4
Kaoline clay (KC)	0.6	0.3	12.2	1.3

From the results presented in Table 2 it is possible to draw the following conclusions. At temperatures of 0-300°C, processes of dehydration, that is, the loss of physically adsorbed on the surface of water molecules (hydrated layer), take place. Compared to BC and KC, DF is characterized by a greater 2-2.5 times the amount of physically adsorbed water. Thus, the thickness of the hydrated surface layer of fillers increases in a row:

$$\text{BC } (0.7\%) < \text{KC } (0.9\%) < \text{DF } (1.8\%). \quad (3)$$

At temperatures of 300-900°C, dehydroxylation processes occur, ie, the loss of chemosorbed water molecules (hydroxyl layer). The surface of DF is characterized by the smallest hydroxyl layer than the surface of the BC and KC. The thickness of the hydroxyl layer of fillers increases in a row:

$$\text{DF } (1.5\%) < \text{BC } (9.3\%) < \text{KC } (13.5\%). \quad (4)$$

Obviously, the established facts substantially affect the acidity of the surface of the fillers. It has previously been established that an increase in the number of physically adsorbed water and the number of hydroxyl groups in the active centers increases the acidity of the surface of inorganic fillers [14]. The results of pK-metric and pH-metric experimental studies of clay fillers are presented in Table 3 and Table 4.

Table 3. Results of study of clay fillers pK-metric method.

Filler	$\sum q_A \cdot 10^{-12}$, 1/sm^2	$\sum q_B \cdot 10^{-12}$, 1/sm^2	$\sum q \cdot 10^{-12}$, 1/sm^2	Q
Bentonite clay (BC)	108.26	119.03	227.29	0.91
Diabase flour (DF)	148.55	185.23	333.78	0.80
Kaoline clay (KC)	78.18	120.93	199.11	0.65

Table 4. Results of study of clay fillers pH-metric method.

Filler	pH_0	pH^{eq}_{susp}	pH_{IIP}	ΔpH	Time to set the equilibrium state, s
Bentonite clay (BC)	7.05	7.44	7.46	0.02	300
Diabase flour (DF)	7.00	8.50	8.50	0	360
Kaoline clay (KC)	7.00	9.11	9.11	0	240

From data Table 3 it follows that the acidity of the surface of clay fillers is significantly dependent on the mineral composition. The mineral composition of the BC and the KC is almost homogeneous and is represented by the dominant mints montmorillonite and kaolinite, respectively. It is obvious that the presence of these minerals has an almost neutral surface of the BC (Q≈0,91) and a weakly basic KC (Q≈0,65). The composition of DF includes several minerals: leucite, orthoclase, montmorillonite and quartz with impurities of limonite. The nature of the surface DF is intermediate and close to the neutral (Q≈0.80). The increase in the alkalinity of the surface DF compared with the BC is due to the presence of minerals leucite and orthoclase, which includes potassium oxide K_2O. In addition, alkalinity may provide impurities of limonite, which includes Fe_2O_3 ferrous oxide. Thus, according to the results of the pK-metric study, the acidity of the surface of clay fillers increases in a row:

$$\text{KC } (Q\approx0{,}65) < \text{DF } (Q\approx0{,}80) < \text{BC } (Q\approx0{,}91). \quad (5)$$

Concerning the acid-base nature of the surface, the results of the pK-metric study are fully confirmed by the data of the pH-metric analysis. So from the data presented in the Table 4 it follows that the acidity of the surface of clay fillers, in accordance with the value of pH_{IIP}, increases in a row:

$$\text{KC } (9.11) < \text{DF } (8.50) < \text{BC } (7.46). \quad (6)$$

Comparing the obtained series 5 and 6 of 4 and 3 it can be concluded that the mineral composition is the determining factor in the acid-base surface properties of the clay fillers. In this case, the thickness of the hydrated and hydroxyl surface layer to a small extent affects the acidity of the surface. This effect, obviously, becomes significant at certain large sizes of the hydroxyl-hydrated layer. On the surface of the clay fillers studied, the thickness of the surface hydroxyl-hydrated layer is not sufficient to significantly affect the acid-base nature of the surface.

From data Table 3 shows that the largest number of active adsorption centers is characterized by the surface DF ($\sum q$ = 333.78 · 10^{12}1/sm^2). This is consistent with the maximum duration of the establishment of the equilibrium state in aqueous suspensions DF (360 s). It should be noted that these indicators are also consistent with the BC and KC: the larger the total number of surface active centers, the greater the duration of the time to establish equilibrium state in aqueous suspensions. According to the results of Table 4 the spectrum of active surface centers of clay fillers is not wide. This conclusion follows from the fact that the values of ΔpH are zero or close to zero. Therefore, most surface active centers of clay fillers have similar values of pK_a.

4 Conclusions

As a result of the research, the following conclusions can be drawn. The mineral composition, surface morphology, dispersion and specific surface of dispersed clay fillers of various chemical natures have been experimentally determined. Using the thermogravimetric (TG) analysis and the differential thermal analysis (DTA), the structure of the hydroxyl-hydrated surface layer of the fillers surface has been studied. The amount and acidic strength of the surface active centers of clay fillers were investigated using pK-metric and pH-metric methods. It was established that the mineral composition is the determining factor in the acid-base surface properties of clay fillers. In this case, the thickness of the hydroxyl-hydrated surface layer affects the surface acidity to a small extent. It has been shown that clay fillers, which include montmorillonite as the dominant minerals, are characterized by higher acidity than the fillers that consist of kaolinite. In the mixed clay fillers, the acid-base properties of the surface depend on the chemical nature of the impurities and related minerals.

References

1. E.F. Kudina, Polym. Mater. and Tech., **3**, **2**, 49-55 (2017)
2. V.A. Andronov, Yu.M. Danchenko, A.V. Skripinets, O.M. Bukhman, Nauk. Visn. Natsion. Hirn. Univ., **6**, 85-91 (2014)
3. A.D. Eselev, Polym. Sci. Ser. D, **2**, **2**, 71-74 (2009)
4. Y-X. Fui, Z-X. He, D-C. Mo, Adhesives Appl. Therm., **66**, 493-498 (2014)
5. Yu. Danchenko, V. Andronov, M. Teslenko, V. Permiakov, E. Rybka, R. Meleshchenko, A. Kosse, Eastern-Europ. Journ. of Enterprise Tech., **1**, **12 (91)**, 9-17 (2018)
6. C. Zapata-Massot, N. Le Boley, Part. Part. Charact., **24**, 339-344 (2007)
7. Yu. Danchenko, V. Andronov, E. Barabash, T. Obigenko, E. Rybka, R. Meleshchenko, A. Romin, Eastern-Europ. Journ. of Enterprise Tech., **6**, **12 (90)**, 4-12 (2017)
8. P. Cardiano, R.C. Ponterio, S. Sergi, S. Lo Schiavo, P. Piraino, Polymer, **46**, **6**, 1857-1864 (2005)
9. E. G. Karayannidou, D.S. Achilias, I.D. Sideridou, Europ. Polym. J., **42**, **12**, 3311-3323 (2006)
10. Yu. Danchenko, M. Kachomanova, Ye. Barabash, Chem. and Chem. Tech., **12**, **2**, 188-195 (2018)
11. Yu. Danchenko, V. Andronov, A. Kariev, V. Lebedev, E. Rybka, R. Meleshchenko, D. Yavorska, Eastern-Europ. Journ. of Enterprise Tech., **5**, **12 (89)**, 20-26 (2017)
12. C. M. Miller, Submitted to the Office of Graduate Studies of Texas A&M University in partial fulfillment of the requirement for the degree of Master of Science, 218 (2010)
13. Yu. M. Danchenko, Polym. Mater. and Tech., **3**, **2**, 56-63 (2017)
14. Yu. Danchenko, V. Andronov, E. Rybka, S. Skliarov, Eastern-Europ. Journ. of Enterprise Tech. **4/12** **(88)**, 17-25 (2017)
15. A. Baldan A., J. of Mater. Sc., **39**, 1-49 (2004)
16. K. Weissenbach, H. Mack, Functional fillers for Plastics, 59-83 (2005)

Evaluating the efficiency of the fluidity retention of ready–mix concrete during the transportation of it to the construction site

Oleksii Kabus*, *Larisa* Butska, *Olga* Makarenko, *Lidiya* Pershina, and *Andriy* Tymoshchenko

Kharkiv National University of Civil Engineering and Architecture, Sumska st. 40, 61002 Kharkiv, Ukraine

Abstract. The work is devoted to the study of changes in the time of mobility of concrete mixtures, which are used for commercial concrete with monolithic construction. The in-situ concrete quality depends on the uniformity, viability and workability of the concrete mix intended for the transportation, unloading and workability at the construction site. One of the techniques used for the solution of this problem can be step-by-step addition of chemical admixtures or introduction of setting retarders whose efficiency is expressed in an increased resiliency of the concrete mix that is extended by two and more hours. Each solution has its advantages and drawbacks; therefore the efficiency can only be estimated under specific conditions taking into account the cost of concrete mix and opportunities for the realization of each individual solution. Were received results showing the presence of a problem of mobility loss of concrete mixtures with chemical additives of high water-reducing action. The use of such technological techniques as increasing the initial mobility and gradual introduction of supplements to superplasticizer proved to be effective technological solutions.

1 Introduction

The in-situ reinforced concrete finds wide application. It is produced using ready-mix concrete that is prepared at special-purpose works and transported to the construction site using truck mixers. State-of-art constructions of any size, shape and purpose are erected using this concrete.

The concrete placing process used for cast-in-place reinforced concrete constructions at the construction site requires the delivery of the mix from the works using the concrete haulers and the structure quality depends on the fluidity retention, uniformity and workability of the mixture intended for the transportation, unloading and laying. However, the hands-on experience shows that there is a danger of the loss of fluidity of concrete mixes especially in the conditions of long-distance transportation and at high ambient temperatures. In large cities with the distance between the concrete plants of 10 to 15 km it

* Corresponding author: calorimetry_centr@ukr.net

usually takes 30 to 40 minutes to transport and place the concrete mix. However, even under such conditions the Customer obtains very often the concrete mix of a low fluidity that requires the water addition.

Many research papers state the availability of the problem related to the loss of fluidity by concrete mixes containing concrete superplasticizer admixtures that happens with the elapse of time [1-4]. This phenomenon can be related to many factors: incompatibility of the admixtures with the cements and with each other, a low concentration of superplasticizers in the concrete mix, high ambient temperature, use of fine quick-hardening cements, availability of flour and clay particles in the aggregates, etc. [1, 4].

If we approach the above problems from the standpoint of directional modifying of the properties of concrete mixes, we will see that different methods are available for their solutions. Cement setting retarders whose efficiency provides a two-hour increase and even more in the resilience retention of concrete mix have been known for ages. [3].

Providing specified parameters of process mixes in the conditions of long-distance transportation without the retardation of hardening processes is a very complicated engineering task indeed for the cast-in-situ construction. One of the technological methods of the solution of such a problem can be a stepwise addition of the admixtures [3].

In the event of stepwise addition one portion of the admixture (40 to 60%) is injected when the concrete mixes are prepared at concrete mortar plant and the other portion is added at the concrete mix placement site right in the structure (usually the admixture solution is poured into the rotary drum of concrete mixer truck). This method enables the renewal of the fluidity of concrete mix adding no water.

The admixture of a different type (plasticizers, superplasticizers, retarders, and accelerators) are combined together and can be introduced individually at different stages depending on the season of the year, arrangement scheme of concrete jobs, the distance to the construction site, taking into consideration the transportation cost and other factors [4].

The analysis done shows that the range of available technological methods of improvement of the properties of concrete mix can be extended from the "maintenance of the fluidity retention of concrete mixes" to "providing the required fluidity at the construction site". However, the most efficient and economic method should be selected in each specific case by way of experiment.

The objective of this research was to study and evaluate the efficiency of different formulation technology-related decisions that the producer of the ready-mix concrete can use to provide specified indices for the concrete mix to insure easy placing at the construction site.

2 Materials and methods of research

The following materials were used to do research:

- cement: Portland cement with the CEM II/ A-S-32.5 (PJSC "Podilsky Cement" CRH Ukraine);
- coarse aggregates: crushed granite Fr.5-10 (OJSC Malokokhnivskyi Granite Pit", Poltava region, city of Kremenchuk);
- fine aggregates: quartz sand with Mk=1.1.(the Bezliudivskyi pit, Kharkiv region)
- chemical admixtures: superplasticizer (SP)-Sika Visco Greate 1020 UA, hardening retarder (C)-Sika Retarder and hydrophobization agent (G) – GKJ -11K (CPC Kremniypolymer")

At the first stage of experimental investigation carried out using mini-cone test we observed the compatibility of added superplasticizer and cement [6]. A plasticizing effect of the action of admixture depending on its concentration and availability of the tendencies to the erroneous hardening or a fast loss of the fluidity by the cement paste were assessed. The test procedure was as follows: the cement paste was prepared with SP admixture concentrations of 0.6; 0.8 and 1.2% of the cement mass and the water-to-cement ratio of 0.4. Then, the paste was placed into the mini-cone and its spread was measured after 5, 30, 60 and 90 minutes. Thereafter, the conclusion was made on the compatibility of SP with the selected cement.

The second stage of experimental investigation included the fabrication of the four series of specimens with different formulation technology decisions aimed at the production of a specified grade of the concrete mix with the fluidity of S4 after 1.5 hours following its preparation. The S4 grade (16-20 cm) was selected, because it is highly desirable for the production of ready-mix concrete and it is usually assumed to be basic for the tests [3, 7]. An opportunity for providing the wanted fluidity grade after 1.5 hours was investigated varying the formulation technology factors:

increasing the SP concentration;

increasing an initial fluidity of the concrete mix (a lower water reduction effect produced by the SP action);

introducing the admixtures retarding the cement hardening;

additional introduction of 25%SP (of the initial value) after 1.5 hours;

stepwise addition of the SP admixture (50+50%).

Table 1 gives the composition of concrete mixes. When the SP admixture was introduced the ratio of cement, crushed stone and sand remained unchanged. The consumption of components was calculated taking into consideration a real density of the concrete mix. The admixtures that retard the hardening were introduced to increase a resiliency of the concrete mix in time. The series of specimens had the following difference in principal:

series 1 -the reference composition had no admixtures;

series 2- the influence of the concentration of SP admixture and an increase in the initial fluidity for the achievement of required fluidity in specified time was investigated;

series 3 - simultaneous introduction with the SP admixture of the admixture that retard the cement hydration;

series 4- stepwise addition and additional introduction of the admixture after 1.5 hours following the preparation;

Concrete mixes were prepared in the laboratory concrete mixer of a forced action and tested according to the standard ISO 4109:1980 [9], in particular we measured the concrete mix fluidity, its retention in time during 1.5 to 2 hours and its density. To define the compression strength of concrete, the concrete specimens of 10x10x10 cm were prepared according to the standard EN 12390-1:2000 [10] and it was measured at an age of 2, 7 and 28 days.

3 The results of research

Table 1 gives the measurement data of the spread of cement pastes with a different concentration of SP using the mini-cone. Introduction of the SP admixture in the amount of 0.6% results in an increased spread of cement pastes ranging from 6.3mm to 16.5 mm; the increase in the concentration to 0.8% results in the increased spread reaching 20 mm, that is not changed later on and it is indicative of the fact that the admixture saturation point has been reached.

The spread of cement pastes with SP after 5, 30, 60 and 90 minutes is not decreased; it is even increased (from 16.5 mm to 19 mm for the composition 2), which is indicative of the rheological compatibility of the admixture and the cement and unavailability of the problems with a long-term maintenance of an appropriate fluidity of concrete mixes.

The research done allows us to assume that it is quite possible to produce concrete mixes with the fluidity stable in time.

Table 1. Cement paste test data.

Item No	Composition	Spread diameter, cm, minute			
		5	30	60	90
1	Reference	6.3	5.3	5.8	5.8
2	0.6% of SP	16.5	17.8	19	19
3	0.8% of SP	20	20	21	20
4	1% of SP	20	20	20.5	20

The test data of concrete mixes and concretes of the four series of specimens are given in Table 2. A decrease in the concrete mix fluidity for the reference series containing no admixtures was equal to 5 cm after 1.5 hours. The concrete mix grade remained at the level of S4 during 60 minutes, which is indicative of its rather high preservation in time. As for the plasticized concrete mixes (series 2) the fluidity change pattern differed significantly; the decrease in the fluidity ranged from 7 to 14 cm after 1.5 hours. For the series I, a decrease in the fluidity had a linear character (1 to 2 cm after 30 minutes) and for the series 2 it had an avalanche -like character, i.e. the slump was decreased by 4 to 11 cm after 30 to 60 minutes. An increase in the admixture concentration failed to result in the mix stabilization in time, though it resulted in an increased water reduction effect and an increased concrete strength. The 0.8 % concentration of SP resulted in the increased strength of 19%, and at 1.2% of SP this value was equal to 61% and at 1.6% of SP it was equal to 87%. We can see a difference in the stability of cement pastes and concrete mixes in time and it can be related to the availability of additional disperse particles in concrete mixes that efficiently adsorb the SP admmixture.

Table 2. Test data of concrete mixes and concretes.

Series No	Consumption of components per 1m^3 of the concrete mix					Slump, cm. minutes				Concrete compression strength, kgf/cm^2, days		
	C, kg	CA, kg	FA, kg	W, kg	Type of the admixture and its concentration	5	30	60	90	2	7	28
1	375	1078	656	246	-	20	18	17	15	70	136	223
2.1	379	1090	664	234	0.8% SP	20	18	15	10	82	168	266
2.2	376	1081	658	242	0.8% SP	25	24	22	18	75	149	233
2.3	390	1113	678	211	1.2% SP	20	13	9	9	99	194	360
2.4	383	1101	670	225	1.2% SP	27	22	18	14	90	181	331
2.5	397	1142	695	194	1.6% SP	20	20	16	8	134	260	416
2.6	389	1119	681	205	1.6% SP	26	25	23	12	116	226	386
3.1	387	1114	678	219	1.2% SP +0.5%C	20	13	8	8	113	230	366
3.2	384	1105	673	208	1.2% SP +0.2%G	20	14	11	10	112	235	362
4.1	390	1121	682	211	1.2%+0.3% SP	20	12	9	20/11*	129	242	380
4.2	386	1110	676	208	0.6%+0.6% SP	9	4	4	20/10*	120	217	355

* after 30 minutes following the admixture introduction

With the increase in mobility to the value of S5 the water reduction effect was reduced, therefore the increase in the strength made up 5, 48 and 73% accordingly (Fig.1). Due to an increase in the initial fluidity for one grade we managed to maintain the fluidity at the level of S4 during 60 minutes and it is a sufficient condition for the most cases of the construction in urban conditions. If we recalculate a required amount of the admixture to be introduced into the mixture with the mobility of S5 in order to reach an equivalent increment in the strength we will see that the increase in the admixture consumption makes up 0.15 to 0.25 %.To increased the concrete strength by 19 % we need to introduce 0.95% of SP instead of 0.8%, to increased the concrete strength by 61% we need to introduce 1.4 % of SP instead of 1.2%, and in the case of 87% we need to introduce 1.85% of SP instead of 1.6 %.

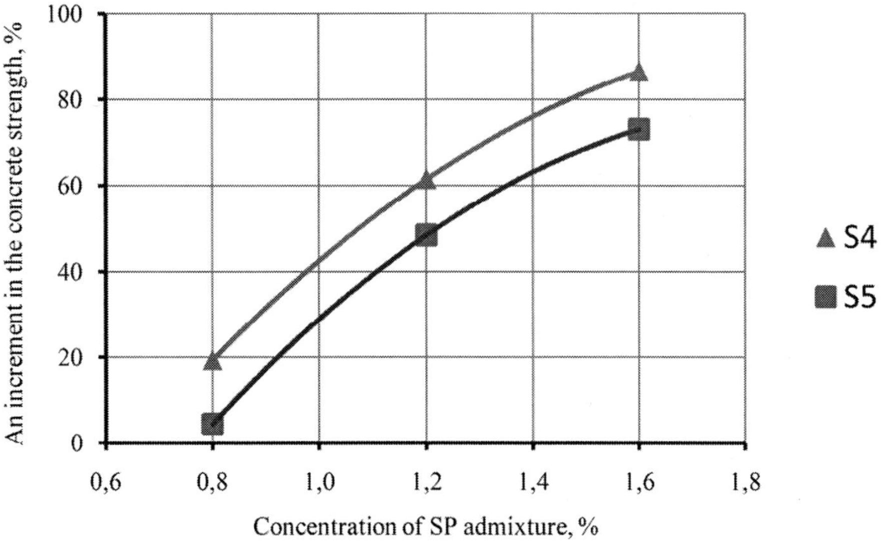

Fig. 1. Relationship of an increase in the concrete strength as a function of the concentration of SP admixture and the initial fluidity.

Addition of the retarding admixtures of cement hydration to the concrete mix (series 3) gave no desired outcome. Neither the hardening retarder (C) nor the hydrophobization agent (G) provided a stable fluidity for the concrete mix. A change in the increment of the strength of specimens for the series 3 is given in Fig.2. It is in the same range as for the specimens of the series 2 with the same amount of SP (series 2.3).

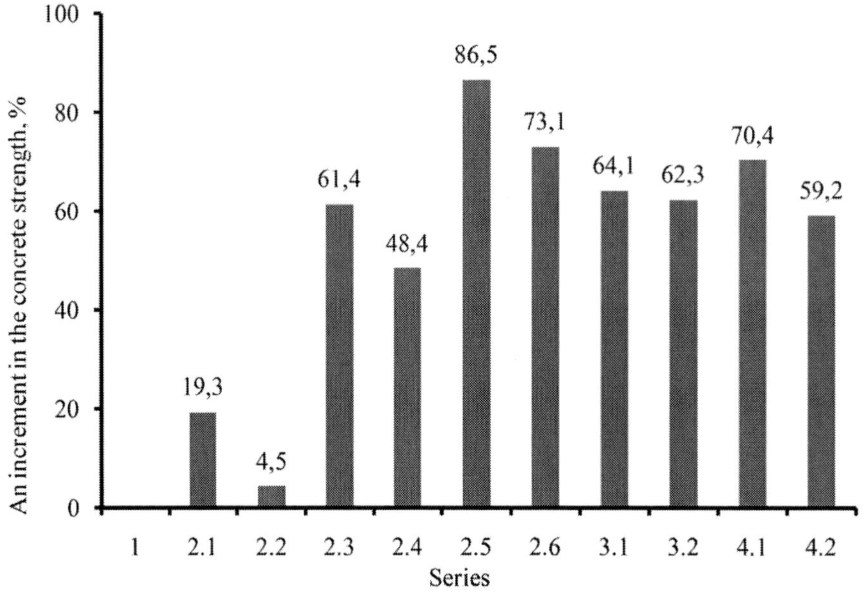

Fig. 2. An increment in the concrete strength for the test series of the specimens.

The last series of the tests (series 4) was related to the verification of the opportunity for the renewal of the fluidity of concrete mix after 1.5 hours following the preparation of it (series 4.1) and the efficiency of the stepwise addition of the admixture (series 4.2). An additional introduction of SP in the amount of 0.3% allowed us to restore completely the initial fluidity of the mix (Fig.3). The renewal of the mix fluidity through the additional introduction of SP indirectly confirms the unavailability of a sufficient amount of the free superplasticizer in the liquid phase that is necessary to reach a high fluidity of concrete mixes [8].

4 Discussion of results

An opportunity for the renewal of the cement fluidity due to the addition of SP is the most appropriate solution for the construction site. Such a necessity can be related to the unforeseen idle time (traffic congestion, equipment failure, etc.); therefore the available stock of the superplasticizer at the construction site can resolve the problem of the need of adding water in extreme cases.

Another method (stepwise addition) that requires the introduction of SP at the construction site is also rather efficient in this case and it requires no additional consumption of SP. However, the use of it can face the problems related to the transportation and technical opportunity for the mixing of the concrete mix of a low mobility (S1) in the concrete truck.

An increment in the strength in both cases was at the level of an ordinary introduction of the admixture and it was 9% higher at additional introduction of the admixture (Fig.2, series 4. 1).

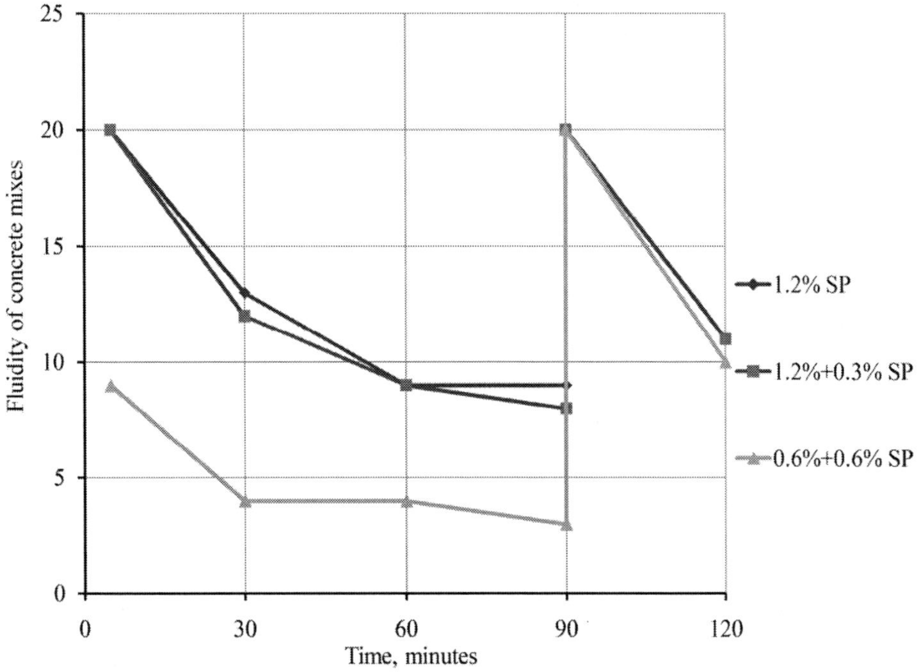

Fig. 3. A change in the fluidity of concrete mixes depending on the method of introduction of the SP admixture.

A drawback of these methods is the need for the time required for the mixing of concrete mix at the construction site and a short-time effect from the repeated admixture introduction. It must not be ruled out that in the case of need an additional amount of the admixture can be introduced several times.

A generalization of the most popular methods used for the maintenance of fluidity of concrete mixes in this research enabled the comparison of them with each other and attention was paid to the need of control of the workability of concrete mixes both at the plant and at the construction site to avoid undesirable phenomena of the degrading properties of concrete mixes and concretes.

5 Conclusions

Three technological techniques turned out to be effective, in particular a two-stage introduction of the admixture though it requires to spend additional time at the construction site and has a short-time effect; an increase in the initial fluidity of mix though it requires an increased admixture consumption in the range of 0.15 to 0.25%; renewal of the mix fluidity due to the additional introduction of 0.3% of the admixture after 1.5 hours following its preparation that includes the drawbacks of two previous techniques, though it can provide the fluidity at a later time. Each solution has its own advantages and drawbacks, therefore their efficiency can only be evaluated under specific conditions taking into consideration the cost of concrete mix and the opportunities for the realization of each solution.

References

1 J. Punkki, ACI Materials Journal, **93** (1996).

2 S.N. Tolmachev, E.A. Belichenko, A.V. Brazhnik. Construction Materials, **5** (2016)
3 A.N. Shakhov. Preservation of Cement Mixes: Contemporary Solutions of Everyday Issues. Concrete Technologies 3-4 (2012)
4 A.V. Usherov-Marshak, V.P. Sopov, Colloid Journal, **56** (**4**), 600-603 (1994)
5 V. Sopov, L. Pershina, L. Butskaya, E. Latorets, O.Makarenko, MATEC Web of Conferences, **116** (2017)
6 Taylor and other. *Identifying Incompatible Combinations of Concrete Materials: Volume I-Final Report*. U.S. department of transport. Federal highway admin (2006)
7 K. Hattori, Journal of the Society of Materials Science, **29**, 318 (1980)
8 A.I. Vovk, Concrete Technology, **5** (2009)
9 ISO 4109:1980 Fresh concrete – Determination of the consistency – Slump test. International Standart. (1980)
10 EN 12390-1:2000 Testing hardened concrete — Part 1 Shape, dimensions and other requirements for specimens and moulds. (2000)

Estimation of corrosion resistance of modifying concrete in a solution of sulfuric acid

Oleksii Kabus, *Yulyia* Kolomiiets*, and *Viktoriia* Lykhohrai

Kharkiv National University of Civil Engineering and Architecture, Sumska st. 40, 61002 Kharkiv, Ukraine

Abstract. This study presents the test results of 8 concrete series in 5% sulfuric acid solution for 28, 56 and 180 days. Each series was made using two mixtures based on CEM II/A-S 32.5 and CEM III/A 32.5 cements. Modification of the properties of concrete was obtained by the introduction of superplasticizer, silica fume, complex of structure modifier and hydrophobic admixture. The water-binder ratio varied from 0.27 to 0.63, the compressive strength was 26 to 83 MPa, and water absorption from 3.4 to 8.4 %. Samples of test cubes (10×10×10 cm) and test beam (16×4×4 cm) were prepared to immersion in the acid solution. The assessment of corrosion resistance was carried out on the weight loss and strength of concrete in relation to the initial value of 28 days. The obtained results showed that the surface layer of concrete in all samples was destroyed after 28 days of exposure. The deterioration of the corrosion resistance of modified concrete in relation to ordinary concrete without additives was observed regardless of the type of cement. An exception was a series of concrete, which had a water-binder ratio of 0.27–0.28 and a compressive strength of 77–83 MPa.

1 Introduction

In Ukraine, sewer tunnels, most of which were constructed 40–50 years ago from ordinary concrete (C16/20, C20/25), are in the bad or emergency condition now. This has been caused by the intensive impact of destructive factors, among which the influence of sulfuric acid of biogenic origin is most significant [1, 2]. Restoration of their operational characteristics to extend the trouble-free life of the service is an expense and difficult problem. It can be solved by modern cement concretes, which are used for repair work, with different technical and economic efficiency. Note that acid corrosion is unavoidable, because the cement-based concrete is alkaline in nature, and decomposes under an acid environment. However, improving the concrete mixture design is still a viable method because it changes the rate of the corrosion reaction process [3].

Modification of properties of concrete with chemical and mineral additives is a common focus of many studies [3-9]. The introduction of chemical admixtures that are aimed at reducing the permeability of concrete according to DSTU B V.2.6–145: 2010 [10] is one of the main ways to reduce the aggressiveness of the environment to concrete. The use of

* Corresponding author: spoonyla@gmail.com

active and pozzolanic mineral additives according to [3, 8, 9] allows to obtain more stable hydrates to sulfuric acid exposure. In the review [3] the authors draw attention to the fact, that the reduced porosity does not guarantee a better acid resistance of concrete and some studies observed opposite results of pozzolan addition effectiveness on acid resistance owing to variations in experimental conditions and the chemical composition of differently sourced pozzolans. Therefore, the chemical compositions of pozzolans should be individually examined, and experimental conditions should be mimicked as close as possible to the real sewer conditions before using it in an actual sewer. As a result of many years of laboratory and field tests, the author [7] also came to the conclusion that concrete additives in highly aggressive environments do not allow to provide the necessary durability of concrete without the use of secondary protection. The investigation of this issue with the testing of concrete in a wide range of their physical and mechanical properties is topical.

One of the simplified methods for assessing the corrosion resistance of concrete for sewer is their laboratory test in solutions of sulfuric acid. This allows, under certain conditions, at the first stage of research to fairly accurate [11] assess the effectiveness of prescription solutions in the development of new compositions of concrete.

The purpose of this study was to evaluate the possibility to improve the corrosion resistance of concrete to the action of solutions of sulfuric acid through the application of complex additives of different principles and mechanisms of action. The total duration of concrete samples exposure under aggressive conditions was 180 days. The obtained results will allow next research directions on the development of effective composite materials for the restoration of sewer structures.

2 Materials and methods

The materials used to produce the 8 series with 16 concrete mixes (Table 1). There are two types of cement (CEM II/A-S 32.5 and CEM III/A 32.5), granite crushed stone (5–10 mm in size), fine quartz sand, tap water and highly effective modifying additives (usually used to reduce porosity and increase the corrosion resistance of concrete) were in concrete mixtures composition. The following were applied:

- superplasticizer on polycarboxylate basis for the production of ready-mix concrete (A1);
- mineral pozzolanic additive – highly dispersed silica fume from waste from a local ferroalloy plant (A2);
- colmatizing additive – portland cement and various active, proprietary chemicals. These active chemicals react with the moisture in fresh concrete and with the by-products of cement hydration to cause a catalytic reaction, which generates a non-soluble crystalline formation throughout the pores and capillary tracts of the concrete (A3);
- hydrophobic waterproofing admixture – water solution of potassium methyl siliconat (A4).

The optimal concentrations of additives were chosen taking into account their effect on concrete hardening and compatibility with cements [12]. The consumption of components is based on the actual density of concrete mixtures. The I–VII series were made from high-workability concrete mixes, and the VIII series from a dry-concrete mix. All series except VII had an approximate consumption of binder 400 kg/m3, and the VII series – 600 kg/m3 of concrete.

According to the requirements of DSTU B V.2.7–213: 2009 [13] and DSTU B V.2.6–181: 2011 [14], the concentration of acid in which the test is conducted is not regulated and should be selected depending on the operation conditions of the structure. In the study of the effect of sulfuric acid on the corrosion rate of cement stone and concrete, its values are taken in the range from 0.01% to 10% [3–5, 15, 16]. Proceeding from the fact that on the concrete surfaces of sewer tunnels the concentration of sulfuric acid can reach several percent, for research purposes its value was selected according to ASTM C267-01 [16] and was equal to 5%.

Table 1. Concrete mix proportions.

Series	Mixes	Cement		Coarse aggregate (kg/m^3)	Fine aggregate (kg/m^3)	Water (kg/m^3)	W/C (W/B)	Slump, cm (Vebe, s)	Admixtures	
		type	kg/m^3						designations	kg/m^3
I	1	CEM II	397	1041	744	247	0.62	20	-	-
	2	CEM III	391	1026	733	245	0.63	20	-	-
II	3	CEM II	397	1043	745	205	0.52	19	A1	6
	4	CEM III	397	1043	745	199	0.5	26	A1	6
III	5	CEM II	403	1057	755	192	0.48	25	A1	8.1
	6	CEM III	404	1061	758	184	0.46	26	A1	8.1
IV	7	CEM II	366	1044	746	175	0.48 (0.44)	21	A1	8
									A2	31.8
	8	CEM III	364	1037	741	168	0.46 (0.42)	22	A1	7.9
									A2	31.6
V	9	CEM II	368	1050	750	178	0.48 (0.44)	23	A1	8
									A2	32
									A3	4
	10	CEM III	363	1037	741	171	0.47 (0.43)	22	A1	7.9
									A2	31.6
									A3	4
VI	11	CEM II	360	1028	734	174	0.48 (0.44)	24	A1	7.8
									A2	31.3
									A3	3.9
									A4	0.8
	12	CEM III	359	1024	731	168	0.47 (0.43)	23	A1	7.8
									A2	31.2
									A3	3.9
									A4	0.8
VII	13	CEM II	558	1062	587	167	0.3 (0.28)	26	A1	12.1
									A2	48.5
									A3	6.1
	14	CEM III	557	1059	585	163	0.29 (0.27)	27	A1	12.1
									A2	48.4
									A3	6.1
VIII	15	CEM II	370	1407	553	117	0.32 (0.29)	0 (30 s)	A1	8
									A2	32.2
									A3	4
	16	CEM III	370	1407	553	117	0.32 (0.29)	0 (30 s)	A1	8
									A2	32.2
									A3	4

Samples of test cubes (10×10×10 cm) and test beam (16×4×4 cm) were made according to the standard method and maintained for 28 days under normal conditions. Then some samples were taken to determine the compressive strength, flexural strength and concrete water absorption by weight. Other samples were immersed in a 5% solution of sulfuric acid (Fig. 1, a) and take out at 28, 56 and 180 days. The concentration of acid solution was controlled by its density. When it was reduced, the required amount of acid was added to reach the initial value of the density of the solution. In each of the required period, two samples were taken from the solution to remove the destroyed concrete layer with a brush (Fig. 1, b). They were then weighed, photographed, tested for strength and found pH. Two different probes sample to determine the pH. First probe selected from the surface and second probe selected from the inside (not less than 1 cm from the edge of the sample), and then pH of the water extract was detected using the indicator.

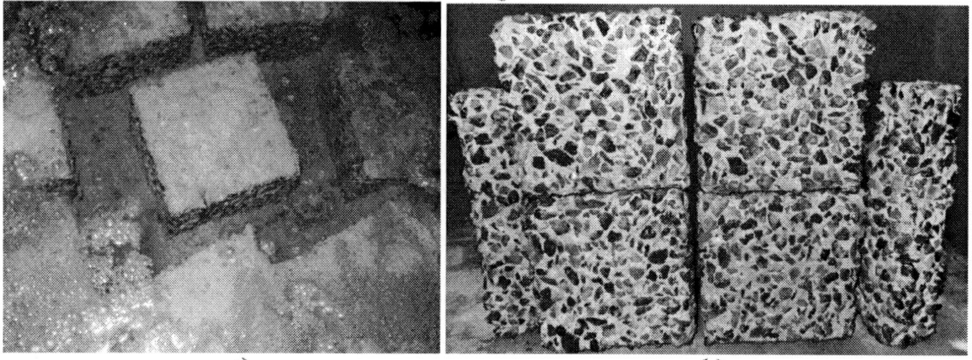

a) b)

Fig. 1. Samples after exposure in a solution of sulfuric acid: a – exposure in acid solution; b – after cleaning the destroyed layer.

3 Results and discussion

Water absorption of concrete samples by weight (Wm) in 28 days is given in Table 2. Additive modifiers have reduced the water absorption by 2–2.5 times. The greatest effect of reducing water absorption was achieved by the introduction of superplasticizer, which allowed reducing the amount of water and water-binder ratio (W/B). All other additives did not have a significant effect on the water absorption of concrete. A more significant decrease in water absorption was observed in concrete made at CEM III/A 32.5.

Table 2. Water absorption of concrete.

Mixes	1	2	3	4	5	6	7	8	9	10	11	12	13	14	15	16
W_m,%	8.1	8.4	7.2	6.9	5.8	5.8	6	5.5	5.6	5.2	5.7	4.8	3.9	3.4	4	3.4

In all samples, irrespective of the type of cement, composition, strength and water absorption, a total destruction of the surface layer of concrete was observed for 28 days. The mass loss of test cube samples ranged from 3.6 to 6.6%, and the test beam samples – 9.2–16.3% (Table 3). In 56 days – 7.2–12.6% and 16.8–32.8%, and in 180 days – 11.8–22.5% and 21.7–50.3%, respectively. The overall dimensions of the specimens decreased by 5–10 mm in 180 days, which corresponds to a corrosion rate of 5–10 mm/year. In the test beams, the mass loss was 2–2.5 times larger than in test cubes, which is explained by a large 1.9 times the coefficient of open surface. Therefore, for accelerated determination of corrosion resistance of concrete, test beams are more appropriate. It should be understood that, in addition to the size of the samples, the conditions of exposure can greatly affect the results obtained. For example, with constant circulation of acid solution, there is an increase in the rate of corrosion [17].

The type of cement did not have a significant effect on the weight loss of the samples, i.e. the increase in the slag additive does not increase the stability of concrete in these conditions. Increasing the strength of samples in 1.5–3 times (Table 4) and reducing water absorption also did not allow to improve the stability of concrete. On the contrary, it was possible to observe somewhat more mass loss in a series of samples with additives. The exception was series VII with high cement content. Its compressive strength was 76.7 – 82.5 MPa and the ratio of W/B was 0.27–0.28. The weight loss of series VII after 56 days of exposure was less than in series I, especially after 180 days. However, it can not be clearly stated that reducing concrete insight and reducing W/B to 0.27 can improve the stability of concrete, because this effect was not observed in series VIII. According to studies [9, 17], the reduction of W/B leads to deterioration of corrosion resistance. In the paper [9], an increase in the mass loss of samples was observed with a decrease in W/B from 0.45 to 0.25, and in [17], the change in W/B from 0.65 to 0.35 resulted in an increase the erosion depth of concrete. This is explained by the fact that concrete with a high water cement ratio has a higher capacity to absorb the expansion of production reaction of gypsum than that with a low water cement ratio [17].

Comparison of series III–VI, which had approximately equal W/B (0.43–0.48), allows us to state that the addition of silica fume, colmatizing and hydrophobic additives at selected concentrations is absolutely ineffective. Thus, the authors' assertion [3] confirms that the effectiveness of mineral additives needs to be checked in each case. In general, it can be concluded that the corrosion rate of concrete is very high for all series of samples, because according to the requirement of DSTU B V.2.7–213: 2009 [13], the weight loss of chemical-resistant concrete in one year of immersion test should not exceed 1%.

Table 3. Loss of concrete samples mass after immersion in 5% solution of sulfuric acid.

Series	Mixes	Mass loss, %					
		test cubes			test beams		
		28 days	56 days	180 days	28 days	56 days	180 days
I	1	3.6	9.0	15.8	10.4	21.9	41.7
	2	3.6	8.6	17.6	9.2	23.8	41.7
II	3	5.0	12.2	20.2	11.3	28.8	42.2
	4	5.0	10.8	20.9	13.2	20.5	43.3
III	5	5.6	12.3	19.0	10.5	28.6	38.8
	6	5.7	9.1	17.6	15.0	26.8	40.6
IV	7	6.3	11.6	21.0	12.0	28.2	43.8
	8	6.6	11.7	21.5	16.3	31.2	48.1
V	9	6.0	12.1	20.9	12.9	29.0	41.9
	10	6.1	11.4	21.0	14.5	32.8	48.9
VI	11	6.3	10.8	21.5	13.5	28.8	44.0
	12	6.7	12.6	22.5	15.8	30.0	50.3
VII	13	4.5	8.5	12.0	10.2	16.8	21.7
	14	4.3	7.2	11.8	12.9	19.3	23.8
VIII	15	5.7	10.0	20.6	12.6	25.1	43.0
	16	5.1	9.2	21.2	13.5	31.5	46.6

The loss of strength of concrete samples as a result of corrosion was more significant than loss of mass. The compressive strength of the samples decreased by 19–46% after 28 days and 39–63% after 56 days of exposure (Table 4). Flexural strength changed differently. There was an increase in strength up to 39% in samples 1, 6, 13, while in other samples, it was reduced up to 62% for 28 days. After 56 days a compressive and flexural strength were reduced on 39–63% and 9–61% respectively. The correlation between compressive strength and flexural strength was not observed, the results were random. For example, sample 13 had a reduction in compressive strength of 38% and an increase in flexural strength by 39%. Proceeding from this, in 180 days, the strength of concrete was not determined. This kind of change in the strength of concrete, in our opinion, is associated with uneven transfer of load on the surface of samples with a bare coarse aggregates (Fig.1, b). Internal concrete did not undergo visual changes (Fig. 2), so all concrete of the samples could not change a strength so much. The absence of corrosion of the internal concrete of the sample holds the value of the pH of the water extract, which on the surface was equal to 3–5, and the inside was 12 for all samples.

Table 4. Change in the strength of concrete samples after immersion in 5% solution of sulfuric acid.

Series	Mixes	Initial strength, MPa		Strength loss, %			
				28 days		56 days	
		compressive strength	flexural strength	compressive strength	flexural strength	compressive strength	flexural strength
I	1	26.4	4.6	19.2	-18.8	38.6	14.9
	2	26.1	5.0	22.6	26.4	43.3	34.5
II	3	38.2	7.0	35.2	19.3	52.3	31.5
	4	39.5	5.7	36.9	6.8	58.3	26.4
III	5	44.1	7.5	36.5	3.1	53.8	50.3
	6	50.0	7.0	37.7	-1	48.7	40.4
IV	7	44.7	7.3	34.5	12.4	53.2	45.1
	8	46.5	6.3	46.1	19.9	60.1	38.2
V	9	48.9	8.4	33.6	10.2	55.1	46
	10	48.2	7.5	35.4	7.7	53.4	43.4
VI	11	45.9	7.8	31.3	26.3	56.4	40.2
	12	51.8	8.6	37.2	316	62.7	46.3
VII	13	76.7	8.2	37.9	-39.4	57.2	10
	14	82.5	8.7	35.8	14.5	48.2	9.4
VIII	15	64.6	9.0	36.6	62	49.0	60.9
	16	62.5	8.7	32.7	32.7	43.6	55.7

According to [15, 17], there are three distinct layers of concrete: the first is a friable surface layer of fully corroded that has no strength, the second – is a thin layer of concrete in a few mm, which has some strength, and the third – the inner concrete in which there is no deep or selective destruction. For a more accurate determination of the concrete strength in the study [9] the surface was aligned with cement mortar of the ratio 1:3. This made it possible to obtain more precise dependencies of concrete durability over time, the value of which for 180 days for various concrete mixes was from 20 to 45%. But, in our opinion, this method of preparing samples should not be considered sufficiently reliable. To obtain accurate results, it is necessary to completely remove the second layer of concrete and prepare the surface of the samples prior to testing in accordance with the standard requirements. An alternative method for analyzing the concrete strength can be to determine the tensile strength when splitting the test cubes (Fig. 2, a.). In this case, the influence of the surface of the samples should not be significant.

a) b)

Fig. 2. The appearance of concrete samples after the tensile test: a – split test of cubes; b – flexural strength test of beams

The obtained results show that the method for determining mass loss is more correct than the method of determining the strength of concrete. Authors [8] weight loss measurement was identified as a proper test method for indicating durability of concrete against sulfuric acid attack. However, authors [9] when testing in a less concentrated solution (1%) believe that mass loss alone is not a reliable index to measure the durability of concrete in acid attack and must be supplemented by strength studies.

4 Conclusions

At present, in the world there is no single method for determining the resistance of concrete to the action of solutions of sulfuric acid, which significantly affect the results obtained by different researchers. The most important factors are the concentration of the acid solution, the immersion test conditions and the size of the samples. Special attention is required to the method of determining the strength of concrete, which has a damaged top layer. The conducted studies have shown that the selected additives do not allow to significantly changes of the resistance of concrete on ordinary cements to the action of 5% solution of sulfuric acid. Increasing the strength, reducing W/B to 0.3 and reducing the absorption of concrete by 2–3 times is not effective. Further tests require concrete with a value of W/B ration of less than 0.25 and more than 0.65.

References

1. A. Garmash, D. Bondarenko, G. Zubko, D. Goncharenko. World Journal of Engineering, **1**, 72 (2016)
2. D. Bondarenko, V. Bulhakov, A. Garmash, D. Goncharenko, S. Pilihram, *Kanalizatsiini tuneli Kharkova: QUO VADIS: monohrafiia* (Kharkiv, Rarytety Ukrainy, 2018)
3. L. Wu, C. Hu, W. Liu, Sustainability, **10(2)**, 517, (Canada, 2018)
4. A. Khodabakhshian, M. Ghalehnovi, J. Brito, E. Shamsabadi, Journal of Cleaner Production, **170**, 42 (2018)
5. N.P. Rajamane, M.C. Nataraja, N. Lakshmanan, J.K. Dattatreya & D. Sabitha, Indian Journal of Engineering & Materials Sciences, 357, (2012)
6. Z. Makhloufia, M. Bederinaa, M.Bouhichaa, E. Kadrib. *Eighth International Conference on Material Sciences. Physics Procedia*, **55**, 329 (2014).

7. N. K. Rozental. Beton i zhelezobeton, **2**, 78 (2011)
8. A.A. Ramezanianpour, A. Zolfagharnasab, F.B. Zadeh, S.H. High, *Tech Concrete: Where Technology and Engineering Meet - Proceedings of the 2017 fib Symposium*, 2290 (2017)
9. S. Goyal, M. Kumar, D. S. Sidhu, Journal of Advanced Concrete Technology, **2**, 273 (2009)
10. *DSTU B V.2.6-145:2010 Zakhyst betonnykh i zalizobetonnykh konstruktsii vid korozii* (2010)
11. B. Huber, H.B. Hilbig, J.E. Drewes, E. Müller, Cement and Concrete Research, **1**, 36 (2017)
12. A.V. Usherov-Marshak, A.V Kabus. Inorganic Materials, **4,** 479 (2016)
13. *DSTU B V.2.7-213:2009. Building materials. Chemically resistant concretes. Test methods* (2009).
14. *DSTU B V.2.6-181:2011. Protection against corrosion of concrete and reinforced concrete constructions. Test methods* (2011).
15. V.M. Moskvyna, *Korrozyia betona v ahressyvnikh*, (Moscow, 1971)
16. *ASTM C267-01. Standard Test Methods for Chemical Resistance of Mortars, Grouts, and Monolithic Surfacings and Polymer Concretes* (2018).
17. K. Kawai, S. Yamaji, T. Shinmi, *10DBMC International Conférence On Durability of Building Materials and Components*, 17 (2005)

Alkali activated cements mix design for concretes application in high corrosive conditions

Oleksandr Kovalchuk[1,*], *Valentina* Grabovchak[2], and *Yaroslav* Govdun[1]

[1]Kyiv National University of construction and Architecture, 03037 Kyiv, Povitroflotsky Avenue 31, Ukraine
[2]National Aviation University, 03058 Kyiv, Kosmonavta Komarov Avenue, 1, Ukraine

Abstract. This paper covers the results of development of corrosion resistant ash alkali-activated cements based on regulation of phase composition of the hydration products through changing the alkali content, content of calcium- containing cement constituents resulting in the increase strength and density of the cement stone. The results of study suggested to conclude that the cement compositions with predominance in the hydration products of weakly soluble low basic hydrosilicates of calcium, hydrogarnets and minerals similar to natural hydroaluminates exhibited the highest corrosion resistance. The results of comparison suggested to draw a conclusion that the alkali-activated cements Types APC III-400 and ACC V-400, according to National Ukrainian Standard DSTU B V.2.7, had high corrosion resistance compared to that of OPC, thus allowing to recommend the developed cements for the concretes intended for use in aggressive environments, inclusive of sodium and magnesium sulphates and others. Coefficients of corrosion resistance of concretes are higher than 1 after even 42 months.

1 Introduction

With the increase in volumes of construction a first-priority issue is how to improve their durability. Durability of concrete structures is dependent upon service conditions and, among many other factors, by characteristics of the cement used for making the concrete. Of special interest is durability of corrosion resistant concretes for special intended use. Until now, there is no unified solution for the concrete mix design which would meet all needs with regard to corrosion resistant concretes. The most important characteristics which determine corrosion resistance of the concretes are: type of cement and its mineralogical composition, composition of hydration products, water to cement ratio (W/C), type of admixtures and additives, and aggregates, pore structure, etc. In the corrosion resistant concrete mix design the following factors should be taken into account: service conditions and aggressive exposure conditions.

* Corresponding author: kovalchuk.oyu@gmail.com

The alkali activated cements based on granulated blast-furnace slags and fly ashes, that have been developed by a scientific school headed by Professor V.D.Glukhovsky, are the cements that can be successfully used for making corrosion resistant concretes [1-3]. The alkali activated cement concretes have high strength properties (40-120 MPa and higher), frost resistance (1000 cycles), water penetration (W10-W50), high corrosion resistance in different mineral and organic environments [4, 5].

Further research work allowed to develop fly ash alkali activated cements in which the fly ash content reached 90% [6-9]. These fly ash alkali activated cements can be an important alternative to portland cements containing fly ash in a quantity of maximum 30%, since in their physical-mechanical properties are similar to the portland cements, and the fly ash alkali activated cement stone is characterized by the higher weather resistance, frost- and corrosion resistance.

2 Raw material and methods of examination

The composition of the concrete without admixtures was constant in all experiments. Low-calcium fly ash class F (under ASTM C 618), ground to a specific surface of 800 m^2/kg by Blaine, sodium carbonate and sodium metasilicate penthahydrate as alkaline activator were used as basic raw materials. Ordinary portland cement (OPC) Type I Grade 52,5 (commercial product with a specific surface of 380 m^2/kg) and blast-furnace slag (specific surface 450 m^2/kg) were taken as additional cement constituents. Chemical composition of raw materials is shown in Table 1.

Table 1. Chemical composition of raw materials.

Material	SiO_2, %	Al_2O_3, %	Fe_2O_3, %	CaO, %	MgO, %	SO_3, %	$Na_2O + K_2O$, %	LOI, %
Fly ash	51.08	24.8	13.67	3.12	1.83	0.08	0.60 + 1.90	1.50
OPC I-500	23.4	5.17	4.12	64.13	0.88	0.55	0.41 + 0.33	0.20
Slag	40.0	5.91	0.32	46.9	5.87	1.62	-	-

The fly ash alkali activated cements under study, they were: alkali activated pozzolanic cement APC III and alkali activated composite cement ACC V [13] were prepared by mixing of the preliminary separately ground fly ash and slag, OPC, and alkaline activator and plasticizer in a ball mill. The slag portland cement M400 (OPC III/A), according to Ukrainian classification [14] was taken as reference. The composition of the fly ash alkali activated cements under study is given in Table 2.

Table 2. Composition of the fly ash alkali activated cements under study.

Cement Type	Composition, % by mass				
	Fly ash	OPC I-500	Slag	Alkaline activator	Plasticizer
APC	66.2	28.4	-	Na_2CO_3-4.7	0.75
ACC	54.2	8.9	26.2	Na_2CO_3-4.7 $Na_2SiO_3 \cdot 5H_2O$-5.3	0.7

Corrosion resistance of the fly ash alkali activated cements under study was determined according to recommendations given in [10] by measuring strength variations of the specimens placed in aggressive environments (10%- and 5%- solutions of sodium sulfate (Na_2SO_4), 4%- and 2%-solutions of magnesium sulfate ($MgSO_4$), concentrated sea water).

The specimens were kept for 3 days in normal conditions, then, for 25 days – in water of technical grade. The test ages were: 30, 60, 90 , 180 and 1300 days.

The corrosion resistance was expressed by a coefficient of corrosion resistance, which is calculated as a ratio between a flexural strength of the specimen after storage for 12 months in aggressive environment and flexural strength of the specimen after storage for 12 months in water. According to above recommendations the cement stone is considered to be a corrosion resistant if the coefficient of corrosion resistance after storage for 12 months is equal or higher than (>)0.8.

3 Results and discussion

At the first stage of the study, main physical-mechanical characteristics of the fly ash alkali activated cements have been determined (Fig.1). The results of test suggested to draw a conclusion that the developed cements fall within the ranges of compressive strength requirements for the alkali activated cements APC and ACC of strength class M400 (their compressive strength was 38.4- 41.3 MPa).

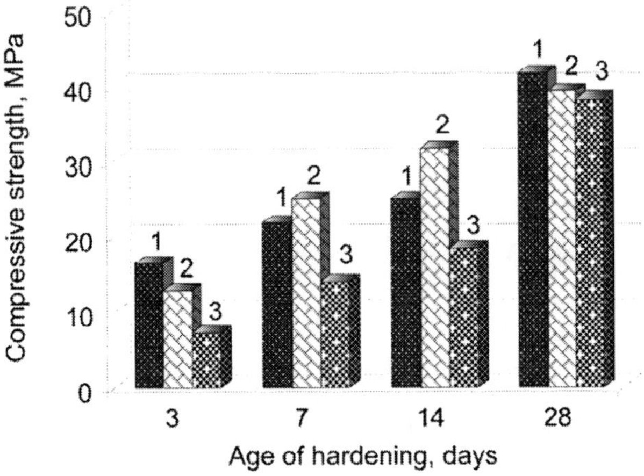

Fig. 1. Compressive strength of the cements under study: 1 – APC; 2 – ACC; 3 – OPC III/A-400.

The results of study of corrosion resistance (Fig.2, a, b, c, d, e, f,) suggested to conclude that all cements under study after long-term storage in aggressive environments at the age of 30 days have increased their strength properties comparing to the reference cement/ This can be attributed to the deepening of the hydration process. However, at the age of 2-3 months strength of the majority of the cements under study tended to lowering. So, at the age of 6 months, the cement OPC III/A-400 has rapidly lost its strength compared to the fly ash alkali activated cements APC and ACC. Thus, the cement OPC III/A-400 could not be considered as corrosion resistant cement (Fig 3).

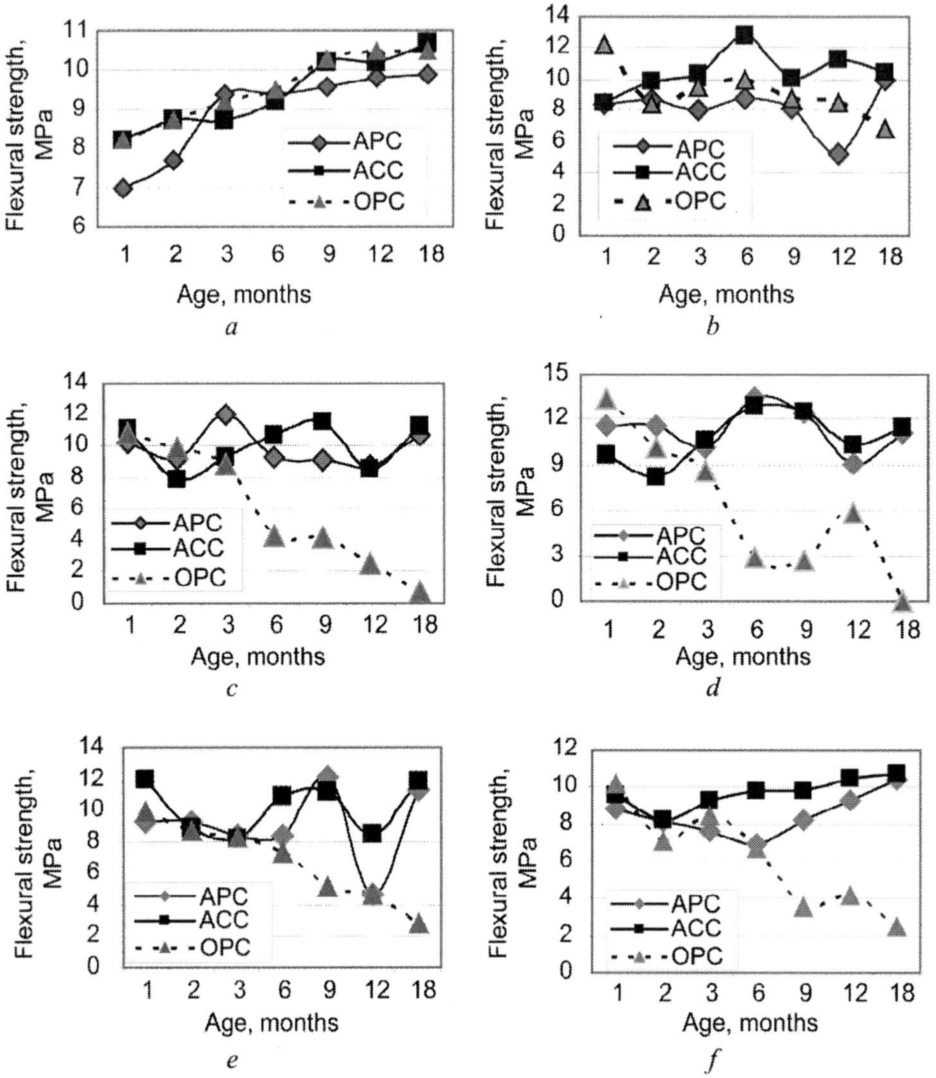

Fig. 2. Flexural strength of the cement under study vs. age (storage in aggressive environment): (a) water of technical grade; (b) sea water; (c) Na_2SO_4 (concentration- 5%); (d) Na_2SO_4 (concentration- 10%); (e) $MgSO_4$ (concentration- 2%); (f) $MgSO_4$ (concentration- 4%).

The calculated coefficients of corrosion resistance (Table 2) suggested to draw a conclusion that the developed fly ash alkali activated cements (APC and ACC) have much better resistance against aggressive environment compared with OPC. So, at the age of 9 months, the coefficient of corrosion resistance of these cements did not lower below 1, and at the age of 12 months the developed composite cement ACC had good flexural strength (the coefficient of corrosion resistance was 1.0 and higher), and the coefficient of corrosion resistance of the developed cement APC declined below critical level only in sea water and magnesium sulfate with concentration of 4%. This result can be explained by a constructive character of the combined alkaline-sulfate activation of the fly ashes [11].

https://doi.org/10.1051/matecconf/201823003007

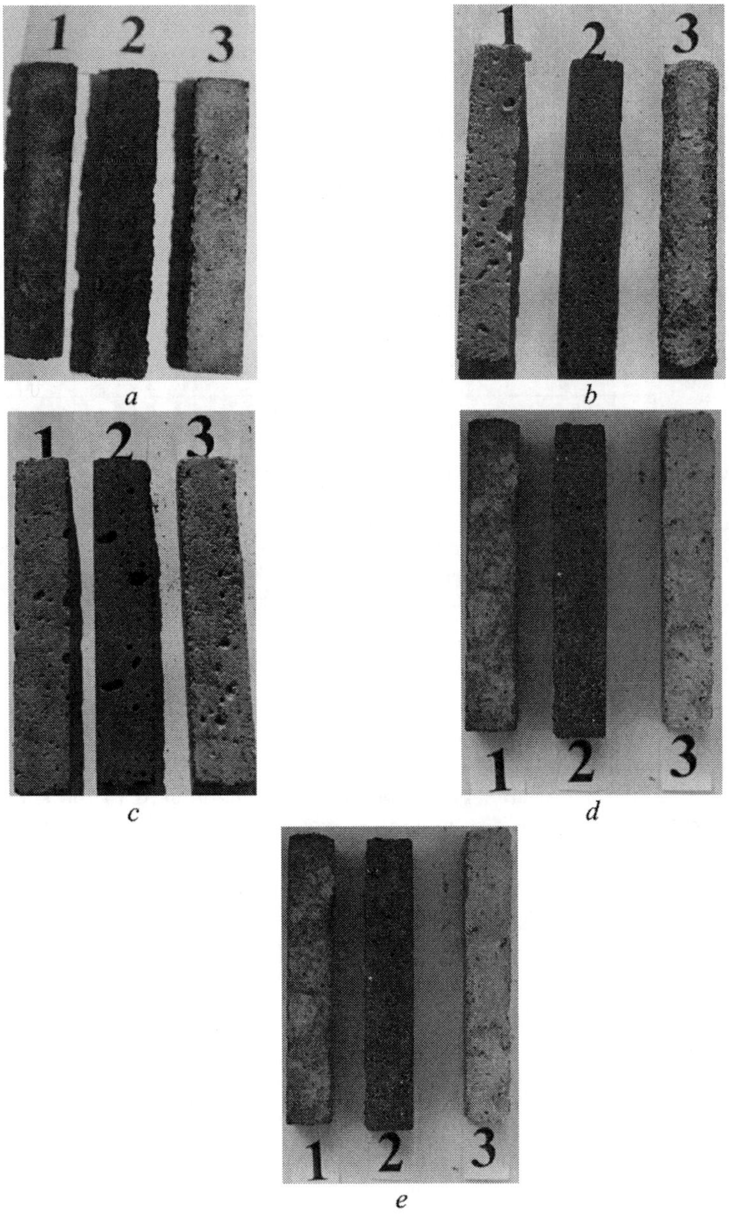

Fig. 3. Photos of the specimens after 1 year-storage in aggressive environment: (a) sea water; (b) 5% Na_2SO_4; (c) 10% Na_2SO_4; (d) 2% $MgSO_4$; (e) 4% $MgSO_4$; 1 – APC; 2 – ACC; 3 – OPC III/A-400.

Analyzing the coefficients of corrosion resistance (Table 3) suggested to draw a conclusion corrosion resistance of the developed fly ash alkali activated cements under study is directly proportional to a quantity of the calcium-containing cement constituents. Thus, with regard to the calcium-containing cement constituents the corrosion resistance decreases as the follows: ACC<APC< OPC III/A-400. With regard of different aggressive environments, the corrosion resistance decreases as follows: sea water < sodium sulfate < magnesium sulfate.

Table 3. Test results of corrosion resistance of the cements under study.

Aggressive environment	Cement type	Coefficient of corrosion resistance (Ks) after storage in aggressive environment, days			
		90	180	365	1300
Sea water	*APC*	0.95	0.84	0.83	0.80
	ACC	1.39	0.99	1.22	1.11
	Slag OPC	1.05	0.85	0.85	0.86
Na_2SO_4 (concentration-5%)	*APC*	1.00	0.95	0.88	1.20
	ACC	1.16	1.13	1.05	1.22
	Slag OPC	0.45	0.45	0.29	0.20
Na_2SO_4 (concentration-10%)	*APC*	1.45	1.29	0.93	1.14
	ACC	1.40	1.23	1.11	1.10
	Slag OPC	0.31	0.26	0.09	0.00
$MgSO_4$ (concentration-2%)	*APC*	0.90	1.27	0.84	1.04
	ACC	1.18	1.10	0.91	1.38
	Slag OPC	0.77	0.50	0.55	0.78
$MgSO_4$ (concentration-4%)	*APC*	0.74	0.85	0.94	0.43
	ACC	1.07	0.96	1.13	0.79
	Slag OPC	0.71	0.35	0.50	0.00

4 Conclusion

The results of study suggested to conclude that in corrosion resistance the developed fly ash alkali activated cements, they are: alkali activated pozzolanic and alkali activated composite cements exhibit the higher corrosion resistance compared to the slag portland cements and can be successfully used in making corrosion resistant concretes for the use in severe aggressive environments.

The results of study suggested to conclude that concretes on the fly ash alkali activated cements APC and ACC exhibit high corrosion resistance (Ks = 1.04...1.39 after 12 months of storage in aggressive environments) and on the the developed composite cement ACC has the higher characteristic. At the age of 42 months slag OPC concretes were totally destroyed, thus alkali activated concretes were showing high corrosion resistance.

References

1. P.V. Krivenko, *Proc. First Intern. Conf. "Alkaline Cements and Concretes*, 11-129 (1994)
2. M. Sanytsky, T. Kropyvnytska, R. Kotiv, AMR, **923**, 42-47 (2014)
3. P.V. Krivenko, *Proc. 10th ICCC*, 4iv046-4iv050 (1997)
4. T. Naik, B.W. Ramme, J.H. Tews, *Proc. 4-th Intern. Conf. on the use of Fly Ash, Silica Fume, Slag and Natural Pozzolans in Concrete*, 200-210 (1992)
5. V.M. Malhotra, *Concrete Admixtures Handbook* (Noyes Publications, 2-nd Edition, 1995)
6. M. Sanytsky, T. Kropyvnytska, T. Kruts, O. Horpynko, I. Geviuk, KEM, **761** (2018)
7. P. Krivenko, G. Kovalchuk, O. Kovalchuk, *Proc. Intern. Conf. "Global Construction: Ultimate Concrete Opportunities"*, 97-104 (2005)
8. P. Krivenko, O. Kovalchuk, O. Petropavlovsky, A. Pasko, S. Lapovska, EEJET, **4/6 (94)**, 6-15 (2018)

9. P.V. Krivenko, G.Yu. Kovalchuk, *Proc. Internatinal Conference Alkali Activated Materials – Research, Production and Utilization*, 349-367 (2007)
10. *DSTU B.V.2.6-145:2010. Protection of concrete and reinforced concrete structures from corrosion*. National standard of ukraine
11. O.S. Kavalerova, E.K. Pushkarova, V.I. Gots, G.YU. Kovalchuk, *Proc. 16th Internat. Conf. "Ibausil"*, 1-0933 – 1-0940 (2006)
12. J. Stark, *Alkali-Kieselsaure-Reaktion* (2007)
13. *DSTU B.V. 2.7-181:2009. Alkaline cements. Specifications.* National standard of ukraine
14. *DSTU B.V. 2.7.-46-96. Building materials. Common cements. Specifications.* National standard of ukraine
15. W. Schroeyers, P. Krivenko, A. Pasko,. JER, **168**, 21–29 (2016)
16. P. Krivenko, O. Kovalchuk, A. Pasko, CBMa, **151**, 819–826 (2017)
17. M.M. Alonso, A. Pasko, C. Gasco, CBM, **159**, 745–754 (2017)
18. P. Krivenko, O. Kovalchuk, A, Pasko, KEM, **761**, 35–38 (2018)
19. P. Kryvenko, H. Cao, O. Petropavlovskyi, L. Weng, O. Kovalchuk, EEJET, **1(6)**, 40-45 (2016)
20. P. Krivenko, O. Kovalchuk, O. Petropavlovsky, A. Pasko, S. Lapovska, EEJET, **4/6 (94)**, 6-15 (2018)
21. P. Krivenko, O. Petropavlovsky, O. Kovalchuk, EEJET, **1/6(91)**, 33-39 (2018)
22. G. Kochetov, T. Prihna, O. Kovalchuk, D. Samchenko, EEJET, **3/6(93)**, 52-60 (2018)
23. R.F. Runova, M.O. Kochevyh, I.I. Rudenko, *Proc. International Conference on Admixtures - Enhancing Concrete Performance*, 149-156 (2005)

Disperse reinforcing of alkaline aluminosilicate binders for protection materials of structures

Pavel Krivenko, and *Volodymyr* Kyrychok*

Scientific Research Institute for Binders and Materials, Kyiv National University of Construction and Architecture, 03037 Povitroflotskyi Avenue, 31, Kyiv, Ukraine

Abstract. The review of the literature and the analysis of the problem of durability and reduction of cracking of binders and materials for the protection of building structures have been carried out, an effective method of improving their cracking resistance during the entire lifetime has been proposed. The influence of reinforcement of the matrix of alkaline aluminosilicate binders on their technological and physical-mechanical properties was investigated. According to the study of influence reinforcing components on the properties of binders, it should be noted that the highest the conditional coefficient of crack resistance is characterized by the contents containing in its composition as a reinforcing component of the basalt scales. The microstructure of the connecting substance and the contact zone "compound substance - the reinforcing component" has been analyzed, the type and the optimum amount of the reinforcing component have been established to provide the technological and operational properties of the materials for the protection of building structures.

1 Introduction

During the operation of building structures in aggressive environments, their durability is sharply reduced due to the diverse nature and origin of corrosion processes.

Building structures constructed in accordance with technical requirements may, for a long time, withstand the negative effects of the atmospheric-climatic and aggressive external environment. In operation, they gradually change their physical, mechanical and structural properties as a result of physico-chemical processes of interaction of cement stone and reinforcement with the external environment [1-3]. Physico-chemical processes passing on the surface of structures in contact with environment, leads to a violation of its functional properties. Therefore, increasing the durability and ensuring the reliable operation of building structures with the help of binders for constructures surfaces is gaining in popularity, and their development is a topical issue of modern materials science.

Protection of building structures by binder on the basis of organic and inorganic substances and their advantages and disadvantages is considered in scientific papers [4-6]. The main drawbacks of organic binders are flammability, aging during the operation, loss of adhesion to the base. The use of modified cements is not effective in view of the low resistance to aggressive media of the matrix, which leads to the destruction of binder.

* Corresponding author: v.kyrychok@gmail.com

Scientific School of V.D. Glukhovsky Scientific Research Institute for Binders and Materials (SRIBM) [7-11] proved the possibility of using alkaline aluminosilicate binders (geocements, geopolymers) to protect building structures from the effects of aggressive environments. Alkaline aluminosilicates are characterized by particularly high reliability indices - corrosion resistance, temperature resistance, fire resistance, which is ensured due to synthesis zeolite-like new formations.

In order to provide the conditions of hardening and structural formation of protective binders, their mortar mixture should have optimum filling, sufficient mobility, high thixotropy and water retention capacity. Cracking of binder at these stages is due to shrinkage deformations brcause of the evaporation of water and its absorption by the base. Therefore, provision of optimal conditions at this stage is ensured by the use of polyfraction composition of fillers and organic and mineral modifiers as a part of binder [12].

Cracking of binders at the stage of their exploitation is a result of stresses that arise under the influence of mechanical loads, temperature or moist gradients. When using buildings or structures in an aggressive environment, cracking resistance of protective binders is a basic requirement. For the protection of building structures, this coverage index is particularly important given a possibility of cracking openings during their operation [4].

According to the data [13, 14], for the evaluation of the resistance to cracking of composite materials, a conditional coefficient of crack resistance is used, that is, the ratio of compressive strength to its tensile strength at bending. Analysis of work on the resistance to cracks in composite materials shows that with the introduction of reinforcing components into their composition, the physico-mechanical parameters are significantly increased, especially the strength of the composite on bending tension. In order to ensure the overlapping of the cracks, reducing the deformability of the binder and increasing its crack resistance, research is conducted in the direction of searching for methods for its reinforcement. Consequently, the reinforcement of the binders can significantly increase their crack resistance during their lifetime [15, 16].

Therefore, the purpose of this work was to study the influence of reinforcing components on the properties of protective alkaline aluminosilicate binders for building structures. To achieve this purpose, the following tasks are set:

- to study the influence of the type of reinforcing components on properties of binders;
- to analyze the formation of reinforced binders structure;
- to determine the optimal number of reinforcing components in the protective binders.

2 Materials and methods

According to the analysis of studies [6, 14], the following types of reinforcing components are used in the production of effective protective materials for their reinforcement: polypropylene fibers (PPF), basalt fibers (BF) and basalt scales (BS) (Figure 1).

a) b) c)

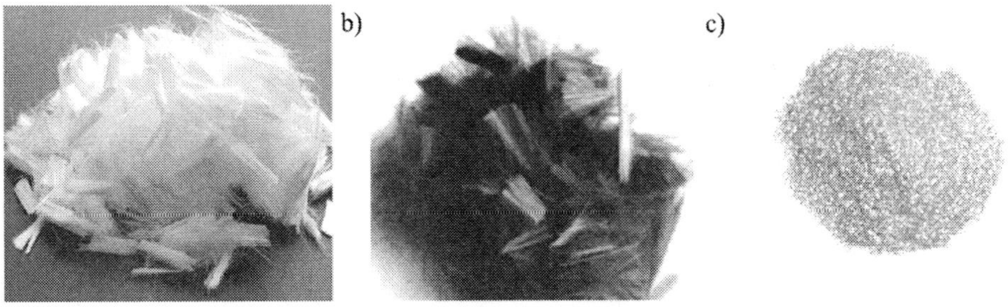

Fig. 1. Reinforcing components for protective binders: a) polypropylene fibers (PPF); b) basalt fibers (BF); c) basalt scales (BS).

The main indicators of evaluation of reinforcing components influence on the formation of binder structure were selected, physical and mechanical characteristics of the composite and the conditional coefficient of crack resistance, C_{cr}, the ratio of flexural R_f to compressive R_c strength. Assessment of binder density was carried out using photographs of composite microstructure, the technological properties of binder were evaluated with dissolution of mortar mixtures.

Before mixing the mortar, the liquid and dry parts of the binder were separately prepared. Optimal composition of the binder was $(0.8Na_2O+0.2K_2O)Al_2O_3{\cdot}4.5SiO_2{\cdot}12H_2O$ – 40%, and sand 0.315-0.63mm – 24.5%, 0-0.315mm – 21.2, fly ash 0-0.16mm – 14.3%. The liquid part included: the alkaline component and its chemical modifiers, to the dry one - the components of the binding, organic additives and fillers. The support components were injected to a dry portion in an amount of 1-3% by weight of the binding agent in the mixture, namely the total mass of liquid alkaline component and dry constituent binders. After preparing the liquid and dry parts of the binder, they were thoroughly mixed together using a HOBART mixer, and the dissolution of the mortar mixture was studied, and samples 40x40x160mm were formed. The hardening of the binder was carried out for 28 days at a temperature of 20 ± 2 °C: the first day in the molds, after which the decomposition and placement of samples in a wet environment were W = 95 ± 5%. After hardening of the binder samples, they were tested for compressive and flexural strength, their microstructure was investigated and the conditional coefficient of crack resistance was calculated.

3 Results and discussion

The results of tests on the influence of reinforcing components on the properties of binders are given in the form of diagrams in Figures 2-4.

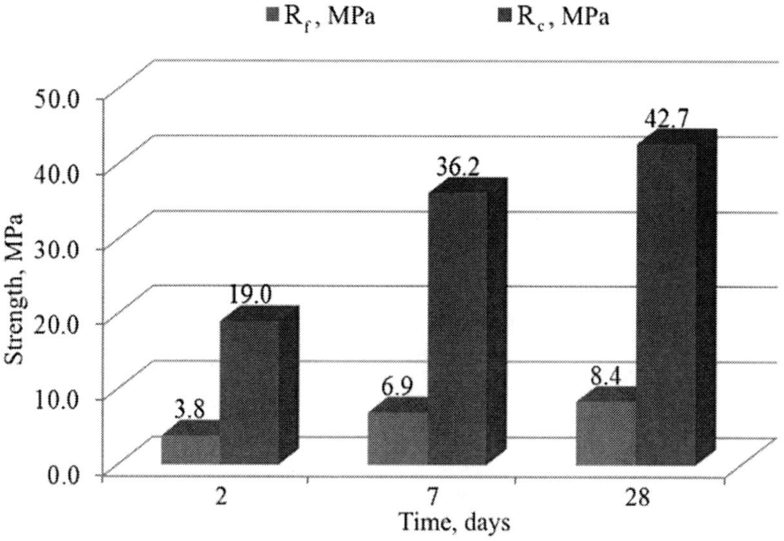

Fig. 2. The effect of adding 1% polypropylene fibers (PPF) to the binder on its physical and mechanical characteristics.

According to the data shown in Fig. 2 the introduction into the binder of 1% polypropylene fibers leads to an increase in flexural strength of 4-8% and a compressive strength of 1 ÷ 3%. However, the maintenance of PPF in the binder is accompanied by a sharp decrease in the flowability of the mortar mixture.

Analyzing the influence of basalt fibers on the physical and mechanical characteristics of the binder of rice. 3, it is worth pointing out that bending strength increases by 8-20% and compressive strength by 9-16%; the addition of 1% BF to the composition also is accompanied by a significant deterioration of the dissolution of the mortar mixture.

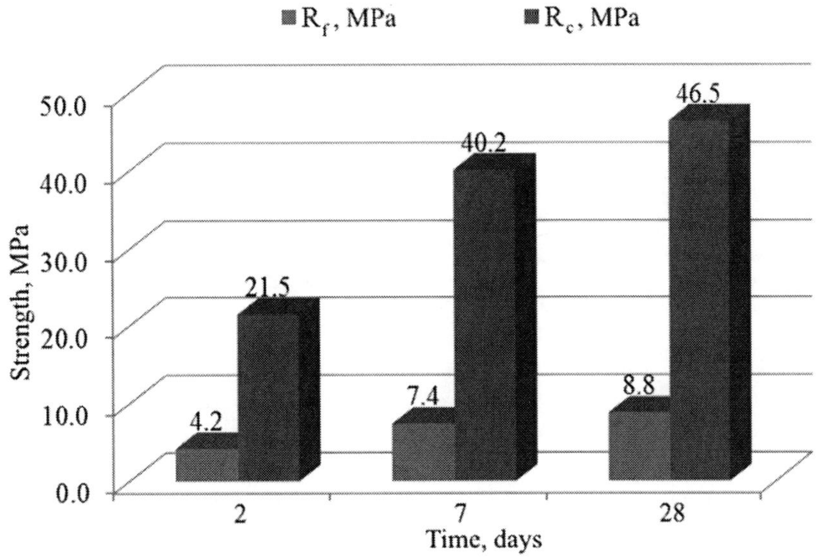

Fig. 3. The effect of adding 1% basalt fibers (BF) to the binder on its physical and mechanical characteristics.

Adding to be deposited basalt scales according to Figure 4 the main way affects the strength of its bending, increasing it to 14-25% and compressive strength increased only by 5-10%. Mortar mixture with contents of 1% BS is characterized by a sufficient index of deformity.

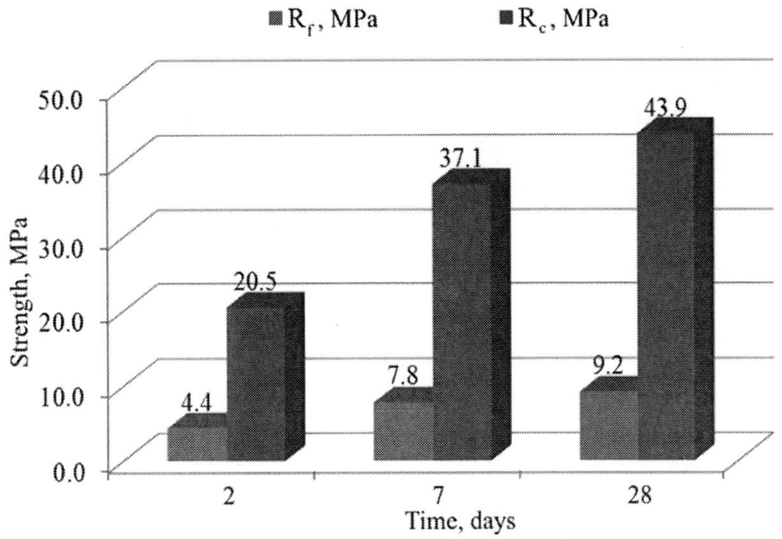

Fig. 4. The effect of adding 1% basalt scales (BS) to the binder on its physical and mechanical characteristics.

For the estimation of the effectiveness of reinforcing components on the crack resistance, calculations of the conditional coefficient of crack resistance (C_{cr}) binders were performed, the results are shown in Figure 5.

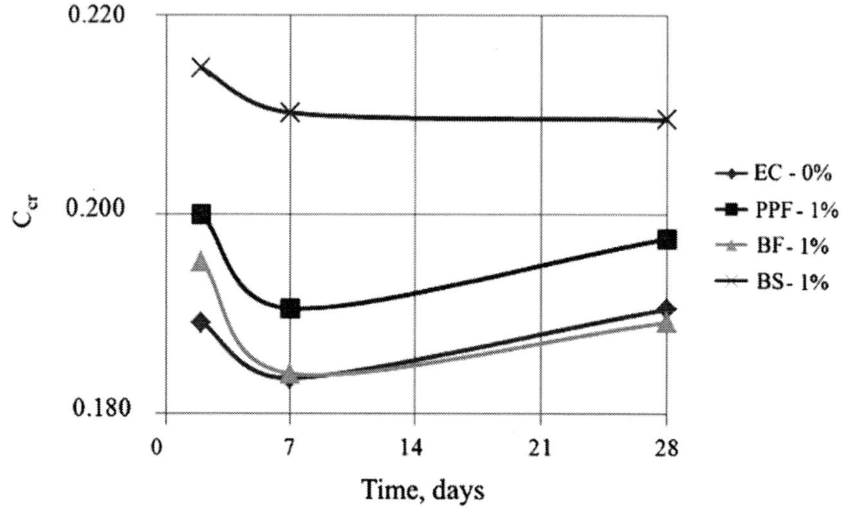

Fig. 5. The effect of adding to the binder of reinforcing components on its coefficient of crack resistance (Ccr).

According to Figure 5 it can be noted that the highest conditional coefficient of crack resistance is characterized by the binder with contents of its composition of basaltic scales, an increase in the coefficient in comparison with the binder without reinforcement of 10-15%. Polypropylene fibers also increase this coefficient by 3-6%; basalt fibers practically do not change the given index, which speaks about its rather low efficiency in relation to increase of cracking resistance of the binder.

For the estimation of the effectiveness of reinforcing components on the crack resistance, calculations of the conditional coefficient of crack resistance (C_{cr}) binders were performed, the results are shown in Figure 5.

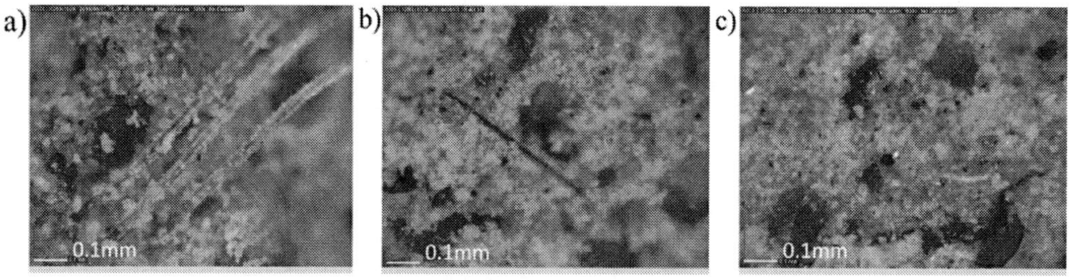

Fig. 6. The Photographs of the microstructure of the binder with 1% reinforcing components : a) PPF; b) BL; c) BS.

In view of the digital photographs of the microstructure of the binders in Figure 6, it can be noted that the polypropylene fibers in the binder acts as an inert structure, no contact was found on the contact zone between the fibers and alkaline aluminum silicate. The analysis of the contact zone between the aluminosilicate and the basalt reinforcing components shows that the contact is more dense and the hydrated aluminosilicate tumors are present on the surface of the fibers and scales. The fibers reinforcement occurs only in one direction, while the basalt scales reinforce the composite in two directions, which explains the higher indexes of the cracking resistance of the protective binder.

Fracture resistance of the binder increases with increasing its ability to withstand tensile stress, while reinforcing the matrix of the binder its strength with bending increases, but deterioration of the mobility of the mortar mixture. Therefore, after analyzing the results of studies on the effect of the type of reinforcing component on the formation of the structure of the binders, it can be noted that the best indicators of physical and mechanical tests and the crack resistance of the binder is characterized by composition with the contents of the basaltic scales. The further research was aimed at determining the effect of the amount of basaltic scales on the flowability of the mortar mixture and the conditional coefficient of crack resistance of the binder, the results of the tests are presented in the graph in Figure 7.

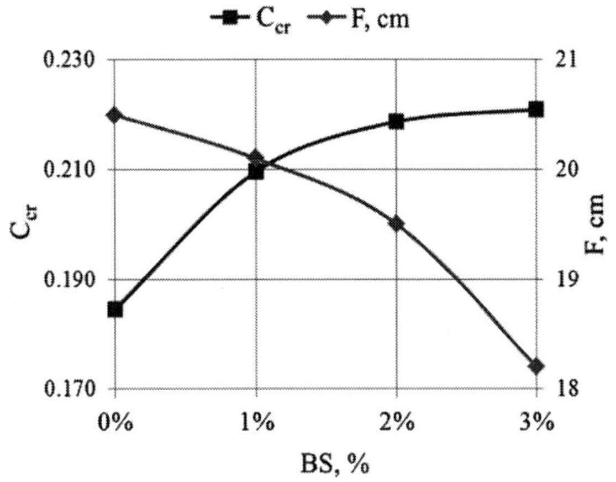

Fig. 7. The effect of the amount of basalt scales (BS) on the flowability (F) of the mortar mixture (F) and the conditional coefficient of cracking resistance of the binder (C_{cr}).

In fig. 7 shows a diagram of the results of the study of effect of the amount of basalt scales on the flowability of the mortar mixture and the crack resistance of the binder. In view of the diagram, it can be noted that a sharp increase in the coefficient of crack resistance occurs when the binder of 1÷2% of the basalt scales is introduced, further increase of its amount leads to decrease of flowability and increase the coefficient of crack resistance of the binder.

4 Conclusions

According to the study of influence of type and quality of reinforcing components on the technological and operational properties of alkaline aluminosilicate binders, it should be noted that:

- the addition to alkaline aluminosilicate binder of reinforcing components contributes to the increase of their strength characteristics and the conditional coefficient of crack resistance.
- the highest the conditional coefficient of crack resistance of the protective binder is characterized by the contents containing in its composition as a reinforcing component of the basalt scales.
- the analysis of protective binders microstructure indicates that the contact zone "aluminosilicate binder - basalt scales" is characterized by the highest density.
- the optimum amount for improving the cracking strength of the binder, ensuring the technological properties of the mortar mixture, is 1÷2% of the basalt scales from the weight of alkaline aluminosilicate binder.

- the further research will focus on the study of operational properties and effectiveness of use of alkaline aluminosilicate binder for the protection of building structures.

The authors express their gratitude to the Ministry of Education and Science of Ukraine for the financial support of work carried out within the framework of budget financing No 1DB-2017, as well as for the development of research topics in the scientific cooperation program COST Action CA15202 "Self Healing building structures: the path to sustainable construction".

References

1. Stark J., Wicht B. *Dauerhaftigkeit von Beton: Der Baustoff als Werkstoff* (Weimar, Baupraxis, 2001).
2. Page C. L., Page M. M. *Durability of Concrete and Cement Composites* (New York, Elsevier, 2007).
3. Mindess, S., Alexander, M., & Bentur, A. *Durability of concrete: Design and Construction* (Boca Raton, CRC Press, 2017).
4. Grantham, M., Grantham, M., & Grantham, M. *Concrete repair: A practical guide* (London, Routledge, 2011).
5. Bassi, R., Roy, S. K. *Handbook of binders for concrete* (Latheronwheel, Whittles, 2002).
6. Swamy, R. N., Tanikawa, S. *Surface binder for sustainable protection and rehabilitation of concrete structures* (Tokyo, Kohbunsha, 2012).
7. Guzii, S., Hela, R., Kirichok, V. Advanced Materials Research, **688**, 107-112, (2013).
8. Krivenko, P. V., Guziy, S. G., Kyrychok, V. I. Advanced Materials Research, **923**, 121-124, (2014).
9. Kyrychok, V., Drochytka, R., Kryvenko, P. Advanced Materials Research, **1122**, 111-114, (2015).
10. Kryvenko, P., Guzii, S., Kovalchuk, O., Kyrychok, V. Materials Science Forum, **865**, 95-106, (2016).
11. Kryvenko, P., Kyrychok, V., Guzii, S. EEJET, **5 (83)**, 40-48, (2016).
12. Schweitzer, P. A. *Paint and binders: Applications and corrosion resistance* (Boca Raton, CRC Press, 2006).
13. Chung, D. D. L. *Composite materials: Science and applications* (London, Springer, 2010).
14. Brigante, D. *New composite materials: Selection, design, and application* (London, Springer, 2014).
15. Omelchuk V., Runova,R., Rudenko I. Key Engineering Materials, **761**, 45-48, (2018).
16. Amziane, Sofiane., & Ferrier, E. *Composite Materials Repairs for Reinforced Concrete Structures* (Iste, Hermes Science Pub, 2016).

Resource-saving technologies for advanced concrete in the Republic of Bashkortostan

Alexander Krot[1,*], *Viktoriya* Ryazanova[2], *Azat* Gabitov[2], *Aleksandr* Salov[2], and *Lyubov* Rolnik[2]

[1]Kharkiv National University of Civil Engineering and Architecture, Sumska st. 40, 61002, Ukraine
[2]Ufa State Petroleum Technological University, Mendeleeva St. 195, Ufa, 450000, Russia

Abstract. Properties of both organo-mineral superplasticized unmodified and modified concretes are considered, impact of workability of concrete and water-binder ratio on strength properties of concrete are analyzed herein. Gonsistent aggregate grading with minimum void, accordingly with maximum packed density under optimum content of dispersed particles in concrete is to be applied to make mortars and concrete with stable properties and less variations in values. The proposed solution enables to simplify proportioning as different combination of aggregates is not required for each proportion. Comparative analysis of various fillers being by-products and materials for making mortars and asphalt-concrete has been made. The results for reinforced steel consumption reducing together with efficient combination of higher grade concrete were considered at amending the "reinforced-concrete structures" section. Computational algorithms made under economic and mathematical methods to improve construction solutions over the reinforced concrete frame are given. Efficient engineering solutions to use high strength concrete and rebar that were applied in some facilities are proposed.

1 Introduction

Concretes with high performance and strength properties and concrete mixes with modifying additives were widely applied in monolithic construction in the Republic of Bashkortostan. Optimization and efficient use of high performance concretes and high-strength concretes especially combined with rebars of new strength grade are currently very important. Tailored composition of the concrete mix, choice of the applicable technology for making, curing and quality maintenance of the concrete blocks and reinforced concrete structures are provided in all advanced concrete manufacturing steps in compliance with strength and operational characteristics [1, 2].

2 Materials and methods

* Corresponding author: my_motherland@ukr.net

Currently applied concrete mixes with ultimate compressive strength of 40MPa (with V30 strength grade) belong to P1 workability group (cone slump up to 1-5cm) and may not be applied in monolithic construction. Concrete mixes of P4-P5 workability group made with organo-mineral modifiers and current superplasticizers for 50-80 strength range (V40-V65 concrete grade) may be basically applied in monolithic construction technology, including using of concrete in thin-walled and densely-reinforced structures. Concrete mixes with advanced chemical additives of P1-P3 workability group enable making V80 concrete and higher (100MPa grade strength and higher) [3, 4].

Proceeding from monolithic construction practice, it may be noted that concretes with compressive strength up to 40MPa (90% of total volume) are largely used. Concreting technology development, improvement of engineering and technological solutions in using concrete and reinforced concrete, optimization in applying high performance concrete and structures made thereof directly depend on testing the Portland cement and its various kinds. So far, the results of certain and our own research in efficient application of modified concrete mixes (Fig.1) are analyzed, classified and summarized.

Current requirements to concrete are different. In certain circumstances, low grade concrete but with higher requirements to special features is required, e.g. BST V15 P4 F75 W6 GOST 7473-2010. To comply with watertightness requirements concrete with higher compressive strength than V15 concrete made by usual technology as the latter will be of lower watertightness grade than W6.

Making and application of modified concrete involve problems not only to attain but to maintain the required strength and estimated performance standards of cement concrete to be accompanied by further researching of structure formation and destruction within cement systems.

To eliminate any possible negative effects of internal stress in high water-reducing systems amorphous microsilica may be efficiently added which reaction when interacting with calcium hydroxide escaping under hydration is accompanied by minimum volume changing. Amorphous microsilica concrete is widely applied in large-scale housing construction nowadays. One of the known efficient ways to improve properties of cementitious composites thereby achieving compositional homogeneity thereof is adding of various mineral addditives to get microfilling effect on nanometric level either [5, 6]. Actually all properties of cementitious composites depend on water-binder ratio, with minor variations being made to water-binder ratio thereby result in significant changing of properties.

Research of impact of general water-binder ratio and filling material on different properties, including concrete strength is to be continued in further wide application of the universal water-binder ratio in concrete making practice.

Developed and proposed original structural and technological concept to apply either general water-binder ratio or equal volume of dispersed part for concrete in wide range of strengths enables:

- to make concrete and concrete mix of wide strength range with required stable and technological properties (workability keeping, watertightness, freeze-thaw durability, rate of strength development, etc.);
- to simplify proportioning of the composition by eliminating the need to specify ratio of the aggregates for each concrete grade and mortar grade by strength by using combination of aggregates with minimum void and optimum content of dispersed particles in concrete and mortars;
- to use high efficiency of plain high grade cement in combination with superplasticizers (increased strength related to particle dispergation effect, keeping of workability under application of retarders, prevention from strength retrogression by increasing aggregate fraction of less than 0.16mm and by adding fillers) in regular concrete and mortars.

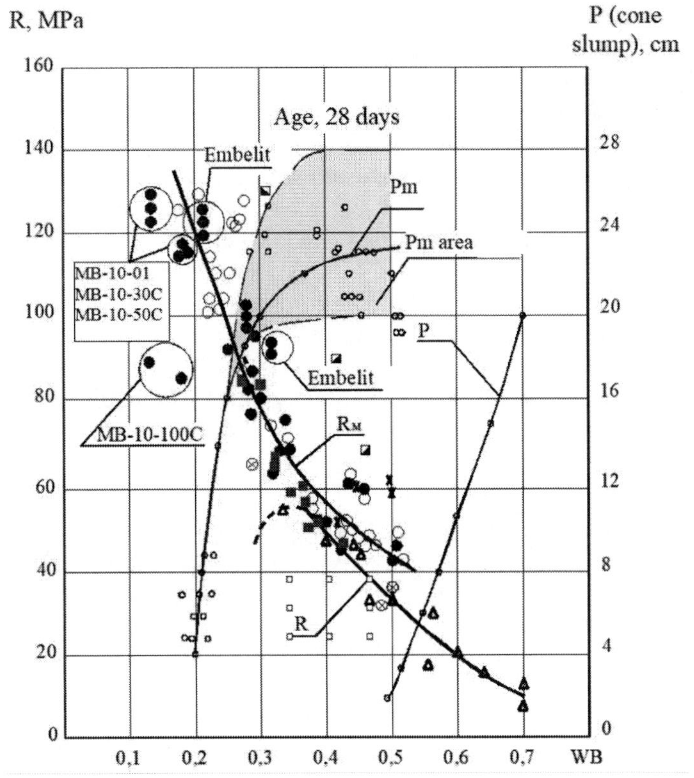

▲ - concrete (unmodified) with no additions;

○ ● ✗ ⊗ ◘ ■ -modified concrete;

○ - V.G.Batrakov, K.G.Sobolev, MB modifier data;

● - S.S.Kaprielov, A.V.Sheinfeld, concrete with MB and Embelit;

✗ - A.T.Kovalyov, I.A.Tsepilova, concrete with SP "Polyplast";

◪, ◩ - V.I.Kalashnikov, concrete with ground meal and MS;
incl. ◩ - reactive powder concrete;

⊗ - As per Biotech data by Lignopan B-1;

■ - Own data;

∘ - Workability of concrete mixes.

Pm – workability of modified concrete mixes, P - workability of unmodified concrete mixes, RM – water-binder dependence for modified concrete, R - water-binder dependence for unmodified concrete

Fig. 1. Dependences of compressive strength of concrete R on concrete mix workability P and water-binder ratio for both unmodified and modified concrete with superplasticizers and organo-mineral additives.

According to analysis of previous and our own results [7, 8] consistent aggregate grading with minimum void, accordingly with maximum packed density under optimum content of dispersed particles in concrete is to be applied to make mortars and concrete with stable properties and less variations in values. Water-binder ratio therewith is to be taken depending on the desired value of the required property, mortar or concrete by substituting part of cement in the cement system with either active or inert mineral dispersed additive (filler) by regulating workability with adding plasticizers.

The proposed solution enables to simplify proportioning as different combination of aggregates is not required for each proportion. Strict requirements are imposed on the filler that should be slightly changing the mortar workability and cementitious composite strength when substituting part of cement. Comparative analysis of various fillers being by-products and materials for making mortars and asphalt-concrete in the Republic of Bashkortostan has been made by us.

Main reasons resulting in instabilities of concrete characteristics, variability of aggregate grain-size distribution and aggregate voidage variation related thereto, variability of fractions less than 0.16mm in the aggregate and silty-clayed particles variation are to be excluded to make modified concrete with the required properties. Adding of mineral additives, such as active ones e.g. amorphous silica, as well as application of cement with the required content and phase relation β-C_2S, C_3S in combination with the appropriate chemical additive are very efficient to prevent from negative destruction processes and retention of the reliable properties in the required life cycle of concrete products and constructions maintenance. R&D and innovation efforts resulted in large-scale manufacture of concrete with superplasticizers [9, 10]. Nowadays in the Republic of Bashkortostan actually all companies making concrete of grade V15 and higher use superplasticizers. Information on some of our projects for making concrete is given below. A batch of prestressed reinforced concrete paving slabs PDNm-AtIV made of concrete with modifier Relamix T-2 with the reduction of cement consumption and time for curing was manufactured in OOO "Zhelezobetonnyi zavod No.1" (Sterlitamak, the Republic of Bashkortostan). A batch of pipe culverts is manufactured by State Unitary Enterprise "Bashkiravtodor" of concrete mixes with organo-mineral modifiers thereby resulting in better quality of the product surface with the reduction of cement consumption [11, 12]. A batch of ready-mix concrete BSG V30 P3 F300 W6 with high performance properties obtained due to application of multicomponent concrete modifier PFM-NLK was prepared by Concrete Batch Plant OOO "Ufimskaya Betonno-Rastvornaya Company" for making flight strip.

The engineered approach and quantitative evaluation for reducing reinforced steel consumption [13, 14] have been applied in engineering of a number of cast-in-place high-rise apartment houses in Ufa. A400 rebar and V25 concrete were substituted by A500S rebar and V40 concrete in one of the facilities, thereby reducing consumption of reinforced steel up to 49.7% (Table 1).

Table 1. Total saving of reinforced steel.

Original grade of concrete	Alternate grade of concrete	Saving of reinforced steel of grades A 400 (μ_0=0,02) /A 500C (μ_0=0,015), %	
		For alternate grade	In relation to previous alternate grade of concrete
V25	V35	10.5 / 9.3	- / -
V25	V40	13.5 / 11.9	2.7 / 2.4
V25	V50	17.9 / 15.8	3.8 / 3.46
V25	V60	20.5 / 18.2	2.3 / 2.1
V25	V70	22.4 / 19.8	1.5 / 1.4

Comparison between two design solutions being developed to construction design stage ("reinforced-concrete structures" section) has revealed that consumption of A500S main rebars is reduced to 25%, while total consumption of A500S rebars (both main and structural one) is reduced to 17% as compared with A400 (AIII) rebars.

The results for reinforced steel consumption reducing together with efficient combination of higher grade concrete were considered at amending the "reinforced-concrete structures" section of the construction design "Ten-storey cast-in-place residential house No.1 "Kaskad" (ICF walls) in Rossiyskaya street, Ufa" and when constructing 25-

storey cast-in-place residential houses Nos.9 & 10 in Ufa in Bakalinsky residential area of Ufa [15, 16].

Feasibility solution for optimization and efficient application of modified high strength concrete was justified and implemented under R&D results at engineering [17, 18], and making concrete at monolithic construction of 25-storey residential houses Nos.9 & 10 in Bakalinsky residential area of Kirovsky District of Ufa.

Consumption of concrete per one storey was 152m^3 for slabs, and 127m^3 per walls according to the original project. After application of the engineered solution the project costs amounted to 152m^3 for slabs, and 85m^3 for walls. Consumption of steel for reinforcing of concrete structures for one storey of residential houses Nos.9 & 10 were as follows: according to the original project with application of V25 concrete and rebars A400 (AIII) for slabs – 43.4t, for walls – 16.1t. Total: 59.5t. After application of the engineered solution for using V35 concrete and rebars A500S consumption of rebars reached: 32.4t for slabs and 12.2t for walls [19, 20, 21]. Total: 44.6t. Total saving of concrete amounted to 42m^3 for one storey and that of steel – 14.9t (25.1%) for one storey by main rebars. Total saving of materials and work costs exceeded RUR 25 mln.

According to amendment made to the "reinforced-concrete structures" section wall thickness of 17 storeys of residential houses Nos.9 & 10 in Bakalinsky residential area was reduced from 300mm (by original project) to 200mm by substituting concrete and rebar grades, weight of the buildings being reduced therewith by several thousands of tons.

3 Results

The results of the feasibility study for efficient application of high-strength modified concrete [22, 23, 24] and of high-strength rebars were adopted to the construction of 25-storey cast-in-place residential houses Nos.9 & 10 in Bakalinsky residential area of Ufa. Actual discrepancy of data between theoretical technical and economical solution and actual expenses for materials at construction amounted to: about 4.5% for concrete and about 6% for rebars.

In accordance with the engineered feasibility solution [25] original AIII rebar grade is substituted to A500S high-strength rebar at efficient combination with the modified high strength concrete. Developed and registered PC program "Calculation of efficient consumption of reinforced steel with respect to cost efficiency for alternate cross-section of the flexible element" [26] was used.

4 Conclusions

The proposed analytical tools engineered and widely applied in the Republic of Bashkortostan enable to define practical areas for applying efficient grades of concrete and reinforced steel in concrete elements accompanied by feasibility study at the design stage and to make quantitative evaluation of efficiency thereof, which is more important when choosing modified concrete and advanced rebars for making cast-in-place concrete facilities.

References

1. V. Braun, *Raskhod armatury v zhelezobetonnykh konstruktsiyakh* (Stroyizdat, Moscow, 1993)

2. A.S. Salov, L.I. Khabibullina, A.I. Gabitov, E.A. Udalova, V.A. Timofeev, A.A. Timofeev, Istoriya nauki i tekhniki, **11**, 37-43 (2017)
3. A.I. Bedov, V.V. Babkov, A.I. Gabitov, A.S. Salov, Vestnik MGSU, **8**, 76-84 (2012)
4. V.I. Vinnichenko, A.N. Ryazanov, MATEC Web of Conferences **116**, 01020 (2017).
5. V.I. Vinnichenko, A.N. Ryazanov, N.Y. Vitsenko, Construction materials, **10**, 76-83 (2016)
6. A.I. Bedov, A.M. Gaysin, A.I. Gabitov, Textile industry technology, **3**(369), 231-236 (2017)
7. A.I. Bedov, A.S. Salov, V.V. Babkov, A.I. Gabitov, I.V. Nedoseko, Current problems for calculation of reinforced concrete constructions, buildings and structures for emergency impact, 503-507 (2016)
8. A.N. Ryazanov, V.I. Vinnichenko, I.V. Nedoseko, V.A. Ryazanova, A.A. Ryazanov, Construction materials **1-2**, 18-22 (2018)
9. S.H. Aminov, I.B. Strugovec, I.V. Nedoseko, V.P. Klimov, V.V. Babkov, Culvert pipes for asphalt concrete roads, **10**, 2 (2003)
10. A.M. Gaisin, V.V. Babkov, Construction materials, **1**, 55-58 (2016)
11. A.G.Tamrazyan, Concrete and reinforced concrete: problems and prospects, **7**, 51-54 (2014)
12. D.N. Sursanov, A.B. Ponomaryov, Bulletin of National Research Perm Polytechnic University Construction and Architecture, **4**, 144-165 (2015)
13. A.I. Bedov, A.I. Gabitov, A.S. Salov, A.M. Gaisin, L.I. Khabibullina, Building and reconstruction, **6(74)**, 3-12 (2017)
14. A.A. Semenov, A.I. Gabitov, Static analysis, **1**, 152 (2005)
15. A.I. Bedov, A.S. Salov, A.I. Gabitov, D.V. Kuznetsov, E.A. Sadykova, International Journal for Computational Civil and Structural Engineering, **13 No. 4**, 37-46 (2017)
16. A.M. Gaisin, S.Y. Samokhodova, I.V. Nedoseko, A.Y. Paymetkina, Residential construction, **5**, 36-40 (2016)
17. V.G. Batrakov, Concrete modifiers: new opportunities and prospects, **10**, 4-7 (2006)
18. S.S. Kaprielov, V.I. Travush, N.I. Karpenko, A.V. Sheynfel'd. Modified concretes of a new generation in the Moscow-City construction sites. Part I, **10**, 13-17 (2006)
19. S.I. Kryshov, I.S. Kurilyuk, Housing construction, **7**, 3-5 (2016)
20. A.M. Gaisin, R.R. Gareev, V.V. Babkov, I.V. Nedoseko, S.Y. Samokhodova, Construction materials, **4**, 82-86 (2015)
21. A.N. Ryazanov, V.I. Vinnichenko, Bulletin of Voronezh State Technological University, **9 No 6-2**, 22-24 (2013)
22. V.I. Kalashnikov, Calculation of high-strength self-compacting concrete compositions, **10**, 4-6 (2008)
23. V.I. Vinnichenko, A.N. Ryazanov. Ecological indices of manufacture of Portland cement clinker and production of the dolomite clinker C. 01020 (2017)
24. A.N. Ryazanov, V.I. Vinnichenko, News of Higher Educational Institutions. North Caucasian Region. Series: Technical Sciences, **1(176)**, 115-117 (2014)
25. D. Balageas, C.P. Fritzen, A. Guemes, *Structural Health Monitoring*, 496 (2006)
26. G.Clark. Challenges for concrete in tall, **7**, 103-112 (2014)

Investigation of the structure of cement stone, obtaining and optimization of high-strength concrete on mechanically activated binder

Liubov Ksonshkevych[1], *Olena* Krantovska[1], *Mykola* Petrov[1], *Sergii* Synii[2,*], and *Anna* Uhl[3]

[1] Odesa State Academy of Civil Engineering and Architecture, Didrihsona st. 4, 65029 Odesa, Ukraine
[2]Lutsk National Technical University, Lvivska st. 75, 43018 Lutsk, Ukraine
[3] Lesya Ukrainka Eastern European National University, Volya Avenue 13, 43025 Lutsk, Ukraine

Annotation. Physico-chemical methods of the study of hydrated cement by mechanically activated binding substance with organo-mineral additive (microsilica + C-3) were presented. It was explored the influence of the additive into the kinetics of formation lowly-basic calcium hydrosilicate. The experiment conducted in accordance with a three-factorial D-optimum plan enabled to establish an influence of blending and technological factors (x_1 – content of microsilica in the binder - 5±5%; x_2 – binder consumption in concrete - 450±100 kg/m^3; x_3 – specific area (S_{sp}) of Portland cement - 400±100 m^2/kg) on physical and mechanical properties of concrete (compression strength, water absorption). It was established that the use of the mechanically activated blended Portland cement of general purpose with an organomineral additive (microsilica+C-3) makes it possible to obtain high-strength concretes which compression strength on the 28th day reaches 120 MPa. Optimization of concrete mixes was achieved on the basis of obtained mathematical models that reflect the influence of blending and technological factors on physical and mechanical properties of a high-strength concrete.

1 Introduction

The amplification of the quality of concrete, more intensive growth of strength, can be achieved through the purposeful change of the structure of the cement stone, as by activating the cement grains[1-3], fillers [4] and microsilica under conditions of intense hydrodynamic influence on them, and due to the modification of their surface area by surfactants [5, 6].

Works by Y.M. Bozhenov [7], R. F. Runova [8] et al. found that the use of active mineral additives provides a real opportunity to obtain concrete with preset physico-mechanical characteristics, with significant savings Portland cement.

*Corresponding author: sergii.synii@gmail.com

2 The purpose of the research

The purpose of the researches was performing X-ray phase analysis and a study of microstructure of hydrated cement, the bonding substance of which was exposed to mechanical activation in specially made triboactivation, as well as increasing the efficiency of the use of ordinary Portland cement in high-strength concrete due to its mechanical activation and modification by organo-mineral additive.

3 Research methods

The research was conducted using of complex of modern methods of physico-chemical analysis with a goal of studying the processes of formation of a phase composition and a microstructure of mechanically activated hydrated cement with the additive of 10 % microsilica [9, 10].

X-ray phase research was conducted by the powder method using the diffractometer DRONE-2.0 with SiKα radiation. Experimental samples are pounded in the agathic mortar until complete passing through the sieve No. 008. The prepared sample was moved on the quartz cuvette with proportional layer, previously greased with Vaseline. The cuvette was installed on the goniometric equipment GUR-5. The diffractograms were recorded in the interval of angles $2\theta = 8 - 50^{\circ}$ by means of detector X-radiation, namely, the scintillation counter with a speed of counting 500 moment/s.

The research of microstructure of prepared samples and photographing the chips of hydrated cement, hydrated by different conditions, were conducted on bit-mapped electron microscope TESLA BS-300 providing possibility of a direct investigation of the surface of solid objects. The bit-mapped electron microscope operates with accelerating tensions to 30 kW. At the same time, the permission 20 nm is achieved. The electron-optical increase is 5-50000 times.

4 Result and discussion

The analysis of diffractograms of hydrated cement in 28 days of maturing is evidence of formation the main characteristic phases of hydrates. During the process of mechanical activation, the processes of hydration are considerably intensified that is confirmed by lowering of density of lines of non-hydrated cement (d/n = 0.26; 0.260; 0.217 nm), and also by increasing of the density of lines of calcium hydroxide (d/n = 0.26; 0.49 nm).

Adding microsilica is accompanied by some regular lowering of density of the lines of the main clinker minerals and by lowering of density of the lines $CaOH_2$ (d/n=0.49; 0.261 nm). It is explained by active pozzolanic reaction between Portland cement and microsilica (Fig. 1). The effectiveness of using the mechanical activation in the complex with additive of microsilica is also proved by the microprobe analysis.

There is a part of compactly packed hexagonal aprons which are related to calcium hydroxide and which are observed on photomicrographies (Fig. 2a).

The photomicrography also shows a grain of microsilica which has not come reacted with $Ca(OH)_2$ (Fig. 2a, b).

At the same time, a major part of added microsilica reacts with $Ca(OH)_2$ producing a considerable quantity of calcium hydrosilicate crystals.

Lowly fundamental C-S-H is crystallized in the shape of needles, contributing to the formation of solid connections between the products of hydration of cement, increasing the density and mechanical durability of stone (Fig. 3b).

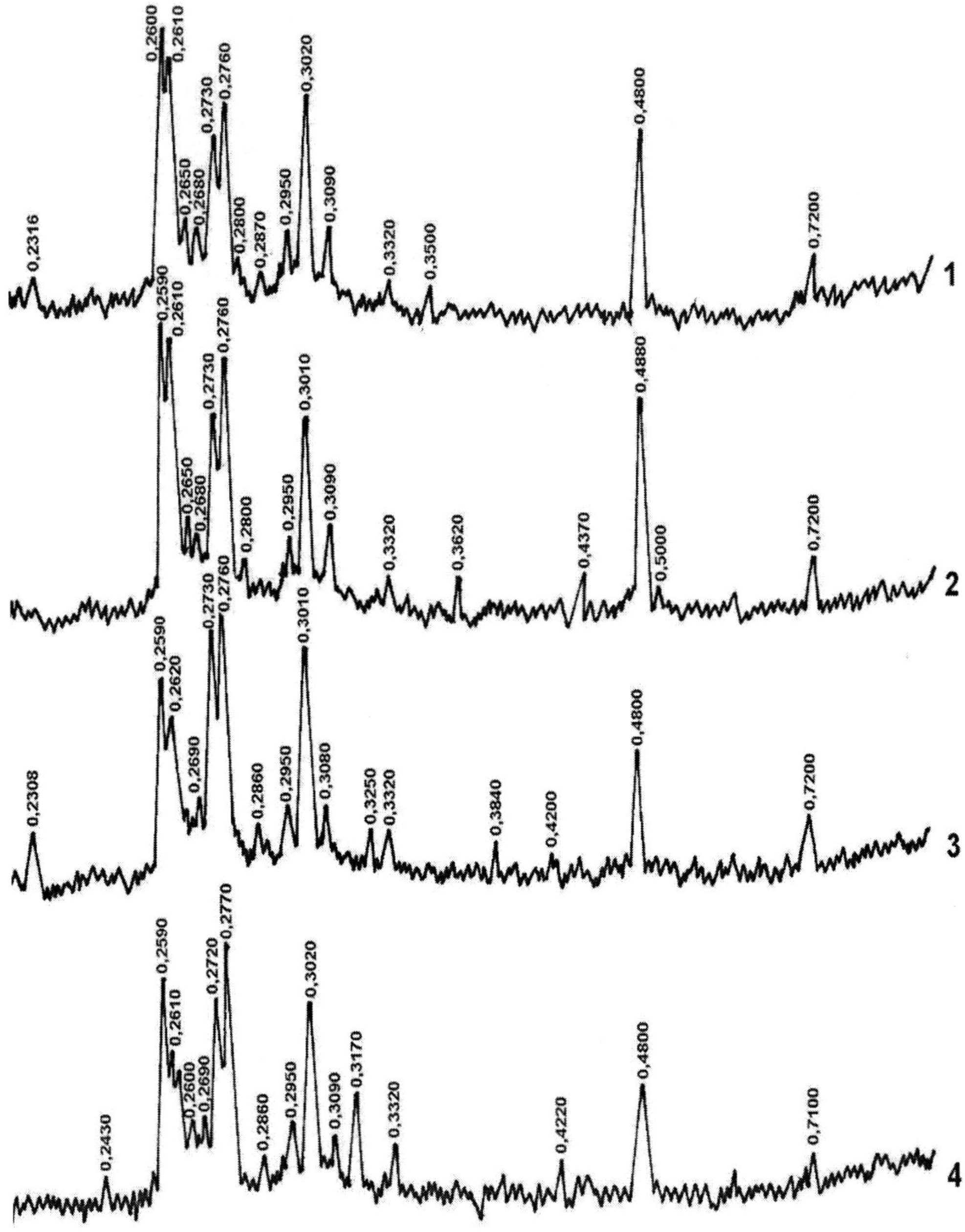

Fig. 1. The diffractograms of Portland cement stone on basis of: 1 - Portland cement without additives; 2 - mechanically activated Portland cement without additives; 3 - Portland cement with 10 % weight of microsilica; 4 - mechanically activated Portland cement with 10 % weight of microsilica.

It was found that adding of organo-mineral additive to Portland cement with simultaneous mechanical activation of bonding substance allows regulating the processes of structure formation. The analysis of the diffractograms of hydrated cement shows that during the process of a mechanical activation the processes of hydration are considerably intensified, and adding 10 % microsilica results in active pozzolanic reaction between Portland cement and microsilica.

a)

b)

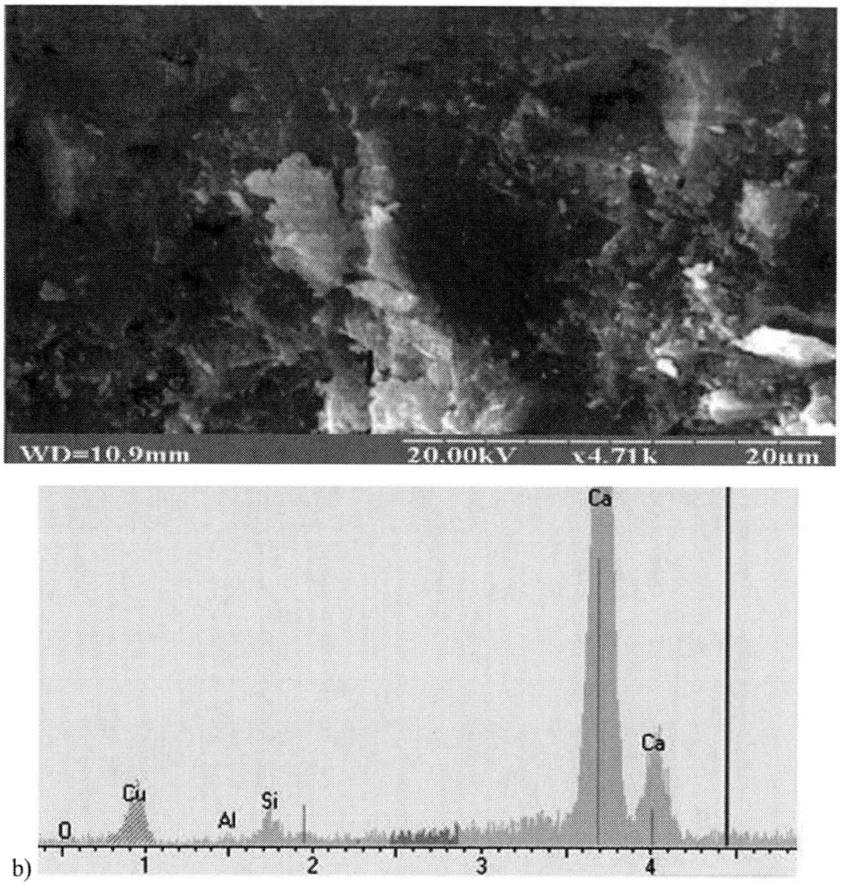

Fig. 2. The microstructure (a) and the spectrum of characteristic X-radiation (b) from a surface of hydrated cement on basis of mechanically activated Portland cement with additive 10 % weight of microsilica.

The studies have shown that the mechanical activation of Portland cement with additive of microsilica changes a qualitative and quantitative composition of new formations that leads to change of conditions of the organization of structure of hydrated cement. For its turn, the change of a character of structure formation excites the change of attributes of solidifying compositions in comparison with cement compositions, received by usual technology [1-3].

It was found that the adding microsilica to Portland cement leads to increasing of durability of hydrated cement in comparison with a control from 26% to 40% (according to terms of maturing), Table 1.

The mechanical activation accelerates a role of microsilica in binding substance. In 28 days' age the durability of hydrated cement by mechanically activated bonding substance with additive of 10% of microsilica is higher in 1.4 in comparison with a control [11]. So that, it may be affirmed that directional structural modification leads to increasing of durability of consolidated cement compositions.

The paper examines the problems associated with production of high-strength concretes by using a general purpose blended Portland cement which is to be mechanically activated and modified with an organomineral additive, and with optimization of concrete mixes taking into account cost-effective use of resources. The research and application of high-strength concrete is an urgent task, especially in road construction [4, 12-13].

a)

b)

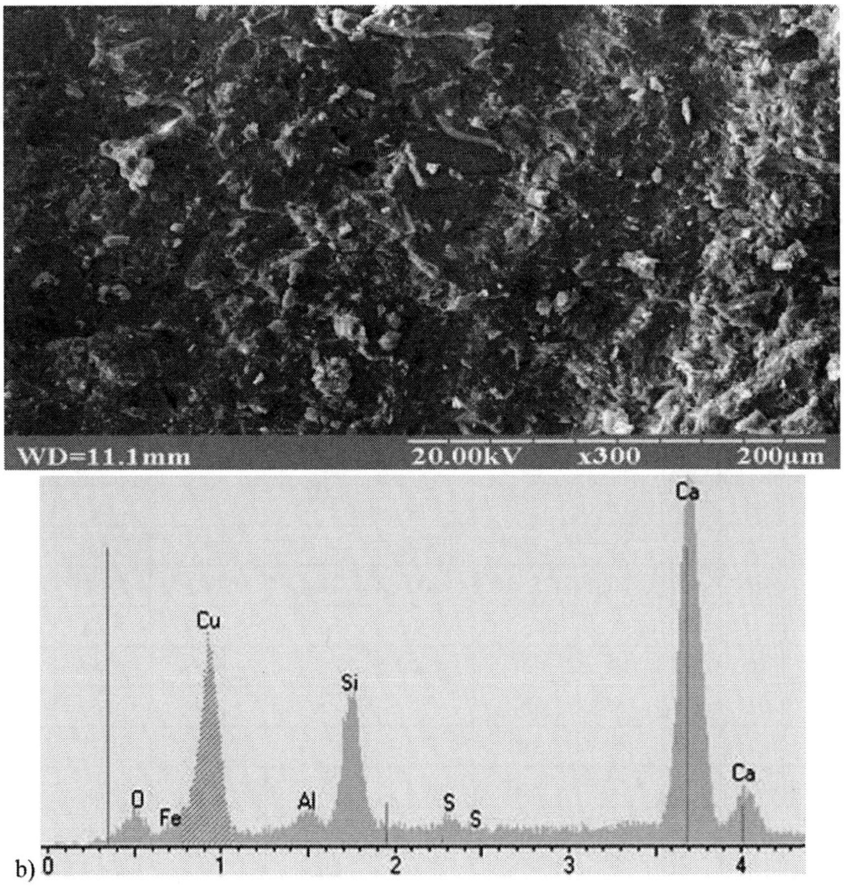

Fig. 3. The microstructure (a) and the spectrum of characteristic X-radiation (b) from the surface of hydrated cement on basis of mechanically activated Portland cement with additive 10 % weight of microsilica.

Table 1. The influence of additive MS on the durability of hydrated cement.

№	The composition of bonding substance, %,		The durability of hydrated cement MPa, at the age					
	Portland-cement (PC)	Micro-silica (MS)	3 days		7 days		28 days	
			$f_{of\ curve}$	$f_{of\ pressure}$	$f_{of\ curve}$	$f_{of\ pressure}$	$f_{of\ curve}$	$f_{of\ pressure}$
1	100	0	5.08 / 4.1	17.1 / 12.2	6.7 / 5.3	25.2 / 18.4	10.9 / 8.1	50.8 / 36.8
2	95	5	6.1 / 4.7	21.2 / 16.6	8.0 / 6.0	28.8 / 21.5	11.6 / 8.51	64.4 / 46.8
3	90	10	7.02 / 5.4	23.4 / 18.2	10.8 / 7.8	34 / 26	14 / 10.1	85.7 / 61.3

Notice: there are significances of durability of hydrated cement by mechanically activated bonding substance above the line; also there is a significance of durability of control samples under the line.

Plan of the experiment and concrete compressive strength value (in the 28 days' age) are summarized in Table 2.

The experiment conducted in accordance with a three-factorial D-optimum plan enabled to establish an influence of blending and technological factors (x_1 – content of microsilica in the binder - 5±5%; x_2 – binder consumption in concrete - 450±100 kg/m^3; x_3 – specific area (S_{sp}) of Portland cement - 400±100 m^2/kg) on physical and mechanical properties of concrete (compression strength, water absorption).

Table 2. Plan of the experiment and mechanical properties of concrete.

№	Variation levels			Micro-silica (MS), %	Consumption of binder, kg/м^3	Specific surface area , м^2/kg	Responses			
	x_1	x_2	x_3				$f^{к}_{cube}$, MPa	$f^{к}_{cube,w}$, MPa	f^{a}_{cube}, MPa	$f^{a}_{cube,w}$, MPa
1	-	-	-	0	350	300	43.6	41.8	59.7	56.4
2	+	-	-	10	350	300	52.3	43.3	73.2	60.6
3	-	+	-	0	550	300	53.9	49.5	75.5	64.4
4	+	+	-	10	550	300	69.5	61.4	97.3	85.9
5	-	-	+	0	350	500	52.7	44.8	68.5	58.2
6	+	-	+	10	350	500	64.3	51.3	90	71.8
7	-	+	+	0	550	500	72.8	69.3	98.3	90.1
8	+	+	+	10	550	500	88.4	71.3	123.8	99.8
9	-	0	0	0	450	400	64.9	55.9	84.4	72.7
10	+	0	0	10	450	400	81.4	67.5	105.8	94.5
1	2	3	4	5	6	7	8	9	10	11
11	0	-	0	5	350	400	61.6	57.6	83.2	77.8
12	0	+	0	5	550	400	75.2	66.1	101	89.2
13	0	0	-	5	450	300	53.9	50.4	72.8	68.0
14	0	0	+	5	450	500	58.1	50.5	78.4	68.2
15	0	0	0	5	450	400	73.4	63.5	99.1	89.8

The choice of several variants of optimal compositions of high-strength concrete was carried out graphically by diagrams constructed according to the corresponding ES-models.

Isolines were placed on the cube displaying the levels of basic physico-mechanical properties of mechanically activated concrete at the 28 days' age, and areas not meeting the limitation criteria were excluded (Fig. 4).

As a common limitation criterion for all three alternatives, the concrete softening coefficient not less than 0.85 ($K_s \geq 0.85$) was adopted. The compression strength of concrete - 80, 90 and 100 MPa was assumed as the second limitation criterion.

Further, optimal compositions were chosen taking into account the required criteria - water absorption and the specific surface area of the binding agent.

A is the consumption of binder 550 kg/m^3, relatively low specific surface area binder (about 320 m^2/kg) and using 10 % microsilica. This composition can be recommended for the basis energy saving to the grinding of the binding agent.

B is the consumption of binder of 550 kg/m^3, the average specific surface area of the binding material (about 400 m^2/kg) and without the microsilica. This composition can be recommended for a reason the absence of the need for an additional technological operation - the introduction of microsilica.

In addition, the composition of B shows less water absorption compared to composition A, which suggests its greater potentially durability.

C is the consumption of binder of 450 kg/m^3, a specific binding surface area of about 380 m^2/kg and in the absence of microsilica. This composition can be recommended for reducing the consumption of binder.

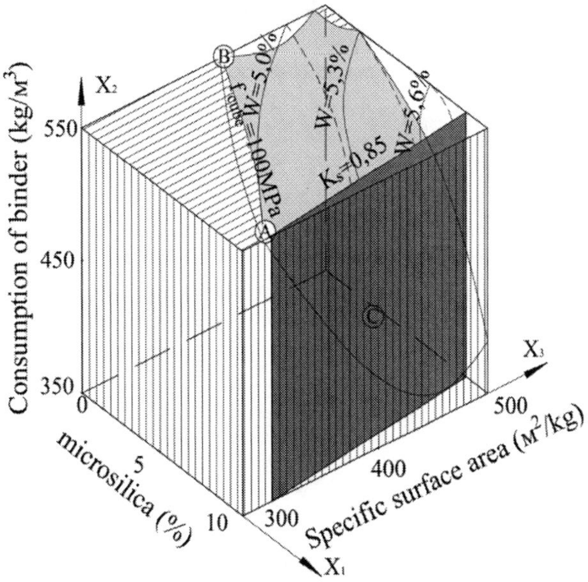

Fig. 4. Selection of optimal composition of concrete in a 3-dimensional diagram.

Solution of the optimization problem enabled to select the optimum compositions of high-strength concretes that can be used in dependence from needed levels for a particular task with due account of the economic efficiency and current prices of energy, equipment and raw materials.

5 Conclusions

The inclusion of microsilica in a composition of binding substance contributes to the formation of strong connections between the products of hydration of cement that is confirmed by the facts on a formation of lowly-basic calcium hydrosilicates.

The mechanical activation of grains of cement and microsilica assures the increase of mechanical characteristic of cement stone in diapason from 26 to 40 %.

It was established experimentally that mechanical activation of blended Portland cement of a general purpose with adding microsilica and C-3 superplasticizer allows to obtain heavy concretes of C35/45...C90/105 classes with consumption of binder from 450 to 550 kg/m^3. The results of the research made it possible to optimize compositions of C70/85...C90/105 high-strength concrete.

References

1. I.V. Barabash, *Mechanochemical Activation of Mineral Weavers*, Training Manual, (Odessa, Astroprint, 2002)
2. S.I. Fedorkin. *Mechanoactivation of secondary raw materials in the production of building materials* (Simferopol, Tavria, 1997)
3. L. Ksonshkevych. *High-strength concrete on the mechanically activated binder*: dis. Cand. tech Sciences: 05.23.05. (Odessa, 2013)

4. O. Andriichuk, V. Babich, I. Yasyuk, S. Uzhehov, MATEC Web of Conferences, **116,** 02001 (2017)
5. V. Ramakrishnan, W. Coyle, S. Pande, Indian Concrete J., **54:1**, 23-26 (1980)
6. *Constitutive modeling o / high strength/ high performance concrete*, FIB Bulletin 42, Sprint-Digital-Druck (Stuttgart, 2008)
7. Y.M. Bazhenov, *Concrete technology* (Moscow, Publishing house ABC, 2003)
8. R.F. Runova, I.I . Rudenko, V.V. Troyan, V.V. Tovstonis, S.P. Shcherbyna, L.D. Pashina, Building materials, products and sanitary equipment, **29**, 91-97 (2008)
9. S.S. Gorelik, L.N. Rastorguev, J.A. Skakov. *Radiographical and electro-optical analysis*. Tutorials for students. – 3 rd ed. comp. alt. (M., MISSIS, 1994)
10. V.M. Melnyk, A.V. Shostak, *Quantitative stereomicrofractography: Monograph* (Lutsk, PVD Tverdinya, 2010)
11. L. Ksonshkevych, I. Barabash, Herald OSACI, Odessa, **40**, 161-167 (2010)
12. O.V. Andriichuk, E.M. Babych, Materials Science, New York, **52:4**, 509-513 (2017)
13. O. Mikulich, V. Shvabyuk, I. Pasternak, O. Andriichuk, Mechanics Research Communications, **91**, 107-111 (2018)

The influence of the content on structure and properties of geopolymer composites on silicate matrix

Yevgen Lutskin[1,*], *Olena* Shynkevych[1], *Igor* Myronenko[2], *Stanislav* Zakabluk[3], and *Oleksiy* Surkov[1]

[1]Odessa State Academy of Civil Engineering and Architecture, 4 Didrikhson St., 65029, Odessa, Ukraine
[2]Odessa National Maritime University, 34 Mechnikov St., 65029, Odessa, Ukraine
[3]TMM Company, 49a Volodymyrska St., 01001, Kyiv, Ukraine

Abstract. The modern nanotechnology techniques allow receiving advanced composites of silicate matrix of thermal-moisture hardening the low-power technology today. A distinctive feature of geopolymer aerated composites is that their properties according to the basic physical and mechanical parameters exceed of the properties of the silicate matrix. Due to the implementation of a complex activation of highly mobile silicate concrete mixture, that is one of the technological features to obtain this type of composites, energy-saving modes of preparation were provided. The composites and products based on them are characterized by low density at high values of strength, water and crack resistance and heat capacity. The technological features of obtaining effective aerated composites on silica matrix of thermal-moisture hardening set out. The comparative analysis of the influence of the tripoli specific surface to the structuring of silicate matrix properties and porous composites modified of alkali containing additives were carried out. Based on this analysis the mechanism of formation of silicate matrix structure and properties was grounded and proposed. It is shown that the tripoli particles promoted "physical" seal structure of silicate matrix and the formation of capillaries' discontinuous structure, by means of their own micro-porosity.

1 Introduction

Traditionally the silicate materials are produced by energy-intensive autoclave technology. The basis of this technology is Michaelis W. development by patent № 14195 05.10.1880.

The modern nanotechnology processes allow today to receive progressive composites on the silicate matrix of thermal-moisture hardening according to the low-power technology. Aerated complex activated composites on the silicate matrix are made by casting technology. A distinctive feature of aerated composites is that their properties

* Corresponding author: lutskin@ukr.net

according to the basic physical and mechanical parameters exceed the properties of the silicate matrix.

At the autoclaves conditions there is activation of crystalline quartz, associated with increased silica solubility. The tendency of minimizing the dissolved lime component in the binder is observed simultaneously. These processes are determined by the peculiarities of chemical thermodynamics of lime and quartz. With higher temperature the solubility of lime decreases, although the rate of dissolution increases and the solubility of crystalline quartz increase [1]. In this case, one of the tasks was to minimize the lime in the composition to reduce the destructive processes. The consequence of these processes was the need to minimize the lime content to its complete binding hydrous calcium to reduce the destructive phenomena.

A large number of research works were devoted to the creation of the desired phase composition of hydrosilicates calcium (GSK). These research works were carried out mainly towards the synthesis of tobermorite, which is formed by minimizing the lime content up to its full bonding. However, the presence of some types of highly basic GSK type hillebrandite, foshagite in the material determines its high frost, air and carbonation resistance under operating conditions [1].

The critical problem of the modern material science is the development and implementation of alternative energy-saving composite binders on nanotechnology and description of the mechanisms of hydration processes at different levels, including nanolevel. The first attempts, to link the mechanical properties of materials with their crystal structure, were made by Griffiths [2].

Many results of works in recent years have shown that for the activated lime-silica composites the presence or absence of any mineral is not exclusive prerequisite for certain physical and mechanical properties [3]. More important factor may be the sizes of the particles of raw mix neoplastic morphology and character of the connection between them. The reactivity mixture, changing of temperature of polymorphic transformations, solubility and kinetics of the displacement in the chemical balance in one direction or another are also connected with the particle size (Le Chatelier). It was found that the properties of ultrafine particles differ in particulate chemical content properties.

The modern nanotechnology receptions allow realizing the effects of the impact of the ultrafine fraction particles. The complex multi-step activation of multicomponent mixtures is used in the research work.

Sufficiently extensive researches of the activation of dispersed particles show the positive effects to the properties of various types of defects, dislocations, etc. It is thermodynamically unstable contacts which can have high strength.

2 Nanotechnology techniques of producing of the complex activated composites on silicate matrix

Developed by the authors the complex activation includes the continuous loop of the different types and methods of activation: mechanical, chemical and thermal [4]. Each type of activation is accompanied by effects which create the conditions for the possibility of subsequent type of activation.

The result of the complex activation is the formation of linear defects, dislocations and point positions and substitutions. Furthermore, there may be angle change between the bonds and the appearance of dangling bonds, which leads to the formation of free radicals in the crystals with covalent bonds and the amorphization for molecular crystals.

Each type of activation will be caused by the prevalence of any type of deformations of the solid phase structure. The differences will be responsible for the nature and kind of dislocations with allowance for the extent and duration of external and internal influences.

Mechanical activation. Mechanical effects in the dispersion medium speed mixer activator provide mechanochemical activation of crystalline quartz surface. The term "mechanochemical reaction" was introduced by W.Ostwald in 1891.

Mechanochemical activation is carried out to the special properties of the newly formed surfaces, especially changing the local chemical and phase content of solids, as well as their aggregate state of under the influence of mechanical effects of high intensity [5].

As nanotechnology reception the mechanochemical activation reduces the viscosity of 3 or more dispersed system containing lime [4, 6]. This effect of viscosity reducing was used to compensate the increased water demand of the mixture introducing the composite porous opal-cristobalite rocks and using the activation of the binder together with the fine aggregate.

Thermal activation. Using as a binder quick lime promotes the "inner activation" of dispersed system under an elevated temperature T= 40÷60°C, the formation of multiple point contacts in the dislocation fields are appeared and the conditions for the hydration hardening unrelated in hydrous calcium lime are created.

External thermal activation occurs in conditions of thermal-moisture treatment. In such conditions at T=85°C contradiction is canceled, which is connected with the increasing the quartz solubility and decreasing lime solubility with increasing its dissolution. In addition, the increase of pH system causes to create favorable conditions for longevity growths GSK during the operational phase when $11.5 \leq pH \leq 12.5$.

Chemical activation. According to E. Avvakumov's works [7] with the presence of water in the inorganic solid-phase system "method of soft mechanochemical synthesis" is happened. To obtain composite oxides from the simple oxide hydroxides are used as the starting components one of which is characterized by acidic, the other one is characterized by basic properties. By means of the neutralization reaction the intensification of the processes takes place. In addition, in such system the conditions for the hydrothermal processes are created.

The acid activation occurs by introducing the amorphous-crystalline silica, alkaline activation – the introduction of increased amounts of lime; it raises pH. The replacement of ground quartz sand in the binder by opal-cristobalite rocks enhances to increase frost and water resistance of the silicate matrix while reducing its density matrix to 20-25%.

Moreover, the presence of opal-cristobalite rock in the dispersion causes formation of nanoscale in the GSK pores of these rocks; in this case they are "nanoreactor"; which walls restrict the growth of neoplasm. Also, the presence of particles of porous rocks with different dispersion binder allows adjusting the speed and reaction kinetics of hydration [8].

The use of quick lime determines the possibility of using high modulus liquid glass as one of the alkali addition, so the temperature rise on the forming stage allows regulating rapid setting of such mixture.

Alkali and alkali-containing additives increase the thermodynamic instability of systems by shifting the equilibrium caused by the formation of additional defects on the surface of silica-containing components [9, 10, 11]. Alkali-containing additives are capable to aeration of mixtures under the certain conditions [12, 13, 14].

In this work, the low-temperature aeration of concrete during introduction of the activation to the mixture for the silicate matrix by liquid glass additives $Na_2O \cdot nSiO_2 + mH_2O$ and sodium hydroxide NaOH was provided.

3 Experiment

In this study, one of the objectives is to identify local laws of formation of structure and properties of the aerated complex activated composites on the silicate matrix of thermal-moisture hardening in order to establish and use in practice the most effective and cost-effective methods of nanotechnology in the production process.

To analyze the possibilities of the regulation of the structure and properties of the aerated composites on the silicate matrix six factorial field experiments according to the 24-point plan, such as "triangles on the cube" type MTQ was carried out [15]. In the plan three mixed factors and three independent ones of the content vary simultaneously. As three mixed factors the surface area of tripoli as the component of lime-silica binder at levels was fixed: $\upsilon_1 - S_{sp1}$=400 m^2/kg, $\upsilon_2 - S_{sp2}$=500 m^2/kg, $\upsilon_3 - S_{sp3}$=600 m^2/kg. As three independent factors in the experiments the content of alkali-containing additives were varied: x_4 – *NaOH* – (0.5÷1) %, x_5 –$Na_2O{\cdot}nSiO_2+mH_2O$ – (1÷5)% and gypsum additives x_6 – $CaSO_4{\cdot}2H_2O$–(2÷4)%.

According to the experiment results ES model is calculated, which allowed to estimate the effect of alkali-containing additives and the specific surface area of tripoli on the properties and characteristics of the structure of aerated composites on silicate matrix.

The computation experiment has showed that the factor which has the greatest influence to the properties is the specific surface area of tripoli. Due to the synergistic action of alkali additives and liquid glass, taken in optimal ratios for each property and given S_{sp} of Tripoli, the aeration of the mixture is achieved, which lowers the density and high values of the properties: compressive strength R_b [MPa], bending strength R_{btb} [MPa], water resistance (coefficient of softening) k_s, crack resistance (critical coefficient of stress intensity) k_{Ic} [MPa·$m^{-0.5}$], heat conductivity (coefficient of heat conductivity) λ [Wt/m·K], frost resistance F [cycles]. The introduction of additives of alkali and liquid glass increases the volume of the mixture to 1.2÷1.4 times. The density of the composite varies from 1300 to 1500 kg/m^3, which is 17÷23% lower than the density of the matrix material and 25÷30% lower than the density of autoclaved silicate brick.

4 Results and discussion

According to the experimental results the experimental-statistical (ES) models were calculated. By ES models of optimal values of quality criteria and compositions of mixtures have been installed. Compressive strength R_b varies from 12 to 18.5 MPa, i.e. 1.5 times, and it is in the range of changing of the matrix material strength. Maximum strength is greater than 18 MPa, it is obtained for compositions which are contain 5% of liquid glass, 0.5% alkali, and 4% gypsum, S_{sp}=400 m^2/kg of tripoli, the same strength of the silicate matrix is obtained on tripoli with S_{sp} = 500 m^2/kg.

According to the ES model (1) the coefficient of heat conductivity λ of aerated composites under the influence of all six factors varies, 1.9 times, from 0.28 to 0.54 Wt/m·K. The minimum value of the t heat conductivity coefficient λ = 0.28 Wt/m·K is obtained on the contents which contain 0.5% NaOH and 1% liquid glass, a mixture of tripoli particles with a specific surface S_{sp}=400 and S_{sp}=600 m^2/kg in equal ratio and gypsum additive 4% (Fig. 1a).

Softening coefficient is varied by k_s 0.81÷1. The softening factor has another impact of additives. The values $k_p \geq 0.95$ were obtained with the content 0.75% NaOH and 5% liquid glass on specific surface of tripoli S_{sp}=400 m^2/kg and the gypsum additive content (Fig. 1b).

Frost resistance of aerated materials is 25-30 cycles that is lay within the changes frost resistance of matrix material – F25-50. The maximum value of the frost resistance of aerated composites were obtained on S_{sp}=400 m^2/kg, and the content of additives gypsum is 4%. The similar values of frost resistance of silicate matrix were obtained in the mixture of tripoli particles S_{sp}=400 and S_{sp}=600 m^2/kg in equal proportions and gypsum content 2.5%.

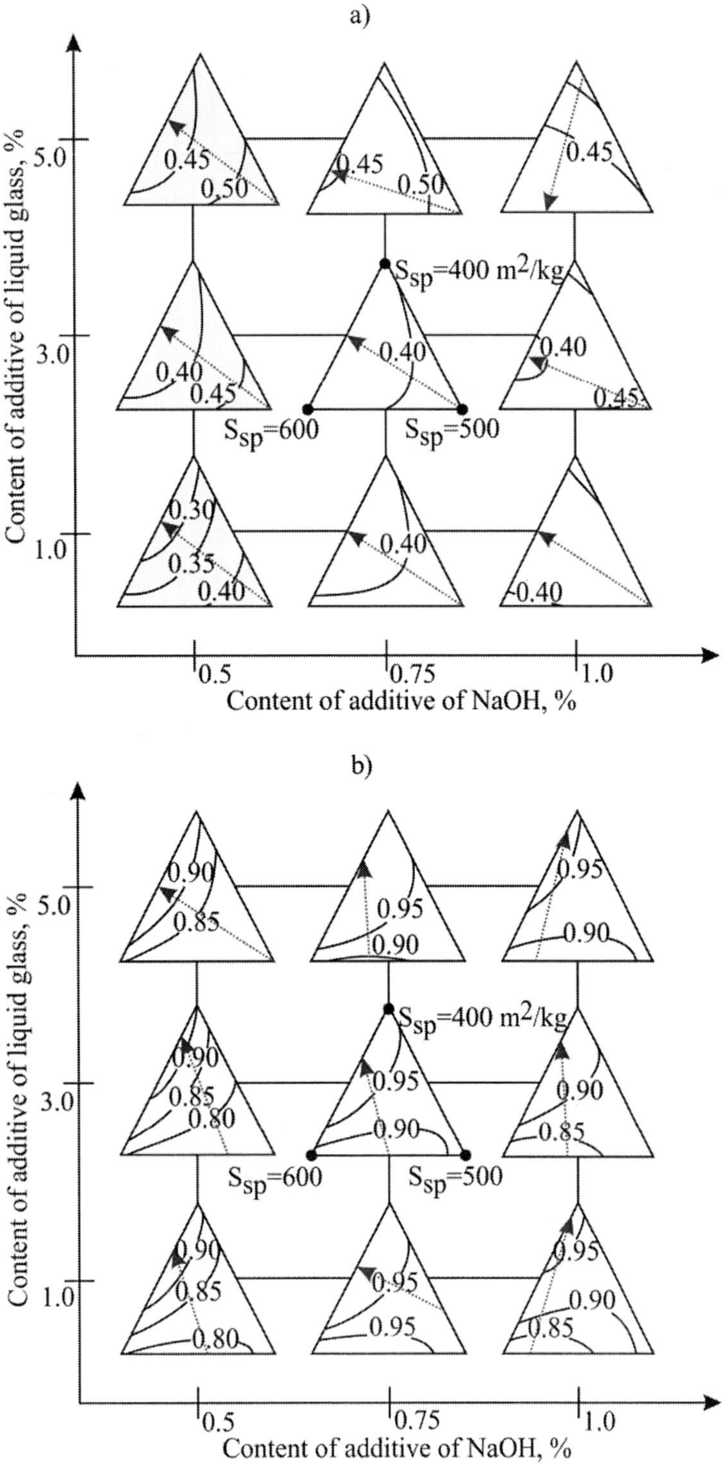

Fig. 1. The influence of additives NaOH and liquid glass on coefficient of heat conductivity λ (a) and softening coefficient k_s (b) an for a fixed value additive of gypsum.

The critical coefficient of stress intensity k_{Ic} under the influence of all factors varies by 1.8 times, from 0.91 to 1.64 MPa·m$^{-0.5}$. S_{sp} tripoli acts on k_{Ic}. According to the ES models, taking into account the quantities of S_{sp} tripoli, which provide maximum and minimum values k_{Ic}, the ratio $\delta k_{Ic} = k_{Ic}^{max}/k_{Ic}^{min}$=1.1÷1.47 time was calculated.

The maximum values of the critical stress intensity factor k_{Ic}=1.64 MPa·m$^{-0.5}$ of aerated composites are equal to k_{Ic} of matrix material and the minimum value k_{Ic} = 0.91 MPa·m$^{-0.5}$ of aerated composites are twice as k_{Ic} of the matrix material. In this case, the maximum and minimum values k_{Ic} for porous composite and the matrix material were obtained at the different values of tripoli S_{sp}. Thus, maximum k_{Ic} for aerated composites was obtained on S_{sp}=600 m^2/kg, and for the matrix material – on the mixture S_{sp}=400 and S_{sp}=600 m^2/kg with an equal ratio, that may be associated with different particles density in the volume and cramped conditions of aerated.

The visualization of ranges of property changes under the influence alkali-containing additives and the specific surface area of tripoli is shown in Fig. 2.

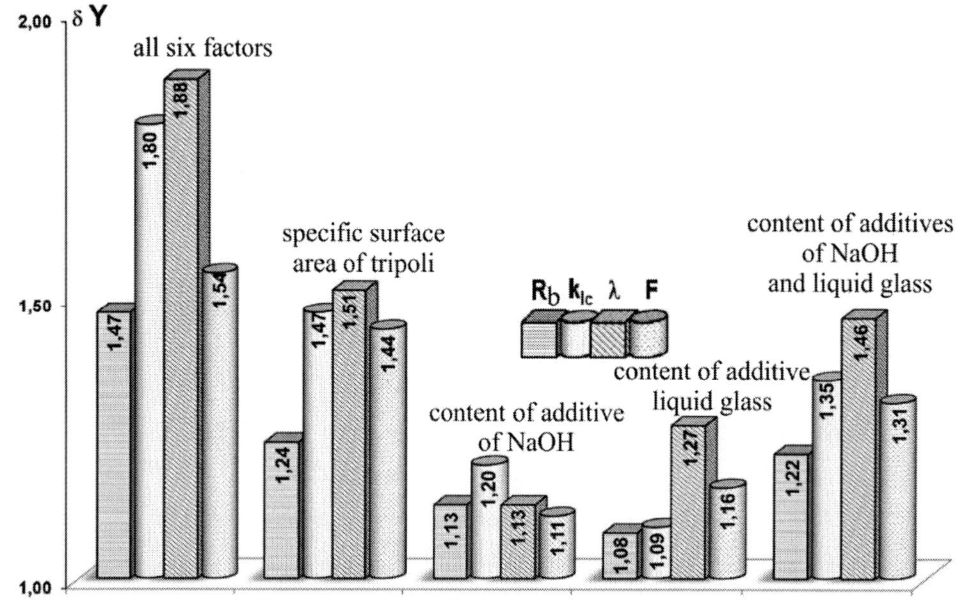

Fig 2. The relative influence of the studied factors and their interactions on the change of properties.

Thus, changed conditions of structure formation by introducing alkali-containing additives promoting aerated, predetermine the introduction of tripoli with a specific surface area, which may be connected, in this case, with the formation of the matrix structure in "straitened circumstances" of interporous partitions. For obtaining optimal compositions of porous composites, introducing alkali-containing additives they condition the necessitate of changes S_{sp} of tripoli and adjustment of the content of gypsum additives compared with the optimal composition of the silicate matrix at constant quantitative content of other components for the mixture and the conditions of their hardening.

This change of the properties is associated with the change of structural parameters of porous composites under the influence of alkali-containing additives and S_{sp} of tripoli.

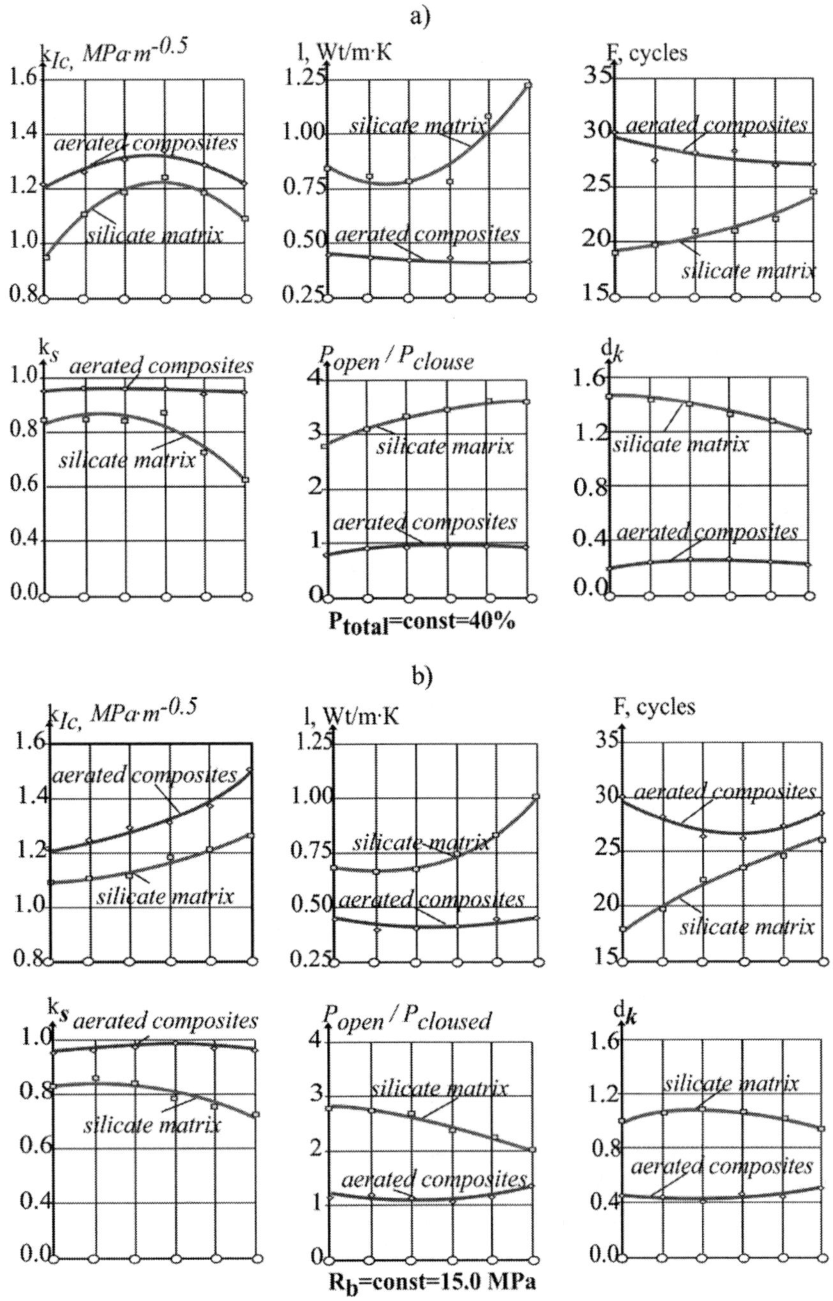

Fig 3. The comparative analysis of the changes in the properties and characteristics of the structure of the silicate matrix and of aerated composites on based it under isoparametric conditions: P_{tot}=const=40% (a) and R_b=const=15.0 MPa (b).

To analyze the impact of the structure characteristics on the properties the ES models of change of total, open and closed porosity and parameters of capillary porosity: d_k and α_k were calculated. The content of closed pores can be increased by reducing the content of open pores 1.7 times.

The total porosity can be increased to 30%. The average size of the capillaries d_k is changed in 4.3 times, the coefficient of distribution uniformity according to their size α_k –

in 2.3 times.

In the next phase of research using the computational experiments [8] according to ES models, the comparative analysis of property changes and the structure characteristics of the silicate matrix and aerated composites based on the isoparametric conditions at the constant total porosity P_{tot}=const=40% was carried out.

It was found that at the constant total porosity P_{tot}=const=40% the aerated composites are characterized by k_{Ic}=1.2÷1.35 MPa·m$^{-0.5}$, that 1.5÷1.7 times higher and the coefficient of heat conductivity is 1.8÷2.8 times lower than at the matrix material, the softening coefficient is $k_s \geq 0.95$ (Fig.3a).

This improvement is connected with change of the structure parameters. Thus, at the aerated composites comparing with the silicate matrix the ratio of the open and closed pores is reduced 3.5 times, the average size of the capillaries relative d_k from 1.2 to 0.35 is reduced more than 3 times.

A similar comparative analysis of property changes and structure characteristics of the silicate matrix and aerated composites based on isoparametric conditions was carried out at R_b=const =15.0 MPa (Fig.3b).

Under these conditions, the levels of the properties and their variation intervals vary: λ is decreased 1.5÷2 times, k_s=1.0, k_{Ic}=1.1÷1.45 MPa·m$^{-0.5}$, it confirms the ambiguous influence P_{tot} on R_b; R_b is determined not only by the total porosity, but also a variety of other characteristics of the structure.

The optimization problem was to obtain the aerated composites on the silicate matrix, which physic-mechanical properties were above the properties of the silicate matrix.

The isoparametric analysis while retaining R_b=const=15MPa and P_{tot}=const=40% showed that the ranges of other properties and structure parameters are significantly reduced: δλ=20%, δR_{btb}=18%, δk_s=6%, δk_{Ic}=7%, δP_{tot} – 10%, $\delta \alpha_k$ – 5% with aerated composites simultaneously with two properties of specified level (R_b и P_{tot}).

Changes of δλ, δR_{btb}, δk_s, δk_{Ic} associated with the presence of a mixture of tripoli different specific surface of tripoli, which confirms its significant influence on the formation of the structure as a silicate matrix and of porous composites based on it.

5 Conclusion

The mechanism of formation structure and properties of complex activated lime-silica mixture, modified by the filler in the form of tripoli on quick lime was proposed.

It is shown that the particles of tripoli contribute to seal structure of silicate matrix and the formation of discontinuous structure of capillaries, including its own microporosity. Furthermore, due to the high sorption capacity the tripoli pores can be the matrix from ultrafine size hydrosilicates, which properties differ from the properties of the calcium hydrosilicates formed in the mixture free space. It contributes to obtain the aerated composites with high physical and mechanical properties.

Thus, the structure modification of the silicate matrix by using quicklime, alkali-containing additives, and tripoli predetermined specific surface area, allows adjusting the levels of the properties and structure parameters of the aerated composites on silicate matrix in a wide range.

According to the results of optimization the contents are recommended which allow to get aerated composites on the silicate matrix given grades of strength, heat conductivity and frost resistance: compressive strength R_b – 10 MPa, 12.5 MPa, density ρ=1300÷1400 kg/m^3, frost resistance F25, heat conductivity λ=0.30÷0.40 Wt/m·K, critical coefficient of stress intensity $k_{Ic} \geq 1$ MPa·m$^{-0.5}$, coefficient of softening $k_s \geq 0.9$.

References

1. V.I. Babushkin, G.M. Matveev, O.P. Mchedlov-Petrosyan, *Thermodynamics of silicates* (Moskow, 1986)
2. P.A. Rebinder, N.B. Uriev, *Surface phenomena in disperse systems. Physico-chemical mechanics. Selected Works of Rebinder P.A.* (Moscow, 1979)
3. E. Shinkevich, E. Lutskin, O. Gnyp, A. Koichev, J. Dotsenko, Proceeding of the 8th Int. Symp. Brittle Matrix Composites 8, 517-525 (2006)
4. O. Shynkevych, Y. Lutskin, O. Koichev, G. Bondarenko, A. Tertychnyi, I. Myronenko, MATEC Web Conf., **116**, 01015 (2017)
5. Y.M. Butt, M.M. Sychev, V.V. Timashev, *Chemical technology of binders* (Moscow, 1980)
6. Y. Lutskin, E. Shinkevich, Proceeding of 14th Int. Congress on the Chemistry of Cement, 2, 632 (2015)
7. E.G. Avvakumov, M. Senna, N. Kosova, W. Soft, *Mechanochemical Synthesis a Basis for New Chemical Technologies* (Netherlands, Kluwer Academic Publishers, 2001)
8. Y. Lutskin, O. Shynkevych, A. Aniskin, Technical Journal, **11 (4-2017)**, 60-165 (2017)
9. O. Borziak, S. Vandolovskyi, V. Chajka, V. Perestiuk, O. Romanenko, Matec Web Conf., **116**, 01001 (2017)
10. O. Pluhin, A. Plugin, D. Plugin, O. Borziak, O. Dudin, Matec Web of Conf., **116**, 01013 (2017)
11. O.I. Demina, A.A. Plugin, E.B. Dedenyova, D.O. Bondarenko, T.A. Kostuk, A.I. Bondarenko, Functional Materials, **24(3),** 415-419 (2017)
12. V.D. Gluhovskij, R.F. Runova, S.E. Maksunov, *Binders and composite materials of contact hardening* (Kiev, 1991)
13. P.V. Kryvenko, E.K. Pushkareva, *Durability of slag-alkaline concrete* (Kiev, 1993)
14. N.R. Rakhimova, R.Z. Rakhimov, Y.S. Lutskin, V.P. Morozov, Y.N. Osin, Romanian Journal of Materials, **48 (2)**, 177–184 (2018)
15. V. Voznesenskiy, T. Lyashenko, J. Ivanov, I. Nikolov, *Computer and optimization of composite materials* (Kiev, 1989)

Research of impact resistance of nanomodified fiberreinforced concrete

*Uliana*Marushchak*, *Myroslav* Sanytsky,*Nazar*Sydor, and *Serhii*Braichenko

Lviv Polytechnic National University, Building Production Department,12 S. Bandera Str., 79013, Lviv, Ukraine

Abstract. Effects of multilevel modification of concrete structure with complex nanomodifier at nano- and microlevel and polypropylene fiber at macro- and mesolevel on mechanical properties and impact resistance of concrete are presented. Nanomodification with complex nanomodifier, which consists of polycarboxylate ether superplasticizer, ultra- and nanofine mineral additives, provides early structure formation, higher strength at early and later ages. The increasing of the static hardness according to Brinell and Mayer of the nanomodified concrete was indicated the high energy of the bond between the structural elements in the surface layer of nanomodified concrete, as well as the ability of the structure to withstand elastic and plastic deformations under uneven compression load. Incorporating of polypropylene fibre to concrete was increased the impact resistance according to parameter of first crack strength and failure energy and changed the failure pattern from brittle to ductile mode.

1Introduction

Advanced construction technologies put forward a new level of requirements for technological and technical properties, as well as durability of concrete, which is associated with their use in structures of civil and military infrastructure, in structures of complex architectural forms, shells, reservoirs, roads, airfields, protective elements.Such structures work in extreme conditions, sothere is a growing trend to enhance their impact and crack resistancefor reliable exploitation throughout the life cycle.There are often high requirements set for the time allocated for construction and repair of building structures and need using of rapid-hardening concrete with necessary technical and technological properties. The disadvantage of high-strength concretes is low impact strength, increasing sensitivity to cracking and brittleness, which manifests itself by reducing plastic deformations under load.In addition, increasing of the loading velocity causes the transition of material to brittle state. Therefore, in the conditions of impact load concrete destructs at lower stresses compared with static loads, which can lead to previous destruction of structures [1].

*Corresponding author: ulmarushchak@ukr.net

The ability of concrete to withstand dynamic load is largely solved by the formation of a multi-level structure of concrete due to complex modification at the nano-, micro- and macro-scale levels with superplasticizers of the new generation, nano- and ultrafine mineral additives, as well as reinforcement by fibres. Herewith structural conditions of braking of cracks, dissipation and absorption of external dynamic impact energy are created. The adding small amounts of different types of fibres into concrete can significantly reduce the damage due to impact load by arresting crack development and bridging of cracks [2, 3].

Changing of fibre reinforced concrete performance is based on the principles of composite development and associated with the joint work of materials with different physical and mechanical properties [4]. The presence of the boundary between the reinforcing elements and the matrix significantly enhances the cracking resistance of the material by increasing of area of load perception and distribution in all directions, which leads to increased ductile characteristics, deformation strengthening [5]. Such fibre reinforced concrete reveals the properties of ultraductile materials.

The improvement of deformation characteristics of cementing materials is associated with the elimination of initial defects at various scale levels, which are the initiators of stress at a load including an impact load. The formation of a low-defective structure on the micro- and nanostructural levels is monitored by the introduction of high effective polycarboxylatesuperplasticizers, energetically active ultra- and nanofine mineral additives [6, 7]. Conditions for the increasing of early strength of concrete due to provide an effect of the filler at the initial period, stimulation of nucleation processes in an intergranular space between cement grains and the early pozzolanic reaction with the formation of nanosized scale fibrous C-S-H phases are created in presence of nanoparticles [8, 9]. Fibrous hydration products provide additional micro- and nanoreinforcement of the cement matrix. The main objective of this paper is to study the impact resistance parameter of nanomodified fibre reinforced concrete.

2 Materials and methods

2.1 Materials

Ordinary Portland cement CEM I 42.5R PJSC Ivano-Frankivsk Cement (Ukraine) according to EN 197-1 standard was used in the investigations. Natural sand of Zhovkva quarry (MF=1.9) and granite crushed stone of 5–20 mm fraction as a coarse aggregate were used for concrete production. The mix proportion was 1:1.35:2.71 with the content of CEM I 42.5R 430 kg per 1 m^3 of concrete. Concrete modified with lignosulfonate plasticizer was used as reference (RC). Methakaoline was used as ultrafine supplementary cementitious materials and nanosilica (Aerosil-380) was used as nanofine admixture for nanomodified concrete (NC). A superplasticizer based on polycarboxylateeter polymer GLENIUM ACE 430 (PCE), which has a powerful dispersing effect on the cement particles, was used as a water reducing admixture for nanomodified concrete. The dosage of polycarboxylate ether superplasticizer which incorporated into nanomodified mixes was 1.0% by weight of cement. For reinforcement at the macro- and the mesoscale levels was used polypropylene fiber (12 mm in length, 18 μm in diameter) to increase an impact resistance of nanomodified concrete. The dosage of fiber in nanomodifiedfiber reinforced concrete (NFRC) was 1.0 wt.%.

2.2 Preparation of the specimens and testing

The coarse aggregate, sand and cement were first mixed in the dry state for one minute before adding the mixing water and plasticizing admixtures. Then the polypropylene fibers were added for prepared fiber reinforced concrete. Mixing was continued for further five minutes to achieve uniform distribution of the fiber. Workability of fresh concrete was determined by the slump test according to EN 12350-2. The slump values of fresh concrete were 180-210 mm. Cubes (100 mm) were prepared from each batch for the compressive strength and hardness test and cubes (70.7 mm) – for impact test. The samples were cured in normal conditions for the hardening of concrete (90-100% RH at 20±2°C). After 1; 2; 7 and 28 days the samples were testing for compressive strength and impact resistance.

The impact test was performed in accordance with the impact testing procedures recommended by ACI Committee 544. The test was carried out by dropping a hammer weighing 20 N from a height of 1000 mm repeatedly on a 30 mm diameter hardened steel ball, which is placed on the top of the centre of a cubical specimen. From this test, impact resistance was estimated as specific energy absorption capacity for initial visible crackand final crack (impact failure energy) of each specimen.

Static indentation tests of concretes at the age of 28 days were carried out by a Brinell testing device with ball diameter of 10 mm. Testing loads of 1000 kg were applied for 10-15 s on the concrete surfaces. Diameters of the residual impressions were measured by a hand microscope of 8× magnification.

3 Result and discussion

The compressive strength of nanomodified concrete NC after 2 and 28 days exceeds the strength of concrete modified with the lignosulfonate admixture - RC 1.8 and 1.4 times respectively (Fig. 1). The adding of fibres causes some decreasing of workability of concrete mixtures, which causes 7.9% increasing of water consumption compared with nanomodified concrete to achieve the required workability and 5-7% decreasing of strength. According to the specific strength parameter f_{cm2}/f_{cm28} = 0.52-0.57, nanomodified concrete and nanomodifiedfiber reinforced concretes are classified as concrete with a rapid strength development. The value of strength after 28 days (f_{cm28} = 98.7-104.5 MPa) of NC and NFRC concretes meet the requirements of high strength concrete (strength class C 80/95).

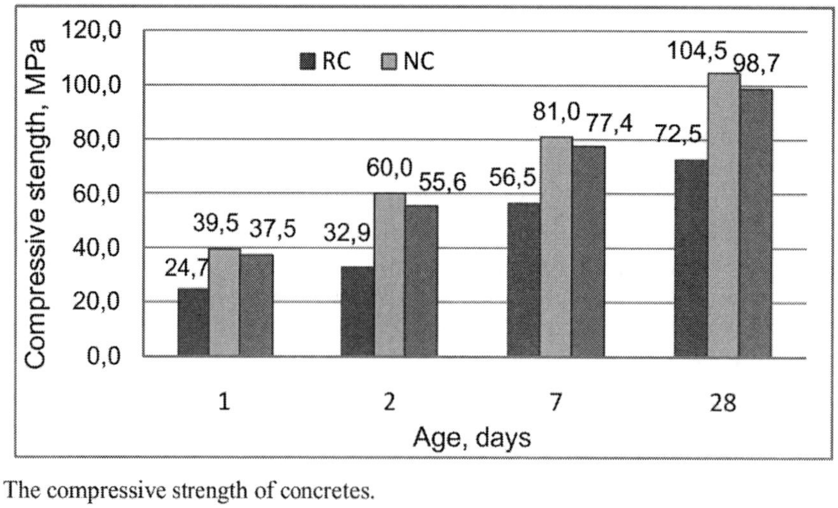

Fig. 1. The compressive strength of concretes.

Impact resistance of concrete is associated with the behaviour of its surface layer under mechanical action on the material, and is characterized by hardness.Brinell(HB) and Mayer (HM) hardness results plotted in Fig. 2 are shown that the use of complex nanomodifierprovides possibility to regulate the structural and energy state of the surface of concrete. The values of the static hardness according to Brinelland Mayerof the nanomodified concrete after 28 days are 812 and 851 N/mm^2 respectively, which is in 1.7-2.2 times exceeds the value of the hardness of reference concrete (390 and 493 N/mm^2). That indicates the high energy of the bond between the structural elements in the surface layer of nanomodified concrete, as well as the ability of the structure to withstand elastic and plastic deformations under uneven compression conditions.The hardness value for nanomodifiedfiberreinforced concrete is lower by 10.3% compared to nanomodified concrete, that indicate the relationship between compressive strength and hardness of concrete and the increasing of plastic deformations when polypropylene fibers are added.

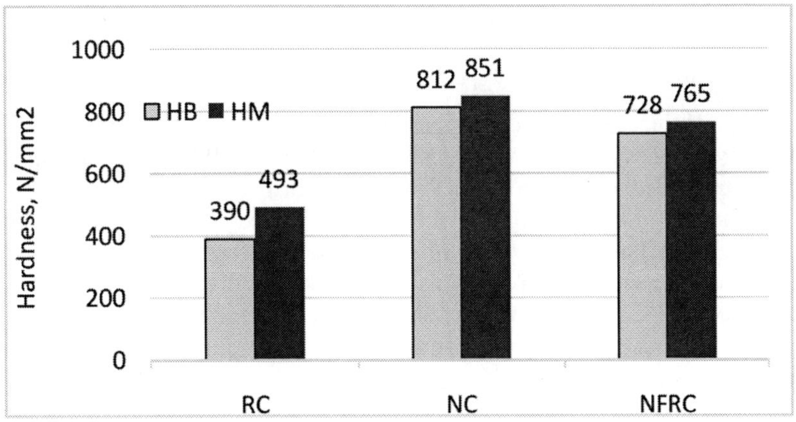

Fig. 2. The Meyer and Brinell hardness of concretes.

The formation of a dense structure of nanomodified concrete NC allowsto increase the impact resistance (specific impact energy to the appearance of the first visible cracks and final crack – impact failure energy) in 2.0-2.6 times compared with reference concrete RC at the early and the later age (Fig. 3).Significant increasing of impact strength occurs when structure of concrete is reinforced with polypropylene fibers. So, after 2 days of hardening, the specific energy of the impact for the first visible crack increases to 2.5 J/cm^3, and the impact failure energy is 3.89 J/cm^3, which exceeds the corresponding indices of nanomodified concrete in 5.0 and 6.7 times respectively.

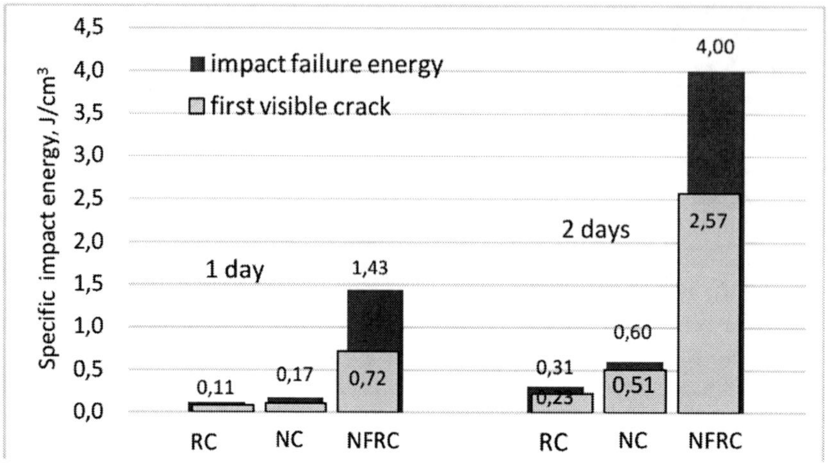

Fig. 3. Impact strength of concrete at early age.

The energy input necessary to cause the visibility of first crack of nanomodified fiber reinforced concrete after 28 days of hardening increases in 10.5 times compared to non-reinforced nanomodifiedconcrete(Fig. 4).The difference between impact energy for final crack and initial crack increased significantlywith adding polypropylene fiber. The energy necessary to cause failure of concrete specimen was increased by 11.8 time over the nanomodified concrete specimen (NC). This proves that the fibers act as an effective crack arrestor in case of NFRC, when an impact load is encountered. Thus the reference concrete exhibits an early brittle failure when compared to fiber reinforced which shows better ductile properties.

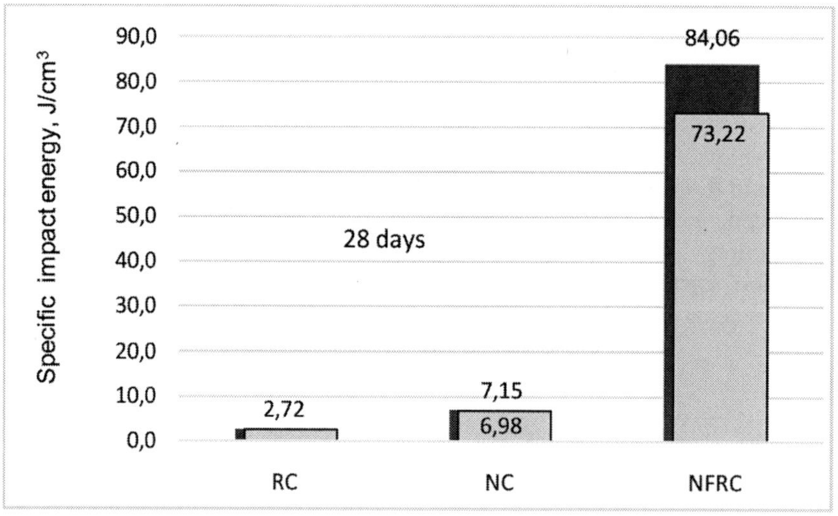

Fig. 4. Impact strength of concrete after 28 days.

The mode of failure of concrete depends upon the matrix strength, aggregate strength and bond strength of fiber with aggregate matrix. Brittle behavior was observed in RC and NFRC concrete specimens with formatiom trunk crackand it was broken into two pieces (Fig. 5, a, b).

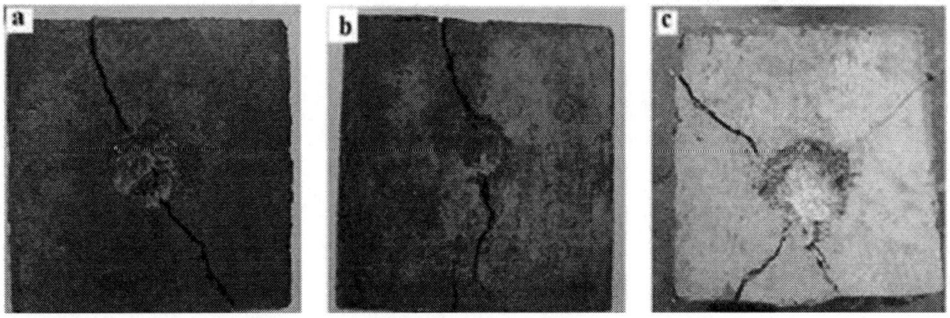

Fig. 5. Failure pattern of specimens after impact test: RC (a), NC (b), NFRC (c).

The NFRC specimens displayed different mode of failure (Fig. 5, c). The mode of failure was changed from brittle to ductile behavior. By incorporating fiber to concrete the cracking pattern was changed from single crack to large number of multiple cracks.When reaching the strength extremum such concrete does not lose integrity and is not accompanied by fragmentation of the material due to the set of reinforcing elements.Further deformation causes expansion of existing microcracks, which involves additional energy.

4 Conclusions

The nanomodification of concrete with organo-mineral additives based on metakaolin, nano-silica and polycarboxylate superplasticizer ensures obtaining nanomodified fiber-reinforced concrete relates to rapid-hardening high strength concretes with spesific strength parameterf_{cm2}/f_{cm28}=0.5-0.52 and strength after 28 days f_{cm28}=98.7 MPa.Reducing the structure defect, forming an increased number of hydrated products in the cement matrix when introducing elements of various scale levels, multilevel reinforcement provides a significant three-dimensional strengthening of nanomodified fibre reinforced concrete, which increases the energy intensity of the destruction process and impact resistance of the composite.By adding polypropylenefiber the energy required to cause the visibility of first crack and failure after 28 days was increased by 26.9 and 30.9 time, respectively over reference concrete, which displays effectiveuse of nanomodified fibre reinforced concretein structural engineering applications.

References

1. S. Fic, *Concrete under the influence of impact loads* (Politechnika Lubelska, Lublin, 2017)
2. A. Badr, A. F. Ashour, A. K. Platten, Int. J. Impact Eng., **32**, 1907 (2006)
3. U. Marushchak, M. Sanytsky, S. Korolko, Yu. Shabatura, N. Sydor. Eastern Eur. J. Enterprise Technologies, **6 (92)**, 34 (2018)
4. L. Dvorkin, O. Dvorkin, V. Zhitkovsky, Y. Ribakov, Mater. Des., **32 (6)**, 3254 (2011)
5. A. Plugin, T. Kostiuk, O. Plugin, D. Bondarenko, Y. Sukhanova, N. Partala, Int. J. Eng. Res. Afr., **31**, 59 (2017)
6. K. Pushkarova, K. Kaverin, D. Kalantaevskiy, Eastern Eur. J. Enterprise Technologies, **5 (5)**, 42 (2015)
7. U. Marushchak, M. Sanytsky, Yu. Olevych, MATEC Web Conf.,**116**, 01008 (2017)

8. O. Borziak, S. Vandolovskyi, V. Chajka, V. Perestiuk, O. Romanenko,MATEC Web Conf.,**116**, 01001 (2017)
9. U. Marushchak, M. Sanytsky, T. Mazurak, Yu. Olevych, Eastern Eur. J. Enterprise Technologies, **6 (84)**, 50 (2016)
10. E. Horszczaruk, E. Mijowska, K. Cendrowski,K. Mijowska, P. Sikora, Cem. Wapno Beton, **1**, 24 (2013)

Improved bearing resistance of soil foundations of buildings with injectable polyurethane composites

Sergii Miroshnichenko[1], *Dmytro* Plugin[1], *Oleg Kalinin*[1], *Alina* Zvierieva[1,*],
and *Ivan* Reznichenko[2]

[1]Ukrainian State University of Railway Transport, Feuerbach sq. 7, 61050 Kharkiv, Ukraine
[2]LLC SPT Ukrayina, Gospitalna str.14, 01601 Kiev, Ukraine

Abstract. The article deals with the analysis of SPT® polyurethane materials and maintenance technologies based on them, and an analytical review of literature on the topic. Solidification with SPT® materials and technologies makes it possible to compact unstable foundation soils of structures and buildings. The solidification is implemened through a guided increase of polymeric material injected into a foundation. As a result, polymeric root-like reinforcing bodies are formed in the foundation, and the water is extruded out of the soil, thus considerably increasing the bearing capacity of the foundation. In order to assess the actual efficiency of the material the authors conducted the full-scale research into the general overhaul technology of the pipe culvert, the static and the dynamic tests on the models of solidified soil was conducted and compared with the results of the tests on the models of non-solidified soil. The authors investigated into the life ratio of SPT® material and the soil solidified with it. By comparing characteristics of SPT® materials and technologies with those of conventional materials, the basic analogies and differences were established. And the possibility to use them for construction, refurbishment and maintenance of rail facilities was determined.

1 Introduction

The use of rail transport facilities on unstable soils, especially watered ones, can lead to irregular soil subsidence, crack development, water resistance failures, initiation of currents, all of which can result in a shorter service life of buildings and possible emergencies [1-3].

In order to improve stability and decrease the water permeability of soils (to create barriers) the following methods are used: mechanical (tamping); drainage (water lowering, electroosmotic solidification due to water electroosmosic and electrophoresis of soil particles when passing direct current); injection; and thermal ones.

Nowadays, the methods based on injection of solidifying materials into soils are becoming more popular. They include cementation (cement solutions), clay grouting (clay

* Corresponding author: farwww@ukr.net

suspension), silicatization (sodium liquid glass as a solidifier: two-solution – liquid glass and solidifier by turn, one-solution two-component – mixture of liquid glass and solidifier, one-solution one-component – liquid glass and soil as solidifier, gas – liquid glass and carbon dioxide as solidifier), electro silicatization (combination of electroosmosis and silicatization), soil tarring (polymerizing tars as a fixing agent and a solidifier), and grouting [1-5].

The polyurethane multicomponent materials and maintenance technologies based on them can be regarded as modern injectable compositions of high efficiency.

2 The analysis of recent research and publications

Since 1980-1990, European countries have used the high pressure cementation (compensational grouting, soilfrac) to avoid subsidence of buildings with underground works beneath or even to elevate buildings already subsided [5-7]. The pioneers of this technique are Keller Grandbau GmhH (Germany) and Soletanche Bachy (France). At present, the high pressure cementation technology used for underground works (compensational grouting) is being automated. Soletanche Bachy (France) conducted the compensation grouting under the building of Alekseevskiy Military School in Moscow during the construction of Lefortovo tunnel underneath [7].

The solidification technique is chosen according to certain requirements (higher capacity, water resistance), type and soil filtration coefficient, structural peculiarities of foundations, etc. In accordance with a specific construction problem the experts choose the type of the device (injectors, wells), determine the form, outlines and size of solidified soil bodies, establish technical requirements for the solidified soils. The design of solidification requires calculation of limiting states in order to avoid subsidence.

The high pressure cementation is injection of a cement suspension through the system of horizontal full-hole pipes under the subsiding building. The full-hole pipes are placed in advance into the wells drilled from the technological shafts in a fan arrangement under the whole subsiding building or territory. The injection runs under the pressure up to 10 MPa which provides the soil layer fracture and elevation of the daylight ground and building.

Subsiding of the building is controlled with automatic theodolites and benchmarks installed on the building. The data from the theodolites are transmitted to the PC-based control system which analyzes it and generates control signals for the injecting equipment.

The jet grouting technique, developed at the beginning of the 1970s in Japan, has become a common practice in Germany, Italy and France (Jet Grouting, Soilcrete) [5, 6, 8]. It involves mounting soil-cement piles and areas with a cutting jet of the cement suspension.

The jet grouting is applied for any soils subject to washout.

In Europe, leaders in manufacturing jet grouting equipment and those who conduct works are Keller Grandbau Gmb, Bauer (Germany) and Rodio, Soilmec (Italy).

In the 1970s, such technologies and equipment were designed by the All-Soviet Union Institute for Mixed Bases and Foundations (VIOS) and Gidrospetspoekt; they were used during construction of Zagorsk Pumped Storage Station where a barrier of 200 m in length and 7 m in depth was built. Nowadays, jet grouting for soils are conducted by such Russian enterprises as Ingeokomspetsstroy (Moscow), Georeconstruktsiya, Geoschit (St.Petersburg), and Region (Perm).

Construction and maintenance of tunnels including underground systems run with the injection of solutions and their treatment [9, 10]. The objective of this injection is filling hollows and voids left after the construction works with solidifying solutions; it also provides interaction with the surrounding soil. Injection of solutions improves the structural behavior, decreases deformations in processing, prevents subsidence of the soil, increases water resistance of the structure and, as a result, lowers the corrosion and extends the service life of constructions.

The injection is conducted in two stages: the initial and the control one. While rating the hydrogeological conditions along the tunnel by the executive documentations (possible elevation of the ground-water level, influences of anthropogenic and other factors), it is a common practice to provide soil packing around the tunnel in order to improve the waterproofness. The division of injections into initial, control and compacting is technologically accepted only for new construction works, and, the operational experience confirms that materials used in wet conditions, as a rule, do not provide needed service life of the barrier.

In order to restore the water resistance behavior and bearing capacity of underground structures, injecting the superplasticized cement-water suspension (SPCWS) in one stage (without dividing into initial and control) is recommended [10]. The SPCWE has the optimal amount of superplasticized additive C-3, Dofen or SL (within 0.2÷0.8 % out of the cement mass, defined by a special technique) which provides its high penetrability at extremely low W/C – 0.33÷0.38. The injection of SPCWS proceeds to a failure under pressure of 0.5÷0.6 MPa.

Since the 1980s, it has been a common practice to inject foam-polymer (usually, polyurethane foam) compositions instead of cement and concrete solutions to provide water resistance for underground structures. The famous Ukrainian polymer composition based on the urethane oligomer Monolit-3 is produced by Monolit-Polymer (Kiev) [11]. It is a three-component composition of resin MA-3, hardener MB-3, and additives – polymerization accelerator systems PAS. Compositions added to cement solutions and concretes in the amount 5÷15% out of the cement create waterproof polymer-cement concretes and solutions.

Though the technology was initially developed for transport tunnels, it is successfully applied for renovation of operational properties of underground structures of various purposes.

Searching for a new material as the hardener and stabilizer of underground structures is required, first of all, due to a number of faults in the above-mentioned techniques, namely:

- high cost of the jet grouting technology as the drilling equipment should inject the solution under high pressure with simultaneous rotation;
- silicatization solutions are hardening for 24 hours;
- under silicatization, there is a high possibility for the material to be washed out;
- soils under the structure cannot be solidified;
- silicatization is not an environment-friendly process; and
- high possibility of damages in the structure due to high pressures and large amount of materials;

All the above-mentioned stipulate the need to use a new material of high durability, stability, manufacturability, water resistance for buildings and structures.

3 The basic part of the research

Solidification with SPT® materials and technologies makes it possible to compact unstable foundation soils of structures and buildings. The solidification is implemented through a guided increase of polymeric material injected into a foundation. And as a result, polymeric root-like reinforcing bodies are formed in the foundation, and the water is extruded out of the soil, thus considerably increasing the bearing capacity of the foundation.

Using SPT® materials and technologies for injection prevents penetration of water inside buildings due to improved waterproofing on the building/soil contact, and also the seam sealing between structures and the sealing of cracks in these structures.

These materials were used for soil solidification under the Main sewage collector in Kiev, under foundations of multistoried buildings, pipes under the railway, soil solidification and slab levelling in Boryspil International Airport and other objects.

The physical and mechanical properties of soils and the bearing capacity of a surface are determined by the soil moisture. If it exceeds the admissible values, deformation processes can be initiated in foundations, often uneven, leading to damages and deterioration to buildings. Therefore, for structures in adverse conditions, such as slopes, under loads from heavy vehicles, with water invasion of the foundations, [11] recommendations are given on constant control of soil moisture; provided it reaches unacceptable values measures for soil draining or solidification should be taken.

In [12, 13] the authors show that stability of disperse systems mostly depends on the interaction between the liquid phase and disperse particles. Thus, for higher efficiency of soil solidification it is reasonable to control the hydrophilic and hydrophobic behavior of disperse particles depending on the injectable materials used. The authors believe that solidification with polar substances requires control of the hydrophilic behavior of particles, and with non-polar ones – the hydrophobic behavior.

In order to assess the actual efficiency of the material the authors conducted the full-scale research into the general overhaul technology of the pipe culvert on the 1216th km of the Kolosivka-Odessa section of the Odessa Railway with injectable polymeric materials. In the framework of the research some models were developed and actual physical and mechanical characteristics of the soils in these models were defined. The static tests on the models of solidified soil were conducted and the results were compared with those on the models of non-solidified soil at various moisture values. The dynamic tests on the models of solidified soil was conducted and compared with the results of the tests on the models of non-solidified soil. The authors investigated into the life ratio of SPT® material and the soil solidified with it (stability to alternating moistening/drying).

In their research the authors used the soils close by their properties to those of engineering structures on Ukrainian railways (soft loamy sand).

The results of the compressive research demonstrated that the deformation modulus of solidified soil 1.4 times higher than the modulus of soil deformation in nature. However, these results describe only changes in the soil structure after compacting and do not consider the influence of the solidified area of soil-polymeric reinforcing bodies on the bearing capacity.

By the results of static stamp tests on models of the soil area, the authors came to the conclusion that soil solidification considerably decreased its stress-strain behavior (an increase of the deformation modulus). The deformation modulus of non-solidified soil in the model by moisture W = 15% was 0.5–0.7 MPa. The soil solidification increased the modulus at W = 10.9 % up to 11.2 MPa and at W = 6.9 % up to 33.4 MPa. Though the deformation modulus increased due to soil dewatering at W = 11.5 % up to 3.9 MPa, at W = 9.5 % up to 33.5 MPa.

The analysis of the research results by deformation characteristics demonstrated that during the static tests on the solidified soil the dependencies of the modulus of elasticity E on moisture W were demonstrated by the diagrams in Fig. 1. These dependencies are approximated by exponential equations with the correlation coefficient 0.99 for non-solidified soil and 0.77 for solidified soil. According to these equations, presented in Fig. 1, the dependency of deformation modulus E on soil moisture W for natural and solidified state was calculated.

The analysis of the calculation results allowed the authors to propose the coefficient of a higher deformation modulus after solidification K_E, the dependency of the deformation modulus of solidified soil on the deformation modulus of the natural soil at equal moisture, in order to evaluate the influence of solidification on the stress-stain behavior of the soil.

The values of these coefficients are given in Fig. 2.

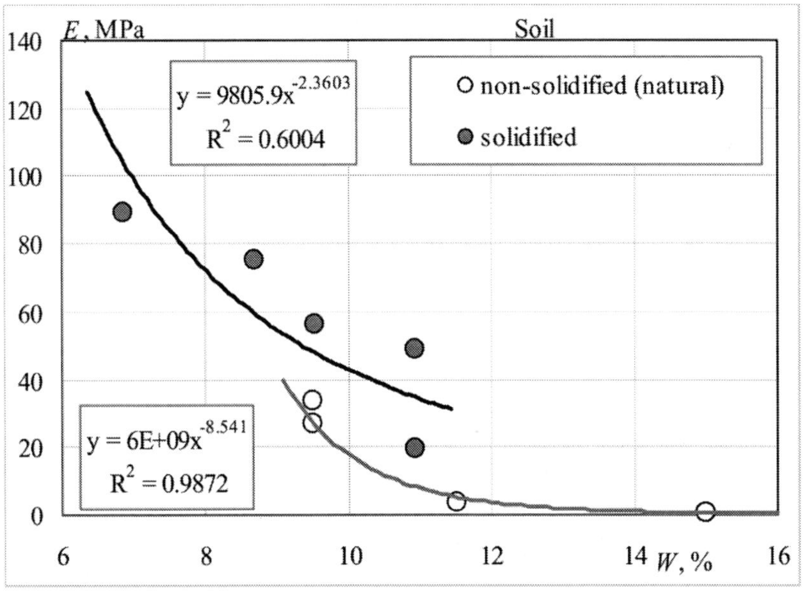

Fig. 1. Dependency of deformation modulus E of natural and solidified soil on moisture W.

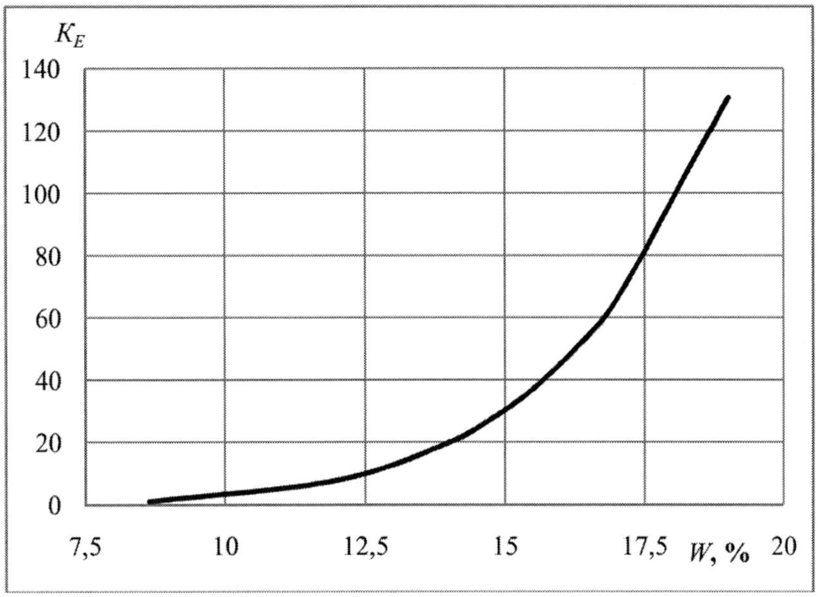

Fig. 2. Dependency of the coefficient of a higher deformation modulus after solidification K_E on soil moisture in nature.

Fig. 2 demonstrates that in dry soils (loamy sand) with moisture less than 8.6 %, solidification does not increase the deformation modulus (does not decrease the stress-strain behavior). For wetter soils, the deformation modulus is increasing due to solidification, at the moisture:

- 10 % – by 2.5 times (from 17.3 to 42.8 MPa);
- 14 % (elasticity limit) – by 20 times (from 0.98 to 19.4 MPa);
- 15 % (middle of the plastic stage limits – by 30 times (from 0.54 to 16.4 MPa);
- 16 % (liquid limit) – from 0 to 12.2 MPa.

The authors believe that these dependencies will cover all loamy and sand-loamy soils. However, it can only be confirmed by the research into soils with other factors of elasticity and liquid limits.

Durability of joint work of the soil/polymer system under the conditions without weathering and insulation will depend on influence of water, especially, on changes in water saturation. The actual foundations of rail transport structures yearly suffer from two-cycle changes in water saturation level. Therefore, in order to evaluate durability of the soil/polymer system the unique research procedure was designed.

Thus, the authors modeled the effect which was definitely destructive for the polymer-soil body (the root-like body taken out of the model and formed by the polymer with soil particles or particles glued with the polymer). This effect was very intensive alternate moistening/drying which eventually ruins virtually all construction composite materials.

By the results of the research into the influence of the alternating moistening/drying of the solidified soil samples on their properties, it was established that the influence ruined the soil, notably led to mass loss both in a dry and water saturated states. After each moistening/drying cycle the difference between the mass in dry and wet states increased, which testified to higher porosity, and, therefore, changes in the stress-strain behavior and soil stability.

Additionally, for the reason of comparison, the similar tests on the soil solidified with one-agent silicatization with silicon hydrofluoric acid as a hardener were made. It is a well-known fact that the durability of such solidification under the conditions of underground waters is 10–15 years.

The research into the influence of the alternating moistening/drying on the samples of solidified soil showed that if solidification was done with SPT® materials and by SPT® technology, the samples lost 5% of their mass in 10 cycles, while silicatization gives a loss of 10% of the mass in 2 cycles. Considering the soil solidification durability by silicatization as 15 years, the life expectancy of solidified soil with SPT® material is no less than 75 years.

The petrographic research demonstrated that unlike traditional chemical soil solidification with silicatization or tarring, which provide even infiltration, SPT® materials and technology provide formation of root-like reinforcing elements in the soil. Some of these roots had the lateral dimension up to 70 mm and consisted of the middle and peripheral zones. The middle zone was a layer of porous well-polymerized material, and the peripheral zone was soil-polymer composite consisting of porous polymerized material and soil particles.

4 Conclusions

In order to assess the way of improvement of the bearing capacity of soil by solidification with SPT® materials and by SPT® technology, the authors propose to use the coefficient of a higher deformation modulus after solidification K_E (the ratio of deformation modulus of the solidified soil to the deformation modulus of natural soil at the same moisture) for calculations.

The research into the influence of alternating moistening/drying on the samples of solidified soil demonstrated that solidification with SPT® materials and by SPT® technology predicts the life expectancy of more than 75 years.

Therefore, the sphere of application for SPT® materials and technology is wider than that of traditional chemical soil solidification with silicatization or tarring (used only for sands and porous foam soils) and can be extended to practically all clay soils. Though, this conclusion can only be confirmed on conducting a full complex of tests on other soils.

References

1. A. Plugin, L. Trykoz, *Vidnovlennya ekspluatatsiynih vlastivostey osnov, fundamentiv, zagliblenih i pidzemnih sporud* (Kharkiv, UkrSURT, 2005)
2. A.M. Plugin, A.A. Plugin, L.V. Trykoz, O.S.Sayapin, O.S. Herasimenko, O.A. Plugin, *Rozrahunki nesuchoyi zdatnosti i tehnologiya zakriplennya osnov budivel i sporud zaliznichnogo transportu* (Kharkiv, UkrSURT, 2012)
3. M.I. Smorodinov, B.S. Fedorov, B.A. Rzhanitsyin i dr., *Osnovaniya i fundamentyi: Spravochnik stroitelya* (Moskva, Stroyizdat, 1983)
4. B.A. Rzhanitsyin, *Himicheskoe zakreplenie gruntov v stroitelstve* (Moskva, Stroyizdat, 1986)
5. *Tehnologii rekonstruktsii fundamentov. Rekonstruktsiya gorodov i geotehnicheskoe stroitelstvo*, www.georec.spb.ru (2003)
6. V.V. Savyovskiy, O.N. Bolotskih, *Remont i rekonstruktsiya grazhdanskih zdaniy* (Kharkiv, Vaterpas, 1999)
7. Zh.-L. Valett, Metro i tonneli, **4**, 16-19 (2002)
8. A.G. Malinin, Metro i tonneli, **4**, 14-15 (2002)
9. *Rekomendatsiyi z gidroizolyatsiyi tuneliv metropolitenu za dopomogoyu tsementnih kompozitsiy* (Kharkiv, Mintrans Ukrayini, Ukrzaliznitsya, 2001)
10. *TsP-0136 Instruktsiya schodo vikonannya robit z nagnitannya rozchiniv za obrobku tuneliv* (M-vo transportu ta zv'yazku Ukrayini, Derzhadministratsiya zaliznichnogo transportu Ukrayini, Ukrzaliznitsya, Golovne upravlinnya koliynogo gospodarstva, 2006)
11. A. Plugin, L. Trykoz, O. Herasymenko, Ant. Pluhin, V. Konev, Diagnostyka. **19 (2)**, 13-23 (2018)
12. A. Plugin, E. Dedeneva, T. Kostyuk, D. Bondarenko, O. Demina, MATEC Web of Conference, **116**, 01010 (2017)
13. A. Plugin, T. Kostiuk, O. Plugin, D. Bondarenko, Yu. Sukhanova, N. Partala, International Journal of Engineering Research in Africa, **31**, 59-68 (2017)

Use of liners made of concrete on quartz aggregates and polyurethane composition to reduce vibration, noise and improve the stability of tramway track

Olena Palant, *Dmytro* Plugin[*], *Andrii* Plugin, *Alexey* Lobiak, and *Oleksii* Pluhin

Ukrainian State University of Railway Transport, Feuerbach sq. 7, 61050 Kharkiv, Ukraine

Abstract. The theoretical basis of reduction of thermal deformations of the ballastless continuous welded track has been developed. The design of the track with integrated rails has been improved by using a special concrete liner of optimal composition with quartz fillers and aggregates and two-component polyurethane. The method of calculating the concrete composition with optimal strength and coefficient of linear thermal expansion together with optimal structural characteristics, To reduce the temperature stresses in the rails, liners should be made of concrete with quartz fillers and aggregates which provide a reduction of the coefficient of linear thermal expansion α_L to 1.1×10^{-6} K^{-1}. The analysis of the stress-strain state of the rails during heating, performed using the finite element method, has shown that the liners made of concrete with quartz aggregates decrease by 10-32% the temperature stresses in rails. The composition of concrete for liners with the compressive strength of above 60 MPa has been substantiated and the polyurethane composition has been selected for insulation of rails from the channel wall with adhesion to concrete of 0.7 MPa in a dry state, 0.43 MPa in a water-saturated state, and adhesion to steel of 1.2 MPa.

1 Introduction

Electric tramway is an environmentally friendly, affordable and convenient type of public transport. However, it also has certain shortcomings, such as hindering movement of other modes of transport, generating noise and vibration, as well as permanent leakage currents from rails which can cause electric corrosion of nearby metal and reinforced concrete structures and pipelines. The use of continuous welded tramway track with insulated rails on the ballastless reinforced concrete base integrated into the roadway cannot completely eliminate the above shortcomings; therefore the rationale for measures to reduce vibration, noise, leakage currents and temperature stresses in such tracks is currently relevant.

[*] Corresponding author: plugin.da@gmail.com

The purpose of the study is theoretical and experimental rationale of measures aimed at reduction of vibration, noise, temperature stresses and leakage currents in tramway tracks with insulated rails on the ballastless reinforced concrete track.

2 Analysis of design features and operating conditions of tramways with insulated rails

The tramway with isolated rails on a ballastless reinforced concrete base integrated into the roadway is becoming increasingly common everywhere in the world [1]. Rails should be installed in railway channels to make sure railheads are on level with the road surface. The rails are mounted on continuous polyurethane rail pads and attached to the concrete walls of the channels and, at the same time, are isolated from them by polyurethane seal. A significant reduction in noise and vibration can thus be achieved.

The disadvantage of this design is high consumption of expensive polyurethane sealer, plastic pipes are laid in the gap between the rail and the channel wall or concrete liners are glued to the rail web to save the sealer.

A characteristic feature of the design being analyzed is the absence of any mechanical fastening of rails to the base. The rails are held in the channel only with the seal which actually acts as a wedge for the rail, and vertical and longitudinal movement is prevented mainly by its adhesion to the concrete surface of the channel walls. As a result, the continuous welded tramway of such design can be unstable, especially when block rails are used, which is evidenced by cases of distortion of the track from temperature stresses [2]. Leakage currents from tramway tracks are known to cause electric corrosion of nearby metal and reinforced concrete structures and pipelines. However, the assumed high protective properties of the analyzed design are still not investigated.

When trams are moving, derailments caused by the wear and tear of the track and rolling stock, speed infringement due to the human factor and traffic accidents occur periodically [2]. Additional factors contributing to such derailments may include electric forces caused by the accumulation of excess electric charges of both natural and anthropogenic origin on the tramway and rolling stock [3, 4].

The tramway design with insulated rails on the ballastless reinforced concrete base has been improved with reference to the tramway in Kharkiv (Ukraine) (Fig. 1) [5].

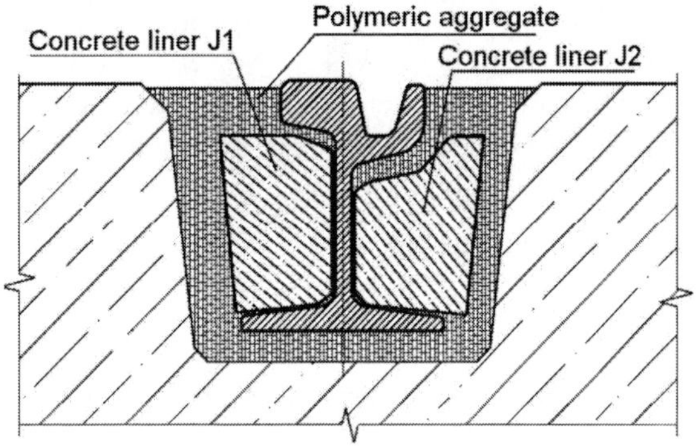

Fig. 1. A rail isolated with a polyurethane composition with concrete liners in the reinforced concrete slab channel.

Concrete liners which are rigidly glued to the rail web only help save polyurethane seal, but also damp vibrations of rails.

3 Main study materials

Prof. A.N. Plugin suggested to use liners (Fig. 1), in particular to reduce the temperature stresses in rails. For this purpose, the minimum possible coefficient of linear thermal expansion α_L should be provided for liners, which would be much lower than that of steel. Since the most volume of concrete is filled with aggregates, aggregates with minimum values of α_L should be used.

Table 1 contains values of the coefficients of linear thermal expansion α_L of the materials of the tramway track shown in Fig. 1

Table 1. Coefficients of linear thermal expansion α_L of the materials of tramway track [7].

Material	α_L, K^{-1}
Steel	13×10^{-6}
Heavy concrete	10×10^{-6}
Granite	7.9×10^{-6}
Quartz	$(0.77\text{-}1.4)\times10^{-6}$
Polyurethane	57.6×10^{-6}

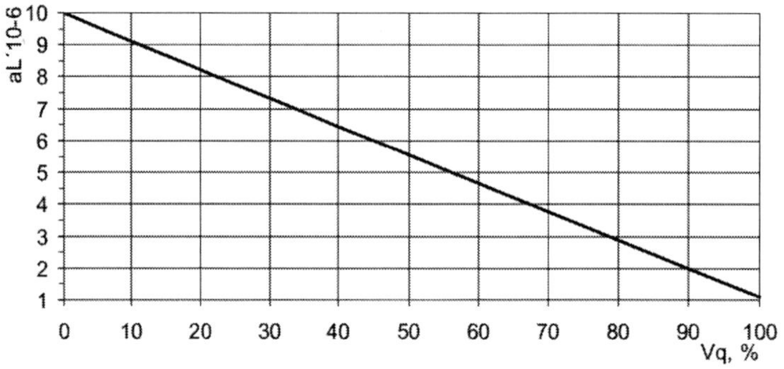

Fig. 2. Dependence of the coefficient of linear thermal expansion of concrete α_L on the volume content of quartz aggregates and fillers V_q in it.

According to Table 1, quartz is characterized with minimum values α_L – in average, 1.1 K^{-1}. Replacement of crushed granite in heavy concrete with crushed quarts, and part of cement (C) with quartz filler (F) will allow decreasing α_L by an order of magnitude – from 10×10^{-6} K^{-1} to about 1.1×10^{-6} K^{-1}. Supposing α_L of cement rock is near to α_L of heavy concrete, and α_L depends inversely on the volume content of quartz (crushed stone (CS), sand (S) and filler) V_q, this dependence will be written as shown in Fig. 2.

3.1 Analysis of the stress-strain state of the rails during heating

The stress-strain state of rails during heating has been analyzed to verify the assumption that the temperature stresses in rails decrease after concrete liners with a lower value of α_L are glued onto it. The stress-strain state of such rails has been analyzed by the finite element method using Lira software complex. Isofields of temperature stresses in the rail without liners and with liners from a temperature increase are presented in Fig. 3, 4, the results of the analysis are given in Table 2.

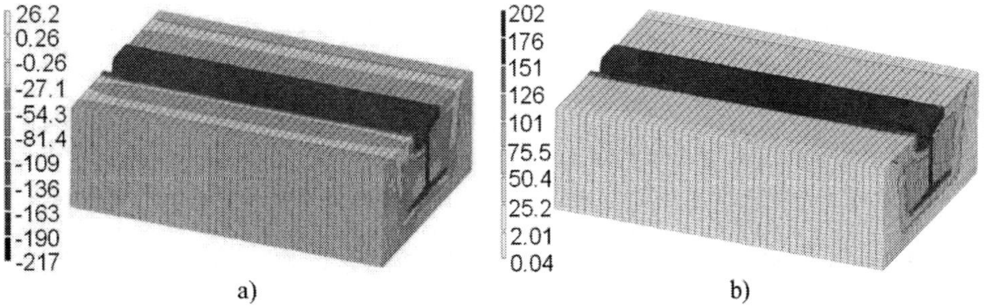

Fig. 3. Stress isofields in the rail with liners when the temperature rises by 80 K: a – main; b – equivalent.

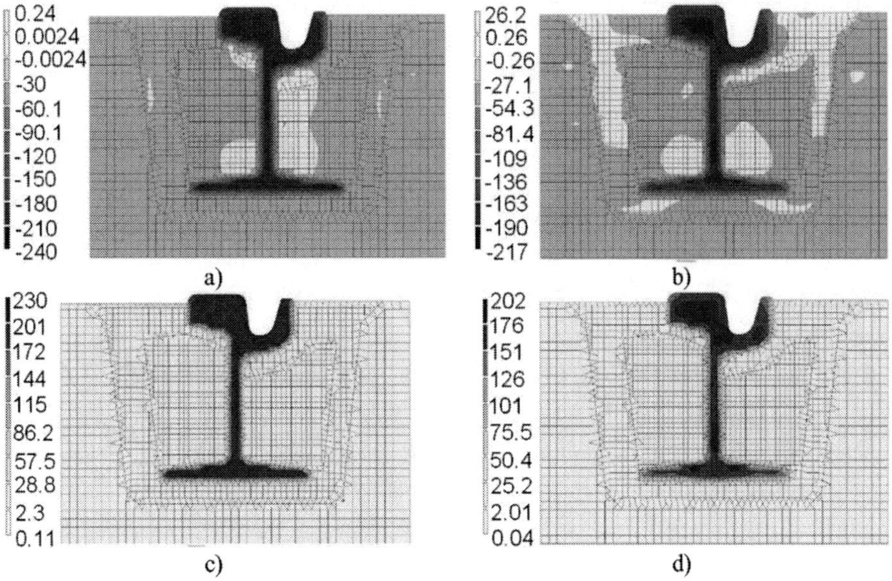

Fig. 4. Stress isofields when the temperature rises by 80 K: a, b – main; c, d – equivalent; a, c – in the rail without liner; c, d – in the rail with liners.

Table 2. Results of the analysis of the stress-strain state of the rail without liners and with liners during heating.

Heating vs. the initial temperature (stress-free temp.)	Stresses	Rail Section area	Stresses in the rail, MPa		Effect of the liners – decrease in stress, %
			without liners	with liners	
by 20 K	main compressive N3	web	57.4	40.7-52.5	12-32
		head		42.0-54.4	10-30
		max		54.4	10
	equivalent		60.1	50.4	12
by 80 K	main compressive N3	web	240	163-195	18-32
		head		175-217	10-27
		max		217	10
	equivalent		230	202	12

Characteristics of the finite element model are as follows:

– length: 1 m, along the ends of the binding rail (along the axis) with finite rigidity being an order of magnitude greater than for all other bonds (maximum displacement of end butts during heating by 80 K is 0.02 mm);

– in the transverse direction along the outer boundary of the model (slab), bonds of finite stiffness are established ensuring the displacement of the boundaries at 80 K not more than by 0.028 mm in horizontal directions and not more than by 0.003 mm vertically downwards;

– slab concrete – C25/30, liner concrete – C32/40, α_L of liners – 1.1×10^{-6} K^{-1};

– rail heating is uniform heightwise, while along other sections – according to the stress diagram from 100% (in the point of contact with the rail) to 40% of the rail temperature from top to bottom, liners – from 90% (in the point of contact with the rail) to 50% of the rail temperature.

Thus, the analysis of the stress-strain state of the rail during heating shows that liners made of concrete on quartz aggregates reduce by 10-32% temperature stresses in rails which cause distortion of the track.

3.2 Rationale of the concrete composition of liners

The liner material is concrete of optimum composition [6], with the maximum strength and low deformability. The properties of such concrete determine its known optimal structural characteristics, such as the coefficients of separation of grains of crushed stone α_{OPT}, sand μ_{OPT}, $(W/C)_{OPT}$ and the optimum amount of the superplasticizer $(SP)_{OPT}$. Optimum coefficients of separation of grains of crushed stone α_{OPT} and sand μ_{OPT} are determined by expressions [8]:

$$\alpha_{OPT} = 2.1\cdot\left(1+\frac{d_S}{d_{CS}}\right)^3 - 1.1, \quad \mu_{OPT} = 2.1\cdot\left(1+\frac{d_C}{d_S}\right)^3 - 1.1 \qquad (1)$$

where d_{CS}, d_S and d_C – average size of crushed stone and sand grains and cement particles, mm.

Crushed stone with fineness 5-10 mm and average grain size d_{CS} 7.5 mm was used. The average sand grain size d_S can be calculated using the known ratios: for sand with fineness modulus FM = 1, d_S = 0.23 mm; FM = 2.5, d_S = 0.49 mm.

Based on the above, an approximate diagram was plotted, Fig. 5, according to which for FM = 1.9 d_S = 0.37 mm.

If we substitute these values to (1), we will get: $\alpha_{OPT} = 2.1\times(1+0.37/7.5)^3-1.1 = 1.32$.

Experimental verification of the value of μ_{OPT} has been performed by making concrete samples using formulations with different values of μ, Table 3, and testing them for compressive strength, Table 4.

This value has been confirmed experimentally.

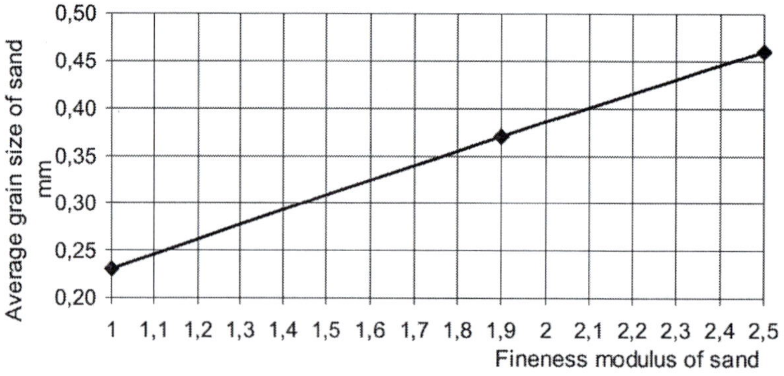

Fig. 5. Approximate dependence of the average sand grain size on its fineness modulus.

Table 3. Experimental concrete compositions and structural coefficients for compressive strength testing of concrete.

Composition No.	S	CS	C	W	C/W	α	μ
1	1.000	900	350	182	1.75	2.56	1.09
2	800	1.100	350	175	1.85	1.83	1.33
3	600	1.300	350	167	2.00	1.32	1.73
4	400	1.500	350	158	2.17	0.95	2.51

Table 4. Compressive strength of concretes, MPa.

Composition No.	without additive			with additive Sika Plast 2508		
	3 days	7 days	28 days	3 days	7 days	28 days
1	23.2	30.0	40.3	29.3	38.2	49.4
2	38.1	48.0	54.0	42.2	51.3	62.1
3	37.4	47.5	53.8	47.3	59.4	69.0
4	33.8	42.1	49.4	39.0	49.7	60.3

In addition, in a concrete mix with a superplasticizer additive, unlike the plain concrete, the like-charged cement particles and sand grains bounce off each other, and the mixture turns into a suspension. Setting of such a suspension begins with the formation of ettringite, which grows with its long side along the normal from the surfaces, similarly to the growth of ettringite in the voids (large air-bubble cavities), Fig. 6.

Judging by the scale of this picture, the length of the ettringite blocks is about 18-20 μm, in average 19 μm. Therefore, in accordance with (1) $\mu_{OPT} = 2.1\times(1+2\times0.019/0.37)^3-1.1 = 1.72$, which is also close to the experimental value $\mu_{OPT} = 1.74$, Fig. 7.

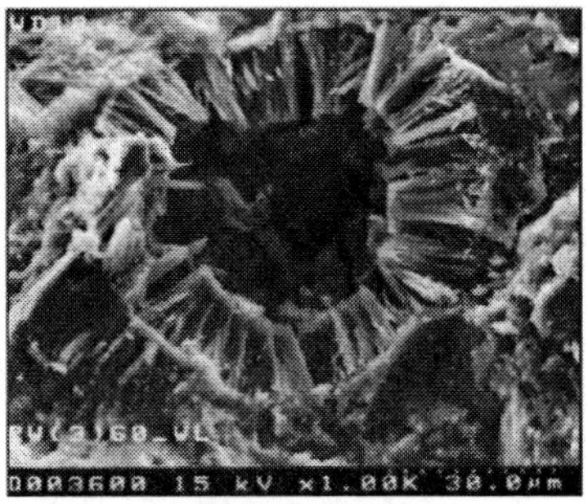

Fig. 6. Ettringite in air voids [9].

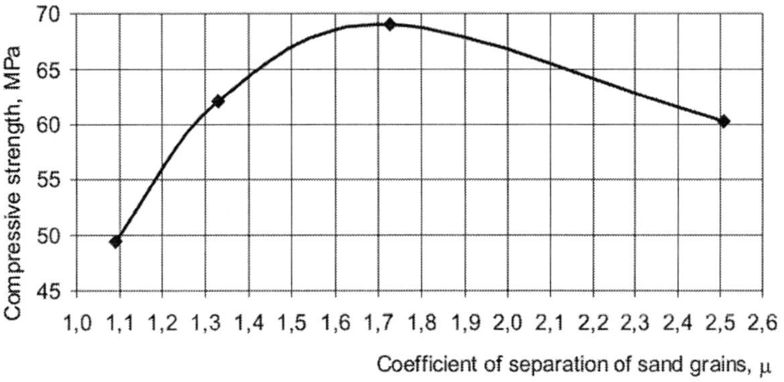

Fig. 7. Dependence of the compressive strength of concrete with the addition of a superplasticizer on the coefficient of separation of sand grains μ and C/W.

Water consumption in the concrete with a superplasticizer additive, according to the experimental data, is 160 kg/m^3 so that a placeable concrete mixture is obtained.

The specific feature of the use of ground quartz as a filler is that its average particle size is 50 μm, similarly to that of cement particles. Thus a uniform, homogeneous mixture of grains can be obtained.

Consumption of crushed stone, sand and water of the liner concrete is determined by the known formulas from [8]. They were, for 1 m^3: CS = 1210 kg, S = 556 kg, W = 167 l.

According to [10], plastic consistency consolidated by vibration SPTsVS is achieved when W/C = 0.32. Accordingly, C = 167/0.32 = 520 kg/m^3.

Absolute volume of the calculated cement is:

$$V_{CP} = \frac{C}{\rho_C} = \frac{520}{3.1} = 168 \quad (l/m^3) \tag{2}$$

Let us assume that absolute volumes of cement and aggregate in the mixture are identical, 0.084 m^3 each. Accordingly, their consumptions will be: C = 0.084×3100 = 260 (kg/m^3); F = 0.084×2.650 = 223 (kg/m^3). Amount of the superplasticizer naphthalene-formaldehyde type additive SP = 0.32×260/100 = 0.83 kg.

The final composition of the concrete of a special rail liner is as follows:

CS = 1210 kg; S = 556 kg; C = 260 kg; F = 223 kg; W = 167 l; SP = 0.83 kg.

The coefficient of separation of the cement mixture and ground quartz $\mu_{OPT \cdot F}$ has its specific characteristics. A mixture of cement and ground quartz with identical particle sizes (50 μm) is placed inside the voids of a mixture of crushed stone and sand. In the presence of a superplasticizer additive, all particles of the mixture of cement and ground quartz have a negative charge, and thus the mixture is a dispersion system. When cement hydration begins, ettringite crystal hydrates are its first products, with the transverse size of 0.3-0.7 μm [11] and an average size of 0.5 μm. When curing begins, they grow from the surface of cement and ground quartz particles as if on the substrate, forming electro-heterogeneous contacts, Fig. 8.

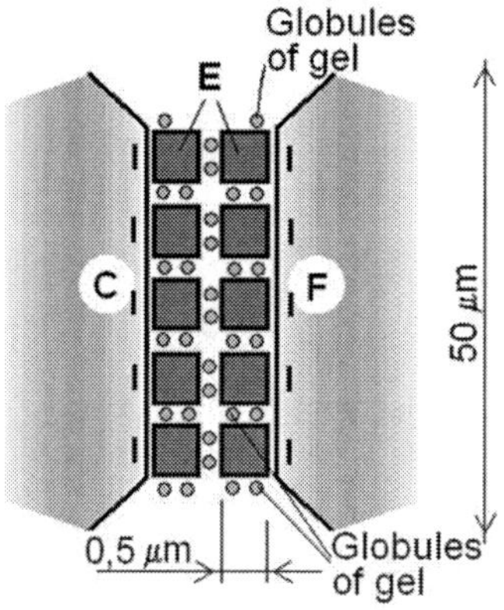

Fig. 8. Electric heterogeneous contact between cement (C) and ground quartz (F) particles via ettringite particles (E).

In accordance with the scheme in Fig. 8, the optimum coefficient of separation of cement particles and ground quartz by the products of cement hydration is $\mu_{C+F} = 2.1 \times (1+2\times0.5/50)^3 - 1.1 = 1.13$.

3.3 Experimental rationale of a polymer composition for rail insulation in the channel

A rail with liners is built into the slab channel, in which the lateral sides extend upward, Fig. 1. This reduces the adhesion of the polyurethane sealing material to the concrete channel, which can draw the rail stack out of it. To prevent this, according to the calculation, adhesion strength of the sealing mass to the steel rail and concrete should be at least 0.7 MPa. The corresponding laboratory tests of two-component polyurethanes Edilon Corkelast® (edilon)(sedra, Netherlands), Icosit KC 340/45 F (Sika, Switzerland-Poland) and new composition "X" (Kharkiv, Ukraine) for adhesion to steel and concrete have been conducted.
The results of the adhesion tests of these polyurethane compositions are summarized in Table 5.

Table 5. Adhesive strength to the surface of steel and concrete of polymeric compositions.

Polyurethane composition	Adhesive strength, MPa		
	To steel	To concrete	
		dry	water-saturated
Edilon Corkelast	>0.8	>0.8	–
Icosit C340/45F	0.88	0.61	0.24
Composition "X"	1.20	0.70	0.43

By the results of these studies, it has been accepted that all investigated polymers can be used as a two-component polyurethane to seal in rails into a tramway slab: Edilon Corkelast (edilon)(sedra, Netherlands), Icosit KC 340/45 F (Sika, Switzerland-Poland) and Composition "X" (Kharkiv, Ukraine). However, the last two in the Ukrainian market are much more economic, than Edilon Corkelast®.

4 Conclusions

1. The design of the tramway track with insulated rails on the ballastless reinforced concrete base, in which the concrete liners that are rigidly glued to the rail web not only help save polyurethane seal, but also damp vibrations of rails and reduce thermal stresses in rails, has been improved.

2. To reduce the temperature stresses in rails, the liners should be made of concrete on quartz aggregates and fillers to ensure a reduction in the coefficient of linear thermal expansion α_L to 1.1×10^{-6} K^{-1}. The analysis of the stress-strain state of rails during heating conducted using the finite element method has shown that the liners made of concrete on quartz aggregates reduce by 10-32% the temperature stresses in rails which cause distortion of the track.

3. The composition of liner concrete with compressive strength over 60 MPa has been substantiated and the polyurethane composition for insulation of rails from the wall of channels with adhesion to dry concrete 0.7 MPa, to water-saturated concrete 0.43 MPa, and to steel 1.2 MPa has been selected.

References

1. T. Shuba, V. Chystiak, V Perestiuk, O. Yeliakina, O.A. Zabiiaka, A.A. Plugin, Collected scientific works Ukrainian State Academy of Railway Transport, **122**, 201-221 (2011)

2. Velkoplošné panely BKV // Prazske tramvaje. Available at: http://www.prazsketramvaje.cz/view.php?cisloclanku=2006041296
3. O.A. Zabiiaka, Collected scientific works Ukrainian State Academy of Railway Transport, **155**, 98-102 (2015)
4. A.M. Plugin, O. Pluhin, O. Palant, O. Konev, A. Nykytynskyj, Matec Web of Conference, **116**, 01011 (2017)
5. O.V. Palant, O.M. Savchenko, D.A. Plugin, A.A. Plugin, Collected scientific works Prydniprovska State Academy of Civil Engineering and Architecture, **104**, 194-199 (2018)
6. D.S. Zakharov, O.V. Palant, S.M. Tolmachiv, D.A. Plugin, P.E. Moisieeva, Collected scientific works Ukrainian State Academy of Railway Transport, **169**, 64-72 (2017)
7. Temperaturnyiy koeffitsient lineynogo rasshireniya // Temperatures.ru. Available at: http://temperatures.ru/pages/temperaturnyi_koefficient_lineinogo_rasshireniya
8. A.N. Plugin, O.A. Kalinin, A.I. Biriukov, Yu.P. Libenko, A.I. Moiseev, Yu.M. Skorodumov, *Sposob opredeleniia sostava betonnoi smesi*: SU1787972[P]. (1993-01-15)
9. J. Stark, K. Bollmann, *Delayed Ettringite Formation in Concrete*. Available at: https://www.imxtechnologies.com/storage/app/media/uploaded-files/ettringite.pdf
10. A.M. Plugin, Art.M. Plugin, O.A. Kalinin, S.V. Miroshnichenko, A.A. Plugin, D.V. Shumyk, D.A. Plugin, A.V. Nykytynskyi, V.A. Liutyi, M.D. Kostiuk. *Superplastyfikovana tsementno-vodiana suspenziia SPTsVS dlia tsementatsii hirskykh porid i budivelnykh konstruktsii*: UA71208[P]. (2008-02-25)
11. A.N. Plugin, A.A. Plugin, O.A. Kalinin, S.V. Miroshnichenko, D.A. Plugin, A.S. Kaganovskiy, Al.A. Plugin, O.V. Gradoboev, *Osnovyi teorii tverdeniya, prochnosti, razrusheniya i dolgovechnosti portlandtsementa, betona i konstruktsiy iz nih* (Naukova dumka, Kyiv, 2012)

Modified Light Concrete of High Strength

Katerina Pushkarova, *Kostiantyn* Kaverin*, and *Danilo* Gadayuchyk

Kyiv National University of Constructioni Architecture, Povitroflotsky avenue, 31, Kiev, Ukraine, 03037

Abstract. The study is devoted to the development of increased strength lightweight concrete based on Portland cement modified with a complex organo-siliceous additive containing a polycarboxylate superplasticizer and an active fine-ground siliceous component. The effect of this complex additive on the physical and chemical characteristics of the hydration and structure formation processes was investigated and it was shown that the high kinetics of the strength gain is provided by directional formation of low-basic calcium hydrosilicates, hydrogranates and plazolite, which are crystallo chemically similar to each other, and provide a dense and strength cement stone matrix. Modification by a complex additive makes possible to obtain concrete mixes with S4 consistency while a Portland cement consumption per 1m3 of not more than 330 kg, with a compressive strength of 55 MPa, watertightness up to W8, frost resistance up to F400, corrosion resistance coefficients Kc in the magnesium sulfate, sodium and ammonium solutions by 10 to 24%. The introduction of a complex additive also helps to reduce the relative shrinkage by 16 to 19% compared to the control composition, which has a positive effect on the durability of the resulting concrete.

1 Introduction

As is already known, lightweight concrete based on expanded clay is used in prefabricated residential, transport and hydraulic construction. Also in the construction of port facilities, in shipbuilding, where, along with the reduction in the mass of structures, other positive properties of expanded clay concrete can be used: increased frost resistance, watertightness, resistance to aggressive exposure conditions.

Research in the field of lightweight expanded clay concrete was carried out by G.A. Buzhevich, A.I. Vaganov, V.M. Virovy, V.G. Dovzhik, V.A. Dorf, O.Sh. Kikawa, L.P. Orentlicher, T.M. Shtolle [1-6].

Abroad in this direction where the works of Rossignolo J. A, Morais J.A., Wee Tiong-Huan, Ergul Yasar, Cengiz Duran Atis, Alaettin Kilic, Hasan Gulsen, Jamal Alduaij, Niyazi Ugur Kockal, Ozturan T., Hugo Costa, Eduardo Julio, Lourenjo J. [7-14].

Fine mineral additives (microfillers) arc an integral part of modern concrete, because their use allows reducing the consumption of cement, increasing the density, strength, resistance of concrete in aggressive environments to guarantee their durability. The positive results that are achieved when active mineral additives are introduced include the binding

* Corresponding author: 1krik.1k1@gmail.com

of portlandite, the formation of low-base hydrosilicates, for example, the C-S-H phase, the reduction of the ratio of capillary and gel micropores in the structure of concrete, the intensification of hydration processes, and the regulation of temperature and volume deformations of cement compositions.

Examples of the practical implementation of additives in the technology of building materials using nanoscale structural elements are given in the papers of Yu.M. Bazhenov, P.G. Komohov, Yu.V. Pukharenko and others. They have showed significant effects achieved by the introduction of various nanoparticles. The trend is based on works by V.G. Batrakov, S.S. Kaprielov, M.A Sanitcky et al. [15-17], related to the modification of the cement stone structure by introducing microsilica and other dispersed mineral additives into the composition of raw mixtures, whose individual fractions can be classified as nanoparticles. The use of nano- and microsilica allows the use of additives along with the superplasticizer to significantly accelerate the processes of hydration and hardening of cement, to provide a high speed of strength gain and reach 115 to 150 MPa by 28 days, increasing the durability of artificial stone by almost 2 to 2.5 times.

However, at the same time, issues related to the study of the replacement of the microsillica with its natural analogues and the study of compatibility with the latest class of superplasticizers based on polycarboxylates, polyacrylates and their esters, need further studying. Also, the issue about the influence of complex organosilicon additives on the processes of structuring high-strength lightweight expanded clay concrete and on their hardened properties still requires clarification.

The prospects for the use of high-strength lightweight concrete are in manufacturing of reinforced concrete products in industrial and high-rise civil engineering, the construction of road bridges, overpasses and interchanges, as well as in the construction of special purpose buildings, due to their positive qualities and advantages vs. heavy concrete. In addition, the reduction in the mass of structural elements of buildings allows solving complicated architectural problems and extends the scope of use of lightweight concrete as structural and insulation material.

Global experience in erection of buildings and structures with the use of structures made of expanded clay (Table 1) confirmed their reliability, durability and high performance of these structures.

Table 1. Comparison of properties of lightweight concrete used in different countries of the world.

Year of implementation	Land	The main characteristics of light concrete		
		Average density of concrete, kg/m^3 (bulk density of expanded clay)	Strength compression for 28 days, MPa	Coefficient of constructive quality (C.C.Q.)
1995	Norway	D1700 to 1950 (M800 to 1000)	60 to 79	35 to 40
1997	USA	D1450 to 1800 (M400 to 600)	18 to 46	12 to 26
2002	Germany	D1800 (M400)	14 to 25	7 to 14
2003	Brazil	D1450 to 1600 (M400 to 600)	40 to 50	27 to 31
2003	Turkey	D1800 to 1860 (M600)	30 to 40	17 to 22
2004	Japan	D1800 to 1850 (M600 to 800)	47 to 54	26 to 29
2007	Russia	D1800 (M600 to 800)	46 to 61	25 to 34

The main precondition for the synthesis and durability of high-strength lightweight concrete is the modification of the cement matrix. In recent years, this has been usually achieved by the use of various complex additives, modifying both binder and concrete mixture.

In this regard, the development of effective high-strength lightweight expanded clay concrete, modified by complex organomineral additives, is a topical task. Consideration of the basic requirements for such additives, namely: component compatibility, availability in the Ukrainian market and the appropriate technical and economic justification for their application will increase the efficiency of using Portland cement. At the same time, the introduction of modified high-strength lightweight expanded clay concrete in monolithic construction requires the study of the peculiarities of the structure of this concrete, taking into account the effect of the compatibility of the action of silica additives of different nature with organic additives.

2 Aim, research objectives, raw materials and research methods

The purpose of the work is to study lightweight expanded clay mixtures based on Portland cement, modified by complex organosilicon additives, and obtaining on this basis high strength lightweight concrete with high physical, mechanical and operational properties.

To achieve this target, the following tasks must be solved:

- to carry out the analysis of already known information sources on the influence of organic and mineral admixtures on the properties of light concrete mixtures and the processes of structure formation during hardening of concrete;
- to develop the composition of a complex organosilicon additive and investigate the compatibility of its components by introducing it into concrete mixtures;
- to optimize the mix design of concrete mixtures and to study the properties of concrete mixtures and concrete.

The subject of the study is high-strength expanded claydite lightweight concrete LC 20/22 to LC 50/55 (DSTU B V.2.7-176: 2008) based on Portland cement modified by a complex organo-silica additive.

The following materials were used as raw materials: Portland cement, CEM I 42,5 R which complies with the requirements of Governmental Standard of Ukraine 2.7-46: 2010 produced by PJSC "Podilsky cement" (Gumentsi village, Ukraine); Dnieper quartz sand according to DSTU B V.2.7-32-95, with Fineness Modulus = 1.28 (Kiev, Ukraine); expanded clay gravel of different grades, the results of which are given in Table 2.

All used silica additives by specific surface area are very finely dispersed, namely: by-products of the Stakhanov Ferroalloy Plant (Blaine = 36,398 cm^2/g), OJSC ArcelorMittal Kryviy Rih (Blaine = 9,478 cm^2/g) and Elkem Microsilica Grade 940-U (Blaine = = 24,795 cm^2/g), SikaFume (Blaine = 8,556 cm^2/g), thin-melting glass "MHS" (Blaine = = 10,485 cm^2/g), trepelle of the Konoplyansky quarry, ground for 2 hours in a ball mill (Blaine = 21,997 cm^2/g).

Polycarboxylate superplasticizers "ViscoCrete-5-600UA", "Plast 555W" (produced by TM Sika, Kiev, Ukraine) and "PowerFlow 3100", "PowerFlow 2695" (produced by TM MC-Bauchemie, Berezan, Ukraine) were used as plasticizers. They meet the requirements of DSTU B V.2.7-69-98, DSTU B V.2.7-171: 2008 (EN 934-2: 2001, NEQ).

The mix design procedure of lightweight concrete was carried out in accordance with DSTU B V.2.7-69-98, DSTU B V.2.7-114-2002, DSTU B V.2.7-176: 2008, DSTU B V.2.7-215: 2009 and mathematical experiment planning methods were used.

The rheological properties of concrete mixtures were determined according to DSTU B V.2.7-114-2002, DSTU B V.2.7-176: 2008. The physical and mechanical properties of the mortar (40 x 40 x 160 mm) and concrete (100 x 100 x 100 mm) were determined according to DSTU B V.2.7-187: 2009 and DSTU B V.2.7-214: 2009, respectively. Open porosity, average density, water absorption and watertightness of concrete samples were determined according to the methods given in DSTU B B.2.7-170: 2009.

Table 2. Characteristics of expanded clay aggregate.

Producer	Bulk dencity, kg/m^3	Compressive strength (cylinder), MPa	Coefficient of thermal conductivity λ, W/m·K	Grade according to DSTU B V. 2.7 –17–95	
				By bulk dencity	By strength
“Plant of expanded clay gravel Novokumol” JSC, Belarus	360	1.65	0.092	M400	S75
“Odessa Keramzite Plant” OJSC	460	1.81	0.1	M500	S75
Khmelnitsky Keramzite Gravel Plant Ltd	600	3.03	0.12	M600	S125

The degree of dispersion of the siliceous additive was estimated using the laser diffraction method on the laser granulometry Mastersizer 2000 Ver. 5.60 and on the Blaine measurement device according to DSTU EN 196-6: 2007. Microhardness of the contact zone “binder – aggregate” of high-strength concrete was determined using a micro-meter PMT-3. The evaluation of the durability of lightweight concrete was performed by studying the kinetics of the strength set, corrosion resistance (DSTU GOST 27677-2011), frost resistance (DSTU B V.2.7-49-96) and deformation of concrete shrinkage according to DSTU B V.2.7-216: 2009.

3 Test results

As previous studies show, the use of a complex organosilicon additive can improve the kinetics of strength gain, and the main regulation factor is the amount of fine-ground tripoli earth in the complex additive [18-20]. Optimization of the composition of lightweight expanded clay concrete, modified with a complex additive, which includes a fine additive based on ground tripoli earth and superplasticizer, was carried out with the use of three-factor three-level mathematical experiment design (Table 2), in which, the consumption of Portland cement (200 to 400 kg) (X1), the content of fine-ground tripoli (5% to 15%) (X2) and the bulk density of expanded clay gravel (M400 to M600) (X3) were used as variable factors.

Table 3. Intervals of variation and values of variable factors of concrete mixtures, modified with a complex organic-mineral additive.

Factors, type		Variation levels			Variation intervals
Natural	Coded	Upper	Middle	High	
Cement dosage, t	X_1	0.4	0.3	0.2	0.1
Tripoli earth dosage, %	X_2	15	10	5	5
Grade of expanded clay by bulk dencity	X_3	600	500	400	100

By analyzing isoparametric diagrams (Fig. 1) of the change in compression strength of lightweight expanded clay concrete, it can be noted that the optimal values of the substitution amount is 10% tripoli of the mass of the binder.

By introducing a complex organic-silica additive with optimal composition and expanded clay M600 and higher makes possible, lightweight expanded clay concrete with compressive strength of 50 to 60 MPa can be produced.

The effectiveness of the modification of Portland cement compositions with a complex organic-silica additive is determined by the nature of the change in the microhardness of the contact zone: "binder – aggregate", the results of which are shown in Fig. 2.

Analysis of the data shows that the microhardness of the samples from unmodified Portland cement to section width of 20 μm from nominal boundary grain filler is gradually increasing from 1,300 MPa (filler) to 2,100 to 2,300 MPa (cement stone), which is 20% higher than microhardness of stone mortar parts of concrete (Fig. 2, mix No. 1).

The modification of the expanded clay concrete by adding the superplasticizer "SikaPlast 555W" (Fig. 2, mix No. 2) increases the microhardness at the border with the aggregate to 2,450 MPa, which is 24% more than the microhardness of the same control mix.

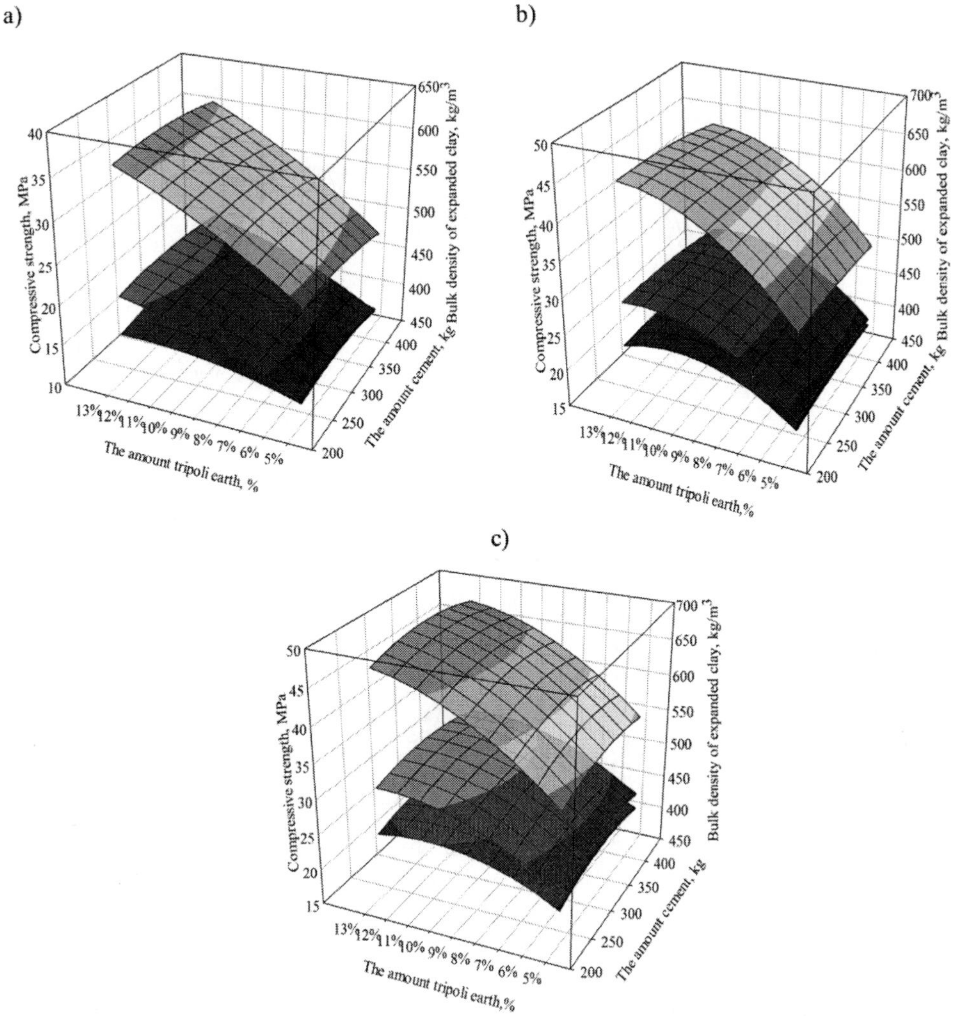

Fig. 1. Isoparametric diagrams of compression strength change for samples concrete on the basis of Portland cement, modified by a complex organo-silica additive based on superplasticizer "SikaPlast 555W" in the amount of 1% and fine-ground tripoli earth (5 to 15%), after hardening in normal conditions during 3 (a), 28 (b) and 365 (c) days.

With introducing a complex organo-silica additive into the lightweight expanded clay concrete (Fig. 2, mix No. 3), the microhardness value on the nominal grain boundary of the aggregate reaches 2,770 MPa, which is 32% more than the microhardness of the cement stone on the boundary of the grain and 19% more than the microhardness of the cement-sand mortar in the intergranular space in comparison with the control mix. The compressive strength of such modified concrete increases by 25 to 30% vs. the composition without additive.

Modification of Portland cement binder systems with a complex organo-silica additive accompanied by the intensive formation of hydrated new substances is represented mainly by low-basic calcium hydrosilicates, which contribute to micro-reinforcement of cement stone, while strengthening its structure and improving adhesion to the aggregate, which is one of the prerequisites for obtaining durable lightweight concrete with increased operational characteristics.

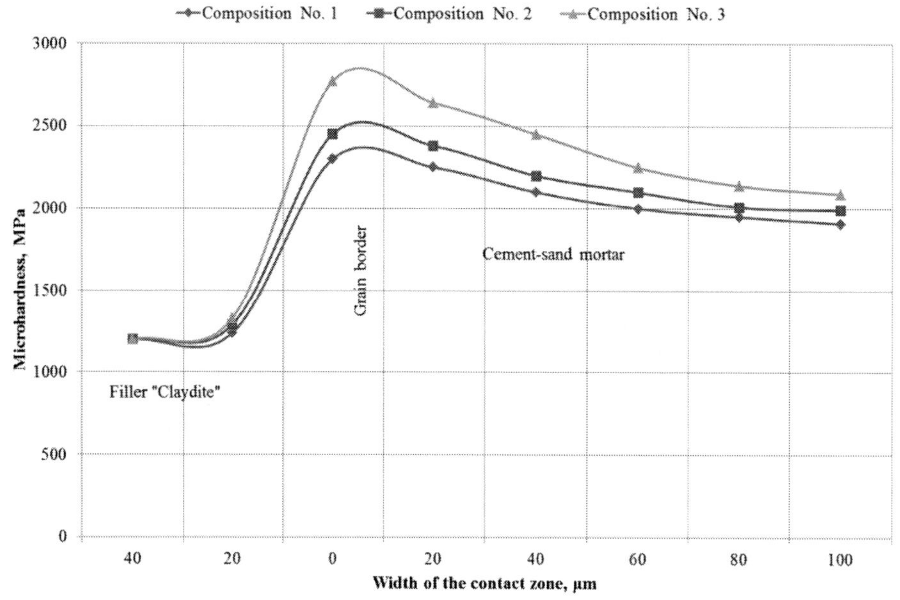

Fig. 2. Change of the microhardness of the contact zone "binding – aggregate" for lightweight concrete: No. 1 - control mix; No. 2 - modified with superplasticizer "SikaPlast 555W"; No. 3 - modified with a complex organo-silica additive on the basis of this superplasticizer and fine-ground tripoli earth.

On the photos of the surface of the artificial stone chip on the basis of Portland cement modified with a complex organo-silica additive fibrous formations and partially recrystallized needle crystals of low-calcium hydrosilicates in the lamellae are clearly visible (Fig. 3), which provides the basis for obtaining high-strength lightweight concrete with a more homogeneous structure.

Thus, when Portland cement modified by a complex additive hardens, due to the reduction of B/C in compressed conditions, there is a directed synthesis of crystallochemically similar compounds and layered reinforcement of the cement matrix with lamellar and needle-like new formation based on low-base calcium hydrosilicates of CSH (I), tobermorite 11.3Å together with platzolite and hydrogarnets that can aggregate and contribute to the formation of a more advanced microstructure of cement stone [21].

The presence in the structure of claydite gravel of isolated closed and open pores promotes penetration of the Portland cement paste into the structure of the expanded clay (due to the self-immolation effect), resulting in interaction between the aluminosilicate component of claydite and the products of hydration of Portland cement, which causes the formation of a stronger contact zone. At the same time, it is likely that in the composition of new formation, in addition to low-basic hydrosilicates of calcium will also be present hydroaluminosilicates, for example, hydrogelinitis.

To assess the performance of lightweight expanded clay concrete, its watertightness should be determined. Water absorption of concrete, as one of the factors of water resistance, can be reduced by increasing the density of the structure of concrete which is achieved by reducing the W/C ratio, as well as the introduction into a concrete mix chemical and mineral additives. Concrete test results are given in Table 4.

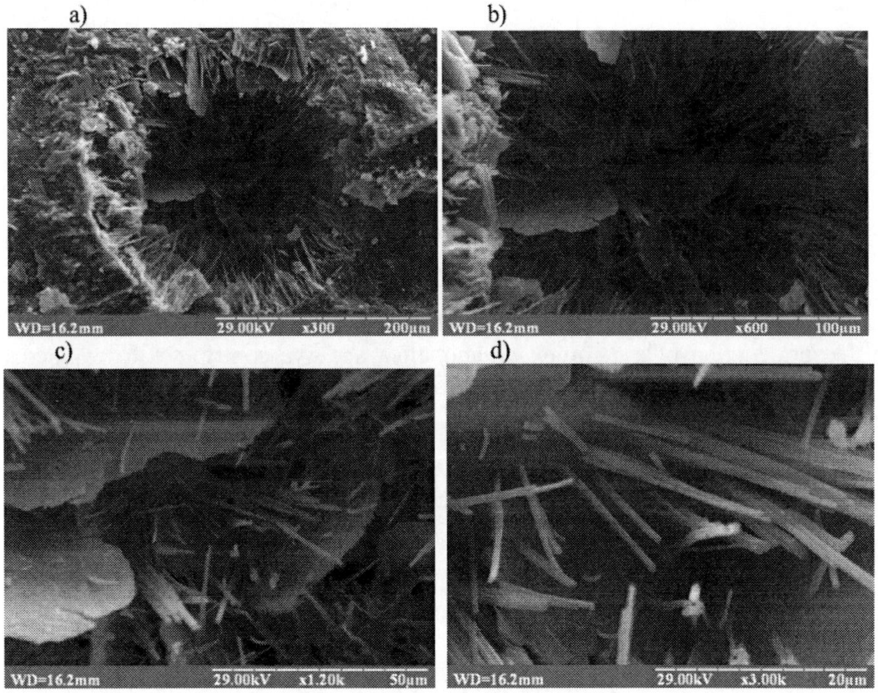

Fig. 3. Photographs of microstructure of cement stone, modified with a complex additive containing polycarboxylate superplasticizer "SikaPlast 555W" and ground shaking tripoli earth after 365 days of hardening: zoom (a) ×300 times; (b) ×600 times; (c) ×1,200 times; (d) ×3,000 times.

The analysis of the data obtained indicates that the complex organo-silica additive developed as a part of lightweight expanded aggregate clay concrete reduces water absorption by 32.6% and reduces volumetric open porosity by 33.5% vs. the additiveless composition, and as a result leads to the formation of optimal structure of concrete.

Waterproofness of concrete plays a significant role in ensuring its water resistance. It is measured by the value of hydrostatic pressure, passing in water through the concrete while it begins to soak. Concrete watertightness is reduced by the use of some technological techniques that increase the density of the structure of both cement stone and concrete in general, which is achieved by lowering the W/C ratio and optimizing granulometric composition. Test results are presented in Table 5.

Table 4. Test results of expanded clay aggregate concrete samples tests for water absorption.

Mix No.	Content of component, wt. %					Average density, kg/m^3	Water absorption, W, %	Open pores, %
	CEM I – 42,5 R	Expanded clay aggregate fraction 5-10, M600	Quartz sand	Superplasticiser «SikaPlast 555W»	Fine-ground tripoli			
1	19.2	23.83	56.97	-	-	1853	3.74	7.04
2	19.14	23.77	56.83	0.26	-	1858	3.2	5.94
3	17.4	23.77	56.83	0.26	1.74	1859	2.52	4.68

In the study of waterptightness of lightweight expanded clay concrete it was established that modification only with superplasticizer provides an increase in watertightness by 13%, while the introduction of a complex organo-silica additive contributes to an increase in watertightness by 37%, compared with the control composition, which corresponds to the grade of watertightness W8 (Table 5).

Table 5. Determination of watertightness of different lightweight expanded clay concrete mixtures.

Mix No.	Content of component, wt. %					Average density, kg/m^3	Compressive strength, MPa	Water pressure, mean, MPa	Watertightness grade
	CEM I – 42,5 R	Expanded clay aggregate fraction 5-10, M600	Quartz sand	Superplasticiser «SikaPlast 555W»	Fine-ground tripoli				
1	19.2	23.83	56.97	-	-	1891	23.9	0.605	W6
2	19.14	23.77	56.83	0.26	-	1897	34.6	0.688	W6
3	17.4	23.77	56.83	0.26	1.74	1898	43.3	0.832	W8

Increasing watertightness can be explained by the predetermined formation of hydration products and reducing the open porosity of clay concrete, which confirms the effectiveness of using this complex additive.

Assessing the physical and mechanical characteristics of modified expanded clay concrete, it is possible to predict that the developed concrete compositions will have improved special properties such as frost resistance, corrosion resistance, as well as reduced shrinkage deformation rates, which to a certain extent depend on water absorption, watertightness and open porosity.

Tests of modified concrete's frost resistance was carried out in accordance with DSTU B V.2.7-49-96 [22], using an accelerated procedure in a 5% sodium chloride solution at a temperature of -50°C. According to the requirements of normative documents, strength loss of specimens that have been under a certain number of cycles of freezing-thawing, should not exceed 5%. The results of studies are presented in Table 6.

The obtained data shows that the control mix of expanded clay concrete on the basis of Portland cement CEM I – 42.5 R provides the frost resistance grade of concrete F200, because the strength loss after 6 cycles of alternating freezing-thawing was 6.82%, exceeding 5%.

Modification of expanded clay concrete with the addition of superplasticizer "SikaPlast 555W" (mix No. 2, Table 6), provides frost resistance F300: strength loss after the 10th cycle of alternating freezing-thawing was 5.61%.

After modification of the expanded clay concrete with a complex additive based on the superplasticizer "SikaPlast 555W" and fine-ground diatomite (mix No. 3, Table 6), there is a rise in frost resistance up to F400, which is confirmed by the results of tests. Loss of strength after the 14th cycle of alternating freezing-thawing was 5%.

The increase of frost resistance of expanded clay concrete on the basis of Portland cement, modified by a complex additive, can be explained by the formation of advanced pore structure and a decrease in the volume of open capillary pores of the material, which provides a smaller amount of ice formed in the pores of concrete at negative temperatures and, accordingly, a decrease in the values of emerging stress forces in the structure of the material.

Table 6. Comparison of frost resistance of expanded clay concrete of different composition.

Mix No.	Content of component, wt. %					Strength loss,% after accelerated methodology freezing-thawing cycles			Frost resistance
	CEM I – 42.5 R	Expanded clay aggregate fraction 5-10, M600	Quartz sand	Superplasticiser «SikaPlast 555W»	Fine-ground tripoli	Number of cycles			
						6	10	14	
1	19.2	23.83	56.97	-	-	-6.82	-	-	F200
2	19.14	23.77	56.83	0.26	-	-	-5.61	-	F300
3	17.4	23.77	56.83	0.26	1.74	-	-	-5.00	F400

The evaluation of the corrosion resistance of concrete was carried out according to the method [23-25]. Samples were made in size 10×10×10 cm, which lasted 24 hours under normal conditions, and then in water. After 28 days of storage, samples were immersed in aggressive environments. Test specimens that were in aggressive environments were carried out after 180 days. Corrosion resistance was determined by the coefficient of corrosion resistance Kc, which is equal to the ratio of the strength of the samples after holding in an aggressive solution (Ragr) to the strength limit of the control samples that were tested at the moment of immersion of the basic samples in the solution (Ro). According to this technique, samples are corrosion-resistant, if, after testing at the age of 180 days, the coefficient of corrosion resistance is $Kc \geq 0.8$.

Control samples of expanded clay concrete and comparative samples were selected for corrosion studies, modified only with superplasticizer and also a complex organo-silica additive. Taking into account possible operating conditions, the corrosion resistance of concrete was investigated in saline solutions, namely in 5% by weight sodium sulfate solution, 5% by weight of ammonium sulfate solution and 1% by weight magnesium sulfate solution.

The residual strength of samples after exposure in aggressive environments for 180 days is given in Table 7, and the results of calculations of the coefficients of corrosion resistance of the investigated samples are presented in Fig. 4.

The analysis of the results suggests that the expanded clay concrete based on the Portland cement binder composition, modified by the additive based on the SikaPlast 555W polycarboxylate superplasticizer and the fine-ground tripoli earth, have rather high coefficients of corrosion resistance compared to the pure Portland cement compositions (Fig. 4, mix No. 3).

The value of K_c of Portland cement clay concrete, modified with a complex additive, in magnesium sulfate solutions increases by 10 to 14% vs. the additiveless composition. In sodium sulfate solutions, this indicator increases by 7.5 to 12.5%, and in ammonium sulphate solutions – by 15 to 24%, compared with the additiveless composition.

The increase in the corrosion resistance of expanded clay concrete can be explained by a decrease in the penetration of aggressive sulfate ions inside the specimens by reducing the open porosity and reducing water absorption, which is achieved by compacting the structure of Portland cement stone and predicted synthesis of increased amount of low-basic calcium hydrosilicates.

Table 7. Change in the strength of expanded clay concrete after storage in aggressive environments.

Mix No.	Content of component, wt. %					Compressive strength, MPa of control samples at 28d	Compressive strength, MPa after aggressive environment storage at 180d		
	CEM I – 42.5 R	Expanded clay aggregate fraction 5-10, M600	Quartz sand	Superplasticiser «SikaPlast 555W»	Fine-ground tripoli		1% solution $MgSO_4$	5% solution Na_2SO_4	5% solution $(NH_4)_2SO_4$
1	19.2	23.83	56.97	-	-	22.9	19.69	18.32	16.03
2	19.14	23.77	56.83	0.26	-	34.9	33.16	30.02	28.27
3	17.4	23.77	56.83	0.26	1.74	38.2	37.44	34.38	33.23

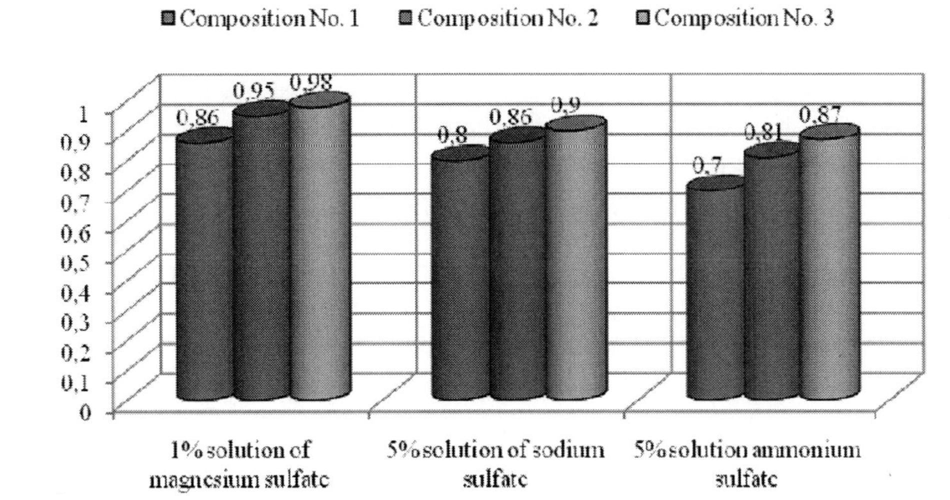

Fig. 4 Comparison of coefficients of corrosion resistance K_s of expanded clay concrete of different composition (according to Table 7) after 180 days aging in aggressive environments: 1%-$MgSO_4$; 5%-Na_2SO_4 and 5%-$(NH_4)_2SO_4$.

An important characteristic of concrete is the value of relative shrinkage, which is defined as the arithmetic mean of the reduction in the size of the sample relative to the initial reference [26]. As known, one of the reasons for shrinkage of Portland cement systems are three-dimensional changes in the gel of a cement stone during drying, which depend on the mineralogical composition of the system, the fineness of cement grinding, conditions and hardening time. At the micro level, the amount of shrinkage depends, first of all, on the ratio of crystalline and gel phases in the products of hydration of the binding composition. The study of the change of relative shrinkage of expanded aggregate clay is given in Table 8 and in Fig. 5.

Table 8. Comparison of relative shrinkage of expanded clay concrete of different composition.

Mix No.	Content of component, wt. %					Kinetics of relative shrinkage of clay concrete, $\varepsilon(t)\cdot10^4$, after days						
	CEM I – 42.5 R	Expanded clay aggregate fraction 5-10, M600	Quartz sand	Superplasticiser «SikaPlast 555W»	Fine-ground tripoli	3	7	14	21	28	60	90
1	19.2	23.83	56.97	-	-	1.49	4.98	7.21	7.71	7.71	7.76	7.76
2	19.14	23.77	56.83	0.26	-	0.99	4.95	6.44	6.93	6.93	6.98	6.98
3	17.4	23.77	56.83	0.26	1.74	0.74	4.46	5.94	6.44	6.44	6.49	6.49

In the study of changes of relative shrinkage of the expanded clay concrete of the above compositions, it was found that the introduction of the admixture of the polycarboxylate superplasticizer "SikaPlast 555W" leads to a reduction of shrinkage by 10 to 12%, while the introduction of a complex organo-silica additive reduces this index by 16 to 19% vs. the control mix.

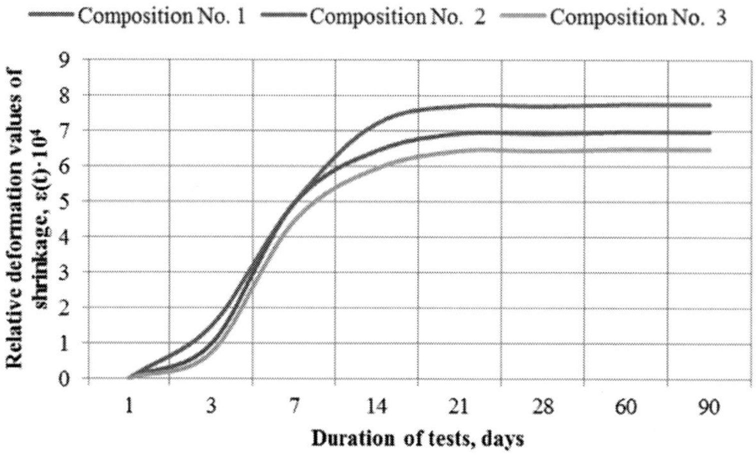

Fig. 5. Comparison of deformations of shrinkage of expanded clay concrete (mixes according to Table 8).

It is known that according to the normative documents for lightweight concrete shrinkage deformation can reach 1 mm/m [26-30]. Thus, the deformation of shrinkage of the additive compositions at the age of 90 days is 0.776 mm/m, whereas modification with a complex additive reduces this index by 16% and is 0.649 mm/m (Table 8 and Fig. 5).

The achieved effect of reducing shrinkage deformation can be explained by the directed formation of low-base calcium hydrosilicates in hydration products.

4 Results discussion

The influence of various grades of claydite gravel on the strength characteristics of lightweight concrete is investigated and it is shown that the use of concrete on the basis of expanded clay aggregate M600 is the most suitable for monolithic construction. It is shown that the highest values of compressive strength increase are associated with concrete mixtures based on expanded clay aggregate (M600), modified with a complex organo-silica based additive based on a polycarboxylate superplasticizer “SikaPlast 555W” in the amount of 1.5% and a fine-groung tripoli earth in an amount of 10% by weight while the intensification of compressive strength gain by 28 days reaches 66%, and after 1 year to 60%, compared with the control mixtures [18-20, 32].

Analyzing the peculiarities of formation of the contact area “aggregate – binder” it was established that when the complex additive is introduced, the composition, gabitius of tumors and the intensity of the process of crystallization in the systems under consideration are changed. The influence of the complex additive on the peculiarities of the formation of the micro-, meso- and macrostructure has been established and a positive role in the synthesis of the strength of crystallochemical low-base calcium hydrosilicates is confirmed, which, in combination with hydroaluminosilicates, which are formed over time, provides a dense bonding and adhesion of the aggregate to the penetrating binder in the pore space of the aggregate grain.

It can also be noted that the use of expanded clay concrete is constantly expanding due to its physical, mechanical and thermal characteristics. For example, Cambio company combines these properties and has created effective expanded clay concrete blocks for the residental construction, which allows you to reduce the cost of construction by 30% and accelerate the construction time by 3 times [31].

According to the results of the study carried out in the batching plant «Budindustriya-1», a production test of total volume of 120 m3 of expanded clay concrete, based on Portland cement modified with a complex organo-silica additive, that was used in the construction of Atlant 2 residential complex, was conducted. The full compliance of the properties of expanded clay aggregate based on Portland cement modified with a complex organo-silica additive, in the actual conditions of production and operation, has been confirmed, which testifies to their high efficiency and functionality in high-rise construction. The economic effect of the implementation of the developed concrete composition was UAH 90.25 for $1m^3$ of concrete.

5 Conclusion

Developed mixes of expanded clay concrete of high strength LC 20/22 to LC 50/55 (DSTU B V.2.7-176: 2008) on the basis of Portland cement, modified with a complex organo-silica additive. It is shown that the introduction of a complex additive provide increasing in durability of expanded clay aggregate concrete at all stages of curing, increases the watertightness of W6 to W8, increases the frost resistance to F400, increases the coefficient of corrosion resistance Ks by 7.5 to 24%, vs. the additiveless composition. It was also shown that the introduction of a complex organo-silica additive reduces the relative shrinkage by 16 to 19%, compared with the control composition, which allows using high strength lightweight expanded clay concrete in residential construction.

References

1. V. H. Batrakov. *Modyfikovani betony. Teoryya y praktyka, yzd. 2-e, pererabotannoe y dopolnennoe* (M., 1998)
2. YU. M. *Bazhenov et al. Modyfytsyrovannye vysokokachestvennye betony* (M, Yzd-vo ASV, 2006)
3. L. I. Dvorkyn, *Tsementni betony z mineralnymy napovnyuvachamy* (Kyyiv, Budivelnyk, 1991)
4. Y. A. Yvanov, *Lehkye betony na shtuchnykh porystykh zapolnytelyakh* (M., Stroyyzdat, 1993)
5. T.M Shtol, *Tekhnolohiya keramzytobetonnykh vyrobiv na horyachomu zapovnyuvachi* (M., Stroyyzdat, 1986)
6. L.P. Orentlykher, *Betony na porystykh napovnyuvachakh u zbirnykh zhelezobetonnykh konstruktsiyakh* (M., Stroyyzdat, 1983)
7. Rossinolo Dzh. A., Ahnesini M.V.S., Morayev YU.A. Tsem Kon Compo., 2003. - №25. - S. 77-82.
8. Vey Tionh-Khuan. *Nedavni podiyi v lehkykh vysokomitsnykh betonakh z ahrehatamy ta bez nykh", Tretye mizhnarodna konferentsiya z budivelnykh materialiv: produktyvnist, innovatsiyi ta strukturni naslidky,* Universytet Brytanskoyi Kolumbiyi, 22-24 (Vankuver, Kanada, 2005)
9. Erhul Yasar, Kenhiz Dyuran Atis, Alaettyn Kilich, Khasan Hulsen, *Sylovi vlastyvosti lehkoho betonu, vyhotovleni z bazaltovoyi pemzy ta lyutserny.* Materialy Lysty, № **57**, 2267-2270 (2003)
10. Dzhamal Alduay, Khalyd Alshalekh, r-n Naseer Khak, Khalid Ellayity *Lehkyy beton u haryachykh pryberezhnykh rayonakh.* Tsementy ta betonni kompozyty, **T. 21**, **5-6**, 453-458 (1999)

11. K H. Al-Khayat, M.N. Khake, Doslidzhennya tsementu ta betonu, **T. 28**, **6**, 859-866 (1998)
12. Padam Shafih, Mokhd Zamin Yumaat, Khilli Makhmud. Budivnytstvo ta budivelni materialy, **T. 25**, **4**, 1848-1853 (2011)
13. Niyazi Uhur Kokal, T. Ozutaran, Materialy ta dyzayn, **T. 32**, **4**, 2396-2403 (2011)
14. Uhu Kosta, Eduardo Khulio, Lourenkho J. Budivelni ta budivelni materialy, **35**, 84-91 (2012)
15. M.A. *Sannitskyy Enerhozberihayuchi tekhnolohiyi v budivnytstvi: navch. Posibnyk.* (Lviv, Vydavnytstvo Lvivskoyi politekhniky, 2012)
16. A. Plugin, E. Dedeneva, T. Kostyuk, D. Bondarenko, O. Demina, Matec Web of Conference, **116**, 01010 (2017)
17. O. Borziak, S. Vandolovskyi, V. Chajka, V. Perestiuk, O. Romanenko, Matec Web of Conference, **116**, 01001 (2017)
18. K.K. Pushkarova, O.A. Honchar, K.O. Kaveryn, Zb. nauk. pratsi «Budivelni materialy, vyroby ta sanitarna tekhnika», **52**, 43-48 (2014)
19. K.O. Kaveryn, Zb. nauk. pratsi «Budivelni materialy, vyroby ta sanitarna tekhnika», **56**, 47-54 (2015)
20. K.K. Pushkarova, K.O. Kaveryn, Zbirnyk naukovykh prats «Resursokomisni materialy, konstruktsiyi, budivli ta sporudy», **33**, 75-83 (2016)
21. K. K. Pushkarova, K. O. Kaverin, D. O. Kalatayevskyy, Skhidnoyevropeyskyy zhurnal pidpryyemstv pidpryyemstva, **5 (77)**, 42-51 (2015)
22. *DSTU B V.2.7-49-96. Budivelni materialy. Betony Metody vyznachennya morozostiykosti. Pryskoreni metody vyznachennya morozostiykosti pry bahatorazovomu zamorozhuvanni ta vidtochenni,* Derzhavnyy standart Ukrayiny (Kyyiv, Derzhkommistobuduvannya Ukrayiny, 1997)
23. M.YU. Leshchynskyy, *Yspytanyya betona* (M., Stroyyzdat, 1980)
24. *DSTU B V.2.6-145: 2011 (HOST 31384: 2008, NEQ). Zakhyst betonnykh i zalizobetonnykh konstruktsiy vid koroziyi. Zahalni tekhnichni vymohy*, Derzhavnyy standart Ukrayiny (Kyyiv, Minrehionbud Ukrayiny, 2010)
25. *DSTU B V.2.6-181: 2011 (HOST 31383: 2008, NEQ). Zakhyst betonnykh i zalizobetonnykh konstruktsiy vid koroziyi. Metody vyprobuvan,* Derzhavnyy standart Ukrayiny (Kyyiv, Minrehionbud Ukrayiny, 2011)
26. *DSTU B V.2.7-216: 2009. Budivelni materialy. Betony Metody vyznachennya deformatsiy usadky ta skorboty*, Derzhavnyy standart Ukrayiny (Kyyiv, Minrehionbud Ukrayiny, 2010)
27. H.S. Solopova, Mizhnarodnyy naukovyy zhurnal «Symvol nauky», **12**, 81-82 (2015)
28. K.S. Karapetyan *Osoblyvosti protsesiv kozatstva ta usadok lehkykh ta inshykh novykh vydiv betoniv. Polzuchest i usadka betonu ta zhelezobetonnykh konstruktsiy* (M., Stroyyzdat, 1976)
29. YU.H. Moskalkova, Nauka ta budivnytstvo, **3**, 40-43 (2017)
30. S.O. Krovyakov, Nauka i budivnytstvo, **3**, 50-58 (2017)
31. V.P. Kholodnyuk, Sbirnyk naukovykh prats «Budivelni materialy», 5-7 (2016)
32. K.K. Pushkarova, Mizhnarodnyy naukovo-vyrobnychyy zhurnal "Keramika: nauka i zhyttya", **1 (34)**, 32-41 (2017)

The efficiency of plasticizing surfactants in alkali-activated cement mortars and concretes

Raisa Runova, *Volodymyr* Gots, *Igor* Rudenko, *Oleksandr* Konstantynovskyi, and *Oles'* Lastivka*

Scientific Research Institute for Binders and Materials, Kyiv National University of Construction and Architecture, 03037 Povitroflotskyi Avenue, 31, Kyiv, Ukraine

Abstract. Functionality of mortar and concrete mixes is regulated by surfactants, which act as plasticizers. The molecular structure of these admixtures can be changed during hydration of alkali-activated cements (AAC). The objective was to determine the chemical nature of plasticizers effective for property modification of mortars and concretes based on AACs with changing content of granulated blast furnace slag from 0 to 100 %. The admixtures without ester links become more effective than polyesters when content of alkaline component increase. The admixtures effective in high alkaline medium were used in dry mixes for anchoring (consistency of mortar 150 mm by Vicat cone; 1 d tensile strength in bending / compressive strength of mortar 6.6 /30.6 MPa) and in ready-mixed concretes (consistency class changed from S1 to S3, S4 with consistency safety during 60 min; 3 d compressive strength of modified concrete was not less than the reference one without admixtures).

1 Indroduction

Application of alkali-activated cements (AAC) allows to provide high quality, functional and durable concretes in the view of efficient use of raw materials and energy. Priority in development of AACs belongs to V.D. Glukhovsky [1, 2]. His scientific school governed by P.V. Krivenko proposed principles of AAC's compositions [3, 4], with involvement of by-products and industrial wastes as well [5, 6]. The mentioned principles underlay the National standard of Ukraine DSTU B V.2.7-181:2009 [7] for practical application. There are five types of AACs depending on composition (% mass of aluminosilicate components): I – granulated blast-furnace slag (GBFS) (90-100), II – portland cement clinker (100), III – fuel fly-ash or basalt (36-80), IV – GBFS (36-89), V – GBFS and fuel fly-ash in total (90-95). Thus the basic factor of AAC composition can be considered as part of GBFS that can change from 0 % (type II) to 100 % (type I) in aluminosilicate component of cement. Compounds of alkaline metals are considered as AAC structurization components, which contents depend on AAC type. Sodium lignosulphonate (LST) is indispensable component of AAC based on GBFS for ensuring of satisfactory setting time and strength of the cements [7, 8].

* Corresponding author: oles.lastivka@gmail.com

It is necessity in chemical admixtures application to obtain modern concretes and mortars. However modification of AAC by admixtures has certain features connected with risk of their molecular structure degradation in hydration medium of AAC due to alkaline hydrolysis (saponification) [9, 10, 11]. The emergence and speed of this process are determined first of all by AAC's composition including nature (group) of alkaline component [12]. For one's turn the efficiency of surfactants as admixtures is caused by their surface activity, depends on molecular mass and adsorption. Compatibility of AACs with aliphatic organic compounds (polyols, polyesters, ethoxylated fat alcohols, salts of carbon acids etc.) as surfactants with plasticizing effect was demonstrated [13, 14, 15]. Surfactants can be used to ensure low water consumption with decrease in the surface tension of porous fluid. It is known that the latter circumstance facilitates migration of porous fluid, respectively, reducing the capillary stress that causes shrinkage [16]. In turn, shrinkage is considered as one of the criteria for autogenous self-healing of cementious materials [17]. Control methods for shrinkage of mortars by changing of AAC composition were considered [18]. To ensure the effective influence of plasticizers on fresh and hardened concretes and mortars based on AACs, the investigations in effective surfactant's nature are reasonable.

The purpose of the paper was to determine the substantiate effectiveness of different chemical types of plasticizers in concretes and mortars while AAC's composition undergoing changes.

2 Experimental program

2.1 Raw Materials

Aluminosilicate components of AACs (% by mass): portland cement clinker (CaO – 66.15; SiO_2 – 22.61; Al_2O_3 – 5.29; Fe_2O_3 – 3.93; MgO – 0.84; Na_2O – 0.15; K_2O – 0.98; SO_3 – 0.50) and GBFS (CaO – 47.30; SiO_2 – 39.00; Al_2O_3 – 5.90; Fe_2O_3 – 0.30; MgO – 5.82; MnO – 0.50; SO_3 – 1.50).

Alkaline components of AACs: sodium silicate pentahydrate $Na_2SiO_3{\cdot}5H_2O$ (SSP) and sodium carbonate ash Na_2CO_3 (SC).

Calcium sulphate (CS) in modification of hemihydrate $CaSO_4{\cdot}0.5H_2O$ was used as setting retarder, density 2.7 g/cm^3.

Plasticizers: «JK-04PP» based on polycarboxylate esters (PC), «JIANKAI» (Chine); sodium lignosulphonate (LST), «Borrespers» (Norway); «Melflux PP100F» based on modified polyethylene glycol (PEG-M), «SKW Polymers» (Germany); PEG-400 based on polyethylene glycol (PEG), «Dow Chemical» (Germany); «Triameen Y12D» based on acyclic polyamine (APM), «AkzoNobel» (Netherlands).

2.2 Experimental Process

The research was realized in two stages.

The first stage. Stability of plasticizers' molecular structure in hydrated medium of AAC with different content of GBFS was investigated by means of infrared spectroscopy (IR) by Specord 75 IR (Germany).

Five compositions of AAC were used:

– № 1 (clinker – 97.6 %, SSP – 2.4 % (0.7 % on Na_2O)) – AAC type II (alkali-activated portland cement, AAPC);

– № 2 (clinker – 67.0 %, GBFS – 30.1 %, SC – 2.9 % (1.7 % by Na_2O)) and № 3 (clinker – 66.7 %, GBFS – 30.0 %, SSP – 3.3 % (1.0 % by Na_2O)) – AAP type IV (alkali-activated slag portland cement, AASPC);

– № 4 (GBFS – 95.2 %, SC – 4.8 % (2.8 % by Na_2O)) and № 5 (GBFS – 94.3 %, SSP – 5.7 % (1.7 % by Na_2O)) – AC type I (alkali-activated slag cement, AASC);

AACs with specific surface 450 m^2/kg (by Blaine) were produced by mixing the components in "HOBART" mixer. The cements were modified by admixture based on ethylhydroxysiloxane polymer (0.04 %) for intensification of grinding and to prevent absorption of water from air as well as for maintaining of AACs performances.

The specimens for IR tests were obtained from cement pastes with one of the mentioned plasticizers (10 % AAC mass) at water-to-cement ratio = 0.4. After 7 d hardening at normal conditions (t= 20±2 °C, RH= 95±5 %) the specimens were dried at t= 40 °C to constant mass and grinded for full passing through sieve 0.08 mm.

The second stage. Dry mixes for anchoring and ready-mixed concretes based on AACs were modified by admixtures most effective in high alkaline medium (see *first stage*).

Slump of fresh concrete mix was determined according to EN 12350-2:1999, compressive strength - EN 12390-3:2001, consistency (flow) of fresh mortar by Vicat cone - EN 196-3:2007, tensile strength in bending and compressive strength of mortars - EN 12190:1999 and EN 196-1:2007, mortar adhesion - EN 1542:1999.

3 Results and discussion

The first stage. Changes of surfactants' molecular structure in AAC medium are shown in fig. 1, 2, 3.

Hardened AAC was characterized by the follow phases (curves 1 - 5, fig. 1, 2, 3): anhydrous calcium silicates (450, 520, 870-880 cm^{-1}), hydrated calcium silicates (920-980, 3400-3450 cm^{-1}), chemically combined and adsorbed water of hydrated calcium silicates (3400-3450 cm^{-1}), molecular water (1620-1650 cm^{-1}), calcium hydroxide (3620-3660 cm^{-1}).

PC admixture was defined by the effect at 1730 cm^{-1} (C=O, ester group) and 1100 cm^{-1} (C-O-C, ether group) (fig. 1a). The admixture undergoes changes in molecular structure due alkaline hydrolysis of ester groups with formation of carboxylate acids and corresponding ethers [19]. The bonds at 1415 cm^{-1} and 1520-1560 cm^{-1} (COO-, carboxylate group) confirm the molecular structure degradation because of alkaline hydrolysis. However hydrolysis was partial that is evidence by effect at 1730 cm^{-1} (curves 1 - 5). Higher destruction of PC admixture took place due to increasing content of alkaline component (by Na_2O) from 0.7 % (curve 1) to 1.7 % (curve 5) that was prescribed by increasing volume of GBFS in AAC. This effect can be confirmed by intensity of bonds at 1415 cm^{-1}, 1520 - 1560 cm^{-1} and decreasing effects at 1730 cm^{-1} and 1100 cm^{-1}.

Thus molecular structure of PC admixture is unstable in alkaline medium. The main chain with carboxylate groups adsorbs on surface of cement particles. Side chains tear from main chain due to hydrolysis of ether bond. Agreeably plasticizing effect of PC admixture due to steric effect of side chains in AAC's medium is significantly less due to increasing of alkaline component content (by Na_2O) from 0.7 % (AAC type II) to 1.7 % (AAC type IV). SSP caused higher degradation of PC admixture than SC, that was fixed for AAC with 1.7 % of alkaline component (by Na_2O) for compositions № 5 (curve 5) and № 2 (curve 2) respectively.

The mentioned effect can be explained by higher pH value of SSP in comparison of SC in the medium, i.e. 12.6 and 11.3 agreeably. Hence stability of PC admixture's molecular structure in hydration medium of AACs depends on the content and nature of alkaline component.

LST admixture unlike to PC admixture characterizes by stable molecular structure in AAC medium. Stability of molecular structure was confirmed by the corresponding effects on all IR curves (fig. 1 b): 620, 650, 780 cm^{-1} (CH_2 bonds); 660, 1200 cm^{-1} (SO_3 groups); 1020 cm^{-1} (OH- groups); 1100 cm^{-1} (OH alcohol groups); 1360 cm^{-1} (CH_3 bonds); 1430, 1500, 1590 cm^{-1} (substituted aromatic ring of lignin); 1630 cm^{-1} (C=C bonds in benzol ring, C=O carbonyl groups and COOH groups); 1710 cm^{-1} (C=O group of aromatic acids and COOH groups); 2910 cm^{-1} (CH_2 bonds); 3380, 3450 cm^{-1} (O-H bonds in hydroxyl, phenol, alcohol and carboxyl groups). The results concerning molecular stability of LST admixture in AAC medium correlate with well known results [20].

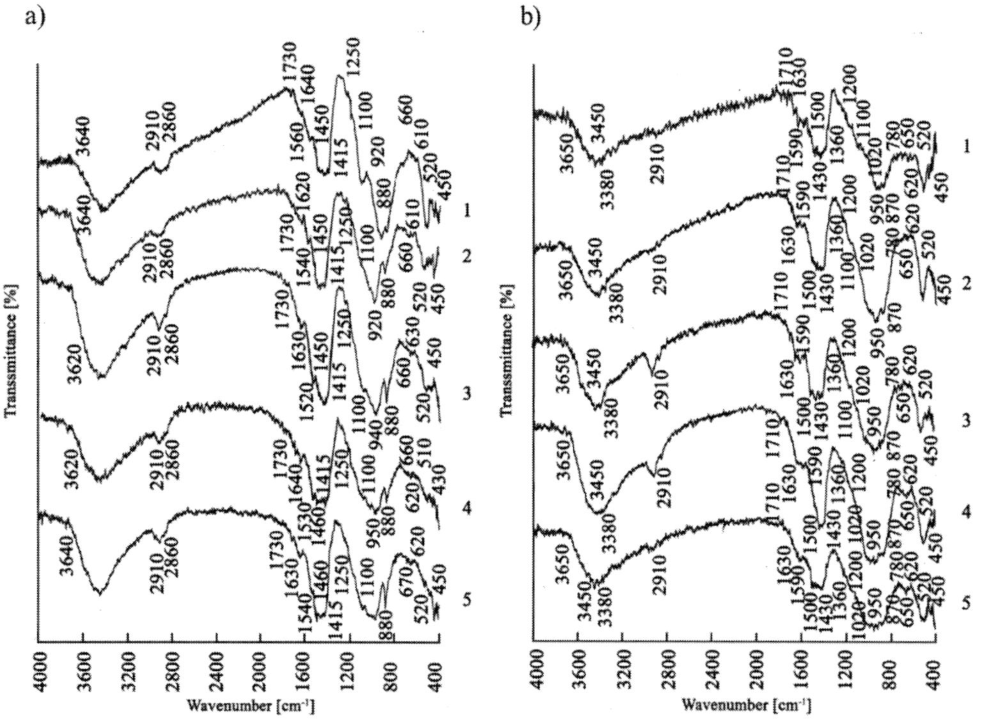

Fig.1. IR spectra of PC (a) and LST (b) in hydration medium of AAC's compositions: 1 – №1; 2 – №2; 3 – №3; 4 – №4; 5 – №5.

PEG admixtures can be also characterized by stable molecular structure in AAC's medium. Characteristic effects (fig. 2 a) are: 710, 2920 cm^{-1} (CH_2 bonds), 1100 cm^{-1} (C-O-C ether group) and 3620 cm^{-1} (OH- group). PEG is nonionic surfactant without dispersible effect. However, this organic substance acts as dispersant after attaching to carboxyl groups (–COOH) of superplasticizer molecules [21]. Such method for increasing adsorption was realized in admixture based on modified polyethylene glycol (PEG-M). Melflux PP100F is copolymer of PEG and acrylic acid, ensure much higher plasticizing effect but is unstable in high alkaline medium. This type of admixture undergoes insignificant changes in medium of AAC type II (fig.2 b, curve 1) only when the content of alkaline component in AAC is low. Effects at 1720 cm^{-1} (C=O, ester groups) confirms such fact. And the presence of effects at 1415 cm^{-1} and 1520 cm^{-1} (COO-, carboxylate group) confirms partial hydrolysis of the admixtures. Increasing content of alkaline component (by Na_2O) from 0.7 % (curve 1) to 1.7 % (curve 5) ensures lack in stability of molecule structure, that can be confirmed by decreasing effect at 1720 cm^{-1} (ether group) with increasing at 1415 cm^{-1} and 1520 cm^{-1}. Destruction of the admixture in presence of SSP is higher in comparison with SC in AAC.

Dodecyl dipropylene triamine (acyclic polyamine, APM) acts as cationic surfactant regarding to AAC and is characterized by stability in high alkaline medium. The effects at 1230 cm^{-1} (C–N bonds, aliphatic amines), 1450, 2860, 2920 cm^{-1} (CH_2, CH_3), 1490 cm^{-1} (C–H, aromatic amines), 1620 cm^{-1} (N–H bond, primary amines), 3320 and 3430 cm^{-1} (two bonds N–H, primary amines) are present on all curves.

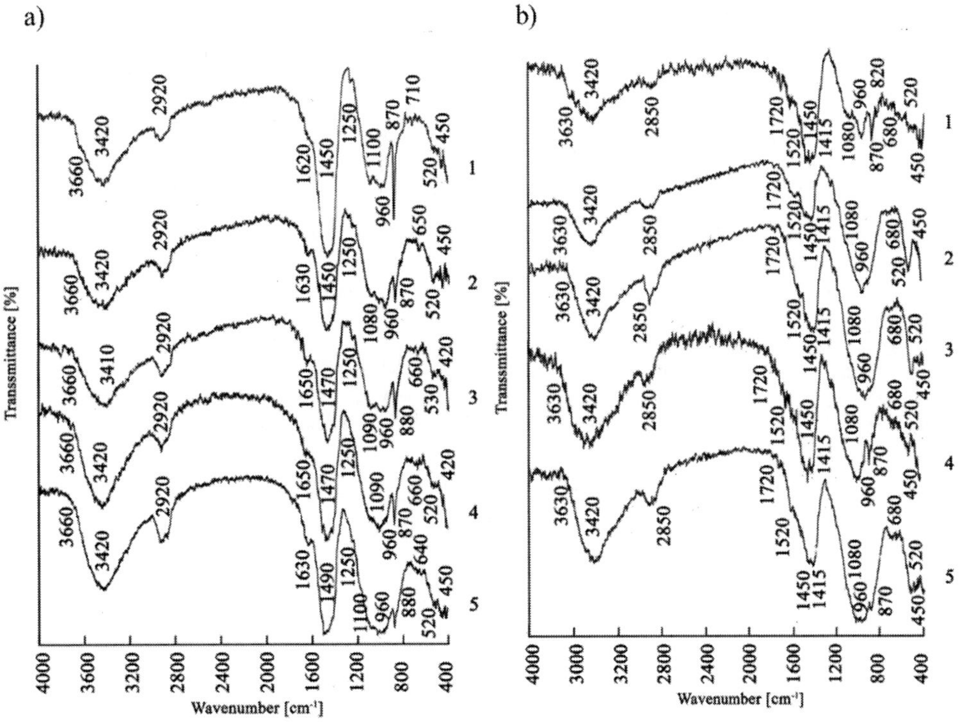

Fig.2. IR spectra of PEG (a) and PEG-M (b) in hydration medium of AAC's compositions: 1 – №1; 2 – №2; 3 – №3; 4 – №4; 5 – №5.

Thus LST, PEG and APM are admixtures with stable molecular structure in AAC medium in all range of alkaline component content's changing. PC and PEG-M admixtures are effective at law content of alkaline component i.e. 0.7 % by Na_2O (AAC type II). Increasing of pH figure that is caused by content and nature of alkaline component ensure significant decreasing in effectiveness of admixtures with ester bands as a result of alkaline hydrolysis.

Second stage. Compositions of dry mixes for anchoring and ready-mixed concretes based on AAC were developed on the grounds of obtained regularities.

AAC type II in system «clinker – SSP – CS» was used in mortar for anchoring. AAC types I and IV in system «clinker – GBFS – alkaline component (SSP, SC) – CS» were used for ready-mixed concretes. LST admixture, 0.8 % mass of aluminosilicate components, was used as necessary component in AAC types I and IV.

CS was used as setting retarder and to increase strength of AAC. The effectiveness of this additive in AAC was already demonstrated [8, 22]. CS was added above 100 % of aluminosilicate component mass: 9.0 % and 6.5 % for AAC type II and AAC type IV agreeable.

PEG-M admixture (Melflux PP100F) was used as water-reducing admixture in mortar for anchoring based on AAC type II. Consistency of mortar was 150 mm. The values of tensile strength in bending / compressive strength of mortars 6.6 /30.6 MPa in 1 d and adhesion 1.1 MPa were obtained.

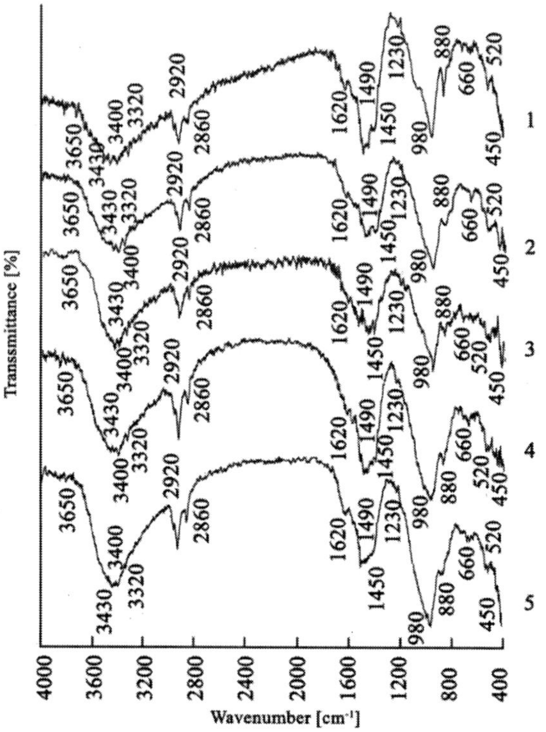

Fig.3. IR spectra of APM in hydration medium of AAC's compositions: 1 – №1; 2 – №2; 3 – №3; 4 – №4; 5 – №5.

4 Conclusions

It was shown that PC and PEG-M admixtures with ester bonds undergo of molecular structure degradation as a result of alkaline hydrolysis at increasing pH values that is caused by content of alkaline component depending on GBFS content and on the nature of alkaline component. LST, PEG and APM admixtures are characterized by stability of molecule regardless of alkaline component content. Effectiveness of admixtures without ester bonds (LST, PEG and APM) is higher at increasing of alkaline component's content agreeably to increasing of GBFS content in AAC type I and IV.

Practicability in application of admixtures according to AAC type was confirmed by properties of mortars and concretes. Thus mortar for anchoring, based on AAC type II modified with PEG-M admixture, was characterized by appropriated consistency (150 mm) and tensile 1 d strength in bending / compressive strength 6.6 /30.6 MPa and adhesion 1.1 MPa. PEG and APM admixtures in presence of LST define changing in consistency of fresh concrete from S1 to S3, S4 and its safety during 60 min; compressive strength of modified concrete was provided at the level of non-modified analog.

References

1. V. Glukhovsky, G. Rostovskaya; G. Rumina, *High strength slag-alkali cement*, 7th Int. Congr. on the Chemistry of Cements, Paris, **III**, 164-168 (1980)
2. V. Gluhovsky: *Alcaline and alcalaine-alkaline-earth hydraulic binders and concretes*, (1979)

3. P. Krivenko, Journal of Ceramic Science and Technology **8(3)**, 323-333 (2017)
4. P. Krivenko, O. Petropavlovskii, H. Vozniuk, S. Lakusta, Procedia Engineering, **195**, 142-146 (2017)
5. T. Croymans, W. Schroeyers, P. Krivenko, O. Kovalchuk, A. Pasko, M. Hult, G. Marissens, G. Lutter & S. Schreurs, Journal of Environmental Radioactivity, **168**, 21-29 (2017)
6. P. Krivenko, O. Kovalchuk, A. Pasko, T. Croymans, M. Hult, G. Lutter, N. Vandevenne, S. Schreurs, & W. Schroeyers, Construction and Building Materials, **151**, 819-826 (2017)
7. *DSTU B V.2.7-181:2009: Alkaline cements. Technical conditions.* National standart of Ukraine
8. O. Lastivka. *Modificated concerts based on alkaline slag portland cement in monolith building*, PhD thesis (Eng), (Kyiv, 2015)
9. M. Collepardi, G. Grossi, M. Pellizon Birelli, G. Ventura. *Influence of D-sorbitol on the properties of binders to immobilize acid nuclear wastes*, 8th CANMET/ACI Int. Conf. on Superplasticizers and Other Chemical Admixtures in Concrete, 525-531 (Sorrento, Italy, 2006)
10. M. Palacios, F. Puertas, Cement and Concrete Research, **35**, 1358-1367, (2005)
11. I. Rudenko, O. Konstantynovskyi, A. Kovalchuk, M. Nikolainko, D. Obremsky, Key Engineering Materials, **761**, 27-30 (2018)
12. P. Krivenko, R. Runova, M. Sanytsky, I. Rudenko I. *Alkaline cements* (Kyiv, 2015)
13. P. Krivenko, R. Runova, I. Rudenko, O. Konstantynovskyi. *Potential of alkali activated portland cement-based dry mixes for anchoring application*, NTCC2014: International Conference on Non-Traditional Cement and Concrete, 123-126 (Brno University of Technology, 2014)
14. I. Rudenko, O. Konstantynovskyi, A. Kovalchuk, V. Marharyta, V. Nikolainko, D. Obremsky, Key Engineering Materials, **761,** 27-30 (2018)
15. P. Krivenko, R. Runova, I. Rudenko, V. Skoryk, V. Omelchuk, EEjET **4(6-88)**, 35-41 (2017)
16. M. Palacios, F. Puertas, Cem Concr Res, **37,** 691-702 (2007)
17. L. Zen, Materiales de Construccion, **64 (316)**, 034 (2014)
18. V. Omelchuk, G. Ye, R. Runova, I. Rudenko, Key Engineering Materials, **761**, 45-48 (2018)
19. M. Palacios, C. Sierra, F. Puertas, Mater. Construcc. **53 (269)**, 89-105 (2003)
20. P. Krivenko, O. Petropavlovskii, G. Vozniuk, V. Pushkar, *Cunstructive properties of the concretes made with alkali-activated cements of new generation*, First Intern. Conf. on Advances of Chemically-activated Materials, 139-146 (Jinan, Shandong, China, 2010)
21. K. Yamada, Cement and Concrete Research, **41**, 793-798 (2011)
22. V. Gots, R. Runova, I. Rudenko, O. Konstantynovskyi, O. Lastivka. *Effect of calcium sulfate modifications on properties of alkaline cements*, 19. Ibausil, Weimar, **1**, 1087-1094 (2015)

Rheological properties of aqueous dispersion of styrene acrylate copolymer incorporating hollow microspheres and AEROSIL®

Natalia Saienko[1,*], *Dmitriy* Demidov[2], *Yuri* Popov[1], *Roman* Bikov[1], and *Vyacheslav* Butskyi[1]

[1]Kharkiv National University of Civil Engineering and Architecture, Sumska st. 40, 61002 Kharkiv, Ukraine
[2]Kharkiv State Auto-transport College, Constitution sq. 28, 61003 Kharkiv, Ukraine

Abstract. The rheological properties of aqueous dispersion of styrene acrylate copolymer incorporating hollow aluminosilicate microspheres and AEROSIL® 300 fumed silica (a high-range water reducing admixture) have been investigated using rotational viscometer. It is established that an increase in the degree of filling with hollow microspheres results in a significant increase in the viscosity of aqueous dispersion leading to the technological difficulties when coating the protected surfaces. AEROSIL® 300 can be dosed in small amounts resulting in a decrease in the structural strength and optimal viscosity of aqueous dispersion of acrylate copolymer, which allows to significantly reduce the viscosity density, while maintaining relatively low water demand and characteristics of highly filled systems, which is crucial in the technology of their application. The combination of such fillers as AEROSIL® 300 and hollow aluminosilicate microspheres in the styrene acrylate copolymer aqueous dispersion allows the combination of a number of positive effects: a reduction in the viscosity of the material during application (without dilution) by applying high shear rate and increasing viscosity (structuring) after application, and also during their storage and transportation.

1 Introduction

Waterborne coatings revealed environmentally friendly features, widely used for the production of protective and decorative coatings on nonmetallic substrates, but have recently been increasingly used in transport construction to protect metal and concrete from corrosion [1].

Such material provides an opportunity to exclude the toxic and fire-hazardous solvents, which are toxic to the environment when exposed to air during hardening. The formation technology and performance characteristics of coatings are determined by the properties of film-forming materials, which shall meet the following requirements: ensuring uniform thin-layer distribution on the surface of the substrate and forming of coatings with

* Corresponding author: natause@ukr.net

predetermined functionality. Such requirements are fully consistent with waterborne coatings [2].

Due to the functional properties and relatively low cost, waterborne coatings based on acrylic binders have become the most widely used [3].

By rational combination of active ingredients: film-forming agent based on aqueous acrylate copolymer dispersion of the pigments, thickeners, targeted additives, mineral fillers and coatings with a given set of properties can be obtained.

Thus, incorporating of high-range water reducing admixtures (talc, kaolin, AEROSIL®, titanium dioxide, a number of organic pigments) into aqueous dispersions of acrylates results in the formation of thixotropic coatings. Due to established structure thixotropic coatings do not flow, but are easily applied to the surface if this structure is damaged.

One of the most common fillers of this type is AEROSIL®, which also serves as a stabilizing agent of waterborne coatings, prevents the settling of pigments and imparts thixotropic properties of the coatings. This is a high-range water reducing admixture of amorphous silicon dioxide, obtained by high-temperature hydrolysis. The presence of silanol groups on the surface of the particles determines the hydrophilic nature of silica obtained according to the said method [4-8].

At present, hollow glass or ceramic microspheres are increasingly used to impart heat-insulating properties to coatings. Microspheres can be extracted or incorporating rarefied air (depending on the conditions of their production) and, thanks to a successful combination of spherical shape, controlled dimensions, low density, relatively high compressive strength; good heat and sound insulation; dielectric properties, are one of the promising technogenic fillers of waterborne coatings [9-12].

Therefore, the study of mineral filler effect on chemical composition of the rheological properties of highly-solid aqueous coatings remains a relevant applied research task, since the filling affects the structural characteristics and predetermines the coating technology respectively.

The objective of this paper was to study the influence of the chemical composition and specific surface of fillers on the acrylic aqueous dispersions properties, such as: optimal dynamic viscosity, structural and rheological properties

2 Experimental program

Acryl-styrene dispersion of ACRONAL® 290 D, produced by BASF (non-volatile compounds content – 50 % wt, PH 7.5…9.0, average particle size about 100 nm, viscosity at 23 °C (ISO 3219, DIN 53019) at shear rate – 100 s^{-1}, 7…15 mPa·s) was used as a paste film thickness.

As the modifiers of the dispersion of ACRONAL® 290 D, cellulosic and acrylic thickeners, a defoamer without mineral oil on a polymer base and dispersing medium were used.

Hollow microspheres (MS) were used as fillers of water-dispersion paint materials (WDCM), which are finely dispersed, free-flowing powders consisting of thin-walled silica-alumina particles of spherical shape with a diameter of 10…100 μm and a specific surface area 0.61 m^2/g.

To regulate the rheological properties of WDCM, a filler AEROSIL® 300 (A-300) with a specific surface area 300 m^2/g was used. The mineral composition of the fillers is presented in Table 1.

Compositions of modifiers and mineral fillers for the study based on the dispersion of ACRONAL® 290 D, were prepared as follows: the components of the composition were dispersed for 20 minutes in a mixer in the form of a dissolver with a paddle stirrer at a rotational speed of 60…120 rpm until a homogeneous mass was obtained.

Table 1. Compound mineral powders

Filler	Surface area, m^2/g	ρ, g/cm^3	pH	Mineral composition, % wt					
				SiO_2	Al_2O_3	MgO	Fe_2O_3	TiO_2	CaO
MS	0.61	0.58…0.69	8.5	52.2…64.3	18.0…39.0	1.0…2.0	1.2…10.0	0.6…1.0	0.1…5.8
A-300	300.00	0.05…0.06	4.5	99.9	0.05	–	0.003	0.03	–

The rheological properties of the compositions were determined using a Reotest-2 rotary viscometer (Germany) with a cylinder-cylinder working assembly using a measuring cylinder H. The flow curves of the compositions were recorded with a change in the speed from 0.1667 to 72.9 sec^{-1} at temperatures: 286, 296 and 306 K. The sample of the prepared compositions was then temperature-controlled for 15 minutes.

To establish the fact that the compositions under study have thixotropic properties (form a hysteresis loop), the flow curves were also measured for forward and reverse travel, that is, for the rise and fall of the shear stress (τ), respectively. Prior to the test, the composition was subjected to a 10-minute deformation at the maximum shear rate (72.9 s^{-1}) and a rest period of 10 minutes during the reverse course of the measurements.

3 Results and discussion

The findings of rheological tests of aqueous dispersion (WDCM) based on styrene acrylate copolymer ACRONAL® 290 D are presented in the form of rheological flow curves (the dependence of the shear stress τ (Pa) on the fillers (initial) and upon the introduction of the AEROSIL® (A-300) fillers and hollow microspheres) on the shear rate Dr (s^{-1}) (Fig. 1a) and in the form of the viscosity change curves η (Pa·s) versus the shear stress τ (Pa) (Fig. 1b).

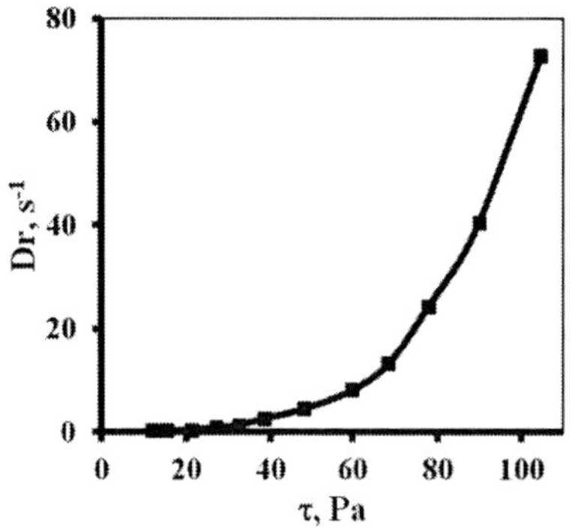

a)

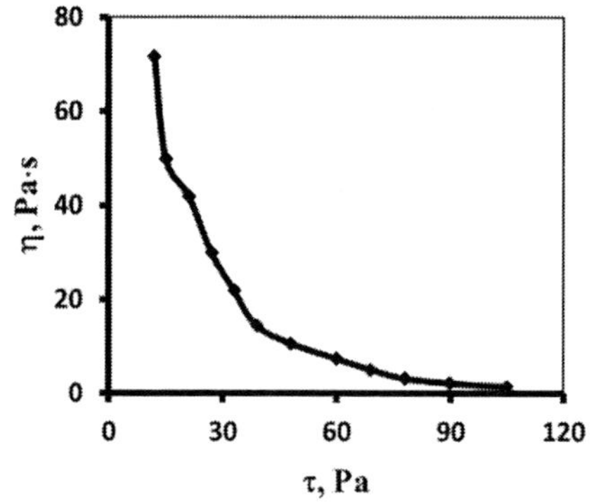

b)

Fig.1. Rheological dependences of the initial WDCM in the form of flow curves (a) and viscosity change curves on the shear stress (b).

First of all, the anomalous nature of the rheological curve (Fig. 1a) as a nonlinear plastic flow of the initial WDCM and the presence of a static limiting shear stress equal to 15.1 Pa (according to Rehbinder, the static static yield point P_{k1} [13]) is of primary interest. This indicates that the dispersed system has a highly developed spatial structure and its flow occurs under the condition that the shear stress value is comparable with the strength of this structure. Consequently, the conditional static yield strength is a measure of the strength of the structural framework. When passing through, the structural grid breaks down, which results in the presence of all nonlinear effects.

The viscosity change in the narrow shear rate is observed on the viscosity change curve from the shear stress of the unfilled WDCM (Fig. 1b), which indicates a significant destruction of its spatial (coagulation) structure.

The character of the rheological curves shown in Fig. 1 testify that WDCM can be attributed to plastic bodies with nonlinear plastic behavior and ultimate shear stress. This conclusion is supported by the fact that rheological curves in the investigated range of shear stresses are well described by the well-known Herschel-Balkley equation [14]:

$$\tau = \tau_s + \eta_{op}\gamma^n, \tag{1}$$

where τ_s is the conditional static yield point, Pa; η_{op} is the optimum water demand (OWD), Pa·s; γ is the shear rate, s^{-1}; n is the the exponent (flow index), which is a measure of the structuredness of systems with a nonlinear plastic pseudoplastic flow pattern $n<1$, and for Newtonian liquids $n=1$.

For the initial (unfilled) WDCM composition, the flow index is 0.355 (n<1), which indicates that it has a spatial structural grid that is destroyed by relatively small applied shear stresses, of the order of 15...20 Pa, and the coagulation structure is easily restored, which indicates the absence of hysteresis loops on the flow curves. It can be assumed that the main contribution to the process of structuring the unfilled WDCM is the interaction between the globules of the latexastirol-acrylate copolymer.

It is known that the introduction of fillers in the polymer dispersion leads to the appearance of a wide range of interactions (from weak physical forces to stronger hydrogen) arising at the polymer-filler-filler-filler interface resulting in the formation of coagulation structures. In structures of the coagulation type, the particles are connected by the intermolecular forces of Van der Waals-London, coagulation and point contacts arise between them. Such structures have a small strength, plastic, elastic and thixotropic properties [15]. The plastic flow is associated with thixotropic phenomena occurring in highly-solid materials. Often, the manifestation of structural viscosity is considered as a positive quality, as the coatings acquire a pasty consistency and in such systems pigments do not settle, materials can be applied thick layers on vertical surfaces without streaks.

Figure 2 shows the rheological flow curves of the WDCM, with a loading level of the MS of 20 % wt and 40 % wt. As can be seen, the introduction of the microcircuit of the flow does not change, but a sharp increase in the strength of the structural grid occurs, as evidenced by the values of the conditional static yield point τ_s, value, which increases more than 3...4 times (see Table 2).

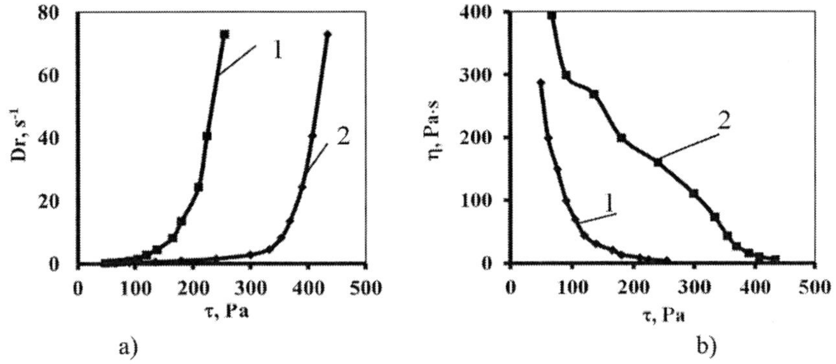

Fig.2. Rheological dependences of WDCM incorporating MS of 20 (1) and 40% (2) by mass in the form of flow curves (a) and viscosity change curves on shear stress (b).

In Fig. 2b shows the dependence of the optimum water demand (OWD) of microsphere-filled WDCM dispersions on the shear stress. It can be seen that the initial values of the optimum water demand (OWD) η_0 increase from 89.8 Pa·s to 287.3 Pa·s with an MS content of 20 % wt and up to 395.1 Pa·s at 40 % wt (see Table 2).

The data presented indicate an increase in the energy between the partial interaction in the resulting structure and the number of contacts with an increase in the content of microspheres, as well as a decrease in the thickness of the interlayers of the dispersion medium. Important in our opinion is that the acidity index of the water extract of the microspheres themselves is pH 8.5 and they, apparently, additionally contribute to the thickening of the styrene acrylate copolymer dispersion due to an increase in the degree of ionization of the carboxyl groups [16]. In addition, for microsphere-filled dispersions of the copolymer, there is also a significant decrease in the optimum water demand (OWD) at the initial sections of the deformation at low shear stresses, which is caused by the rupture at the initial stage of less strong contacts between dispersion globules, between mineral particles and polymeric globules, and also the orientation of partially destroyed aggregates in the direction of shear. Despite the considerable hardening of the structure incorporating WDCM microspheres, the coagulation structure is easily reconstructed camping, as evidenced by the absence of flow curves of the hysteresis loop.

The specificity of the interaction of AEROSIL® with aqueous dispersion of the polymer lies in the fact that a relatively low mass fraction of the solid phase creates an anomalously high concentration of particles – 10^{15}-10^{16} per cm^3 of the dispersion volume [17]. Thus, the investigated AEROSIL® dispersions can be attributed to the so-called highly concentrated dispersive systems, for which the influence of the concentration factor on the properties is decisive.

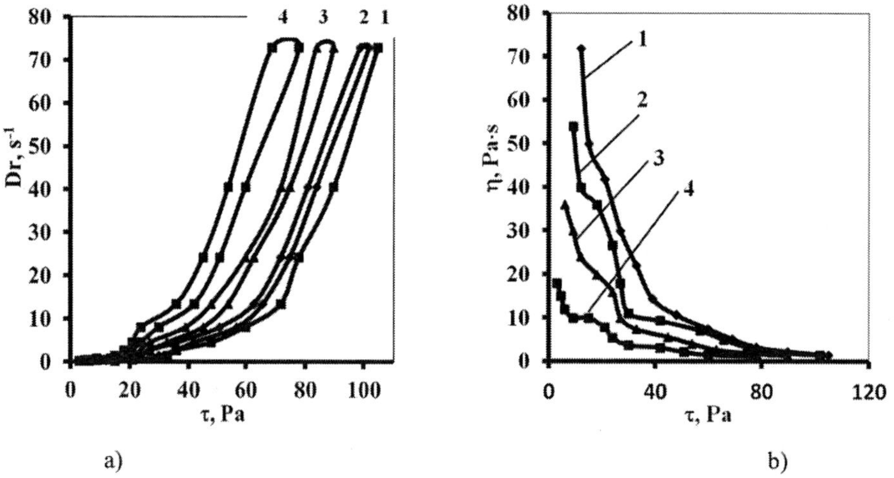

Fig. 3. Rheological dependencies of WDCM A-300 content in the form of flow curves (a) and viscosity change curves versus shear stress (b): 1 – WDCM; 2 – A-300 0.5 % wt; 3 – A-300 1.0 % wt; 4 – A-300 1.5 % wt.

In Fig. 3 a, rheology-impenetrable dispersions of WDCM incorporating AEROSIL® A-300 in an amount of 0.5, 1.0 and 1.5% by weight are presented. It can be seen that the nature of the flow curves does not change with the introduction of A-300, but the strength of the structural grid decreases, as evidenced by the values of the conditional static yield point, the value of which decreases from 15.1 Pa for WDCM to 6.1, 5.4 and 3.1 Pa at the introduction of A-300 in an amount of 0.5, 1.0 and 1.5 % wt, Respectively (see Table 2). At the same time, the values of the initial optimum water demand (OWD) η_0 decrease by more than 2…4 times (see Table 2). This is most likely due to the fact that when the A-300 is introduced, the alkalinity of the WDCM itself decreases (the pH value of aqueous extract of AEROSIL® is pH = 4.1) resulting in a decrease in the viscosity of WDCM due to a decrease in the degree of ionization of carboxyl groups *(-COOH)* particles of latex.

The analysis of the obtained results shows that the rheological behavior of aqueous acrylic-styrene dispersions incorporating AEROSIL® has all the characteristics typical for highly concentrated disperse systems with a coagulation-type structure. A sign of this is the nonlinear character of rheological curves and incomplete display of the rheological behavior of dispersions under dynamic conditions. This is manifested in the appearance of a hysteresis loop, whose dimensions increase noticeably with increasing AEROSIL® content, and the rheological curves themselves are shifted to the region of lower shear stresses.

At the same time, enhancement of packing density hollow microspheres up to 40 % wt Significantly increases the viscosity of the composition (by 4 times) resulting in technological difficulties in the application of heat-insulating coatings to the protected surfaces. These difficulties are associated with the need to avoid the flow of the film in the gravitational field. Such a "flow" of the material can lead to the appearance of an inhomogeneity in its film thickness. Therefore, considering the fact that the introduction of A-300 leads to a certain decrease in the structural strength and optimum water demand (OWD) of aqueous dispersion of the acrylate copolymer, the rheological properties of WDCM filled simultaneously with hollow microspheres and A-300 were studied.

In Fig. 3a, the flow curves of WDCM incorporating hollow microspheres (40 % wt) and additionally with AEROSIL® (0.5, 1.0 and 1.5 % wt) are presented. It can be seen that the nature of the flow curves does not change with the addition of the A-300 to the filled system, but the strength of the structural grid decreases, as evidenced by the values of the conditional static yield point (τ_s), which decreases from 65.8 Pa to 59.9, 52.4 and 44.9 Pa with the introduction of A-300 in an amount of 0.5, 1.0 and 1.5 % wt, respectively (see Table 2). At the same time, the values of the initial optimum water demand (OWD) (η_0) are reduced from 395.1 Pa·s to 359.1, 314.2 and 269.3 Pa·s (see Table 2), and the shear stress interval of the WDCM, filled simultaneously with microspheres and AEROSIL®. This system is also characterized by the nonlinear behavior of rheological curves and the appearance of hysteresis loops, indicating an incomplete reconstruction of the structure of the complexly filled WDCM in a dynamic mode with forward and reverse deformation of shear stresses.

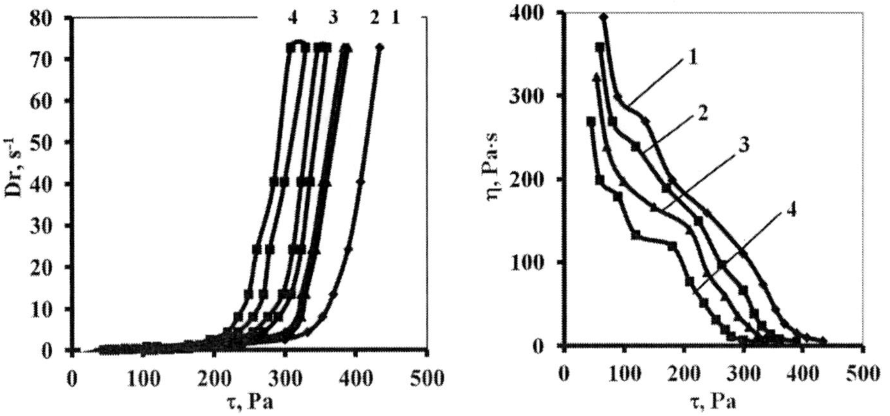

a) b)

Fig. 4. Rheological dependences of WDCM on the content of MS and A-300 in the form of flow curves (a) and viscosity change curves on shear stress (b): 1 – MS 40 % wt; 2 – MS 40+A-300 0.5 % wt; 3 – MS 40+A-300 1.0 % wt; 4 – MS 40+A-300 1.5 % wt.

To assess the degree of structuring and the strength of intermolecular interactions in the WCDMA, structural and rheological properties allow for structural and rheological properties. Table 2 shows the structural and rheological properties: the conditionally static yield strength (τ_s), the viscosity at the minimum rate, the onset of failure (initial optimum water demand (OWD) (η_0), the minimum viscosity at the maximum strain rate (η_{min}), and n – the exponent in the Herschel-Balkley equation.

Table 2. Structuraal and rgeological characteristics.

Compositions, % wt	τ_s, **Pa**	η_0, **Pa·s**	η_{min}, **Pa·s**	n
WDCM	15.1	89.8	5.3	0.355
WDCM+A-300 (0.5)	6.1	35.9	4.9	0.393
WDCM+A-300 (1.0)	5.4	32.3	4.1	0.433
WDCM+A-300 (1.5)	3.1	18.1	3.1	0.534
WDCM+MC (20)	47.9	287.3	13.3	0.269
WDCM+MC (40)	65.8	395.1	27.3	0.296
WDCM+MC (40)+A-300 (0.5)	59.9	359.1	24.4	0.292
WDCM+MC (40)+A-300 (1.0)	52.4	314.2	22.8	0.303
WDCM+MC (40)+A-300 (1.5)	44.9	269.3	19.9	0.315

4 Conclusions

The experimental studies on the rheological properties of aqueous dispersion of styrene acrylate copolymer incorporating aluminosilicate microspheres (MS) and high-range water reducing AEROSIL® 300 fumed silica (A-300) determined the viscosity and strain rate of WDCM versus the shear stress for the fillers: MC from 20 up to 40 % wt and A-300 from 0.5 to 1.5 % wt.

By the nature of the rheological curves, the unfilled WDCM refers to plastic bodies with nonlinear plastic behavior and limiting shear stress, has a spatial structural grid that is destroyed by relatively small applied shear stresses, of the order of 15-20 Pa and easily restores the coagulation structure, as evidenced by the absence on the curves of the flow of the hysteresis loop.

The degree of thixotropy of the WDCM incorporating A-300 is characterized by the area of hysteresis loop and the absolute shear stress at low shear rate and increases with the content of the filler, since the coagulation interaction of the particles has a major role in the formation of spatial structures. The introduction of small additives A-300 (0.5…1.5 % wt) results in a certain decrease in the structural strength and optimum water demand (OWD) of aqueous dispersion of the acrylate copolymer, which allows to significantly reduce the viscosity and regulate the rheological properties of the WCDMA application.

Enhancement of packing density of hollow microspheres up to 40 % wt significantly increases the viscosity of the composition resulting in technological difficulties when applying heat-insulating coatings to the protected surfaces. These difficulties are associated with the need to avoid the flow of the film in the gravitational field. Such a "flow" of the material can lead to the appearance of an inhomogeneity in its film thickness.

Incorporating of AEROSIL® to WCDMA allows to regulate the rheological properties at the stage of production of coatings or before using them. That allows to combine a number of positive effects simultaneously: decrease in the viscosity of the material during application (without dilution) due to application of high shear rate and viscosity increase (structuring) after application, as well as during storage and transportation.

References

1 V. Lobkovsky, Industrial coatings, **4**, 28-31 (2016)

2 E.E. Kazakova, *Water-dispersions acrylic paint and varnish materials for construction purposes*, (Moscow, 2003)

3 N. Saienko, D. Demidov, M. Myagkikh, Scientific Bulletin of Civil Engineering, **86** (4), 154-157 (2016)

4 G. Michael, Russian Coatings Journal, **6**, 42-47 (2007)

5 M. Dolz, F. González, J. Delegido, M. Hernández and J. Pellicer, Journal of pharmaceutical sciences, **89** (6), 790-797 (2000)

6 N. Saienko, R. Bukov, D. Demidov and D. Kovalenko, International Scientific Conference "Transport technologies and infrastructure", 356-358 (2018)

7 A. S. Makarov, I. A. Andreeva, V. Y. Tretinnik, Colloid Journal, **63** (6), 731-737 (2001)

8 V. Katnov, S. Stepin, R. Katnova, R. Mingaliyeva and P. Grishin, Herald of Kazan Technological University, **15**, (7), 95-96 (2012)

9 H. S. Katz, J.V. Milewski, *Handbook of fillers and reinforcements for plastics*, (New York: Van Nostrand Reinhold Co, 1978)

10 L. Vakhitova, A. Zavertatnyy, F+ S: fare and security, **45** (3), 64-66 (2010)

11 A. Inozemtsev, Ye. Korolev, Industrial and Civil Engineering, **10**, 80-83 (2013)

12 V.S. Drozhzhin, I.V. Piculin, J. Moscow Phys. Soc., **9**, 209-214 (1999)

13 P. V. Mikhailov, P.A. Rebinder, Colloid Journal, **18** (2), 107-119 (1955)

14 V.N. Matveenko, Kirsanov E.A. MSU Vestnik, Series 2. Chemistry, **52** (4), 243-276 (2011)

15 A.Y. Malkin A. Y. *Rheology in the processes of formation and transformation of po*lymers, Moscow, (1985)

16 V.A. Epishkina, R.N. Celms, V.K. Vasiliev and A.M. Kiselev, Proceedings of high schools. Technology of textile industry, **6**, 86-92 (2006)

17 V. Zakordonsky, B. Lazaruk, Visnyk of tne Lviv University. Series Chemistry, 55 (2), 472-485 (2014)

Study of influence of the alkaline component on the physico-mechanical properties of the low clinker and clinkerless waterproof compositions

Yuliya Savchuk*, *Andrii* Plugin, *Vitalii* Lyuty, *Oleksii* Pluhin, and *Olga* Borziak

Ukrainian State University of Railway Transport, Feuerbach sq. 7, 61050 Kharkiv, Ukraine

Abstract. The influence of the alkaline component on the physico-mechanical properties and the structure of the clinkerless waterproof composition on the basis of Kryvyi Rih blast furnace slag and an alkaline component – sodium hydroxide – have been investigated. High and very high correlation between the compressive strength *f* of the composition and the alkali – sodium hydroxide – strength *A/S* has been established. The dependence *f* on *A/S* has an extreme wave-like pattern with maxima at *A/S* = 1.5% and 15-20%, which is explained by the formation of one and two series of hydration products between slag particles similarly to Portland cement-based compositions. It has been established that the hydration products of the composition are predominantly gel-like one, such as hydrosilicate gel and zeolite-like sodium calcium hydroalumosilicates; fewer crystalline hydration products include prismatic and needle-like crystals of complex salts of aluminate phases. The pore size in the composition structure mostly does not exceed 2 microns, although single pores up to 10 microns in size also occur.

1 Introduction

During the operation, concrete structures of engineering installations suffer damage not only from mechanical loads, but also temperature and humid influences and deformations, the impact of aggressive media, electrical influences – leakage currents and wandering currents, electrical potentials and charges, etc. This requires their repair and protection against aggressive action of chemically and biologically active media, electrical currents, potentials and charges. Cement mortars, dry mixtures, Portland cement clinker-based plasters are mainly used for repair and protection [1-3]. An effective variety of such materials is the Portland cement-based penetrating compositions which compact the surface concrete layers of the damaged installation [4, 5]. However, Portland cement-based materials are often insufficiently stable in many operating conditions. For example, at metallurgical plants, they are quickly broken when affected by acid gases, such as sulphur dioxide, carbon dioxide (Fig. 1) [6, 7]. In such conditions, alkaline and slag-alkali binding materials produced from blast-furnace granulated slag and characterized by high corrosion resistance in many environments are more resistant [8, 9]. Therefore, the development of

* Corresponding author:uliasavhyk83@gmail.com

penetrating compositions based on slag-alkali binding materials that would be resistant to acid gases and could be made from available raw materials is a relevant task. The creation of such compositions will also contribute to solving the environmental and economic problems of industrial waste utilization.

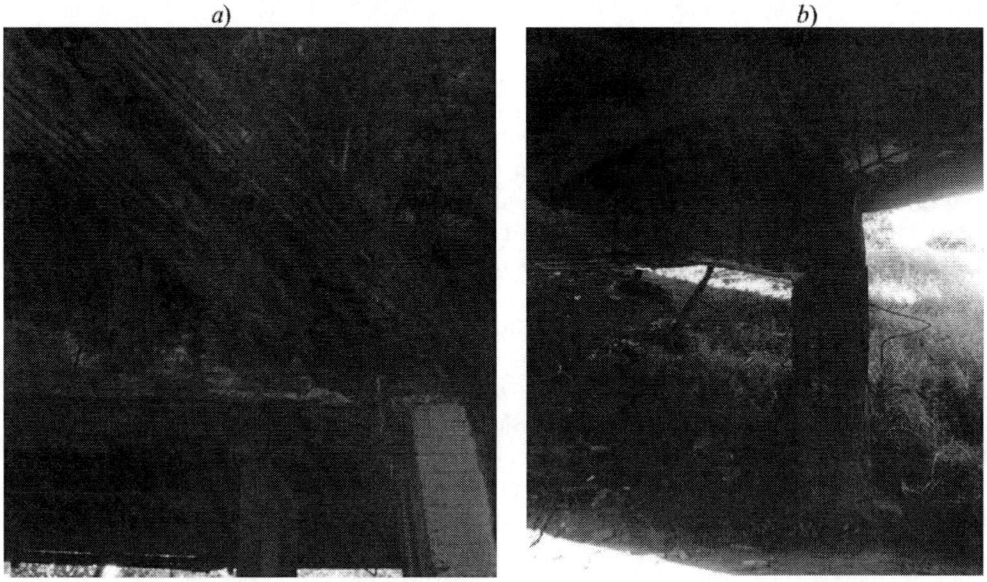

Fig. 1. Damage of reinforced concrete girders (a, b) and supports (b) of the pipeline within the territory of PJSC MMK Azovstal (Mariupol) in the area with high content of sulphur dioxide and carbon dioxide in the air.

Penetrating action, adhesion to the repaired surface, protective properties and water resistance of the penetrating compositions depend on their structure [1, 2, 10, 11] and are achieved by adding chemical [4] and mineral admixtures [2, 12-15], disperse reinforcement [16-19]. A complex of such additives, including nitrates, carbonates, sulphates, sodium and calcium chlorides, are proposed in [4, 16, 18], but the effect of this complex additive has been tested only for Portland cement and its varieties.

The authors investigate the effect of the complex chemical additive and the alkaline component on the physico-mechanical properties of artificial stone on the basis of a slag-alkali binding material [6, 7]. It has been established in [7], that its physic-mechanical properties in the investigated range are unaffected by the content of the complex chemical additive, but are extremely dependent on the content of sodium hydroxide. The maximum compressive and bending strengths have been recorded when the content of sodium hydroxide was about 5% of the slag content by weight, but the study was conducted only on the slag of one metallurgical plant – PJSC MMK Azovstal (Mariupol), and the region of the low content of sodium hydroxide less than 1 % remained unexplored.

The purpose of the work is determination of the influence of the alkaline component on the physic-mechanical parameters and the structure of the slag-alkali composition intended for use as a penetrating waterproof compound.

Tasks of the study: investigation of the dependences of compressive strength of a slag-based composition on the content of sodium hydroxide; conducting electron microscopic studies of the structural features of the solidified composition

2 Research methods and materials

The quartz sand composition with a grain fineness 1.1, blast-furnace granulated slag of PJSC MMK Azovstal (Mariupol), OJSC Mittal Steel Kryviy Rih (Kryvyi Rih), ground so that at least 85% pass through a 008 sieve, sodium hydroxide *NaOH* was investigated. A complex chemical additive that provides penetrating action and water resistance of the composition contained nitrates, carbonates, sulphates, sodium and calcium chlorides in the ratio established in [4, 16, 18]. A dry mixture was prepared from all components, which was mixed with water immediately before use. Dependence of average consistency, water absorption, bending and compressive strength, water resistance (softening factor) of the composition on the content of sodium hydroxide and the complex chemical additive was studied. The relationship between them was estimated by the coefficient of correlation by the Chaddock scale.

For the study, mixtures with different ratios of components were prepared. The ratio in the base composition was: sand - 60%, slag - 35%, sodium hydroxide - 5% by weight. The amount of the complex chemical additive was 1% of the slag weight. The amount of sodium hydroxide varied from 0 to 40% by slag weight. The amounts of sodium hydroxide and the complex chemical additive were increased, while the sand to slag ratio 60:35 remained unchanged. For the determination of bending and compressive strength and average consistency, water absorption, water resistance (softening factor), samples shaped as small beams 40×10×10 mm in size were made from the mixture, which were tested by standard (conventional) methods after a fixed natural cure period. The softening factor was determined as the ratio of the compressive strength of the sample in the water-saturated to a constant mass state to the compressive strength in the air-dry state.

Electron microscopic analysis of the artificial stone structure was conducted at its spalls at 15 kV using a scanning electron microscope Philips XL 30 ESEM-FEG by F.A.Finger-Institut für Baustoffkunde of Bauhaus-Universität Weimar.

3 Results of experimental studies

The dependences of the compressive strength *f* of the Kryvyi Rih slag-based composition on the alkali – sodium hydroxide – strength *A/S* are shown in Fig. 2, 3. Electronic microscopic images of artificial stone on Mariupol slag containing 5% sodium hydroxide and a complex chemical additive of 1.5% of the slag weight are shown in Fig. 4.

4 Discussion of results obtained in the study

The analysis of Figs. 2 and 3 showed high and very high Chaddock correlation between the compressive strength of the composition and the content of sodium hydroxide – the squared correlation R^2 reaches 0.62-0.92, the correlation coefficient R is 0.79-0.96. The dependence of strength *f* on the content of alkali – sodium hydroxide – *A/S* for the Kryviy Rih slag-based composition, as well as for the Mariupol slag-based composition [7] has an extreme character, but unlike it – wave-like with maxima at *A/S* = 0.015 (1.5 %) and 0.15-0.20 (20%). This can be explained by the formation of one and two series of hydration products between slag particles and/or slag particles between sand grains similarly to cement concrete and compositions [19]. Evidently, for practical purposes, a dosage of alkali – sodium hydroxide – should be 1.5%. The obtained strength values are not very high – up to 10 MPa, which can be explained by the features of slag selection for the study and the method of determination. However, such values are quite sufficient for repair and protective plasters, since, as known, the strength and modulus of elasticity of the plaster must be less than those of the material of the repaired surface.

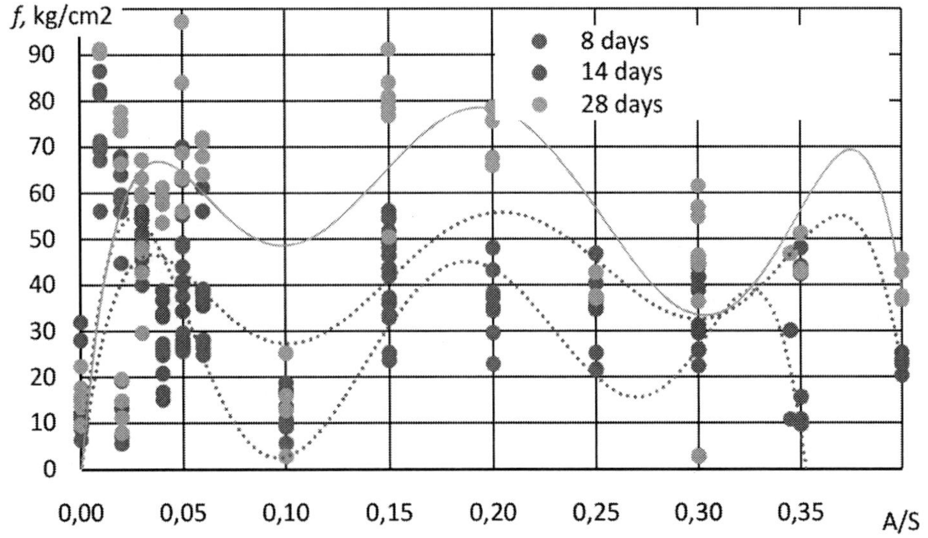

Fig. 2. Dependence of compressive strength *f* of the Kryviy Rih slag-based composition on the content of alkaline NaOH *A/S* after 8, 14 and 28 days of natural curing.

On the basis of the analysis of electron microscopic images shown in Fig. 4 and others is was established as follows. Fig. 4, *a* shows a fraction of sand (right) and slag particles (left) covered with hydration products. The hydration products are mostly gel-like (Fig. 4, *b-d*), probably, hydrosilic gel and zeolite-like sodium-calcium hydroaluminosilicates. Fewer crystalline hydration products are also observed – prismatic and needle-like crystals up to 2 μm in length (Fig. 4, *a, c, d*), probably, complex salts of aluminate *AFm* and *Aft* phases [20-23]. No hexagonal plates of Portlandite are observed, which may be the evidence of a general deficit of crystalline hydration products and electroheterogeneous contacts and, as a result, low strength. Most pores are up to 2 microns in size between the hydration products (Fig. 4, *d*), the largest pore size is up to 10 microns (a single pore in the contact zone of sand grain, Fig. 4, *a*).

5 Conclusions

As a result of the research carried out:

1. High and very high correlation between the compressive strength *f* of the Kryviy Rih slag-based composition and the alkali – sodium hydroxide – strength *A/S* has been established with the correlation coefficient of 0.79-0.96. Dependence of *f* on *A/S* has an extreme wave-like character with maxima at *A/S* = 0.015 (1.5%) and 0.15-0.20 (20%). This is explained by the formation of one and two series of hydration products between the slag particles similarly to Portland cement-based compositions.

2. Hydration products of the composition are predominantly gel-like, apparently, hydrosilicidal gel and zeolite-like sodium calcium hydroaluminosilicates, fewer crystalline hydration products include prismatic and needle-like crystals of complex salts of aluminate phases.

3. Physicochemical studies of the composition of the hydration products of the composition should be conducted in order to confirm the results of electron microscopic studies.

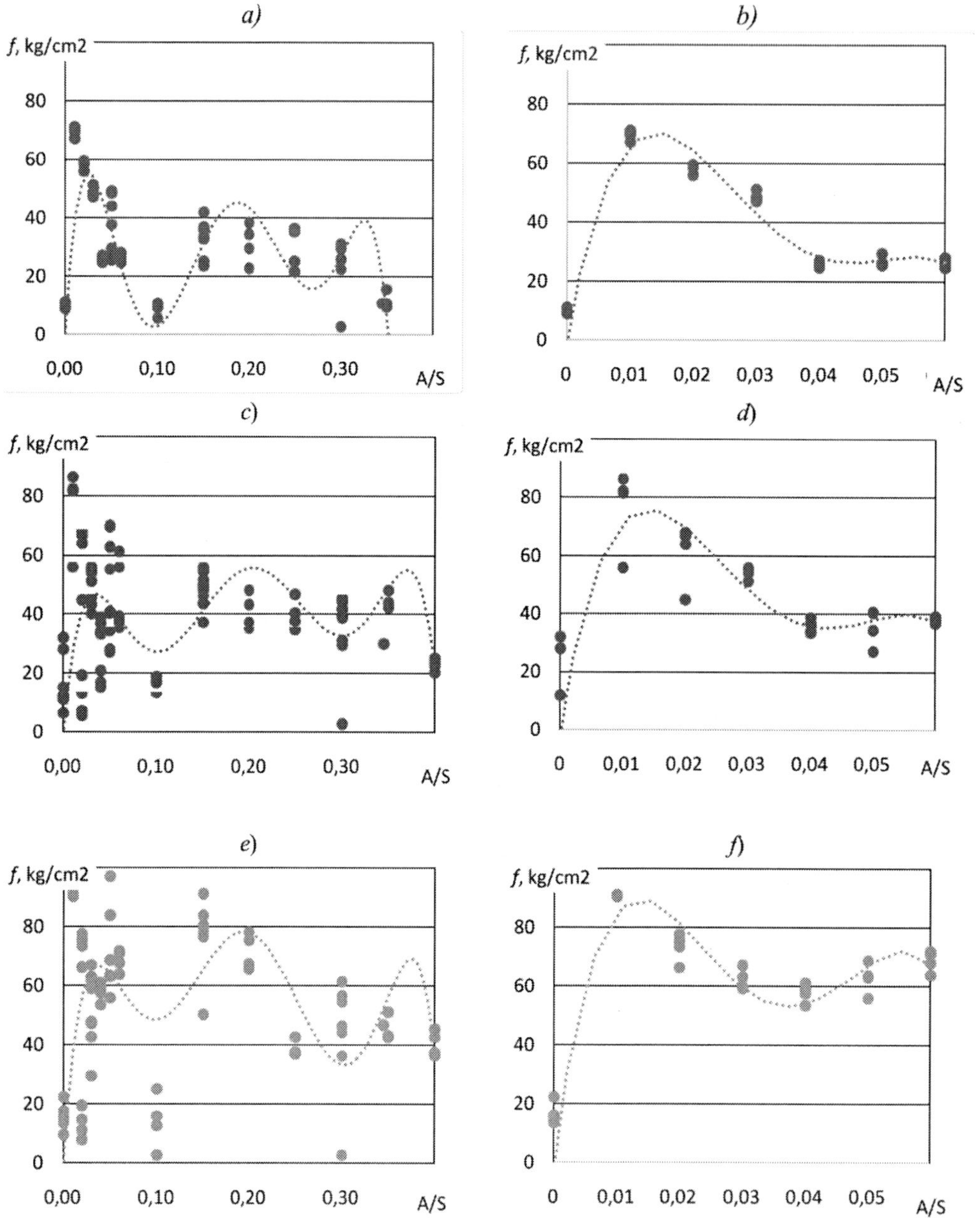

Fig. 3. Dependence of compressive strength *f* of the composition on the content of alkaline *NaOH A/S* after 8 (a, b), 14 (c, d) and 28 (e, f) days of natural curing: a, c, e – at *A/S* range from 0 to 0.4; b, d, e - at *A/S* range from 0 to 0.06.

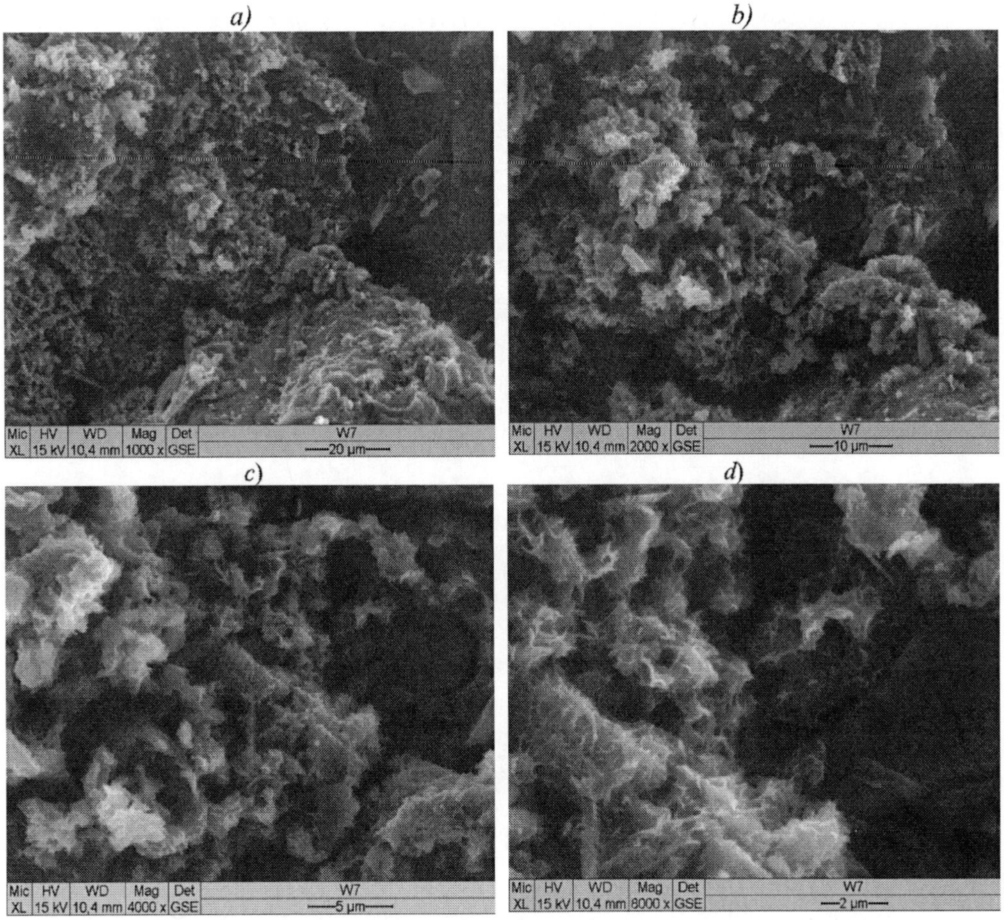

Fig. 4. Zoomed electron microscopic pictures of the composition: a - × 1000; b - × 2000; c - × 4000; d - × 8000. Each subsequent picture is a zoomed fragment of the previous one.

References

1 M. Sanytsky, T. Kropyvnytska, R. Kotiv, Advanced Materials Research, **923**, 42-47 (2014)

2 O. Pluhin, A. Plugin, D. Plugin, O. Borziak, O. Dudin, Matec Web of Conference, **116**, 01013 (2017)

3 P. V. Krivenko, M. Sanytsky, T. Kropyvnytska, R. Kotiv, Advanced Materials Research, **897**, 45-48 (2014)

4 O.I. Demina, A.A. Plugin, E.B. Dedenyova, D.O. Bondarenko, T.A. Kostuk, A.I. Bondarenko, Functional Materials, **24 (3)**, 415-419 (2017)

5 D.O. Bondarenko, R.A. Yakovleva, T.O. Kostiuk, O.Yu. Proshin, *30th SAMPE Europe International Jubilee Conference,* 563-569 (Paris, 2009)

6 A.A. Plugin, V.A. Liutyi, Yu.Iu. Savchuk, Zb. nauk. prats UkrDUZT, **171**, 19-29 (2017)

7 A.A. Plugin, Y.Y. Savchuk, V.A. Liutyi, T.O. Kostiuk, D.O. Bondarenko *Penetrating Fine-Clinker and Clinkerless Cement-Based Waterproof Compounds* (19'Ibausil, Weimar, 2018)

8 P.V. Kryvenko, E.K. Pushkareva *Dolhovechnost shlakoshchelochnoho betona [Durability slag-alkali cement]* (Kyiv, Budivelnyk, 1993)

9 Pavel V. Krivenko, Myroslav Sanytsky, Tetiana Kropyvnytska, Key Engineering Materials, 761, 193-196 (2018)

10 E. Lutskin, E. Shinkevich, Technical Journal, **9 (1-2015)**, 27-33 (2015)

11 E. Shinkevich, Y. Zaytsev, E. Lutskin, G. Bondarenko, A. Tymnyak, *Proceeding of 2nd International conference on Microstructural-related Durability of Cementitious Composites,* 1462-1468 (RILEM Publications SARL, 2017)

12 U. Marushchak, M. Sanytsky, T. Mazurak, Y. Olevych, Eastern-European Journal of Enterprise Technologies, **6/6 (84)**, 50–57 (2016)

13 O. Borziak, S. Vandolovskyi, V. Chajka, V. Perestiuk, O. Romanenko, Matec Web of Conference, **116**, 01001 (2017)

14 M.H. Salyia, T.A. Kostiuk, Yu.A. Spyryn, A.A. Plugin, Zb. nauk. prats UkrDAZT, **130**, 49-56 (2012)

15 U. Maruchchak, M. Sanytsky, S. Korolko, Y. Shabatura, N. Sydor, Eastern-European Journal of Enterprise Technologies, **2/6 (92)**, 34-41 (2018)

16 A.A. Plugin, T.O. Kostiuk, O.A. Plugin,D.O. Bondarenko, Yu.A. Sukhanova, N.N. Partala, International Journal of Engineering Research in Africa, **31**, 59-68 (2017)

17 A.A. Plugin, T.A. Kostiuk, D.A. Bondarenko, M.H. Salyia, Naukovyi visnyk budivnytstva, **59**, 157-162 (2010)

18 T.A. Kostiuk, A.A. Plugin, Al.A. Plugin, Zb. nauk. prats UkrDAZT, **148(2)**, 32-38 (2014)

19 A.N. Plugin, A.A. Plugin et al. *Osnovy teorii tverdeniya, prochnosti, razrusheniya i dolgovechnosti portlandtsementa, betona i konstruktsiy iz nikh* (Naukova dumka, Kyiv, 2012)

20 T. Matschei, B. Lothenbach, F.P. Glasser, Cem. Concr. Res, **37 (2)**, 118-130 (2007)

21 M. Balonis, B. Lothenbach, G. Le Saout, F.P. Glasser, Cem. Concr. Res., **40(7)**, 1009-1022 (2010)

22 Balonis M., Medala M., Glasser F.P. Cem. Res. **23(3)** (2011)

23 A.A. Plugin, R.F. Runova, International Journal of Engineering Research in Africa, **36**, 69-73 (2018)

Investigation of the bottom ash slags influence on the heavy concrete frost resistance

Volodymyr Shulgin, *Dmytro* Yermolenko, *Heorhii* Durachenko, Oleksandr Petrash, and *Oksana* Demchenko*

Poltava National Technical Yuri Kondratyuk University, Pershotravnevyi Avenue, 24, Poltava, 36011, Ukraine

Abstract. This paper addresses the issue of the influence of cement consumption, plasticizing additive, and compaction time on the strength and frost resistance of concrete manufactured with bottom ash slags from thermal power plants. The investigation was carried out using the three-factor experiment with variables varied at three levels. There were received response functions, which characterize the correlation between frost resistance and variables: cement consumption, plasticizer additives and compaction time. An analysis of the surface of the response function, where the frost resistance of the concrete was accepted as a response, revealed the optimal correlation between the components of the mixture and the compaction time of the seal. By using the STATISTICA 12 software complex, the values of the factors were specified, when the optimal components ratio with GLENIUM 51 superplasticizer is achieved for the production of concrete with high frost resistance that is greater than F300. The conclusions quantify the results of the investigation.

1 Introduction

During combustion of solid fuels for the production of heat and electricity at the thermal power plants waste is generated in the form of slag and ashes. Ash and slag are the large-scale waste.

One of the ways to safely dispose of such waste is to use it in the construction industry for the manufacture of heavy concrete. The optimum content of ash in heavy concretes is determined by the selection of specific concrete mixtures under the condition that the necessary indicators of their quality are provided. This paper addresses the issue of the influence of cement consumption [1]. Concrete mixtures with ashes have a higher viscosity, better mobility, less water drainage and stratification, which makes such mixtures ideal for use in monolithic concrete. According to the sources [2], replacement of a part of cement by ash, as a rule, leads to lower frost resistance. This is explained, firstly, by an increase in the actual water-cement ratio and the formation of an additional volume of capillary pores. The reduction of frost resistance is also facilitated by the reduction of the entrained air content when the ash is introduced and the presence of ash in organic residues that swell in water and capable of forming compounds that destroy the binder.

*Corresponding author: homenko_81@ukr.net

Organic residues are usually contained in a coarse fraction of ash. Grinding of ash leads to the destruction of organic particles and their uniform distribution throughout the material, which positively affects the frost resistance of concrete.

A significant increase of the frost resistance of concrete with ash can be reached by the introduction of surface-active additives of hydrophobic-plasticizing type into concrete. The plasticizing component of such additives reduces the water consumption of the concrete mixture, and hence the volume of capillary pores. [3-6]. The hydrophobic component, in turn, affects the walls of pores and capillaries in concrete and thus prevents the water from entering them. Thereby, after 150 cycles of freezing and thawing of concrete with the replacement of 20% of cement by ground molded ash, combined with a complex hydrophobic-plasticizing additive, increased strength by 10% compared with concrete without ash. The use of a hydrophobic-plasticizer additive when replacing 25% of cement by fine ash allowed to obtain concrete that is not inferior to the frost-resistance of concrete without ash and capable of withstanding up to 200 freezing and thawing cycles [7]. However, the properties of such concrete are still remaining not investigated and the effect of ash slags on the frost resistance of concrete [8].

2 Investigation results

2.1 Materials for concrete with bottom ash slags

The main purpose of these investigations is to determine the effect of the composition of concrete mixtures with ash slags on the frost resistance of concrete, and the selection of the most optimal concrete with a strength class of C20/25 using bottom ash slags. The planning of the experiment and the results of research on concrete for strength using ash-slag are given in [9].

For this investigation, a mixture of the following materials was chosen: cement CEM I 42.5N which is characterized by high early strength, high content of C_3S in clinker and low content of C_3A [10].

The ash slag (bottom ash) from Starobesheivska TPP; true density is 2.6 g/cm^3; fill density is 1.206 g/cm^3; fineness modulus 1; water consumption 24%. Granite gravel of 5-10 mm fraction is used as a coarse aggregate, voidness of gravel is 47 %, fill density is $\rho_н$ = 1350 kg/m^3, true density is ρ = 2.71 g/cm^3. As an additive, a polycarboxylate superplasticizer is used GLENIUM 51.

2.2 Mathematical planning of experiments

The research was planned on the basis of using a three-factor experiment in three levels. As variable factors, there were selected: the consumption of cement, the compaction time of the concrete mixture, and the plasticizer additive consumption.

In order to ensure the appropriate level of porosity and taking into account the features of the source materials structure, the water-cement ratio (W/C) of 0.4 was taken. In the research, control samples were made on quartz sand with strength of 31 N/mm^2.

2.3 Regression analysis of the concrete frost resistance investigation

Frost resistance was determined using a "Concrete-Frost" device. After determining all the results were processed on the PC using the device programs, which allows determination of water absorption by volume of samples and obtain graphs of change in samples volume in time. In accordance with the composition of concrete for a series of samples, certain

patterns are noticed. Namely, samples that were made with the maximum expands of plasticizer were distinguished by low frost resistance, as well as those containing the least amount of cement. Below is a detailed analysis of these phenomena.
Using STATISTICA software, the coefficients of the regression equation (Table 1) are derived as well as the exact equation (1).

Table 1. Results of calculation of regression coefficients.

Factor	Regressn - Coeff.	Std.Err.	t(8)	p	-95,% - Cnf.Limt	+95,% - Cnf.Limt
Mean/Interc.	-7075.50	4743.786	-1.49153	0.179449	-18292.8	4141.773
(1)Cement consumption. kg(L)	24.25	25.697	0.94368	0.376756	-36.5	85.013
Cement consumption. kg(Q)	-0.02	0.034	-0.64231	0.541128	-0.1	0.058
(2)Compaction time, min.(L)	510.41	295.180	1.72915	0.127408	-187.6	1208.399
Compaction time, min.(Q)	-16.92	30.228	-0.55961	0.593195	-88.4	54.561
(3)Plasticizer consumption. l(L)	443.28	166.262	2.66616	0.032178	50.1	836.431
Plasticizer consumption. l(Q)	-15.85	7.557	-2.09794	0.074097	-33.7	2.015
1L by 2L	-0.90	0.583	-1.53633	0.168338	-2.3	0.483
1L by 3L	-0.45	0.292	-1.53633	0.168338	-1.1	0.241
2L by 3L	-9.19	8.746	-1.05042	0.328431	-29.9	11.495

$$y = -7075.5 + 24.25\ x_1 - 0{,}02\ x_1^2 + 510.41 x_2 - 16.92\ x_2^2 + 443.28\ x_3 - 15.85\ x_3^2 \quad (1)$$

The diagram of the marginal average dependence of frost resistance on the variation factors was constructed.

The same dependence is noticed in the investigation of the dependence of frost resistance on the cement quantity (Fig. 1): the greater the quantity of cement, the less frost resistance. For the quantity of cement at 350 kg/m^3, sufficient compaction could not be provided.

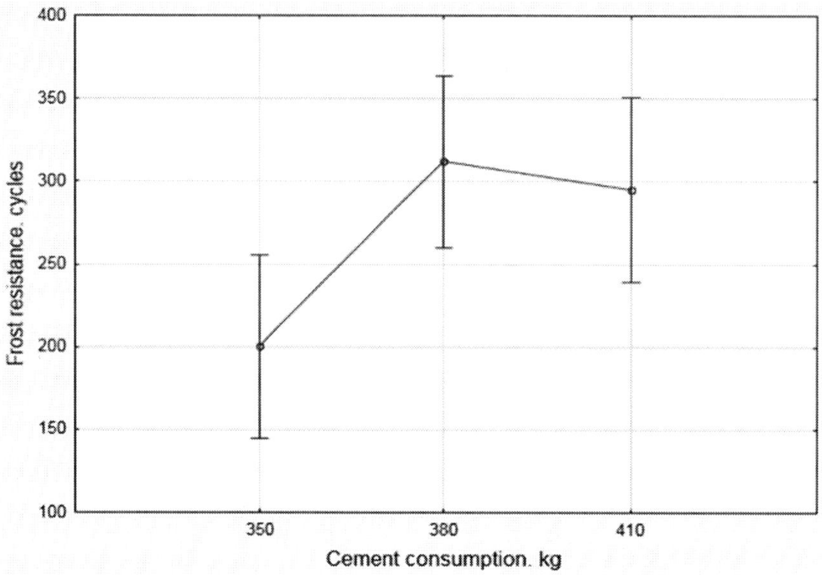

Fig. 1 Diagram of marginal average dependence of frost resistance on cement consumption.

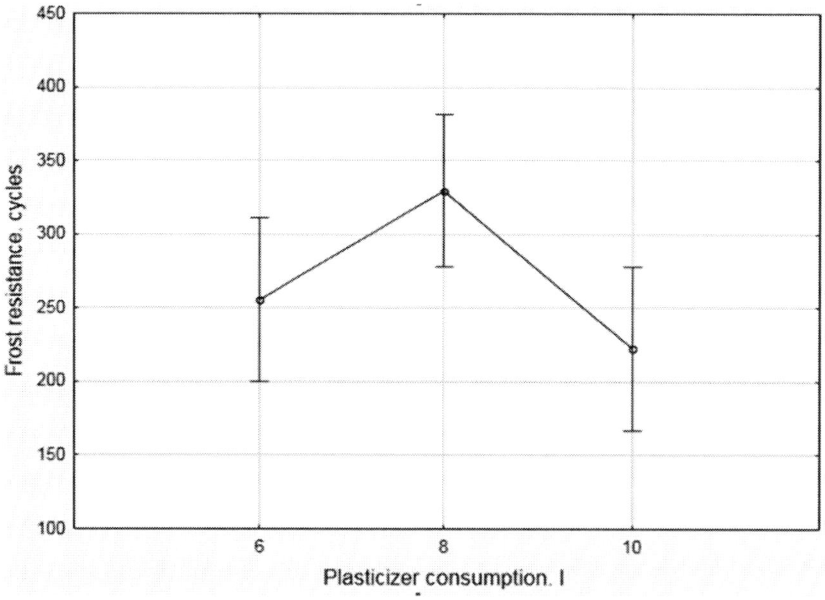

Fig. 2. Diagram of marginal average frost resistance dependence on plasticizer consumption.

The high consumption of plasticizer has an adverse effect on the frost resistance of samples as a chemically active filler (ash-slag) interacts with the organic components of the super-plasticizer additive, the result of which is the emergence of inclusions, which have a negative effect on the frost resistance of the samples.

Low consumption of plasticizer also has a adverse effect on frost resistance, since under such conditions the plasticization action of the supplement is not sufficiently manifested, it is not possible to provide an appropriate level of porosity, resulting in low frost resistance.

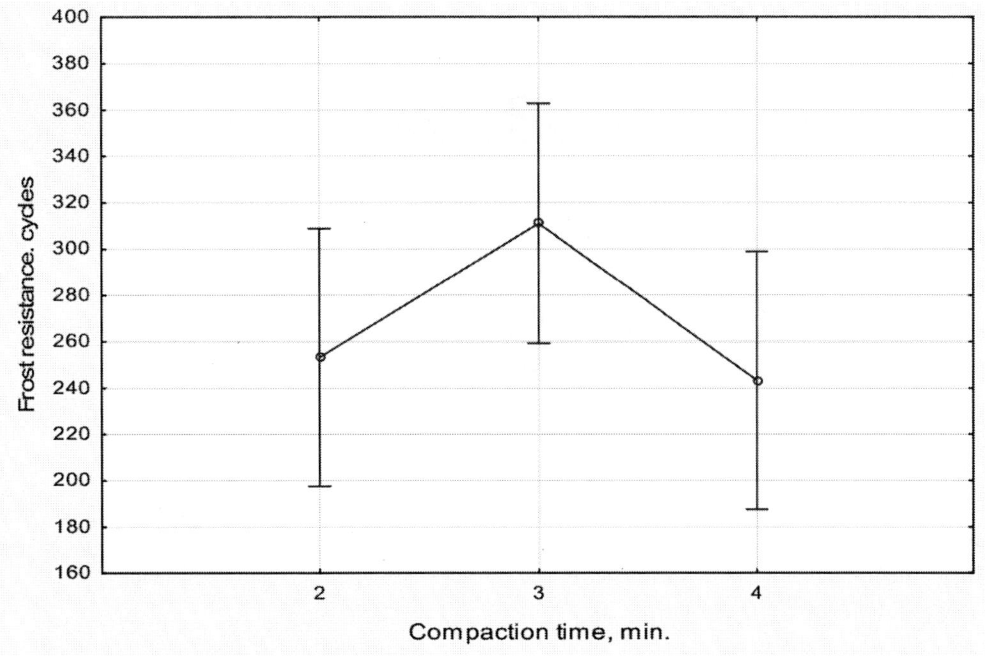

Fig. 3. Diagram of marginal mean frost resistance dependence from compaction time.

As a result of the experiment, it was possible to determine the optimal sample compaction time, which is 3 minutes.

Interconnection of frost resistance and variation factors is obtained. For this purpose, the correlation coefficients have been calculated.

From Table 2 it is getting obvious that plasticizer consumption has the greatest influence on the frost resistance.

Table 2. Results of calculation of correlation coefficients.

Factor	Frost resistance, cycles
(1) Cement consumption, kg (L)	0.51
Cement consumption, kg (Q)	0.43
(2) Plasticizer consumption, % (L)	-0.18
Plasticizer consumption, %(Q)	0.60
(3) Compaction time, min (L)	-0.05
Compaction time, min (Q)	0.42

The surfaces of influence are constructed for a clear reflection of the effect of variation factors on the frost resistance using STATISTICA software system.

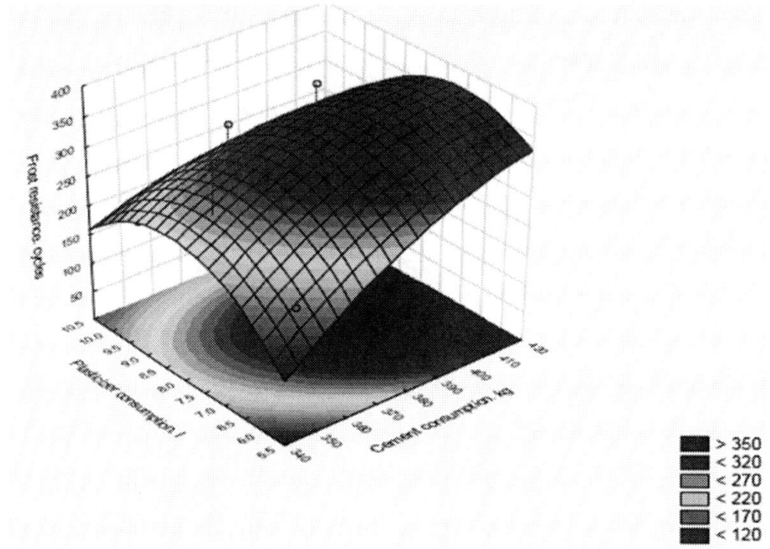

Fig. 4. Surface of the effect of the plasticizer and cement consumptions on frost resistance.

From Fig. 4 it is clear that the highest frost resistance is showed by samples for which the cement consumption is about 400 kg/m^3, and the consumption of plasticiser – 8 liters.

Fig. 5 implies that the optimal compaction time is 3 minutes.

Fig. 5 and 6 also suggest that the impact of the compaction time on frost resistance is much less than that of plasticizer and cement.

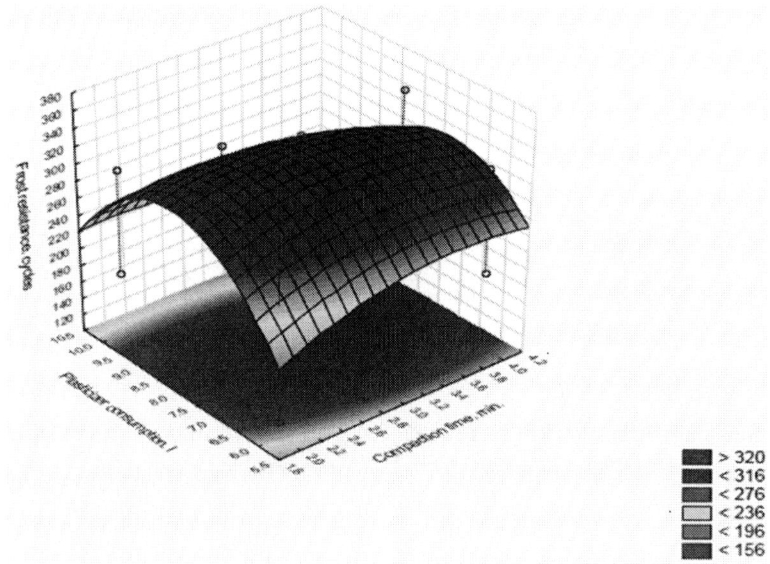

Fig5. The surface of the effect of the plasticizer and the compaction time on frost resistance.

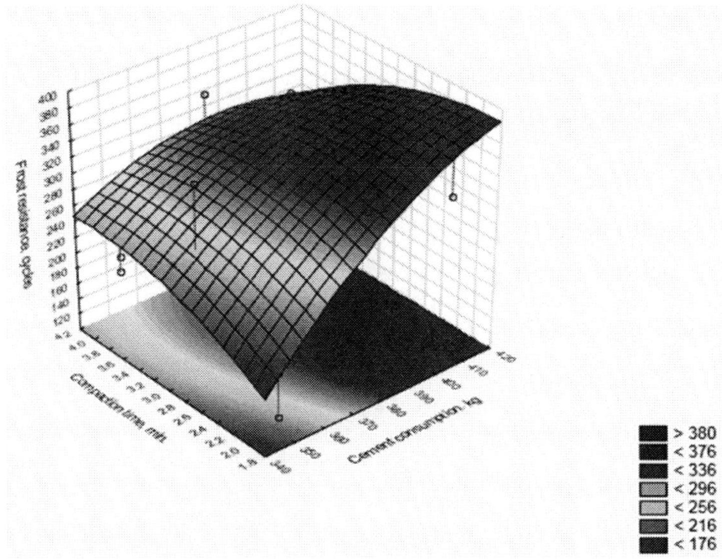

Fig. 6. The surface of the impact of cement consumption and compaction time on frost resistance.

Visual analysis of graphs and surfaces gives only approximate values of optimal component ratios. However, the software complex STATISTICA allows determining the degree of the factors at which the researched values reaches its maximum magnitude (Table 3).

Table 3. Critical variables.

Factor	Observed - Minimum	Critical - Values	Observed - Maximum
Cement consumption, kg	350.0000	478.0123	410.0000
Compaction time, min.	2.0000	2.5065	4.0000
Plasticizer consumption. l	6.0000	7.0809	10.0000

The predicted value of the response (frost resistance) for the above critical values of the input factors is of the order of 377 cycles.

3 Main conclusions

Utilizing the bottom ash slag as a fine filler allowed obtaining the concrete of the desired frost resistance class of F300. It should be noted that some samples manufactured using bottom ash-slag had a higher level of frost resistance than ones from the control group. The greatest frost resistance (the average value from the sample series is 317 cycles) was achieved for the cement consumption at 380 kg/m^3. The frost resistance is negatively impacted by the increased (10 l) consumption of plasticizer while the average frost resistance of the samples was 222 cycles. Applying the bottom ash slag as a fine filler allows manufacturing the concrete with a strength of 36 N/mm^2, that corresponds to C 20/25 class of concrete and F 300 frost resistance index. Using the bottom ash slag as a fine filler instead of sand in concrete mixture ensures not only the waste utilization but manufacturing the concrete for road construction as well.

References

1. Yogesh Aggarwal, Rafat Siddique CONSTR BUILD MATER, **54**, 210-223 (2014)
2. Dvorkin L.I. *Building Materials from Industrial Waste,* (Rostov-on-Don, 2007)

3. P.V. Krivenko, E.K. Pushkareva, V.I. Gotz, G. Y. *Kovalchuk, Tsementy i betony na osnovi palyvnykh zol ta shlakiv* (Kiev, 2012)
4. Nurul Izzati Raihan Ramzi Hannan, Shahiron Shahidan, Noorwirdawati Ali, Mohamad Zulkhairi Maarof, MATEC WEB CONF, **97** (2017)
5. V.M. Malhotra, P.K. Mehta, High-Performance, High-Volume Fly Ash Concrete, Inc., (Canada, 2002)
6. N. Zaichenko, A. Serduk, Visnyk DNABiA, **1(99)**, 137-144 (2013)
7. Seyoon Yoon, P.J.M. Monteiro, D.E. Macphee, F.P. Glasser, Mohammed Salah-Eldin Imbabi, CONSTR BUILD MATER, **54**, 432-442 (2014)
8. V.M. Malhotra, High-Performance, High-Volume Fly Ash Concrete, Concrete International **24(7)**, 30-34 (2002)
9. Victor Bondar, Volodymyr Shulgin, Oksana Demchenko, Ludmila Bondar, MATEC WEB CONF, **116** (2017)
10. P. K. Mehta, High-performance, high-volume fly ash concrete for sustainable Development, International Workshop on Sustainable Development and Concrete Technology, 3-14 (2004)

MATEC Web of Conferences **230**, 03020 (2018) https://doi.org/10.1051/matecconf/201823003020
Transbud-2018

Construction binders and environmental indicators of their production

Varvara Vinnichenko[1,*], *Alexander* Ryazanov[2], and *Olga* Krot[1]

[1]Kharkiv National University of Civil Engineering and Architecture, Sumska st. 40, 61002, Kharkiv, Ukraine
[2] Ufa State Petroleum Technological University, Mendeleeva st. 195, 450000, Ufa, Russia

Abstract. A comparative analysis of carbon dioxide emissions during the burning of Portland cement clinker, lime and gypsum was performed. The interrelation between energy indices and the emission of carbon dioxide is shown. The quantitative indices of atmospheric pollution in the manufacture of binders are given. Ecological expediency of production of gypsum and cement clinker in comparison with lime is shown. The article is made rough calculation of environmental pollution by carbon dioxide, which is produced and released into the atmosphere during the production of building binders. It was determined the directions of reducing the formation of gases during firing of the substances to reduce environmental pollution: decrease in endothermy of chemical transformations; reduction of the content of calcium and magnesium carbonates in the feedstock mixture; reduction of raw material moisture; reduction of fuel consumption for the process; increasing the technological efficiency of the process. It was determined the directions of reducing the formation of gases during firing of the substances to reduce environmental pollution. It is shown that carbon dioxide is released into the Earth's atmosphere in the amount of more than three billion metric tons per year by enterprises that produce cement, lime, and gypsum.

1 Introduction

Currently, environmental protection issues are addressed from industrial emissions, most often through the use of gas cleaning and dust removal technologies designed to reduce toxic emissions through their utilization or capture. Cleaning of exhaust gases from dust is carried out, mainly, by means of electric filters, fabric filters, scrubbers and cyclones. And reducing the emission of gaseous pollutants in the building materials industry is almost not drawn to attention.

The process of burning the building viscosity is accompanied by significant emissions of carbon dioxide, nitrogen oxides, water vapour, carbon monoxide, sulfur oxides and dust into the atmosphere [1]. The increase in the concentration of carbon dioxide in the atmosphere is a large-scale problem that can change the climate of the planet. Because carbon dioxide absorbs long-wave radiation reflected from the surface of the Earth, its

*Corresponding author: vvinnichenko@ukr.net

presence in the atmosphere acts as a protective shield that reduces heat losses of the planet. This phenomenon is called the greenhouse effect, which is accompanied by an increase in temperature. According to the available estimates, the warming of the climate increases the rate of increase of the world ocean level by 5-25 times due to the melting of glaciers and the thermal expansion of sea water [2]. The problem of the climate on the Earth in the future is closely linked with the ecological balance between the two main factors - the concentration of aerosols and carbon dioxide in the atmosphere. In previous years, carbon dioxide emissions have not received sufficient attention. But the climate change, which began in recent years, indicates the importance of this problem.

2 Results and discussion

2.1 Theory of the question

Ecologists warn that if i Carbon dioxide during the firing of clinker is released in the decomposition of carbonates of calcium and magnesium and in the combustion of carbon-containing components of fuels. Therefore, reducing the content of calcium carbonate and magnesium carbonate in the raw material mixture at the clinker firing is one way to reduce carbon dioxide emissions into the atmosphere. The content of carbon dioxide in the composition of the exhaust gases also decreases with a smaller amount of combustible fuel.

The most common binders are cement, lime and gypsum. Cement raw materials are calcium and magnesium carbonates, as well as aluminosilicates. Lime is made only of carbonates. Therefore, the formation of carbon dioxide in the burning of lime is greater than that of the firing of cement clinker. Construction gypsum (calcium sulfate semihydrate) is formed as a result of the decomposition of dihydrogen gypsum (calcium sulphate dihydrate) on semi-water gypsum and water at certain temperatures. In other words, there is no carbon dioxide in the reaction products. As a result of the reaction, only water is released.

Total amount of gases released during roasting:

$$G = G_g \cdot F + G_m \cdot M_{CO_2}\left(\frac{CaCO_3}{M_{CaCO_3}} + \frac{MgCO_3}{M_{MgCO_3}}\right)\cdot 0.01 +$$
$$+\frac{G_m \cdot W}{100-W} + G_m\left[\frac{n \cdot M_{H_2O} \cdot ASH}{M_{ASH}} + \frac{2 \cdot M_{H_2O}(CaSO_4 \cdot 2H_2O)}{M_{CaSO_4 \cdot 2H_2O}}\right]\cdot 0.01, \quad (1)$$

where $M_{CaCO_3}, M_{MgCO_3}, M_{H_2O}, M_{ASH}, M_{CaSO_4 \cdot 2H_2O}$ - the molecuflar weight of carbon dioxide, calcium carbonate, magnesium carbonate, water, aluminosilicate minerals and calcium sulfate dihydrate, kg, respectively,

n - the number of water molecules in the aluminosilicate,

G_g –volume of gases released during the combustion of fuel, m^3/m^3 of fuel,

$CaCO_3$, $MgCO_3$, ASH, $CaSO_4$ $2H_2O$ - amount of calcium carbonate, magnesium carbonate, aluminosilicate minerals and calcium sulfate dihydrate in the feed, % of the weight of the feed mixture,

G_m - consumption of dry raw materials, kg/kg of the binder,

W - the moisture content of the raw mix, % of the mass of the raw mix,

F - fuel consumption for the finished product.

$$F = \frac{q_{en} - q_{ex}}{\eta \cdot Q_r^n}, \quad (2)$$

where q_{en}, q_{ex} – enthalpy of endothermic and exothermic reactions in the preparation of binders,

η - efficiency of the firing process,

Q_r^n - calorific value of the fuel.

In the formula, the first term shows the evolution of gases as a result of fuel combustion, the second - the release of carbon dioxide in the decomposition of carbonates, the third - the release of water vapour from the raw materials, the fourth - the liberation of hydrated moisture from aluminosilicates and calcium sulfate dihydrate.

As follows from the above formulas, an increase in the content of calcium and magnesium carbonates in the raw materials contributes to an increase in the formation of carbon dioxide and an increase in emissions into the atmosphere. Emissions of pollutants are also interrelated with energy costs [4-6]: the more fuel consumption for the firing process, the more the combustion products are released. When calcining cement clinker, the carbonate content in the raw material is approximately 80%. The starting materials for the production of lime are 100% carbonates. A building gypsum is obtained from calcium sulfate hemihydrate. From the dependence obtained, it is possible to determine the directions of reducing the formation of gases during firing of the substances to reduce environmental pollution:

- decrease in endothermy of chemical transformations;
- reduction of the content of calcium and magnesium carbonates in the feedstock mixture;
- reduction of raw material moisture;
- reduction of fuel consumption for the process;
- increasing the technological efficiency of the process.

And as a result of the reactions, only water is released.

2.2 Comparative indicators for the release of carbon dioxide in the manufacture of binders

It is easy to calculate the amount of carbon dioxide that enters the atmosphere only from the work of cement kilns.

If we take as a basis the heat engineering calculations of the rotary kiln for the firing of cement clinker wet production, the output of carbon dioxide from the raw material mixture- 0.57 kg/kg of clinker, and the formation of carbon dioxide from fuel combustion - 0.6 kg/kg of clinker [3] . The total carbon dioxide emissions are $0.57 + 0.6 \approx 1.2$ kg/kg of clinker. With a dry method of production, this figure will be 0.8 kg/kg clinker. Comparative indicators of carbon dioxide release in the production of one ton of binders, tons of CO_2 / ton of binder, are shown in the diagram (Fig. 1). Global cement production is expected to increase from 3.27 billion metric tons in 2010 to 4.83 billion metric tons in 2030 [7]. In 2017, world cement production amounted to about 4.1 billion metric tons [8].

In total, approximately 350 million metric tons of lime and limestone were produced in 2015 [9]. In 2016, world total production of gypsum was estimated to be 263 million metric tons [10].

In the production of building binders, the annual emissions of carbon dioxide on the planet are approximately - for the production of cement $0.8{\cdot}0.8{\cdot}4.1+1.2{\cdot}0.35+0.06{\cdot}0.263 =$ $= 3.060$ billion metric tons. And this is only CO_2!

0.06 is the formation of carbon dioxide in the combustion of fuel consumed for the firing of gypsum. The figures speak for themselves. From the rotary kiln for burning cement clinker leaves 2-4 tons of gas per ton of clinker [11]. This figure varies depending on the moisture content of the source material, the amount of carbonates in the raw materials, the fuel consumption for firing and the type of fuel. Hence, all polluting gases are released into the environment two or more times more than the binder is produced.

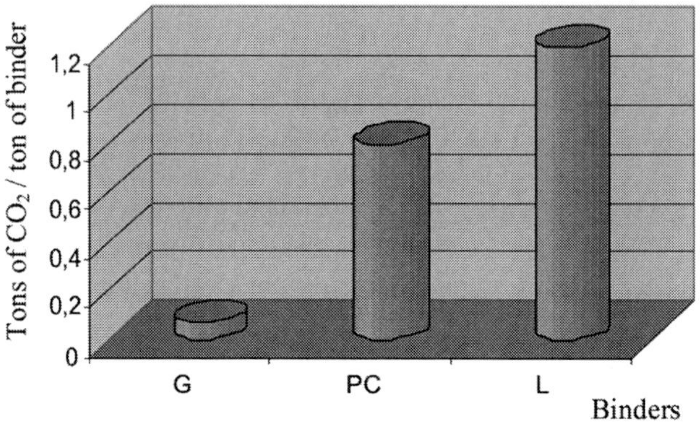

Fig. 1. Emissions of carbon dioxide in the production of gypsum (G), Portland cement clinker (PC) and lime (L)

The production of low-clinker and zero-clinker cements is noteworthy. They can be obtained from the waste [12-15] containing calcium oxides. The decarbonization process was realized in the previous production.

3 Conclusions

It is shown that approximately 3.06 billion metric tons of carbon dioxide is released annually into the atmosphere only from the production of binders: cement, lime, and gypsum. The formation of carbon dioxide in the production of construction binders may be reduced in the following areas: the use of raw materials with reduced carbonate content, the introduction of dry production methods and other economical technologies, the use of energy sources that do not contain carbon and its compounds. In addition, the urgency of this problem requires the creation of new and the introduction of known technologies for capturing carbon dioxide in the building materials industry.

References

1. Bruce Vigon with contributions from Tiffany Brunetti, Vinay Gadkari, Scott Butner, and Jill Engel-Cox. Toward a Sustainable Cement Industry. *Industrial Ecology in the Cement Industry* . **9 (9)**, 3-33 (2002).
2. IPCC. Technical summary. In: Climate Change 2013: The Physical Science Basis. Working Group I Contribution to the Fifth Assessment Report of the Intergovernmental Panel on Climate Change (Stocker TF, et al., eds.). Cambridge, United Kingdom and New York, NY:Cambridge University Press (2014). Available: http://www.climatechange2013.org [accessed 21 May 2015]
3. KH.S. Vorob'yov, D.YA. Mazurov, *Teplotekhnicheskiye raschety tsementnykh pechey i apparatov* (Vysshaya shkola, Moskva, 1962)
4. V. Vinnichenko, A. Ryazanov, MATEC Web of Conferences **116**, 01020 (2017)
5. A. Ryazanov, V. Vinnichenko, A. Plugin, Proc. scientific. works, UkrSART, **138**, 77 (2013)

6. A. Plugin, V.Vinnichenko, O. Borzjak, Proc. scientific. works, UkrSART, **143**, 87 (2014)
7. Global cement production from 1990 to 2030. https://www.statista.com/statistics/373845/global-cement-production-forecast/
8. Cement production globally and in the U.S. from 2010 to 2017. https://www.statista.com/statistics/219343/cement-production-worldwide/
9. Lime and limestone production by country in 2015. https://www.statista.com/statistics/657049/production-of-lime-worldwide/
10. Prepared by Robert D. Crangle, Jr. GYPSUM. *U.S. Geological Survey*. 2 Mineral Commodity Summaries, 76-77 (January 2017) https://minerals.usgs.gov/minerals/pubs/commodity/gypsum/mcs-2017-gypsu.pdf
11. E. Hodorov, Uspekhi tsementnoy promyshlennosti**, 432** (1968)
12. Y.L. Wang, S.P. Cui, G.P. Tian, M.Z. Lan, Z.H. Wang. Materials Science Forum, **814**, 564-568 (2015)
13. Zur Strassen H. *Zement Kalk Gips*, **10**.1, 1-12 (1957)
14. W.C.Liu, Z.L.Wu, Y.J.Wang, Y.L. Zhao, Y.Yu. *Ind. Eng. Chem. Res.*, **52**, 13761-13767 (2013)
15. V. Vinnichenko, A. Ryazanov. *Ecology and industry*, **2,** 111 (2013)

Resistance of concrete and expanded clay concrete under periodic external influences

Valery Vyrovoy, *Oksana* Korobko*, *Vladimir* Sukhanov, and *Yuri* Zakorchemny

Odessa State Academy Civil Engineering and Architecture, Didrihsona st. 4, 65029 Odessa, Ukraine

Abstract. The purpose of the work is to increase resistance of concrete in conditions of multiple freezing-thawing and wetting-drying cycles. An important property of concrete is its ability to adapt to the periodical effect of external factors by self-organization of the structure. It assumes the presence at the material of active elements capable of changing their parameters adequately to impacts. Such elements include cracks and inner surfaces of partition as integral elements of the structure of any material. Sets of active elements can be to obtained by directing the structural organization of concrete. The operating factor of structure formation can be a change in the ratio of adhesion and cohesive bonding forces at the surface of partition between the matrix material and the fillers. Concrete the structure of which was initiated by selective adhesion of a matrix to the surface of fillers differed in the raised values of properties. Structural potential has determined resistance of concrete at freezing-thawing and wetting-drying. Sets of active elements of concrete structure are the factors of safe functioning of building constructions in conditions of influence of the operating environment.

1 Introduction

At operation concrete as a material of building constructions undergoes influence of periodic changes of temperature and humidity of the environmental and reacts to them by spontaneous reorganization of the structure. Structural adaptation of the material provides stability of initial level of properties of concrete under the influence of changing stress. This is, to a large extent, determined by changes the parameters of technological cracks and inner surfaces of partition as active elements of the structure, capable in a timely manner of responding to external influences [1]. Poly-structure of concretes implies mutual influence and interaction of the structural levels at all periods of their life cycle. Therefore, parameters of level of interaction of a matrix material with fillers can be used as the factor of regulation of processes of microstructure formation and conditions of initiation and development of cracks which arise in technological period of production of a construction.

Change of a ratio of adhesive and cohesive bonding forces at the surface of partition between the matrix and the surface of fillers influences technological damage of concrete as the total quantity of cracks and inner surfaces of partition in material volume. Technological cracks pass into rank operational cracks under the first influence of the external environment. Sets of technological cracks determine dynamics of change of damage of concretes in period their functioning as realization of adaptation effects. Material properties remain at the predetermined level as result of structure self-organization. On the basis of the above, the purpose of the study is to increase the resistance of concrete under freezing-thawing and wetting-drying at the expense of the directed organization their structure by changing the ratio of adhesive and cohesive bonding forces at the surface of partition between the matrix material and the fillers.

2 Analysis of recent research

In structural materials science the structural approach, which is based on the determining role of the structure in ensuring the properties of construction as system of a certain type, is becoming increasingly important [1, 2, 3, 4, 5]. This is consistent with the poly-structural approach to building composites, which takes into account the interdependence of different scale structural levels [6]. There are new possibilities for the directed organization of the structure of concrete as a complex dynamic open system by changing the initial conditions for the structure formation of individual subsystems - the levels of structural heterogeneities [7, 8, 9, 10]. It is shown in the works [1, 9] that cracks and inner surfaces of partition as integral elements of the structure determine structural changes in the periods of production and operation of materials. Adaptation of composites as open systems to external influences mainly depends on their damage by technological and operational cracks and inner surfaces of partition [11]. This does not exclude the certain effect of conservative and meta-stable structural elements (relict zone cement particles, pores, capillaries and other) on the behavior of materials [12, 13, 14]. The sets of cracks and inner surfaces of partition can be received at different structural levels given their characteristic elements. The change of conditions of the interaction of the matrix component with fillers allows providing damage of concrete under which the effects of adaptation favorable for resistance of a material will occur. This, in turn, will contribute to the safety of the functioning of constructions under the influence of temperature and humidity of the environment.

3 Object and methods of the research

In order to achieve the objective of the research, the following tasks were set: - to analyze the dynamics of changes of damage as a structure characteristic at low-cycle fatigue of the concrete; - to determine the coefficients of resistance of concrete and expanded clay concrete in conditions of repeated external exposures.

The object of researches is the structure of concrete as open self-organizing system.

Researches were carried out on samples-cubes with the size of 100×100×100 mm, made of concrete and expanded clay concrete.

The surface of the grain part of granite rubble and expanded clay gravel was processed by hydrophobic agent. It has provided various ratios of adhesive and cohesive bonding forces at the surface of partition between the cement matrix and the surface of fillers: $R_A>R_C$; $R_A<R_K$; $R_A=R_C$ (R_A is the adhesion value of the matrix to the surface of the fillers; R_C is the cohesive strength of the matrix material.

Controlled properties of concrete: technological and operational damage (K_d), average density (ρ_ω, kg/m^3), the volume of open capillary pores (P_o, %), water absorption (W, %), modulus of elasticity (E, mPa·10^{-3}), strength ($f_{ck.cube}$, mPa), weight (m, gr), resistance (K_{res}) of concrete and expanded clay concrete under periodic freezing-thawing and wetting-drying.

Concrete damage was determined using the coefficient of damage [11] as the ratio:

- $K_d=\Sigma L/S$, cm/cm^2, ΣL_i, cm, is the total length of crack and inner surfaces of partition boundaries on the sample surface; S, cm^2, is the fixed surface area of the sample;

- $K_d=/\Sigma L_i/L_i$, ΣL_i, cm, is the total length of crack and inner surfaces of partition boundaries on the sample surface (the actual crack); L_i, cm, is the geodesic line (the shortest distance between the exit points of the actual crack at end faces of the sample).

Measurements were made on 5 sides of each sample, except for the top, calculating K_d as the arithmetic mean of the values obtained.

Coefficients of resistance of concrete and expanded clay concrete after 500 and 100 cycles of alternating freezing-thawing and wetting-drying were calculated on the basis of the obtained values of properties.

The coefficients of resistance of concrete were determined as: $K_f=Q^{Ncycles}/Q^{28}$: $Q^{Ncycles}$ – the value of property of the samples after a certain quantity of exposure cycles; Q^{28} – the value of property of the samples after 28 days of normal hardening.

4 Results of the research

In the case of a statistically arbitrary arrangement of grains in the volume of concrete, the groups of fillers form the ordered cells, restricting this or that part of the matrix material. Even with the same concrete mix, cells of different shapes and sizes coexist at the level of the macrostructure with different ratios of adhesion and cohesive bonding forces at the surfaces of partition between the matrix and the surface of the fillers. In each cell, the individual initial conditions for the structure formation of the cement matrix are provided [15]. This leads to different periods of formation of the structure and manifestation of the volume deformations of the matrix material and the properties of the cement stone in local volumes of concrete. Concretes of industrial mixes include sets of structural cells with multi-variant geometrical parameters at given conditions of interaction of the matrix with the surface of fillers. To increase the variety of cells can be by changing the ratio of the values of cohesion and adhesion on the surfaces of partition between the matrix and the fillers. Cells are relatively autonomous interconnected concrete parts, in which sets of different structural elements are provided. The interaction of these elements determines the structure potential and the level of properties of the entire material prior to operation.

The researches carried out on the models of structural cells showed that the values of the damage of the cement stone in different cells varied up to 2,5 times, depending on the ratio R_A/R_C. The value of K_d changed on 15-45 % at variation of the form of cells, at variation of orientation of the fillers in the cells by 18-40%. Distinction of the value of matrix damage was 22-34% at various ratios of R_A/R_C.

Difference of the structure of the cement matrix in individual concrete cells was confirmed by a wide range of quantitative values of the strength and speed of propagation of ultrasonic waves in local zones of concrete samples. At the wide variety of cells of the cells, the greatest range of the mechanical and acoustic characteristics in the sample material is ensured. It testifies to coexistence in the concrete of various density sections of the matrix material. The values of strength of a cement stone in various cells of the same sample differed by 13-33%. At the wide variety R_A/R_C ratios, the range of the values of the properties was in 2 times more, than in samples with complete or imperfect adhesion of the matrix to the surface of the fillers. The range of cement matrix properties in the local volumes of concrete samples at $R_A>R_C$ and $R_A<R_C$ can be explained by the influence of the geometrical parameters of the cells on the structure formation of the matrix material and the properties of concrete. A more varied set of cells in concretes is formed by selective adhesion of the cement matrix to the surface of fillers ($R_A=R_C$).

Depending on the conditions of interaction of the matrix material with the surface of the fillers, the concrete mixes were corrected for the W/C values.

Concrete grade C25/30 at $R_A<R_C$ (mix *I*); C30/35 at $R_A>R_C$ (mix *II*); C32/40 at $R_A=R_C$ (mix *III*) were obtained in conditions of constant mobility of concrete mixture.

Expanded clay concrete grade LC12/13 at $R_A>R_C$ (mix *I*) and grade LC16/18 at $R_A<R_C$ (mix *II*) and $R_A=R_C$ (mix *III*) was obtained at the various ratios R_A/R_C with corresponding values of properties.

The formation of structural blocks at all levels of heterogeneity is an equifinal result of diverse processes and phenomena of concrete structure organization [16, 17, 18, 19, 20]. It initiates the spontaneous generation and self-development of the surfaces of partition between the individual blocks. These elements can be transformed into cracks and inner surfaces of partition under the influence of own deformations of the material. In turn, cracks and inner surfaces of partition are integral elements of the structure that determine the heterogeneity of concrete before the beginning of functioning of the construction. Cracks and inner surfaces of partition provide the possibility of manifesting various mechanisms of structure adaptation to the negative influences of the external environment. A feature of these elements is the ability to react instantly to virtually any external influences, timely changing own parameters and initiate action of structural elements with slow responses to the rapidly changing operating conditions of material. The coexistence of cracks and inner surfaces of partition causes the possibility of self-organization of the material that prevents the destruction of the construction [21, 22]. The manifestation of adaptation effects connected with self-development of cracks can be realized as the transformation of cracks into inner surfaces of partition. It stops the growth of cracks, increases the discreteness of composites and creates preconditions for the localization of deformations and stresses in individual blocks of material. It contributes to the dissipation of excess energy on the banks of various cracks and inner surfaces of partition. The ability of cracks to redistribute deformations between blocks prevents their local accumulation that reduces the general stress-strain state of the material. In the local volumes of macrostructure of concrete, a variety of sets of active elements are formed, which at the level of the material structure integrative determine its functional potential and the conditions of interaction with the environment.

In the conditions of the experiment, the best properties were of concretes, the structure of which was initiated by the selective adhesion of the matrix to the surface of the fillers.

At $R_A=R_C$, the structure is formed, which leads to a decrease values of water absorption and porosity of concrete and expanded clay concrete at increasing values of their average density, technological damage, strength and modulus of elasticity. The change in the ratio of adhesive and cohesive bonding forces at the surface of partition between the matrix material and fillers makes it possible to increase the concrete strength up to 26% and expanded clay concrete strength up to 10% after 28 days of normal hardening, at the age of 720 days - up to 24% and 12%, respectively. The modulus of elasticity increased by 12-21%, the decrease values of water absorption was 8-14%. The volume of the open capillary pores decreased to 18% at the organization of structure of concrete in conditions of selective adhesion of the matrix to the fillers. The speed of passage of ultrasonic waves through concrete has decreased from U=3400 m/s to U=3200 m/s, through expanded clay concrete - from U= 3200 m/s to U=2900 m/s. Increasing the density of materials should lead to increase of their resistance under action of external factors.

Analysis of the results showed that the change of ratio of R_A and R_C values determines the technological (initial) damage of concrete, the values of K_d of which, under the experimental conditions, differ by 11-15%. It, in turn, predetermined the dynamics of changes of material operational damage under repeated external exposures, fig.1.

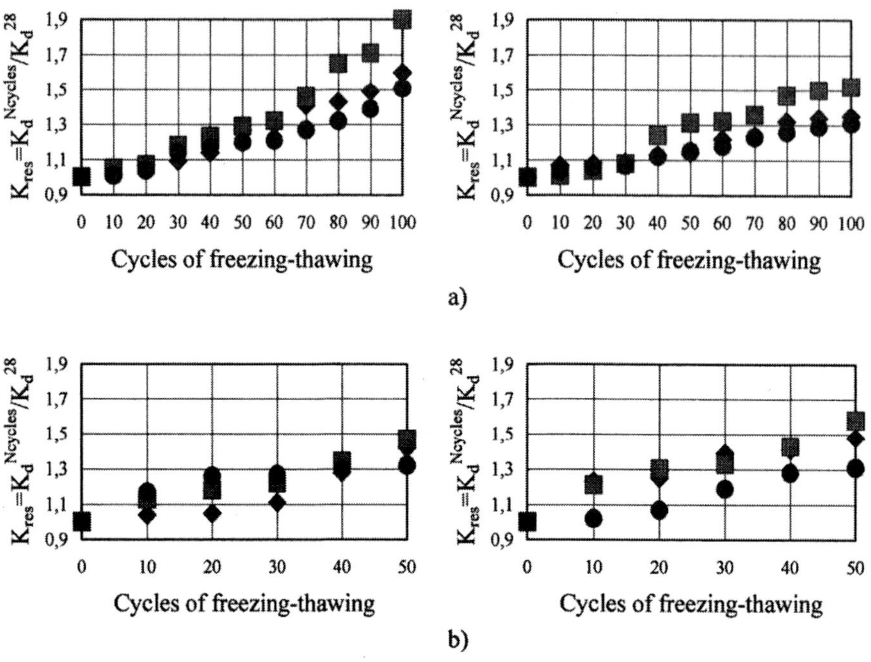

Fig. 1. Coefficients of resistance, K_d, of concrete (a) and expanded clay concrete (b) at low-cycle fatigue. $K_d=K_d^{Ncycles}/K_d^{28}$: $K_d^{Ncycles}$ – the value of damage of the samples after a certain quantity of exposure cycles; K_d^{28} – the value of damage of the samples after 28 days of normal hardening.
♦ – $R_A<R_C$; ■ – $R_A>R_C$; ● – $R_A=R_C$.

As a result of periodic external exposures, damage of concrete increased by 26-48%, expanded clay concrete to 24-37%. Values *Kd* of concretes with the structure formed under the influence of selective adhesion of the matrix to the surface of fillers were 20-25% less than the values of concrete damage at other R_A/R_C ratios. It can be explained by the fact that the cracks were closed in the structural blocks of the cement matrix, which did not allow them to grow to the dimensions dangerous for this level.

The development of cracks in materials with block structure is a discrete-continuous process of incremental growth of cracks after each cycle of repeated influences. The following structural transformations depend on the previous changes of the parameters of cracks and inner surfaces of partition due to the fact that new conditions are constantly created for the redistribution of deformations and stresses on the boundaries of cracks.

The change of the parameters of cracks and inner surfaces of partition is reflected on the values of the coefficient of concrete resistance and expanded clay concrete in case of low cycle fatigue caused by cycles of periodic changes of temperature (*T*) and humidity (*W*) of the environment (Table 1). Increased values of the properties provided by the structure of concrete under the experimental conditions are highlighted in the color.

The analysis of the coefficients of resistance of concrete and expanded clay concrete showed that the concretes, whose structure was organized at the initial condition $R_A=R_C$, were more adapted to the influence of alternating external factors.

After 100 cycles of periodic exposures, the strength of concrete of mix *III* practically did not change, the modulus of elasticity of concrete increased by 11%, the modulus of elasticity of expanded clay concrete has decreased by 10%. The volume of open capillary pores has decreased by 7-10%, water absorption by 0-5% (concrete) and 14-16% (expanded clay concrete), the mass loss was not more than 2%.

Table 1. Resistance of concrete and expanded clay concrete at low cycle fatigue.

N^{cycle}	$K_{res}=f_{ck}^{Ncycle}/f_{ck}^{28}$			$K_{res}=E^{Ncycle}/E^{28}$			$K_{res}=W^{Ncycle}/W^{28}$			$K_{res}=\Pi_o^{Ncycle}/\Pi_o^{28}$			$K_{res}=m^{Ncycle}/m^{28}$		
	Mixes			Mixes			Mixes			Mixes			Mixes		
	I	II	III	I	II	III	I	II	III	I	II	III	I	II	III
Concrete															
T=0	1	1	1	1	1	1	1	1	1	1	1	1	1	1	1
T=50	0.97	0.98	0.99	1.04	0.96	1.24	0.85	0.93	0.76	0.84	0.90	0.77	0.99	0.98	0.99
T=100	0.96	0.96	0.98	1.02	0.94	1.12	0.97	0.99	0.95	0.92	0.94	0.93	0.95	0.95	0.97
W=0	1	1	1	1	1	1	1	1	1	1	1	1	1	1	1
W=50	1.02	1	1	1.28	1.11	1.32	0.90	0.97	0.88	0.88	0.93	0.88	0.98	0.98	0.99
W=100	1	0.98	0.99	1.22	1.02	1.29	1.04	1.03	1.0	0.99	0.98	1.0	0.95	0.95	0.98
Expanded clay concrete															
T=0	1	1	1	1	1	1	1	1	1	1	1	1	1	1	1
T=50	0.97	0.98	1	0.86	0.95	0.93	0.78	0.82	0.73	0.76	0.81	0.71	0.98	0.98	1
T=100	0.95	0.95	0.98	0.82	0.9	0.89	0.89	0.92	0.83	0.85	0.87	0.81	0.95	0.95	0.97
W=0	1	1	1	1	1	1	1	1	1	1	1	1	1	1	1
W=50	1	1.01	1.03	1.11	1.17	1.03	0.92	1.01	0.96	0.91	0.96	0.98	0.99	0.97	0.99
W=100	0.97	0.99	0.99	1.07	1.12	0.99	1.12	1.15	1.16	1.06	1.08	1.16	0.96	0.95	0.96

For concretes of mixes *I* and *II*, the strength values decreased to 5% at change of the modulus of elastic modulus of up to 18%, an increase in water absorption by 3-13% and the open porosity by 11-14%, loss of weight reached 5 %.

Difference of initial conditions has provided structures which potential allows the use of building products on the basis of concrete and expanded clay concrete at periodic variations of humidity and temperature of the environment.

5 Conclusions

The carried out researches and the analysis of results allow concluding, that increase of resistance of concrete at alternating external influences is provided by the directed organization of their multilevel structures. The initial conditions for the formation of the structure can be determined by changing the ratio of adhesion-cohesive bonding forces at the surface of partition between the matrix material and the surface of the fillers. The reorganization of the concrete structure determines the resistance of concrete at low-cycle fatigue and the preservation of the product properties level. The dynamics of change of damage as a manifestation of the effects of adaptation of concrete during the operation of the product depends on the structural potential in the form of sets of technological cracks and inner surfaces of partition. Concretes with a structure formed by selective adhesion of the cement matrix to the fillers had the raised parameters of quality and were better adapted to low-cycle fatigue stress. The variety of the initial conditions of interfacial interactions has provided increase strength of concrete to 26% and strength of expanded clay lightweight concrete to 10%, modulus of elasticity to 12-21% at reduction of total porosity by 18% and water absorptions by 8-14%. Interference and interaction of the active elements of the structure is a factor of the safe functioning of building constructions.

References

1. V. Vyrovoy, V. Dorofeev, V. Sukhanov, *Composite building materials and constructions. Structure, self-organization, properties* (TES, Odessa, 2010)
2. V. Babkov, V. Mokhov, S. Kapitonov, P. Komokhov, *Structure formation and destruction of cement concrete* (Public administration of the press, Ufa, 2002)

3. Yu. Bazhenov, A. Danilov, I. Garkina, etc., *System approach to development and quality management of building materials* (PALEOTYPE, Moscow, 2006)
4. L. Sheynich, E. Pushkareva, *Processes of self-organization of the structure of building composites* (Gamma Print, Kyiv, 2009)
5. F. Campbell, *Structural Composite Materials* (ASM International Press, Ohio, 2010)
6. V. Solomatov, The development of the polystructural theory of composite building materials, Proceedings of high schools. Construction and architectur, **8**, 58-64 (1985)
7. V. Mogilevsky, *Methodology of systems* (Economics, Moscow, 1999)
8. Prangishvili, *System approach and system-wide regularities* (SINTEG, Moscow, 2000)
9. V. Vyrovoy, O. Korobko, V. Sukhanov, E. Rogniuk, collection of scientific works of OSACEA, **55**, 57-63 (2014)
10. V. Vyrovoy, A. Gerega, V. Dorofeev, V. Sukhanov, J. Modern concrete. Prospects of development, 15-17 (2010)
11. V. Dorofeev, V. Vyrovoy, *Technological damage of building materials and constructions* (Gorod masterov, Odessa, 1998)
12. V. Chernyavsky, *Adaptation of abiotic systems: concrete and reinforced concrete* (DNYRT, Dnepropetrovsk, 2008)
13. J. Stark, V. Bernd, *Durability of Concrete* (Oranta, Kiev, 2004)
14. *Durability of Concrete and Cement Composites* by C.L. Page, M.M. Page (CRC Press, 2007).
15. V. Vyrovoy, O. Korobko, V. Panasyuk, J. Tech. of conc., **5-6**, 16-18 (2016)
16. L. Shpinova, V. Chikh, M. Sanitsky, etc, *Physico-chemical basis for the formation of the structure of cement stone* (Vishcha school, Lviv, 1981)
17. Pascenco, E. Myasnikov, V. Vimen, *Cement Theory* (Budivelnik, Kiev, 1991)
18. J. Stark, B. Wicht, *Cement and lime* (Oranta Kiev, 2008)
19. Y. Ballim, P. Graham, Cem Concr Compos., **26 (5)**, 417–426 (2004)
20. P. Metha, P. Monteiro, *Concrete: Microstructure, Properties and Materials* (McGraw-Hill, New York, 2006.
21. O. Korobko, V. Sukhanov, V. Vyrovoy, V. Tofanilo, Collection of scientific works of KSABU, **3 (29)**,114-121 (2014)
22. V. Shevchenko, Yu. Antonyuk, Collection of scientific works of conf. “Structure formation, strength and destruction of composite building materials and constructions”, 149-152 (2018)

Casting of concrete and reinforced concrete pipes by vibro-vacuum technique

Basheer Younis*, *Ibragim* Kazimahomedov, *Saad* Salem, *Tatiana* Kostuk and *Elena* Dedenyova

Kharkiv National University of Civil Engineering and Architecture, 61002 Sumska st, 40, Kharkiv, Ukraine

Abstract. This article deals with the casting of concrete and reinforced concrete pipes by vibro-vacuum technique. The existing techniques are investigated, the shortcomings are revealed and the solutions to major issues with the improved techniques were offered. Improvement of a technological process of concrete and reinforced concrete pipes production is an important and current task. Development of such technologies should ensure the required supporting capacity reducing the cost of pipeline construction thereof. In this article the existing ways of molding concrete and reinforced concrete pipes have been evaluated. Their engineering deficiencies in terms of labour-intensity and obtaining the products with increased physical and mechanical characteristics have been identified. The offered solution addressing the problems has been identified, namely, the casting of concrete pipes by vibro-vacuum technique. Pilot studies have confirmed that this method makes it possible to achieve the sufficiently high physico-mechanical characteristics of the products. Tensile strength of vibro-vacuum pipes is 40% higher than that of pipes obtained by layer-by-layer axial compaction.

1 Introduction

The modern requirements to the quality of construction products are not inferior to the demands for the newest architectural elements [1, 2]. In addition to the high requirements for physico-mechanical characteristics and durability of concrete and reinforced concrete pipes, the requirements for the quality and casting techniques are set. The casting process consists of assembling molds, fixture installation, laying and compaction of concrete mix.

The most useful casting technique is determined in accordance with design features of the concrete products, technological production flow diagram, degree and type of reinforcement, mass and dimensions of the products, quality requirements for casting and compacting of concrete mix, quality requirements for the surface of the product, availability and volume of production [1-6].

Currently, in the construction market there is a steady trend towards automated production. For laying and compacting of concrete mix the following techniques are

*Corresponding author: docbasheer01@gmail.com

applied: vibrating, vibro-stamping, vibro-rolling, vibro-vacuum, centrifugation, etc., as well as the combined ones, such as: concrete vibrating and pressing, centrifuging, rolling, etc. [7-10].

The purpose of the article is to investigate the current requirements for construction products, the casting techiques and their relevance so far, to identify the negative aspects of existing techniques and to suggest the ways to solve them in terms of operation and product quality.

2 Analysis of publications on the research topic

The article is based on studies in the field of building materials and casting techiques for pipelines by Yu.M.Bazhenov [6], A.G. Komar B.G., Skramtaev [7], A.E. Desov et al. The analysis of these works allows considering the casting techiques for the pipelines, in particular, vibro-vacuum, but they are not enough for an integrated approach to pipeline systems upgrading. Analysis of these works allows considering the casting techiques for the pipelines, to determine their efficiency, accompanied by an integrated approach to improving physical and mechanical characteristics of concrete and reinforced concrete pipes . The article is based on studies in the field of production of concrete pipes by I.N. Akhverdova V.I. Babushkin, A.G. Vandolovsky, V.N. Ladyzhensky, O.P. Mchedlov-Petrosyan, A.E. Shmurnov, B.B. Khasanov, A.N. Popov, P.A. Makarov et al. The two methods of compacting a concrete mix with a core rotated inside the mold have proliferated in the production of concrete pipes: rotational layer-by-layer pressing and radial pressing.

The rotational pressing is made by pressing a curved blade sliding over the compacted concrete, while it falls under sliding friction and wears out pretty quickly.

For the layer-by-layer radial pressing a sealing core with a roller system on the pressing core instead of the blade increasing the service life has been developed. An important condition is the rotation of the rollers during compaction without sliding which causes certain technological difficulties.

The proposed vibro-vacuum sealing technique in comparison with previous methods makes it possible to obtain the pipes and shaped elements of increased strength characteristics, resulting in the increased the service life of pipelines.

3 The basic content of the article

There is a casting technique for dry-concrete mixes in molds with one open surface by feeding a concrete mix along a number of identical, parallel, freely rotating blades on the axes contacting the exposed surface of the molded product and moving relative to the open surface of the mold in the transverse and longitudinal directions. Moreover, the concrete mix is fed under the blades by a layer exceeding the thickness of the molded article under the blade, and the speed of the blades moving along their axes is selected, depending on the speed of the layer of the concrete mix being squeezed out from under the front ends of the blades along the axis of their movement. However, such techniques do not guarantee the high quality of the molded articles. In the same pipe, areas of the highest quality of compaction may be along with insufficiently compacted ones. The reason for this is insufficient control and the impossibility of maintaining a constant layer of loose concrete mix over the blade head.

The closest to the claimed technical solution is the precast-concrete pipes casting technique where the concrete mix is moistened in the space between the mold and the core and compact the mix due to excessive pressure in the core with simultaneous vibration. The excessive pressure is created by water injection between the rubber sleeve and the metal core from the inside. Preserving the excess pressure of the rubber stocking on the concrete mixture, the mold is directed to heat treatment. The shortcomings of the above method include the injection of concrete mix into the interstice space from below due to excess pressure resulting in additional labor costs and other problems.

The task is to increase the productivity and quality of molded concrete and reinforced concrete pipes. To solve the existing problem, it is proposed to use the vibro-vacuum technique in casting of concrete and reinforced concrete pipes in the building industry.

Vibro-vacuuming is a casting technique where vibration of a concrete mix, carried out by one of the known techniques, is combined with the removal of air and excess water from it. Vibro-vacuuming is a complex way of casting products, including vacuum dewatering and repeated short-term vibration. Casting of products is carried out in vertical and horizontal facilities, providing compaction of the concrete mix, immediate stripping and unattended (on the sucker) removal of freshly molded products.

In connection with the fact that apart from the pipes there is a need for branch pipe parts (Y-branch fittings) in pipelines a technological solution for casting of complex shaped products from high-flow concrete mixes was developed.

Close to the developed technical solution is the precast-concrete pipes casting technique where the concrete mix is moistened in the space between the mold and the core and compact the mix due to excessive pressure in the core with simultaneous vibration. The excessive pressure is created by water injection between the rubber sleeve and the metal core. In the production of pipes this way the upper fed high-flow concrete mixes the mold are used.

Fig. 1. Laboratory facility for the production of concrete and reinforced concrete pipes by vibro-vacuum technique.

It has been established by experimental means that the highest compaction quality of the pipes is achieved using the vibro-vacuum casting technique where up to 30% of excessive water is removed from freshly placed concrete mix, as a result of this treatment, the concrete mix is compacted, which in turn improves its properties, namely, tensile strength. The figures obtained are presented in Table 1.

Table 1. Tensile strength of concrete pipes

Tensile strength of concrete pipes (R_{bt}) MPa		W/C
Samples of pipes obtained by layer-by-layer axial compaction	Samples of pipes obtained by vibro-vacuum technique	
4.20	6.32	0.5
3.93	6.80	0.55
3.12	7.41	0.65

The tensile strength of vibro-vacuum pipes is 40% higher than that of pipes obtained by layer-by-layer axial compaction. Simultaneously with vacuum treatment the concrete mix is compacted by vibration to increase its strength.

4 Conclusion

The study highlighted the main problems of quality and operation of concrete and reinforced concrete pipelines with existing casting techniques. The offered casting technique improves the characteristics of the products, in particular tensile strength. This technique is intended for casting of concrete and reinforced concrete pipes, rings, pillars of lighting supports and overhead contact system of concrete mixes.

References

1. R.F. Runova, V.I. Gotz, M.A. Sanytsky, *Design materials of the new generation and technology in the future in the future* (2008)
2. Z. Jamrozy, Concrete and its technologies (Scientific publishing house Pwn., Warsaw, 2000)
3. N. Zaichenko, *High-strength fine-grained concrete with a complex microstructure of the modified* (2009)
4. Yu. Bazhenov, *Vysokoprochnyy melkozernistyy beton dlya armotsementnykh konstruktsiy* (Moscow, 1963)
5. Akhverdov, *Osnovy fiziki betona* (Moscow, 1981)
6. Yu.M. Bazhenov, *Technology of concrete and reinforced concrete products* (Moscow, 1984)
7. B.G. Skramtaev, Building industry, **3**, 64-72 (1938)
8. A.G. Vandolovsky, E.B. Dedeneva, Scientific Herald of Construction, **47**, 192-195 (Kharkiv, 2008)
9. E. Dedeneva, I. Kazimagomedov, Saad Salem, T. Kostyuk, Basheer N. Younis, M. Yakimenko, MATEC Web Conf., **116** (2017)
10. B.N. Younis, E.B. Dedeneva, T.O. Kostiuk, , Proceedings of III International scientific conference, 28-31 (Morrisville ,2017)

AUTHOR INDEX

AUTHOR INDEX

AUTHOR INDEX